T0328260

Microalgae Cultivation for Biofuels Production

Microalgae Cultivation for Biofuels Production

Edited by

ABU YOUSUF
Chemical Engineering & Polymer Science
Shahjalal University of Science and Technology
Sylhet, Bangladesh

ELSEVIER

ACADEMIC PRESS
An imprint of Elsevier

Academic Press is an imprint of Elsevier
125 London Wall, London EC2Y 5AS, United Kingdom
525 B Street, Suite 1650, San Diego, CA 92101, United States
50 Hampshire Street, 5th Floor, Cambridge, MA 02139, United States
The Boulevard, Langford Lane, Kidlington, Oxford OX5 1GB, United Kingdom

Notices
Knowledge and best practice in this field are constantly changing. As new research and experience broaden our
understanding, changes in research methods, professional practices, or medical treatment may become
necessary.

Practitioners and researchers must always rely on their own experience and knowledge in evaluating and using
any information, methods, compounds, or experiments described herein. In using such information or methods
they should be mindful of their own safety and the safety of others, including parties for whom they have a
professional responsibility.

To the fullest extent of the law, neither the Publisher nor the authors, contributors, or editors, assume any
liability for any injury and/or damage to persons or property as a matter of products liability, negligence or
otherwise, or from any use or operation of any methods, products, instructions, or ideas contained in the
material herein.

Library of Congress Cataloging-in-Publication Data
A catalog record for this book is available from the Library of Congress

British Library Cataloguing-in-Publication Data
A catalogue record for this book is available from the British Library

ISBN: 978-0-12-817536-1

For information on all Academic Press publications visit our website at
https://www.elsevier.com/books-and-journals

Publisher: Brian Romer
Acquisition Editor: Raquel Zanol
Editorial Project Manager: Peter Adamson
Production Project Manager: Poulouse Joseph
Cover Designer: Alan Studholme

Typeset by TNQ Technologies

Working together
to grow libraries in
developing countries

www.elsevier.com • www.bookaid.org

My parents and wife, who's encouragement were the driving force.......

List of Contributors

Iftikhar Ahmed
Energy Research Centre
COMSATS University Islamabad
Lahore Campus, Lahore, Pakistan

M. Amirul Islam
Laboratory for Quantum Semiconductors and
    Photon-Based BioNanotechnology
Department of Electrical and Computer Engineering
Faculty of Engineering
Université de Sherbrooke
Sherbrooke, QC, Canada

M. Anand
Department of Chemistry
Science and Humanities Kingston Engineering College
Vellore, Tamil Nadu, India

Ambreen Aslam
Department of Environmental Sciences
The University of Lahore
Lahore, Punjab, Pakistan

Muhammad Roil Bilad
Chemical Engineering Department
Universiti Teknologi PETRONAS
Bandar Seri Iskandar, Perak Darul Ridzuan, Malaysia

Rolando Chamy
Laboratorio de Biotecnología Ambiental
Escuela de Ingeniería Bioquímica
Facultad de Ingeniería
Pontificia Universidad Católica de Valparaíso
Valparaíso, Chile

Benjamas Cheirsilp
Biotechnology for Bioresource Utilization Laboratory
Department of Industrial Biotechnology
Faculty of Agro-Industry
Prince of Songkla University
Hat Yai, Songkhla, Thailand

Min-Yee Choo
Institute of Biological Sciences
Faculty of Science
University of Malaya
Kuala Lumpur, Malaysia

Olivia Córdova
Laboratorio de Biotecnología Ambiental
Escuela de Ingeniería Bioquímica
Facultad de Ingeniería
Pontificia Universidad Católica de Valparaíso
Valparaíso, Chile

Tahir Fazal
Biorefinery Engineering And Microfluidic (BEAM)
    Research Group
Department of Chemical Engineering
COMSATS University Islamabad
Lahore Campus
Lahore, Punjab, Pakistan

Che Ku Mohammad Faizal
Faculty of Engineering Technology
Universiti Malaysia Pahang
Gambang, Pahang, Malaysia

M. Franco Morgado
Instituto de Ingeniería
Universidad Nacional Autónoma de México (UNAM)
Ciudad de México, México

D. García
Center for Research in Aquatic Resources of Nicaragua
National Autonomous University of Nicaragua
Managua, Nicaragua

A. González-Sánchez
Instituto de Ingeniería
Universidad Nacional Autónoma de México (UNAM)
Ciudad de México, México

Mohd Dzul Hakim Wirzal
Chemical Engineering Department
Universiti Teknologi PETRONAS
Bandar Seri Iskandar, Perak Darul Ridzuan, Malaysia

Pei Han
School of Resources, Environmental & Chemical
    Engineering and Key Laboratory of Poyang Lake
    Environment and Resource Utilization
Ministry of Education
Nanchang University
Nanchang, Jiangxi, China

Javed Iqbal
Department of Chemical Engineering
Khwaja Fareed University of Engineering and
    Information Technology
Rahim Yar Khan, Punjab, Pakistan

Kailin Jiao
College of Energy
Xiamen University
Xiamen, Fujian, China

Keju Jing
Department of Chemical and Biochemical Engineering
College of Chemistry and Chemical Engineering
The Key Lab for Synthetic Biotechnology of Xiamen
    City Xiamen University
Xiamen, Fujian, China

Joon Ching Juan
Nanotechnology and Catalysis Research Center
    (NANOCAT)
University of Malaya
Kuala Lumpur, Malaysia

Ahasanul Karim
Faculty of Engineering Technology
Universiti Malaysia Pahang
Gambang, Pahang, Malaysia

Mohd Asyraf Kassim
Bioprocess Technology Programme
School of Industrial Technology
Universiti Sains Malaysia (USM)
Gelugor, Penang, Malaysia

Zaied Bin Khalid
Faculty of Engineering Technology
Universiti Malaysia Pahang
Gambang, Pahang, Malaysia

Md Maksudur Rahman Khan
Faculty of Chemical and Natural Resources Engineering
Universiti Malaysia Pahang
Gambang, Pahang, Malaysia

Michael Kornaros
Laboratory of Biochemical Engineering and
    Environmental Technology (LBEET)
Department of Chemical Engineering
University of Patras
Patras, Greece

Eleni Koutra
Laboratory of Biochemical Engineering and
    Environmental Technology (LBEET)
Department of Chemical Engineering
University of Patras
Patras, Greece

Shishir Kumar
Nanomaterials and Bioprocessing Laboratory
    (NABLAB)
Dan F. Smith Department of Chemical Engineering
Lamar University
Beaumont, TX, United States

Jia Jia Leam
Chemical Engineering Department
Universiti Teknologi PETRONAS
Bandar Seri Iskandar, Perak Darul Ridzuan, Malaysia

Hwei Voon Lee
Nanotechnology and Catalysis Research Center
    (NANOCAT)
University of Malaya
Kuala Lumpur, Malaysia

Jingjing Li
School of Resources, Environmental & Chemical
    Engineering and Key Laboratory of Poyang Lake
    Environment and Resource Utilization
Ministry of Education
Nanchang University
Nanchang, Jiangxi, China

Tau Chuan Ling
Institute of Biological Sciences
Faculty of Science
University of Malaya
Kuala Lumpur, Malaysia

Lu Lin
College of Energy
Xiamen University
Xiamen, Fujian, People of Republic of China

Qian Lu
School of Resources
Environmental & Chemical Engineering and Key
    Laboratory of Poyang Lake Environment and
    Resource Utilization
Ministry of Education
Nanchang University
Nanchang, Jiangxi, China

Yinghua Lu
Department of Chemical and Biochemical Engineering
College of Chemistry and Chemical Engineering
The Key Lab for Synthetic Biotechnology of Xiamen
    City
Xiamen University
Xiamen, Fujian, China

Yohanis Irenius Mandik
Department of Chemistry
Faculty of Mathematics and Natural Sciences
University of Cenderawasih
Jayapura, Indonesia

Emmanuel Manirafasha
Department of Chemical and Biochemical Engineering
College of Chemistry and Chemical Engineering
The Key Lab for Synthetic Biotechnology of Xiamen
    City
Xiamen University
Xiamen, Fujian, China

H.O. Méndez-Acosta
Centro Universitario de Ciencias Exactas e Ingenierías
Universidad de Guadalajara
Guadalajara-Jalisco, México

Sandhya Mishra
Academy of Scientific & Innovative Research (AcSIR)
CSIR-Central Salt and Marine Chemicals Research
    Institute
Bhavnagar, Gujarat, India

Dongyan Mu
School of Environmental and Sustainability Sciences
Kean University
Union, NJ, United States

Theophile Murwanashyaka
Department of Chemical and Biochemical Engineering
College of Chemistry and Chemical Engineering
The Key Lab for Synthetic Biotechnology of Xiamen
    City
Xiamen University
Xiamen, Fujian, China

Tabish Nawaz
Environmental Science and Engineering Department
    (CESE)
Indian Institute of Technology Bombay
Mumbai, Maharashtra, India

Center for Advances in Water and Air Quality
Lamar University
Beaumont, TX, United States

Theoneste Ndikubwimana
Kibogora Polytechnic
Rusizi, Rwanda

Eng-Poh Ng
School of Chemical Sciences
Universiti Sains Malaysia
Penang, Malaysia

Lee Eng Oi
Nanotechnology and Catalysis Research Center
    (NANOCAT)
University of Malaya
Kuala Lumpur, Malaysia

Imran Pancha
Academy of Scientific & Innovative Research (AcSIR)
CSIR-Central Salt and Marine Chemicals Research
    Institute
Bhavnagar, Gujarat, India

Ashok Pandey
Distinguished Scientist
CSIR-Indian Institute for Toxicology Research
Lucknow, Uttar Pradesh, India

Won-Kun Park
Department of Chemistry & Energy Engineering
Sangmyung University
Seoul, Republic of Korea

Antonino Pollio
Department of Biology
University of Naples Federico II
University Complex of Monte Sant'Angelo
Naples, Italy

Ashiqur Rahman
Nanomaterials and Bioprocessing Laboratory
    (NABLAB)
Dan F. Smith Department of Chemical Engineering
Lamar University
Beaumont, TX, United States

Shristi Ram
Academy of Scientific & Innovative Research (AcSIR)
CSIR-Central Salt and Marine Chemicals Research
    Institute
Bhavnagar, Gujarat, India

J. Ranjitha
CO_2 Research and Green Technologies Centre
Vellore Institute of Technology
Vellore, Tamil Nadu, India

Naim Rashid
Department of Chemical Engineering
COMSATS University Islamabad
Lahore, Punjab, Pakistan

Fahad Rehman
Biorefinery Engineering and Microfluidic (BEAM)
    Research Group
Department of Chemical Engineering
COMSATS University Islamabad
Lahore Campus
Lahore, Punjab, Pakistan

Dianbandhu Sahoo
Institute of Bioresources and Sustainable Development
A National Institute under Department of
    Biotechnology
Govt. of India
Takyelpat, Imphal, Manipur, India

Muhammad Saif Ur Rehman
Department of Chemical Engineering
Khwaja Fareed University of Engineering and
    Information Technology
Rahim Yar Khan, Pakistan

Myrsini Sakarika
Center for Microbial Ecology and Technology (CMET)
Ghent University
Gent, Belgium

Thinesh Selvaratnam
Department of Civil and Environmental Engineering
Lamar University
Beaumont, TX, United States

M.L. Serejo
Centro Universitario de Ciencias Exactas e Ingenierías
Universidad de Guadalajara
Guadalajara-Jalisco, México

Ali Shan
Department of Environmental Sciences
The University of Lahore
Lahore, Punjab, Pakistan

Sirasit Srinuanpan
Biotechnology for Bioresource Utilization Laboratory
Department of Industrial Biotechnology
Faculty of Agro-Industry
Prince of Songkla University
Hat Yai, Songkhla, Thailand

Jean-Philippe Steyer
Laboratory of Environmental Biotechnology
National Institute of Agronomic Research
University of Montpellier
Narbonne, France

Yong Sun
College of Energy
Xiamen University
Xiamen, Fujian, China

Tan Kean Meng
Bioprocess Technology Programme
School of Industrial Technology
Universiti Sains Malaysia (USM)
Gelugor, Penang, Malaysia

Xing Tang
University of Rwanda-College of Education
Kigali, Rwanda

Fujian Engineering and Research Center of Clean and
    High-Valued Technologies for Biomass
Xiamen Key Laboratory of Clean and High-Valued
    Utilization for Biomass
Xiamen University
Xiamen, Fujian, China

Antonia Terpou
Laboratory of Biochemical Engineering and
    Environmental Technology (LBEET)
Department of Chemical Engineering
University of Patras
Patras, Greece

A. Toledo-Cervantes
Centro Universitario de Ciencias Exactas e Ingenierías
Universidad de Guadalajara
Guadalajara-Jalisco, México

Panagiota Tsafrakidou
Laboratory of Biochemical Engineering and
    Environmental Technology (LBEET)
Department of Chemical Engineering
University of Patras
Patras, Greece

Sabeela Beevi Ummalyma
Institute of Bioresources and Sustainable Development
A National Institute under Department of
    Biotechnology
Govt. of India
Sikkim Centre
Tadong, Gangtok, Sikkim, India

S.V. Vamsi Bharadwaj
Academy of Scientific & Innovative Research (AcSIR)
CSIR-Central Salt and Marine Chemicals Research
    Institute
Bhavnagar, Gujarat, India

S. Vijayalakshmi
CO_2 Research and Green Technologies Centre
Vellore Institute of Technology
Vellore, Tamil Nadu, India

Yusuf Wibisono
Bioprocess Engineering Department
Universitas Brawijaya
Malang, Indonesia

Chunhua Xin
School of Management
China University of Mining and Technology (Beijing
    Campus)
Beijing, China

Yuanchao Xu
College of Biologic Engineering
Qi Lu University of Technology
Jinan, Shandong, China

Abu Yousuf
Department of Chemical Engineering & Polymer
    Science
Shahjalal University of Science and Technology
Sylhet, Bangladesh

Qamar uz Zaman
Department of Environmental Sciences
The University of Lahore
Lahore, Punjab, Pakistan

Xianhai Zeng
College of Energy
Xiamen University
Xiamen, Fujian, China

Wenguang Zhou
School of Resources
Environmental & Chemical Engineering and Key
    Laboratory of Poyang Lake Environment and
    Resource Utilization
Ministry of Education
Nanchang University
Nanchang, Jiangxi, China

Gaetano Zuccaro
Laboratory of Environmental Biotechnology
National Institute of Agronomic Research
University of Montpellier
Narbonne, France

About the Editor

Abu Yousuf holds a PhD in Chemical Engineering from the University of Naples Federico II, Italy. His primary research interests include biorefinery, bioenergy, bioremediation, and waste-to-energy. He has published more than 50 papers in reputed ISI and Scopus indexed journals and 5 book chapters. He has been editing 3 books, will be published by Elsevier, and has been serving as an editorial board member of several reputed journals. He won the UNESCO Prize on E-learning course of "Energy for sustainable development in Asia," Jakarta, Indonesia, 2011. He attended the "BIOVISION.Next Fellowship Programme 2013" at Lyon, France, after a selection based on scientific excellence, mobility, and involvement in civil society. He also received several awards for his research outcomes from UK, Malaysia and Bangladesh. He successfully accomplished 10 research project including the grants provided by The World Academy of Science (TWAS), Italy; Ministry of Higher Education, Malaysia; and Ministry of Science and Technology, Bangladesh. Dr. Yousuf is a member of IChemE, American Chemical Society (ACS), American Association for Science and Technology (AASCIT), and Bangladesh Chemical Society (BCS). He has 12 years' experience in teaching at the undergraduate and postgraduate levels, having very good remarks from the students. Currently, Dr. Yousuf is serving as a Professor in Chemical Engineering and Polymer Science, Shahjalal University of Science and Technology, Bangladesh. Previously he held the position of Senior Lecturer at Faculty of Engineering Technology, Universiti Malaysia Pahang, Malaysia. He has presented his research work in Germany, France, Italy, India, Vietnam, Malaysia and Bangladesh. He can be contacted at ayousufcep@yahoo.com/ayousufcep@sust.edu.

Preface

Bio-based industries are drawing attention because of their major role in energy security and climate change mitigation during the 21st century. At the same time, microalgae have come into sight as a potential nonedible feedstock with advantages over traditional land crops, such as high productivity, continuous harvesting throughout the year, and minimal problems regarding land use and race with food. Therefore, understanding of microalgae biorefinery is a significant part of the growing demand for energy, fuels, chemicals, and materials worldwide.

In this context, the book is intended to compile significant technological challenges to produce economically competitive algae-derived biofuel, such as efficient methods for cultivation; improvement of harvesting and lipid extraction techniques; optimization of conversion/ production processes of fuels and co-products; integration of microalgae biorefineries to several industries; environmental resilience by microalgae; and techno-economic and life cycle analysis of production chain to get the maximum benefits from microalgae biorefineries.

The book starts (Chapter 1) with basic information of microalgae, like biology of microalgae, their lipid content, mass cultivation systems with their comparison, and influential growth factors. Chapters 2—7 address mainly microalgae cultivation, harvesting, nutrients metabolism, and bioreactor design. They also discuss the modification of microalgae cultivation via engineering approaches such as photobioreactor development and mode of cultivation to improve microalgae biomass production.

Chapters 8—10 describe the current technologies published in recent studies on downstream processes of microalgal biomass into bioproducts including biodiesel, biogas, hydrogen, pigments, and fatty acids as end products. These chapters further explain how to cut down the cost of microalgae in downstream processes particularly in the stages of harvesting and dewatering, extraction, fractionation, and conversion.

Chapters 11—13 address the challenges of microalgal cultivation and exploitation of different algal metabolites for various industrial products (fuel and nonfuel) in

the biorefinery concept to solve the issues associated with microalgal biomass and its future perspectives. In this regard, the chapters highlight fuel-based products such as biodiesel, biogas, biohydrogen, and bioethanol, and nonfuel products such as fertilizer, forage, food, feed, and bioactive compounds. These chapters also provide a perspective to improve their yield, and outline a strategy to integrate biorefinery with current environmental techniques.

Chapter 14 introduces contemporary genetic engineering techniques and tools to improve biofuel and biochemical production in various microalgae. They also discuss various system biology tools like genomics, transcriptomics, etc., to improve the final product yield using various microalgae.

Chapter 15 presents microbiological communities associated with anaerobic sludge for the biomethanization of a microalgal biomass that should be considered for the optimal development of anaerobic digestion to produce biogas.

Chapter 16 describes the various types of nanobiocatalysts used for microalgae-based biodiesel production using transesterification process and their merits in the current industrial application.

Chapter 17 focuses on a green chemistry-based approach on the biosynthesis of nanomaterials using algae. Special attention is placed on AgNP synthesis and microwave-assisted synthesis methods. Various factors that affect nanomaterial synthesis such as pH, temperature, incubation time, ionic strength, light intensity, and algal biomass have been extensively discussed.

Chapter 18 provides a critical review of the existing methodology and application of life cycle assessment (LCA) and techno-economic assessment (TEA) on algae cultivation and algal biofuel production.

Chapters 19—21 emphasize on the direct benefits of environmental bioremediation through microalgae culture. These chapters provide the viability of environmental resilience of some microalgae strains for treating highly polluted wastewaters. The chapters also show the pathway how to achieve major breakthrough by qualifying microalgae as a sustainable biorefinery feedstock and by resource recovery from waste streams through their culture.

Researchers and academicians with strong scientific knowledge and practical experiences have shared their thoughts in this book. We trust the book will enrich the foresight of current and future researchers who are dedicatedly working on microalgae-based products.

Abu Yousuf
Sylhet, Bangladesh

Acknowledgment

We are expressing our gratitude to all distinguished authors for their thoughtful contribution to make this book project a success. Their patience and diligence in revising the first draft of the chapters after assimilating the suggestions and comments of reviewers are highly appreciated.

We would like to acknowledge the solicitous contributions of all the reviewers who spent their valuable time in a constructive and professional manner to improve the quality of the book.

We are grateful to the staff at Elsevier, particularly Ms. Raquel Zanol (Acquisitions Editor), who supported us tremendously throughout the project and for her great and encouraging mind. We also acknowledge Dr. Peter W. Adamson (Editorial Project Manager), Poulouse Joseph (Senior Project Manager) and Swapna Praveen (Copyrights Coordinator) for their cordial handling of the book.

Contents

1 **Fundamentals of Microalgae Cultivation**, 1
Abu Yousuf

2 **Microalgae Cultivation Systems**, 11
Gaetano Zuccaro, Abu Yousuf, Antonino Pollio
and Jean-Philippe Steyer

3 **Microalgae Cultivation and Photobioreactor Design**, 31
Wenguang Zhou, Qian Lu, Pei Han and
Jingjing Li

4 **Mixotrophic Cultivation: Biomass and Biochemical Biosynthesis for Biofuel Production**, 51
Tan Kean Meng, Mohd Asyraf Kassim and
Benjamas Cheirsilp

5 **Oleaginous Microalgae Cultivation for Biogas Upgrading and Phytoremediation of Wastewater**, 69
Sirasit Srinuanpan, Benjamas Cheirsilp and
Mohd Asyraf Kassim

6 **Efficient Harvesting of Microalgal biomass and Direct Conversion of Microalgal Lipids into Biodiesel**, 83
Benjamas Cheirsilp, Sirasit Srinuanpan and
Yohanis Irenius Mandik

7 **Membrane Technology for Microalgae Harvesting**, 97
Jia Jia Leam, Muhammad Roil Bilad,
Yusuf Wibisono, Mohd Dzul Hakim Wirzal and
Iftikhar Ahmed

8 **Processing of Microalgae to Biofuels**, 111
Emmanuel Manirafasha, Kailin Jiao,
Xianhai Zeng, Yuanchao Xu, Xing Tang,
Yong Sun, Lu Lin, Theophile Murwanashyaka,
Theoneste Ndikubwimana, Keju Jing and
Yinghua Lu

9 **Microalgal Cell Disruption and Lipid Extraction Techniques for Potential Biofuel Production**, 129
Ahasanul Karim, M. Amirul Islam,
Zaied Bin Khalid, Che Ku Mohammad Faizal,
Md. Maksudur Rahman Khan and
Abu Yousuf

10 **Conversion of Microalgae Biomass to Biofuels**, 149
Min-Yee Choo, Lee Eng Oi, Tau Chuan Ling,
Eng-Poh Ng, Hwei Voon Lee and
Joon Ching Juan

11 **Microalgal Biorefinery**, 163
Eleni Koutra, Panagiota Tsafrakidou,
Myrsini Sakarika and Michael Kornaros

12 **Microalgal Biorefineries for Industrial Products**, 187
Sabeela Beevi Ummalyma, Dinabandhu Sahoo
and Ashok Pandey

13 **Biorefinery of Microalgae for Nonfuel Products**, 197
Ambreen Aslam, Tahir Fazal, Qamar uz Zaman,
Ali Shan, Fahad Rehman, Javed Iqbal,
Naim Rashid and Muhammad Saif Ur Rehman

14 **Recent Trends in Strain Improvement for Production of Biofuels From Microalgae**, *211*
S.V. Vamsi Bharadwaj, Shristi Ram, Imran Pancha and Sandhya Mishra

15 **Microalgae to Biogas: Microbiological Communities Involved**, *227*
Olivia Córdova and Rolando Chamy

16 **Microalgae-Based Biofuel Production Using Low-Cost Nanobiocatalysts**, *251*
S. Vijayalakshmi, M. Anand and J. Ranjitha

17 **Biosynthesis of Nanomaterials Using Algae**, *265*
Ashiqur Rahman, Shishir Kumar and Tabish Nawaz

18 **Life Cycle Assessment and Techno-Economic Analysis of Algal Biofuel Production**, *281*
Dongyan Mu, Chunhua Xin and Wenguang Zhou

19 **Environmental Resilience by Microalgae**, *293*
M.L. Serejo, M. Franco Morgado, D. García, A. González-Sánchez, H.O. Méndez-Acosta and A. Toledo-Cervantes

20 **Microalgae-based Remediation of Wastewaters**, *317*
Myrsini Sakarika, Eleni Koutra, Panagiota Tsafrakidou, Antonia Terpou and Michael Kornaros

21 **Resource Recovery From Waste Streams Using Microalgae: Opportunities and Threats**, *337*
Naim Rashid, Thinesh Selvaratnam and Won-Kun Park

INDEX, *353*

CHAPTER 1

Fundamentals of Microalgae Cultivation

ABU YOUSUF

INTRODUCTION

Microalgae or microphytes are microscopic eukaryotic organisms, usually found in freshwater and marine systems, use solar energy to produce ATP, which is converted to lipid, carbohydrate, and proteins by its metabolic function. Microalgae offer great promise in contributing to renewable bioenergy, as they have high lipid content, rapid growth rate, and aquatic growth environment using solar energy and also avoid the food versus fuel debate. Although microalgae have a wide range of applications such as food supplements, lipids, enzymes, biomass, polymers, toxins, pigments, tertiary wastewater treatment, and "green energy," in the past few decades, it has been considered as highly potential source of biofuels.

Algae are having both plant-like and animal-like characteristics. Plant-like algae are typically found in aquatic environments that contain chloroplasts and are capable of photosynthesis. Unlike higher plants, algae lack vascular tissue and do not possess roots, stems, leaves, or flowers. Animal-like algae are having of flagella and centrioles and are capable of feeding on organic material in their metabolism. The size of algae varies from single cell to giant multicellular species, and they can survive or grow in multivariate environments such as in fresh water, salt water, or wet soil or on moist rocks. They can be reproduced by sexually or asexually or by a combination of both processes through alternation of generations.

Algae have been used in the past as a way to recycle some of the nutrients from wastewater sources and also as a step in industrial wastewater treatment. As long as the world's population continues to grow, the source of wastewater only gets greater and greater. Utilizing this harmful product as a nutrient source, microalgae are capable of producing lipid that is considered as a potential source of biodiesel. However, three aspects of microalgae production that will strongly influence the future sustainability of algal biofuel production are the energy and carbon balance, environmental impacts, and production cost.

To consider microalgae as a viable feedstock, overall energy and carbon balance must be favorable.

Comparing biomass production system in terms of net energy ratio (NER), a positive energy balance should be required. When NER is defined as the sum of the energy used for cultivation, harvesting, and drying, divided by the energy content of the dry biomass, then, if it is less than unity, the process produces more energy than it consumes. In realistic case, most systems have the NER value > 1. In the systems that have NER<1, if drying, dewatering processes, and lipid extraction are considered, the NER changes from 0.05−0.1 to 0.5−0.75 [1].

Recovery and preparation of microalgal biomass for transesterification reaction requires more than 20% −30% of the total cost of biofuel preparation. Floatation, centrifugation, coagulation-flocculation, and filtration are some of the mostly used ways for separation; however, no single best method for harvesting is not yet involved. Most methods used to extract the oil from algal biomass rely on a dry biomass product. Lyophilization, oven drying, or forced air drying also causes an additional cost. For these reasons, commercial-grade algal oil production is still not cost-effective. However, microalgae are given priority over other energy crops because of the following:

- They can be cultivated in nonarable land.
- They can be grown on ponds and photobioreactors.
- They can tolerate wide range of pH, salinity, and temperature.
- They are continually cultivable, not seasonally harvested.
- They can mitigate CO_2 from industrial and atmospheric sources.
- Complete usage of biomass (proteins, lipids, and carbohydrates).
- They are not finite resource, i.e., renewable and sustainable.

Therefore, the aim of the chapter is to address the basic issues of microalgae cultivation, which should be considered to make them sustainable for microalgae-based industries.

Microalgae Cultivation for Biofuels Production. https://doi.org/10.1016/B978-0-12-817536-1.00001-1

FUNDAMENTAL CHARACTERISTICS OF ALGAE

Algae are broadly classified into two categories: macroalgae and microalgae. Up to June 2012, from 30,000 to more than 1 million species have been estimated, of which 44,000 names have probably been published, and 33,248 names have been processed [2].

Biology of Microalgae

Microalgae is a general term for the algae that form a multicellular thallium at least in one stage of the life cycle. It can also show differentiation between tissues and reproductive system than unicellular microalgae. The largest multicellular algae are called seaweed, which can be well over 25 m in length. The oil content of macroalgae usually ranges between 10% and 30%.

Microalgae are simple microscopic heterotrophic or autotrophic photosynthetic organisms, also called phytoplanktons by biologists ranging between 1 and 50 μm in diameters. These are usually found in pond/damp places or aquatic environment. Most species contain chlorophyll, which uses photonic energy (light), carbon dioxide (CO_2) and water to synthesize carbohydrates (energy storage) and make biomass (algae growth).

In a batch culture system, algal growth experiences five different phases. In this system, inoculum with a fixed volume of nutrient media is charged into the system. As the microalgae replicate, the nutrients depleted and final product obtained. The phases are as follows. (1) Lag phase: Initial period of slow growth where microalgae take time for adaptation into the new environment. The length of the lag phase depends on the size of the inoculum as well as shock of the environment. (2) Exponential: Rapid growth and often cell division occurs when there is an appreciable amount of cell and microalgae grow very rapidly. (3) Declining relative growth: This phase occurs when a growth requirement for cell division is limiting. (4) Stationary: Cell division slows due to the lack of resources necessary for growth. (5) Death/lysis: Cells begin to die due to lack of nutrients [3,4] (Fig. 1.1).

Microalgae are easy to cultivate and can tolerate broad range of pH, salinity, and temperature. They commonly double every 24 h. During the peak growth phase, some microalgae can double every 3.5 h. Oil contents of microalgae are usually between 20% and 50% (dry weight), while some strains can reach as high as 80%. It has the ability to survive much more extreme condition, which triggers this to generate higher lipid content [5].

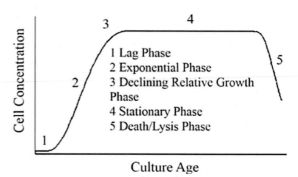

FIG. 1.1 Schematic growth curve of a microalgal batch culture system.

Lipid Content of Microalgae

Although several kinds of microorganisms may accumulate oils, such as microalgae, bacillus, fungi, and yeast, not all of them are suitable for biodiesel production (Table 1.1). Microalgae offer high lipid contents, but they need a larger acreage to be cultured and longer fermentation periods than bacteria. Although bacteria accumulate lower lipid than microalgae, they offer higher growth rates (only 12−24 h is needed for the maximum biomass concentration to be reached) and easy culture methods. Besides this, the most efficient oleaginous yeasts *Cryptococcus curvatus* can accumulate storage lipids up to >60% on a dry weight basis. In addition, when the *C. curvatus* grow under N-limiting conditions, these lipids usually consist of single oil cell (SOC) 90% w/w triacylglycerol with a percentage of saturated fatty acids (% SFAs) of about 44%, which is similar to many plant seed oils [6].

ALGAE MASS CULTIVATION SYSTEM

At present, approximately 5000 tons of dry algae are produced per year worldwide. It is considered as green gold for future [4]. Several cultivation techniques are currently employed for the large-scale production of microalgae biomass; microalgae can be normally cultivated in the open-culture systems (lakes or ponds) due to cheap and easy process and also in the closed-culture systems called photobioreactors (PBRs) due to higher yield of biomass and better control opportunity. Recently, many researchers try to combine these two to obtain the best case [5].

Photobioreactor

A PBR is a closed (or mostly closed) vessel for phototrophic production where energy is supplied via electric lights [8]. PBRs can be located indoors or outdoors depending upon the light collection and distribution

TABLE 1.1
Lipid Content of Different Classes of Microorganisms [7].

Microorganisms	Lipid Content, % dry wt	Microorganisms	Lipid Content, % dry wt
Microalgae		Yeast	
Botryococcus braunii	25–75	*Candida curvata*	58
Crypthecodinium cohnii	16–37	*Cryptococcus albidus*	65
Nitzschia sp.	45–47	*Lipomyces starkeyi*	64
Schizochytrium sp.	50–77	*Rhodotorula glutinis*	72
Bacterium		Fungi	
Arthrobacter sp.	>40	*Aspergillus oryzae*	57
Acinetobacter calcoaceticus	27–38	*Mortierella isabellina*	86
Rhodococcus opacus	24–25	*Humicola lanuginosa*	75
Bacillus alcalophilus	18–24	*Mortierella vinacea*	66

systems and their commercial feasibility. In PBRs, the culture medium is enclosed in a transparent array of tubes or plates, and the microalgal broth is circulated from a central reservoir [9]. PBR can be different kind (Table 1.2), like polyethylene bags, glass fibers cylinder, flat modular photobioreactor, tubular inclined, segmented glass plate, and annular photobioreactor. The main objective of any PBR is the reduction of biomass production costs. To achieve the goal, numerous studies have been done on catalysts improvement, shaping of the PBR, controlling environmental parameters during cultivation, and aseptic designs. The controlling of operational parameters such as pH, temperature, and gas diffusion is also a vital issue in PBR [10].

Open Pond

Open pond cultivation system typically consists of simple water tank or bigger earthen-bank ponds where nutrients are added from outsource. Natural light plays a role in photosynthesis, and CO_2 comes from atmosphere. The pond is usually designed in a raceway or track configuration, in which a paddlewheel provides circulation and mixing of the algal cells and nutrients. The raceways are typically made from poured concrete, or they are simply dug into the earth and lined with a plastic liner to prevent the ground from soaking up the liquid. Baffles in the channel guide the flow around bends that also minimize space and loss. Medium is added in front of the paddlewheel, and algal broth is harvested behind the paddlewheel, after it has circulated through the loop. A comparison between open pond and closed-system PBR is summarized in Table 1.3.

Heterotrophic and Autotrophic Microalgal Culture System

Algae that require light source as energy to run photosynthesis are called autotrophic. Some species of algae are heterotrophic, meaning that they use organic carbon substrates such as glucose or acetate for energy and do not require sunlight to grow. There are some kinds of algae that can operate in both an autotrophic and heterotrophic manner depending on the available resources. These algae species are referred to as mixotrophic [13].

Heterotrophic and mixotrophic algae are very interesting to some researchers, as they grow 24 h a day compared with the autotrophic algae that grow 12 h a day. Heterotrophic algae are independent of light source, which makes their growth condition energy intensive and their surface to volume ratio lower than PBRs and open pond.

This makes growing heterotrophic algae easier to scale up. The growth systems are relatively simple and cheap to set up and can provide high yields of algae per volume. However, the systems require a source of organic carbon, unlike autotrophic systems. Some researchers suggest that heterotrophic growth systems may be better for producing high biomass and high lipid contents.

ALGAE GROWTH FACTOR

Major key components for algal growth are a growth medium with proper nutrients, a light source for photosynthesis, and CO_2 or air flow. All of these growth factors must be specified for successful microalgae cultivation for a specific purpose, which can vary from

TABLE 1.2
Reactors and Their Specification [11,12].

Microorganism	Reactor type	Advantages	Limitations
Microalgae	Open pond photobioreactors	- Relatively cheap - Easy to clean - Utilizes nonagricultural land - Low energy inputs - Easy maintenance - Good for mass cultivation	- Poor biomass productivity - Large area of land required - Poor mixing, light and CO_2 utilization - Cultures are easily contaminated - Difficulty in growing algal cultures for long periods
	Tubular photobioreactors	- Large illumination surface area - Suitable for outdoor cultures - Relatively cheap - Good biomass productivities	- Some degree of wall growth - Fouling - Requires large land space - Gradients of pH, dissolved oxygen, and CO_2 along the tubes
	Flat plate-airlift reactors	- High biomass productivities - Easy to sterilize - Low oxygen buildup - Readily tempered - Good light path - Relatively cheap - Easy to clean up - good for immobilization of algae - Large illumination surface area - Suitable for outdoor cultures	- Require many compartments and support materials - Difficult temperature control - Small degree of hydrodynamic stress - Some degree of wall growth
Closed photobioreactors	Column photobioreactor	- Compact - High mass transfer - Low energy consumption - Good mixing with low shear stress - Easy to sterilize - High potentials for scalability - Readily tempered - Good for immobilization of algae - Reduced photoinhibition and photooxidation	- Small illumination area - Expensive compared with open ponds - Shear stress - Sophisticated construction - Decrease of illumination surface area upon scale-up

species to species [14]. These factors can be divided into three categories (Table 1.4):

Environmental Factors

pH

pH measures the level of acidity or alkalinity that a body of water has. Most algae have pH optima for growth generally 6–8 and photosynthesis ability in the neutral to alkaline pH range. Variation of pH affects the growth in a number of ways. pH increases during daytime caused by photosynthetic CO_2 assimilation by the algae followed by decrease in pH at night due to the respiratory process of the community [8]. There is a relationship between CO_2 concentration and pH of microalgal culture medium. It is related to the chemical equilibrium among chemical species such as CO_2, H_2CO_3, HCO_3^-, and CO_3^{2-}. The equilibrium of these species is pH dependent when CO_2 is predominant at pH below 7.0 and CO_3^{2-} predominant above pH 10. Microalgae have been shown to cause a rise in pH to

TABLE 1.3
Comparison Between Open and Closed Algal Cultivation System.

Parameters	Open Pond (Raceway Pond)	Photobioreactor (Closed system)
Contamination risk	Extremely high	Low
Space required	High	Low
Water losses	Extremely high	Almost none
CO_2 losses	High	Almost none
Biomass quality	Not susceptible	Susceptible
Variability as to cultivated species	Not given, cultivation possibilities is limited to a few algal variety	Nearly all microalgae varieties
Flexibility of production	Change of production between the possible varieties nearly impossible	Change of production without any problem
Reproducibility of production parameters	Dependent on exterior condition	Possible within certain tolerances
Process control	Not given	Given
Standardization	Not possible	Possible
Weather dependence	Absolute, production impossible during rain	Insignificant, because closed configurations allow production also during bad weather
Biomass concentration during production	Low. Approximately 0.1−0.2 gm/L	High. Approximately 2−8 gm/L
Efficiency of treatment process	Low. Time-consuming, lower mass per volume	Comparatively higher mass per volume

10−11 because of CO_2 uptake photosynthetically to convert its biomass [15]. Increasing CO_2 can lead to higher biomass accumulation but leads to lower the pH value, which has adverse effect on microalgal physiology. However, when pH increases too high, photosynthesis can be limited due to scarcity of CO_2

Nutrients

To grow algae, macronutrients should contain nitrogen and phosphorus mainly (silicon is also required for salt-water algae). In addition, trace metals, such as Fe, Mg, Mn, B, Mo, K, Co, and Zn, are also needed. CO_2 fixation is also necessary to build up a balanced medium for optimum growth. Microalgae biomass usually consists of around 40%−50% carbon, 4%−8% nitrogen, and 0.1% phosphate by dry weight. In nature, some of these can be found easily from mineral source, and some can be found by bacterial metabolism. Nitrogen is abundant in nature because it fixes by bacteria continuously, while phosphorus is limited, which is effectively bound as orthophosphate in sediment. Nutrient levels, especially nitrogen and phosphorus combinedly, are very important in the production of lipids. Many articles

have shown that a nitrogen-deficient growth medium triggers the algae to produce higher levels of lipids. Under phosphorus limitation, Xin et al. [16] reported lipid content of 53% for *Scenedesmus* sp. With the same species, nitrogen limitation only induced a lipid content of 30%. Nutrient deficiencies and excess nutrients, both, can cause physiological and morphological changes in microalgae, since they can inhibit some of the vital

TABLE 1.4
Growth Factors of Microalgae.

Categories of Factors	Factors/parameters
Environmental factors	• pH • Nutrients • Temperature
Processing parameters	• Mixing • Light intensity
Biotic factors	• Invasive species and predators

metabolic pathways. Recently, researchers are focused upon two-phase growth system, where in the first phase, algae are grown in nutrient-abundant medium and then transferred to a nutrient-deficient medium where lipid accumulation is boosted up [17].

Temperature

Most algae of interest for lipid production have a temperature tolerance between 15 and 40°C. Researchers [15] have shown that many of the oil-producing algae species grow best between 25 and 30°C. The optimal growth temperature varies by species and the desired algae response. But controlling the temperature at outdoor condition is tough and expensive. Not only the extreme temperature but also the evaporation of media, overheating and cooling at outdoor condition, and lipid composition are also major issues for algal growth. Chaisutyakorn et al. [18] found that higher temperature is responsible for saturated lipid accumulation and lower temperature for unsaturated lipid accumulation. Contrarily, many researchers [19,20] found there is no effect of temperature upon lipid accumulation. So, there exist contradictory views about this issue.

Processing Parameters
Mixing

Mixing is important to prevent sedimentation of algae and to move the algae between the light and dark regions of the pond/reactor. Without any forced mixing, algae at the surface absorb all the available light and can become photoinhibited, while algae deeper in the media are light deprived. It also helps to maintain a homogeneous cell concentration in medium. Mixing can be provided in several ways such as open ponds use a mechanical stirrer (a paddlewheel in raceways) and bubbling in gas (air, CO_2) to provide mixing. PBRs use pumps and bubbling in gas for mixing [8]. It is also noted that many microalgae species cannot tolerate vigorous mixing.

Light intensity

When algae are cultivated photosynthetically, the efficiency of photosynthesis is a crucial determinant in their productivity since it affects the growth rate, biomass production, and lipid accumulation. Effect of light intensity depends on depth of the culture medium and density of the algal biomass. If the depth and cell concentration of the culture is higher, light intensity must be increased to penetrate through the medium. On the other hand, direct sunlight or high-density artificial light may act as photoinhibitor. Overheating caused by both natural and artificial illumination is also unexpected and should be avoided. It is suggested that light intensity of 1000 lux is suitable for the culture in Erlenmeyer flasks; 5000−10,000 is required for larger volumes. A light/dark system is required for the efficient photosynthesis of microalgae because light is needed for photochemical phase to produce ATP and NADPH and dark for biochemical phase to synthesize essential biomolecules for microbial growth.

Biotic Factors
Predators

Invasive species and predators can be any kind of living organism that is totally unexpected in the microalgae culture area because they inhibit microalgae growth, pollute the culture medium, and deficit the nutrient. Predators may be fungus, bacteria, insect, and even unwanted microalgae species. To avoid any potential issues from invasive and predator species, industrial algae growth is mostly limited to extremophile algae species, which can grow in extreme environments in which competing species are unable to survive. Open ponds are susceptible to invasion by low oil−producing algae strains, while PBRs prevent this by keeping the algae contained from the outside environment.

TRADITIONAL SOURCES OF BIODIESEL AND THEIR LIMITATIONS

To deal with deteriorated situation of the whole world energy supply, energy environment, and energy security, renewable biofuels are receiving considerable attention as substitute [21]. One of the most prominent renewable energy resources is biodiesel, which is produced from renewable biomass by transesterification of triacylglycerols, yielding monoalkyl esters of long-chain fatty acids with short-chain alcohols, for example, fatty acid methyl esters and fatty acid ethyl esters. Traditionally, biodiesel is obtained from vegetable oils such as soybean oil [22,23], jatropha oil [24], rapeseed oil, palm oil, sunflower oil, corn oil, peanut oil, canola oil, and cottonseed oil [25]. Apart from vegetable oils, biodiesel can also be produced from other sources such as animal fat (beef tallow, lard), waste cooking oil, greases (trap grease, float grease), and algae. In South East Asia, Europe, the United States, and China, palm oil, rapeseed oil, transgenic soybeans, and wasting oil, respectively, were used to produce biodiesel. However, all these plant oil materials require energy and acreage for sufficient production of oilseed crops. Likewise, animal fat oils need to feed these animals. In spite of the favorable impacts that its commercialization could provide, the economic aspect of biodiesel production has been restricted by the cost of oil raw materials. If plant oil was used for biodiesel production, the cost of source has accounted to 70%—85% of the whole production cost [26]. Therefore, taking into account of these inhibition factors, exploring ways to reduce the high cost of biodiesel is of much interest in recent research, especially for those methods concentrating on lowering the cost of oil raw material. Microorganisms (microalgae, yeast) have often been considered for the production of oils and fats as an alternative to agricultural and animal sources. Synthesis of microbial lipids or single cell oil have some unique advantages over vegetable oils, such as they can be cultivated on degraded and nonagricultural lands that avoid use of high-value lands and crop-producing areas and can utilize salt and wastewater streams, thereby greatly reducing freshwater use [27].

Apart from these conditions, a comparison between first-, second-, and third-generation biodiesels is presented in Table 1.5.

MICROALGAE AS AN ALTERNATIVE

The composition of the biomass is useful for characterizing the best use of microalgae species. The biomass of microalgae contains a number of compounds such as proteins and lipids, which make up the organelles. Algal biomass contains three main components: carbohydrates, proteins, and lipids/natural oil. For example, with the knowledge that biodiesel is made from oils, a microalga with a very high protein content and low lipid content would not be useful as a biofuel feedstock (Table 1.6). It is found that the bulk of the natural oil made by microalgae is in the form of triacylglycerides

---

TABLE 1.5
Review of the Specification of the First-, Second-, and Third-Generation Biofuels.

Fuel Generation	Feedstock	Limitations	Advantages
First generation Produced directly from food crops	Grains, sugar cane, wheat and vegetable oils, animal fats [11,28]	- Effects of the environment and climate change - Changes in ecological balance and biodiversity - Suppressed the food production - Releasing more carbon in their production than their feedstock capture in their growth [28,29]	- Can offer some CO_2 benefits - Can help to improve domestic energy security. [28]
Second generation Produced from nonfood crops	Wood, organic waste, food crop waste and specific biomass crops, oil and sugar crops such as *Jatropha*, cassava, or *Miscanthus* [11];	- Requires costly technologies - Involves pretreatment with special enzymes [11] - Burning the crop residues means decreasing of future organic matter in agricultural soils [29]	- Do not compete with food production - Significantly reduce CO_2 production - More comparable with standard petrol and diesel and would be most cost-effective route to renewable, low carbon energy for road transport. [11,28].
Third generation Produced from microorganisms	Algae, yeast, seaweeds, microbes. [11].	- Require high-oil content species, which is usually associated with lower yields and is more prone to contamination - Dewatering or drying is energy intensive - Easily being contaminated	- No arable land use - Not depend on climate or season - Able to produce 15—300 times more oil than traditional crops on an area basis - Microalgae have a very short harvesting cycle (≈ 1—10 days depending on the process) [11]

TABLE 1.6
Oil Content and Lipid Productivity of AAlgae Species of Interest for Making Biodiesel [30,31].

Microalgae	Oil Content (% dry wt)	Lipid Productivity (mg/L/d)
Botryococcus braunii	25–75	–
Chlamydomonas sp.	16.6–25.25*	–
Chlorella emersonii	25–63	10.3–50
Chlorella protothecoides	14.6–57.8	1214
Chlorella sp.	5–58	11.2–40
Crypthecodinium cohnii	20–51.1	–
Cylindrotheca sp.	16–37	–
Dunaliella primolecta	23.1	–
Dunaliella salina	6–25	116
Isochrysis sp.	7.1–33	37.8
Monallanthus salina	20–22	–
Nannochloris sp.	20–56	60.9–76.5
Nannochloropsis sp.	12–68	37.6–90
Neochloris oleoabundans	29–65	90–134
Nitzschia sp.	16–47	–
Phaeodactylum tricornutum	18–57	44.8
Scenedesmus sp.	11–55	–
Schizochytrium sp.	50–77	–
Tetraselmis sueica	8.5–23	–

(TAGs), which is the right kind of oil for producing biodiesel. The fatty acids attached to the TAG within the algal cells can be both short- and long-chain hydrocarbons. The shorter chain length acids are ideal for the creation of biodiesel, and some of the longer ones can have other beneficial uses. Some microalgae species with high lipid content are listed in Table 1.6.

CONCLUSION AND RECOMMENDATION
Microalgae provide us oxygen we need to breathe by photosynthesis. For this reason, we owe them our lives, but they will also improve our lifestyle. Since they are rich in proteins, carbohydrates, and lipids, which are the source of many beneficial products for mankind, such as human nutrition, feed, agriculture, aquaculture, cosmetics, medicinal, and others, the scientific community agrees that, in the near future, microalgae will competitively generate clean energy and third-generation biofuels, contributing therefore to a sustainable development in both environmental and circular economy. Furthermore, recent research and interest upon bioenergy sector has renewed again; it may also

be found that algal biofuels will play a vital role in future, and attention should be given on the following:
- Optimization of the culture parameters necessary for high yields of biomass and total lipid content
- Significant reductions in complexity of cultivation and costs
- Adoption of dual-benefits large-scale industrial processes, such as wastewater treatment and biofuels production
- Development of the mathematical modeling-based process, before industrial trial and scale-up.

REFERENCES
[1] R. Slade, A. Bauen, Micro-algae cultivation for biofuels: cost, energy balance, environmental impacts and future prospects, Biomass Bioenergy 53 (2013) 29–38.
[2] H.C. Greenwell, L. Laurens, R. Shields, R. Lovitt, K. Flynn, Placing microalgae on the biofuels priority list: a review of the technological challenges, J. R. Soc. Interface 7 (2009) 703–726.
[3] R. Halim, M.K. Danquah, P.A. Webley, Extraction of oil from microalgae for biodiesel production: a review, Biotechnol. Adv. 30 (2012) 709–732.

[4] G. Huang, F. Chen, D. Wei, X. Zhang, G. Chen, Biodiesel production by microalgal biotechnology, Appl. Energy 87 (2010) 38−46.

[5] M. Hung, J. Liu, Microfiltration for separation of green algae from water, Colloids Surf. B Biointerfaces 51 (2006) 157−164.

[6] C. Ratledge, Single cell oils—have they a biotechnological future? Trends Biotechnol. 11 (1993) 278−284.

[7] X. Meng, J. Yang, X. Xu, L. Zhang, Q. Nie, M. Xian, Biodiesel production from oleaginous microorganisms, Renew. Energy 34 (2009) 1−5.

[8] M.B. Johnson, Microalgal biodiesel production through a novel attached culture system and conversion parameters, Virginia Tech (2009).

[9] B.J. Krohn, C.V. McNeff, B. Yan, D. Nowlan, Production of algae-based biodiesel using the continuous catalytic Mcgyan® process, Bioresour. Technol. 102 (2011) 94−100.

[10] O. Perez-Garcia, F.M. Escalante, L.E. de-Bashan, Y. Bashan, Heterotrophic cultures of microalgae: metabolism and potential products, Water Res. 45 (2011) 11−36.

[11] G. Dragone, B.D. Fernandes, A.A. Vicente, J.A. Teixeira, Third Generation Biofuels from Microalgae, 2010.

[12] A. Singh, P.S. Nigam, J.D. Murphy, Mechanism and challenges in commercialisation of algal biofuels, Bioresour. Technol. 102 (2011) 26−34.

[13] R. Maceiras, M. Rodrı, A. Cancela, S. Urréjola, A. Sánchez, Macroalgae: raw material for biodiesel production, Appl. Energy 88 (2011) 3318−3323.

[14] E.M. Grima, E.-H. Belarbi, F.A. Fernández, A.R. Medina, Y. Chisti, Recovery of microalgal biomass and metabolites: process options and economics, Biotechnol. Adv. 20 (2003) 491−515.

[15] N. Pragya, K.K. Pandey, P. Sahoo, A review on harvesting, oil extraction and biofuels production technologies from microalgae, Renew. Sustain. Energy Rev. 24 (2013) 159−171.

[16] L. Xin, H. Hong-Ying, G. Ke, S. Ying-Xue, Effects of different nitrogen and phosphorus concentrations on the growth, nutrient uptake, and lipid accumulation of a freshwater microalga *Scenedesmus* sp, Bioresour. Technol. 101 (2010) 5494−5500.

[17] P. Mercer, R.E. Armenta, Developments in oil extraction from microalgae, Eur. J. Lipid Sci. Technol. 113 (2011) 539−547.

[18] P. Chaisutyakorn, J. Praiboon, C. Kaewsuralikhit, The effect of temperature on growth and lipid and fatty acid composition on marine microalgae used for biodiesel production, J. Appl. Phycol. 30 (2018) 37−45.

[19] Q. Zhang, J.-j. Zhan, Y. Hong, The effects of temperature on the growth, lipid accumulation and nutrient removal characteristics of *Chlorella* sp. HQ, Desalin Water Treat 57 (2016) 10403−10408.

[20] M.Y. Roleda, S.P. Slocombe, R.J. Leakey, J.G. Day, E.M. Bell, M.S. Stanley, Effects of temperature and nutrient regimes on biomass and lipid production by six oleaginous microalgae in batch culture employing a two-phase cultivation strategy, Bioresour. Technol. 129 (2013) 439−449.

[21] J. Hill, E. Nelson, D. Tilman, S. Polasky, D. Tiffany, Environmental, economic, and energetic costs and benefits of biodiesel and ethanol biofuels, Proc. Natl. Acad. Sci. USA 103 (2006) 11206−11210.

[22] T. Samukawa, M. Kaieda, T. Matsumoto, K. Ban, A. Kondo, Y. Shimada, H. Noda, H. Fukuda, Pretreatment of immobilized Candida Antarctica lipase for biodiesel fuel production from plant oil, J. Biosci. Bioeng. 90 (2000) 180−183.

[23] W. Du, Y. Xu, D. Liu, J. Zeng, Comparative study on lipase-catalyzed transformation of soybean oil for biodiesel production with different acyl acceptors, J. Mol. Catal. B Enzym. 30 (2004) 125−129.

[24] S.M.A. Sujan, M. Jamal, M.N. Haque, M.Y. Miah, Performance of different catalysts on biodiesel production from Jatropha curcas oil through transesterification, Bangladesh J. Sci. Ind. Res. 45 (2010) 85−90.

[25] S. Al-Zuhair, Production of biodiesel: possibilities and challenges, Biofuel Bioprod Bioref 1 (2007) 57−66.

[26] X. Miao, Q. Wu, Biodiesel production from heterotrophic microalgal oil, Bioresour. Technol. 97 (2006) 841−846.

[27] P.M. Schenk, S.R. Thomas-Hall, E. Stephens, U.C. Marx, J.H. Mussgnug, C. Posten, O. Kruse, B. Hankamer, Second generation biofuels: high-efficiency microalgae for biodiesel production, Bioenergy Res. 1 (2008) 20−43.

[28] S. Naik, V.V. Goud, P.K. Rout, A.K. Dalai, Production of first and second generation biofuels: a comprehensive review, Renew. Sustain. Energy Rev. 14 (2010) 578−597.

[29] S. Tanner, Biofuels of the Third Generation, 2009.

[30] Y. Chisti, Biodiesel from microalgae, Biotechnol. Adv. 25 (2007) 294−306.

[31] T.M. Mata, A.A. Martins, N.S. Caetano, Microalgae for biodiesel production and other applications: a review, Renew. Sustain. Energy Rev. 14 (2010) 217−232.

CHAPTER 2

Microalgae Cultivation Systems

GAETANO ZUCCARO • ABU YOUSUF • ANTONINO POLLIO •
JEAN-PHILIPPE STEYER

INTRODUCTION

Microalgae are prokaryotic and eukaryotic cellular organisms with a size from a few μm to *ca.* 200 μm, able to produce around 50% of all O_2 on Earth via photosynthesis [1], a wide range of bioproducts including polysaccharides, lipids, pigments, proteins, vitamins, bioactive compounds, and antioxidants. It has been estimated that the number of available species is 2-8 $\times$ 10^5, but only 4-5 $\times$ 10^4 are described in literature [2] and classified as *Chlorophyceae* (green algae), *Cyanophyceae* (blue-green algae), *Chrysophyceae* (golden algae), *Rhodophyceae* (red algae), *Phaeophyceae* (brown algae), and *Bacillariophyceae* (diatoms).

The potential of microalgae has attracted increasing interest in research and industrial fields because of their extensive applications such as renewable energies, pharmaceuticals, and nutraceuticals. Indeed, they are different in some aspects because it is possible to cultivate them in the vicinity of already existing industrial sites as they do not depend on the availability of fertile lands like in the case of bioprocesses that use feedstocks, like lignocellulosic biomasses, and need the transport to the biorefinery plants. This factor represents a way to facilitate the pretreatment processes and to reduce the overall production costs and the environmental concerns.

Microalgae cultivation requires specific environmental conditions including temperature ranges, light intensities, mixing conditions, nutrient composition, and gas exchange. They can be cultured using different metabolic pathways (photoautotrophic, heterotrophic, and mixotrophic) and by using different cultivations systems, commonly classified as open and closed systems. This chapter will provide an overview of the fundamentals to determine the future role of microalgae in biorefinery applications.

GROWTH FACTOR AFFECTING MICROALGAL CULTURES

The growth of microalgae is affected by not only several factors such as temperature, light, nutrient availability, gas exchange, salinity, cell density, pH but also operational parameters such as fluid and hydrodynamic stress, mixing, culture depth, dilution rate, and harvest frequency [3].

Temperature

Physiological and morphological responses of microalgal growth, including photosynthesis and carbon fixation, are strictly related to the temperature [4]. Each species has its own optimal range of growth temperature for which it is possible to classify the algae as psychrophiles (<15 °C), mesophiles (<50 °C), and thermophiles (>50 °C) (Table 2.1).

The upper temperature for eukaryotic algae is 62°C [23], and no photosynthetic organisms have been reported able to grow beyond 75°C, due to chlorophyll instability.

The importance of the optimal temperature range is also related to the carbon fixation. Indeed, higher temperatures enhance CO_2 absorption and fixation but represent an inhibitor factor for the respiration metabolism and for the photosynthetic proteins as they disturb the energy balance in cells. High temperatures also reduce cell size and microbial biomass growth (due to the affinity of ribulose for CO_2), particularly in outdoor culture systems.

The effect of moderately elevated temperatures has also been found in the activation state of ribulose-1,5-bisphosphate (Rubisco), an enzyme able to act as an oxygenase or as carboxylase, depending on the relative amount of O_2 and CO_2 present in the chloroplast. The CO_2 fixation activity of Rubisco enzyme is affected by the rising temperature up to a certain level and then

Microalgae Cultivation for Biofuels Production. https://doi.org/10.1016/B978-0-12-817536-1.00002-3

TABLE 2.1
Temperature Classification for Microalgal Life.

Strain	Class	Classification	References
Entemoneis kufferatii	Bacillariophyceae	Psychrophile	[5]
Fragilariopsis cylindrus	Bacillariophyceae	Psychrophile	[6]
Chlamydomonas nivalis	Chlorophyceae	Psychrophile	[7]
Chloromonas nivalis	Chlorophyceae	Psychrophile	[8]
Gloeocapsa ralfsiana	Cyanophyceae	Psychrophile	[5]
Nostoc commune	Cyanophyceae	Psychrophile	[9]
Phormidium murrayi	Cyanophyceae	Psychrophile	[10]
Chlamydomonas raudensis	Chlorophyceae	Psychrophile	[5]
Euglena gracilis	Euglenoidea	Mesophile	[11]
Synechocystis sp.	Cyanophyceae	Mesophile	[12]
Chlorella spp.	Trebouxiophyceae	Mesophile	[13]
Scenedesmus spp.	Chlorophyceae	Mesophile	[14]
Chlorogleopsis sp.	Cyanophyceae	Mesophile	[15]
Desmodesmus sp.	Chlorophyceae	Mesophile	[16]
Chlorella sorokiniana	Chlorophyceae	Mesophile	[13]
Arthrospira platensis	Cyanophyceae	Mesophile	[17]
Thermosynecocchus elongatus	Cyanophyceae	Thermophile	[18]
Oscillatoria terebriformis	Cyanophyceae	Thermophile	[19]
Cyanidium caldarium	Rhodophyta	Thermophile	[20]
Galdieria sulphuraria	Rhodophyta	Thermophile	[21]
Synecocchus sp.	Cyanophyceae	Thermophile	[22]

declines [24]. Temperature can also be used as a stress treatment to induce the production of valuable metabolites. Because the temperature has an evident impact on the photosynthesis, it is difficult to evaluate the kinetic equation able to describe this phenomenon.

Bechet et al. [25] studied different approaches, that can be summarized as "uncoupled" and "coupled," where μ (h^{-1}), specific growth rate of photosynthesis is the product of two distinct functions of light intensity (Monod function) and temperature (Arrhenius equation) [Eq. 2.1]:

$$\mu - \mu_{m,0} \, exp\left(-\frac{E_a}{KT}\right) \cdot \frac{I_{av}}{K + I_{av}} \quad (2.1)$$

where μ is the specific growth rate (h^{-1}), $\mu_{m,0}$ is the maximum specific growth rate (h^{-1}), E_a is the activation energy for photosynthesis (J), k is the Boltzmann constant (J K^{-1}), T is the temperature (K), I_{av} is the average

light intensity in the culture broth (μmol m^{-2}s^{-1}), and K is a light constant (μmol m^{-2}s^{-1}).

The potential interdependence of light and temperature is defined according to Eq. (2.2):

$$\mu = 2\mu_m \, (T)(1 + \beta_I) \, \frac{I/I_{opt}(T)}{1 + 2\beta_I \, I/I_{opt} \, (T) + \left(I/I_{opt} \, (T)\right)^2} \quad (2.2)$$

where $\mu_m(T)$ is the maximum specific growth rate (h^{-1}) at the temperature T (°C), β_I is a constant, I is the light intensity (μmol/m^{2}s), and $I_{opt}(T)$ is the optimum light intensity for photosynthesis (μmol m^{-2}s^{-1}) at the temperature T (°C).

Light

There is a direct relationship between microalgae growth, light intensity, and cultivation duration because the variation of the latter ones is able to directly affect photosynthesis and the related biochemical composition as well as biomass yield [26].

There is a relationship between light intensity (I) and rate of photosynthesis (P) as shown in Fig. 2.1 [25], where is possible to observe three different light regimes:

1. $I < I_k$ (the rate of photosynthesis is proportional to low light intensities and the photosynthesis is limited by the rate of photon capture);
2. $I_k < I < I_{inhib}$ (the rate of photosynthesis is usually maximal and independent from light intensity. This regime is light saturated, it means that the rate of photosynthesis is limited by rate of reactions following the photon capture); and
3. $I > I_{inhib}$ (the rate of photosynthesis starts to decrease with the light intensity which, above a certain level can damage light receptors, such as key proteins, in the chloroplasts. This regime is known as photoinhibition).

The optimum level of light intensities for most of microalgae species are about $200-400 \ \mu mol \ m^{-2} \ s^{-1}$ [27], and for some microalgal species, it could also be equal to $100 \ \mu mol \ m^{-2} \ s^{-1}$.

Light penetrates through the cell medium, but it is also attenuated according to Lambert–Beer's law [Eq. 2.3].

$$I(l) = I_0 \exp(-\sigma Xl) \qquad (2.3)$$

where $I(l)$ is the local light intensity at a distance l from the external surface; I_0, the incident light intensity; σ, the extinction coefficient and X, the cell concentration.

Light reactions provide the conversion of light energy in chemical energy in the form of short-term energy storing molecules. This is possible because chlorophyll molecules absorb photons, which induce the excitation of a pair of electrons, leading indirectly to ATP and NADPH production. However, when the light intensity surpasses a limit, it can represent the cause of the photoinhibition and photooxidation, which negatively influences cell density growth [28]. Moreover, the phenomenon of photoinhibition is also related to the preexposure to high or low irradiances. Indeed, Beardall and Morris [29] demonstrated a dynamic effect known as "hysteresis," where a lower productivity of algae cells preexposed to high light intensity was observed if compared to productivity of cells preexposed to a dark regime, that can be caused to overestimate productivity, for example, during outdoor cultivation.

On the other hand, optimal light intensity needs to be experimentally evaluated in each case to maximize CO_2 assimilation with a minimum rate of photorespiration and as little photoinhibition as possible [25]. The photoinhibition is associated to self or mutual shading effects, in which cells close to the bottom are shaded from the cells on the surface. One of the solutions identified to solve this problem is, for example, using of LED lights instead of fluorescent tubes that are commonly used [30].

Microalgal photosynthesis is adapted to perform in a spectrum range of solar radiation. However, an increase of solar UV radiation, caused by the depletion of the stratospheric ozone layer may result in cell damage, that

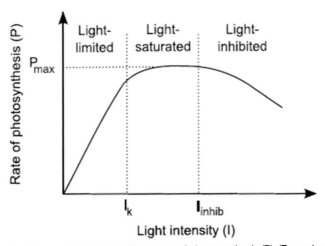

FIG. 2.1 Relationship between light intensity (I) and rate of photosynthesis (P). (Reproduced from Q. Béchet, A Shilton, B Guieysse, Modeling the effects of light and temperature on algae growth: State of the art and critical assessment for productivity prediction during outdoor cultivation, Biotechnology Advances 31 (2013) 1648–1663.)

it can be avoid because of the presence of some pigments able to increase the photosynthetic rate by resonance electron transfer with Chlorophyll A [31] (Fig. 2.2).

Anyway microalgae have physiological strategies to cope with excess of solar radiation including antioxidants such as β-carotene, astaxanthin, α-tocopherol, or mechanisms for scavenging reactive oxygen species, arising when there are insufficient electron sinks available [32].

Nutrients

Microalgal cultures require essential macronutrients, vitamins, and trace elements in adequate quantities and in bioavailable chemical forms. In the literature, the recipes of the cultivation media are frequently considered fixed and attributable to the Redfield C:N:P ratio of 106:16:1. Nevertheless, the experience of cultivating microalgae in various types of media with different nutrient compositions, diverging from the canonical Redfield ratio, suggested that the cultivation media could be flexible and adapted to microalgal metabolic needs, as well as specific environmental conditions [33].

Carbon

Carbon is the main microalgal biomass element, equal to 65% of dry weight, but in some species and culture conditions, it does not exceed 18%. The majority of species contains about 50%, but its increase is strictly related to the limitation of other nutrients, such as nitrogen and phosphorus. It is mainly taken up in its inorganic form through photosynthesis according to the following equilibrium equation [Eq. 2.4] depending on the pH of the solution.

$$CO_{2(aq)} + H_2O \leftrightarrow H_2CO_3{}^* \leftrightarrow HCO_3{}^- + H^+ \leftrightarrow CO_3{}^{2-} + 2H^+$$

$$(2.4)$$

When pH is lower than 6.5, H_2CO_3 is predominant; from 6.5 to 10, the dominant form is $HCO_3{}^-$. At pH higher than 10, more $CO_3{}^{2-}$ becomes predominant.

Inorganic carbon is fixed inside the microalgal cells through Calvin cycle. The assimilation of CO_2 is catalyzed by Ribulose-1,5-bisphosphate carboxylase oxygenase (Rubisco). In CO_2-concentrating mechanisms, the intracellular concentration of CO_2 around Rubisco is high compared with its extracellular concentration, and it is used as substrate for its fixation [34], avoiding the presence of a respiration reaction, that is present when intracellular CO_2 is low and O_2 concentration is high. In this case, Rubisco enzyme will react with O_2 rather than CO_2.

The inorganic carbon uptake is performed passively through membrane diffusion, inserting the cells as free CO_2, or actively through membrane pumps. The

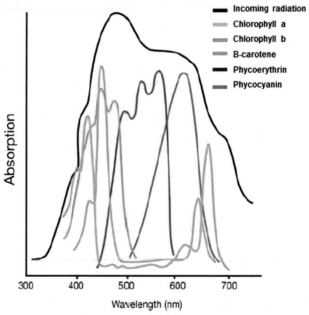

FIG. 2.2 Absorption range of pigments and quantum photosynthesis rate of photosynthesis in microalgae. (Reproduced from P.J. le B. Williams and L.M.L. Laurens, Microalgae as biodiesel & biomass feedstocks: Review & analysis of the biochemistry, energetics & economics, Energy Environ. Sci. 3 (2010) 554–590.)

pH of the cultivation medium affects the uptake of inorganic carbon. In alkaline environments, carbon is metabolized mainly by active transportation than diffusion, excreting H^+ ions, able to react with HCO_3^- to give CO_2. In the same environmental reaction, microalgal cells can regulate their intracellular pH by homeostasis. Calcification is another pathway used to obtain CO_2, if the pH is high. The calcification process consists of $CaCO_3$ precipitation starting from Ca^{2+}, which is used to prevent intracellular toxicity and as a way to enhance nutrient uptake [35].

Photosynthetic activity, in particular the extracellular and intracellular conversion of bicarbonate to carbon dioxide is obtained according to Eq. (2.5):

$$HCO_3^- \rightarrow CO_2 + OH^- \qquad (2.5)$$

The passive uptake of CO_2 does not alkalize the medium because of the absence of hydroxylic group formation during its fixation. When CO_2 is supplied as gas, the solubility and the mass transfer represent a critical parameter. CO_2 solubility in water at 25°C and 1 atm is 1.5 g/L, but it is strictly related to several parameters such as temperature and pressure as shown in Fig. 2.3 [36].

The mass transfer rate from the gaseous to the liquid phase also influences the CO_2 fixation. The specific contact area and the concentration gradient between the two phases affects the transfer. The mass transfer of CO_2 in solution is too low to compensate microalgal uptake. For this reason, the medium is sparged with CO_2-rich gas, which has different dissolutions related to different pH. Indeed, at high pH, the mass transfer of CO_2 is faster than in low pH values, because of the chemical reaction of CO_2 and OH^- faster than the

hydration of CO_2 to H_2CO_3 [35]. The mass transfer rate of CO_2 (M_{CO_2}) is evaluated according to Eq. (2.6) [35]:

$$M_{CO_2} = K_L\,\alpha(C^*_{CO_2} - C_{CO_2}) \qquad (2.6)$$

where, K_L is the liquid phase mass transfer coefficient, α is the contact area, $C^*_{CO_2}$ is CO_2 concentration in the liquid phase at surface equilibrium, or equal to partial pressure in gas phase, and C_{CO_2} is the concentration of CO_2 in the liquid.

The production costs to supply CO_2 to microalgae cultivation is estimated to be about 50% compared with the cost for biomass production. A low-cost option is represented by the biogas, as product of anaerobic digestion, which contains 20%–40% of CO_2, or the CO_2, as product of alcoholic fermentation. An option could be represented by the flue gas derived from coal-fired plants, or cement production plants, or natural gas combustion, which contains 10%–25% of CO_2, but also a concentration of inhibitors such as NOx, SOx, eliminated through purifying processes such as the reaction to form sodium or ammonium bicarbonate and urea. The main drawback is the increase of pH and the increase of ionic strength but also an advantage related to the increase of solubility if compared with CO_2. Na_2HCO_3 solubility is > 90 g/L at 25°C, that make these types of processes attractive, except for the economic impact that is three times if compared with the use of CO_2 [37,38]. Chisti et al. [39] studied $NaHCO_3$ tolerance of some strain such as *Synechocystis* sp. PCC6803, *Cyanothece* sp., *Chlorella sorokiniana*, *Dunaliella salina*, *Dunaliella viridis*, *Dunaliella primolecta*, and *Arthrospira platensis*. The challenge is to screen and find strain able to grow with high pH, alkalinity, and ion strength.

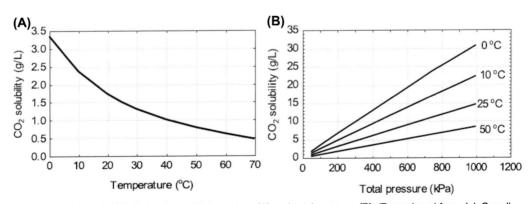

FIG. 2.3 CO_2 solubility in function of temperature **(A)** and total pressure **(B)**. (Reproduced from J.J. Carroll, J.D. Slupsky, A.E. Mather, The solubility of carbon dioxide in water at low pressure. Phys. Chem. Ref. Data 20 (1991) 1201–1209.)

Ideally, the mass transfer rate of CO_2 and the rate of its assimilation should match the microalgal activity and simultaneously the regulation and adjustment of pH. However, the theoretical equilibrium is affected by the partial pressure and the alkalinity [40].

Finally, there are many species that can grow on organic substrates according to a heterotrophic or mixotrophic metabolism. The main ways of organic carbon uptake into the cells are diffusion, active transportation, and phosphorylation [41]. The main organic compounds are monosaccharides, volatile fatty acids, glycerol, and urea. Moreover, in mixotrophic regime, the growth rate is higher than that in autotrophic regime, but an increase of the cost related to the organic carbon source supply has to be considered, even if in some cases, for example, the support of different processes such as anaerobic digestion could represent a help beceause bacteria could convert organic matter in acetate or some other volatile fatty acids suitable for the microalgal growth [42].

Nitrogen

Nitrogen is the second most abundant element in microalgal biomass, with a concentration that is typically around 1%–14% of dry weight [43], and it is an essential biochemical compound present in DNA, RNA, proteins, and pigments (chlorophylls and phycocyanin). Glutamine synthetase enzyme system is the main pathway used to metabolize nitrogen. Glutamate reacts with ammonium to form amino acid glutamine [42]. Nitrogen is supplied in inorganic form as NO_3^-, NO_2^-, NO, NH_4^+ but also in organic form as urea or amino acids.

Nitrate (NO_3^-) is frequently supplied as $NaNO_3$. Microalgal cells can tolerate concentration up to 100 mM of nitrate, but higher concentrations negatively affect microbial growth, probably because of an increase of nitrate reductase activity and a simultaneous increase in nitrite and ammonium concentration in a range that could be toxic for the cells [44]. Nitrite (NO_2^-) is considered an intermediate of the nitrification processes due to bacteria, that is the oxidation of ammonia to nitrate, but it can be also an intracellular intermediate product by the nitrate reductase reaction. It is assimilated through active transportation and diffusion. Yang et al. [45] observed an increase of lag phase in cultures of *Botryococcus braunii* equal to 10 days in the presence of 4 mM in nitrite concentration and a total inhibition of microalgal growth with 8 mM in nitrite concentration. Taziki et al. [46], on the other hand, observed that *Chlorella vulgaris* is able to efficiently remove and utilize both nitrate and nitrite from culture medium even at high concentrations,

producing simultaneously algal biomass for several applications and a promising candidate for wastewater treatment. High CO_2 concentrations positively affect nitrite uptake, as the nitrite reductase activity increases, enhancing the nitrite assimilation [47].

Nitrite oxide (NO) is a small and nonpolar molecule, able to diffuse directly in the cell. It is a free radical that can have detrimental effects on the cells, and its low solubility can be another limiting factor. To increase NO solubility, ferrous-complexed EDTA can be added to the cultivation medium [48]. Dissolved NO is oxidized to nitrite or nitrate in the presence of dissolved oxygen, both taken up by microalgae [49].

However, the preferred nitrogen source for microalgae is ammonium/ammonia due to less energy consumption for its assimilation if compared with the other nitrogen sources [41]. Ammonia is volatile, and it has a very high solubility, which means that it is frequently in liquid solution. Anyway, ammonia reacts in the presence of water forming a system ammonia/ammonium, which is in equilibrium in dependence of pH (Fig. 2.4) according Eq. (2.7):

$$NH_4^+ + OH^- \leftrightarrow NH_3 + H_2O \qquad (2.7)$$

The toxicity of ammonium ions is less than of free ammonia because it is related to pH of cultivation medium. Indeed, when it is supplied as nitrogen source, the pH drops because of H^+ release. Ammonium is assimilated by transportation mechanisms, which can positively control intracellular concentrations; on contrast, since ammonia is metabolized through passive diffusion into cells, the control of intracellular concentration is more difficult. Moreover, free ammonia affects the photosynthetic system and in particular induces photo damage of photosystem II [50]. To avoid the negative effects due to the presence of ammonia, the pH control is one strategy. Another strategy is to feed gradually ammonia to the culture medium. An additional constraint of using ammonia for nitrogen application for microalgae growth is that it can be lost from the cultivation media due to volatilization, especially at higher pH values [51]. Ammonia can react with CO_2 according to Eqs. (2.8) and (2.9) to produce bicarbonate or urea, respectively; which can be used as nitrogen source for microalgal growth.

$$CO_2 + NH_3 + H_2O \leftrightarrow NH_4CO_3 \qquad (2.8)$$

$$CO_2 + 2NH_3 \leftrightarrow H_2NCOONH_4 \leftrightarrow (NH_2)_2CO + H_2O \quad (2.9)$$

Dinitrogen (N_2) can be assimilated as nitrogen source by the reduction of N_2 to NH_4^+ using nitrogenase enzyme complex in microorganisms, mainly

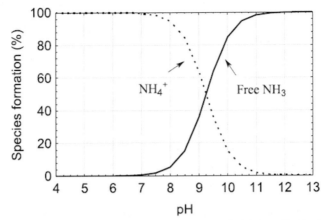

FIG. 2.4 Effect of pH on species formation. (Reproduced from G. Markou, D. Vandamme, K. Muylaert, Microalgal and cyanobacterial cultivation: The supply of nutrients, Water Res. 65 (2014) 186–202.)

cyanobacteria and diatoms, called diazotrophic [52]. However, the nitrogen-fixing process is a very energy-costly process, and it consumes 16 ATPs for the generation of two NH_3 according to Eq. (2.10) [53]:

$$N_2 + 8H^+ + 8F_{red} + 16ATP \rightarrow 2NH_3 + H_2 + 16ADP + 8F_{dox} + 16P_i$$

$$(2.10)$$

where F_{red} and F_{dox} are the reduced and oxidized form of ferredoxin, respectively, and P_i is inorganic phosphate.

Organic nitrogen can also be metabolized from microalgae in form of urea or amino acids transported actively into the cells. Generally, urea, which is the most significant organic nitrogen source, is hydrolyzed to ammonia and carbonic acid (Eq. 2.11), both utilized by microalgae.

$$(NH_2)_2CO + H_2O \rightarrow NH_3 + H_2CO_3 \qquad (2.11)$$

Phosphorus

Phosphorus is another important nutrient for microalgal growth, since its content in biomass is 0.05%−3.3% [43] and it is a limiting nutrient for microalgae.

Organic molecules such as RNA, DNA, membrane phospholipids, and ATP contain phosphorus, which mainly come from potassium, ammonium, and sodium phosphate produced from phosphate rocks. Microalgal cells absorb phosphorus in orthophosphate form by the action of intracellular, extracellular enzymes or directly attached to cell wall. It also can be metabolized as dissolved organic phosphorus by using an extracellular mineralization by phosphate enzymes

[54]. The uptake rate of phosphorus is affected by available light, pH, temperature, ionic strength, and available ions (K^+, Na^+ and Mg^{2+}) that, in their part, promote the precipitation with phosphate [55]. Microalgae can accumulate intracellular phosphorus reserves as polyphosphate granules used as phosphorus source, according to a behavior known as "luxury" [56]. Because of this property to store excess of phosphorus, it is used to remove phosphorus from wastewater.

Other nutrients

The cultivation medium requires the presence of several other micronutrients, such as magnesium (Mg) that is mainly present in aqueous solution as Mg^{2+} and provided as $MgSO_4$ or $MgCl_2$, in the range of 0.35% −0.7%. It is considered an activator for several enzymes, a constituent of photosynthetic apparatus, that is, chlorophylls, and participates in processes such as ATP formation [42]. When the pH is higher than 11, it precipitates as magnesium phosphate or magnesium hydroxide, able to promote flocculation of microalgal biomass.

Sulfur (S) is also a significant micronutrient. Its content in microalgal biomass is included in the range of 0.15%−1.6% [43] and in form of amino acid, sulfur-based lipid, vitamin, and different sulfur-containing secondary metabolite [57]. Sulfur is mainly assimilated as sulfate (SO_4^{2-}), whereas in the form of sulfite, it is toxic [58].

Calcium (Ca) content in microalgal biomass varies in the range 0.1%−1.4%, but it also can reach 8% [43]. It is present in form of Ca^{2+}, frequently added as $CaCl_2$, and able to affect the cell division and the overall morphogenesis. When intracellular Ca^{2+} of cultivation medium reaches high value, it has negative effects on the microalgal growth.

High calcium concentrations in the cultivation medium along with high pH values results to the formation of $CaCO_3$ and various other calcium salts which precipitate, decreasing the alkalinity of the medium and the concentration of some minerals such as iron and phosphorus [59].

Iron (Fe) is involved in fundamental enzymatic processes such as oxygen metabolism, electron transfer, nitrogen assimilation, and chlorophyll synthesis [42] and supplied as chelated complexes to increase its bioavailability.

pH

Microalgal metabolism is strictly related to pH because it regulates the uptake of ions, enzymatic activity, phosphorus availability, inorganic carbon availability, and ammonia toxicity [60]. Two practical pH ranges are 7.9−8.3 (marine water) and 6.0−8.0 (fresh water). Most microalgal species are pH sensitive, and few of them can reach the range tolerated by C. vulgaris [61]. In a closed system, the pH can increase up to pH 10. This rising in pH value can be controlled flushing CO_2 or using inorganic or organic acids [43]. An option is represented by combustion flue gas with high CO_2 concentrations, that can drop the pH to 5, but, at stronger acidity level, the photosynthetic growth is limited. However, the simultaneous degradation of metabolites, as well as the release of several organic acids is a factor able to affect pH values. In contrast, in alkaline solutions, the excess of OH^- reacts with CO_2 to form HCO_3^-, resulting to a higher bicarbonate−carbonate alkalinity and consequently a higher total carbon availability [62]. It seems that CO_2 storage is unavoidable not only to control the pH but also to maintain the overall carbon balance. Finally, some studies mentioned that high pH stress inhibits the cell cycle and triggers the lipid accumulation [63].

METABOLIC PRODUCTION SYSTEMS

In general, microalgae grow fixing inorganic carbons as CO_2 or as sodium bicarbonate, and absorbing light as energy source. This route is called autotrophic. At same time, they can utilize organic substrate as energy and C source, in heterotrophic regime. Finally, they can grow mixotrophically, when microalgae perform photosynthesis as principal energy source but both organic and inorganic (i.e., CO_2) are essential.

The selection of a microalgae cultivation mode is important. Table 2.2 shows the advantage and the drawbacks of the three major modes of microalgae cultivation.

Autotrophic Cultivation

Autotrophic cultivation is considered the most common growth system for microalgae. The light can be supplied in form of sunlight, which represents an additional advantage even if the availability of solar light is a limiting factor in zones where sunlight supply is not constant and intense enough to support algae growth, or in form of artificial light, such as light-emitting diodes (LEDs), which is a source of energy developed because of its low energy consumption and its range lights (e.g., red LED, 624−634 nm; green LED, 515−525 nm; blue LED; 460−465 nm) to enhance the production of specific molecules [64]. Autotrophic cultivation is considered technically and commercial scaling-up typically at outdoor environment, especially to enhance lipid productivity (by using 2% CO_2 in air) and also to recycle industrial CO_2 [64,65]. However, the enhancement of autotrophic microalgal growth requires to overcome the problem related to the self-shading due to the photoacclimation and, simultaneously, increasing the net activity of Rubisco [66]. Flynn and Raven [67] observed such an increase in microalgal growth rate appears implausible without a de facto artificial replacement for Rubisco, which is the single most important and abundant enzyme on Earth.

Heterotrophic Cultivation

This process is a solution to the problem of light availability because several microalgae can grow under dark conditions. In this case, carbon and energy sources are supplied as organic carbon substrate. The main advantages of heterotrophic cultivation are a high growth control and high productivity, performed exploiting two-stage cultivation in a fed-batch or a heterotrophic-photoinduction regime [68]. The use of cheap carbon sources (glucose, acetate, glycerol) commonly used in fermentation processes and derived from low cost feedstocks or wastewater. Moreover, the high attainable cell density allows to reduce the costs associated with dewatering during the harvesting step. A constrained number of microalgae strains can be cultivated under these conditions, such as C. vulgaris, Chlorella protothecoides, Crypthecodinium cohnii, and Schizochytrium limacinum [63,65]. On the other hand, the presence of organic substrates and water are potential microbial contamination of culture media [69], and dark conditions in heterotrophic conditions can inhibit pigmentation and secondary metabolite production.

TABLE 2.2
The Overall Differences Among the Growth Modes of Microalgae Cultivation.

Cultivation Mode	Carbon Source	Energy Supply	Light Availability Requirements	Advantage	Disadvantage
Autotrophic	CO_2	Light	Obligatory	Low cost	Low growth rate
				High production of pigments	Low biomass productivity
Heterotrophic	Organic carbon source	Organic carbon source	No requirements	High growth rate	Carbon sources costs
				High biomass productivity	Problems of contamination by other microoganisms
				Easy scaling-up	
				Potential to remove C (organic), N and P from wastewater	
Mixotrophic	CO_2 and organic carbon source	Light and organic carbon source	No obligatory	High growth rate	High cost
				High biomass density	Problems of contamination
				Prolonged exponential growth phase	
				Reduction of lost biomass from respiration during dark regime	
				Reduction of photo-inhibitory effect	

Mixotrophic Cultivation

In mixotrophic conditions, microalgae can ingest both exogenous organic compounds (cheese whey permeate, sodium acetate, fruit peel, glucose, fructose, glycerol) and CO_2 as carbon source. Light and inorganic carbon are used in the photosynthetic metabolism, while the organic carbon is used during the respiration, where CO_2 released is, in its part, used during photosynthesis. Mixotrophy combines the advantages of autotrophy and heterotrophy, and it overcomes the disadvantages of autotrophy [70]. Another advantage is related to the flexible usage of light, which is not an absolutely limiting factor. On the contrary, it could allow a reduction of photoinhibition phenomena, if compared with autotrophic and heterotrophic cultures and if integrated with dark/light cycle, which allows the growth of autotrophy under their optimum conditions, the biomass

and lipid content of the mixotrophic cultivation will be not merely a sum of autotrophy and heterotrophy [70]. Typical mixotrophic metabolisms can be found in *Spirulina platensis*, *Chlamydomonas reinhardtii*, *Chlorella sorokiniana*, *Scenedesmus obliquus*, and *C. vulgaris* [65]. Different from heterotrophy, the valuable pigments and photosynthetic carotenoids such as β-carotene could be preserved in illuminated conditions under mixotrophic cultivation mode [71], although the mixotrophic energy conversion efficiency is lower because of photosynthetic losses.

CULTURE SYSTEMS

Microalgae can be cultivated in systems mainly classified into open and closed. An open system exploits sunlight, and it is widely exposed to the environment, resulting in an important advantage due to the usage of free natural resources. The closed systems or photobioreactors (PBRs) can be classified in tubular, column, membrane, flat plate. There are advantages and drawbacks for each culture system.

Open Ponds

Open ponds could be further classified in unstirred, raceway, paddlewheel open ponds depending of mixing and cultivation. Generally, they require low initial investments if compared with the closed systems and are generally constructed in concrete or compacted earth, in various shapes, usually able to respect basic criteria as to provide enough sunlight, an optimal hydrodynamic force and a closed loop channel mixed to uniform the cells. They do not require arable land, but a large amount of water, which represents still a drawback.

Fig. 2.5 shows a typical configuration of a raceway open pond, where the surface area (A) and the working volume (V_L) can be estimated according to Eqs. (2.12) and (2.13) [72]:

$$A = \frac{\pi q^2}{4} + pq \qquad (2.12)$$

$$V_L = Ah \qquad (2.13)$$

where p and q are the length and the width of the pond, respectively. Instead, h is the depth of the culture medium. A lower depth increases the surface-to-volume ratio, and this improves light penetration, but in a large pond, up to 1 ha in area, the depth is in the range of 20–30 cm. Ostwald [73] studied the relationship between the concentration of algae, C_x (mg L^{-1}), and the light penetration depth, h (cm), determined according to Eq. (2.14):

$$h = \frac{6000}{C_x} \qquad (2.14)$$

Therefore, a continuously mixed outdoor culture of algae at a depth of 30 cm should achieve an average light-limited algal concentration of about 300 mg L^{-1} dry weight. Higher photon flux densities increase this only slightly because the penetration of light is proportional to the log of its intensity [74]. Another parameter that can increase the efficiency of raceway open pond is the turbulence, calculated on the basis of Re value (Eq. 2.15), without damaging the algal cells, but simultaneously able to keep the cells in suspension preventing

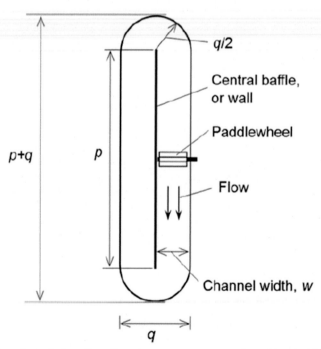

FIG. 2.5 A top view of a typical configuration of a raceway open pond used for algal biomass production. (Reproduced from Y. Chisti, Large-scale production of algal biomass: raceway ponds. in: F. Bux, Y. Chisti (Eds.), Algae biotechnology, Springer, New York, NY, 2016, pp. 21–40.)

thermal stratification and facilitating removal of oxygen generated by photosynthesis.

$$Re = \frac{\rho v d_h}{\mu} \qquad (2.15)$$

where ρ is the density of the culture broth, v is its average flow velocity, d_h is the hydraulic diameter of the flow conduit, and μ is the viscosity of the algal broth. The density and viscosity of algal broth are assumed like a water solution at the operating conditions. The hydraulic diameter is calculated according to Eq. (2.16):

$$d_h = \frac{4wh}{w + 2h} \qquad (2.16)$$

where w is the channel width as shown in Fig. 2.5 and h is the average of the depth of the broth. A Re (Reynolds number) that exceeds 4000 is generally considered right to define a turbulent regime, not in this case where the value of Re expected has to be about 8000. In ponds with semicircular ends, curved baffles or flow deflectors are installed to minimize the formation of death zones [72].

The paddlewheel sited in a depression of the pond is able to generate a flow and assure the mixing. The hydraulic power (P, kW) is calculated according to Eq. (2.17):

$$P = \frac{Qw \cdot d}{102e} \qquad (2.17)$$

where Q is the quantity of water in motion ($m^3\ s^{-1}$), w the specific weight of water ($kg\ m^{-3}$) and e the efficiency of the paddlewheel; 102 is the conversion factor required to convert $m\ kg\ s^{-1}$ to kW [74].

Engineering design including some parameters discussed above is connected with several environmental factors, process, and biological parameters that affect the biomass productivity. Indeed, the success of cultivation is based on right estimations of light input and water temperature. Another important aspect to take in account is the uncontrolled change in ionic composition of the broth due to the evaporation and changes of seasonal conditions that can affect the productivity. Finally, the biomass productivity can be influenced by the evolution of microbial community because the algal species grow in the presence of microorganisms such as fungus, bacteria, virus, rotifers, Cladocerans (e.g., Daphnia), Amebae, Cyclopoid copepods, Ciliates, and Chironomid midges. This is the reason of moving toward extremophile microalgae, which always have less chances of having predators during outdoor mass cultivation.

On the other hand, the interactions of microbial community in open ponds are a mutual advantage for populations, for example, from bacteria and microalgal symbiosis, benefit extracellular products. Moreover, N and P removal from wastewaters and simultaneous growth can be improved by using a cocultivation systems of microalgae and bacteria [75,76]. Globally, more than 80% of algal biomass is generated in open ponds, mainly due to the lower investments required for these systems. However, the use of closed PBRs is projected to grow by 2024 in terms of demand and sales because of the advantages related to these cultivation systems [77].

Photobioreactors

The choice of PBRs depends on several factors, including the productivity of microalgal biomass and the final products. However, this cultivation mode requires additional costs in light illumination, carbon dioxide, and cultivation feedings, but it is simpler to control if compared with the open systems and able to minimize the problems related to the contamination. However, PBRs suffer from several limitation related, for example, the biofilm formation that leads to oxygen accumulation in the culture, which can have possible toxic effects on photosynthetic growth and the light that remains the main limiting factor. These problems can be overcome in part by appropriate engineering solutions and consequently different types of PBRs, most often, more expansive and complex, but it is not the reason to reduce the interest to the potential utilization of this technology in industrial scale.

The basic criteria to build a PBR are still related to the role of light, the role of circulation, the mass transfer, the materials used for the construction, and the temperature.

The light penetration in the culture medium is a limiting factor because it can be absorbed or scattered. The traditional light radiative equation (Eq. 2.18) contains the factors of spatial position (r) and the angular direction (s) to validate this dependency [78].

$$\frac{dI(r,s)}{ds} = -aI(r,s) - \sigma_s I(r,s) + \frac{\sigma_s}{4\pi} \int_0^{4\pi} I(r,s')\Phi(s \cdot s')d\Omega' \qquad (2.18)$$

where r, s, and s' are the position vector, direction vector, and scattering direction vector, respectively; s, a, σ_s, and I are the path length, absorption coefficient, scattering coefficient, and light intensity, respectively; and Φ and Ω' are the phase function and solid angle, respectively. In Eq. (2.18), the first term on the right-hand side accounts for the loss of photons due to light absorption, and the second term represents the loss of radiation due to light out-scattering by the cells. The last term indicates the profit of the radiation induced by light in-scattering, and the fraction of the radiation intensity, $I(r, s')$, in the direction s' scattered in direction s is determined by the phase function, $\Phi(s \cdot s')$.

The total incident intensity, $G(r)$, can be computed according to Eqs. (2.19) and (2.20) for a monochromatic and polychromatic radiation, respectively:

$$G(r) = \int_{\Omega=0}^{\Omega=4\pi} I(r, s)d\Omega \qquad (2.19)$$

$$G(r) = \sum_{\lambda_{min}}^{\lambda_{max}} G_\lambda(r) \qquad (2.20)$$

where λ_{min} and λ_{max} are the minimum and maximum PAR wavelengths (400−700 nm) of the light source, respectively; and $G_\lambda(r)$ is the total incident intensity at the wavelength λ [79].

Column photobioreactors (bubble and airlift)

Airlift and bubble column PBRs are simple cylinder devices with a radius, which should not exceed 0.2 m to avoid problems related to the light availability in the center of PBR and a height limitation of about 4 m for structural reasons, due to the strength of the transparent materials employed and to avoid shading effects. They are commonly used in bioprocessing and wastewater treatment because of their low cost. They have a low shear forces, absence of wall growth, high mass transfer, consequently high efficiency of CO_2 use. The operating efficiency and the maximum biomass productivity are strictly related to the column dimensions, to the specific growth rate or algae strains doubling time, intensity of light and surface area. High cell density, which could reach 10^9 cells, can affect light penetration because of the mutual shading effects between different cells. To overcome these problems and the sedimentation of microalgal cells, an adequate mixing, accomplished by aeration, is needed, simultaneously able to ensure the uniform exposure to light and nutrients and facilitate heat transfer and gas exchange. Based on aeration mode, column PBRs can be divided into bubble column and airlift reactor (Fig. 2.6). However, the bubble size is a function of several factors such as the properties of sparger, liquid and gas phase physical properties and the ratio H/D of the column. It has to take in account also phenomena of bubble coalescence or breakage and the possibility of clogging effects due to the presence of micron size algae, more common at high

biomass concentration and high pressure drop. The bubble size distribution in the column is very important because the size of gas bubbles at the top decides the down comer gas holdup, which then results in specific liquid circulation velocity and the light/dark cycles [80].

Flat-plate photobioreactors

Flat-plate PBRs are cuboidal-shaped reactors patented by Huang et al. [81] and made in glass, polycarbonate, or other transparent materials with a high ratio S/V. This type of PBR combines many advantages because it has a high ratio of illuminated area to volume and an easy way to control the temperature by spraying water onto the irradiated surface or by submerging the bottom of the PBR in a water pool. Finally, it has low mechanical forces on the cells, a high gas−liquid mass transfer rate, a good mixing provided by means air bubbling or mechanical rotation.

There are different configurations of flat panel PBRs as shown in Fig. 2.7 characterized by thickness larger than 7 mm, but Gifuni et al. [82] developed a new ultra-flat PBR to reduce optical path of 3 mm and simultaneously increase microalgal concentration and productivity.

Flat-plate PBRs are considered as one of the most promising reactor types for mass cultivation of photosynthetic microorganisms.

Tubular photobioreactors

Tubular PBRs (Fig. 2.8) are mainly used for outdoor mass cultivation and commonly made up in glass or plastic. They can be arranged in different orientations, for example, horizontal, inclined, vertical, or helical, to maximize the sunlight capture. The microalgal culture is pumped into the reactors, which have generally a diameter from 10 to 60 mm and the length that can be also 100 m. A small diameter of 10 mm may be designed when the goal is the production of high cell concentrations [83].

They have several drawbacks such as the limitation in the photosynthetic efficiency, frequently in outdoor cultivations, for the oxygen build up and also the very high energy consumption when compared with bubble column and flat plate PBRs. A possible inhibiting concentration due to oxygen toxicity can occur after only 1 min in a tube without gas exchange [83]. Moreover, due to the photolimitation and problems related to the mass transfer, the growth of the cells in the center of the tube is restricted. Therefore, in theory, the width of the tubes should be kept as short as possible depending on the potential O_2 accumulation and CO_2

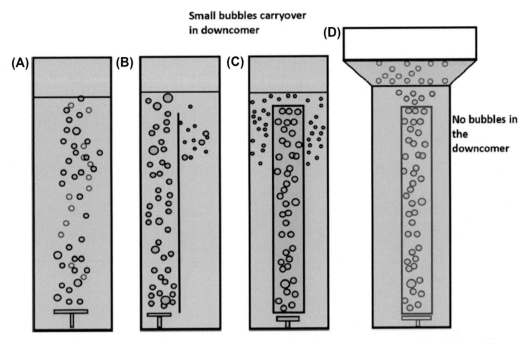

FIG. 2.6 Schematic different photobioreactors; **(A)** bubble column, **(B)** split column airlift, **(C)** internal loop airlift, and **(D)** internal loop airlift with gas separator. (Reproduced from S.B. Pawar, Process engineering aspects of vertical column photobioreactors for mass production of microalgae, ChemBioEng Rev 3 (2016) 101−115.)

depletion. It is understandable as CO_2, O_2 gradients, and pH between the medium inlet and the outlet should be carefully considered [79]. Another drawback of these systems is the uncontrolled grow of pathogenic microorganisms in the inner walls and the formation of biofilms, which influence the mass transfer of reagents because it was proven that the presence of external mass transfer resistance at biofilm surface is able to create profile switches of reagent concentration inside the biofilm [84]. A novel tubular PBR with the outer

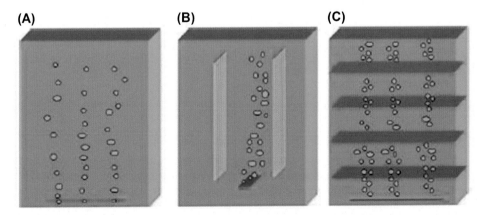

FIG. 2.7 Flat plate photobioreactors; **(A)** simple flat panel, **(B)** flat panel with vertical baffles, and **(C)** flat panel with horizontal baffles. (Reproduced from S.B. Pawar, Process engineering aspects of vertical column photobioreactors for mass production of microalgae, ChemBioEng Rev 3 (2016) 101−115.)

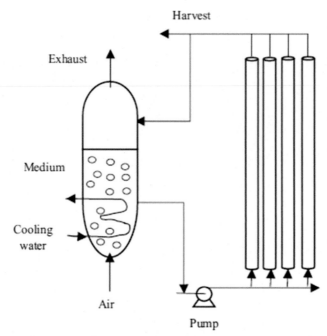

FIG. 2.8 Tubular photobioreactor. (Reproduced from L. Moreno-Garcia, K. Adjalle, S. Barnabé, G.S.V. Raghavan, Microalgae biomass production for a biorefinery system: Recent advances and the way toward sustainability, Renewable and Sustainable Energy Reviews 76 (2017) 493–506.)

surface periodically shaded was shown to enhance the photosynthetic efficiency by increasing the biomass productivity by 21.6% compared with conventional PBRs. The principle explaining this increase is the creation of an artificial light/dark cycle that favors the microalgae growth [65,85].

In addition, a high temperature in the culture may occur because of limited volume resulting from small diameters of tubular PBRs. Finally, cleaning the walls of tubular PBRs is an annoying problem that is closely connected to light permeability. At present, mechanical cleaning is the procedure most often used to do this washing [86].

Membrane photobioreactors

The membrane photobioreactors (MPBRs) (Fig. 2.9) offer an alternative to the conventional PBR technology. Indeed, the most important limitation of PBRs is related to biomass wash-out due to short residence time, resulting in a harvesting rate that is higher than growth rate [87]. Membrane PBRs consist of an additional filtration tank where a membrane provides the retention of microalgal cells, preventing the wash out and increasing biomass concentration, while the medium passes as permeate. MPBRs can be operated at higher dilution

and higher growth rates if compared with PBRs [88], but a recycling of remaining nutrients in the permeate is one of the key issues for the development of large scale cultures because it is possible to minimize the water and nutrients consumption. Indeed, some metabolites produced by microalgae and known as algogenic organic matter, and no assimilated nutrients can be accumulated.

Plastic bag photobioreactors

Plastic bag PBRs (Fig. 2.10) received increasing attention because of their low cost. These bags can be arranged in different patterns according to their respective volume. Abomohra et al. [89] performed experiments in a pilot scale, including 20 bags (20 cm in width, 2 m in length, and 0.2 mm in material thickness, for a volume of 16 L) made in polyethylene.

Plastic bag PBRs with greater volumes can be immersed in a water pool to reduce cost and to control the temperature in summer. Plastic bag PBRs can even be designed to produce cells in the ocean; mixing and mass transfer are improved by the efficient utilization of ocean waves, which can reduce the cost substantially [79,90]. However, they have many disadvantages related to the inadequate mixing, which influences

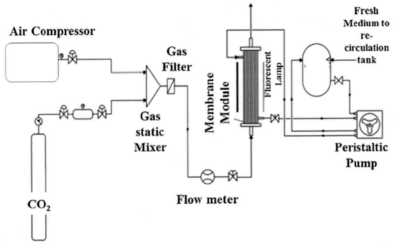

FIG. 2.9 Schematic diagram of membrane photobioreactor. (Reproduced from V. Mortezaeikia, R. Yegani, O. Tavakoli, Membrane-sparger versus membrane contactor as a photobioreactors for carbon dioxide biofixation of *Synechoccus elongates* in batch and in semi-continuous mode, Journal of CO_2 Utilization 16 (2016) 23—31.)

cell growth in some zones. In addition, photolimitation can occur due to the shape of bags. They are fragile, and leakages happen more than occasionally, making the life of these systems too short [79].

Solid-state photobioreactors
Attached cultivation or solid-state PBRs (Fig. 2.11) are studied to improve the productivity and to reduce the usage of water. Algal cells are attached onto a supporting material containing a small volume of culture medium to provide nutrients to the attached cells. It could be arranged in multiple layers to increase the

FIG. 2.10 Plastic bag photobioreactor. (Reproduced from Q. Huang, F. Jiang, L. Wang, C. Yang, Design of photobioreactors for mass cultivation of photosynthetic organisms, Engineering 3 (2017) 318—329.)

illuminated area and to promote the photoinduction of algal cells [91]. This innovative culture method has been successfully used in indoor and outdoor operations, and several fresh water and marine species have been grown in this type of systems showing high biomass productivities and photosynthetic efficiencies.

The attached cultivation system has been found for having the highest biomass productivities and the easiest harvesting procedures at a lab scale. These systems are also likely to scale up with a low risk of contamination and high gas exchange rates. However, the biomass production costs are higher compared with the suspended cultures, which claims for a need to increase the productivities and light efficiencies and find the best commitment between production cost and biomass yield, composition, and productivities.

CONCLUDING REMARKS
Microalgae offer interesting features to qualify them as alternative feedstocks for various applications including environmental and industrial. However, efforts are required to win different challenges such as controlling culture conditions by selecting favorable conditions able to improve the productivity and choosing systems designed specifically to meet needs such as CO_2 mitigation or wastewater treatment or photoinhibition problems or the cost of carbon substrates.

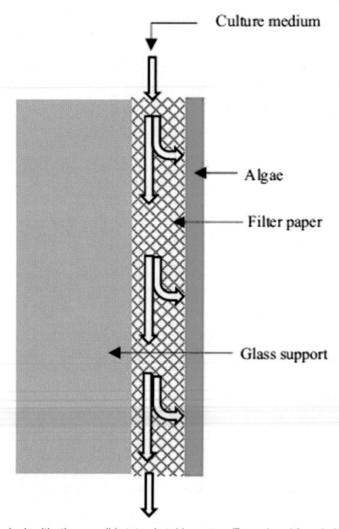

FIG. 2.11 Attached cultivation or solid-state photobioreactor. (Reproduced from L. Moreno-Garcia, K. Adjalle, S. Barnabé, G.S.V. Raghavan, Microalgae biomass production for a biorefinery system: Recent advances and the way toward sustainability, Renewable and Sustainable Energy Reviews 76 (2017) 493–506.)

This chapter has also outlined several reactor conditions to highlight the positive and negative aspects. Open pond plants undoubtedly represent a better and low-cost culture systems able to overcome the problems associated with closed systems such as PBRs. The problem related to the use of water remains to be solved, for example, by innovative systems in a solid-state regime. Therefore, reconciling cultural problems and the possibility of application on an industrial scale are certainly challenges that cannot be overlooked in this particular phase of growing interest in microalgae cultures.

REFERENCES

[1] M.I. Khan, J.H. Shin, J.D. Kim, The promising future of microalgae: current status, challenges, and optimization of a sustainable and renewable industry for biofuels, feed, and other products, Microb. Cell Fact. 17 (2018) 36–57.
[2] T. Suganya, M. Varman, H. Masjuki, S. Renganathan, Macroalgae and microalgae as a potential source for commercial applications along with biofuels production: a biorefinery approach, Renew. Sustain. Energy Rev. 55 (2016) 909–941.

[3] Y. Bartosh, C.J. Banks, Algal growth response and survival in a range of light and temperature conditions: implications for non-steady-state conditions in waste stabilisation ponds, Water Sci. Technol. 55 (2007) 211–218.

[4] S.V. Mohan, M. Rohit, P. Chiranjeevi, R. Chandra, B. Navaneeth, Heterotrophic microalgae cultivation to synergize biodiesel production with waste remediation: progress and perspectives, Bioresour. Technol. 184 (2015) 169–178.

[5] J. Seckbach, D. Chapman, D. Garbary, A. Oren, W. Reisser, Oxygenic photosynthetic microorganisms in extreme environments: possibilities and limitations, in: J. Seckbach (Ed.), Algae and Cyanobacteria in Extreme Environments, Springer, Netherlands, 2007, pp. 3–25.

[6] T. Mock, K. Valentin, Photosynthesis and cold acclimation: molecular evidence from a polar diatom, J. Phycol. 40 (2004) 732–741.

[7] D. Remias, U. Lütz-Meindl, C. Lütz, Photosynthesis, pigments and ultrastructure of the alpine snow alga *Chlamydomonas nivalis*, Eur. J. Phycol. 40 (2005) 259–268.

[8] D. Remias, U. Karsten, C. Lütz, T. Leya, Physiological and morphological processes in the Alpine snow alga *Chloromonas nivalis* (Chlorophyceae) during cyst formation, Protoplasma 243 (2010) 73–86.

[9] W.F. Vincent, A. Quesada, Ultraviolet radiation effects on cyanobacteria: implications for Antarctic microbial ecosystems, in: C.S. Weiler, P.A. Penhale (Eds.), Ultraviolet Radiation in Antarctica: Measurements and Biological Effects, Antarctic Research Series, American Geophysical Union, Washington, D. C., 1994, pp. 111–124.

[10] A. Quesada, J. Mouget, W.F. Vincent, Growth of antarctic cyanobacteria under ultraviolet radiation: UVA counteracts UVB inhibition, J. Phycol. 31 (1995) 242–248.

[11] Y. Wang, T. SeppaÈnen-Laakso, H. Rischer, M.G. Wiebe, *Euglena gracilis* growth and cell composition under different temperature, light and trophic conditions, PLoS One 13 (2018) 1–17.

[12] Y. Yu, L. You, D. Liu, W. Hollinshead, Y.J. Tang, F. Zhang, Development of *Synechocystis* sp. PCC 6803 as a phototrophic cell factory, Mar. Drugs 11 (2013) 2894–2916, 2013.

[13] E. Kessler, Upper limits of temperature for growth in *Chlorella* (Chlorophyceae), Plant Syst. Eval. 151 (1985) 67–71.

[14] N. Hanagata, T. Takeuchi, Y. Fukuju, D.J. Barnes, I. Karube, Tolerance of microalgae to high CO_2 and high temperature, Phytochemistry 31 (1992) 3345–3348.

[15] E. Ono, J.L. Cuello, Carbon dioxide mitigation using thermophilic cyanobacteria, Biosyst. Eng. 96 (2007) 129–134.

[16] Y. Xie, S.-H. Ho, C.-N.N. Chen, C.-Y. Chen, I.-S. Ng, K.-J. Jing, J.-S. Chang, Y. Lu, Phototrophic cultivation of a thermo-tolerant *Desmodesmus* sp. for lutein production: effects of nitrate concentration, light intensity and fed-batch operation, Bioresour. Technol. 144 (2013) 435–444.

[17] F. Delrue, E. Alaux, L. Moudjaoui, C. Grignard, G. Fleury, A. Perilhou, P. Richaud, M. Petitjean, J.-F. Sassi, Optimization of *Arthorospira platensis* (Spirulina) growth: from laboratory scale to pilot scale, Fermentation 3 (2017) 59–73.

[18] J.-Y. Leu, T. H. Lin, M.J.P. Selvamani, H.-C. Chen, J.-Z. Liang, K.-M. Pan, Characterization of a novel thermophilic cyanobacterial strain from Taian hot springs in Taiwan for high CO_2 mitigation and C-phycocyanin extraction, Process Biochem. 48 (2013) 41–48.

[19] R.W. Castenholz, The behavior of *Oscillatoria terebriformis* in hot springs, J. Phycol. 4 (1968) 132–139.

[20] R. Ikan, J. Seckbach, Lipids of the thermophilic alga *Cyanidium calderium*, Phytochemistry 11 (1972) 1077–1082.

[21] C. Ciniglia, E.C. Yang, A. Pollio, G. Pinto, M. Iovinella, L. Vitale, H.S. Yoon, Cyanidiophyceae in Iceland: plastid rbcL gene elucidates origin and dispersal of extremophilic *Galdieria sulphuraria* and *G. maxima* (Galdieriaceae, Rhodophyta), Phycologia 6 (2014) 542–551.

[22] D. Pedersen, S.R. Miller, Photosynthetic temperature adaptation during niche diversification of the thermophilic cyanobacterium *Synechococcus* A/B clade, ISME J. 11 (2017) 1053–1057.

[23] L.J. Rothschild, R.L. Mancinelli, Life in extreme environments, Nature 409 (2001) 1092–1101.

[24] M.E. Salvucci, S.J. Crafts-Brandner, Relationship between the heat tolerance of photosynthesis and the thermal stability of rubisco activase in plants from contrasting thermal environments, Plant Physiol. 134 (2004) 1460–1470.

[25] Q. Béchet, A. Shilton, B. Guieysse, Modeling the effects of light and temperature on algae growth: state of the art and critical assessment for productivity prediction during outdoor cultivation, Biotechnol. Adv. 31 (2013) 1648–1663.

[26] I. Krzemińska, B. Pawlik-Skowrońska, M. Trzcińska, J. Tys, Influence of photoperiods on the growth rate and biomass productivity of green microalgae, Bioprocess Biosyst. Eng. 37 (2014) 735–741.

[27] R.M. Schuurmans, P. van Alphen, J.M. Schuurmans, H.C.P. Matthijs, K.J. Hellingwerf, Comparison of the photosynthetic yield of cyanobacteria and green algae: different methods give different answers, PLoS One 10 (2015) e0139061.

[28] S.A. Razzak, M.M. Hossain, R.A. Lucky, A.S. Bassi, H. de Lasa, Integrated CO_2 capture, wastewater treatment and biofuel production by microalgae culturing – a review, Renew. Sustain. Energy Rev. 27 (2013) 622–653.

[29] J. Beardall, I. Morris, The concept of light intensity adaptation in marine phytoplankton: some experiments with *Phaeodactylum tricornutum*, Mar. Biol. 37 (1976) 377–387.

[30] H. Wu, Effect of different light qualities on growth, pigment content, chlorophyll fluorescence, and antioxidant enzyme activity in the red alga *Pyropia haitanensis* (Bangiales, Rhodophyta), BioMed Res. Int. 2016 (2016) 1–8.

[31] M. Hildebrand, R.M. Abbriano, J.E. Polle, J.C. Traller, E.M. Trentacoste, S.R. Smith, et al., Metabolic and cellular organization in evolutionarily diverse microalgae as related to biofuels production, Curr. Opin. Chem. Biol. 17 (2013) 506–514.

[32] J.A. Raven, P.J. Ralph, Enhanced biofuel production using optimality, pathway modification and waste minimization, J. Appl. Phycol. 27 (2015) 1–31.

[33] K.R. Arrigo, Marine microorganisms and global nutrient cycles, Nature 437 (2005) 349–355.

[34] G.D. Price, M.R. Badger, F.J. Woodger, B.M. Long, Advances in understanding the cyanobacterial CO_2-concentrating mechanism (CCM): functional components, C_i transporters, diversity, genetic regulation and prospects for engineering into plants, J. Exp. Bot. 59 (2008) 1441–1461.

[35] S. Van Den Hende, H. Vervaeren, N. Boon, Flue gas compounds and microalgae: (Bio-) chemical interactions leading to biotechnological opportunities, Biotechnol. Adv. 30 (2012) 1405–1424.

[36] J.J. Carroll, J.D. Slupsky, A.E. Mather, The solubility of carbon dioxide in water at low pressure, J. Phys. Chem. Ref. Data 20 (1991) 1201–1209.

[37] I. Suh, C.-G. Lee, Photobioreactor engineering: design and performance, Biotechnol. Bioprocess Eng. 8 (2003) 313–321.

[38] Z. Chi, F. Elloy, Y. Xie, Y. Hu, S. Chen, Selection of microalgae and cyanobacteria strains for bicarbonate-based integrated carbon capture and algae production system, Appl. Biochem. Biotechnol. (2013) 1–11.

[39] Y. Chisti, Constraints to commercialization of algal fuels, J. Biotechnol. 167 (2013) 201–214.

[40] M. Olaizola, Microalgal removal of CO_2 from flue gases: changes in medium pH and flue gas composition do not appear to affect the photochemical yield of microalgal cultures, Biotechnol. Bioprocess Eng. 8 (2003) 360–367.

[41] O. Perez-Garcia, F.M.E. Escalante, L.E. de-Bashan, Y. Bashan, Heterotrophic cultures of microalgae: metabolism and potential products, Water Res. 45 (2011) 11–36.

[42] G. Markou, D. Vandamme, K. Muylaert, Microalgal and cyanobacterial cultivation: the supply of nutrients, Water Res. 65 (2014) 186–202.

[43] J.U. Grobbelaar, in: A. Richmond (Ed.), Handobook of Microalgal Culture: Biotechnology and Applied Phycology, Blackwell Publishing Ltd., Oxford, 2004, pp. 97–115.

[44] J. Jeanfils, M. Canisius, N. Burlion, Effect of high nitrate concentrations on growth and nitrate uptake by free-living and immobilized Chlorella vulgaris cells, J. Appl. Phycol. 5 (1993) 369–374.

[45] S. Yang, J. Wang, W. Cong, Z. Cai, F. Ouyang, Utilization of nitrite as a nitrogen source by Botryococcus braunii, Biotechnol. Lett. 26 (2004) 239–243.

[46] M. Taziki, H. Ahmadzadeh, M.A. Murry, Growth of Chlorella vulgaris in high concentrations of nitrate and nitrite for wastewater treatment, Current Biotechnol. 4 (2015).

[47] E. Flores, A. Herrero, G. Miguel, Nitrite uptake and its regulation in the cyanobacterium Anacystis nidulans, Biochim. Biophys. Acta (BBA)-Biomembranes 896 (1987) 103–108.

[48] D.E. Santiago, H.-F. Jin, K. Lee, The influence of ferrous complexes EDTA as a solubilization agent and its autoregeneration on the removal of nitric oxide gas through the culture of green alga Scenedesmus sp, Pro. Biochem. 45 (2010) 1949–1953.

[49] H. Nagase, K.-i. Yoshihara, K. Eguchi, Y. Okamoto, S. Murasaki, R. Yamashita, K. Hirata, K. Miyamoto, Uptake pathway and continuous removal of nitric oxide from flue gas using microalgae, Biochem. Eng. J. 7 (2001) 241–246.

[50] M. Drath, N. Kloft, A. Batschauer, K. Marin, J. Novak, K. Forchhammer, Ammonia triggers photodamage of photosystem II in the cyanobacterium Synechocystis sp. Strain PCC 6803, Plant Physiol. 147 (2008) 206–215.

[51] G. Markou, D. Vandamme, K. Muylaert, Ammonia inhibition on Arthrospira platensis in relation to the initial biomass density and pH, Bioresour. Technol. 166 (2014) 259–265.

[52] J. Peccia, B. Haznedaroglu, J. Gutierrez, J.B. Zimmerman, Nitrogen supply is an important driver of sustainable microalgae biofuel production, Trends Biotechnol. 31 (2013) 134–138.

[53] T. Großkopf, J. LaRoche, Direct and indirect costs of dinitrogen fixation in Crocosphaera watsonii WH8501 and possible implications for the nitrogen cycle, Front. Microbiol. 3 (2012).

[54] S.T. Dyhrman, K.C. Ruttenberg, Presence and regulation of alkaline phosphatase activity in eukaryotic phytoplankton from the coastal ocean: implications for dissolved organic phosphorus remineralization, Limnol. Oceanogr. 51 (2006) 1381–1390.

[55] A.D. Cembella, N.J. Antia, P.J. Harrison, The utilization of inorganic and organic phosphorous compounds as nutrients by eukaryotic microalgae: a multidisciplinary perspective: part I, Crit. Rev. Microbiol. 10 (1982) 317–391.

[56] N. Powell, A. Shilton, Y. Chisti, S. Pratt, Towards a luxury uptake process via microalgae e defining the polyphosphate dynamics, Water Res. 43 (2009) 4207–4213.

[57] A. Melis, H.-C. Chen, Chloroplast sulfate transport in green algae e genes, proteins and effects, Photosynth. Res. 86 (2005) 299–307.

[58] A. Oren, E. Padan, S. Malkin, Sulfide inhibition on photosystem II in cyanobacteria (blue-green algae) and tobacco chloroplasts, Biochim. Biophys. Acta (BBA)-Bioenergetics 546 (1979) 270–279.

[59] H. Shimamatsu, in: P. Ang Jr. (Ed.), Asian Pacific Phycology in the 21st Century: Prospects and Challenges, Springer, Netherlands, 2004, pp. 39–44.

[60] I. Havlik, T. Scheper, K.F. Reardon, Monitoring of microalgal processes, in: C. Posten, S. Feng Chen (Eds.), Microalgae Biotechnology, Springer International Publishing, Cham, 2016, pp. 89–142.

[61] M.K. Lam, K.T. Lee, Potential of using organic fertilizer to cultivate *Chlorella vulgaris* for biodiesel production, Appl. Energy 94 (2012) 303–308.

[62] R. Münkel, U. Schmid-Staiger, A. Werner, T. Hirth, Optimization of outdoor cultivation in flat panel airlift reactors for lipid production by *Chlorella vulgaris*, Biotechnol. Bioeng. 110 (2013) 2882–2893.

[63] A.K. Vuppaladadiyam, P. Prinsen, A. Raheem, R. Luque, M. Zhao, Microalgae cultivation and metabolites production: a comprehensive review, Biofuels Bioprod. Bioref. 12 (2018) 304–324.

[64] S.K. Saha, P. Murray, Exploitation of microalgae species for nutraceutical purposes: cultivation aspects, Fermentation 4 (2018) 46–63.

[65] L. Moreno-Garcia, K. Adjalle, S. Barnabé, G.S.V. Raghavan, Microalgae biomass production for a biorefinery system: recent advances and the way towards sustainability, Renew. Sust. Energ. Rev. 76 (2017) 493–506.

[66] P. Kenny, K.J. Flynn, Physiology limits commercially viable photoautotrophic production of microalgal biofuels, J. Appl. Phycol. 29 (2017) 2713–2727.

[67] K.J. Flynn, J.A. Raven, What is the limit for photoautotrophic plankton growth rates? J. Plankton Res. 39 (2017) 13–22.

[68] Y. Li, H. Xu, F. Han, J. Mu, D. Chen, B. Feng, H. Zeng, Regulation of lipid metabolism in the green microalga *Chlorella protothecoides* by heterotrophy-photoinduction cultivation regime, Biores. Technol. 192 (2015) 781–791.

[69] J. Lowrey, R.E. Armenta, M.S. Brooks, Nutrient and media reciclyng in heterotrophic microalgae cultures, Appl. Microbiol. Biotechnol. 100 (2016) 1061–1075.

[70] J. Zhan, J. Rong, Q. Wang, Mixotrophic cultivation, a preferable microalgae cultivation mode for biomass/bioenergy production and bioremediation, advances and prospect, Int. J. Hydrog. Energy 42 (2017) 8505–8517.

[71] Y. Alkhamis, J.G. Qin, Comparison of pigment and proximate compositions of *Tisochrysis lutea* in phototrophic and mixotrophic cultures, J. Appl. Phycol. 28 (2015) 35–42.

[72] Y. Chisti, Large-scale production of algal biomass: raceway ponds, in: F. Bux, Y. Chisti (Eds.), Algae Biotechnology, Springer, New York, NY, 2016, pp. 21–40.

[73] W.J. Oswald, Large-scale algal culture systems (engineering aspects), in: M.A. Borowitzka, L.J. Borowitzka (Eds.), Micro-Algal Biotechnology, Cambridge University, Cambridge, 1988, pp. 357–394.

[74] M.A. Borowitzka, Culturing microalgae in outdoor ponds, in: R.A. Andersen (Ed.), Algal Culturing Techniques, Elsevier, UK, 2005, pp. 205–219.

[75] K. Kumar, S.K. Mishra, A. Shrivastav, M.S. Park, J.W. Yang, Recent trends in the mass cultivation of algae in raceway ponds, Renew. Sust. Energ. Rev. 51 (2015) 875–885.

[76] B.-H. Kim, R. Ramanan, D.-H. Cho, H.-M. Oh, H.-S. Kim, Role of Rhizobium, a plant growth promoting bacterium, in enhancing algal biomass through mutualistic interaction, Biomass Bioenergy 69 (2014) 95–105.

[77] Transparency Market Research, Algae Fuel Market - Global Industry Analysis, Market, Size, Share, Growth, Trends and Forecast 2015–2023, 2015. http://www.transparencymarketresearch.com/algae-fuel-market.html.

[78] Q. Huang, T. Liu, J. Yang, L. Yao, L. Gao, Evaluation of radiative transfer using the finite volume method in cylindrical photobioreactors, Chem. Eng. Sci. 66 (2011) 3930–3940.

[79] Q. Huang, F. Jiang, L. Wang, C. Yang, Design of photobioreactors for mass cultivation of photosynthetic organisms, Engineering 3 (2017) 318–329.

[80] S.B. Pawar, Process engineering aspects of vertical column photobioreactors for mass production of microalgae, Chem. Bio. Eng. Rev. 3 (2016) 101–115.

[81] Q. Huang, L. Yao, X. Guo, A device consisted of airlift flat-panel photobioreactors for the cultivation of photosynthetic microorganism, 2015. China patent, Appl. Num. 201510478717.1.

[82] I. Gifuni, A. Pollio, A. Marzocchella, G. Olivieri, New ultra-flat photobioreactor for intensive microalgal production : the effect of light irradiance, Algal Res. 34 (2018) 134–142.

[83] C. Posten, Design principles of photo-bioreactors for cultivation of microalgae, Eng. Life Sci. 9 (2009) 165–177.

[84] S. Skoneczny, B. Tabiś, The method for steady states determination in tubular biofilm reactors, Chem. Eng. Sci. 137 (2015) 178–187.

[85] Q. Liao, L. Li, R. Chen, X. Zhu, A novel photobioreactor generating the light/dark cycle to improve microalgae cultivation, Bioresour. Technol. 161 (2014) 186–191.

[86] J. Zhu, J. Rong, B. Zong, Factors in mass cultivation of microalgae for biodiesel, Chin. J. Catal. 34 (2013) 80–100.

[87] V. Discart, M.R. Bilad, L. Marbelia, I.F.J. Vankelcom, Impact of changes in broth composition on Chlorella vulgaris cultivation in a membrane photobioreactor (MPBR) with permeate recycle, Bioresour. Technol. 152 (2014) 321–328.

[88] L. Marbelia, M.R. Bilad, V. Passaris, V. Discart, D. Vandamme, A. Beuckels, K. Muylaert, I.F.J. Vankelecom, Membrane photobioreactors for integrated microalgae cultivation and nutrient remediation of membrane bioreactors effluent, Bioresour. Technol. 163 (2014) 228–235.

[89] A.E.F. Abomohra, M. El-Sheekh, D. Hanelt, Pilot cultivation of the chlorophyte microalga *Scenedesmus obliquus* as a promising feedstock for biofuel, Biomass Bioenerg. 64 (2014) 237–244.

[90] Z.H. Kim, H. Park, S.J. Hong, S.M. Lim, C.G. Lee, Development of a floating photobioreactor with internal partitions for efficient utilization of ocean wave into improved mass transfer and algal culture mixing, Bioprocess. Biosyst. Eng. 39 (2016) 713–723.

[91] J. Shi, B. Podola, M. Melkonian, Application of a prototype-scale twin-layer photbioreactor for effective N and P removal from different process stages of municipal wastewater by immobilized microalgae, Bioresour. Technol. 154 (2014) 260–266.

Microalgae Cultivation and Photobioreactor Design

WENGUANG ZHOU • QIAN LU • PEI HAN • JINGJING LI

INTRODUCTION

The never-ending energy demand and the fast-increasing CO_2 emission have prompted scientists on a quest to find sustainable resources for fuel production and CO_2 fixation [1]. Microalgae with the potential of being exploited for biofuel, animal feed, and food supplement have been considered as a promising resource to alleviate aforementioned problems [2]. Cultivation and harvesting, determining the cost and the value of biomass, are critical steps in the exploitation of microalgae for biofuel and other value-added products. Nutrients metabolism and photobioreactor design, in which both fundamental research and applied research have been widely conducted, are two important concerns in microalgae cultivation as well.

This chapter, aiming at providing readers with knowledge of microalgae cultivation, harvesting, nutrients metabolism, and photobioreactor design, consists of four parts. Firstly, nutrients assimilation by microalgae cells was discussed. Secondly, basic cultivation modes, including autotrophic cultivation, heterotrophic cultivation, and mixotrophic cultivation and synthesis pathways of some value-added compositions in microalgae were discussed. Thirdly, details of some widely used photobioreactors for large-scale microalgae cultivation were provided. Fourthly, a couple of conventional and novel harvesting technologies used for further biomass refining were introduced.

NUTRIENTS METABOLISMS

Carbon Assimilation

As an essential element of protein, lipid, and carbohydrate, carbon plays a critical role in microalgae growth. Deficiency of carbon would reduce the biomass productivity and limit the lipid accumulation in microalgae cells. Also, the syntheses of some value-added compositions in microalgae cells are also limited by carbon deficiency in culture medium [3]. Carbon resources commonly used for microalgae cultivation include carbon dioxide, carbonate, bicarbonate, and organic carbon, such as glucose, acetic acid, and glycerol. For the purpose of producing microalgae-based biodiesel and other value-added products, it is important to provide sufficient carbon at sensible cost and promote the carbon assimilation by microalgae cells.

Utilization of inorganic carbon

Inorganic carbon sources commonly used for microalgae cultivation include carbon dioxide, bicarbonate, and carbonate. Since carbon dioxide could be obtained from air at very low cost by pumping air into culture medium, carbon dioxide is a very common carbon source employed for microalgae growth [4]. To increase the assimilation efficiency of carbon dioxide, photosynthesis of microalgae cell could be enhanced by increasing light intensity, adjusting pH value of culture medium, and reducing the turbidity of culture medium [5,6].

To efficiently utilize the low percentage (about 0.04%) of carbon dioxide in air, almost all microalgae have developed a carbon-concentrating mechanism that enhances photosynthetic efficiency and thus permits high growth rates of microalgae at low CO_2 concentrations. Absorbed carbon dioxide is converted to glucose for short-term energy storage by photosynthesis in microalgae cells. Glucose in microalgae cells is transformed into pyruvate, a precursor for the synthesis of fatty acids, carotene, and other compositions, through glycolysis pathway [7]. Compared with organic carbon sources, carbon dioxide obtained from air, which is more cost-saving, is preferred by some users in microalgae cultivation.

Microalgae-based CO_2 fixation has a great potential in various fields. In recent years, more and more researchers are trying to use microalgae to fix carbon dioxide in air and improve the air quality [8]. Zhao et al. [4] even proposed a concept of mitigating greenhouse effects by growing microalgae. Some researchers proposed a concept of using microalgae in spacecraft to recycle carbon dioxide produced by astronauts [9].

Utilization of organic carbon

Organic carbon sources in wastewater are very complex since industries have different production procedures, which impact the forms of organic carbon in effluent [10]. Common organic carbon sources that could be efficiently utilized by algal cells include soluble saccharides, volatile fatty acids (VFAs), glycerol, and so on [11,12]. Not all the organic carbon sources could be assimilated by microalgae cells, and some of the organic carbon sources even have hazardous effects on algae growth. The study of Luet al. [13] showed that some solid particles in wastewater also cannot be transported through cell membrane into intracellular environment. Besides the insoluble organic carbon, some soluble organic carbon could not be utilized by algae due to the lack of certain metabolic pathways in algal cells. Algae with different enzyme activities and metabolic pathways have selectivity on metabolizing organic carbon source [14]. For example, *Scenedesmus* sp. had high biomass yields in medium with glucose and acetate, while this could not efficiently utilize sucrose and maltose [15]. As shown in Table 3.1, microalgae could utilize various organic carbon sources for biomass synthesis.

Carbon from waste resources

In a real-world application, to reduce the cost of microalgae production, carbon for microalgae cultivation could be obtained from waste resources, such as flue gas and waste stream [19]. Flue gas from industry always contains high percentage (10%−25%) of carbon

dioxide. Without appropriate treatment, the emission of flue gas will accelerate greenhouse effects. Waste streams, obtained from industrial section, agricultural section, and municipal section, which are rich with carbon, could be exploited as culture media for microalgae growth [20]. The recovery of carbon from waste resources not only reduces the production cost of microalgae biomass but also contributes to the environmental protection. The progress of exploiting waste streams for microalgae cultivation is discussed as follows.

As the greenhouse effects of carbon dioxide are inducing serious climate crisis, flue gas with high percentage of carbon dioxide should be treated before emission. Conventional technologies for the removal of carbon dioxide in flue gas include ethanolamine-based absorption, membrane separation, and so on [21]. However, these technologies with high investment cost would bring heavy financial burden to the industries. The concept of using microalgae-treated flue gas was proposed to convert carbon dioxide to valuable microalgae biomass for further refining [22]. The core technology of this concept is that catalyzed by carbonic anhydrase, carbon dioxide is converted to bicarbonate, which could be assimilated by microalgae cells in a more efficient way. Some examples of using flue gas for microalgae cultivation are shown in Table 3.2.

One of the main technical problems associated with the microalgae-based carbon dioxide fixation and flue gas treatment is the toxicity caused by high concentration of carbon dioxide [28]. High concentration of dissolved carbon dioxide would reduce the pH value of culture medium and create an unfavorable condition for microalgae growth. Another technical problem is that in most cases, flue gas with high temperature (over 70°C) would cause thermal damage to microalgae cells [28]. In a real-world application, to promote the

TABLE 3.1
Organic Carbon Sources for Microalgae Growth.

Carbon Source	Microalgal Strain	Biomass Yield (g/L)	Growth Period (day)	References
Glycerol	*Chlorella vulgaris*	2.8	5	[16]
Ethanol and glycerol	*Chlorella protothecoides*	12.73	6	[17]
Acetic acid	*Scenedesmus* sp.	1.9	6	[15]
Glucose	*Scenedesmus* sp.	4.3	6	[15]
Glycerol and glucose	*Chlorella vulgaris*	2.62	4	[18]
Propionate	*Scenedesmus* sp.	0.5	6	[15]

TABLE 3.2
Examples of Using Flue Gas as Carbon Source for Microalgae Cultivation.

Microalgal Strain	Percentage of Carbon Dioxide in Flue Gas (%)	Removal Efficiency of CO_2 (%)	Biomass Yield (g/L)	References
Chlorella sp.	10	46	2.25	[23]
Chlorella sp.	10	63	5.15	[24]
Chlorella sp.	15	85.6	0.95	[25]
Spirulina sp.	6	53.29	3.40	[26]
Scenedesmus sp. and *Ankistrodesmus* sp.	5	59.80	4.90	[27]
Spirulina sp.	12	45.61	3.50	[26]

use of inorganic carbon in flue gas for microalgae cultivation, these technical problems should be overcome.

Waste stream is another important resource to provide carbon at low cost for microalgae cultivation. Different from the flue gas, waste streams, which contain various organics, such as soluble saccharides and VFAs, mainly provide organic carbon [29]. As shown in Table 3.3, depending on the raw materials and processing steps, waste streams from industries have very different nutrients profiles. For example, waste streams from sugar process plant are rich with glucose and other forms of saccharides, while waste streams from manure pretreatment plant contain more VFAs. Since microalgal species have preferences on the carbon sources, in the practice, to use waste streams for microalgae production, robust algal species should be selected in advance [30]. In addition, some pretreatment could be conducted to convert organics in wastewater to nutrients, which could be used by microalgae for biomass. The study of Hu et al. [29] showed that through anaerobic fermentation, organic carbon in swine manure could be converted to VFAs, which could be assimilated by microalgae cells efficiently.

Nitrogen Assimilation
Nitrogen sources and assimilation pathways
In algal cells, nitrogen is an essential element for the synthesis of protein, genetic compositions (DNA and RNA), and energy-storing molecules (e.g., ATP/ADP) [33]. Nitrogen sources available for microalgae cells include nitrate, ammonium, and organic nitrogen. As shown in Fig. 3.1, driven by various enzymes, including nitrate reductase, nitrite reductase, and urease, absorbed nitrogen is converted into NH_3-N for further assimilation process through glutamine synthetase-glutamine oxoglutarate aminotransferase (GS-GOGAT) pathway or glutamate dehydrogenase (GDH) pathway [33]. Nitrogen profile in wastewater depends on how the wastewater is generated in specific industries. For example,

TABLE 3.3
Exploitation of Some Waste Streams for Microalgae Cultivation.

Wastewater	Major Carbon Source	Microalgal Strain	Biomass Yield (g/L)	References
Dairy processing wastewater	Saccharides	*Chlorella* sp.	3.2	[6]
Cane molasses and algae residue	Saccharides	*Schizochytrium* sp.	52.33	[31]
Fermented swine manure	Volatile fatty acids	*Chlorella* sp.	1.6	[29]
Meat-processing wastewater	Protein decomposition	*Chlorella* sp.	1.54	[13]
Oil crop residue	Oil and protein	*Scenedesmus* sp.	4.5	[20]
Anaerobic digested dairy manure	Volatile fatty acids	*Chlorella* sp.	3.5	[32]

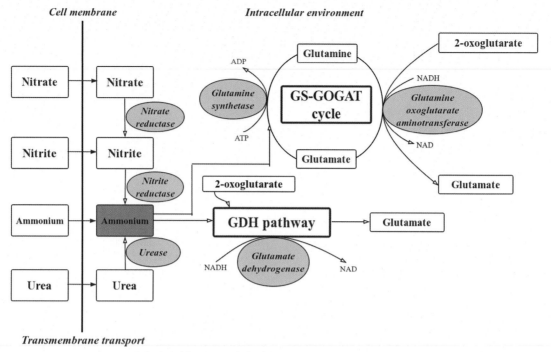

FIG. 3.1 Nitrogen assimilation pathways in microalgae cells.

excessive nitrogen in food industry wastewater is mainly caused by the protein degradation products, while in coking wastewater, it is the heterocyclic compounds that contribute to high nitrogen content [13,34]. Hence, maximizing the nitrogen absorption and assimilation in algal cells would not only increase the protein content in biomass but also enhance the intracellular metabolisms and promote algae growth.

Ammonium toxicity

Although ammonium plays an important role in microalgae metabolisms, high concentration of ammonium will cause toxicity and prohibit microalgae growth. In culture medium or wastewater, ammonium toxicity is caused by the effects of both ionized ammonium (NH_4^+) and unionized ammonia (NH_3). Generally, as an uncharged and lipid soluble chemical, ammonia, which can pass through the plasma membrane efficiently, is more toxic to microalgae cells [35]. Possible mechanisms of ammonium toxicity to microalgae include (1) high concentration of ammonium could change the osmotic pressure in culture medium or wastewater, further causing the dehydration of microalgae cells [36] and (2) oxidative stress caused by high concentration of ammonium would accelerate the

degradation of chlorophyll in microalgae cells and prohibit the photosynthesis [37].

Microalgae have very different tolerance levels to ammonium toxicity. For example, the tolerance level of Chlorophyceae to ammonia toxicity reached 60 mM, while Dinophyceae only tolerate about 2.5 mM ammonium [35]. As reported by previous work, tolerance levels of microalgae species are closely related with the metabolic mechanisms of cells and the growth conditions [7]. Hence, to grow microalgae in culture medium or wastewater with high concentrations of ammonia, selection of robust microalgal strain with high tolerance level is an essential step. For example, Zhao et al. [4] isolated a microalgal strain from landfill and found that it could tolerate high concentration of ammonium (over 500 mg/L). In addition, it is a feasible way to alleviate ammonia toxicity by creating a favorable growth condition. According to the study of Lu et al. [7]; carbon source in culture medium would directly impact the assimilation process of nitrogen. In terms of the alleviation of ammonia toxicity, glucose is much better than citric acid.

In a real-world application, to alleviate ammonia toxicity and promote the microalgae growth, a couple of strategies have been developed [4]. First, dilution is the simplest strategy to reduce the ammonia toxicity. Previous studies have mitigated ammonia toxicity in food-processing wastewater, municipal wastewater, and manure by using this strategy [38,39]. The disadvantage of this strategy is that a large amount of freshwater will be consumed. Second, ammonia stripping is another simple way to remove excessive ammonia in culture medium and alleviation toxicity. Park et al [39] developed vacuum-driven ammonia stripping system to pretreat the manure with high concentration of ammonia for microalgae cultivation. Ammonia removal by pumping air into wastewater or culture medium has also been proven to be an effective pretreatment strategy to alleviate ammonia toxicity.

Strategies to promote nitrogen utilization

Based on the above discussion, it is believed that three critical points in nitrogen assimilation should be considered to promote the nitrogen absorption and assimilation. The first point is the membrane transport protein, which is critical to the absorption of nitrogen source in wastewater. The second one is nitrate reductase, nitrite reductase, and urease whose activities are critical to the formation of NH_3-N. The third point is the GS-GOGAT pathway and GDH pathway, which are critical to the conversion of intracellular ammonia to glutamate in algal cells [33]. Issues with any of these critical points could result in inadequate nitrogen absorption or excessive ammonia in algae cells, posing lethal threats to the well-being of cells. In practice, both endogenous and exogenous approaches have been applied to accelerate nitrogen absorption and assimilation. Sun et al. [33] established the transformation system in the green alga *Dunaliella viridis* and demonstrated the function of nitrate reductase for promoting the conversion of NO_3-N into NH_3-N by genetic engineering. Many strategies, including balancing the ratio of C/N in wastewater, applying nitrogen starvation pretreatment, and optimizing the cultivation temperature, were also widely used [40,41].

Of these exogenous approaches discussed above, balancing the ratio of C/N is one of the most efficient and simple ways to promote nitrogen assimilation. According to the theory of Redfield stoichiometry, the ratio of C/N in culture medium should be around 6:1. This theory was confirmed by the study of Ma et al. [40] in which the highest algal biomass yield (3.4 g/L) was obtained when the ratio of C/N in molasses wastewater was 5:1. The main mechanism for this theory is that nitrogen assimilation is directly associated with carbon assimilation in algal cells. Some organic carbon entering Krebs cycle would produce an intermediate product, 2-oxoglutarate, which is the essential substrate for nitrogen assimilation by both GS-GOGAT pathway and GDH pathway (Fig. 3.1). In addition, some carbon metabolism processes, particularly glycolysis and Krebs cycle, supporting the generation of ATP and NADH, which drive the biochemical reactions of GS-GOGAT and GDH pathways, would promote the nitrogen assimilation in algal cells. Therefore, the increase of carbon content in wastewater could potentially promote the accumulation of intracellular 2-oxoglutarate for nitrogen assimilation.

Phosphorus Assimilation

Since phosphorus is not a major component of carbohydrates, fatty acids, or amino acids, its content in algal biomass is low. However, phosphorus plays an indispensable role in the regulation of algal metabolisms [42]. Ionized phosphorus in wastewater is absorbed by algal cells through either high-affinity Pi transporter or low-affinity Pi transporter on plasma membrane [43]. In algal cells, the form of absorbed phosphorus changes with the intracellular pH value. The pKs for the dissociation of H_3PO_4 to $H_2PO_4^-$ and then to HPO_4^{2-} are 2.1 and 7.2, respectively [44]. Absorbed phosphorus is assimilated to synthesize ATP through photophosphorylation, substrate-level phosphorylation, and oxidative phosphorylation [45]. Assimilated phosphorus participates in the metabolisms of algal cells in the forms of intermediate products of Calvin cycle, energy-storing molecules (ATP/ADP), nucleic acids, and phospholipid.

Phosphorus plays an important role in the metabolisms of microalgal cells. Firstly, phosphorus in the form of ribulose-1, 5-bisphosphate (RuBP) is the substrate for carbon dioxide assimilation through Calvin cycle, and all the intermediate products in Calvin cycle contain phosphorus. So assimilated phosphorus regulates the fixation of carbon dioxide through Calvin cycle in algal cells. Secondly, phosphorus is used to synthesize energy-storing molecules (ATP/ADP), which transport chemical energy in algal cells for metabolisms. Phosphorus deficiency would reduce the contents of ATP/ADP, further inhibiting all the cellular functions, including respiration, protein synthesis, primary active transport on plasma membrane, and lipid synthesis, driven by ATP [45,46]. Thirdly, phosphorus is an essential substrate for the synthesis of nucleic acids, which further regulate the expression of proteins in algal cells. Wurch et al. [42] demonstrated that phosphorus deficiency would disturb the gene transcripts and protein expression profiles. Fourthly, phospholipid bilayer, which is another composition with phosphorus, provides a barrier around algal cells and regulates the transport of some nutrients into intracellular environment.

THREE CULTIVATION MODES AND MICROALGAL COMPONENTS SYNTHESIS
Introduction of Three Cultivation Modes
According to the nutritional modes, microalgae cultivation strategies could be classified into three categories, namely, autotrophic cultivation, heterotrophic cultivation, and mixotrophic cultivation. Inorganic carbon is the main carbon source, and light or solar energy is the main energy source for microalgae growth in autotrophic cultivation, while microalgae in heterotrophic cultivation mainly use organic carbon and energy from Krebs cycle. In mixotrophic cultivation, carbon sources for microalgae growth include both inorganic carbon and organic carbon. The application and advantages/disadvantages of each cultivation strategy are discussed as follows.

Autotrophic cultivation
As shown in Fig. 3.2, photosynthesis, which converts carbon dioxide or bicarbonate to saccharides in microalgae cells, plays an essential role in autotrophic cultivation. The biomass yield and growth rate of microalgae are partly attributed to photosynthesis rate. Hence, it is a critical step to screen microalgal species, such as *Chlorella* sp., with high photosynthesis rate for autotrophic cultivation. Some critical control points, such as light intensity and carbon dioxide content, which determine the photosynthesis rate, should be strictly controlled in autotrophic cultivation. First, to ensure that CO_2 emission is less than CO_2 fixation by microalgae cells, light intensity should be kept higher than the compensation point. However, if the light intensity is over the threshold, microalgae growth will be limited by the oxidative stress caused by the light irradiation [47]. Second, CO_2 concentration in culture medium should be kept higher than the CO_2 compensation point. On the contrary, fixed carbon will not be sufficient to support the microalgae growth and biomass production [48].

In autotrophic cultivation mode, the risk of bacterial or fungal contamination is very low due to the lack of organic carbon in culture medium. Also the intensive light plays an important role in the prevention of bacterial contamination. Since the cultivation process could be performed in ambient environment, instead of sterile environment, autotrophic cultivation is a cost-saving mode for microalgae production. In addition, carbon dioxide for microalgae growth could be obtained from air at very low cost. Although autotrophic cultivation is cost-saving, it has very low biomass productivity, ranging from 0.02 to 0.20 mg/L/day [48,49].

Heterotrophic cultivation
Heterotrophic cultivation refers to the growth mode in which microalgae use organic carbon as the carbon source. The heterotrophic microalgal strains require an organic carbon source and can be grown devoid of illumination. Some microalgal species could utilize glucose, VFAs, glycerol, acetate, and other organic carbon for biomass synthesis in the conditions without any light [50]. In terms of biomass yield or productivity,

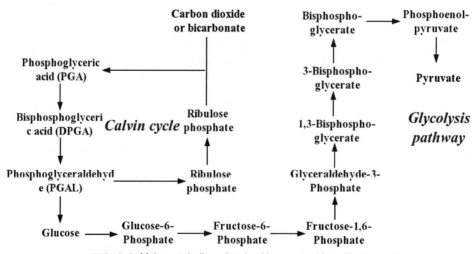

FIG. 3.2 Main metabolisms involved in autotrophic cultivation.

heterotrophic cultivation is much better than autotrophic cultivation. However, compared with carbon dioxide, organic carbon sources, such as glucose, VFAs, and acetate, are much more expensive. In a real-world application, to control the production cost of microalgae biomass by heterotrophic cultivation, organic carbon is mainly obtained from waste streams [51,52].

Different from the artificial culture medium, some waste streams may contain a lot of organics, which could not be effectively utilized by microalgae cells [53]. For example, it was observed that in swine manure, some of the insoluble organics could not be absorbed by microalgae cells. As a result, only a certain amount of organic carbon could be utilized by microalgae cells in heterotrophic cultivation mode. To improve the utilization efficiency of organic carbon by microalgae cells, Hu et al. [29] pretreated swine manure by anaerobic digestion, through which the solid organics were converted to VFAs. Thus, the utilization efficiency of organic carbon by microalgae cells in heterotrophic cultivation mode could be improved.

Although heterotrophic cultivation is considered as a promising way to achieve high biomass productivity, high investment cost and high risk of bacterial contamination seriously limit its application in microalgae production under some conditions. For example, heterotrophic cultivation is not feasible in the open pond system for microalgae biomass production. In addition, potential CO_2 emission by microalgae in heterotrophic cultivation mode is another problem criticized by some researchers and environmentalists. The potential CO_2 emission in heterotrophic cultivation mainly occurs in the Krebs cycle. In the practice, the cultivation modes for microalgae growth are selected according to the actual conditions.

Mixotrophic cultivation

To overcome the technical and economic disadvantages of heterotrophic and autotrophic cultivation modes, some studies used mixotrophic cultivation for microalgae biomass production [54,55]. In the mixotrophic cultivation, microalgae cells not only utilize inorganic carbon through photosynthesis but also assimilate organic carbon. Generally, in terms of carbon assimilation, microalgae growth in mixotrophic cultivation mode could be divided into two stages: organic carbon assimilation and inorganic carbon assimilation. In the first stage, microalgae growth could be described as heterotrophy due to high concentration of initial organic carbon in culture medium. When the organic carbon is consumed by microalgae cells and reduced to a certain level, photosynthesis is induced in microalgae

cells and the inorganic carbon sources, such as carbon dioxide and bicarbonate, indicating the onset of the second stage [56].

Compared with autotrophic cultivation and heterotrophic cultivation, mixotrophic cultivation has the following advantages. First, under mixotrophic cultivation mode, microalgae cells not only grow heterotrophically by assimilating organic carbon, increasing biomass yield, and enhancing lipid synthesis but also consume inorganic carbon (carbon dioxide and bicarbonate) and produce oxygen through photosynthesis reactions, making the overall CO_2 emission lower than that of heterotrophic cultivation mode [57]. In most cases, biomass yield and lipid content of microalgae under mixotrophic cultivation mode are not merely a sum of autotrophy and heterotrophy, suggesting that mixotrophic cultivation has favorable effects on microalgae growth and intracellular composition synthesis [58]. Second, some value-added compositions in microalgae cells could be preserved under mixotrophic cultivation mode. For example, the synthesis of photosynthetic pigments is enhanced by illumination under mixotrophic cultivation mode while prohibited under heterotrophic cultivation mode without any light source [59,60]. Third, illumination in mixotrophic cultivation mode could partly contribute to the control of bacterial contamination in culture medium with high concentration of organic carbon.

Synthesis of Value-Added Compositions in Microalgae Cells

To use microalgae biomass for the production of biofuel, animal feed, fertilizer, or food, the synthesis of value-added compositions, such as protein, lipid, and some bioactive compounds, should be enhanced. In this section, relation between nutrients assimilation and value-added compositions synthesis was assessed. Also, strategies to enhance the synthesis of value-added compositions were discussed.

Synthesis of lipid

A couple of years ago, studies on the microalgae lipid were conducted with the purpose of producing high-quality microalgae biodiesel. Some microalgal strains with high lipid yield are shown in Table 3.4 cited from the studies of Lowrey et al. [58] and Nascimento et al. (2013). Recently, the recession in global oil market and the decrease of oil price seriously prohibited the microalgae-based biodiesel. Thus, more and more studies are focusing on the value-added lipid compositions, particularly polyunsaturated fatty acids, for food or animal feed use. As shown in Fig. 3.2, since pyruvate is a precursor of fatty acid synthesis in microalgae cells,

TABLE 3.4
Microalgal Strains with High Lipid Productivity.

Microalgal Strain	Biomass Productivity (g/L/day)	Lipid Content (%)	Lipid Productivity (mg/L/day)
Ankistrodesmus falcatus	0.34	16.49	56.07
Ankistrodesmus fusiformis	0.24	20.66	49.58
Scenedesmus sp.	0.217	–	20.65
Kirchneriella lunaris	0.14	17.30	24.22
Chlamydomonas sp.	0.24	15.07	36.17
Chlamydocapsa bacillus	0.32	13.52	43.26
Chlorella sp.	0.104	–	6.91
Coelastrum microporum	0.11	20.55	22.61
Scenedesmus obliquus	0.16	16.73	26.77
Pseudokirchneriella subcapitata	0.08	28.43	22.74

carbon supply plays a critical role in lipid synthesis and accumulation in microalgae cells. In microalgal metabolisms, pyruvate is mainly from the glycolysis process, of which saccharide is one of the reaction substrates. Hence, to promote the lipid synthesis, it is necessary to provide sufficient organic carbon in culture medium or enhance the photosynthesis of microalgae cells. Besides carbon supply, some enzymes play essential roles in the synthesis of polyunsaturated fatty acids. The synthesis of polyunsaturated fatty acid is catalyzed by a couple of desaturases and elongases, so the performance of these essential enzymes will directly determine the fatty acid profile of microalgae biomass.

To produce microalgae biomass with high contents of polyunsaturated fatty acids, three strategies have been applied by previous studies. First, screening microalgal strains that are rich with polyunsaturated fatty acids is an effective strategy. Reported microalgae that could be exploited for the production of polyunsaturated fatty acids include Bacillariophyceae, Haptophyceae, Raphidophyceae, and so on [61]. Second, optimization of growth condition is another strategy to promote the synthesis of polyunsaturated fatty acids in microalgae cells. Third, molecular technology could be applied to modify the gene of microalgae for the synthesis of polyunsaturated fatty acids. However, since the safety issues of gene-modified (GM) food or feed are criticized by many researchers, the use of GM microalgae for polyunsaturated fatty acids has not been widely commercialized [62]. Thus, the first two strategies, which have been widely studied by previous research, are introduced as follows.

Most of the microalgal strains with high contents of polyunsaturated fatty acids are living in the ocean [63]. On the contrary, freshwater microalgae contain much lower contents of polyunsaturated fatty acids. Such a phenomenon is partly determined by the growth conditions of freshwater microalgae and marine microalgae in the evolutionary history. To survive in the marine environment, microalgae have to tolerate the low temperature and the high light intensity. Since polyunsaturated fatty acids could increase the fluidity of plasma membrane and decrease its freezing point, microalgae with high content of polyunsaturated fatty acids are more likely to survive in the environment with low temperature [64]. Also, polyunsaturated fatty acids, as antioxidants, could alleviate the oxidation process and improve the survival of microalgae in the environment with intensive irradiation.

Besides screening microalgal strains, optimizing the growth conditions, particularly decreasing the temperature, could produce biomass with high percentage of polyunsaturated fatty acids. Under the conditions with low temperature, driven by self-protection mechanisms, microalgae cells would decrease the freezing point of plasma membrane by accumulating polyunsaturated fatty acids. Previous studies have successfully used this strategy to produce biomass with high concentrations of polyunsaturated fatty acids by growing freshwater microalgae in culture medium or waste streams [65,66]. In a real-world application, considering the low biomass productivity of microalgae at low temperature, microalgae are always grown through two stages for polyunsaturated fatty acids accumulation. At the first stage, microalgae are grown at higher temperature (over 25°C) for biomass production. When the biomass yield reached peak value, the microalgae cultivation is shifted to the second stage, in which the temperature is controlled below 10°C [67]. Thus, polyunsaturated fatty acids will not be induced at the expense of biomass production.

Synthesis of bioactive compounds

Previous studies have fully documented the bioactive compounds, such as astaxanthin, β-carotene, polysaccharide, dietary fiber, peptide, and so forth, in microalgae biomass. Compared with microalgae protein and lipid, these bioactive compounds have wider application range in the industry. For example, astaxanthin could be used as a raw material for the production of animal feed, food supplement, and even some medical products. Considering the large market demand and high price of the bioactive compounds, researchers are paying more attention to bioactive compounds in microalgae. In this section, synthesis of astaxanthin in microalgae is introduced.

Astaxanthin (3,3'-dihydroxy-β, β-carotene-4,4'-dione), which is a ketocarotenoid with superior antioxidative activity, is an essential component in some fishery feeds. Ambati et al. (2014) reported that the addition of astaxanthin in the salmon could improve the meat quality. In terms of astaxanthin production, compared with conventional resources, such as fish and shrimp, microalgae have three advantages as follows: first, supported by modern microalgae cultivation technology, effects of extreme weather on the production of astaxanthin could be avoided [39]. Second, microalgae-based astaxanthin production can prevent the contamination of poisonous chemicals and ensure the safety of astaxanthin. Third, biomass productivities of microalgae are very high [39]. Because of these great advantages, in the market, astaxanthin from microalgae has replaced astaxanthin extracted from fish and shrimp. In the practice, to strictly control the potential toxic chemicals from wastewater, food processing effluent without any toxicity is preferred in the production of pigments for animal feed.

In nature, astaxanthin has three main geometric isomers, namely, all-*trans* astaxanthin, 9-*cis* astaxanthin, and 13-*cis* astaxanthin. Astaxanthin has a hydroxyl group, which can react with acid, such as a fatty acid, on each terminal ring. Sometimes, the hydroxyl group is "free," namely reacting with no acid. The reaction with acids can impact the hydrophobicity. In order of hydrophobicity, it is observed that diesters astaxanthin > monoesters astaxanthin > free

astaxanthin [68]. To promote the dissolution of astaxanthin in oil product, diesters astaxanthin is preferred.

In different algal strains, biosynthesis pathways of astaxanthin are different. The astaxanthin synthesis pathways in *Haematococcus pluvialis*, *Chlorella zofingiensis*, and *Scenedesmus obliquus* have been reported by previous studies [68,69]. Same as the polysaturated fatty acid, astaxanthin is accumulated by microalgae cells under specific conditions. As shown in Table 3.5, specific conditions to induce astaxanthin synthesis in microalgae cells include nitrogen deficiency, high salinity, and high irradiation. The research of Li et al. [68] reported that high concentration of glucose in culture medium is another critical factor to astaxanthin synthesis.

Synthesis of protein

Microalgae are considered as a viable source of protein for food industry or feed industry. Synthesis pathways of protein in microalgae cells have been fully documented by previous publications [70,71]. Some species of microalgae are known to contain protein contents similar to those of traditional protein sources, such as egg, meat, and soybean. Compared with these traditional protein sources, microalgae have a couple of benefits in resource utilization, nutritional value, and productivity. First, microalgae cultivation does not compete with traditional crops for agricultural resources, such as arable lands. The cultivation of marine microalgae even does not require freshwater [72]. Thus, microalgae protein will not be produced at the expense of other agricultural crops. Second, some microalgal species that are rich in essential amino acids have high nutritional values in food or feed industry [71]. Third, microalgae have much higher biomass productivity than traditional protein crops. It was reported that average productivity of microalgae protein could reach 4−15 tons/Ha/year, while terrestrial crops, such as soybean and pulse legumes, only produce protein at the productivity of 0.5−2.0 tons/Ha/year. The productivity of animal protein is much lower. Considering these benefits, microalgae-based protein production is attracting more attentions in both academic research and industry.

DESIGN OF PHOTOBIOREACTORS AND MICROALGAE HARVESTING TECHNOLOGIES

Examples of Photobioreactors for Microalgae Production

The selection and specific design of photobioreactors for microalgae production include a series of decisions

TABLE 3.5
Astaxanthin Yields (*Haematococcus pluvialis*) in Different Growth Conditions

Medium	Major Inducers	Cultivation Model	Biomass Yield or Cell Density	Astaxanthin Yield	References
Basal medium (sodium nitrate)	Nitrogen deficiency	One-stage cultivation	0.45 g/L	2 mg/L	[47]
Basal medium (calcium nitrate)	Nitrogen deficiency	One-stage cultivation	0.7 g/L	7.5 mg/L	[47]
BG-11 medium	High salinity (0.8% NaCl); high irradiance (350 $\mu E/m^2$ s)	One-stage cultivation	–	0.8% DW (dry weight)	[48]
OHM medium	Nitrogen deficiency; high irradiance	Two-stage cultivation	–	14 mg/L	[50]
Special heterotrophic medium	Acetate; high irradiance (950 $\mu mol/m^2$ s)	Two-stage cultivation	6 g/L	114 mg/L	[70]
OHM medium	Nitrogen deficiency; high irradiance; acetate	Two-stage cultivation	–	25 mg/L	[50]
OHM medium	High irradiance (240 mol photon $m^{-2}s^{-1}$)	Two-stage cultivation	6×10^5 cells/mL	50 mg/L	[49]
BG-11 medium	High salinity (0.8% NaCl); high irradiance (350 $\mu E/m^2$ s)	Two-stage cultivation	–	4.0% DW (dry weight)	[48]

on matters ranging from light source to bioreactor structure. Parameters that should be considered in the design of photobioreactors include the ratio of surface area to volume (s/v), light angle, accumulation of oxygen, mixing strategy, carbon dioxide supply, and temperature. To cultivate microalgae at sensible cost, various photobioreactors, including open photobioreactor and closed photobioreactor, have been designed. As open photobioreactors (having direct contact with the environment), raceway ponds, tanks, and inclined-surface platforms are widely used. As closed photobioreactors (having no direct contact between culture and environment), column bioreactor, tubular bioreactor, and flat plate bioreactor are typically used. Recently, to further reduce the cultivation cost or simplify the biomass harvesting procedure, some novel photo-bioreactors, such as revolving algal biofilm bioreactor and OMEGA (Offshore Membrane Enclosures for Growing Algae) system, were developed.

Open raceway bioreactor

Open raceway bioreactor was firstly reported by Ostwald in 1960s and had been revised and improved in the past decades [5]. A shown in Fig. 3.3, a typical open raceway bioreactor consists of two channels connected by bends. One or two paddle wheels are always employed to recirculate the culture medium and prevent the precipitation of microalgae (Fig. 3.3). In the practice, light source and gas supplier could be combined with the bioreactor to accelerate microalgae growth. The energy consumption of open raceway bioreactor is as low as 3.72 W/m^3, while the biomass yield could reach 0.4 g/L [73]. This type of bioreactor, with low energy consumption and low operation cost, is one of the most frequently used systems for microalgae cultivation. At industrial scale, the size of open raceway bioreactor could range between 1000 and 5000 m^2.

Many microalgal strains, including marine microalgae and freshwater microalgae, have been grown in open raceway bioreactors for either biomass production or wastewater remediation [74,75]. For the purpose of accelerating microalgae growth and promoting nutrients recovery, a couple of critical parameters, such as temperature and light intensity, have been optimized by previous work [76,77].

To further commercialize the open raceway bioreactor in microalgae industry, some technical bottlenecks should be overcome. Firstly, bacterial contamination caused by the direct contact between

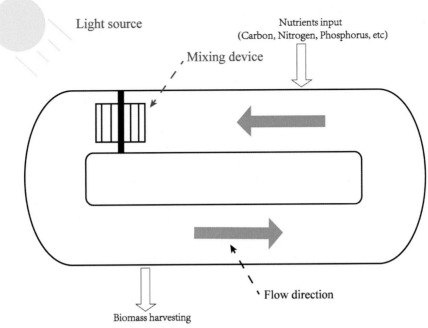

FIG. 3.3 Schematic drawing of open raceway bioreactor.

culture medium and environment may reduce the biomass yield and even cause the failure of microalgae cultivation. Although antibiotics could be applied to control the bacteria, overuse of antibiotics would increase the bacterial resistance and reduce the safety level of microalgae biomass. For example, microalgae biomass with the excessive residual of antibiotics could not be used for food or animal feed production. Secondly, compared with the vertical photobioreactors, horizontal open raceway bioreactors have much lower space utilization ratio. Thus, in a real-world application, most of the open raceway bioreactors are constructed on the nonarable land to alleviate the competition between microalgae cultivation and traditional crops production. Third, low biomass productivity caused by the temperature fluctuation, insufficient illumination, and nutrients deficiency in open raceway system is another technical bottleneck.

Tubular/flat plate/column bioreactor

Closed photobioreactors, including tubular bioreactor, flat plate bioreactor, and column bioreactor, have been developed to reduce the risks of bacterial contamination and produce value-added microalgae biomass [78].

In the past decades, the designs of these closed bioreactors have been improved to create a much better condition for microalgae growth. For example, the closed bioreactor, which could automatically change the illumination incident angle according to the position of sun, was used for microalgae cultivation [79]. As a result, the surface of microalgae cultivation bioreactor could be toward the light source all the time. Besides the light angle, light quality is another parameter that is studied by researchers in the design of closed bioreactors. Effects of light wavelength and light intensity on microalgae growth and nutrients value of biomass have been widely studied in previous work [80]. Atta et al. [80] proposed a novel concept of using light-emitting diode (LED) lamp to induce the synthesis of valuable nutrients, such as antioxidants and natural pigments, in *Spirulina* sp. grown in a closed column bioreactor. Since bacterial contamination could be effectively prevented in the use of closed photobioreactor, culture media or waste streams with high contents of organics could be used for microalgae cultivation [81]. Generally, in terms of microalgae biomass production, these closed bioreactors are much better than open raceway systems.

Revolving algal biofilm bioreactor

Although aforementioned bioreactors have been studied for many years, low algal cell density and high biomass harvesting cost seriously limit their application in the industry. Revolving algal biofilm bioreactor was firstly designed by the lab of Prof. Zhiyou Wen at Iowa State University to increase the cell density and reduce harvesting cost by constructing a condensed microalgae film [82]. According to the report of Gross et al. [82]; microalgae biomass density on the biofilm could reach 40 g/m^2. Thus, the biomass yield of 1 m^2 biofilm is equal to that of 50 L culture medium under autotrophic cultivation mode (about 0.8 g/L microalgae biomass). In this way, the size of bioreactor could be significantly reduced, and the harvesting process could be simplified. With the effects of other research teams, now this novel biofilm bioreactor has been successfully commercialized in some wastewater treatment plants.

According to previous studies, the critical control points of this biofilm bioreactor are listed as follows. First, selection of appropriate film for microalgae attachment is an essential step in the construction of this biofilm bioreactor [83]. Gross and Wen [83] discovered that cotton duct is a good material for the construction of biofilm by comparing a couple of materials, such as cotton denim, corduroy, and cotton rag. Second, it is necessary to screen microalgal strains that could attach on the film easily and assimilate nutrients in culture medium efficiently. Now, the microalgal strains used for the construction of biofilm include *Chlorella* sp. and *Scenedesmus* sp. Third, the design of cost-saving and energy-saving harvesting device is another critical control point in the use of revolving algal biofilm bioreactor.

Although revolving algal biofilm bioreactor has been commercialized in microalgae cultivation, some studies should be conducted in the coming future to further improve its performance in biomass production and wastewater remediation. First, the interaction between microalgae and medium-borne bacteria or fungi in the revolving algal biofilm bioreactor should be studied. Theoretically, the activities of some bacteria or fungi in this bioreactor could promote the attachment of microalgae on the film. Second, technologies to extend the usage period of film materials should be developed to reduce the replacement frequency of microalgal biofilm.

OMEGA bioreactor

OMEGA project was led by National Aeronautics and Space Administration (NASA) of The United States during 2009–12, aiming at using Offshore Membrane Enclosures for Growing Algae (OMEGA) system for microalgae cultivation, waste stream remediation, and carbon dioxide capture. The OMEGA system mainly consists of large flexible plastic tubes. According to the experimental results, biomass productivity of OMEGA system could reach 4.5 g/m^2 day [84].

In this case, the mixing process, which usually consumes a lot of energy, is driven by the wave. Thus, the energy consumption for microalgae cultivation could be reduced to a lower level. In addition, the offshore placement eliminates the need for terrestrial resources and reduces the investment cost of microalgae cultivation. Another advantage of OMEGA system is that structural supports of the whole system are provided by ocean, instead of expensive mechanical components typically used by traditional photobioreactors. Previous studies further reduced the cost of OMEGA system by using wastewater, instead of artificial culture medium, for microalgae cultivation, and treated wastewater could be directly discharged to the ocean [84]. In this way, not only the microalgae production cost but also the wastewater treatment cost could be reduced to a very low level. The limitation of OMEGA bioreactor is that this bioreactor is only applicable in the areas with ocean, lake, or river.

Negative pressure–driven bioreactor

In most of the conventional photobioreactors, for the purpose of accelerating the mixing of microalgae with culture medium and promoting the nutrients assimilation, energy-intensive power units, such as pumps (over 1 kW), have to be employed to overcome the air resistance in the system. Thus, the energy consumption caused by air resistance increases the production cost of microalgae biomass. To reduce energy consumption and operation cost, the research team of Prof. Wenguang Zhou at Nanchang University (Jiangxi, China) designed a novel negative pressure–driven tubular photobioreactor (Fig. 3.4).

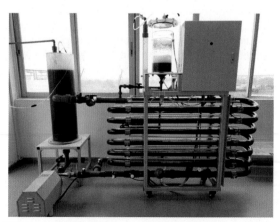

FIG. 3.4 Negative pressure–driven tubular photobioreactor.

In this negative pressure—driven tubular photobioreactor, culture medium is lifted to a top container by a vacuum pump with very low energy consumption (about 7W). According to the experimental results, compared with conventional tubular photobioreactor, this novel negative pressure—driven photobioreactor reduced the energy consumption by 99.3%. Due to the low energy consumption, it has been successfully commercialized for both industrial microalgae cultivation and household food-grade microalgae production.

Aquaponics system for microalgae production

Besides the design of photobioreactors mentioned above, recently, a novel concept of integrating photobioreactor with other agricultural systems to reduce the cost of microalgae biomass or improve the recovery efficiency of agricultural residuals is attracting researchers' attention. The use of aquaponics system for microalgae cultivation, which was firstly studied by the researchers from the University of Minnesota [8], is a good example of such a novel concept. As shown in Fig. 3.5, wastewater from the fish tank goes through the open pond for microalgae growth and then pump into the vegetable cultivation system. Harvested microalgae biomass could be directly used as feed for fish growth.

Since some of the vegetables have very low tolerance levels of ammonia, organics excreted by fish might threaten the growth of vegetables. Furthermore, in conventional aquaponics system, final products, such as vegetables, could not partly reduce the input of raw materials, particularly fish feed. Hence, such a system does not meet the basic criteria of circular economy. The benefits of integrating aquaponics system with microalgae production include the following (1) microalgae with great capacity of removing ammonia in aquaculture wastewater could effectively assimilate ammonia accumulated in fish tank and alleviate ammonia toxicity, creating a better condition for fish growth, and (2) the use of valuable microalgae biomass as fish feed would reduce the cost and achieve the goals of circular economy.

Recently, the research team of Prof. Wenguang Zhou from Nanchang University (Jiangxi, China) is trying to integrate the symbiosis of microalgae and probiotics with the aquaponics system. Probiotics would convert solids in wastewater to soluble nutrients, which could be utilized by microalgae cells, by releasing various extracellular enzymes. At the same time, the probiotics could accumulate highly valuable compositions, such as vitamins, pigments, and polypeptides [86]. As a result, harvested biomass consisted of microalgae, and probiotics would increase the immunity of fish and prevent the fishery diseases. The success of this project will prohibit the overuse of antibiotics in fish feed and promote further development of eco-friendly agriculture.

Harvesting Technologies

Efficient harvesting of biomass from culture medium at low cost is essential for industrial use of microalgae. The techniques that have been widely used to harvest microalgae include gravity sedimentation, flocculation,

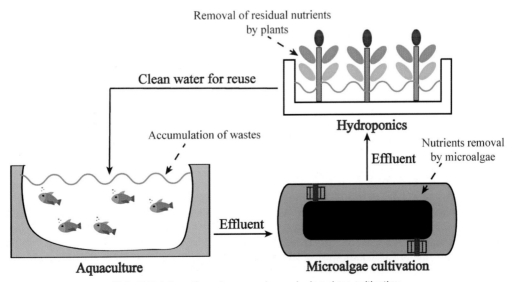

FIG. 3.5 Integration of aquaponics and microalgae cultivation.

centrifugation, filtration, flotation, and electrocoagulation [87]. The cost and energy consumption of microalgae harvesting are very high, since the biomass density in culture medium is generally low and most of the microalgae cells carry negative charge that keeps the stability of cells in a dispersed state. It was reported that the cost of harvesting process even accounts for 30% of the total cost of microalgae biomass production [88]. Therefore, it is a feasible way to reduce the production cost of microalgae biomass by improving the harvesting technologies.

The selection of harvesting technique is dependent on the properties of microalgae and the value of the biomass products. Generally, the harvesting process could be divided into two steps, namely, bulk harvesting and thickening. The purpose of bulk harvesting technologies, including flocculation, flotation, and gravity sedimentation, is to separate biomass from the culture medium. Thickening, with much higher energy consumption than bulk harvesting, includes filtration and centrifugation. Specific technologies for microalgae harvesting are listed as follows.

Sedimentation/centrifugation/flocculation

Gravity sedimentation is the most simple and energy-saving method for algae harvesting. However, in some cases, this harvesting method is extremely time-consuming and inefficient. Griffiths et al. [89] reported that applied gravity sedimentation to harvest 11 algal species demonstrated that the maximum sedimentation rate of *Spirulina platensis* and *Cylindrotheca fusiformis* reached 11 g/(L·h), while that of some other algal species was even lower than 0.008 g/(L·h) [89]. Therefore, gravity sedimentation is only applicable in the harvesting of algal species with large sizes and high biomass densities. In addition, in algae-based wastewater treatment, it is impossible to separate the algal biomass and solid in wastewater by gravity.

Although centrifugation-based strategy has high harvesting efficiency and brings no additional contaminants into algal biomass, high energy consumption and cost make it unfeasible in practice [90]. Milledge and Heaven [90] estimated that the energy available in the algal biodiesel (approximately 9.09 kW h/L or 10.33 kW h/kg) is only one-fourth of the energy consumed by centrifugation. Dassey and Theegala [91] claimed that the centrifugation harvesting cost of algae with high oil content (65% by dry mass) could be reduced by 82% through conducting multiple stage harvesting and increasing the flow rate of algae culture medium. Even under the ideal conditions of this improved harvesting method, the estimate energy

required to produce algal oil and the price of oil were higher than 9.6 kW h/L and \$0.86/L, respectively [91]. Therefore, centrifugation is mainly applied in the harvesting of algae with highly valuable compositions [92].

Chemical flocculation with low energy consumption is a simple and efficient way for algae harvesting [93]. Chemical flocculants include aluminum salt, ferric salt, and synthetic polyacrylamide polymers. Chatsungnoen and Chisti [93] reported that aluminum sulfate and ferric chloride were used as flocculants, and harvesting efficiency reached 95% after 62 min of sedimentation. However, the usage of excessive dosages of flocculants, particularly aluminum salts and some toxic polymers, in algae harvesting would contaminate the algal biomass and prohibit its use in food, feed, and fertilizer industries. Filtration, which is carried out commonly on membranes with assistance of a suction pump, is particularly suitable for the collection of algae of low densities. However, the solid particles in wastewaters would lead to the eventual clogging of the filter and reduce the harvesting efficiency. Therefore, filtration is not feasible in the harvesting of algae grown on most wastewaters with solids.

Fungi-assisted sedimentation

Filamentous fungi with the ability of forming large pellets provide an alternative way for algae harvesting. Impacted by growth conditions, fungi have various morphologies, such as dispersed hyphae, denser spherical aggregates, and loose hyphal aggregates [94]. According to previous studies, only denser spherical aggregates, also called pelleted fungi, have been effectively applied in algae harvesting. Therefore, the formation of fungal pellet, which is driven by electrostatic force, hydrophobicity, and specific interaction of fungal cell wall compositions, is significant to algae harvesting. The main mechanism of fungi-assisted sedimentation in algae harvesting is that algal cells are copelletized into the large fungal pellets, which could be harvested through sieve, much easier than suspended cells. Compared with centrifugation and electric flocculation, fungi-assisted sedimentation consumes much less energy in the harvesting process. In addition, many filamentous fungal products are certificated as "generally regarded as safe" by the Food and Drug Administration (FDA) in the United States. Considering the nontoxicity of those fungal species, this method would not prohibit the use of harvested biomass in some industries. In this way, the purposes of refining algal biomass and improving its safety-level could be achieved in the harvesting process. Besides, due to the different sizes, algae biomass entrapped in fungal pellets (2−10 mm) could be separated from solids (<0.2 mm) in wastewaters by a simple way [95]. This point is significant to the harvesting of algae grown on wastewaters. Because of these advantages, it is of great interest to apply the fungi-assisted sedimentation in algae harvesting.

Recent studies have successfully used various filamentous fungi, such as *Aspergillus* sp., *Mucor* sp., *Rhizopus* sp., and *Penicillium* sp., to harvest algae [96,97]. Under the optimum conditions, algae harvesting efficiencies of fungi-assisted sedimentation ranged between 90% and 100% [94]. The structure and formation of fungal pellets, which determine the harvesting efficiency, are impacted by various factors, such as fungal species, carbon nitrogen ratio (C/N ratio), cell wall composition, inoculation ratio, temperature, agitation, pH, suspended solids in medium, trace metals, and so on [95]. Four cultivation parameters with decisive effects on harvesting efficiencies are discussed as follows. Firstly, effect of pH on fungal pellet formation is mainly through a change in the surface properties of fungi. Due to the different surface properties, fungal species have different responses to the pH values of culture medium. Zhou et al. [97] reported that acidic conditions were favorable to the formation of *Aspergillus oryzae* pellets, while Liu et al. [96] did not observe significant difference in the formation of *Rhizopus oryzae* pellets with a pH range of 2.5–7.0. Secondly, agitation speed is a key parameter that determines the size of fungal pellets. An inverse relationship between agitation speed and pellet formation was observed in previous studies. Zhou et al. [97] demonstrated that diameters of pellets formed in flasks agitated at 50 and 150 rpm were about 3 and 8 mm, respectively. As shown in Table 3.4, agitation speed for fungal pelletization is usually controlled below 190 rpm. Thirdly, temperature plays an important role in the formation of fungal pellets [98]. Fourthly, nutrients, particularly carbon sources, in culture medium also impact the fungal pelletization. Glucose is the most common carbon source, while glycerol and acetate could also be used for the formation of fungal pellets. In practice, to make this harvesting method feasible, food wastes, including glycerol and acetate, and some food processing wastewaters with high content of glucose could be used as cheap culture media for the fungal pelletization.

Nontoxic components-based flocculation
To avoid toxic contamination, some studies modified chemical flocculation methods by using nontoxic components, particularly naturally occurring bioflocculants produced by bacteria without toxicity [99]. Most bioflocculants are biopolymers with various compositions, such as carbohydrates and proteins. Elkady et al. [99] purified bioflocculant from *Bacillus mojavensis* and detected the bioflocculant consisting of 17 types of amino acids and five types of sugars. Due to the different metabolisms in microbial cells, compositions of bioflocculants vary from one to another. Cultivation condition could also impact the compositions of bioflocculants.

Due to the complicated compositions, various hypotheses have been proposed to explain the bioflocculation mechanisms. Firstly, based on the study of bioflocculant (97.54% polysaccharides and 2.46% protein) produced by *Bacillus mucilaginosus*, Lian et al. (2008) claimed that the flocculation takes place mainly through charge neutralization and bridging mechanism. It was discovered that driven by electrostatic force, bioflocculants were absorbed onto negatively charged kaolin particles, thereby bridging particles. When the gravity of floc exceeded buoyant force, aggregated floc settled down and precipitated. A theory based on cations mediated bridging was also reported by studying the bioflocculants produced by *Halomonas* sp. and *Pseudoalteromonas* sp., respectively. According to this theory, cations could neutralize negative charges of both bioflocculants and suspended particles. The connection between bioflocculants and particles is established by such a cations-mediated bridge. Therefore, addition of metal cations could enhance the performance of negatively charged bioflocculants.

Besides the advantage of nontoxicity, high harvesting efficiency is also an advantage of bioflocculants. The use of bioflocculants in algae harvesting demonstrated that under the optimized conditions, harvesting efficiencies of many algal species reached 90%. In the harvesting process, to maximize the harvesting efficiency, two main parameters should be taken into consideration. Firstly, pH value could determine the bioflocculation process by impacting the charge density of flocculants. Zhang et al. [100] demonstrated that the increase of pH value neutralized the negative charge of functional groups on some bioflocculants. The changes of charge density would further impact the attraction and connection between bioflocculants and algal cells in the harvesting process. Secondly, metal ions play a critical role in the formation of flocs connected by cations-mediated bridge. Therefore, presence of metal ions in wastewater could enhance the harvesting performance of negatively charged bio-flocculants while having no significant effects on positively charged bioflocculants [100]. Besides pH value and metal ions, some other parameters, including temperatures, mixing speeds, and concentrations of bioflocculants, were explored by previous studies as well. Before the use of certain bioflocculant, its biochemical properties and physical properties should be measured to identify the appropriate flocculation conditions.

Flotation

Flotation, of which the mechanism for algae harvesting is that microbubbles generated physically or chemically adhere to algal cells and cause them to float onto the surface of culture medium for further collection, has been applied in algae production since the 1960s [101]. However, high energy consumption of compressor and saturator in traditional flotation (also called "dissolved flotation") system is a serious concern. In addition, algal cells may block the nozzles for bubbles generations, increasing the maintenance cost of traditional flotation system.

In recent years, some modified flotation techniques have attracted attentions of researchers. Wiley et al. [101] developed a suspended flotation system that used surfactant to chemically generate microbubbles for algae harvesting. It was proven that suspended flotation method is much more cost-saving than dissolved flotation method. However, until now, research on the suspended flotation method has not been widely conducted. To verify the feasibility and applicability of suspended flotation method in algae harvesting, further studies are needed. Another emerging harvesting technology is electroflotation, which generates microbubbles, including hydrogen gas and oxygen gas, by electrolysis. At the same time, electrolysis would neutralize the negative charge on the surface of algal cells, causing coagulation that is favorable to algae harvesting as well. According to the principles of electroflotation, some harvesting facilities have been designed [102]. However, considering the potential risks of hydrogen gas explosion caused by electrolysis, electroflotation method should be modified before wide implementation. One of the greatest advantages of these new flotation methods is that the nozzle block caused by algal cells is effectively prevented by avoiding the use of traditional bubbles generation systems. As a result, the maintenance cost of harvesting systems would be reduced significantly.

CONCLUSIONS

Technologies, including the exploitation of waste streams as culture media, the design of photobioreactor, and the biomass harvesting techniques, associated with microalgae cultivation have been significantly improved. To further promote the commercialization

of microalgae cultivation, some issues still need to be addressed. First, production cost of microalgae biomass should be reduced by improving the photosynthesis rate, reducing cost of culture medium, simplifying the harvesting procedure, and so forth. Second, the value-added compositions in microalgae biomass should be exploited. Considering the price fluctuation in global oil market, microalgae-based biodiesel may not be commercialized in some cases. Hence, it is necessary to further extend the use of microalgae biomass in other fields, such as food supplement, animal feed, and organic fertilizer. Third, techniques to improve the nutrients value of microalgae biomass are required to promote the commercialization of microalgae cultivation. For example, strategies to enhance the synthesis of value-added compositions, such as polyunsaturated fatty acids, essential amino acids, and natural pigments, should be further developed in the coming future. Advances in these areas will allow the achievement of a relevant scale and establishment of a valuable industry based on microalgae cultivation.

REFERENCES

[1] Y. Shan, J. Liu, Z. Liu, X. Xu, S. Shao, P. Wang, D. Guan, New provincial CO_2 emission inventories in China based on apparent energy consumption data and updated emission factors, Appl. Energy 184 (2016) 742−750.

[2] W. Zhou, J. Wang, P. Chen, C. Ji, Q. Kang, B. Lu, K. Li, J. Liu, R. Ruan, Bio-mitigation of carbon dioxide using microalgal systems: advances and perspectives, Renew. Sustain. Energy Rev. 76 (2017) 1163−1175.

[3] T. Suganya, M. Varman, H. Masjuki, S. Renganathan, Macroalgae and microalgae as a potential source for commercial applications along with biofuels production: a biorefinery approach, Renew. Sustain. Energy Rev. 55 (2016) 909−941.

[4] X. Zhao, Y. Zhou, S. Huang, D. Qiu, L. Schideman, X. Chai, Y. Zhao, Characterization of microalgae-bacteria consortium cultured in landfill leachate for carbon fixation and lipid production, Bioresour. Technol. 156 (2014) 322−328.

[5] F. Acién, E. Molina, A. Reis, G. Torzillo, G. Zittelli, C. Sepúlveda, J. Masojídek, Photobioreactors for the Production of Microalgae, Microalgae-Based Biofuels and Bioproducts, Elsevier, 2018, pp. 1−44.

[6] Q. Lu, W. Zhou, M. Min, X. Ma, Y. Ma, P. Chen, H. Zheng, Y.T. Doan, H. Liu, C. Chen, Mitigating ammonia nitrogen deficiency in dairy wastewaters for algae cultivation, Bioresour. Technol. 201 (2016) 33−40.

[7] Q. Lu, P. Chen, M. Addy, R. Zhang, X. Deng, Y. Ma, Y. Cheng, F. Hussain, C. Chen, Y. Liu, Carbon-dependent alleviation of ammonia toxicity for algae

cultivation and associated mechanisms exploration, Bioresour. Technol. 249 (2018) 99−107.

[8] J. Ouwehand, E. Van Eynde, E. De Canck, S. Lenaerts, A. Verberckmoes, P. Van Der Voort, Titania-functionalized diatom frustules as photocatalyst for indoor air purification, Appl. Catal. B Environ. 226 (2018) 303−310.

[9] T. Niederwieser, P. Kociolek, D. Klaus, Spacecraft cabin environment effects on the growth and behavior of *Chlorella vulgaris* for life support applications, Life. Sci. Space. Res. (2017).

[10] D.-H. Cho, R. Ramanan, J. Heo, Z. Kang, B.-H. Kim, C.-Y. Ahn, H.-M. Oh, H.-S. Kim, Organic carbon, influent microbial diversity and temperature strongly influence algal diversity and biomass in raceway ponds treating raw municipal wastewater, Bioresour. Technol. 191 (2015) 481−487.

[11] T.M. Sobczuk, M.I. González, E.M. Grima, Y. Chisti, Forward osmosis with waste glycerol for concentrating microalgae slurries, Algal Res. 8 (2015) 168−173.

[12] P. Tsapekos, P. Kougias, M. Alvarado-Morales, A. Kovalovszki, M. Corbière, I. Angelidaki, Energy recovery from wastewater microalgae through anaerobic digestion process: methane potential, continuous reactor operation and modelling aspects, Biochem. Eng. J. 139 (2018) 1−7.

[13] Q. Lu, W. Zhou, M. Min, X. Ma, C. Chandra, Y.T. Doan, Y. Ma, H. Zheng, S. Cheng, R. Griffith, Growing *Chlorella* sp. on meat processing wastewater for nutrient removal and biomass production, Bioresour. Technol. 198 (2015) 189−197.

[14] T. Moriyama, K. Sakurai, K. Sekine, N. Sato, Subcellular distribution of central carbohydrate metabolism pathways in the red alga *Cyanidioschyzon merolae*, Planta 240 (3) (2014) 585−598.

[15] H.-Y. Ren, B.-F. Liu, C. Ma, L. Zhao, N.-Q. Ren, A new lipid-rich microalga Scenedesmus sp. strain R-16 isolated using Nile red staining: effects of carbon and nitrogen sources and initial pH on the biomass and lipid production, Biotechnol. Biofuels 6 (1) (2013) 143.

[16] B. Cheirsilp, S. Kitcha, S. Torpee, Co-culture of an oleaginous yeast Rhodotorula glutinis and a microalga *Chlorella vulgaris* for biomass and lipid production using pure and crude glycerol as a sole carbon source, Ann. Microbiol. 62 (3) (2012) 987−993.

[17] X. Feng, T.H. Walker, W.C. Bridges, C. Thornton, K. Gopalakrishnan, Biomass and lipid production of Chlorella protothecoides under heterotrophic cultivation on a mixed waste substrate of brewer fermentation and crude glycerol, Bioresour. Technol. 166 (2014) 17−23.

[18] W.-B. Kong, H. Yang, Y.-T. Cao, H. Song, S.-F. Hua, C.-G. Xia, Effect of glycerol and glucose on the enhancement of biomass, lipid and soluble carbohydrate production by *Chlorella vulgaris* in mixotrophic culture, Food Technol. Biotechnol. 51 (1) (2013) 62.

[19] J. Pires, M. Alvim-Ferraz, F. Martins, M. Simões, Carbon dioxide capture from flue gases using microalgae: engineering aspects and biorefinery concept, Renew. Sustain. Energy Rev. 16 (5) (2012) 3043−3053.

[20] Q. Lu, J. Li, J. Wang, K. Li, J. Li, P. Han, P. Chen, W. Zhou, Exploration of a mechanism for the production of highly unsaturated fatty acids in *Scenedesmus* sp. at low temperature grown on oil crop residue based medium, Bioresour. Technol. 244 (2017) 542−551.

[21] M. Hasib-ur-Rahman, F. Larachi, Prospects of using room-temperature ionic liquids as corrosion inhibitors in aqueous ethanolamine-based CO_2 capture solvents, Ind. Eng. Chem. Res. 52 (49) (2013) 17682−17685.

[22] S.A. Razzak, M.M. Hossain, R.A. Lucky, A.S. Bassi, H. de Lasa, Integrated CO_2 capture, wastewater treatment and biofuel production by microalgae culturing—a review, Renew. Sustain. Energy Rev. 27 (2013) 622−653.

[23] M.S.A. Rahaman, L.-H. Cheng, X.-H. Xu, L. Zhang, H.-L. Chen, A review of carbon dioxide capture and utilization by membrane integrated microalgal cultivation processes, Renew. Sustain. Energy Rev. 15 (8) (2011) 4002−4012.

[24] S.-Y. Chiu, C.-Y. Kao, M.-T. Tsai, S.-C. Ong, C.-H. Chen, C.-S. Lin, Lipid accumulation and CO_2 utilization of *Nannochloropsis oculata* in response to CO_2 aeration, Bioresour. Technol. 100 (2) (2009) 833−838.

[25] J. Cheng, Y. Huang, J. Feng, J. Sun, J. Zhou, K. Cen, Improving CO_2 fixation efficiency by optimizing Chlorella PY-ZU1 culture conditions in sequential bioreactors, Bioresour. Technol. 144 (2013) 321−327.

[26] M.G. De Morais, J.A.V. Costa, Biofixation of carbon dioxide by *Spirulina* sp. and *Scenedesmus obliquus* cultivated in a three-stage serial tubular photobioreactor, J. Biotechnol. 129 (3) (2007) 439−445.

[27] W.Y. Cheah, P.L. Show, J.-S. Chang, T.C. Ling, J.C. Juan, Biosequestration of atmospheric CO_2 and flue gas-containing CO_2 by microalgae, Bioresour. Technol. 184 (2015) 190−201.

[28] J.A. Lara-Gil, M.M. Álvarez, A. Pacheco, Toxicity of flue gas components from cement plants in microalgae CO_2 mitigation systems, J. Appl. Phycol. 26 (1) (2014) 357−368.

[29] B. Hu, M. Min, W. Zhou, Z. Du, M. Mohr, P. Chen, J. Zhu, Y. Liu, R. Ruan, Enhanced mixotrophic growth of microalga *Chlorella* sp. on pretreated swine manure for simultaneous biofuel feedstock production and nutrient removal, Bioresour. Technol. 126 (2012) 71−79.

[30] A.F. Talebi, S.K. Mohtashami, M. Tabatabaei, M. Tohidfar, A. Bagheri, M. Zeinalabedini, H.H. Mirzaei, M. Mirzajanzadeh, S.M. Shafaroudi, S. Bakhtiari, Fatty acids profiling: a selective criterion for screening microalgae strains for biodiesel production, Algal Res. 2 (3) (2013) 258−267.

[31] F.-W. Yin, S.-Y. Zhu, D.-S. Guo, L.-J. Ren, X.-J. Ji, H. Huang, Z. Gao, Development of a strategy for the production of docosahexaenoic acid by *Schizochytrium* sp. from cane molasses and algae-residue, Bioresour. Technol. 217 (2019) 118−124.

[32] B. Molinuevo-Salces, A. Mahdy, M. Ballesteros, C. González-Fernández, From piggery wastewater nutrients to biogas: microalgae biomass revalorization through anaerobic digestion, Renew. Energy 96 (2016) 1103−1110.

[33] Y. Sun, X. Gao, Q. Li, Q. Zhang, Z. Xu, Functional complementation of a nitrate reductase defective mutant of a green alga *Dunaliella viridis* by introducing the nitrate reductase gene, Gene 377 (2006) 140−149.

[34] S. Shi, Y. Qu, F. Ma, J. Zhou, Bioremediation of coking wastewater containing carbazole, dibenzofuran, dibenzothiophene and naphthalene by a naphthalene-cultivated *Arthrobacter* sp. W1, Bioresour. Technol. 164 (2014) 28−33.

[35] Y. Collos, P.J. Harrison, Acclimation and toxicity of high ammonium concentrations to unicellular algae, Mar. Pollut. Bull. 80 (1−2) (2014) 8−23.

[36] P. Buckwalter, T. Embaye, S. Gormly, J.D. Trent, Dewatering microalgae by forward osmosis, Desalination 312 (2013) 19−22.

[37] J. Nimptsch, S. Pflugmacher, Ammonia triggers the promotion of oxidative stress in the aquatic macrophyte Myriophyllum mattogrossense, Chemosphere 66 (4) (2007) 708−714.

[38] Y. He, R. Wang, G. Liviu, Q. Lu, An integrated algal-bacterial system for the bio-conversion of wheat bran and treatment of rural domestic effluent, J. Clean. Prod. 165 (2017) 458−467.

[39] J. Park, H.-F. Jin, B.-R. Lim, K.-Y. Park, K. Lee, Ammonia removal from anaerobic digestion effluent of livestock waste using green alga *Scenedesmus* sp. Bioresour. Technol. 101 (22) (2010) 8649−8657.

[40] C. Ma, H. Wen, D. Xing, X. Pei, J. Zhu, N. Ren, B. Liu, Molasses wastewater treatment and lipid production at low temperature conditions by a microalgal mutant *Scenedesmus* sp. Z-4, Biotechnol. Biofuels 10 (1) (2017) 111.

[41] J. Wang, W. Zhou, H. Yang, F. Wang, R. Ruan, Trophic mode conversion and nitrogen deprivation of microalgae for high ammonium removal from synthetic wastewater, Bioresour. Technol. 196 (2015) 668−676.

[42] L.L. Wurch, E.M. Bertrand, M.A. Saito, B.A. Van Mooy, S.T. Dyhrman, Proteome changes driven by phosphorus deficiency and recovery in the brown tide-forming alga Aureococcus anophagefferens, PLoS One 6 (12) (2011) e28949.

[43] S. Lin, R.W. Litaker, W.G. Sunda, Phosphorus physiological ecology and molecular mechanisms in marine phytoplankton, J. Phycol. 52 (1) (2016) 10−36.

[44] P. Lobit, P. Soing, M. Génard, R. Habib, Theoretical analysis of relationships between composition, pH, and titratable acidity of peach fruit, J. Plant Nutr. 25 (12) (2002) 2775−2792.

[45] G. Kiss, C. Konrad, I. Pour-Ghaz, J.J. Mansour, B. Németh, A.A. Starkov, V. Adam-Vizi, C. Chinopoulos, Mitochondrial diaphorases as NAD+ donors to segments of the citric acid cycle that support substrate-level phosphorylation yielding ATP during respiratory inhibition, FASEB J. 28 (4) (2014) 1682−1697.

[46] Y. Wang, S.-H. Ho, C.-L. Cheng, W.-Q. Guo, D. Nagarajan, N.-Q. Ren, D.-J. Lee, J.-S. Chang, Perspectives on the feasibility of using microalgae for industrial wastewater treatment, Bioresour. Technol. 222 (2016) 485−497.

[47] S. Wahidin, A. Idris, S.R.M. Shaleh, The influence of light intensity and photoperiod on the growth and lipid content of microalgae *Nannochloropsis* sp, Bioresour. Technol. 129 (2013) 7−11.

[48] A. Kumar, S. Ergas, X. Yuan, A. Sahu, Q. Zhang, J. Dewulf, F.X. Malcata, H. Van Langenhove, Enhanced CO_2 fixation and biofuel production via microalgae: recent developments and future directions, Trends Biotechnol. 28 (7) (2010) 371−380.

[49] A. Demirbas, M.F. Demirbas, Importance of algae oil as a source of biodiesel, Energ. Convers. Manage. 52 (1) (2011) 163−170.

[50] O. Perez-Garcia, F.M. Escalante, L.E. de-Bashan, Y. Bashan, Heterotrophic cultures of microalgae: metabolism and potential products, Water Res. 45 (1) (2011) 11−36.

[51] T. Cai, S.Y. Park, Y. Li, Nutrient recovery from wastewater streams by microalgae: status and prospects, Renew. Sustain. Energy Rev. 19 (2013) 360−369.

[52] M.P. Devi, G.V. Subhash, S.V. Mohan, Heterotrophic cultivation of mixed microalgae for lipid accumulation and wastewater treatment during sequential growth and starvation phases: effect of nutrient supplementation, Renew. Energy 43 (2012) 276−283.

[53] T.-Y. Zhang, Y.-H. Wu, S.-f. Zhu, F.-M. Li, H.-Y. Hu, Isolation and heterotrophic cultivation of mixotrophic microalgae strains for domestic wastewater treatment and lipid production under dark condition, Bioresour. Technol. 149 (2013) 586−589.

[54] S. Kim, J.-e. Park, Y.-B. Cho, S.-J. Hwang, Growth rate, organic carbon and nutrient removal rates of *Chlorella sorokiniana* in autotrophic, heterotrophic and mixotrophic conditions, Bioresour. Technol. 144 (2013) 8−13.

[55] S.V. Mohan, M.P. Devi, Salinity stress induced lipid synthesis to harness biodiesel during dual mode cultivation of mixotrophic microalgae, Bioresour. Technol. 165 (2014) 288−294.

[56] W. Farooq, Y.-C. Lee, B.-G. Ryu, B.-H. Kim, H.-S. Kim, Y.-E. Choi, J.-W. Yang, Two-stage cultivation of two *Chlorella* sp. strains by simultaneous treatment of brewery wastewater and maximizing lipid productivity, Bioresour. Technol. 132 (2013) 230−238.

[57] S.-H. Ho, C.-Y. Chen, D.-J. Lee, J.-S. Chang, Perspectives on microalgal CO_2-emission mitigation systems—a review, Biotechnol. Adv. 29 (2) (2011) 189−198.

[58] J. Lowrey, M.S. Brooks, P.J. McGinn, Heterotrophic and mixotrophic cultivation of microalgae for biodiesel production in agricultural wastewaters and associated challenges—a critical review, J. Appl. Phycol. 27 (4) (2015) 1485−1498.

[59] B. Cheirsilp, S. Torpee, Enhanced growth and lipid production of microalgae under mixotrophic culture condition: effect of light intensity, glucose concentration and fed-batch cultivation, Bioresour. Technol. 110 (2012) 510−516.

[60] G. Markou, E. Nerantzis, Microalgae for high-value compounds and biofuels production: a review with focus on cultivation under stress conditions, Biotechnol. Adv. 31 (8) (2013) 1532−1542.

[61] I. Lang, L. Hodac, T. Friedl, I. Feussner, Fatty acid profiles and their distribution patterns in microalgae: a comprehensive analysis of more than 2000 strains from the SAG culture collection, BMC Plant Biol. 11 (1) (2011) 124.

[62] H.-M. Lam, J. Remais, M.-C. Fung, L. Xu, S.S.-M. Sun, Food supply and food safety issues in China, Lancet 381 (9882) (2013) 2044−2053.

[63] Y.-F. Niu, M.-H. Zhang, D.-W. Li, W.-D. Yang, J.-S. Liu, W.-B. Bai, H.-Y. Li, Improvement of neutral lipid and polyunsaturated fatty acid biosynthesis by overexpressing a type 2 diacylglycerol acyltransferase in marine diatom Phaeodactylum tricornutum, Mar. Drugs 11 (11) (2013) 4558−4569.

[64] A.C. Guedes, H.M. Amaro, C.R. Barbosa, R.D. Pereira, F.X. Malcata, Fatty acid composition of several wild microalgae and cyanobacteria, with a focus on eicosapentaenoic, docosahexaenoic and α-linolenic acids for eventual dietary uses, Food Res. Int. 44 (9) (2011) 2721−2729.

[65] Y. Li, Y.-F. Chen, P. Chen, M. Min, W. Zhou, B. Martinez, J. Zhu, R. Ruan, Characterization of a microalga Chlorella sp. well adapted to highly concentrated municipal wastewater for nutrient removal and biodiesel production, Bioresour. Technol. 102 (8) (2011) 5138−5144.

[66] W. Mulbry, S. Kondrad, J. Buyer, Treatment of dairy and swine manure effluents using freshwater algae: fatty acid content and composition of algal biomass at different manure loading rates, J. Appl. Phycol. 20 (6) (2008) 1079−1085.

[67] W. Zhou, Y. Li, M. Min, B. Hu, P. Chen, R. Ruan, Local bioprospecting for high-lipid producing microalgal strains to be grown on concentrated municipal wastewater for biofuel production, Bioresour. Technol. 102 (13) (2011) 6909−6919.

[68] Y. Li, J. Huang, G. Sandmann, F. Chen, Glucose sensing and the mitochondrial alternative pathway are involved in the regulation of astaxanthin biosynthesis in the dark-grown Chlorella zofingiensis (Chlorophyceae), Planta 228 (5) (2008) 735−743.

[69] N. Sun, Y. Wang, Y.-T. Li, J.-C. Huang, F. Chen, Sugar-based growth, astaxanthin accumulation and carotenogenic transcription of heterotrophic Chlorella zofingiensis (Chlorophyta), Process Biochem. 43 (11) (2008) 1288−1292.

[70] G. Potvin, Z. Zhang, Strategies for high-level recombinant protein expression in transgenic microalgae: a review, Biotechnol. Adv. 28 (6) (2010) 910−918.

[71] A. Barka, C. Blecker, Microalgae as a Potential Source of Single-Cell Proteins, A Review, Base, 2016.

[72] W. Zhou, Y. Cheng, Y. Li, Y. Wan, Y. Liu, X. Lin, R. Ruan, Novel fungal pelletization-assisted technology for algae harvesting and wastewater treatment, Appl. Biochem. Biotechnol. 167 (2) (2012) 214−228.

[73] O. Jorquera, A. Kiperstok, E.A. Sales, M.L. Embirucu, M.L. Ghirardi, Comparative energy life-cycle analyses of microalgal biomass production in open ponds and photobioreactors, Bioresour. Technol. 101 (4) (2010) 1406−1413.

[74] S.F. Sing, A. Isdepsky, M. Borowitzka, D. Lewis, Pilot-scale continuous recycling of growth medium for the mass culture of a halotolerant Tetraselmis sp. in raceway ponds under increasing salinity: a novel protocol for commercial microalgal biomass production, Bioresour. Technol. 161 (2014) 47−54.

[75] B. Zhu, F. Sun, M. Yang, L. Lu, G. Yang, K. Pan, Large-scale biodiesel production using flue gas from coal-fired power plants with Nannochloropsis microalgal biomass in open raceway ponds, Bioresour. Technol. 174 (2014) 53−59.

[76] Y. Sun, Y. Huang, Q. Liao, A. Xia, Q. Fu, X. Zhu, J. Fu, Boosting Nannochloropsis oculata growth and lipid accumulation in a lab-scale open raceway pond characterized by improved light distributions employing built-in planar waveguide modules, Bioresour. Technol. 249 (2018) 880−889.

[77] Q. Zhang, S. Xue, C. Yan, X. Wu, S. Wen, W. Cong, Installation of flow deflectors and wing baffles to reduce dead zone and enhance flashing light effect in an open raceway pond, Bioresour. Technol. 198 (2015) 150−156.

[78] A.P. Carvalho, S.O. Silva, J.M. Baptista, F.X. Malcata, Light requirements in microalgal photobioreactors: an overview of biophotonic aspects, Appl. Microbiol. Biotechnol. 89 (5) (2011) 1275−1288.

[79] A. Souliès, J. Legrand, H. Marec, J. Pruvost, C. Castelain, T. Burghelea, J.F. Cornet, Investigation and modeling of the effects of light spectrum and incident angle on the growth of Chlorella vulgaris in photobioreactors, Biotechnol. Prog. 32 (2) (2016) 247−261.

[80] M. Atta, A. Idris, A. Bukhari, S. Wahidin, Intensity of blue LED light: a potential stimulus for biomass and lipid content in fresh water microalgae Chlorella vulgaris, Bioresour. Technol. 148 (2013) 373−378.

[81] J. Liu, C. Yuan, G. Hu, F. Li, Effects of light intensity on the growth and lipid accumulation of microalga Scenedesmus sp. 11-1 under nitrogen limitation, Appl. Biochem. Biotechnol. 166 (8) (2012) 2127−2137.

[82] M. Gross, D. Jarboe, Z. Wen, Biofilm-based algal cultivation systems, Appl. Microbiol. Biotechnol. 99 (14) (2015) 5781−5789.

[83] M. Gross, Z. Wen, Yearlong evaluation of performance and durability of a pilot-scale revolving algal biofilm (RAB) cultivation system, Bioresour. Technol. 171 (2014) 50−58.

[84] P.E. Wiley, Microalgae Cultivation Using Offshore Membrane Enclosures for Growing Algae (OMEGA), UC Merced, 2013.

[85] M.M. Addy, F. Kabir, R. Zhang, Q. Lu, X. Deng, D. Current, R. Griffith, Y. Ma, W. Zhou, P. Chen, Co-cultivation of microalgae in aquaponic systems, Bioresour. Technol. 245 (2017) 27–34.

[86] J.Y. Yoo, S.S. Kim, Probiotics and prebiotics: present status and future perspectives on metabolic disorders, Nutrients 8 (3) (2016) 173.

[87] D. Vandamme, I. Foubert, K. Muylaert, Flocculation as a low-cost method for harvesting microalgae for bulk biomass production, Trends Biotechnol. 31 (4) (2013) 233–239.

[88] J. Zhang, B. Hu, A novel method to harvest microalgae via co-culture of filamentous fungi to form cell pellets, Bioresour. Technol. 114 (2012) 529–535.

[89] M.J. Griffiths, R.P. van Hille, S.T. Harrison, Lipid productivity, settling potential and fatty acid profile of 11 microalgal species grown under nitrogen replete and limited conditions, J. Appl. Phycol. 24 (5) (2012) 989–1001.

[90] J.J. Milledge, S. Heaven, A review of the harvesting of micro-algae for biofuel production, Rev. Environ. Sci. Biotechnol. 12 (2) (2013) 165–178.

[91] A.J. Dassey, C.S. Theegala, Harvesting economics and strategies using centrifugation for cost effective separation of microalgae cells for biodiesel applications, Bioresour. Technol. 128 (2013) 241–245.

[92] A.I. Barros, A.L. Gonçalves, M. Simões, J.C. Pires, Harvesting techniques applied to microalgae: a review, Renew. Sustain. Energy Rev. 41 (2015) 1489–1500.

[93] T. Chatsungnoen, Y. Chisti, Harvesting microalgae by flocculation–sedimentation, Algal Res. 13 (2016) 271–283.

[94] S.O. Gultom, B. Hu, Review of microalgae harvesting via co-pelletization with filamentous fungus, Energies 6 (11) (2013) 5921–5939.

[95] L.G. Chan, J.L. Cohen, J.M.L.N. de Moura Bell, Conversion of agricultural streams and food-processing by-products to value-added compounds using filamentous fungi, Annu. Rev. Food. Sci. T. 9 (2018) 503–523.

[96] Y. Liu, W. Liao, S. Chen, Study of pellet formation of filamentous fungi *Rhizopus oryzae* using a multiple logistic regression model, Biotechnol. Bioeng. 99 (1) (2008) 117–128.

[97] W. Zhou, M. Min, B. Hu, X. Ma, Y. Liu, Q. Wang, J. Shi, P. Chen, R. Ruan, Filamentous fungi assisted bio-flocculation: a novel alternative technique for harvesting heterotrophic and autotrophic microalgal cells, Separ. Purif. Technol. 107 (2013) 158–165.

[98] J. Nyman, M.G. Lacintra, J.O. Westman, M. Berglin, M. Lundin, P.R. Lennartsson, M.J. Taherzadeh, Pellet formation of zygomycetes and immobilization of yeast, New Biotechnol. 30 (5) (2013) 516–522.

[99] M. Elkady, S. Farag, S. Zaki, G. Abu-Elreesh, D. Abd-El-Haleem, Bacillus mojavensis strain 32A, a bioflocculant-producing bacterium isolated from an Egyptian salt production pond, Bioresour. Technol. 102 (17) (2011) 8143–8151.

[100] Z. Zhang, S. Xia, J. Zhao, J. Zhang, Characterization and flocculation mechanism of high efficiency microbial flocculant TJ-F1 from *Proteus mirabilis*, Colliads Surface B. 75 (1) (2010) 247–251.

[101] P.E. Wiley, K.J. Brenneman, A.E. Jacobson, Improved algal harvesting using suspended air flotation, Water Environ. Res. 81 (7) (2009) 702–708.

[102] H. Shin, K. Kim, J.-Y. Jung, S.C. Bai, Y.K. Chang, J.-I. Han, Harvesting of *Scenedesmus obliquus* cultivated in seawater using electro-flotation, Korean J. Chem. Eng. 34 (1) (2017) 62–65.

CHAPTER 4

Mixotrophic Cultivation: Biomass and Biochemical Biosynthesis for Biofuel Production

TAN KEAN MENG • MOHD ASYRAF KASSIM • BENJAMAS CHEIRSILP

INTRODUCTION

In recent years, the production of biofuel and fine chemicals from renewable biomass such as microalgae has gained significant attention. Biofuel from microalgae is believed to be alternative renewable energy to partially replace the energy that being produced by fossil fuel. In 2050, it is estimated that 20% of the total energy that currently being produced by fossil fuel will be replaced with renewable energy from biomass [1].

The utilization of microalgae as biofuel feedstock and the current focus such as microalgae growth rate is contributed by various factors. It was reported that microalgae exhibited several advantages over the terrestrial plant. For instance, the microalgae cultivation can be carried out using waste such as wastewater and carbon dioxide (CO_2), which is generated from various industries. Also, these microalgae can grow in different types of environment and show higher growth rate as compared with other terrestrial plant.

On the other hand, microalgae are the photosynthetic microorganisms that carry out photosynthesis and respiration using CO_2 for the production of its biomass. During this biochemical reaction process, the additional value of biochemical compounds such as protein, carbohydrate, and lipid is also produced. It is known that these chemical compounds were a good platform and can be converted into a various types of value-added products such as chemicals and biofuel.

To ensure the feasibility of biofuel production, achieving high productivity of biomass, lipid, and carbohydrate is very important. To date, many investigations have been explored in either engineering or molecular aspect of microalgae growth. The modification of microalgae cultivation via engineering approaches such as photobioreactor development and mode of cultivation are among the focus in the improvement study for microalgae biomass production. There are three major cultivation modes, namely, autotrophic, heterotrophic, and mixotrophic cultivation strategies. As per to date, there is limited information on the microalgae cultivation via mixotrophic. Thus, this chapter reviews and discusses on the mixotrophic cultivation system along with the metabolisms that occur during cultivation process. The effect of cultivation condition and mode of cultivation strategy will be also discussed in this chapter.

MIXOTROPHIC MICROALGAE CULTIVATION
Mode of Microalgal Cultivation

To ensure the feasibility of biorefinery production, microalgae biomass production and its biochemical components should be taken into consideration. The growth characteristic and biochemical accumulation are significantly affected by the mode of cultivation conditions [2]. Generally, the mode of microalgae cultivation can be divided into four main types: photoautotrophic, heterotrophic, mixotrophic, and photoheterotrophic cultivations. However, the cultivation mode based on photoautotrophy, heterotrophy, and photoheterotrophy has observed major limitations, in which the microalgae were highly dependent on weather condition and higher cost expenses (carbon sources), respectively [3]. Subsequently, this leads to unfeasibility of biofuel-based biorefinery process. The mixotrophic cultivation is a good strategy to overcome the limitation from previous cultivation modes and combine their advantages to promote high growth rate and biochemical accumulation at the same time.

Mixotrophic Cultivation

Under mixotrophic cultivation, microalgae are able to grow using light as the energy source as well as CO_2 and organic carbon as the carbon sources [2]. Generally, mixotrophic cultivation is a technique that combines the advantages from autotrophic and heterotrophic cultivations while overcoming the disadvantages from both cultivations. The advantages of this cultivation mode are in agreement with the previous study, which showed that *Chlorella vulgaris* is able to utilize glycerol as a sole carbon substrate and enhance the algal growth rate simultaneously in mixotrophic culture medium [3]. The utilization of complex organic carbon substrate can stimulate the biosynthesis of lipids and soluble carbohydrates as the raw materials for biodiesel and bioethanol production and reduce the anabolism of photosynthetic pigments and proteins. Apart from that, the major advantages of mixotrophic cultivation also include the following:

I. Higher growth rate and biomass productivity
II. Higher lipid and carbohydrate productivity
III. Easy to scaled-up process
IV. Prolonged exponential growth phase
V. The possibility of manipulating biomass composition

In earlier 1977, Endo et al. [4] reported that the *C. vulgaris* growth rate was almost the summation of autotrophy and heterotrophy when cultivated under mixotrophic mode supplied with light source and acetate as the carbon source. Under this cultivation mode, microalgae are able to exhibit dual functions. They can grow heterotrophically with organic carbon, increase in growth rate, enhancement in biomass production, primary metabolites including carbohydrate and lipid accumulation, as well as able to recycle the inorganic carbon (CO_2) and utilize sunlight to produce oxygen through photosynthesis. Thus, the overall CO_2 emission is lower under the mixotrophic cultivation mode, unlike heterotrophic cultivation mode [5]. Mixotrophic cultivation of microalgae can be well defined with both sunlight and organic carbon, which are not the limiting factors for growth. This statement was in agreement with the study done by Liang et al. [6] who demonstrated the cultivation of *C. vulgaris* exhibited the maximum biomass density up to 2 g/L in the presence of 1% (w/w) of glucose. A similar result was observed by Marquez et al. [7] who demonstrated the microalgae cultivation in mixotrophic condition (presence of glucose) generated higher growth rate as compared with photoautotrophy and heterotrophy conditions. Therefore, it can be concluded that the microalgae growth rate is highly affected with the

presence of organic substance especially the carbon source. Glucose and its analogs (3-O-methyl glucose or 6-deoxyglucose) could contribute to high biomass under the presence of light, which results in the increased growth rate and biomass production [8,9].

To ensure the feasibility of biofuel production, high lipid and biomass productivity are the crucial parameters in mixotrophic cultivation mode. Previous study showed that the enhancement of lipid up to 57% was observed for mixotrophic cultivation of oleaginous *Chlorella protothecoides* [10]. The result coincided with another study that showed the mixotrophic cultivation of *C. protothecoides* was able to accumulate high lipid content up to $21.3 \pm 2.5\%$ in the presence of glycerol as the carbon source [11]. The effectiveness of mixotrophic mode was further evaluated by Kong et al. [3], who showed that oleaginous microalgae *C. vulgaris* with 0.098 g/h of lipid productivity can be achieved using a mixture of glycerol and glucose as the substrates under this cultivation mode. This study found that the effect of glycerol and glucose addition caused the metabolic changes and subsequently led to more lipid content of microalgae.

Apart from the synthesized lipid in microalgae, carbohydrate is also one of the components that is produced by microalgae through the fixation from atmospheric CO_2 or organic carbon source. Carbohydrate was converted through a series of biochemical pathway. A study done by Perez-Garcia and Bashan [5] proved that mixotrophic cultivation of microalgae is a promising strategy to accumulate carbohydrate in the microalgae cells. Previous study showed that cultivation of *Asterarcys quadricellulare* under mixotrophic condition supplemented with 0.1 g/L glucose was able to accumulate up to 36.6% of carbohydrate content [12]. This result coincided with the previous study that used mixotrophic cultivation the *C. vulgaris*, which accumulated higher carbohydrate content and productivity that were 8.74% and 57.26 mg/L d, respectively, using a mixture of glycerol and glucose when compared with autotrophic control condition [3]. This phenomenon might be contributed by the shifting of algal cell metabolic pathway when supplemented with organic carbon source and energy (light and glucose). This was noticed by a previous study who illustrated that the protein content reduction in mixotrophic *C. vulgaris* UAM 101 cells was compensated by an increase in carbohydrate content [13]. Therefore, the mixotrophic cultivation was a good strategy to obtain a large biomass high lipid and carbohydrate content to ensure the feasibility of biorefinery processes. The difference between each cultivation modes is summarized in Table 4.1.

TABLE 4.1
Different Cultivation Modes of Microalgae and its Advantages and Disadvantages.

Cultivation Mode	Carbon Source	Energy Supply	Light Availability Requirement	Advantages	Disadvantages
Photoautotrophic	Inorganic (CO_2)	Light	Obligatory	(1) Low cost (2) Low energy consumption	(1) Low growth rate and biomass production (2) Need special bioreactor (closed photobioreactor) (3) High dependence on weather condition
Heterotrophic	Organic carbon source (e.g., glucose, glycerol, acetate)	Organic carbon source	No requirement	1) High biomass productivity and lipid accumulation 2) Bioreactors design with little constraints 3) Easier scaling-up process 4) Possible to change the biomass composition by changing the culture medium organic substrates	1) Limited microalgae species that can grow under heterotrophically 2) Higher cost (expenses of organic substance and energy are required) 3) Need for sterile media and easy to be contaminated 4) Maybe inhibit growth due to excess of organic substance 5) Inability to produce light-induced metabolites
Mixotrophic	Inorganic (CO_2) and organic carbon source	Light and organic carbon source	No obligatory	1) Higher growth rate, higher biomass, carbohydrate, and lipid accumulation 2) Prolonged exponential growth phase 3) Reduction of lost biomass from respiration during dark hour 4) Reduction or stopping of photoinhibitory effect 5) Flexible to switch between photoautotroph and heterotroph regimens at will	1) Limited microalgae species that can grow under mixotrophically 2) Higher cost (energy) 3) Need for sterile media and easy to be contaminated 4) Indirect use of arable land
Photoheterotrophic	Organic carbon	Light	Obligatory	1) High biomass productivity	1) High cost (equipment and substrate) 2) Low carbohydrate and lipid content 3) Require both light and organic carbon at the same time 4) Contamination

METABOLISMS OF MIXOTROPHIC CULTIVATION
Carbohydrate Metabolism

The carbohydrate serves two main purposes for microalgae. It serves as the structural components in cell walls and storage components inside the microalgae cell [14]. Carbohydrate that acts as the structural components in cell walls is important in terms of protection, structure, and support, whereas the storage compounds in the cell are used to provide energy needed for the metabolic processes of organisms. Generally, the carbohydrate can be categorized into mono-, di-, oligo-, and polysaccharides. In this context, we mainly focused on fermentable sugars since it possesses the advantage for downstream processing by producing a wide array of valuable by-products such as chemicals and biofuel. The reserved carbohydrates compositions vary, which mostly depend on the microalgae strains as well as nutritional and environmental conditions.

The study on carbohydrate biosynthesis pathway in microalgae becomes crucial to determine the key enzyme and protein that is responsible for carbohydrate synthesis. Carbohydrate metabolism occurs through the reversal of glycolysis pathway that took place in the cytoplasm microalgae [5]. The details on carbohydrate metabolism pathway in microalgae are illustrated in Fig. 4.1.

It is also important to study the gene that controlled the related enzyme for carbohydrate synthesis. In glycolysis pathway, fructokinase, glyceraldehyde 3-phosphate dehydrogenase, phosphoglycerate mutase, and pyruvate kinase are downregulated in mixotrophy, indicating the crucial role of this metabolic process in the absence of acetate, where sugar is produced in the dark phase of photosynthesis. This will subsequently produce ATP and reducing power for upregulated process [15]. In upregulated glycolysis, it involves several important enzymes to ensure the production of glucose or starch such as diphosphate-fructose-6-phosphate-1-phosphotransferase (PFP), nicotinamide adenine dinucleotide phosphate ($NADP^+$), $NADP^+$-dependent glyceraldehyde-3-phosphate dehydrogenase (GapN), diphosphate—fructose-6-phosphate 1-phosphotransferase (PFP), and phosphofructokinase (PFK). The enzymes involved in increasing the rate of gluconeogenesis consistently with the increased starch accumulation observed in mixotrophy [15]. Table 4.2 summarizes the enzymes and their function in regulating the carbohydrate metabolism in microalgae.

CULTIVATION PARAMETERS

Cell growth rate, biomass production, carbohydrate, and lipid productivity are the critical factors to ensure the feasibility for microalgae large-scale production and its biorefinery process. Hence, the study of cultivation methods for high biomass production is important to reduce the cost of microalgae utilization. Based on the mechanism and characteristic of the four cultivation modes (photoautotrophy, heterotrophy, mixotrophy, and photoheterotrophy) discussed previously, mixotrophic cultivation of microalgae is the best in terms of biomass production as well as carbohydrate and lipid content in microalgal biomass when compared with other cultivation modes. Hence, this subtopic will further discuss about the virtues of mixotrophic cultivation with the manipulation of various environmental and nutrient factors. Successful manipulation of these factors will lead to maximized productivity. Details of the cultivation parameters and its by-products are illustrated in Fig. 4.2.

ENVIRONMENTAL FACTORS
Light Intensity

Light is the main energy provider for photosynthetic organisms, including microalgae. Generally, the majority of microalgae are light-saturated under the light intensity of $200-400$ $\mu mol/m^2 s$ [7]. The cultivation of microalgae under appropriate light intensity will significantly affect the microalgae growth, chlorophyll production, and primary metabolites accumulation such as carbohydrate and lipid [7,16]. However, further increase of light source will lead to photoinhibition and resulted in the slow growth rate of microalgae [17].

According to the previous study done by Rai et al. [16], the cultivation of *Chlorella* sp. microalgae exhibited higher dry weight up to 1.05 g/L under 2700 lux illumination, unlike under 2100 lux illumination, with only 0.5426 g/L. The similar result was observed by Li et al. [18] who showed that increment in biomass production of *Chlorella kessleri* microalgae corresponds to the increased light intensity up to 120 $\mu mol/m^2 s$. It can be explained by the stimulation microalgae with higher light intensity, which results in synthesizing higher photochemical phase to produce adenosine triphosphate (ATP) and nicotinamide adenine dinucleotide phosphate-oxidase (NADPH). They are needed by dark reaction for biochemical phase that synthesizes essential molecules for growth [19]. Another study also showed that *C. protothecoides* will reduce the lag phase in cultivation period with the presence of light source, thus increasing its growth rate [18]. However, further exposure to high light intensity would lead to drastic decrease of cell biomass. Based on the previous study, it showed that cultivation of *Chlorella* sp. microalgae under high light intensity up to

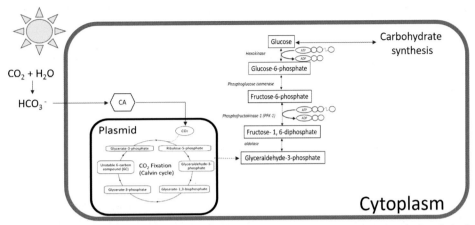

FIG. 4.1 Simplified overview of the metabolites and representative pathways in microalgal carbohydrate biosynthesis.

TABLE 4.2
Types of Enzymes Involved in Glycolysis Pathway and Their Functions.

Regulation Location	Types of Enzymes	Function
Downregulated enzymes	Fructokinase	Controls the use of inorganic pyrophosphate and makes the reaction reversible
	Glyceraldehyde 3-phosphate dehydrogenase	Final enzyme that produces glyceraldehyde 3-phosphate in the Calvin Benson cycle
	Phosphoglycerate mutase	Catalyzes the internal transfer of a phosphate group from C-3 to C-2, resulting in the conversion of 3-phosphoglycerate (3PG) to 2-phosphoglycerate (2PG) through a 2,3-bisphosphoglycerate intermediate
	Pyruvate kinase	Catalyzes the transfer of a phosphoryl group from PEP to ADP
Upregulated enzymes	Diphosphate-fructose-6-phosphate-1-phosphotransferase (PFP)	Catalyzes the reversible interconversion of fructose 6-phosphate and fructose 1,6-bisphosphate using inorganic pyrophosphate as the phosphoryl donor
	Nicotinamide adenine dinucleotide phosphate ($NADP^+$)	Is a cofactor used in anabolic reactions and at the same time requires NADPH as a reducing agent
	$NADP^+$-dependent glyceraldehyde-3-phosphate dehydrogenase	Catalyzes an alternative reaction to produce NADPH and glycerate 3-phosphate directly from glyceraldehyde-3-phosphate, without producing the intermediate glyceraldehyde-1,3-bisphosphate
	Diphosphate-fructose-6-phosphate phosphotransferase	Catalyzes the formation of fructose 1,6-bisphosphate from fructose 6-phosphate using inorganic pyrophosphate as the phosphoryl donor, rather than ATP
	Phosphofructokinase	The use of inorganic pyrophosphate makes the reaction reversible, increasing the rate of gluconeogenesis

3300 lux would lead to decrease of cell dry weight up to 35.78%. This result was in concordance with the findings by Xie et al. [19] who showed that the specific growth rate of *Platymonas subcordiformis* was remarkably decreased when illuminated using 126 μmol/m² s under mixotrophic cultivation. The same result was observed by Li et al. [18] who indicated that under high light intensity, e.g., 200 μmol/s, this would lead

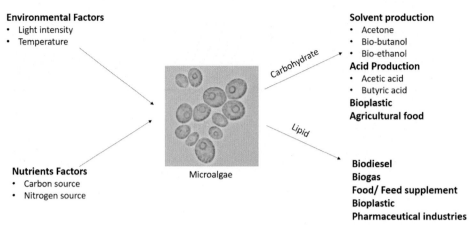

Environmental Factors
- Light intensity
- Temperature

Nutrients Factors
- Carbon source
- Nitrogen source

Microalgae

Carbohydrate

Lipid

Solvent production
- Acetone
- Bio-butanol
- Bio-ethanol

Acid Production
- Acetic acid
- Butyric acid

Bioplastic
Agricultural food

Biodiesel
Biogas
Food/ Feed supplement
Bioplastic
Pharmaceutical industries

FIG. 4.2 Effect of different cultivation parameters toward microalgae and their by-products.

to the effect of photoinhibition. This result was further agreed by another study who demonstrated that *Chlorella* sp. microalgae exhibit a slower growth rate under high light intensity (10,000 lux). High light intensity would rapidly damage the photosystem II and subsequently lead to a decrease in bioproductivity, which is known as photoinhibition [20].

Light intensity also influences the carbohydrate content in microalgae biomass. Generally, it was known that high light intensity results in an increase of carbohydrate content [21]. Based on the previous study, it showed the threefold increase in carbohydrate content of *Porphyridium aerugineum* when cultivated under high proton flux [22]. It can be explained by the previous study that explained that the luminous intensity was an important factor to stimulate the production of carbon compounds including carbohydrate from organic substances through microalgae metabolism pathway [23]. However, the different result was found by Gim et al. [24] who demonstrated that increasing light intensity up to 150 $\mu mol/m^2$ s had increased the total lipid content in microalgae biomass. Another study was done by Cheirsilp and Torpee [25], which found that lipid content of marine *Chlorella* sp. was increased up to 397.8 ± 7.9 mg/L when microalgae were cultivated under 8000 lux of light intensity. This result was in agreement with the another study that found increase in lipid content up to 63.71% when the light intensity was increased from 2100 to 2700 lux [16]. Light intensity changed the lipid metabolism pathway and enhanced the lipid content in microalgae. It can be further explained by Gim et al. [24] who showed that the enzymes involved in fatty acid biosynthesis were stimulated, particularly the acetyl-CoA carboxylase, desaturase, acyl-carrier protein

synthase, ATP/citrate lyase, and membrane-bound glucose permease in the presence of high light intensity. This was further agreed by a study that showed that the excessive light energy was converted and stored as fatty acids through CO_2 fixation by Calvin cycle and lipogenesis and thus increased the lipid accumulation of microalgae.

TEMPERATURE

There is a correlation between the temperature and light intensity on the biomass composition. Generally, low temperature was resulted from the less exposure toward light intensity and contributing to low biomass production of microalgae, whereas the effect of high temperature was observed to be vice versa. Previous study showed that the cell density would inactivate below its optimum temperature, and increase in temperature above the saturating condition will limit the cell density due to photoinhibition [26]. This owned to the disruption of chloroplast lamellae that caused by high light intensity [27]. This result was in concordance with the study done by Ogbonda et al. [28] who showed the increment of biomass concentration of *Spirulina* sp. corresponding to the increment of temperature up to 30 °C. The similar result was also observed by Venkata Subhash et al. [29] who showed the increment of biomass production that corresponded with the increment of temperature up to 30°C, wheras the reduction of biomass production was observed when the microalgae were cultured under 35°C with domestic wastewater in mixotrophic condition as shown in Fig. 4.3.

One of the most commonly observed changes with temperature shift is the alteration on the level of unsaturation of fatty acids in the lipid membrane. Previous

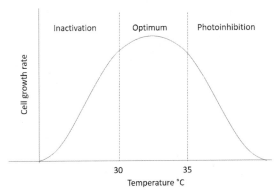

FIG. 4.3 Typical microalgal cell growth rate profile at a different temperature.

study reported the increment of fatty acids (14:0) from ~4% at 10°C to > 20% at 25°C using marine plankton species [30]. However, different result was observed by Lynch and Thompson [31] who reported that *Dunaliella salina* (UTEX 1644) had shown a considerable increase in unsaturated fatty acid in response to temperature drops of 30–12°C. The result was further agreed by Converti et al. [32] who showed decrease in lipid content of *C. vulgaris* from 14.71% to 5.90% when the temperature was increased from 25 to 30°C. This phenomenon was most probably due to the low temperature that reduced the fluidity of cell membrane and subsequently compensated by increasing the level of unsaturated fatty acids [33].

Temperature was also reported to impact the carbohydrate content in microalgae biomass. Generally, an increase in temperature will lead to the degradation of starch produced in microalgae. According to Nakamura and Miyachi [34], their study demonstrated that increase in temperature up to 40°C will lead to the degradation of ^{14}C starch. This finding can be explained by the degradation of starch contributed by the degeneration of starch-controlling enzymes such as α-amylase and α-glucan phosphorylase under high temperature [35]. Therefore, the temperature is one of the key factors that needed to be considered for cultivation.

NUTRIENTS FACTOR
Carbon Source
Carbon compound is one of the essential elements that can induce the cell growth rate. Apart from that, the type of carbon source also greatly affects the accumulation of biomass, carbohydrate, and lipid in microalgal biomass. Since both CO_2 and organic carbon utilization coexist under mixotrophic condition, the CO_2

concentration and organic compounds supplied need to be finely optimized to achieve the best productivities in mixotrophic condition.

CO_2 is found to be a primary contributor for microalgae growth, and it strongly stimulates the photosystem and enhances the photosynthetic productivity under appropriate range. For mixotrophic cultivation condition, the microalgae can recycle the inorganic carbon source such as CO_2 for photosynthesis. The influence of exogenous CO_2 on the mixotrophic cultivation of *Auxenochlorella protothecoides* was studied by B. Hu et al. [36] who showed that CO_2 supply (5% CO_2 (v/v)) enhanced the biomass production after 4–5 days. The same result was observed by the another study who demonstrated that cultivation of *C. vulgaris* ARC1 exhibits higher growth rate when supplied with 6% CO_2 concentration as compared with ambient air [37]. Also, the CO_2 concentration has significantly affected the microalgae carbohydrate accumulation. Generally, by increasing the supply of CO_2, it enhanced the carbohydrate accumulation in microalgae biomass. Based on the previous study done by Xia and Gao [38], the increase in dissolved CO_2 concentration from 3 to 186 μmol/L was found during the cultivation of *Chlorella pyrenoidosa* and *Chlamydomonas reinhardtii* with elevated carbohydrate content from 9.30% to 21.0% and 3.19%–7.40% (w/w), respectively. Besides, another study also showed an increase in carbohydrate productivity of *Scenedesmus obliquus* up to 15.8 mg/L d when supplied with 14.1% CO_2 concentration when compared with other conditions [39]. This can be explained by the fact that excess energy such as ATP was left out after cellular maintenance during dark phase, which could be used for energy storage either in the form of starch or lipid [40]. However, the different result was reported by Thyssen et al. [41] who showed the decrease in CO_2 concentration resulted in an increase of carbohydrate content in the microalgae biomass. A study was done by Izumo et al. [42] who showed an increase in carbohydrate productivity up to 2.5-fold when the amount of CO_2 concentration decreased from 3% to 0.04%. This was due to an extracellular carbonic enzyme that located in CO_2 concentrating mechanisms (CCMs), enabling microalgae to acquire and concentrate inorganic carbon from the extracellular environment under low CO_2 conditions and causing the efficient CO_2 utilization during photosynthesis [42].

Organic carbon source also contributes to the microalgae growth under mixotrophic culture condition. Among the organic carbon sources such as glucose, xylose, rhamnose, fructose, sucrose, and galactose, glucose was the ideal carbon source for microalgae in terms of biomass and lipid production under mixotrophic cultivation [43]. This result was further in agreement with the previous study who determined the effect of organic carbon sources such as glucose, glycerol, and ethanol on *C. kessleri*. An interesting result indicated that glucose was the most effective carbon source for cell growth and fatty acid accumulation [44]. This is because of the other carbon sources involved with complicated interconversion metabolic process in microalgae metabolism to furnish energy for cell growth as well as bioenergy production when compared with glucose [45]. The utilization of glucose was further shown by Kong et al. [3] who demonstrated that the high biomass and lipid content could be achieved in *C. vulgaris* microalgae using the mixture of carbon source, glucose, and glycerol.

Besides, acetate is also a commonly used organic carbon in mixotrophy culture. It was found that sodium acetate is a potential carbon source that could help to enhance both microalgae growth and lipid biosynthesis. A previous study done by Heredia-Arroyo et al. [46] showed that the cultivation of *C. vulgaris* could exhibit a high growth rate and lipid content using sodium acetate as the sole carbon source. The same findings also indicated by Xiuxia et al. [47] who used sodium acetate as an organic nutrient to stimulate high growth rate and lipid production for *Brachiomonas submarina* microalgae under mixotrophic cultivation. In addition, Haiying et al. [48] also showed that cultivation of *Phaeodactylum tricornutum* supplied with glucose and acetate sodium could enhance the growth and lipid content, while no such enhancement was exhibited by starch. However, different result was observed by the other study. It showed that *C. vulgaris* can accumulate more carbohydrates using glycerol or glucose as the carbon source under mixotrophic cultivation [49]. Hence, it can be concluded that carbon source is the important factor for microalgae growth productivity and its biochemical accumulation depends on the microalgae strain and its metabolism.

NITROGEN SOURCE

Nitrogen is one of the essential nutrients that contributed a crucial role for microalgae growth. It plays an important role in the formation of vital compounds such as DNA, proteins, and pigments in microalgae metabolism process. Nitrogen starvation is a common strategy to improve the formation of biochemical compounds in microalgae. Several common nitrogens used in previous studies include urea, sodium nitrate

($NaNO_3$), and potassium nitrate (KNO_3) [50,51]. Nitrogen starvation in microalgae will affect the photosynthesis systems, mainly the PSII with a negative impact on the synthesis of proteins participating in both photosystems (PSI and PSII) [45]. Under nitrogen starvation, the flow of photosynthetically fixed carbon changed from the protein synthesis metabolic pathway to the lipid or carbohydrate synthesis pathway and resulted in the accumulation of either carbohydrates or lipids in microalgae [51].

Nitrogen starvation is a common strategy in green algae to stimulate the lipid accumulation, and some studies also reported that it caused the carbohydrates accumulation. The previous study done by Ratnapuram et al. [52] demonstrated that higher lipid productivity of *C. pyrenoidosa* can be achieved up to 284 g/kg and neutral lipids up to 154.3 g/kg when cultivation was performed under the stress phase condition. This result was further in agreement with Paranjape et al. [53] who showed the enhancement in lipid content of *Chlorella sorokiniana* (PCH02) and *C. vulgaris* (PCH05) under nitrogen limitation. The synthesis of lipid under stress condition was due to the metabolic pathway changes in microalgae system. It was supported by Guschina and Harwood [54], who showed that the metabolic pathway changes from lipid metabolism, specifically membrane lipid synthesis to neutral lipid storage. This, in turn, increases the total lipid content of green algae under nutrient starvation. By using the proteomic analysis, Shen [55] proved that the activity of tricarboxylic acid cycle was improved in mixotrophic cultivation of microalgae, which leads to more fatty acid synthesis in *S. obliquus* under nutrient starvation or limitation conditions.

Apart from that, a nutrient limitation can also enhance the carbohydrates accumulation in microalgae biomass. Based on the previous study, freshwater alga, *Chlorella* sp., can accumulate up to 38% of starch content when cultivated in the absence of urea as the nitrogen-limited condition [50]. This result was further in agreement with Dragone et al. [56] who demonstrated that the high carbohydrate accumulation up to > 40% was obtained using *C. vulgaris* P12 under nitrogen depletion in the two-stage cultivation process. This result was further ratified by previous findings of Behrens et al. [57], where the starch accumulation in *C. vulgaris* increased up to 55% under nitrogen-limited conditions. High carbohydrate content was obtained from the nitrogen-limited or deficiency condition because the available nitrogen was utilized for enzyme synthesis and essential cell structures in microalgae. Hence, the available CO_2 subsequently was fixed and converted into carbohydrate or lipid rather than protein [58].

ACTIVATOR AND INHIBITORS

Microalgae growth and chemical accumulation under mixotrophic cultivation also can be improved by the addition of other components in the medium during the cultivation period such as activators or inhibitors.

Activator

Activator is normally used to further improve the biomass production as well as carbohydrate and lipid accumulation in microalgae biomass. Fig. 4.4 shows the chemical structures for common activator used to enhance microalgal growth and biochemical accumulation in the cell. The presence of activator in the cultivation medium was found which could contribute the pH value of the cultivation medium, liquid-liquid viscosity and gas-liquid transfer coefficient that largely affected the cell metabolism process. This subsequently could specifically target the key enzymes or metabolites for the desired [5]. Based on the previous result exhibited by Cui et al. [59], the addition of low-cost antioxidant sodium erythorbate ($C_6H_7NaO_6 \cdot H_2O$, purity $\geq$ 98.0%) as a yellowish powder (8 g/L) significantly accelerated the growth of microalgae in the first stage with air aeration. This activator will shorten the duration of lag phase for the microalgae culture and hence increase the microalgae growth rate.

Apart from that, another study reported the enhancement in C. pyrenoidosa biomass productivity up to 6-fold (121 mg/m^3 d) and 32-fold in lipid productivity (16.6 mg/m^3 d) in the presence of sodium acetate and glycerol as the activator [60]. The effect of acetate was further evaluated by Cecchin et al. [15] who showed that the cultivation of C. sorokiniana under mixotrophic condition could enhance the biomass production up to 42% in airlift photobioreactors when compared with autotrophic conditions with the presence of acetate as the activator. This result was further in agreement by other studies, which indicated that the combination of 4 g/L sodium acetate supplementation with nitrogen and phosphorous deficiency condition can increase total fatty acid yield by 93.0% and 150.1%, when compared with nutrient-depleted and

normal culture conditions, respectively [61]. Hence, it can be explained that the acetate compound functions as a carbon precursor, undergoes a series of chemical reactions, and involves in the Kennedy pathway for the enhancement of triacylglyceride synthesis [62]. Generally, this will provide a better way to produce biofuel at reduced overall cost for biodiesel production.

The auxin family such as α-naphthalene acetic acid (NAA) also has been reported to stimulate higher growth rate on microalgae cultivation. Hunt et al. [63] reported that NAA at 5 ppm as a growth stimulant enhanced the C. sorokiniana biomass productivity up to 0.042 g/L d as compared with control condition, which only exhibited growth rate of 0.018 g/L d on day 10. The effect of NAA was further evaluated by Czerpak et al. [64] who showed that using plant regulator, NAA with the concentration of 5×10^2 M could exhibit the highest stimulatory effect on the fresh weight of C. pyrenoidosa culture up to 8.14 ± 0.56 g/dm^3 or 270.2% when compared with the control culture. The increment of cell dry weight was due to the growth activator, auxins that can display an inhibitory effect on the process of oxidation and degeneration of chlorophylls and carotenoids, thus delaying the algal senescence [64]. Besides, NAA is also a natural auxin that is commonly found in vascular plant and exhibited much structural homology to indole-3-acetic acid (IAA), which is the growth-promoting enzyme in microalgae [65].

The effect of auxin was further observed by the previous researcher who found that the brassinosteroids cooperated synergistically with auxins in stimulating cell proliferation and endogenous accumulation of monosaccharides in C. vulgaris [66]. Coapplication of 50 μM IAA with 1 nM brassinolide (BR) exhibited the highest increase in cell number up to 321% upon 48 h cultivation as compared with the control condition. Under the same concentration of IAA and BR, it also stimulated the sugar accumulation content in C. vulgaris culture up to 273% as compared with the control culture. The increasing of monosaccharide in microalgae was initiated by the increased chlorophyll

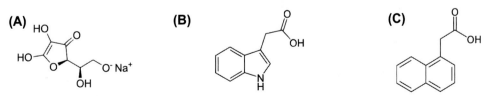

FIG. 4.4 Chemical structure of common activators used to enhance biomass and biochemical biosynthesis in microalgal cell: **(A)** sodium erythorbate, **(B)** auxins, and **(C)** α-naphthalene acetic acid (NAA).

content in response to brassinosteroids, which subsequently enhanced the photosynthesis rate and contributed to the monosaccharide production [66].

Apart from that, Tiburcio et al. [67] reported that polyamines act as the polycation nitrogen provided upon growth and monosaccharides accumulation in *C. vulgaris* Beijerinck (Chlorophyceae). These regulator polyamines (spermine) contain three amine groups that can enhance the number of *C. vulgaris* cells when compared with the control condition. This polyamine (spermidine) also contains four amine groups that displayed the greatest simulative activity in monosaccharide accumulation up to 22×10^{-8} μg/cell during the third day of cultivation under the same condition.

Inhibitor

Another compound that can be used to manipulate the microalgae growth and chemical compounds accumulation during mixotrophic cultivation is the inhibitor. The presence of inhibitors during cultivation was proved to inhibit or retard the cell growth and biochemical components in microalgae [68]. Several types of inhibitors have been reported such as dimethyl sulfoxide (DMSO) and brassinosteroids that affect the growth and chemical production of microalgae [68]. Pelekis et al. [69] reported that the presence of 2,4-dichlorophenoxyacetic acid (2,4-D) with the concentration of 2×10^{-3} M could significantly decrease up to 85% microalgae cell number in the mature phase, 69% of fresh weight, and 60% of dry weight when compared with control media. Apart from that, the 2,4-D inhibitor also could lead to the decrease in starch content of growing cultures such as *Polytoma uvella*. Therefore, it could be suggested that the presence of herbicide will most likely interfere with metabolic processes and photosynthetic algae, which lead to reduced microalgae growth rate and starch accumulation.

Other studies have also been reported on the effect of 2,4-D on the oxidative phosphorylation of microorganism [70]. This will lead to a reduction in the amount of energy available for normal metabolic activity, therefore affecting the growth rate. This result was further agreed by the other study that used the plant growth regulator 2,4-D with the concentrations of up to 10^{-3}. It stimulated marked changes in the phosphorus metabolism by inhibiting the incorporation of P^{32} into nucleotides as ATP and ADP (uncoupling phosphorylation) in the cells. This decreases the oxygen uptake by *C. pyrenoidosa* Chick and results in the reduction of microalgae growth rate [71]. However, there was still limited information about the inhibitors that could affect microalgae growth rate and their chemical composition. Hence, future study should be considered to provide comprehensive information.

CULTIVATION MODE
Two-Stage Cultivation

Two-stage cultivation is one of the promising approaches for microalgae cultivation to achieve high biomass productivity and biochemical compound accumulation at the same time [72]. Generally, during the first-stage cultivation, the microalgae were first grown under photoautotroph condition, in which sunlight was provided as the energy source for the culture to maximize the microalgae biomass productivity. Next, when the maximum biomass concentration was reached, the microalgae cultivation system was converted into mixotrophic cultivation mode as shown in Fig. 4.5. The maximum biomass concentration can be indicated with self-shading effect because light becomes the limiting factor. Under mixotrophic cultivation mode, the sufficient carbon source was provided to promote accumulation of biochemical components in microalgae biomass [59,73].

Several investigations on the potential of two-stage mixotrophic cultivation microalgae were evaluated by few research groups [74]. Yen and Chang [72] showed that the two-stage cultivation of *C. vulgaris* achieved higher biomass up to 7.4 g/L and 57.1% of linoleic acid (C18:2) using glycerol as a carbon source when compared with photoautotrophic cultivation. This result was further in agreement with another study who showed the cultivation using *Nannochloropsis* sp. (MT-I5), resulting in higher desired saturated fatty acid up to 48% when two-stage cultivation was applied during the cultivation process [71]. Besides, another study also showed that the cultivation of *C. vulgaris* (UTEX-265) under photomixotrophic condition could generate a significant result in terms of high biomass concentration and lipid content up to 3.20 g/L and 24%, respectively [75]. The high lipid content was due to nitrogen-deficient cultivation media that result in the cessation of microalgae growth and animate a metabolic pathway favorable to the accumulation of reserve lipids [76]. Apart from that, a better performance of microalga growth and lipid accumulation could be attributed to the environmental factor during the cultivation [77,78].

Two-stage cultivation could also further enhance the carbohydrate productivity. A study done by Ratnapuram et al. [52] demonstrated that the cultivation of *C. pyrenoidosa* could be achieved up to 2.85 g/L biomass production and accumulate up to 110 mg/g of carbohydrate concentration. Hence, the increment of starch

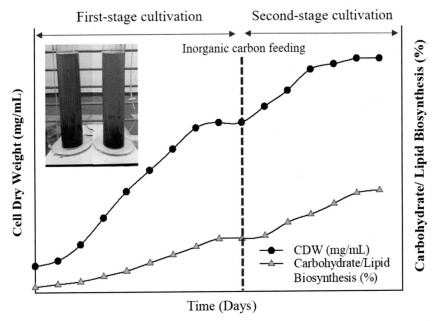

FIG. 4.5 Two-stage cultivation of microalgae for cell dry weight and carbohydrate/lipid biosynthesis.

content can be explained with the high light intensity, resulting in enhanced production of polysaccharides in microalgae [56].

Therefore, the cultivation using mixotrophic cultivation is an attractive concept for the purpose of enhancing the final concentration of algal biomass, carbohydrate, and lipid yields. By adopting a two-stage cultivation process, it could also reduce the potential contamination. Therefore, it might be a suitable application for the scale-up cultivation of algae.

Fed-Batch Cultivation

Fed-batch cultivation is a technique in microorganism cultivation processes, where one or more nutrients are supplied in the cultivation vessel and the products produced during the process remain in the reactor until the process is complete. This technique offers several advantages in microalgae cultivation in comparison with batch cultivation system. The advantages of this approach include (1) producing higher biomass concentration from a well-established condition and (2) controlling by-product and metabolite production by providing a specific substrate that is solely required for final product formation. The detail of fed-batch cultivation of microalgae is illustrated in Fig. 4.6.

Several studies were reported on fed-batch cultivation of various microalgae under mixotrophic condition [79–81]. Cheirslip et al. [25] investigated the effect of fed-batch cultivation toward the biomass and lipid

productivity of *Chlorella* sp. and *Nannochloropsis* sp. Stepwise fed-batch cultivation was conducted by adding acetic acid and nutrient during the cultivation process, and it was found that the lipid production of both microalgae species was increased twice, higher than those in batch cultivation system. A similar observation was also reported on the mixotrophic fed-batch cultivation of *Nannochloropsis oculata*, which found that addition of vitamin and nutrient solution using this strategy enhanced 1.25% of biomass density when compared with those in batch cultivation system [82]. Martinez-macias et al. [83] also reported that the highest biomass and lipid production was obtained from the cultivation using fed-batch strategy. The study also suggested that cultivation condition could contribute significant effect on the biomass production. On the other hand, the study also suggested the fed-batch cultivation usage for the biomass production followed with continuous feeding system to enhance the accumulation of lipid content in the cell. Moreover, another study indicated that the fed-batch cultivation strategy was able to promote the growth of *N. oculata* in two different cultivation media, namely, F/2 Guillard medium and algal medium.

BIOFUEL PRODUCTION UNDER MIXOTROPHIC CULTIVATION CONDITION

Nowadays, the main focus of microalgae biofuel research is the optimization of conditions for the

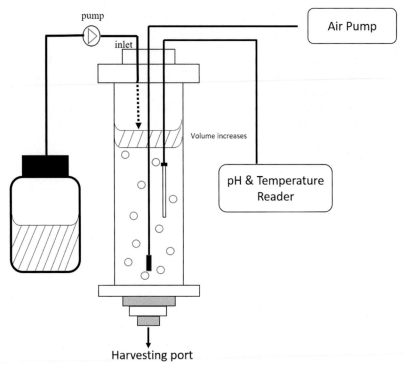

FIG. 4.6 Schematic design representation of the typical fed-batch microalgae cultivation.

production of lipid-rich microalgae. Besides the microalgae lipids for biodiesel production, carbohydrates are also a preferable feedstock for several biological biomass conversion technologies and especially in the technology of sugar fermentation for the production of bioethanol.

With the information that discussed previously, it showed that the biofuel production, especially biodiesel and bioethanol using mixotrophic cultivation system, is one of the promising methods to reduce the cost of microalgae biomass production. This could be because the mixotrophic condition provides dual activities that enable the usage of CO_2 and organic carbons simultaneously, and both respiratory and photosynthetic metabolisms happen concurrently.

Based on the result showed by Cheirsilp and Torpee [25], it was demonstrated that the cultivations of marine *Chlorella* sp. and *Nannochloropsis* sp. were the ideal candidates for biodiesel production due to their high content of lipid up to 369.2 ⊥ 2.7 and 305.8 ± 5.5 mg/L, respectively, under mixotrophic cultivation. From the lipid content showed in this study, it was largely contributed by 72.04% and 70.18% of palmitic acid (C16:0) as well as 13.39% and 13.91% of stearic acid (C18:0) for both microalgae. These results are consistent with those of

Wang et al. [84] who reported that palmitic acid (C 16:0), linoleic acid (C 18:2), and linolenic acid (C 18: 3) were the abundant fatty acids in green microalgae, *C. pyrenoidosa*. Another study by *Cerón-García* et al. [85] demonstrated that the cultivation of *C. protothecoides* was able to generate a high amount of oleic acid methyl ester up to 66% under mixotrophic cultivation condition. Oils with a high oleic acid content are considered to be the most suitable fuels due to the greater stability of their alkyl esters and their better characteristics as fuels [86]. On the other hand, the cultivation under mixotrophic condition also could accumulate high amount of starch for bioethanol production. According to the previous study, the highest amount of carbohydrate content and productivity could be obtained up to 8.74% and 57.26 mg/L d when cultivation of *C. vulgaris* was carried out in the presence of 10 g/L of glycerol and 2 g/L of glucose [3]. Table 4.3 tabulates some of the examples of potential microalgae that can be used for biofuel production.

To sum up, this chapter provides a promising niche for reducing the overall cost of biodiesel and bioethanol production from microalgae. It investigates the by-products of algal biodiesel production and algal cell hydrolysis as the possible raw materials (lipids and carbohydrates) and organic carbon substrates (soluble

TABLE 4.3
Potential Microalgae Be Used for Biofuel Production.

Name of Microalgae	Biofuel	Sources	Production	References
Chlorella vulgaris	Bioethanol	Plantain peel extract	10.82%	[87]
Spirulina platensis	Bioethanol	Glucose	0.04%	[7]
Chlorella sp.	Biodiesel	Glucose	72.04%	[25]
Nannochloropsis sp.	Biodiesel	Glucose	70.18%	[25]
Scenedesmus obliquus	Biodiesel	Glucose	75.00%	[88]
C. vulgaris	Biodiesel	Dry-grind ethanol thin stillage	52.20%	[89]

carbohydrates and glycerol) for the mixotrophic cultivation of microalgae.

Hence, this chapter provides a promising niche for reducing the overall cost of biodiesel and bioethanol production from microalgae as it investigates the by-products of algal biodiesel production and algal cell hydrolysis as possible raw materials (lipids and carbohydrates) and organic carbon substrates (soluble carbohydrates and glycerol) for mixotrophic cultivation of microalgae.

CONCLUSION AND FUTURE PROSPECT

Production of microalgal biomass for biofuel production via mixotrophic condition is believed to be one of the alternatives to enhance biomass production and biofuel productivity. Further research and development should be considered to ensure the feasibility of biofuel production.

The most critical part is high-cost production due to the additional organic source to enhance and achieve satisfactory microalgal biomass production. To ensure the biomass production cost-effectiveness and reduce the dependency of using the fresh organic source, implementing integrated cultivation system using wastewater and CO_2 from flue gas emitted from various industries as a carbon source for biomass production is one of the alternative approaches to overcome this issue. Using waste generated from the industry not only reduces production cost but could also reduce pollution caused by the industry.

Another future research in this area includes the development of innovative photobioreactor for microalgal cultivation. Auto smart monitoring medium supply control for microalgae is considered the advanced biotechnology in this field. This technology is one of the most promising methods to accurate tracking of multiple culture parameters such as temperature, dissolved oxygen, flow rate, pH, and nutrient concentrations, meanwhile to correctly reflect the process dynamics with implementing the appropriate actions to enhance the microalgal cellular growth and its metabolism [90]. Therefore, this application could achieve the optimal microalgae growth rate and better process performance. Similar application has been adopted by Marxen et al. [91] who applied the computer-based for *Synechocystis* sp. PCC6803 microalgae cultivation under controlled physiological conditions with the aided of feedback loop in the cultivation system. Another study also reported that computer-based controlled microalgae in outdoor tubular photobioreactor can enhance microalgae growth rate and productivity with the simulated different control growth environment conditions [90]. This advanced technology is still remaining unclear, and more effort and information should be explored in the future.

In terms of fundamental research, isolation and characterization of extremophile microalgae that could stand the fluctuating cultivation conditions are required to give maximum biomass productivity when the cultivation is performed in an uncontrolled condition. On the other hand, the further development should also focus on various by-products produced from single cultivation via biorefinery approach. Apart from biofuel production, full utilization of the microalgal biomass for production of various types of products could enhance the value and give more profit to the microalgal biotechnology industry.

The understanding of metabolisms at cellular levels such as biomass production and biochemical biosynthesis pathway is a critical aspect to manipulate the cultivation and therefore, further development is essential to enhance microalgal biomass and chemical production.

ACKNOWLEDGMENTS

The authors wish to thank School of Industrial Technology, Universiti Sains Malaysia, and the Ministry of Higher Education, Malaysia (USM). The project and manuscript preparation were financially supported by the RUI Grant 1001/PTEKIND/8011043 and USM Fellowship RU(1001/CIPS/AUPE001) from Universiti Sains Malaysia.

REFERENCES

[1] N.K. Patel, S.N. Shah, in: S. Ahuja (Ed.), Biodiesel from Plant Oils, Food, Energy, and Water, Elsevier, Boston, 2015, pp. 277−307.

[2] K. Chojnacka, F.-J. Marquez-Rocha, Kinetic and stoichiometric relationships of the energy and carbon metabolism in the culture of microalgae, Biotechnology 3 (2004) 21−34.

[3] W.-B. Kong, H. Yang, Y.-T. Cao, H. Song, S.-F. Hua, C.-G. Xia, Effect of glycerol and glucose on the enhancement of biomass, lipid and soluble carbohydrate production by *Chlorella vulgaris* in mixotrophic culture, Food Technol. Biotechnol. 51 (2013) 62−69.

[4] H. Endo, S. Hiroshi, K. Nakajima, Studies on *Chlorella regularis*, heterotrophic fast-growing strain II. Mixotrophic growth in relation to light intensity and acetate concentration, Plant Cell Physiol. 18 (1977) 199−205.

[5] O. Perez-Garcia, Y. Bashan, O. Perez-Garcia, Y. Bashan, Microalgal Heterotrophic and Mixotrophic Culturing for Bio-Refining: From Metabolic Routes to Techno-Economics, Algal Biorefineries, 2015, pp. 61−131.

[6] Y. Liang, N. Sarkany, Y. Cui, Biomass and lipid productivities of *Chlorella vulgaris* under autotrophic, heterotrophic and mixotrophic growth conditions, Biotechnol. Lett. 31 (2009) 1043−1049.

[7] F.J. Marquez, K. Sasaki, T. Kakizono, N. Nishio, S. Nagai, Growth characteristics of *Spirulina platensis* in mixotrophic and heterotrophic conditions, J. Ferment. Bioeng. 76 (1993) 408−410.

[8] E. Komor, W. Tanneb, The hexose-proton symport system of *Chlorella vulguris* specificity, stoichiometry and energetics of sugar-induced proton uptake, Eur. J. Biochem. 44 (1974) 219−223.

[9] W. Tanner, Light-driven active uptake of 3-O-methylglucose via an inducible hexose uptake system of *Chlorella*, Biochem. Biophys. Res. Commun. 36 (1969) 278−283.

[10] Y. Wang, H. Rischer, N.T. Eriksen, M.G. Wiebe, Mixotrophic continuous flow cultivation of *Chlorella prototheccides* for lipids, Bioresour. Technol. 144 (2013) 608−614.

[11] E. Sforza, R. Cipriani, T. Morosinotto, A. Bertucco, G.M. Giacometti, Excess CO2 supply inhibits mixotrophic growth of *Chlorella prototheccides* and *Nannochloropsis salina*, Bioresour. Technol. 104 (2012) 523−529.

[12] O. Oliveira, S. Gianesella, V. Silva, T. Mata, N. Caetano, Lipid and carbohydrate profile of a microalga isolated from wastewater, Energy Procedia 136 (2017) 468−473.

[13] M.I. Orús, E. Marco, F. Martínez, Suitability of *Chlorella vulgaris* UAM 101 for heterotrophic biomass production, Bioresour. Technol. 38 (1991) 179−184.

[14] G. Markou, I. Angelidaki, D. Georgakakis, Microalgal carbohydrates: an overview of the factors influencing carbohydrates production, and of main bioconversion technologies for production of biofuels, Appl. Microbiol. Biotechnol. 96 (2012) 631−645.

[15] M. Cecchin, S. Benfatto, F. Griggio, A. Mori, S. Cazzaniga, N. Vitulo, M. Delledonne, M. Ballottari, Molecular basis of autotrophic vs mixotrophic growth in *Chlorella sorokiniana*, Sci. Rep. 8 (2018) 6465.

[16] M.P. Rai, T. Gautom, N. Sharma, Effect of Salinity, pH, light intensity on growth and lipid production of microalgae for bioenergy application, Online J. Biol. Sci. 15 (4) (2015) 260−267.

[17] A. Kamiya, W. Kowallik, Photoinhibition of glucose uptake in *Chlorella*, Plant Cell Physiol. 28 (1987) 611−619.

[18] Y. Li, W. Zhou, B. Hu, M. Min, P. Chen, R.R. Ruan, Effect of light intensity on algal biomass accumulation and biodiesel production for mixotrophic strains *Chlorella kessleri* and *Chlorella prototheccide* cultivated in highly concentrated municipal wastewater, Biotechnol. Bioeng. 109 (2012) 2222−2229.

[19] J. Xie, Y. Zhang, Y. Li, Y. Wang, Mixotrophic cultivation of *Platymonas subcordiformis*, J. Appl. Phycol. 13 (2001) 343−347.

[20] A.P. Carvalho, S.O. Silva, J.M. Baptista, F.X. Malcata, Light requirements in microalgal photobioreactors: an overview of biophotonic aspects, Appl. Microbiol. Biotechnol. 89 (2011) 1275−1288.

[21] Z. Amini Khoeyi, J. Seyfabadi, Z. Ramezanpour, Effect of light intensity and photoperiod on biomass and fatty acid composition of the microalgae, *Chlorella vulgaris*, Aquacult. Int. 20 (2012) 41−49.

[22] O. Friedman, Z. Dubinsky, S. (Malis) Arad, Effect of light intensity on growth and polysaccharide production in red and blue-green *Rhodophyta* unicells, Bioresour. Technol. 38 (1991) 105−110.

[23] H.-X. Chang, Y. Huang, Q. Fu, Q. Liao, X. Zhu, Kinetic characteristics and modeling of microalgae *Chlorella vulgaris* growth and CO_2 biofixation considering the coupled effects of light intensity and dissolved inorganic carbon, Bioresour. Technol. 206 (2016) 231−238.

[24] G.H. Gim, J. Ryu, M.J. Kim, P. Il Kim, S.W. Kim, Effects of carbon source and light intensity on the growth and total lipid production of three microalgae under different culture conditions, J. Ind. Microbiol. Biotechnol. 43 (2016) 605−616.

[25] B. Cheirsilp, S. Torpee, Enhanced growth and lipid production of microalgae under mixotrophic culture condition: effect of light intensity, glucose concentration and fed-batch cultivation, Bioresour. Technol. 110 (2012) 510−516.

[26] A. Juneja, R. Ceballos, G. Murthy, A. Juneja, R.M. Ceballos, G.S. Murthy, Effects of environmental factors and nutrient availability on the biochemical

composition of algae for biofuels production: a review, Energies 6 (2013) 4607—4638.

[27] M. Brody, A.E. Vatter, Observations on cellular structures of *Porphyridium cruentum*, J. Biophys. Biochem. Cytol. 5 (1959) 289—294.

[28] K.H. Ogbonda, R.E. Aminigo, G.O. Abu, Influence of temperature and pH on biomass production and protein biosynthesis in a putative *Spirulina* sp, Bioresour. Technol. 98 (2007) 2207—2211.

[29] G. Venkata Subhash, M.V. Rohit, M.P. Devi, Y.V. Swamy, S. Venkata Mohan, Temperature induced stress influence on biodiesel productivity during mixotrophic microalgae cultivation with wastewater, Bioresour. Technol. 169 (2014) 789—793.

[30] P.A. Thompson, M. Guo, P.J. Harrison, Effects of variation in temperature. i. on the biochemical composition of eight species of marine phytoplankton1, J. Phycol. 28 (1992) 481—488.

[31] D. V Lynch, G.A. Thompson, Low temperature-induced alterations in the chloroplast and microsomal membranes of *Dunaliella salina*, Plant Physiol. 69 (1982) 1369—1375.

[32] A. Converti, A.A. Casazza, E.Y. Ortiz, P. Perego, M. Del Borghi, Effect of temperature and nitrogen concentration on the growth and lipid content of *Nannochloropsis oculata* and *Chlorella vulgaris* for biodiesel production, Chem. Eng. Process 48 (2009) 1146—1151.

[33] I. Nishida, N. Murata, Chilling sensitivity in plants and cyanobacteria: the crucial contribution of membrane lipids, Annu. Rev. Plant Physiol. Plant Mol. Biol. 47 (1996) 541—568.

[34] Y. Nakamura, S. Miyachi, Effect of temperature on starch degradation in *Chlorella vulgaris* 11h cells, Plant Cell Physiol. 23 (1982) 333—341.

[35] Y. Nakamura, Change in molecular weight distribution in starch when degraded at different temperatures in *Chlorella vulgaris*, Plant Sci. Lett. 30 (1983) 259—265.

[36] B. Hu, M. Min, W. Zhou, Y. Li, M. Mohr, Y. Cheng, H. Lei, Y. Liu, X. Lin, P. Chen, R. Ruan, Influence of Exogenous CO_2 on biomass and lipid accumulation of microalgae *Auxenochlorella protothecoides* cultivated in concentrated municipal wastewater, Appl. Biochem. Biotech. 166 (2012) 1661—1673.

[37] S. Chinnasamy, B. Ramakrishnan, A. Bhatnagar, K.C. Das, Biomass production potential of a wastewater alga *Chlorella vulgaris* ARC 1 under elevated levels of CO_2 and temperature, Int. J. Mol. Sci. 10 (2009) 518—532.

[38] J.-R. Xia, K.-S. Gao, Impacts of elevated CO_2 concentration on biochemical composition, carbonic anhydrase, and nitrate reductase activity of freshwater green algae, J. Integr. Plant Biol. 47 (2005) 668—675.

[39] M.-K. Ji, H.-S. Yun, Y.-T. Park, A.N. Kabra, I.-H. Oh, J. Choi, Mixotrophic cultivation of a microalga *Scenedesmus obliquus* in municipal wastewater supplemented with food wastewater and flue gas CO_2 for biomass production, J. Environ. Manage. 159 (2015) 115—120.

[40] C. Yang, Q. Hua, K. Shimizu, Energetics and carbon metabolism during growth of microalgal cells under photoautotrophic, mixotrophic and cyclic light-autotrophic/dark-heterotrophic conditions, Biochem. Eng. J. 6 (2000) 87—102.

[41] C. Thyssen, R. Schlichting, C. Giersch, The CO_2 -concentrating mechanism in the physiological context: lowering the CO_2 supply diminishes culture growth and economises starch utilisation in *Chlamydomonas reinhardtii*, Planta 213 (2001) 629—639.

[42] A. Izumo, S. Fujiwara, Y. Oyama, A. Satoh, N. Fujita, Y. Nakamura, M. Tsuzuki, Physicochemical properties of starch in *Chlorella* change depending on the CO_2 concentration during growth: comparison of structure and properties of pyrenoid and stroma starch, Plant Sci. 172 (2007) 1138—1147.

[43] G.H. Gim, J.K. Kim, H.S. Kim, M.N. Kathiravan, H. Yang, S.-H. Jeong, S.W. Kim, Comparison of biomass production and total lipid content of freshwater green microalgae cultivated under various culture conditions, Bioprocess Biosyst. Eng. 37 (2014) 99—106.

[44] Y. Wang, T. Chen, S. Qin, Differential fatty acid profiles of *Chlorella kessleri* grown with organic materials, J. Chem. Technol Biot. 88 (2013) 651—657.

[45] J. Zhan, J. Rong, Q. Wang, Mixotrophic cultivation, a preferable microalgae cultivation mode for biomass/bioenergy production, and bioremediation, advances and prospect, Int. J. Hydrogen Energ. 42 (2017) 8505—8517.

[46] T. Heredia-Arroyo, W. Wei, R. Ruan, B. Hu, Mixotrophic cultivation of *Chlorella vulgaris* and its potential application for the oil accumulation from non-sugar materials, Biomass Bioenergy 35 (2011) 2245—2253.

[47] Y. Xiuxia, Z. Xiaoqi, J. Yupeng, L. Qun, Effects of sodium nitrate and sodium acetate concentrations on the growth and fatty acid composition of *Brachiomonas submarina*, J. Ocean Univ. Qingdao 2 (2003) 75—78.

[48] W. Haiying, F. Ru, P. Guofeng, A study on lipid production of the mixotrophic microalgae *Phaeodactylum tricornutum* on various carbon sources, Afr. J. Microbiol. Res. 6 (2012) 1041—1047.

[49] A.P. Abreu, B. Fernandes, A.A. Vicente, J. Teixeira, G. Dragone, Mixotrophic cultivation of *Chlorella vulgaris* using industrial dairy waste as organic carbon source, Bioresour. Technol. 118 (2012) 61—66.

[50] I. Brányiková, B. Maršálková, J. Doucha, T. Brányik, K. Bišová, V. Zachleder, M. Vítová, Microalgae-novel highly efficient starch producers, Biotechnol. Bioeng. 108 (2011) 766—776.

[51] [51a] Y.-M. Zhang, H. Chen, C.-L. He, Q. Wang, Nitrogen starvation induced oxidative stress in an oil-producing green alga *Chlorella sorokiniana* C3, PLoS One 8 (7) (2013);
[51b] A. Banerjee, S.K. Maiti, C. Guria, C. Banerjee, Metabolic pathways for lipid synthesis under nitrogen stress in *Chlamydomonas* and *Nannochloropsis*, Biotechnol. Lett. 39 (2017) 1—11.

[52] H.P. Ratnapuram, S.S. Vutukuru, R. Yadavalli, Mixotrophic transition induced lipid productivity in *Chlorella pyrenoidosa* under stress conditions for biodiesel production, Heliyon 4 (1) (2018).

[53] K. Paranjape, G.B. Leite, P.C. Hallenbeck, Effect of nitrogen regime on microalgal lipid production during mixotrophic growth with glycerol, Bioresour. Technol. 214 (2016) 778–786.

[54] I.A. Guschina, J.L. Harwood, Lipids and lipid metabolism in eukaryotic algae, Prog. Lipid Res. 45 (2006) 160–186.

[55] X. Shen, Heterotrophic and Mixotrophic Cultivation of Microalgae under Nitrogen Starvation for Biodiesel Production, City University of Hong Kong, 2017.

[56] G. Dragone, B.D. Fernandes, A.P. Abreu, A.A. Vicente, J.A. Teixeira, Nutrient limitation as a strategy for increasing starch accumulation in microalgae, Appl. Energ. 88 (2011) 3331–3335.

[57] P.W. Behrens, S.E. Bingham, S.D. Hoeksema, D.L. Cohoon, J.C. Cox, Studies on the incorporation of CO_2 into starch by *Chlorella vulgaris*, J. Appl. Phycol. 1 (1989) 123–130.

[58] B. Richardson, D.M. Orcutit, H.A. Schwertner, C.L. Martinez, H.E. Wickline, Effects of nitrogen limitation on the growth and composition of unicellular algae in continuous culture, Appl. Microbiol. 18 (2) (1969) 245–250.

[59] H. Cui, F. Meng, F. Li, Y. Wang, W. Duan, Y. Lin, Two-stage mixotrophic cultivation for enhancing the biomass and lipid productivity of *Chlorella vulgaris*, Amb. Express 7 (2017) 187.

[60] M.P. Rai, S. Nigam, R. Sharma, Response of growth and fatty acid compositions of *Chlorella pyrenoidosa* under mixotrophic cultivation with acetate and glycerol for bioenergy application, Biomass Bioenergy 58 (2013) 251–257.

[61] L. Yang, J. Chen, S. Qin, M. Zeng, Y. Jiang, L. Hu, P. Xiao, W. Hao, Z. Hu, A. Lei, J. Wang, Growth and lipid accumulation by different nutrients in the microalga *Chlamydomonas reinhardtii*, Biotechnol. Biofuels 11 (2018) 40.

[62] R. Ramanan, B.-H. Kim, D.-H. Cho, S.-R. Ko, H.-M. Oh, H.-S. Kim, Lipid droplet synthesis is limited by acetate availability in starchless mutant of *Chlamydomonas reinhardtii*, FEBS Lett. 587 (2013) 370–377.

[63] R.W. Hunt, S. Chinnasamy, A. Bhatnagar, K.C. Das, Effect of biochemical stimulants on biomass productivity and metabolite content of the microalga, *Chlorella sorokiniana*, Appl. Biochem. Biotechnol. 162 (2010) 2400–2414.

[64] R. Czerpak, P. Dobrzyń, A. Krotke, E. Kicińska, The effect of auxins and salicylic acid on chlorophyll and carotenoid contents in *Wolffia arrhiza* (l.) Wimm. (Lemnaceae) growing on media of various trophicities, Pol. J. Eviron. Stud. 11 (3) (2002) 231–235.

[65] Y. Zhao, Auxin biosynthesis and its role in plant development, Annu. Rev. Plant Biol. 61 (2010) 49–64.

[66] A. Bajguz, A. Piotrowska-Niczyporuk, Synergistic effect of auxins and brassinosteroids on the growth and regulation of metabolite content in the green alga *Chlorella vulgaris* (Trebouxiophyceae), Plant Physiol. Biochem. 71 (2013) 290–297.

[67] A.F. Tiburcio, T. Altabella, A. Borrell, C. Masgrau, Polyamine metabolism and its regulation, Physiol. Plant. 100 (1997) 664–674.

[68] M.M. Allaf, L. Rehmann, M. Allaf, Effect of Plant Hormones on the Production of Biomass and Lipid in Microalgae, Western University, 2013.

[69] M.L. Pelekis, B.S. Mangat, K. Krishnan, Influence of 2,4-dichlorophenoxyacetic acid on the growth and stored polyglucan content of three species of heterotrophic algae, Pestic. Biochem. Physiol. 28 (1987) 349–353.

[70] G.B. Pinchot, The mechanism of uncoupling of oxidative phosphorylation by 2,4-Dinitrophenol*, J. Biol. Chem. 242 (20) (1967) 4577–4583.

[71] R.T. Wedding, M. Kay Black, Uncoupling of phosphorylation in *Chlorella* by 2,4-dichlorophenoxyacetic acid, Plant Soil 9 (1961) 242–248.

[72] H.-W. Yen, J.-T. Chang, A two-stage cultivation process for the growth enhancement of *Chlorella vulgaris*, Bioprocess Biosyst. Eng. 36 (2013) 1797–1801.

[73] E. Bardone, M. Bravi, T. Keshavarz, F. Di Caprio, A. Visca, P. Altimari, L. Toro, B. Masciocchi, G. Iaquaniello, F. Pagnanelli, Two stage process of microalgae cultivation for starch and carotenoid production, Chem. Eng. Trans. 49 (2016).

[74] Y.T.T. Doan, J.P. Obbard, Two-stage cultivation of a *Nannochloropsis* mutant for biodiesel feedstock, J. Appl. Phycol. 27 (2015) 2203–2208.

[75] W. Farooq, Y.-C. Lee, B.-G. Ryu, B.-H. Kim, H.-S. Kim, Y.-E. Choi, J.-W. Yang, Two-stage cultivation of two *Chlorella* sp. strains by simultaneous treatment of brewery wastewater and maximizing lipid productivity, Bioresour. Technol. 132 (2013) 230–238.

[76] A.T. Lombardi, P.J. Wangersky, Influence of phosphorus and silicon on lipid class production by the marine diatom *Chaetoceros gracilis* grown in turbidostat cage cultures, Mar. Ecol. Prog. Ser. 77 (1991) 39–47.

[77] G.A. Dunstan', J.K. Volkman, S.M. Barrett, C.D. Garland, Changes in the lipid composition and maximisation of the polyunsaturated fatty acid content of three microalgae grown in mass culture, J. Appl. Phycol. 5 (1) (1993) 71–83.

[78] J. Fábregas, A. Maseda, A. Domínguez, M. Ferreira, A. Otero, Changes in the cell composition of the marine microalga, *Nannochloropsis gaditana*, during a light:dark cycle, Biotechnol. Lett. 24 (20) (2002) 1699–1703.

[79] F.J. Fields, J.T. Ostrand, S.P. Mayfield, Fed-batch mixotrophic cultivation of *Chlamydomonas reinhardtii* for high-density cultures, Algal Res. 33 (2018) 109–117.

[80] F. Chen, Y. Zhang, High cell density mixotrophic culture of *Spirulina platensis* on glucose for phycocyanin production using a fed-batch system, Enzyme Microb. Technol. 20 (1997) 221–224.

[81] Y. Xie, Y. Jin, X. Zeng, J. Chen, Y. Lu, K. Jing, Fed-batch strategy for enhancing cell growth and C-phycocyanin production of *Arthrospira (Spirulina) platensis* under phototrophic cultivation, Bioresour. Technol. 180 (2015) 281–287.

[82] T.-H. Wang, S.-H. Chu, Y.-Y. Tsai, F.-C. Lin, W.-C. Lee, Influence of inoculum cell density and carbon dioxide concentration on fed-batch cultivation of *Nannochloropsis oculata*, Biomass Bioenergy 77 (2015) 9–15.

[83] R. Martínez-Macías, E. Meza-Escalante, D. Serrano-Palacios, P. Gortáres-Moroyoqui, P.E. Ruíz-Ruíz,

G. Ulloa-Mercado, Effect of fed-batch and semicontinuous regimen on *Nannochloropsis oculata* grown in different culture media to high-value products, J. Chem. Technol. Biotechnol. 93 (2018) 585–590.

[84] H. Wang, H. Xiong, Z. Hui, X. Zeng, Mixotrophic cultivation of *Chlorella pyrenoidosa* with diluted primary piggery wastewater to produce lipids, Bioresour. Technol. 104 (2012) 215–220.

[85] M.C. Cerón-García, M.D. Macías-Sánchez, A. Sánchez-Mirón, F. García-Camacho, E. Molina-Grima, A process for biodiesel production involving the heterotrophic fermentation of *Chlorella protothecoides* with glycerol as the carbon source, Appl. Energy 103 (2013) 341–349.

[86] A. Robles-Medina, P.A. González-Moreno, L. Esteban-Cerdán, E. Molina-Grima, Biocatalysis: towards ever greener biodiesel production, Biotechnol. Adv. 27 (2009) 398–408.

[87] O.K. Agwa, I.G. Nwosu, G.O. Abu, Bioethanol production from *Chlorella vulgaris* biomass cultivated with plantain (*Musa Paradisiaca*) peels extract, Adv. Biosci. Biotechnol. 8 (2017) 478–490.

[88] S. Mandal, N. Mallick, Microalga *Scenedesmus obliquus* as a potential source for biodiesel production, Appl. Microbiol. Biotechnol. 84 (2009) 281–291.

[89] D. Mitra, J. (Hans) van Leeuwen, B. Lamsal, Heterotrophic/mixotrophic cultivation of oleaginous *Chlorella vulgaris* on industrial co-products, Algal Res. 1 (2012) 40–48.

[90] R. Dormido, J. Sánchez, N. Duro, S. Dormido-Canto, M. Guinaldo, S. Dormido, R. Dormido, J. Sánchez, N. Duro, S. Dormido-Canto, M. Guinaldo, S. Dormido, An interactive tool for outdoor computer controlled cultivation of microalgae in a tubular photobioreactor system, Sensors 14 (2014) 4466–4483.

[91] K. Marxen, K.H. Vanselow, S. Lippemeier, R. Hintze, A. Ruser, U.-P. Hansen, A photobioreactor system for computer controlled cultivation of microalgae, J. Appl. Phycol. 17 (2005) 535–549.

CHAPTER 5

Oleaginous Microalgae Cultivation for Biogas Upgrading and Phytoremediation of Wastewater

SIRASIT SRINUANPAN • BENJAMAS CHEIRSILP • MOHD ASYRAF KASSIM

INTRODUCTION

Biogas produced from anaerobic digestion of organic wastes is considered as a renewable and sustainable bioenergy. The main component of crude biogas is methane (CH_4) at 50%−80% and carbon dioxide (CO_2) at 35%−40%. It also contains trace amount of water (H_2O), oxygen (O_2), and hydrogen sulfide (H_2S) depending on the composition of the organic wastes used. The CH_4 content in biogas is the most important factor for its ignition quality in engines and electricity energy generation. More than 90% of CH_4 content in biogas is required for effective burning and high specific calorific value. There are several physical/chemical methods to remove CO_2 and increase CH_4 content in biogas. However, most of these technologies consume huge amount of energy, ancillary equipment, and harmful additives and additionally produce pollutants. Therefore, the biological method for removing CO_2 in biogas through microalgae cultivation has been attempted to improve the biogas quality. Due to high CO_2 fixation ability of microalgae, a large number of researches have shown that CH_4 content in biogas could be upgraded by microalgae to be more than 90%. Moreover, through this strategy microalgal biomass with high-value products could also be generated [1].

The use of specific microalgae with high lipid content, namely oleaginous microalgae for removing CO_2 from biogas not only can purify the biogas but also produce biodiesel feedstocks. This integrated process is very attractive to offset the production costs of biofuels and environmentally friendly mitigate CO_2 into microalgal biomass. As commercial media are not practical for microalgae cultivation at industrial level, many researchers in this field have paid increasing attention on the alternative use of industrial wastewater as good nutrient sources for microalgae cultivation. The concepts of microalgae cultivation and its cocultivation with other microorganisms for simultaneously removing CO_2 from biogas, phytoremediation of industrial wastewater, and low-cost production of biofuels may contribute greatly to the sustainable and environmental friendly production of renewable energy. Fig. 5.1 shows the microalgae cultivation scheme for simultaneous biogas upgrading, production of biodiesel feedstock, and phytoremediation of wastewater. This chapter aims to summarize the recent research on microalgae cultivation for biogas upgrading, production of biodiesel feedstocks, and phytoremediation of wastewater.

BIOGAS UPGRADING AND LIPID PRODUCTION

Several microalgae have been studied for CO_2 removal from biogas, such as *Chlorella* spp. [2−7], *Leptolyngbya* spp. [8], *Scenedesmus* spp. [1,7−10], *Nannochloropsis* sp. [11], *Selenastrum* spp. [7,9], and *Anabaena* spp. [7]. According to these reports, CO_2 removal efficiency by the microalgae could be as high as >90% and the CH_4 content in biogas could be upgraded up to >90%, which meets the standard level of biogas used in combustion (Table 5.1). However, only few oleaginous microalgae have been evaluated for their ability to accumulate high lipid content during biogas upgrading process. *Chlorella* spp. [6]. and *Scenedesmus* spp. [1,10,12]. have been examined for their ability to upgrade biogas and accumulate lipids, and it was found that more than 90% of their fatty acid compositions are palmitic acid (C16:0), oleic acid (C18:1), linoleic acid (C18:2), and linolenic acid (C18:3), which are similar to those of plant oils and suitable as biodiesel feedstocks.

Microalgae Cultivation for Biofuels Production. https://doi.org/10.1016/B978-0-12-817536-1.00005-9

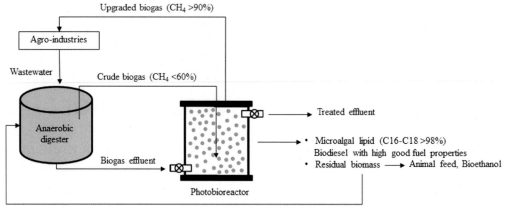

FIG. 5.1 Microalgae cultivation scheme for simultaneous biogas upgrading, production of biodiesel feedstock, and phytoremediation of wastewater.

FACTORS AFFECTING BIOGAS UPGRADING AND LIPID PRODUCTION

Biogas Composition

The cultivation of microalgae using CO_2 from biogas as an inorganic carbon source could be limited by the variation in concentration of CO_2, CH_4, and H_2S in biogas [13–15]. Several studies have evaluated the effect of CO_2 concentration on the microalgae growth [16–18]. Some microalgae exhibit excellent tolerance ability to high CO_2 concentration. The genus *Chlorella* and *Scenedesmus*, growing in the wide range of high CO_2 concentration, are considered as suitable strains for studying the possibility of CO_2 removal from biogas [1,6]. At a low concentration of CO_2, the growth of microalgae may be limited, whereas a high CO_2 concentration may cause a reduction in pH and inhibit the growth of microalgae and finally reduce the CO_2 fixation ability. Wang et al. [19] found that the microalgal biomass of *Chlorella vulgaris* increased with increasing CO_2 concentration from 25% to 55%. Yan et al. [20] suggested that CO_2 in biogas above 55% was harmful to the microalgal cells and inhibited the growth of *Chlorella* sp. The determination of microalgae growth kinetics follows the well-known Monod model. The growth rate relates to the initial dissolved inorganic carbon, CO_2, can be expressed as

$$\mu - \frac{\mu_{max}S}{K_s + S} \qquad (5.1)$$

where μ is specific growth rate, μ_{max} is maximum specific growth rate, K_S is half-saturation constant, and S could be either nutrient concentration or CO_2 concentration. The μ_{max} and K_S of microalgae growth have

been reported to be in the range of 0.02–2.45 day and 0.3–1.2 mg/L, respectively [21].

The percent of CH_4 content in biogas is one of crucial factors for the performance of microalgae to grow and remove CO_2 from biogas. Meier et al. [11] have reported that 50%–100% of CH_4 content in biogas did not affect the growth of *Nannochloropsis gaditana*, while Kao et al. [2] claimed that *Chlorella* sp. MM-2 grew less when the CH_4 content in biogas was increased from 20% up to 80%. Yan et al. [20] have found that the acceptable range of CH_4 content in biogas for cultivation of *Chlorella* sp. was 45%–55% (v/v). Srinuanpan et al. [1] compared the cultivation of two microalgae marine *Chlorella* sp. and *Scenedesmus* sp. and found that both microalgae were able to grow similarly either using 40% CO_2 in air or using 40% CO_2 in biogas. It was explained that CH_4 in biogas might be insoluble in culture medium at identical condition and then did not negatively affect the microalgae growth.

In addition to CO_2 and CH_4 content in biogas, H_2S content in biogas also greatly influences microalgae growth and their CO_2 fixation ability. Depending on the organic wastes used in anaerobic digester, the H_2S content in biogas varied from 0.005% to 2% [13]. Only few researchers reported the effect of H_2S content in biogas on the performance of microalgae. *Chlorella* sp. has been reported that it presented high biomass production and exhibited its well-being during its cultivation using CO_2 from biogas containing H_2S less than 150 ppm [2]. In one study, the microalgae seemed to convert H_2S to sulfate to reduce the toxicity and the generated sulfate was assimilated by the microalgae as a growth-limiting substrate [22]. But this ability is strongly pH dependent, and low pH is more suitable

TABLE 5.1
Microalgae Cultivation for Biogas Upgrading.

Microalgae Strain	Biogas Composition	CH$_4$ Content after Upgrading (%v/v)	References
Mutant *Chlorella* sp.	Desulfurized biogas containing 70% CH$_4$, 20% CO$_2$, and 8% N$_2$	84–87	Kao et al. [2]
	Desulfurized biogas containing 69% CH$_4$, 20% CO$_2$, and <50 ppm H$_2$S	85–90	Kao et al. [3]
Chlorella sp.	Crude biogas containing 70.65% CH$_4$, 26.14% CO$_2$, 0.23% O$_2$, 3.11% H$_2$O, and <0.005% H$_2$S	92.16	Yan and Zheng [4]
	Crude biogas containing 67.35% CH$_4$, 28.41% CO$_2$, 0.73% O$_2$, 3.48% H$_2$O, and <0.005% H$_2$S	92.74	Zhao et al. [5]
	Synthetic biogas containing 50% CH$_4$ and 50% CO$_2$	94.7	Tongprawhan et al. [6]
	Crude biogas containing 61.38% CH$_4$, 32.57% CO$_2$, 0.54% O$_2$, 5.52% H$_2$O, and <0.005% H$_2$S	84.21	Wang et al. [7]
Leptolyngbya sp.	Synthetic biogas containing 25% CO$_2$ and 75% CH$_4$	99.56	Choix et al. [8]
Scenedesmus obliquus	Crude biogas containing 61.38% CH$_4$, 32.57% CO$_2$, 0.54% O$_2$, 5.52% H$_2$O, and <0.005% H$_2$S	84.28	Wang et al. [7]
	Crude biogas containing 61.75% CH$_4$, 32.28% CO$_2$, 0.31% O$_2$, 2.68% H$_2$O, and <0.005% H$_2$S	94.41	Ouyang et al. [9]
	Synthetic biogas containing 25% CO$_2$ and 75% CH$_4$	96.5	Choix et al. [12]
Scenedesmus sp.	Synthetic biogas containing 40% CO$_2$ and 60% CH$_4$	>90	Srinuanpan et al. [1]
	Synthetic biogas containing 40% CO$_2$ and 60% CH$_4$	>98	Srinuanpan et al. [10]
Anabaena spiroides	Crude biogas containing 61.38% CH$_4$, 32.57% CO$_2$, 0.54% O$_2$, 5.52% H$_2$O, and <0.005% H$_2$S	81	Wang et al. [7]
Selenastrum capricornutum	Crude biogas containing 61.38% CH$_4$, 32.57% CO$_2$, 0.54% O$_2$, 5.52% H$_2$O, and <0.005% H$_2$S	78	Wang et al. [7]
Selenastrum bibraianum	Crude biogas containing 61.75% CH$_4$, 32.28% CO$_2$, 0.31% O$_2$, 2.68% H$_2$O, and <0.005% H$_2$S	79	Ouyang et al. [9]
Nannochloropsis gaditana	Synthetic biogas containing 28% CO$_2$ and 72% CH$_4$	98.6	Meier et al. [11]

for better H$_2$S uptake rate. Other types of sulfur, that is, thiosulfate, tetrathionate, thiocyanate, and elemental sulfur are less beneficial for the microalgae growth [13]. González-Sánchez and Posten [13] proposed the growth kinetics of *Chlorella* sp. as the function of H$_2$S concentration. They found that the H$_2$S started inhibiting the microalgae growth at a concentration >200 ppm. The sulfate was generated and later adsorbed by the microalgae.

It should be noted that the CO$_2$ content also affected the lipid content and fatty acid compositions of the microalgae. The CO$_2$ content higher than 10% could induce the accumulation of lipids in *Chaetoceros muelleri* to be more than 40% of their dry biomass, and this high CO_2 content also induced the synthesis of unsaturated fatty acids rather than saturated fatty acids, especially the polyunsaturated fatty acids [16,23]. Only Tongprawhan et al. [6] and Srinuanpan et al. [1,10] reported the effect of using biogas on the fatty acid compositions of microalgal lipids. They found that the biogas with 40% CO_2 could stimulate the production of unsaturated fatty acids such oleic acid (C18:1) and linolenic acid (C18:3) in the microalgae.

Gas Flow Rate

Gas flow rate is the key factor to maximize CO_2 removal rate and biomass productivity by the microalgae. Appropriate flow rate of gas could help in good mixing and adequate supply of CO_2 for microalgae growth [4]. Increasing gas flow rate can lead to an increase in microalgal biomass productivity and CO_2 removal efficiency. However, in case the supply of CO_2 is much higher than the ability of the microalgae to consume, the culture pH would become acidic. The acidic pH negatively affects solubility and availability of nutrients, activity of enzymes, nutrient transport across plasma membrane, and electron transport in cell metabolisms, especially respiration and photosynthesis [16]. Therefore, too high gas flow rate with excess supply of CO_2 decreased the performance of the microalgae [1,10]. Another possible reason has been described that as the flow rate increases, the size of gas bubble also increases and the specific surface area per gas volume decreases [2]. In addition, the evaporation rate of the culture medium also increases when the gas flow rate increases [24].

Srinuanpan et al. [1] suggested that using a low flow rate of biogas the microalgae could completely remove CO_2 from biogas, but the biogas would be treated at a low rate and resulted in low CO_2 removal rate and low productivity of purified biogas. Using higher gas flow rate means that higher amount of biogas could be purified. As both percent CO_2 removal and CO_2 removal rate are important for purifying biogas, the strategy to achieve both targets was developed by Srinuanpan et al. [10]. The stepwise-increasing of biogas flow rate was attempted to improve both percent CO_2 removal and CO_2 removal rate. Through this strategy, the biogas was initially supplied at a low biogas flow rate which is appropriate for supporting the growth of microalgae cells at low cell density. As the microalgae grew and their biomass increased, the biogas flow rate was increased to supply more CO_2. This strategy did enhance the overall process productivity. Interestingly, the gas flow rate also affected the lipid content and fatty

acid composition of the microalgae. Several researchers reported that the lipid content in microalgal cells increased with increasing gas flow rate [25,26]. It is possibly due to the high gas flow rate could constantly supply high carbon-to-nitrogen ratio, which has been proven to play an important role in elevating the lipid content in microalgal cells [24]. Carvalho and Malcata [27] have reported that the concentration of palmitoleic acid in the flagellate *Pavlova lutheri* increased with increasing gas flow rate.

Light Energy

Light intensity and photoperiods

Light is an important energy source for the photosynthesis by the microalgae. When the light energy is insufficient, the microalgae growth is restricted. The biomass productivity is low and the microalgae even consumed the storage energy (carbohydrates/lipids) during photorespiration [5]. Therefore, during low light intensity period, an increase in photoperiod (i.e., longer lighting time) could support the better microalgae growth [4,28]. On the other hand, with the high light intensity above saturation level the microalgae could not grow well due to the light absorb limitation. Moreover, an excessive light intensity also destroyed the light adapter system, that is, photosystems I and II [4,29]. Thus, several reports have suggested decreasing photoperiod (i.e., shorter lighting time) during high light intensity illumination to avoid this photoinhibition effect [4,28].

In addition to nutrients, the metabolism of microalgae depends on photosynthesis and photosynthesis exclusively requires light energy to drive it. Therefore, the light energy is one of the crucial factors for the balanced growth of microalgae. The light energy can be expressed as photosynthetic photon flux density (PPFD). There have been many efforts developing equations for describing the relationship between specific growth rate (μ) and PPFD (I).

$$\mu = \mu_{max}.I/(K_E + I) \tag{5.2}$$

where μ_{max} is the maximum specific growth rate and K_E is the half-saturation constant for light, which is analogous to the Monod equation. The reported μ_{max} and K_E were in the range of 0.06–0.14 h and 95–178 $\mu mol/m^2$ s, respectively [21,30].

The effects of light intensity on CO_2 removal from biogas by microalgae have been reported. Some microalgae required moderate light intensity (150–350 $\mu mol/m^2$ s) and moderate photoperiod (12–14 h a day) for efficient biogas upgrading [4,9]. It should be noted that the optimal light intensity and photoperiod was controversial with different microalgal strains. In fact, the optimal light intensity would change upon the change in density of the microalgal cells. During an initial stage of cultivation, as the microalgae cell density is low the light intensity should not exceed the saturation level to the microalgal cells otherwise photoinhibition may occur and damage the microalgal cells. Moreover, using low light intensity during the initial stage of indoor cultivation could save electricity consumption. On the contrary, as the microalgae grew and the high density of microalgal cells causes mutual cell shading and resulted in lower light penetration. Therefore, the higher density of microalgae cells required the higher light intensity to avoid the light limitation phenomenon [31−33]. Srinuanpan et al. [10] suggested the stepwise-increasing of light intensity during the biogas upgrading based on the growth curve of the microalgae. This strategy successfully enhanced the microalgae growth and their ability to remove CO_2 from biogas.

Although several reports suggested that high light intensity could enhance the microalgae growth, it reduced lipid content of the microalgae [34]. This was because the microalgae used light energy for cell division rather than lipid accumulation [35,36]. However, many researchers reported different trends of results [37−39]. They explained that under high light intensity the saturation of light could be occurred and resulted in low photosynthesis activity; overall anabolic reaction flux was harshly limited, more photosynthetic flow of carbon and energy from carbohydrate and protein turning into the lipids synthesis.

Light wavelength
Light wavelength is also one of important parameters for photosynthesis in the microalgae. The light source should contain the absorption bands of chlorophyll pigments or comprise growth-efficient light spectra [40]. Generally, the microalgae absorb light wavelength at 400−700 nm for supporting their photosynthesis through chlorophyll a (450−475 nm), chlorophyll b (630−675 nm), and carotenoids (400−500 nm). The wavelengths absorbed by microalgae also differ depending on the species. Recently, it has been proved that using narrow band light-emitting diode (LED) as light source could reduce operation cost of indoor cultivation more than using ordinary fluorescent lamps. Several researchers found that the optimum light wavelength for supporting microalgae growth and removing CO_2 from biogas was the red light wavelength (600−700 nm) [40−42]. It was because the chlorophylls in microalgae cells effectively absorb the red light wavelength better than other wavelengths [5]. Moreover, the shorter wavelengths, that is, blue light

has a higher probability of striking the light-harvesting complex and inhibiting the photosynthesis of the microalgae [43]. Zhao et al. [5] and Yan et al. [40] suggested that using a longer wavelength or combining red light with other lights could avoid light saturation and photoinhibition effects. They also reported that combining red and blue light wavelengths gave higher microalgae growth rate than the use of single light wavelength, irrespective of the combining ratio.

The lipid composition of the microalgae is also affected by light wavelength [44,45]. The cellular fatty acid content of microalga *Nannochloropsis* sp. was highest when exposed to green light wavelength [30]. It has been reported that green light wavelength could stimulus the increase of thylakoid membranes and fatty acid contents in plants [46−48]. However, these reports were controversial with those of Teo et al. [44] who reported that cultivating *Tetraselmis* sp. and *Nannochloropsis* sp. under blue light wavelength gave the highest microalgae biomass production and also lipid content than other light wavelengths. They explained that both ribulose bisphosphate carboxylase/oxygenase (Rubisco) and carbonic anhydrase enzymes affect the utilization of carbon dioxide in microalgae cells. Both enzymes are principally active under the blue light wavelength. Interestingly, the two-stage light wavelength has been used to enhance the lipid accumulation in the microalgae [46,49].

Nitrogen Source and its Concentration
Nitrogen source involves in the biochemical synthesis of nucleic acids, lipids, and proteins in the microalgae cells. Higher nitrogen source concentration could support better microalgae growth [50]. However, the excessive level of nitrogen could inhibit microalgae cells [50,51]. Several nitrogen sources including KNO_3, $NaNO_3$, Urea, $CaNO_3$, NH_4NO_3, and NH_4Cl have been used for cultivation of microalga *Scenedesmus* [50]. Among these nitrogen sources, KNO_3 most supported the microalgae growth due to the availability of both nitrogen and potassium. The optimal KNO_3 concentration was in the range of 0.5−1.0 g/L. It has been reported that the biogas CO_2 removal efficiency and lipid productivity of two microalgae *Chlorella* sp. and *Scenedesmus* sp. were enhanced with the optimal KNO_3 concentration of 0.62−0.8 g/L [1,6]. As nitrogen insufficiency may limit and excess nitrogen concentration may inhibit the microalgae growth and result in a low CO_2 removal efficiency, the stepwise addition of nitrogen source has been proposed by Srinuanpan et al. [10]. In their study, the KNO_3 concentration at 0.2 g/L was fed every 48 h, and this strategy successfully

enhanced the microalgae growth and the CO_2 removal efficiency from biogas but with a low lipid content. Although cultivating microalgae under nitrogen deficiency was effective for lipid accumulation [52], the nitrogen deficiency did limit the microalgae growth and thereby limit the CO_2 removal efficiency.

COMBINING BIOGAS UPGRADING AND WASTEWATER TREATMENT

Recently, a number of studies have reported the simultaneous biogas upgrading and nutrients removal by microalgae. Table 5.2 summarizes microalgae cultivation for biogas upgrading and wastewater treatment. Cultivating *Chlorella* spp. in domestic digestate effluent could upgrade biogas with the CH_4 content >90%, while removing about >80% of chemical oxygen demand (COD), >70% of total nitrogen (TN), and >70% of total phosphorus (TP) from wastewater. *Scenedesmus* spp. upgraded the CH_4 content in biogas >70% when they were grown in swine wastewater [40,53–55]. Meanwhile, COD, TN, and TP were effectively removed >60%, >60%, and >70%, respectively. Zhao et al. [18] found that *Nannochloropsis oleoabundans* could also upgrade CH_4 content in biogas to be nearly 80% when it was grown in biogas effluent. Not all types of wastewater could be directly used for microalgae cultivation. The dark color of the wastewater also had poor light penetration and limited the photosynthesis of the microalgae. Hence, both high concentration of inhibitory compounds and low light penetration caused poor microalgae growth and then resulted in low CO_2 removal efficiency. Therefore, those wastewaters should be pretreated or diluted before use. Xu et al. [54] suggested that the COD below 1600 mg/L was suitable for culturing the microalgae to upgrade biogas.

Most researches focused on combination of two processes, biogas upgrading with wastewater treatment by the microalgae, while the combination of three processes, that is, biogas upgrading, wastewater treatment, and production of biodiesel feedstocks has been rarely reported. Only Khan et al. [56] evaluated the feasibility of combining microalgae cultivation for biogas upgrading and production of biodiesel feedstock using digestate as low-cost medium. The microalgae increased the CH_4 content in biogas up to 81.6% and accumulated lipid at 26% based on its dry biomass. The produced lipids have high potential as biodiesel feedstock due to their similar fatty acid methyl ester profile to those of plant oils. The idea of integrating three processes of biogas upgrading, wastewater treatment, and lipid production by oleaginous microalgae would contribute greatly to the industrialization of microalgae-based biofuels.

RECENT CONCEPTS FOR MICROALGAE CULTIVATION

Recently, the cocultivations of microalgae with other microorganisms have been proposed and their applications have been reported. Table 5.3 summarizes the performance of biogas upgrading and phytoremediation of wastewater by various types of cocultivation. Fig. 5.2 shows cocultivation scheme of microalgae with filamentous fungi through copelletization (A) and pellet immobilization (B) and the aggregation of microalgae with activated sludge (C), and the coimmobilization of microalgae with bacterial cells (D). *C. vulgaris* has been cocultivated with several filamentous fungi including *Pleurotus geesteranus*, *Ganoderma lucidum*, and *Pleurotus ostreatu* and used for biogas upgrading and pollutant removal from the effluent [57]. *C. vulgaris—G. lucidum* pellets most efficiently purified biogas with percent CO_2 removal of 75.61%. The CH_4 content in biogas was increased up to 90%. COD, TN, and TP removal were >75%. *G. lucidum* fungus has been reported that it could rapidly grow in swine wastewater and efficiently removed pollutants. The microalgal cells could also be immobilized with fungi pellets through absorption mechanism [58]. The intergrowth of microalgae and fungi could enhance the specific surface area of algae-fungi symbionts and the potentiality of nutrient uptake.

Activated sludge is generated during the aerobic treatment of wastewater. They have been used adsorptive materials for contaminants due to their high absorbing surface area and absorbing property. They are also potentially used as supporter materials for microalgae [59]. The cocultured microalgae-activated sludge system has shown high CO_2 removal efficiency from biogas and high pollutant removal from anaerobic digester effluent possibly due to the symbiosis between microalgae and indigenous bacteria in activated sludge [58]. In this consortia system, the oxygen was generated by microalgae photosynthesis, and then was used as oxygen source by bacteria while the bacteria could produce CO_2 through respiration in their cells and serve as carbon source for microalgae [60]. The imaging SEM characterization of the microalgae—sludge pellets are shown in the report of Anbalagan et al. [61]. Sun et al. [62] evaluated the CO_2 removal from biogas using three microalgal strains including *C. vulgaris*, *Scenedesmus obliquus*, and *Neochloris*

TABLE 5.2
Microalgae Cultivation for Biogas Upgrading and Wastewater Treatment.

Microalgae Strain	Biogas Composition	Wastewater	CH4 Content After Upgrading (%v/v)	REMOVAL EFFICIENCY			References
				COD Removal (%)	TN Removal (%)	TP Removal (%)	
Chlorella sp.	Crude biogas containing 67.35% CH_4, 28.41% CO_2, 0.73% O_2, 3.48% H_2O, and <0.005% H_2S	Biogas effluent pH = 6.77 COD = 986.05 mg/L TN = 357.41 mg/L- TP = 37.24 mg/L	92.16	88.74	83.94	80.43	Yan and Zheng [4]
	Crude biogas containing 67.35% CH_4, 28.41% CO_2, 0.73% O_2, 3.48% H_2O, and <0.005% H_2S	Biogas effluent - pH = 6.50 - COD = 1203.61 mg - TN = 492.02 mg/L - TP = 53.28 mg/L	92.74	85.35	77.98	73.03	Zhao et al. [5]
	Crude biogas containing 61.38% CH_4, 32.57% CO_2, 0.54% O_2, 5.52% H_2O, and <0.005% H_2S	Biogas effluent - pH = 6.99 - COD = 789.13 mg/L - TN = 189.15 mg/L - TP = 15.53 mg/L	84.21	64.76	55.67	53.84	Wang et al. [7]
	Synthetic biogas containing 70% CH_4, 29.5% CO_2, and 0.5% H_2S	Wastewaters - pH = 6.43 - COD = 592 mg/L - TN = 580 mg/L - TP = 34 mg/L	94	70	80	85	Posadas et al. [66]
Chlorella vulgaris	Crude biogas containing 67.32% CH_4, 34.45% CO_2, 0.62% O_2, 3.66% H_2O, and <0.005% H_2S	Biogas effluent - pH = 6.84 - COD = 1013.87 mg/L - TN = 308.75 mg/L - TP = 9.93 mg/L	80.04	59.27	51.32	63.22	Zhao et al. [18]
	Synthetic biogas containing 70% CH_4, 29.5% CO_2, and 0.5% H_2S	Vinasse wastewater - pH = 7.84 - COD = 306 mg/L - TN = 71 mg/L - TP = 3.3 mg/L	93.9	51	37	86	Serejo et al. [64]
Chlorella minutissima	Synthetic biogas containing 70% CH_4, 29.5% CO_2, and 0.5% H_2S	Biogas effluent - COD = 1745 mg/L - TN = 1815 mg/L - TP = 48 mg/L	96	85	85	100	Toledo-Cervantes et al. [67]

Continued

TABLE 5.2
Microalgae Cultivation for Biogas Upgrading and Wastewater Treatment.—cont'd

Microalgae Strain	Biogas Composition	Wastewater	CH4 Content After Upgrading (%v/v)	REMOVAL EFFICIENCY			References
				COD Removal (%)	TN Removal (%)	TP Removal (%)	
	Biogas containing 63.53% CH_4 and 36.8% CO_2	Wastewater	86.4	–	–	–	Khan et al. [56]
Chlorella pyrenoidosa	Crude biogas containing 63.84% CH_4, 31.02% CO_2, 0.56% O_2, 3.59% H_2O, and <0.005% H_2S	Swine biogas effluent - pH = 6.79 - COD = 962.41 mg/L - TN = 360.28 mg/L - TP = 31.46 mg/L	92.87	92.67	80.87	79.33	Yan et al. [39]
Scenedesmus spp.	Crude biogas containing 61.75% CH_4, 32.28% CO_2, 0.31% O_2, 2.68% H_2O, and <0.005% H_2S	Biogas effluent - pH = 7.14 - COD = 1294.32 mg/L - TN = 419.49 mg/L - TP = 14.84 mg/L	94.41	93.03	84.12	86.76	Ouyang et al. [9]
	Crude biogas containing 67.32% CH_4, 34.45% CO_2, 0.62% O_2, 3.66% H_2O, and <0.005% H_2S	Biogas effluent - pH = 6.84 - COD = 1013.87 mg/L - TN = 308.75 mg/L - TP = 9.93 mg/L	82.64	63.12	53.04	59.14	Zhao et al. [18]
	Crude biogas containing 58.67% CH_4, 37.54% CO_2, 0.79% O_2, 3.01% H_2O, and <0.005% H_2S	Diluted swine biogas effluent - pH = 6.43 - COD = 1600 mg/L - TN = 60 mg/L - TP = 65 mg/L	88.25	75.29	62.54	88.79	Xu et al. [54]
	Raw swine biogas containing 70.7% CH_4, 26.1% CO_2, 0.23% O_2, $H_2S \cong 1550$ ppm	Swine biogas effluent - pH = 7.9 - $N-NH_3$ = 120 mg/L - $P-PO_4^{3-}$ = 90 mg/L	70	–	83	78	Prandini et al. [55]
Nannochloropsis oleoabundans	Crude biogas containing 67.32% CH_4, 34.45% CO_2, 0.62% O_2, 3.66% H_2O, and <0.005% H_2S	Biogas effluent - pH = 6.84 - COD = 1013.87 mg/L - TN = 308.75 mg/L - TP = 9.93 mg/L	80.06	61.03	51.74	54.33	Zhao et al. [18]

COD, Chemical oxygen demand; TN, Total nitrogen; TP, Total phosphorus; -, Not available.

TABLE 5.3
Co−cultivation of Microalgae With Other Microorganisms for Biogas Upgrading and Wastewater Treatment.

Co−Culture Technology	Biogas Composition	Wastewater	CH_4 Content after Upgrading (%v/v)	REMOVAL EFFICIENCY			References
				COD Removal (%)	TN Removal (%)	TP Removal (%)	
Co−cultivation of microalgae with fungi							
Microalgae *Chlorella vulgaris*—fungi *Ganoderma lucidum*	Synthetic biogas containing 64.92% CH_4 and 35.08% CO_2	Biogas effluent - pH = 7.15 - COD = 1041 mg/L - TN = 288 mg/L - TP = 12 mg/L	93	86	86	86	Wang et al. [19]
	Synthetic biogas containing 62.38% CH_4 and 31.19% CO_2	Biogas effluent - pH = 6.97 - COD = 1496 mg/L - TN = 278 mg/L - TP = 28 mg/L	92	70	76	78	Cao et al. [57]
	Crude biogas containing 64.58% CH_4, 31.72% CO_2, 0.54% O_2, 3.15% H_2O, and <0.005% H_2S	Biogas effluent - pH = 6.95 - COD = 1213 mg/L - TN = 239 mg/L - TP = 13 mg/L	90	74	76	77	Zhang et al. [58]
	Crude biogas containing 64.59% CH_4, 33.79% CO_2, 0.38% O_2, 1.23% H_2O, and <0.01% H_2S	Biogas effluent - pH = 6.97 - COD = 1496 mg/L - TN = 278 mg/L - TP = 28 mg/L	86	83	81	84	Guo et al. [63]
Co−cultivation of microalgae with activated sludge/bacteria							
Microalgae chlorella vulgaris—activated sludge	Synthetic biogas containing 64.92% CH_4 and 35.08% CO_2	Biogas effluent - pH = 7.15 - COD = 1041 mg/L - TN = 288 mg/L - TP = 12 mg/L	88	75	74	74	Wang et al. [17]
	Crude biogas containing 64.58% CH_4, 31.72% CO_2, 0.54% O_2, 3.15% H_2O, and <0.005% H_2S	Biogas effluent - pH = 6.95 - COD = 1213 mg/L - TN = 239 mg/L	93	72	75	74	Zhang et al. [58]

Continued

TABLE 5.3
Co—cultivation of Microalgae With Other Microorganisms for Biogas Upgrading and Wastewater Treatment.—cont'd

Co—Culture Technology	Biogas Composition	Wastewater	CH$_4$ Content after Upgrading (%v/v)	REMOVAL EFFICIENCY			References
				COD Removal (%)	TN Removal (%)	TP Removal (%)	
	Crude biogas containing 64.59% CH$_4$, 33.79% CO$_2$, 0.38% O$_2$, 1.23% H$_2$O, and <0.01% H$_2$S	- TP = 13 mg/L Biogas effluent - pH = 6.97 - COD = 1496 mg/L - TN = 278 mg/L - TP = 28 mg/L	83	81	83	83	Guo et al. [63]
Microalgae Scenedesmus—activated sludge	Synthetic biogas containing 54.5% CH$_4$, 45% CO$_2$, and 0.5% H$_2$S	Biogas effluent - pH = 7.43 - COD = 999 mg/L - TN = 257 mg/L - TP = 14 mg/L	93	65	63	74	Sun et al. [62]
Microalgae *Picochlorum* sp. and *Halospirulina* sp.—bacteria	Synthetic biogas containing 69.5% CH$_4$, 30% CO$_2$, and 0.5% H$_2$S	Mineral salt medium - pH = 9.3	>96	—	52–55	12–29	Franco-Morgado et al. [65]
Unknown microalgae species—bacteria	Synthetic biogas containing 70% CH$_4$, 29.5% CO$_2$, and 0.5% H$_2$S	Domestic wastewater - pH = 7.15 - IC = 1500 mg/L - TN = 1719 mg/L	76–99	—	—	—	Rodero et al. [68]
	Synthetic biogas containing 70% CH$_4$, 29.5% CO$_2$, and 0.5% H$_2$S	Domestic wastewater - pH = 9 - IC = 1663 mg/L - TN = 336 mg/L - TP = 25 mg/L	97.9	—	—	—	Marin et al. [69]
	Synthetic biogas containing 70% CH$_4$, 29.5% CO$_2$, and 0.5% H$_2$S	Domestic wastewater - pH = 9 - TOC = 16–523 mg/L - IC = 450–600 mg/L - TN = 374–718 mg/L - TP = 26–135 mg/L	99.6	—	—	—	Marin et al. [70]

—, Not available.

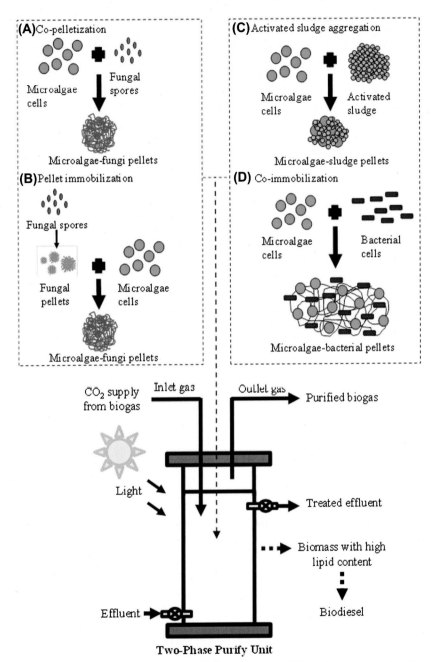

FIG. 5.2 Co–cultivation scheme of microalgae with filamentous fungi through co–pelletization **(A)** and pellet immobilization **(B)** and the aggregation of microalgae with activated sludge **(C)**, and the coimmobilization of microalgae with bacterial cells **(D)**.

oleoabundans mixed with activated sludge and used biogas slurry as nutrient medium. They found that the cocultivation of *S. obliquus* with activated sludge pellets could effectively treat biogas slurry and removal CO_2

>75% and the CH_4 content in biogas increased >90%. The nutrient removal including COD, TN, and TP was more than 50%. The performance of these cocultivation systems for biogas upgrading and pollutant

removal were superior to the use of pure microalgae [19,58,63]. The lipids in the generated microalgae-activated sludge biomass were in the range of 10% −40% [62−65]. Srinuanpan et al. [60] found that the lipid content in microalga−fungi biomass cultured in secondary effluent fed with air was 15.83%. The extracted lipids mainly contained long-chain fatty acids of 16 and 18 carbon atoms (>83%) which was suitable for being used as a feedstock of biodiesel.

CONCLUSIONS

Biogas upgrading and production of biodiesel feedstock by microalgae are highly potential for environmental friendly and renewable energy production. The integration of microalgae cultivation process for removing CO_2 from biogas and treating of wastewater is a promising strategy for economic viability of the microalgae-based biofuel production. Co−cultivation of microalgae with other microorganisms and their applications for biogas upgrading and treating of wastewater would also give high impact on process efficiency. As microalgae are also renewable and sustainable sources of value-added products, the biorefinery approach for microalgae biomass should be considered and implemented. The combined technoeconomic and environmental impact assessment would be useful tool for policy makers and producers in identifying potential bottlenecks and choosing suitable implementations.

REFERENCES

[1] S. Srinuanpan, B. Cheirsilp, W. Kitcha, P. Prasertsan, Strategies to improve methane content in biogas by cultivation of oleaginous microalgae and the evaluation of fuel properties of the microalgal lipids, Renew. Energy 113 (2017) 1229−1241.

[2] C.Y. Kao, S.Y. Chiu, T.T. Huang, L. Dai, G.H. Wanga, C.P. Tseng, C.H. Chen, C.S. Lin, A mutant strain of microalga *Chlorella* sp. for the carbon dioxide capture from biogas, Biomass Bioenerg. 36 (2012) 132−140.

[3] C.Y. Kao, S.Y. Chiu, T.T. Huang, L. Dai, L.K. Hsu, C.S. Lin, Ability of a mutant strain of the microalga *Chlorella* sp. to capture carbon dioxide for biogas upgrading, Appl. Energy 93 (2012) 176−183.

[4] C. Yan, Z. Zheng, Performance of photoperiod and light intensity on biogas upgrade and biogas effluent nutrient reduction by the microalgae *Chlorella* sp, Bioresour. Technol. 139 (2013) 292−299.

[5] Y. Zhao, J. Wang, H. Zhang, C. Yan, Y. Zhang, Effects of various LED light wavelengths and intensities on microalgae-based simultaneous biogas upgrading and digestate nutrient reduction process, Bioresour. Technol. 136 (2013) 461−468.

[6] W. Tongprawhan, S. Srinuanpan, B. Cheirsilp, Biocapture of CO_2 from biogas by oleaginous microalgae for improving methane content and simultaneously producing lipid, Bioresour. Technol. 170 (2014) 90−99.

[7] Z. Wang, Y. Zhao, Z. Ge, H. Zhang, S. Sun, Selection of microalgae for simultaneous biogas upgrading and biogas slurry nutrient reduction under various photoperiods, J. Chem. Technol. Biotechnol. 97 (2015) 1982−1989.

[8] F.J. Choix, R. Snell-Castro, J. Arreola-Vargas, A. Carbajal-López, H.O. Méndez-Acosta, CO_2 removal from biogas by cyanobacterium *Leptolyngbya* sp. CChF1 isolated from the Lake Chapala, Mexico: optimization of the temperature and light intensity, Appl. Biochem. Biotechnol. 183 (2017) 1304−1322.

[9] Y. Ouyang, Y. Zhao, S. Sun, C. Hu, L. Ping, Effect of light intensity on the capability of different microalgae species for simultaneous biogas upgrading and biogas slurry nutrient reduction, Int. Biodeterior. Biodegrad. 104 (2015) 157−163.

[10] S. Srinuanpan, B. Cheirsilp, P. Prasertsan, Effective biogas upgrading and production of biodiesel feedstocks by strategic cultivation of oleaginous microalgae, Energy 148 (2018), 766−764.

[11] L. Meier, R. Pérez, L. Azócar, M. Rivas, D. Jeison, Photosynthetic CO_2 uptake by microalgae: an attractive tool for biogas upgrading, Biomass Bioenerg. 7 (2015) 102−109.

[12] F.J. Choix, E. Polster, R.I. Corona-González, R. Snell-Castro, H.O. Méndez-Acosta, Nutrient composition of culture media induces different patterns of CO_2 fixation from biogas and biomass production by the microalga *Scenedesmus obliquus* U169, Bioprocess Biosyst. Eng. 40 (2017) 1733−1742.

[13] A. González-Sánchez, C. Posten, Fate of H_2S during the cultivation of *Chlorella* sp. deployed for biogas upgrading, J. Environ. Manag. 191 (2017) 252−257.

[14] X.W. Wang, J.R. Liang, C.S. Luo, C.P. Chen, Y.H. Gao, Biomass, total lipid production, and fatty acid composition of the marine diatom *Chaetoceros muelleri* in response to different CO_2 levels, Bioresour. Technol. 161 (2014) 124−130.

[15] H.M. Amaro, A.C. Guedes, F.X. Malcata, Advances and perspectives in using microalgae to produce biodiesel, Appl. Energy 88 (2011) 3402−3410.

[16] D. Tang, W. Han, P. Li, X. Miao, J. Zhong, CO_2 biofixation and fatty acid composition of *Scenedesmus obliquus* and *Chlorella pyrenoidosa* in response to different CO_2 levels, Bioresour. Technol. 102 (2011) 3071−3076.

[17] T. Thawechai, B. Cheirsilp, Y. Louhasakul, P. Boonsawang, P. Prasertsan, Mitigation of carbon dioxide by oleaginous microalgae for lipids and pigments production: effect of light illumination and carbon dioxide feeding strategies, Bioresour. Technol. 219 (2016) 139−149.

[18] Y. Zhao, S. Sun, C. Hu, H. Zhang, J. Xu, L. Ping, Performance of three microalgal strains in biogas slurry purification and biogas upgrade in response to various mixed light-emitting diode light wavelengths, Bioresour. Technol. 187 (2015) 338−345.

[19] X. Wang, S. Gao, Y. Zhang, Y. Zhao, W. Cao, Performance of different microalgae-based technologies in biogas slurry nutrient removal and biogas upgrading in response to various initial CO_2 concentration and mixed light-emitting diode light wavelength treatments, J. Clean. Prod. 166 (2017) 408–416.

[20] C. Yan, L. Zhang, X. Luo, Z. Zheng, Influence of influent methane concentration on biogas upgrading and biogas slurry purification under various LED (light-emitting diode) light wavelengths using *Chlorella* sp, Energy 69 (2014) 419–426.

[21] E. Lee, M. Jalalizadeh, Q. Zhang, Growth kinetic models for microalgae cultivation: a review, Algal Res. 12 (2015) 497–512.

[22] R. Mera, E. Torres, J. Abalde, Influence of sulphate on the reduction of cadmium toxicity in the microalga *Chlamydomonas moewusii*, Ecotoxicol. Environ. Saf. 128 (2016) 236–245.

[23] M.A. Kassim, T.K. Meng, Carbon dioxide (CO_2) biofixation by microalgae and its potential for biorefinery and biofuel production, Sci. Total Environ. 584–585 (2017) 1121–1129.

[24] Y. Su, K. Song, P. Zhang, Y. Su, J. Cheng, X. Chen, Progress of microalgae biofuel's commercialization, Renew. Sustain. Energy Rev. 74 (2017) 402–411.

[25] A. Widjaja, C.C. Chien, Y.H. Ju, Study of increasing lipid production from fresh water microalgae *Chlorella vulgaris*, J. Taiwan. Inst. Chem. Eng. 40 (2009) 13–20.

[26] P. Binnal, P.N. Babu, Statistical optimization of parameters affecting lipid productivity of microalga *Chlorella protothecoides* cultivated in photobioreactor under nitrogen starvation, S. Afr. J. Chem. Eng. 23 (2017) 26–37.

[27] A.P. Carvalho, F.X. Malcata, Optimization of omega-3 fatty acid production by microalgae: crossover effects of CO_2 and light intensity under batch and continuous cultivation modes, Mar. Biotechnol. 7 (2005) 381–388.

[28] S. Xue, Z. Su, W. Cong, Growth of *Spirulina platensis* enhanced under intermittent illumination, J. Biotechnol. 151 (2011) 271–277.

[29] H. Jeong, J. Lee, M. Cha, Energy efficient growth control of microalgae using photobiological methods, Renew. Energy 54 (2013) 161–165.

[30] N. Kurano1, S. Miyachi, Selection of microalgal growth model for describing specific growth rate-light response using extended information criterion, J. Biosci. Bioeng. 100 (2005) 403–408.

[31] P. Das, W. Lei, S.S. Aziz, J.P. Obbard, Enhanced algae growth in both phototrophic and mixotrophic culture under blue light, Bioresour. Technol. 102 (2011) 3883–3887.

[32] L. Pilon, H. Berberoğlu, R. Kandilian, Radiation transfer in photobiological carbon dioxide fixation and fuel production by microalgae, J. Quant. Spectrosc. Radiat. Transfer 112 (2011) 2639–2660.

[33] C. Yan, L. Zhang, X. Luo, Z. Zheng, Effects of various LED light wavelengths and intensities on the performance of purifying synthetic domestic sewage by microalgae at different influent C/N ratios, Ecol. Eng. 51 (2013) 24–32.

[34] B. Cheirsilp, S. Torpee, Enhanced growth and lipid production of microalgae under mixotrophic culture condition: effect of light intensity, glucose concentration and fed-batch cultivation, Bioresour. Technol. 110 (2012) 510–516.

[35] C. Yeesang, B. Cheirsilp, Effect of nitrogen, salt, and iron content in the growth medium and light intensity on lipid production by microalgae isolated from freshwater sources in Thailand, Bioresour. Technol. 102 (2011) 3034–3040.

[36] B. George, I. Pancha, C. Desai, K. Chokshi, C. Paliwal, T. Ghosh, S. Mishra, Effects of different media composition, light intensity and photoperiod on morphology and physiology of freshwater microalgae *Ankistrodesmus falcatus*-a potential strain for bio-fuel production, Bioresour. Technol. 171 (2014) 367–374.

[37] J. Liu, C. Yuan, G. Hu, F. Li, Effects of light intensity on the growth and lipid accumulation of microalga *Scenedesmus* sp. 11–1 under nitrogen limitation, Appl. Biochem. Biotechnol. 166 (2012) 2127–2137.

[38] A. Difusa, J. Talukdar, M.C. Kalita, K. Mohanty, V.V. Goud, Effect of light intensity and pH condition on the growth, biomass and lipid content of microalgae *Scenedesmus species*, Biofuels 6 (2015) 37–44.

[39] Q. He, H. Yang, L. Wu, C. Hu, Effect of light intensity on physiological changes, carbon allocation and neutral lipid accumulation in oleaginous microalgae, Bioresour. Technol. 191 (2015) 219–228.

[40] C. Yan, R. Muñoz, L. Zhu, Y. Wang, The effects of various LED (light emitting diode) lighting strategies on simultaneous biogas upgrading and biogas slurry nutrient reduction by using of microalgae *Chlorella* sp. Energy 106 (2016) 554–561.

[41] T.H. Kim, Y. Lee, S.H. Han, S.J. Hwang, The effects of wavelength and wavelength mixing ratios on microalgae growth and nitrogen, phosphorus removal using *Scenedesmus* sp. for wastewater treatment, Bioresour. Technol. 130 (2013) 75–80.

[42] S. Ho, C. Chen, D. Lee, J. Chang, Perspectives on microalgal CO_2-emission mitigation systems – a review, Biotechnol. Adv. 29 (2011) 189–198.

[43] C. Yan, L. Zhu, Y. Wang, Photosynthetic CO_2 uptake by microalgae for biogas upgrading and simultaneously biogas slurry decontamination by using of microalgae photobioreactor under various light wavelengths, light intensities, and photoperiods, Appl. Energy 178 (2016) 9–18.

[44] C.L. Teo, M. Atta, A. Bukhari, M. Taisir, A.M. Yusuf, A. Idris, Enhancing growth and lipid production of marine microalgae for biodiesel production via the use of different LED wavelengths, Bioresour. Technol. 162 (2014) 38–44.

[45] S. Wahidin, A. Idris, S.R.M. Shaleh, The influence of light intensity and photoperiod on the growth and lipid content of microalgae *Nannochloropsis* sp. Bioresour. Technol. 129 (2013) 7–11.

[46] C.H. Ra, C.H. Kang, J.H. Jung, G.T. Jeong, S.K. Kim, Effects of light-emitting diodes (LEDs) on the accumulation of lipid content using a two-phase culture process with three microalgae, Bioresour. Technol. 212 (2016) 254–261.

[47] J.M. Anderson, P. Horton, E.H. Kim, W.S. Chow, Towards elucidation of dynamic structural changes of plant

thylakoid architecture, Philos. Trans. R. Soc. Lond. B. Biol. Sci. 367 (2012) 3515−3524.

[48] Y. Yamamoto, Quality control of photosystem II: the mechanisms for avoidance and tolerance of light and heat stresses are closely linked to membrane fluidity of the thylakoids, Front. Plant Sci. 7 (2016) 1136.

[49] D.G. Kim, C. Lee, S.M. Park, Y.E. Choi, Manipulation of light wavelength at appropriate growth stage to enhance biomass productivity and fatty acid methyl ester yield using *Chlorella vulgaris*, Bioresour. Technol. 159 (2014) 240−248.

[50] M. Arumugam, A. Agarwal, M.C. Chandra-Arya, Z. Ahmed, Influence of nitrogen sources on biomass productivity of microalgae *Scenedesmus bijugatus*, Bioresour. Technol. 131 (2013) 246−249.

[51] Y. Li, M. Horsman, B. Wang, N. Wu, C.Q. Lan, Effects of nitrogen sources on cell growth and lipid accumulation of green alga *Neochloris oleoabundans*, Appl. Microbiol. Biotechnol. 81 (2008) 629−636.

[52] M. Siaut, S. Cuine, C. Cagnon, B. Fessler, M. Nguyen, P. Carrier, A. Beyly, F. Beisson, C. Triantaphylides, Y.H. Li-Beisson, G. Peltier, Oil accumulation in the model green alga *Chlamydomonas reinhardtii*: characterization, variability between common laboratory strains and relationship with starch reserves, BMC Biotechnol. 11 (2011) 7.

[53] B. Cheirsilp, T. Thawechai, P. Prasertsan, Immobilized oleaginous microalgae for production of lipid and phytoremediation of secondary effluent from palm oil mill in fluidized bed photobioreactor, Bioresour. Technol. 241 (2017) 787−794.

[54] J. Xu, Y. Zhao, G. Zhao, H. Zhang, Nutrient removal and biogas upgrading by integrating freshwater algae cultivation with piggery anaerobic digestate liquid treatment, Appl. Microbiol. Biotechnol. 99 (2015) 6493−6501.

[55] J.M. Prandini, M.L.B. da Silva, M.P. Mezzari, M. Pirolli, W. Michelon, H.M. Soares, Enhancement of nutrient removal from swine wastewater digestate coupled to biogas purification by microalgae *Scenedesmus* spp. Bioresour. Technol. 202 (2016) 67−75.

[56] S.A. Khan, F.A. Malla, L.C. Malav, N. Gupta, A. Kumar, Potential of wastewater treating *Chlorella minutissima*, for methane enrichment and CO_2, sequestration of biogas and producing lipids, Energy 150 (2018) 153−163.

[57] W. Cao, X. Wang, S. Sun, C. Hu, Y. Zhao, Simultaneously upgrading biogas and purifying biogas slurry using co-cultivation of *Chlorella vulgaris* and three different fungi under various mixed light wavelength and photoperiods, Bioresour. Technol. 241 (2017) 701−709.

[58] Y. Zhang, K. Bao, J. Wang, Y. Zhao, C. Hu, Performance of mixed LED light wavelengths on nutrient removal and biogas upgrading by different microalgal-based treatment technologies, Energy 130 (2017) 392−401.

[59] D. Fytili, A. Zabaniotou, Utilization of sewage sludge in EU application of old and new methods: a review, Renew. Sustain. Energy Rev. 12 (2018) 116−140.

[60] S. Srinuanpan, A. Chawpraknoi, S. Chantarit, B. Cheirsilp, P. Prasertsan, A rapid method for harvesting and immobilization of oleaginous microalgae using pellet-forming filamentous fungi and the application in phytoremediation of secondary effluent, Int. J. Phytoremediat. 20 (2018) 1017−1024.

[61] A. Anbalagan, S. Schwede, C.F. Lindberg, E. Nehrenheim, Influence of iron precipitated condition and light intensity on microalgae activated sludge based wastewater remediation, Chemosphere 168 (2017) 1523−1530.

[62] S. Sun, Z. Ge, Y. Zhao, C. Hu, H. Zhang, L. Ping, Performance of CO_2 concentrations on nutrient removal and biogas upgrading by integrating microalgal strains cultivation with activated sludge, Energy 97 (2016) 229−237.

[63] P. Guo, Y. Zhang, Y. Zhao, Biocapture of CO_2 by different microalgal-based technologies for biogas upgrading and simultaneous biogas slurry purification under various light intensities and photoperiods, Int. J. Environ. Res. Public Health 15 (2018) 528.

[64] M.L. Serejo, E. Posadas, M.A. Boncz, S. Blanco, P. García-Encina, R. Muñoz, Influence of biogas flow rate on biomass composition during the optimization of biogas upgrading in microalgal-bacterial processes, Environ. Sci. Technol. 49 (2015) 3228−3236.

[65] M. Franco-Morgado, A. González-Sánchez, A. Toledo-Cervantes, R. Lebrero, R. Muñoz, Integral (VOCs, CO_2, mercaptans and H_2S) photosynthetic biogas upgrading using innovative biogas and digestate supply strategies, Chem. Eng. J. 354 (2018) 363−369.

[66] E. Posadas, D. Marín, S. Blanco, R. Lebrero, R. Muñoz, Simultaneous biogas upgrading and centrate treatment in an outdoors pilot scale high rate algal pond, Bioresour. Technol. 232 (2017) 133−141.

[67] A. Toledo-Cervantes, C. Madrid-Chirinos, S. Cantera, R. Lebrero, R. Muñoz, Influence of the gas-liquid flow configuration in the absorption column on photosynthetic biogas upgrading in algal-bacterial photobioreactors, Bioresour. Technol. 225 (2017) 336−342.

[68] R. Rodero, E. Posadas, A. Toledo-Cervantes, R. Lebrero, R. Muñoz, Influence of alkalinity and temperature on photosynthetic biogas upgrading efficiency in high rate algal ponds, Algal Res. 33 (2017) 284−290.

[69] D. Marín, E. Posadas, P. Cano, V. Pérez, S. Blanco, R. Lebrero, R. Muñoz, Seasonal variation of biogas upgrading coupled with digestate treatment in an outdoors pilot scale algal-bacterial photobioreactor, Bioresour. Technol. 263 (2018) 58−66.

[70] D. Marín, E. Posadas, P. Cano, V. Pérez, R. Lebrero, R. Muñoz, Influence of the seasonal variation of environmental conditions on biogas upgrading in an outdoors pilot scale high rate algal pond, Bioresour. Technol. 255 (2018) 354−358.

FURTHER READING

[1] G.T. Ding, Z. Yaakob, M.S. Takriff, J. Salihon, M.S. Abd Rahaman, Biomass production and nutrients removal by a newly-isolated microalgal strain *Chlamydomonas* sp. in palm oil mill effluent (POME), Int. J. Hydrogen Energy 41 (2016) 4888−4895.

Efficient Harvesting of Microalgal biomass and Direct Conversion of Microalgal Lipids into Biodiesel

BENJAMAS CHEIRSILP • SIRASIT SRINUANPAN • YOHANIS IRENIUS MANDIK

INTRODUCTION

Biofuel and biorefinery have been expected to mitigate the problems from energy security and climate change at least to some extent. Nowadays, three generations of biofuels have been developed. Biofuel that is produced from sugar/starch/oil crops is the first-generation, that is produced from lignocellulosic biomass is the second-generation, and that is produced from microalgae and macroalgae is the third-generation biofuel. Microalgae can use sunlight energy to fix carbon dioxide into biomass and produce oxygen through photosynthesis. It is well known that photosynthesis is the most cost-effective way to mitigate anthropogenic carbon dioxide. In addition, the microalgae have been considered as fast-growing species, and therefore, their carbon dioxide fixation rates are much higher than those of the plants [1,2]. It is interesting that several biofuels can be produced by and from microalgae. These include hydrogen (directly produced by specific microalgae), biodiesel (produced via transesterification reaction of microalgal lipids), bioethanol, and biogas (produced via fermentation of microalgal biomass) [3,4,5,6]. Oleaginous microalgae can utilize both inorganic carbon (carbon dioxide) and organic carbon sources (glucose, acetate, etc.) to synthesize and accumulate lipids that can be converted into biodiesel. These approaches can then be the solutions for two major problems: (1) air pollution from carbon dioxide evolution and (2) energy crises due to the shortage of energy sources [6,7].

Microalgae as sources for biofuel production have advantages over the crops in terms of higher growth rate, higher productivity of biomass, ability to grow both in freshwater and sea water environments, not competing with food, and lower production cost. Some species of microalgae can adapt to grow in various environments. It has been reported that microalgae need only 49–132 times of smaller land and averagely need only 8–10 days for cultivation [7,8,9,10,11]. The productivity of bioethanol from microalgae could be more than 100 times of that from cassava at the same harvesting time [12]. The productivity of biodiesel from microalgae also could reach 120,000 kg/Ha year. This value is more 20 times higher than that of the palm oil (5800 kg/Ha year) and 80 times higher than that of the jatropha oil (1500 kg/Ha year) [11]. Moreover, the utilization of microalgae as biodiesel feedstocks could avoid competition with food resources. Microalgae are then promising organisms that have high potentials as biodiesel feedstocks [13,14,15,16]. However, many costly and energy-consuming steps in the downstream process, including harvesting of low-density microalgal biomass, biomass drying, extraction of lipid, and transesterification of microalgal lipids, have jeopardized interests in biodiesel production from microalgae [17,18,19]. Therefore, to be more sustainable, the effective downstream process for microalgae-based biodiesel should be investigated. This chapter aims to give an overview of fundamental and recently developed harvesting and conversion of microalgal biomass into biodiesel.

MICROALGAL BIOMASS

Due to the very high diversity of microalgae, microalgal biomass are rich sources of many biochemical products that can be used in a number of industries including food, fuel, nutritional, cosmetic, and pharmaceutical industries. Many species of microalgae can produce specific products such as, fatty acids, sterols, peptides, polysacharides, enzymes, carotenoids, antioxidants, and some toxins either inside or outside the cells

Microalgae Cultivation for Biofuels Production. https://doi.org/10.1016/B978-0-12-817536-1.00006-0

[20,21,22]. Approximately 35,000 species of microalgal species have already been identified, such as *Spirulina* spp., *Chlorella* spp., *Botryococcus* spp., *Nitzschia* spp., *Nannochloropsis* spp., *Dunaliella primolecta, Tetraselmis suecia*, etc. Microalgal biomass are mainly composed of lipids, carbohydrates, and proteins. The medium composition and the culture conditions are the crucial factors that affected the composition of microalgal biomass. The production of these products by microalgae depends on species and culture conditions. Therefore, there are opportunities to recover microalgal biomass and their biochemical products with a certain amount by selecting the suitable species and manipulating the medium components and culture conditions. The main factors could be carbon dioxide availability, light intensity and photoperiod, culture temperature and pH, salts, and other nutrients. In general, oleaginous microalgae contain lipids more than 20% of their biomass. Some of those microalgae may contain lipids as high as 50% of their biomass and also had very high growth rate [22,23,24], while the carbohydrate content of microalgae could be around 30% of their biomass, which is likely higher than cassava's (23% of the biomass). Protein is one of the main components in microalgal biomass. Its content in specific microalgae may be as high as 50% of the biomass. Based on the amino acid profiles in microalgal protein, it has been considered as one of the alternative protein sources in foods and also plays an important role in microalgae biorefinery [25,26]. Table 6.1 shows the percent compositions of carbohydrate, lipid, and protein in some microalgae.

HARVESTING OF MICROALGAL BIOMASS

Microalgal biomass cultivated in either open ponds or closed photobioreactors commonly reach relatively low concentrations in the range of 0.5–5 g/L [27]. The small size and low concentration of microalgal biomass make their harvesting process challenging. Fig. 6.1 shows physical/chemical and biological harvesting methods of microalgal biomass. The physical/chemical methods include gravity sedimentation, centrifugation, filtration, flocculation, or combination of these [28,29]. The harvesting cost of microalgae biomass has been estimated to contribute 20%–30% of the production cost [30]. Most commercial systems use centrifugation method to harvest microalgal biomass, but it consumes large amount of energy [30,31]. Filtration is alternatively used for harvesting of microalgal biomass, but the membranes are easily fouled if the medium is directly filtered [32,33,34]. The use of vacuum filtration

is suitable to harvest some microalgae cells with large size. But these methods are not suitable to harvest small-size microalgae like *Chlorella* spp. and *Dunaliella* spp. [34]. Based on life cycle analysis, the use of centrifugation or filtration for harvesting of microalgal biomass needs intensive maintenance and large amount of energy [35,36,37]. Therefore, other low-cost and low-energy-consumed gravity sedimentation and flocculation would be more convenient and effective for harvesting of microalgal biomass in large scale [38,39,40]. However, the methods used for microalgae harvesting must be energy- and cost-efficient for viable biofuel production. Both common and recently developed methods for harvesting of microalgal biomass are described in detail.

Gravity sedimentation

Gravity sedimentation is considered at the cheapest and most conventional method, which allows discarding liquid at least 90% prior to further processing. It has been commonly used in treatment of wastewater and sludge due to the capacity to handle large volumes and low commercial value of the biomass formed [39,40]. The liquids or solids having different densities can be separated from suspensions by sedimentation, and mostly clear effluents are produced. These processes require low energy consumption, low design cost, and less skilled operators. However, their disadvantages include the slow sedimentation rate and the requirement of consequent large footprint. The cell motility, cell density, cell size, and water turbulence influence the sedimentation rate [40]. However, this method is effective only when the densities of the particles are different enough. Chen et al. [28] suggested that the gravity sedimentation is suitable for separation of large-size microalgae such as *Spirulina* spp. To increase the recovery of cells via gravity sedimentation, the aeration should be stopped, and this could facilitate the autoflocculation. Flocculation is usually used to increase the efficiency of sedimentation. In addition, the autoflocculation could be enhanced by adjusting the culture pH that is suitable for cell aggregation [28,29,42,43]. Chatsungnoen and Chisti [44] have optimized harvesting process of microalgal biomass using aluminum sulfate and ferric chloride as coagulants in a flocculation-sedimentation process and could recover microalgal biomass >95%.

Centrifugation

Centrifugation is one of the rapid and effective methods to harvest microalgal biomass but requires high energy input. The high gravitational and strong shear forces may also cause cell disruption and structural damage

TABLE 6.1
Percent (% w/w) Carbohydrate, Lipid, and Protein Contents of Microalgae.

Microalgae species	Lipid	Protein	Carbohydrate	References
Scenedesmus obtusiusculus	52	9	25	Toledo-Cervantes et al. [146]
Botryococcus braunii	37	32	26	Pérez-Mora et al. [147]
Phaeodactylum tricornutum	18	40	25	Tibbetts et al. [148]
Nannochloropsis granulata	48	28	17	Tibbetts et al. [148]
Tetraselmis chuii	13	47	25	Tibbetts et al. [148]
Arthrospira (Spirulina) platensis	23	28	41	Rodrigues et al. [149]
Chlorella pyrenoidosa	13	63	13	Zhang et al. [150]
Micractinium reisseri	25	40	19	Srinuanpan et al. [29]
Scenedesmus obliquus	33	36	19	Srinuanpan et al. [29]
Synechococcus sp.	11	63	15	Becker [151]
Chaetoceros sp.	17	37	6	Renaud et al. [152]
Cryptomonas sp.	22	47	4	Renaud et al. [152]
Rhodomonas sp.	19	29	9	Renaud et al. [152]
Skeletonema costatum	14	28	12	Renaud et al. [152]

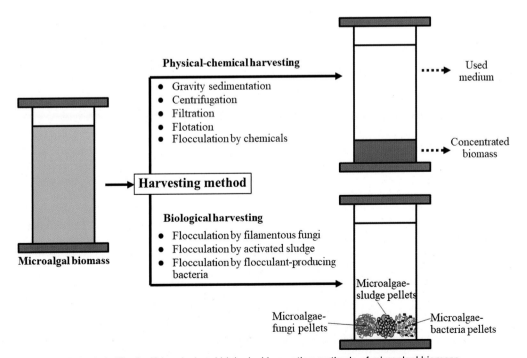

FIG. 6.1 Physical/chemical and biological harvesting methods of microalgal biomass.

and limits the application for recovering high-value products, such as pharmaceuticals, polyunsaturated fatty acids, and other commodities. The small size of microalgae cells needs longer centrifugation time to be effectively harvested. Dassey and Theegala [30] reported high cell recovery efficiency (96%) but with a low volume flow rate (0.94 L/min) implied a 20 kWh/m^3 of energy input by using continuous-flow centrifuge, whereas only 0.80 kWh/m^3 was needed for 17% recovery at a volume flow rate of 23 L/min. Several centrifugation systems, including tubular centrifuge, nozzle-type centrifuge, solid-bowl decanter centrifuge, and solid-ejecting disc centrifuge, have been examined for microalgae separation. Grima et al. [41] reported that the disc-stack centrifuge and nozzle-type centrifuge could give high efficiency of microalgae harvesting followed by nozzle discharge centrifuge and hydrocyclone, respectively.

Filtration

Filtration is the process to separate microalgal biomass from liquid medium by causing the latter to pass through the filter under gravity or vacuum pressure to hold back microalgae on the filter. The filtration process can either be performed continuously or discontinuously [45]. The benefits of using filtration include no chemical additives, simplicity in operation, and low energy consumption [46]. Several filtration methods, including pressure filtration, vacuum filtration, and cross-flow filtration, have been used to recover microalgal biomass [45,47,48]. The membrane materials could be either typical polymeric [41,49,50,53] or ceramic [51,52]. Compared with polymeric membranes, ceramic membranes offer better performances in terms of high flow rate and high reproducibility. But their prices are also higher than that of the polymeric membranes [54]. Recently, cheap polymeric materials such as acrylonitrile butadiene styrene polymer have been applied in the dewatering step for microalgae biorefining [41,53]. The development of the cheaper membranes could contribute to a significant cost reduction for harvesting process of microalgal biomass through filtration.

Flotation

In flotation process, the gas or air bubbles are introduced through culture suspension, and the microalgal biomass get attached to gaseous molecules and accumulated on the liquid surface. This method is particularly effective for thin microalgae suspension that could be simply gravity thickening [38]. The basic variations of this process are dispersed air flotation, dissolved air flotation,

electroflotation, and ozone flotation [55,56,57]. The ratio of gaseous molecules to microalgae is one of the most important factors affecting the performance of the flotation efficiency. Several researchers have confirmed that ozone flotation was more effective than other methods [58,59]. Also, ozoflotation could improve lipid recovery yields and modify fatty acid methyl ester (FAME) profiles. The ozone flotation could increase the cell flotation efficiency by modifying the cell wall surface and/or releasing the active agents from microalgal cells [60]. Moreover, the ozone flotation can also improve the quality of water by lowering the turbidity and organic contents of the effluent [58]. Flotation separation efficiency relates to bubble size [61]. Smaller size of gas bubbles has lower rise velocity and higher surface area to volume ratio. This enables their longer retention time and better attachment efficiency with the microalgae cells and leads to the increasing in harvesting efficiency by floatation [64]. Thus, one of the most efficient ways of achieving maximum attachment is by generating as many small bubbles as possible [61,62,63]. Combinations of flocculation with flotation have been also used to increase the harvesting efficiency [64,65,66].

Flocculation

Flocculation is cheap, simple, and feasible to harvest large volume of microalgae culture [44,67,68,69]. There are several flocculation types such as physical, chemical, and biological methods. Generally, flocculation is applied as a preconcentration step for harvesting of microalgal cells. Through this preconcentration step, the overall energy consumption and cost could be substantially reduced [70,71,72]. It has been reported that more than 95% of total energy consumption in centrifugation step could be reduced by using chitosan-based flocculations as preconcentration step for harvesting of microalgal cells [71]. Moreover, larger amount of microalgal biomass aggregations led to the greater flocculation efficiency of microalgal cells than other methods including filtration and centrifugation [44,73]. The combination of flocculation with other methods could also effectively preconcentrate a large volume of microalgal culture with low cell density.

Harvesting of microalgal cells through flocculation has been performed using either physical/chemical method or biological method. Microalgal cells that have negative charge surface can form a stable suspension [29]. The adjustment of pH by changing ratio of H^+/OH^- ions may induce the autoflocculation. Mg^{2+} ions can also affect the electrostatic interactions between microalgal cells and induce flocculation. Microalgal flocculation could occur under alkaline conditions (pH > 10) by coprecipitation with Mg^{2+} and/or Ca^{2+} [73], while inorganic flocculants, i.e., iron chloride and aluminum sulfate, need acidic conditions for effective microalgal flocculation [44,74]. Flocculation of microalgal cells by addition of inorganic salts has been widely used and known as the most promising method because this type of flocculation can be easily scaled up and applied for various species of microalgal cells [74]. Mandik et al. [68] have reported that the flocculation of microalgal biomass in mixotrophic culture did require lower amount of aluminum ion than those in photoautotrophic and heterotrophic cultures. However, the flocculants should be nontoxic, inexpensive, and effective at a low concentration. Organic cationic polymers such as chitosan, polyacrylamide, tannin, and starch are nontoxic flocculating agents that have been widely used in wastewater treatment and in the food industry [29,73,74,75,76,77].

Bioflocculation by filamentous fungi

Microalgae harvesting via biological methods is an alternative way to separate the microalgal biomass from liquid medium. As this method requires low energy inputs and no addition of chemicals, it is therefore suitable as a new insight for harvesting of microalgal cells. Recently, coculturing of microalgae with filamentous fungi is considered to be potential method for microalgal harvesting [69]. The fungal-assisted bioflocculation could be performed either by adding fungal pellets or spores (Fig. 6.1). The fungal pellet-assisted bioflocculation needs two steps of pelletization of fungal cells prior to their addition into microalgae culture broth. This technique could rapidly harvest the microalgae cells but required extra cultivation medium and time for fungal pellet formation [81,82,83]. The second technique using fungal spores, namely copelletization, allows the fungi to grow and simultaneously flocculate microalgal cells in the pellets. This technique could reduce the operating step and increase the overall productivity. However, not all fungal species could be used in the copelletization. Srinuanpan et al. [69] reported that filamentous fungus *Trichoderma reesei* could not form pellets with microalgal cells, but the pure fungal pellets did effectively harvest the microalgal cells with 10 min. Other fungal pellets used for the bioflocculation of microalgal cells were the pellets from *Mucor circinelloides*, *Penicillium expansum*, and *Rhizopus oryzae* [79,80]. Zhou et al. [78] optimized the conditions for copelletization of fungus *Aspergillus oryzae* with *Chlorella vulgaris* UMN235. The optimal spore concentration, glucose concentration, and pH for harvesting of microalgal cells were 1.2×10^4 spores/mL,

10 g/L, and 4.0–5.0, respectively. Muradov et al. [83] indicated that the fungi might neutralize negative charge on the microalgal cell surface and make the microalgal cells get attached to the fungal cells. It was also possible that the fungal cells might secrete sticky extracellular polymeric substances and enable the cell-to-cell attachment [69,82,84].

Bioflocculation by flocculant-producing bacteria

Recently, the innovative biological flocculations of microalgae using bacterial bioflocculants and also the bacterial cells themselves have been proposed [38]. The flocculation efficiency of *C. vulgaris* and *Chlorella prototothecoides* using commercial γ-PGA bioflocculant produced by *Bacillus subtilis* was as high as 95% [85]. Several novel flocculant-producing bacteria isolated from various wastewaters have been reported and optimized for their bioflocculant production with high yields [86,87,88]. To reduce the production cost, various industrial wastewater, such as phenol-containing wastewater [89], livestock wastewater [86], and palm oil mill effluent [90], have been used as low-cost cultivation media for the bacterial growth and bioflocculant production. Most of them contained enough nutrition for bacterial growth, but their imbalanced nutrients are likely to limit the bacteria growth and bioflocculant production [88,91,92].

The cocultivation of microalgae with flocculant-producing bacteria could reduce step and time for bacteria cultivation (Fig. 6.1). Bacterial cells likely produced extracellular polymeric substances that could act as bioflocculants and enable the flocculation of microalgal cells. It was also possible that both cells might produce extracellular polymeric substances that are different from each other and get to the cell-to cell contact prior to coflocculation [93,94]. The bacteria from genera *Terrimonas*, *Sphingobacterium*, and *Flavobacterium*, which are associated with microalgal growth in nature, have been applied for harvesting of *C. vulgaris* [94]. Ji et al. [95] reported that *Bacillus licheniformis* cells got attached to the cell surface of *C. vulgaris* through the cocultivation in synthetic wastewater. Adjusting pH to alkaline could strengthen the adhesion of *B. licheniformis* cells to microalgal surface and also the energy exchanges between microalgae and bacteria [96]. Another studies reported that the addition of the bacterial culture broth to the microalgal culture in a later growth stage showed great flocculation efficiency of 83% within 24 h. It was indicated that both bacterial cells and their extracellular metabolites played important roles in the flocculation process

[97,98]. The symbiotic effect between bacterial and microalgae cells has also been reported. As the bacterial cells are aerobes and the microalgae are oxygen producers, the gas exchange between both cells could occur [98]. The cocultivation of microalgae with activated sludge, which is called activated microalgae granules or activated microalgae floc, presents an alternative microalgae-bacteria system. This strategy was proposed to eliminate the economic obstacle of microalgae-harvesting processes [99] and also perform the phytoremediation of wastewater [100].

CONVERSION OF MICROALGAL LIPIDS INTO BIODIESEL

Microalgal Lipids

High lipid/oil content of some microalgal species has been attractive as feedstocks for biodiesel production. Green algae (*Chlorophyta*) and diatoms (*Bacillariophyta*) are mostly considered as feedstocks for biodiesel due to their high growth rates and high lipid contents. Microalgae can use both inorganic carbon (CO_2) and organic carbon sources (glucose, acetate, etc.) for lipid accumulation. The types of lipids are classified into polar lipids (e.g., phospholipids) and neutral lipids (triglycerides). Neutral lipids are the most important materials for biodiesel production [101]. Lipid contents in microalgae are in the range of 5%–77% biomass, and the fatty acid compositions vary from species to species depending on microalgae life cycle and culture conditions, i.e., medium composition, culture temperature, light intensity, photoperiod, and air flow rate [102–108]. It has been reported that the microalgal cells during stationary phase had lower content of polar lipids compared with those during growth phase. The lipid contents of some microalgae possibly increased from ~10 wt% to almost 20 wt% of their biomass under oxygen deprivation [107]. In addition, microalgae have been commonly reported that they likely respond to the nutrient starvations by intensifying their metabolisms toward neutral lipid synthesis [29].

The microalgal lipids are composed of C12 to C22 fatty acids that are either unsaturated or saturated types. The double bonds in these fatty acid chains are never more than six and the unsaturated fatty acids are almost *cis*-isomers [109]. The main fatty acids (C16:0, C18:1, and C18:2) in microalgal lipids are potential feedstocks for high-quality biodiesel [110]. Table 6.2 summarizes the biodiesel properties of microalgal lipids including saponification number: SN; iodine number: IN; cetane number: CN; degree of unsaturation: DU; long-chain

saturation factor: LCSF; and cold filter plugging point: CFPP. These values can be calculated from their fatty acid profiles. Several researches confirmed that the biodiesel properties of microalgal lipids meet the biodiesel fuel specifications given by the regulatory international standard of ASTM D6751 and EN-14214 [111,112].

Lipid Extraction From Microalgal Biomass

There are several procedures for lipid extraction from microalgal biomass. These include milling, homogenization, mechanical pressing, enzymatic extraction, ultrasonic-assisted extraction, solvent extraction, osmotic shock, and supercritical fluid extraction. These methods have their own advantages and disadvantages. Milling method uses grinding small beads and agitation to disrupt cell walls and extract the lipids out of the cells. Pressures from homogenization and mechanical pressing are used to rupture cell walls and recover the lipids from the cells. Enzymes can hydrolyze cell walls and release lipids. These methods are usually combined with either solvent extraction or ultrasonic-assisted extraction. Supercritical fluid extractions aimed to utilize chemical compounds having properties of both liquids and gases at high temperatures and pressures. The advantage of this method is no chemical residues left after the reaction. The combination of these methods with sonication showed faster extractions and higher yields [113–116]. The sonication-assisted solvent extraction is widely applied for lipid extraction from microalgal biomass. This combination requires less toxic solvent mixture and can minimize energy consumption, reaction time, and volume of solvents, while this maximizes lipid yield [116,117].

Biodiesel Production From Microalgal Lipids via Transesterification Reaction

Traditionally, biodiesel is produced via transesterification reaction of lipids and oils. The microalgal biomass have to be dewatered by drying before extraction of lipids. Consequently, the lipids in microalgal biomass have to be extracted before transesterification. This is called three-step method. Alternatively, the wet microalgal biomass can be used in lipid extraction step without drying step. Recently, the direct transesterification reactions of wet microalgal biomass into biodiesel have been developed to reduce the number of unit operations and overall production cost of microalgae-based biodiesel. The transesterification reaction can be catalyzed either by homogeneous or heterogeneous catalysts, and these catalysts are acid, alkaline, and enzymes [118–123]. Fig. 6.2 shows biodiesel

TABLE 6.2
Prospect Biodiesel Fuel Properties of Lipids From Several Microalgae.

	BIODIESEL FUEL PROPERTIES						
Microalgae	Saponification number	Iodine number	Cetane number	Degree of unsaturation	Long-chain saturation factor	Cold filter plugging point	References
Scenedesmus sp.	207.4	111.2	47.9	101.5	7.6	7.4	Srinuanpan et al. [69]
Coelastrella sp.	194.5	84.9	55.5	91.9	19.3	44.3	Karpagam et al. [111]
Micractinium sp.	190.6	53.4	62.9	57.8	3.9	3	Karpagam et al. [111]
Anabaena sphaerica	228.6	32.5	62.8	87.1	7	5.7	Anahas and Muralitharan [153]
Nostoc calcicola	201.7	64.6	52.7	63.5	9.1	12.2	Anahas and Muralitharan [153]
Chlorella minutissima	199	40.14	96	63	5	0.6	Arora et al. [154]
EN-14214	–	≤120	≥51	–	–	≤5/−20	–
ASTM D6751	–	–	≥47	–	–	6/18	–

(A) Extraction and transesterification

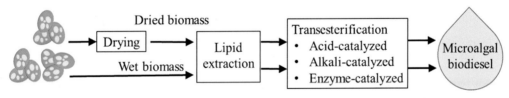

(B) Direct transesterification

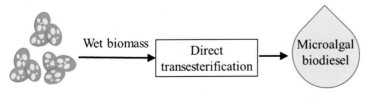

FIG. 6.2 Biodiesel production from microalgal lipids via conventional and direct transesterification reactions.

production from microalgal lipids via conventional and direct transesterification reactions.

Acid-catalyzed biodiesel production

Several homogeneous acid catalysts such as H_2SO_4, HCl, H_3PO_4, BF_3, and organic sulfonic acids have been used in transesterification of microalgal lipids into biodiesel. Among these acid catalysts, sulfuric and sulfonic acid are more preferred due to their higher conversion rates [124]. Kim et al. [125] and Macías-Sánchez et al. [126] used sulfuric acid as catalyst to convert *Nannochloropsis gaditana* lipids into biodiesel and achieved relatively high biodiesel yields of 80% and 87%, respectively. The conversion yield of biodiesel from *Nannochloropsis oceanica* lipids was also as high as 91% when using sulfuric acid as catalyst [127]. Although the transesterification using acid catalysts could produce biodiesel at high yields, the rates of reaction are very slow compared with those using alkaline catalysts. However, the transesterifications of cheap oil feedstocks such as used cooking oil and animal oil that contained high free fatty acid content (>40%) require acid catalysts to prevent the formation of soap from saponification reaction between free fatty acids and alkaline catalysts. It has been reported that the use of excess acid catalysts could increase the conversion rate and yield of lipids into biodiesel [128,129].

Alkali-catalyzed biodiesel production

Several alkaline catalysts have been widely used in biodiesel production. These include sodium hydroxide and potassium hydroxide due to their high conversion rate >98% within a short reaction time about 30 min while using very low loading amount (0.5 mol%). However, the free fatty acids in the feedstocks significantly affect this alkali-catalyzed process because the alkaline catalysts react with free fatty acids to form soap that would reduce the efficiency and yield of the reaction and also increase the viscosity of the reaction mixture [124,130]. NaOH was used as catalyst for transesterification of *C. protothecoides* lipids and gave the maximal biodiesel yield of 98% [131]. Similarly, KOH has been used for transesterification of lipids from microalgae *Spirulina maxima* [132] and *C. vulgaris* [133] and gave the maximum biodiesel yield of 86.1% and 85%, respectively.

Enzyme-catalyzed biodiesel production

Enzyme-catalyzed biodiesel production requires lipases as biocatalysts to transesterify lipids into biodiesel. In general, lipases can catalyze various bioconversion reactions and also has advantages over acid and alkaline catalysts including easier removal of by-products, easier recovery of glycerol, no side reactions, and more environmentally friendly [122,123]. Several researchers have studied lipase-catalyzed biodiesel production from microalgal lipids [134–139]. Teo et al. [134] conducted the biodiesel production from *Tetraselmis* sp. lipids via lipase-catalyzed transesterification and found that the biodiesel yield from the lipase-catalyzed reaction was sevenfold higher than that from alkaline-catalyzed reaction. More than 94% of *N. gaditana* lipids were transformed to biodiesel via lipase-catalyzed transesterification [135]. Moreover, the immobilized lipases were also attempted for conversion of microalgal lipids into biodiesel to reduce the production cost due to their high stability and reusability. The immobilized lipase from yeast *Rhodotorula mucilaginosa* could be reused for 10 cycles without significant loss of activity [140]. Guldhe et al. [138] reported that immobilized *Aspergillus niger* lipase could be used for two cycles without significant loss in conversion efficiency, and more than 80% of biodiesel yield was achieved.

Simultaneous Lipid Extraction and Transesterification of Microalgal Lipids Into Biodiesel

The traditional process for biodiesel production from microalgal lipids requires three steps of microalgae dewatering or drying, lipid extraction, and transesterification. Among these three steps, the energy consumption in drying step accounts for 84% of total energy consumption [141]. Furthermore, this process often requires 0.5–3 h in lipid extraction and subsequent transesterification of lipids into biodiesel. To reduce energy consumption in microalgae drying step and to simplify the process for biodiesel production, several researchers have proposed the direct transesterification that combine lipid extraction step and transesterification step into a single process [142–145]. Wahidin et al. [145] evaluated the effectiveness of the simultaneous microwave-assisted extraction and transesterification. However, the biodiesel yield from this process was only 42.22%. The acid-catalyzed direct transesterification of *N. gaditana* biomass achieved higher biodiesel yield of 64.98% [142]. Shirazi et al. [143] achieved the maximum biodiesel yield of 99.32% when using mixed solvents in direct transesterification process. The use of cosolvent system (ethanol/chloroform) in direct transesterification of wet *N. gaditana* biomass could yield biodiesel at 80% [144]. The use of ethyl acetate in direct transesterification contributes not only increasing biodiesel yield up to 97.1% and eliminating additional cosolvent but also reducing the overall production cost. Therefore, the direct transesterification with optimized solvents has high potential for disrupting microalgae cell wall and directly converting extracted lipids into biodiesel.

CONCLUSIONS

Conventional and recently developed harvesting methods for microalgal biomass and conversion of microalgal lipids into biodiesel have been summarized. Biological methods have been introduced as an alternative way that requires low energy consumption and no chemicals use. The direct transesterification of microalgal lipids into biodiesel presents as promising technique with high efficiency and low energy consumption. This review provides a new sight for the technical, economic, and environmentally friendly harvesting of microalgal biomass and direct conversion of microalgal lipids into biodiesel.

REFERENCES

[1] A.C. Eloka-Eboka, F.L. Inambao, Effects of CO_2 sequestration on lipid and biomass productivity in microalgal biomass production, Appl. Energy. 195 (2017) 1100−1111.

[2] X. Miao, Q. Wu, Biodiesel production from heterotrophic microalgal oil, Bioresour. Technol. 97 (2006) 841−846.

[3] S.H. Ho, S.W. Huang, C.Y. Chen, T. Hasunuma, A. Kondo, J.S. Chang, Bioethanol production using carbohydrate-rich microalgae biomass as feedstock, Bioresour. Technol. 135 (2013) 191−198.

[4] D. Sengmee, B. Cheirsilp, T.T. Suksaroge, P. Prasertsan, Biophotolysis-based hydrogen and lipid production by oleaginous microalgae using crude glycerol as exogenous carbon source, Int. J. Hydrogen Energy. 42 (2017) 1970−1976.

[5] Y. Zhang, X. Kong, Z. Wang, Y. Sun, S. Zhu, L. Li, P. Lv, Optimization of enzymatic hydrolysis for effective lipid extraction from microalgae Scenedesmus sp, Renew. Energy. 125 (2018) 1049−1057.

[6] S. Srinuanpan, B. Cheirsilp, P. Prasertsan, Effective biogas upgrading and production of biodiesel feedstocks by strategic cultivation of oleaginous microalgae, Energy 148 (2018), 766−764.

[7] Y. Chisti, Biodiesel from microalgae, Biotechnol. Adv. 25 (2007) 294−306.

[8] T.M. Mata, A.A. Martins, N.S. Caetano, Microalgae for biodiesel production and other applications: A review, Renew. Sustain. Energy Rev. 14 (2010) 217−232.

[9] B. Cheirsilp, S. Torpee, Enhanced growth and lipid production of microalgae under mixotrophic culture condition: effect of light intensity, glucose concentration and fed-batch cultivation, Bioresour. Technol. 110 (2012) 510−516.

[10] M. Daroch, S. Geng, G. Wang, Recent advances in liquid biofuel production from algal feedstocks, Appl. Energy. 102 (2013) 1371−1381.

[11] M.M. Teresa, A.M. Antonio, N.S. Caetano, Microalgae for biodiesel production and other applications: a review, Renew. Sustain. Energy. 14 (2010) 217−232.

[12] Ansyori, Ethanol sebagai Bahan Bakar Alternatif, Erlangga, Jakarta, 2004.

[13] X. Deng, Y. Li, X. Fei, Microalgae: A promising feedstock for biodiesel, Afr. J. Microbiol. Res. 3 (2009) 1008−1014.

[14] Y. Li, M. Horsman, N. Wu, C. Lan, N. Dubois-Calero, Biofuels from microalgae, Biotech. Progress. 24 (2008) 815−820.

[15] L. Gouveia, A.C. Oliveira, Microalgae as a raw material for biofuels production, J. Indust. Microbiol. Biotechnol. 36 (2009) 269−274.

[16] C.K. Phwan, H.C. Ong, W.H. Chen, T.C. Ling, E.P. Ng, P.L. Show, Overview: Comparison of pretreatment technologies and fermentation processes of bioethanol from microalgae, Energy Convers. Manag. 173 (2018) 81−94.

[17] Z. Yang, R. Guo, X. Xu, X. Fan, X. Li, Enhanced hydrogen production from lipid-extracted microalgal biomass residues through pretreatment, Int. J. Hydrogen Energy. 35 (2010) 9618−9623.

[18] H.L. Zheng, Z. Gao, J.W. Yin, X.J. Ji, H. Huang, Lipid production of Chlorella vulgaris from lipid-extracted microalgal biomass residues through two-step enzymatic hydrolysis, Bioresour. Technol. 117 (2012) 1−6.

[19] T. Mathimani, L. Uma, D. Prabaharan, Formulation of low-cost seawater medium for high cell density and high lipid content of Chlorella vulgaris BDUG 91771 using central composite design in biodiesel perspective, J. Clean Prod. 198 (2018) 575−586.

[20] M. Olaizola, Commercial development of microalgal biotechnology: from the test tube to the marketplace, Biomol. Eng. 20 (2003) 459−466.

[21] K.W. Chew, J.Y. Yap, P.L. Show, N.H. Suan, J.C. Juan, T.C. Ling, D.J. Lee, J.S. Chang, Microalgae biorefinery: high value products perspectives, Bioresour. Technol. 229 (2017) 53−62.

[22] A.B.M. Hossain, A. Salleh, A.N. Boyce, P. Chowdhurry, M. Naqiuddin, Biodiesel fuel production from algae as renewable energy, Am. J. Biochem. Biotech. 4 (2008) 250−254.

[23] Q. Hu, M. Sommerfeld, E. Jarvis, M. Ghirardi, M. Posewitz, M. Seibert, A. Darzins, Microalgal triacylglycerols as feedstocks for biofuel production: perspectives and advances, Plant J. 54 (2008) 621−639.

[24] M.J. Massinggil (Ed.), 15 Years of experience producing microalgae feedstock and resulting co-products, Kent Bioenergy Corporation, San Diego, 2009.

[25] P.J.L.B. Williams, L.M.L. Laurens, Microalgae as biodiesel and biomass feedstocks: review and analysis of the biochemistry, energetics and economics, Energy Environ. Sci. 3 (2010) 554−590.

[26] P. Spolaore, C. Joannis-Cassan, E. Duran, A. Isambert, Commercial applications of microalgae, J. Biosci. Bioeng. 101 (2006) 87–96.

[27] M.K. Lam, K.T. Lee, Microalgae biofuels: a critical review of tissues, problems and the way forward, Biotechnol. Adv. 30 (2012) 673–690.

[28] C.Y. Chen, K.L. Yeh, R. Aisyah, D.J. Lee, J.S. Chang, Cultivation, photobioreactor design and harvesting of microalgae for biodiesel production: a critical review, Bioresour. Technol. 102 (2011) 71–81.

[29] S. Srinuanpan, B. Cheirsilp, P. Prasertsan, Y. Kato, Y. Asano, Strategies to increase the potential use of oleaginous microalgae as biodiesel feedstocks: nutrient starvations and cost-effective harvesting process, Renew. Energy. 113 (2018) 1229–1241.

[30] A.J. Dassey, C.S. Theegala, Harvesting economics and strategies using centrifugation for cost effective separation of microalgae cells for biodiesel applications, Bioresour. Technol. 128 (2013) 241–245.

[31] R. Divakaran, V.N.P. Pillai, Flocculation of algae using chitosan, J. Appl. Phycol. 14 (2002) 419–422.

[32] S. Babel, S. Takizawa, Microfiltration membrane fouling and cake behavior during algal filtration, Desalination 261 (2010) 46–51.

[33] A. Sandip, V.H. Smith, T.N. Faddis, An experimental investigation of microalgal dewatering efficiency of belt filter system, Energy Rep. 1 (2015) 169–174.

[34] R. Harun, M.K. Danquah, G.M. Forde, Microalgal biomass as a fermentation feedstock for bioethanol production, J. Chem. Technol. Biotechnol. 85 (2010) 199–203.

[35] K. Sander, G. Murthy, Life cycle analysis of algae biodiesel, Int. J. Life Cycle Ass. 5 (2010) 704–714.

[36] F. Fasaei, J.H. Bitter, P.M. Slegers, A.J.B. van Boxtel, Techno-economic evaluation of microalgae harvesting and dewatering systems, Algal Res. 31 (2018) 347–362.

[37] Z. Wu, Y. Zhu, W. Huang, C. Zhang, T. Li, Y. Zhang, A. Li, Evaluation of flocculation induced by pH increase for harvesting microalgae and reuse of flocculated medium, Bioresour. Technol. 110 (2012) 496–502.

[38] T. Ndikubwimana, X. Zeng, N. He, Z. Xiao, Y. Xie, J.S. Chang, L. Lin, Y. Lu, Microalgae biomass harvesting by bioflocculation-interpretation by classical DLVO theory, Biochem. Eng. J. 101 (2015) 160–167.

[39] R. Munoz, B. Guieysse, Algal–bacterial processes for the treatment of hazardous contaminants: a review, Water Res. 40 (2006) 2799–2815.

[40] M. Hapońska, E. Clavero, J. Salvadó, X. Farriol, C. Torras, Pilot scale dewatering of *Chlorella sorokiniana* and *Dunaliella tertiolecta* by sedimentation followed by dynamic filtration, Algal Res. 33 (2018) 118–124.

[41] E.M. Grima, E.H. Belarbi, F.G.A. Fernandez, A.R. Medina, Y. Chisti, Recovery of microalgal biomass and metabolites: process options and economics, Biotechnol. Adv. 20 (2003) 491–515.

[42] S. Şirin, E. Clavero, J. Salvadó, Potential preconcentration methods for Nannochloropsis gaditana

and a comparative study of pre-concentrated sample properties, Bioresour. Technol. 132 (2013) 293–304.

[43] A.S. Japar, M.S. Takriff, N.H.M. Yasin, Harvesting microalgal biomass and lipid extraction for potential biofuel production: a review, J. Environ. Chem. Eng. 5 (2017) 555–563.

[44] T. Chatsungnoen, Y. Chisti, Harvesting microalgae by flocculation-sedimentation, Algal Res. 13 (2016) 271–283.

[45] G. Singh, S.K. Patidar, Microalgae harvesting techniques: a review, J. Environ. Manage. 217 (2018) 499–508.

[46] G. Kang, Y. Cao, Application and modification of poly (vinylidene fluoride) (PVDF) membranes-a review, J. Membr. Sci. 463 (2014) 145–165.

[47] J.J. Milledge, S. Heaven, A review of the harvesting of micro-algae for biofuel production, Rev. Environ. Sci. Biotechnol. 12 (2013) 165–178.

[48] N. Pragya, K.K. Pandey, P.K. Sahoo, A review on harvesting, oil extraction and biofuels production technologies from microalgae, Renew. Sustain. Energy Rev. 24 (2013) 159–171.

[49] T. Baerdemaeker, B. Lemmens, C. Dotremont, J. Fret, L. Roef, K. Goiris, L. Diels, Benchmark study on algae harvesting with backwashable submerged flat panel membranes, Bioresour. Technol. 129 (2013) 582–591.

[50] I.L. Drexler, D.H. Yeh, Membrane applications for microalgae cultivation and harvesting: a review, Rev. Environ. Sci. Bio/Technol. 13 (2014) 487–504.

[51] W. Mo, L. Soh, J.R. Werber, M. Elimelech, J.B. Zimmerman, Application of membrane dewatering for algal biofuel, Algal Res 11 (2015) 1–12.

[52] D. Mancinelli, C. Halle, Nano-filtration and ultra-filtration ceramic membranes for food processing: a mini review, J. Membr. Sci. Technol. 5 (2015) 140.

[53] C. Nurra, E. Clavero, J. Salvado, C. Torras, Vibrating membrane filtration as improved technology for microalgae dewatering, Bioresour. Technol. 157 (2014) 247–253.

[54] S.D. Rios, J. Salvado, X. Farriol, C. Torras, Antifouling microfiltration strategies to harvest microalgae for biofuel, Bioresour. Technol. 119 (2012) 406–418.

[55] M. Niaghi, M. Mahdavi, R. Gheshlaghi, Optimization of dissolved air flotation technique in harvesting microalgae from treated wastewater without flocculants addition, J. Renew. Sustain. Energy. 7 (2015) 013130.

[56] W. Phoochinda, D. White, B. Briscoe, An algal removal using a combination of flocculation and flotation processes, Environ. Technol. 25 (2015) 1385–1395.

[57] Y.L. Cheng, Y.C. Juang, G.Y. Liao, S.H. Ho, K. L Yeh, C.H. Chen, J.S. Chang, J.C. Liu, D.J. Lee, Dispersed ozone flotation of *Chlorella vulgaris*, Bioresour. Technol. 101 (2010) 9092–9096.

[58] Y.S. Cheng, Y. Zheng, J.M. Labavitch, J.S. Vander Gheynst, The impact of cell wall carbohydrate composition on the chitosan flocculation of *Chlorella*, Process Biochem. 46 (2011) 1927–1933.

[59] M.V. González, I. Monje-Ramírez, M.O. Ledesma, J.G. Fadrique, S. Velásquez-Orta, Harvesting microalgae

using ozoflotation releases surfactant proteins, facilitates biomass recovery and lipid extraction, Biomass Bioenerg. 95 (2016) 109−115.

[60] X. Jin, P. Jin, X. Wang, A study on the effects of ozone dosage on dissolved-ozone flotation (DOF) process performance, Water Sci. Technol. 71 (2006) 1423−1428.

[61] Z. Dai, S. Dukhin, J. Ralston, The inertial hydrodynamic interaction of particles and rising bubbles with mobile surfaces, J. Colloid Interf. Sci. 197 (1998) 275−292.

[62] Z. Dai, D. Fornasiero, J. Ralston, Particle−bubble collision models-a review, Adv. Colloid Interf. Sci. 85 (2000) 231−256.

[63] J. Hanotu, B.H.C. Hemaka, W.B. Zimmerman, Microflotation performance for algal separation, Biotechnol. Bioeng. 109 (2012) 1663−1673.

[64] S. Garg, L. Wang, P.M. Schenk, Flotation separation of marine microalgae from aqueous medium, Sep. Purif. Technol. 156 (2015) 636−641.

[65] H.A. Kurniawati, S. Ismadji, J.C. Liu, Microalgae harvesting by flotation using natural saponin and chitosan, Bioresour. Technol. 166 (2014) 429−434.

[66] N.R.H. Rao, R. Yap, M. Whittaker, R.M. Stuetz, B. Jefferson, W.L. Peirson, A.M. Granville, R.K. Henderson, The role of algal organic matter in the separation of algae and cyanobacteria using the novel "Posi"-Dissolved air flotation process, Sep. Purif. Technol. 130 (2018) 20−30.

[67] F.F. Bauer, P. Govender, M. Bester, Yeast flocculation and its biotechnological relevance, Appl. Microbiol. Biot. 88 (2010) 31−39.

[68] Y.I. Mandik, B. Cheirsilp, P. Boonsawang, P. Prasertsan, Optimization of flocculation efficiency of lipid-rich marine Chlorella sp. biomass and evaluation of its composition in different cultivation modes, Bioresour. Technol. 182 (2015) 89−97.

[69] S. Srinuanpan, A. Chawpraknoi, S. Chantarit, B. Cheirsilp, P. Prasertsan, A rapid method for harvesting and immobilization of oleaginous microalgae using pellet-forming filamentous fungi and the application in phytoremediation of secondary effluent, Int. J. Phytoremediat. 20 (2018) 1017−1024.

[70] L.B. Brentner, M.J. Eckelman, J.B. Zimmerman, Combinatorial life cycle assessment to inform process design of industrial production of algal biodiesel, Environ. Sci. Technol. 45 (2011) 7060−7067.

[71] Y. Xu, S. Purton, F. Baganz, Chitosan flocculation to aid the harvesting of the microalga Chlorella sorokiniana, Bioresour. Technol. 129 (2013) 296−301.

[72] S. Salim, M.H. Vermue, R.H. Wijffels, Ratio between autoflocculating and target microalgae affects the energy-efficient harvesting by bio-flocculation, Bioresour. Technol. 118 (2012) 49−55.

[73] P.V. Brady, P.I. Pohl, J.C. Hewson, A coordination chemistry model of algal autoflocculation, Algal Res. 5 (2014) 226−230.

[74] N. Uduman, Y. Qi, M.K. Danquah, G.M. Forde, A. Hoadley, Dewatering of microalgal cultures: a major bottleneck to algae-based fuels, J. Renew. Sustain. Energy 2 (2010) 0127011−0127015.

[75] J.C.M. Pires, M.C.M. Alvim-Ferraz, F.G. Martins, M. Simões, Carbon dioxide capture from flue gases using microalgae: Engineering aspects and biorefinery concept, Renew. Sust. Energ. Rev. 16 (2012) 3043−3063.

[76] F. Bleeke, G. Quante, D. Winckelmann, G. Klöck, Effect of voltage and electrode material on electroflocculation of Scenedesmus acuminatus, Bioresour. Bioprocess. 2 (2015) 36.

[77] D. Vandamme, I. Foubert, K. Muylaert, Flocculation as a low-cost method for harvesting microalgae for bulk biomass production, Trends Biotechnol. 31 (2013) 233−239.

[78] W. Zhou, M. Min, B. Hu, X. Ma, Y. Liu, Q. Wang, J. Shi, P. Chen, R. Ruan, Filamentous fungi assisted bioflocculation: a novel alternative technique for harvesting heterotrophic and autotrophic microalgal cells, Sep. Purif. Technol. 107 (2013) 158−165.

[79] S.O. Gultom, B. Hu, Review of microalgae harvesting via co-pelletization with filamentous fungus, Energies 6 (2013) 5921−5939.

[80] J. Chen, L. Leng, C. Ye, Q. Lu, M. Addy, J. Wang, J. Liu, P. Chen, R. Ruan, W. Zhou, A comparative study between fungal pellet- and spore-assisted microalgae harvesting methods for algae bioflocculation, Bioresour. Technol. 259 (2018) 181−190.

[81] D. Wrede, M. Taha, A.F. Miranda, K. Kadali, T. Stevenson, A.S. Ball, A. Mouradov, Co-cultivation of fungal and microalgal cells as an efficient system for harvesting of microalgal cells, lipid production and wastewater treatment, PLoS One 9 (2014) e113497.

[82] A.F. Miranda, M. Taha, D. Wrede, P. Morrison, A.S. Ball, T. Stevenson, A. Mouradov, Lipid production in association of filamentous fungi with genetically modified cyanobacterial cells, Biotechnol. Biofuels. 8 (2015) 179.

[83] N. Muradov, M. Taha, A.F. Miranda, D. Wrede, K. Kadali, A. Gujar, T. Stevenson, A.S. Ball, A. Mouradov, Fungal-assisted algal flocculation: application in wastewater treatment and biofuel production, Biotechnol. Biofuels. 8 (2015) 1.

[84] L. Selbmann, F. Stingele, M. Petruccioli, Exopolysaccharide production by filamentous fungi: the example of Botryosphaeria rhodina, Anton Leeuw. Int. J. G. 84 (2003) 135−145.

[85] H. Zheng, Z. Gao, J. Yin, X. Tang, X. Ji, H. Huang, Harvesting of microalgae by flocculation with poly (γ-glutamic acid), Bioresour. Technol. 112 (2012) 212−220.

[86] F.Q. Peng, G.G. Ying, B. Yang, S. Liu, H.J. Lai, Y.S. Liu, Z.F. Chen, G.J. Zhou, Biotransformation of progesterone and norgestrel by two freshwater microalgae (Scenedesmus obliquus and Chlorella pyrenoidosa): transformation kinetics and products identification, Chemosphere 95 (2014) 581−588.

[87] C. Liu, K. Wang, J.H. Jiang, W.J. Liu, J.Y. Wang, A novel bioflocculant produced by a salt-tolerant, alkaliphilic and biofilm-forming strain Bacillus agaradhaerens C9

and its application in harvesting *Chlorella minutissima* UTEX2341, Biochem. Eng. J. 93 (2015) 166–172.

[88] H. Guo, C. Hong, C. Zhang, B. Zheng, D. Jiang, W. Qin, Bioflocculants' production from a cellulase-free xylanase-producing *Pseudomonas boreopolis* G22 by degrading biomass and its application in cost-effective harvest of microalgae, Bioresour. Technol. 255 (2018) 171–179.

[89] H. Chen, C. Zhong, H. Berkhouse, Y. Zhang, Y. Lv, W. Lu, Y. Yang, J. Zhou, Removal of cadmium by bioflocculant produced by *Stenotrophomonas maltophilia* using phenol-containing wastewater, Chemosphere 155 (2016) 163–169.

[90] W. Chaisorn, P. Prasertsan, S. O-Thong, P. Methacanon, Production and characterization of biopolymer as bioflocculant from thermotolerant *Bacillus subtilis* WD161 in palm oil mill effluent, Int. J. Hydrogen Energy 41 (2016) 21657–21664.

[91] G.E. Adebami, B.C. Adebayo-Tayo, Comparative effect of medium composition on bioflocculant production by microorganisms isolated from wastewater samples, Rep. Opin. 5 (2013) 46–53.

[92] S.J. Ramsden, S. Ghosh, L.J. Bohl, T.M. Lapara, Phenotypic and genotypic analysis of bacteria isolated from three municipal wastewater treatment plants on tetracycline-amended and ciprofloxacin-amended growth media, J. Appl. Microbiol. 109 (2010) 1609–1618.

[93] A.K. Lee, D.M. Lewis, P.J. Ashman, Microbial flocculation, a potentially low-cost harvesting technique for marine microalgae for the production of biodiesel, J. Appl. Phycol. 21 (2009) 559–567.

[94] S.B. Ummalyma, E. Gnansounou, R.K. Sukumaran, R. Sindhu, A. Pandey, D. Sahoo, Bioflocculation: An alternative strategy for harvesting of microalgae – An overview, Bioresour. Technol. 242 (2017) 227–235.

[95] X. Ji, M. Jiang, J. Zhang, X. Jiang, Z. Zheng, The interactions of algae-bacteria symbiotic system and its effects on nutrients removal from synthetic wastewater, Bioresour. Technol. 247 (2018) 44–50.

[96] Z. Liang, Y. Liu, F. Ge, N. Liu, M. Wong, A pH-dependent enhancement effect of co-cultured bacillus licheniformis on nutrient removal by *Chlorella vulgaris*, Ecol. Eng. 75 (2015) 258–263.

[97] J. Lee, D.H. Cho, R. Ramanan, B.H. Kim, H.M. Oh, H.S. Kim, Microalgae-associated bacteria play a key role in the flocculation of *Chlorella vulgaris*, Bioresour. Technol. 131 (2013) 195–201.

[98] A.A.K. Das, J. Bovill, M. Ayesh, S.D. Stoyanov, V.N. Paunov, Fabrication of living soft matter by symbiotic growth of unicellular microorganisms, J. Mater. Chem. B. 4 (2016) 3685–3694.

[99] O. Tiron, C. Bumbac, E. Manea, M. Stefanescu, M.N. Lazar, Overcoming microalgae harvesting barrier by activated algae granules, Sci. Rep. 7 (2017) 4646.

[100] D.S. Wágner, M. Radovici, B.S. Smets, I. Angelidaki, B. Valverde-Pérez, B.G. Plósz, Harvesting microalgae using activated sludge can decrease polymer dosing and enhance methane production via co-digestion in a bacterial-microalgal process, Algal Res. 20 (2016) 197–204.

[101] G. Bagnato, A. Iulianelli, A. Sanna, A. Basile, Glycerol production and transformation: a critical review with particular emphasis on glycerol reforming reaction for producing hydrogen in conventional and membrane reactors, Membranes 7 (2017) 17.

[102] M.R. Brown, S.W. Jeffrey, J.K. Volkman, G.A. Dunstan, Nutritional properties of microalgae for mariculture, Aquaculture 151 (1997) 315–331.

[103] J.M. Lv, L.H. Cheng, X.H. Xu, L. Zhang, H.L. Chen, Enhanced lipid production of *Chlorella vulgaris* by adjustment of cultivation conditions, Bioresour. Technol. 101 (2010) 6797–6804.

[104] R.S. Gour, M. Bairagi, V.K. Garlapati, A. Kant, Enhanced microalgal lipid production with media engineering of potassium nitrate as a nitrogen source, Bioengineered 9 (2018) 98–107.

[105] P. Chaisutyakorn, J. Praiboon, C. Kaewsuralikhit, The effect of temperature on growth and lipid and fatty acid composition on marine microalgae used for biodiesel production, J. Appl. Phycol. 30 (2017) 37–45.

[106] P. Binnal, P.N. Babu, Statistical optimization of parameters affecting lipid productivity of microalga *Chlorella protothecoides* cultivated in photobioreactor under nitrogen starvation, S. Afr. J. Chem. Eng. 23 (2017) 26–37.

[107] G.A. Dunstan, J.K. Volkman, S.M. Barrett, C.D. Garland, Changes in the lipid composition and maximisation of the polyunsaturated fatty acid content of three microalgae grown in mass culture, J. Appl. Phycol. 5 (1993) 71–83.

[108] R. Halim, M.K. Danquah, P.A. Webley, Extraction of oil from microalgae for biodiesel production: A review, Biotechnol. Adv. 30 (2012) 709–732.

[109] A.R. Medina, E.M. Grima, A.G. Gimenez, M.J. Ibanez, Downstream processing of algal polyunsaturated fatty acids, Biotechnol. Adv. 16 (1998) 517–580.

[110] F.J. Zendejas, P.I. Benke, P.D. Lane, B.A. Simmons, T.W. Lane, Characterization of the acylglycerols and resulting biodiesel derived from vegetable oil and microalgae (*Thalassiosira pseudonana* and *Phaeodactylum tricornutum*), Biotechnol. Bioeng. 109 (2012) 1146–1154.

[111] R. Karpagam, K.J. Raj, B. Ashokkumar, P. Varalakshmi, Characterization and fatty acid profiling in two fresh water microalgae for biodiesel production: lipid enhancement methods and media optimization using response surface methodology, Bioresour. Technol. 188 (2015) 177–184.

[112] R. Karpagam, R. Preeti, K.J. Raj, S. Saranya, B. Ashokkumar, P. Varalakshmi, Fatty acid biosynthesis from a new isolate *Meyerella* sp. N4: molecular characterization, nutrient starvation, and fatty acid profiling for lipid enhancement, Energy Fuel 29 (2014) 143–149.

[113] G. Cravotto, L. Boffa, S. Mantegna, P. Perego, M. Avogadro, P. Cintas, Improved extraction of vegetable oils under high-intensity ultrasound and/or microwaves, Ultrason. Sonochem. 15 (2008) 898–902.

[114] F. Wei, G.Z. Gao, X.F. Wang, X.Y. Dong, P.P. Li, W. Hua, X. Wang, X.M. Wu, H. Chen, Quantitative determination of oil content in small quantity of oilseed rape by ultrasound-assisted extraction combined with gas chromatography, Ultrason. Sonochem. 15 (2008) 938–942.

[115] A.M.P. Neto, R.A. Sotana de Souza, A.D. Leon-Nino, J.D.A. da Costa, R.S. Tiburcio, T.A. Nunes, T.C.S. de Mello, F.T. Kanemoto, F.M.P. Saldanha-Corrêa, S.M.F. Gianesella, Improvement in microalgae lipid extraction using a sonication-assisted method, Renew. Energy. 55 (2013) 525–531.

[116] A.L. Ido, M.D.G. de Luna, S.C. Capareda, A.L. Maglinao, Application of central composite design in the optimization of lipid yield from *Scenedesmus obliquus* microalgae by ultrasonic-assisted solvent extraction, Energy 157 (2018) 949–956.

[117] B. Singh, A. Guldhe, I. Rawat, F. Bux, Towards a sustainable approach for development of biodiesel from plant and microalgae, Renew. Sustain. Energy Rev. 29 (2014) 216–245.

[118] G. Pahl (Ed.), Biodiesel: Growing a New Energy Economy, 2nd ed., CGPC, Vermont, TX, 2008.

[119] B. Sialve, N. Bernet, O. Bernard, Anaerobic digestion of microalgae as a necessary step to make microalgal biodiesel sustainable, Biotechnol. Adv. 27 (2009) 409–416.

[120] P.J. Deuss, K. Barta, J.G. de Vries, Homogeneous catalysis for the conversion of biomass and biomass-derived platform chemicals, Catal. Sci. Technol. 4 (2014) 1174–1196.

[121] A. Guldhe, P. Singh, F.A. Ansari, B. Singh, F. Bux, Biodiesel synthesis from microalgal lipids using tungstated zirconia as a heterogeneous acid catalyst and its comparison with homogeneous acid and enzyme catalysts, Fuel 187 (2017) 180–188.

[122] S.B. Velasquez-Orta, J.G.M. Lee, A.P. Harvey, Evaluation of FAME production from wet marine and freshwater microalgae by in situ transesterification, Biochem. Eng. J. 76 (2013) 83–89.

[123] B. Bharathiraja, M. Chakravarthy, R.R. Kumar, D. Yuvaraj, J. Jayamuthunagai, R.P. Kumar, S. Palani, Biodiesel production using chemical and biological methods-A review of process, catalyst, acyl acceptor, source and process variables, Renew. Sustain. Energy Rev. 38 (2014) 368–382.

[124] J.M. Marchetti, V.U. Miguel, A.F. Errazu, Possible methods for biodiesel production, Renew. Sustain. Energy Rev. 11 (2007) 1300–1311.

[125] B. Kim, H. Im, J.W. Lee, In situ transesterification of highly wet microalgae usinghydrochloric acid, Bioresour. Technol. 185 (2015) 421–425.

[126] M.D. Macías-Sánchez, A. Robles-Medina, M.J. Jiménez-Callejón, E. Hita-Peña, L. Estéban-Cerdán, P.A. González-Moreno, E. Navarro-López, E. Molina-Grima, Optimization of biodiesel production from wet microalgal biomass by direct transesterification using the surface response methodology, Renew. Energ. 129 (2018) 141–149.

[127] H.J. Im, H.S. Lee, M.S. Park, J.W. Yang, J.W. Lee, Concurrent extraction and reaction for the production of biodiesel from wet microalgae, Bioresour. Technol. 152 (2014) 534–537.

[128] V.B. Borugadda, V.V. Goud, Biodiesel production from renewable and sustainable feedstocks: status and opportunities, Renew. Sustain. Energy Rev. 16 (2012) 4763–4784.

[129] K.G. Mangesh, D.K. Ajay, M.L. Charan, N.S. Narayan, Transesterification of karanja (*Pongamia pinnata*) oil by solid basic catalysts, Eur. J. Lipid Sci. Technol. 108 (2006) 389–397.

[130] M. Veillette, A. Giroir-Fendler, N. Faucheux, M. Heitz, A biodiesel production process catalyzed by the leaching of alkaline metal earths in methanol: from a model oil to microalgae lipids, J. Chem. Technol. Biot. 92 (2016) 1094–1103.

[131] V. Makareviciene, S. Lebedevas, P. Rapalis, M. Gumbyte, V. Skorupskaite, J. Zaglinskis, Performance and emission characteristics of diesel fuel containing microalgae oil methyl esters, Fuel 120 (2014) 233–239.

[132] M.A. Rahman, M.A. Aziz, R.A. Al-khulaidi, N. Sakib, M. Islam, Biodiesel production from microalgae *Spirulina maxima* by two step process: Optimization of process variable, J. Radiat. Res. Appl. Sci. 10 (2017) 140–147.

[133] A.P. Cercado, F. Ballesteros, S. Capareda, Ultrasound assisted transesterification of microalgae using synthesized novel catalyst, Sustain. Environ. Res. 28 (2018) 234–239.

[134] C.L. Teo, H. Jamaluddin, N.A.M. Zain, A. Idris, Biodiesel production via lipase catalysed transesterification of microalgae lipids from *Tetraselmis* sp, Renew. Energ. 68 (2014) 1–5.

[135] E.N. López, A.R. Medina, P.A.G. Moreno, M.J.J. Callejón, L.E. Cerdán, L.M. Valverde, B.C. Lopez, E.M. Grima, Enzymatic production of biodiesel from *Nannochloropsis gaditana* lipids: influence of operational variables and polar lipid content, Bioresour. Technol. 187 (2015) 346–353.

[136] J. Huang, J. Xia, W. Jiang, Y. Li, J. Li, Biodiesel production from microalgae oil catalyzed by a recombinant lipase, Bioresour. Technol. 180 (2015) 47–53.

[137] A. Guldhe, B. Singh, I. Rawat, K. Permaul, F. Bux, Biocatalytic conversion of lipids from microalgae *Scenedesmus obliquus* to biodiesel using Pseudomonas fluorescens lipase, Fuel 147 (2015) 117–124.

[138] A. Guldhe, P. Singh, S. Kumari, I. Rawat, K. Permaul, F. Bux, Biodiesel synthesis from microalgae using immobilized *Aspergillus niger* whole cell lipase biocatalyst, Renew. Energy. 85 (2016) 1002–1010.

[139] J. Amoah, S.H. Ho, S. Hama, A. Yoshida, A. Nakanishi, T. Hasunuma, C. Ogino, A. Kondo, Conversion of *Chlamydomonas* sp. JSC4 lipids to biodiesel using Fusarium heterosporum lipase-expressing *Aspergillus oryzae* whole-cell as biocatalyst, Algal Res 28 (2017) 16–23.

[140] D. Surendhiran, M. Vijay, A.R. Sirajunnisa, Biodiesel production from marine microalga *Chlorella salina* using

whole cell yeast immobilized on sugarcane bagasse, J. Environ. Chem. Eng. 2 (2014) 1294−1300.

[141] P. Patil, H. Reddy, T. Muppaneni, S. Deng, P. Cooke, P. Lammers, N. Khandan, F.O. Holguin, T. Schuab, Power dissipation in microwave-enhanced in-situ transesterification of algal biomass, Green Chem 14 (2012) 809−818.

[142] S. Torres, G. Acien, F. García-Cuadra, R. Navia, Direct transesterification of microalgae biomass and biodiesel refining with vacuum distillation, Algal Res 28 (2017) 30−38.

[143] H.M. Shirazi, J. Karimi-Sabet, C. Ghotbi, Biodiesel production from *Spirulina* microlagae feedstock using direct transesterification near supercritical methanol condition, Bioresour. Technol. 239 (2017) 378−386.

[144] J. Park, B. Kim, Y.K. Chang, J.W. Lee, Wet in situ transesterification of microalgae using ethyl acetate as a cosolvent and reactant, Bioresour. Technol. 230 (2017) 8−14.

[145] S. Wahidin, A. Idris, N.M. Yusof, N.H.H. Kamis, S.R.M. Shaleh, Optimization of the ionic liquid-microwave assisted one-step biodiesel production process from wet microalgal biomass, Energy Convers. Manag. 54 (2018) 1−6.

[146] A. Toledo-Cervantes, G.G. Solórzano, J.E. Campos, M. Martínez-Garcíac, M. Morales, Characterization of *Scenedesmus obtusiusculus* AT-UAM for high-energy molecules accumulation: deeper insight into biotechnological potential of strains of the same species, Biotechnol. Reports 17 (2018) 16−23.

[147] L.S. Pérez-Mora, M.C. Matsudo, E.A. Cezare, G.J.C.M. Carvalho, An investigation into producing *Botryococcus braunii* in a tubular photobioreactor, J. Chem. Technol. Biotechnol. 91 (2016) 3053−3060.

[148] S.M. Tibbetts, W.J. Bjornsson, P.J. McGinn, Biochemical composition and amino acid profiles of Nannochloropsis granulata algal biomass before and after supercritical fluid CO_2 extraction at two processing temperatures, Animal Feed Sci. Technol. 204 (2015) 62−71.

[149] M.S. Rodrigues, Influence of ammonium sulphate feeding time on fed-batch *Arthrospira (Spirulina) platensis* cultivation and biomass composition with and without pH control, Bioresour. Technol. 102 (2011) 6587−6592.

[150] L. Zhang, M. He, J. Liu, The enhancement mechanism of hydrogen photoproduction in *Chlorella protothecoides* under nitrogen limitation and sulfur deprivation, Int. J. Hydrogen Energy. 39 (17) (2014) 8969−8976.

[151] E.W. Becker, Microalgae: Biotechnology and microbiology, Cambridge University Press, Cambridge, Great Britain, 1994, p. 293.

[152] S.M. Renaud, L.V. Thinh, D.L. Parry, The gross composition and fatty acids composition of 18 species of tropical Australian microalgae for possible use in mariculture, Aquacult 170 (1999) 147−159.

[153] A.M.P. Anahas, G. Muralitharan, Central composite design (CCD) optimization of phytohormones supplementation for enhanced cyanobacterial biodiesel production, Renew. Energy 130 (2019) 749−761.

[154] N. Arora, A. Patel, P.A. Pruthi, K.M. Poluri, V. Pruthi, Utilization of stagnant non-potable pond water for cultivating oleaginous microalga *Chlorella minutissima* for biodiesel production, Renew. Energy 126 (2018) 30−37.

CHAPTER 7

Membrane Technology for Microalgae Harvesting

JIA JIA LEAM • MUHAMMAD ROIL BILAD • YUSUF WIBISONO •
MOHD DZUL HAKIM WIRZAL • IFTIKHAR AHMED

INTRODUCTION

Increasing world population is accompanied by an increase in global energy consumption, which brings concern to the energy security. Researchers have been exploring options for utilization of more renewable energy source to replace the depleting and polluting traditional fossil fuels from coal, natural gas, and crude oil. Currently, renewable energy still only covers 13% of the world energy demand, with bioenergy taking up 10% [1]. In terms of global renewable electricity, 24% of world electricity generation is derived from renewable resources, with biomass contributes only 1.9%.

Feedstock materials derived from biomass for renewable energy are classified into first, second, and third generations. The first generation constitutes of food crops, which may disrupt global food security if they are expansively exploited. This feedstock also only yields a low net energy gain [2]. The second generation of feedstock constitutes of the nonfood crops, which are lower in quantity and thus considered as very low in sustainability and demand sophisticated processing technologies [3]. This brings the interest toward microalgae as the third-generation feedstock.

Microalgae biomass is very interesting feedstock. Microalgae are simple multicellular organisms that can grow rapidly even in harsh conditions [4]. Microalgae cells contain high amount of lipids (triacylglycerides), a promising biodiesel feedstock, besides sugars and carbohydrates that can be fermented into bioethanol [5]. Microalgae uptake inorganic carbon source, typically CO_2, for cell building block, and obtain energy for metabolism through photosynthesis to form chemical energy [2]. They can be grown intensively and thus require a small land foot-print for cultivation. Many microalgae species can be cultivated using saline water,

hence they do not compete for other uses of the fresh water that increasingly depleted [5]. The then can be seen as a very promising feedstock for biofuel production. Their fast growing rates independent of the climate also allow them to be harvested all year round [2].

To produce microalgae biomass under controlled environment, the microalgae are initially grown in a photobioreactor (PBR). The cultivation stage can only reach a relatively low final biomass concentration of 0.02%–0.5% [6]. To obtain the processable biomass from the diluted suspension, the biomass broth undergoes dewatering process until reaching concentration range of 15%–20% prior drying, extraction, conversion, and other downstream processes [7]. Basic illustration of microalgae biomass production and processing is shown in Fig. 7.1.

Harvesting of microalgae biomass is a challenge due to their small cell size (typically a few micrometers), good suspension stability (their density is close to that of water), and very diluted nature. Therefore, harvesting processes are often energy intensive and potentially requiring net energy exceeding the chemical energy content of the produced biofuel [8–10], which make them untenable for energy generation purposes. Therefore, to be economically attractive for biofuel production, the production costs and the energy input during harvesting must be substantially reduced [11,12].

In this chapter, recent developments of membrane technology for microalgae harvesting are discussed. It includes two main membrane processes: pressure driven and osmotically driven. The discussion covers material and system developments as well as membrane fouling fundamentals and control. Lastly, future prospects of membrane technology and potential research area are briefly elaborated.

Microalgae Cultivation for Biofuels Production. https://doi.org/10.1016/B978-0-12-817536-1.00007-2

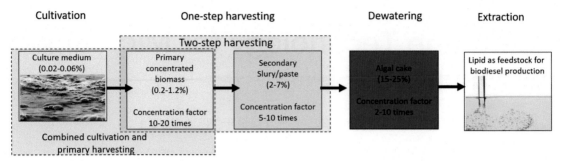

FIG. 7.1 Illustration of production and processing stages of microalgae biomass.

PRESSURE-DRIVEN MEMBRANE PROCESS

Pressure-driven membrane processes, or simply active filtration processes [13], use hydraulic pressure to force mass transport through a semipermeable membrane to separate or concentrate the feed streams [14]. Common pressure-driven membranes include microfiltration (MF), ultrafiltration (UF), nanofiltration (NF), and reverse osmosis (RO) [15]. For application of microalgae harvesting, only the first two are mostly explored.

Implementation

Two harvesting strategies have been reported for microalgae biomass harvesting from a diluted broth medium as illustrated in Fig. 7.2. In the first strategy, the pressure-driven membrane process is used to concentrate the biomass to lower its water content and convert the broth from diluted to more concentrated solution. The membrane facilitates uptake of water as the permeate stream. The concentration method is relatively simple but would still require further concentration process to produce a paste of biomass for drying. In the second strategy, the membrane is used as a medium to develop a biomass cake, which later allows scrapping of the attached biomass from the membrane surface [16]. During the filtration, the biomass is retained and accumulated on the membrane surface. When the cultivating medium is drained, a paste of microalgae biomass can then be scrapped off. This method yields a highly concentrated paste of biomass cake, but the scrapping process makes the operation cycle rather complex.

Another unconventional method is in the form of membrane PBR. This approach combines cultivation and harvesting by introducing a membrane filtration integrated in the cultivation system. The membrane enables to an extent decoupling of the solid retention time and hydraulic retention time and avoids washout. The harvesting step can be assigned in this process by its ability to maintain the microalgae biomass concentration at a high level in the cultivation broth [17,18]. A small increment of biomass concentration in the growth medium would lead to significant reduction of treated volume for the next concentration step (i.e., centrifugation). For example, to obtain 5 wt% biomass paste from 0.01 wt% feed biomass, it requires $500\times$ concentration factor. It decreases to only $250\times$ concentration factor if the feed biomass is at 0.02 wt% concentration.

Microalgae biofilm membrane photobioreactor (BMPBR) has also been explored. BMPBR consists of submerged media in which the microalgae cells are grown and attached. This system exhibits higher volume of microalgae production and nutrients removal rate besides producing high-quality effluent [18]. However, irrespective of the approach, high biomass concentration increases membrane fouling propensity of the feed. Under this condition, the permeate flux decreases as microalgae cake layer builds up at a faster rate, promoting membrane fouling [19]. This aspect is discussed in detail in Section 4.

Both dead end and cross-flow modes have been tested for microalgae concentration with promising results. Dead-end mode works well for feed with low concentration because high concentrations lead to rapid membrane fouling. This process requires as little energy as flocculation or vacuum filtration but still higher than the cross-flow one. The cross-flow filtration is seen as a more promising mode and economically more attractive for initial dewatering due to its ability to achieve higher concentration factors. In the cross-flow filtration, the integrity of microalgae cells can be preserved with minimum damages in motility features and reproductive capabilities, allowing large concentration factors (5−40 times) [13]. On the other hand, increasing cross-flow velocity increases the filtration flux as it is difficult for microalgae to accumulate on the membrane surface due to good local mixing [20].

In the cross-flow mode, transmembrane pressure (TMP) is a sensitive parameter. An increase in TMP

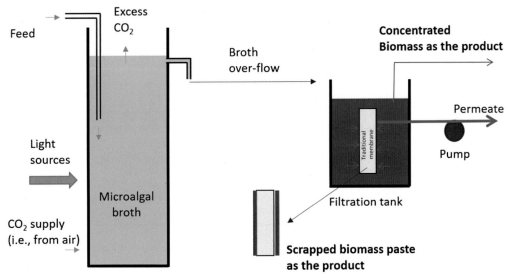

FIG. 7.2 Illustration of pressure-driven membrane process for microalgae biomass harvesting. In the first method, the product is in the form of concentrated biomass, while in the second method, the product is in the form of scrapped biomass cake accumulating on the membrane surface.

also increases the filtration flux due to the high turbulence from higher cross-flow velocities and shear stress, hindering deposition of microalgae on membrane surface. However, further increase in TMP results in a permeance drop due to cake layer formation [21]. Permeance increment can only be achieved by increasing TMP up to certain extent of threshold value. Beyond that threshold value, it does not affect the filtration flux due to excessive exopolymeric organic matters secreted by the microalgae cells at high TMP that worsen the membrane fouling [20].

UF process is more desirable than MF in microalgae harvesting. The larger pores in MF expose them easier to pore blocking. MF is normally attractive for large microalgae, because bigger microalgae size brings more advantage as they are easier to be filtered [13]. The use of MF is more preferred when effective membrane fouling control is available. This way, the advantage of higher permeability of the MF membrane can be maximized.

Developments in Membrane Materials

Selection of membrane materials is important for both MF and UF processes. The criteria for selection include biomass concentration, species characteristics, surface charge, hydrophobicity, and flow parameters of the feed [13]. Some polymer materials have been explored for microalgae harvesting. They include polyvinylidene fluoride (PVDF), polyacrylonitrile (PAN), polyether sulfone (PES), and polysulfone (PS) for MF/UF

and cellulose acetate (CA) and polyamide for NF/RO [7]. In a membrane material screening study for polysulfone (PS), fluoropolymer (PVDF), and regenerated cellulose acetate (RCA), PVDF shows the greatest increase in flux with an increase in applied cross-flow velocity. For all membranes, fouling materials attach weaker on the membrane surface at high applied cross-flow velocity, and PVDF is the most sensitive to the flow hydrodynamic. On the other hand, RCA shows the lowest fouling tendency compared with PS and PVDF due to the hydrophilicity of RCA membranes [21].

Polymeric phase inversion membranes are the most common materials in the pressure-driven processes. Detailed information on these kinds of materials can be found elsewhere [7]. However, recent reports show the emergence of ceramic-based membrane as summarized in Table 7.1. A new type of membrane material in the form of diatomite dynamic membrane (DDM) has recently been developed [22]. It consists of a slime, algae, and diatomite layers formed on the underlying primary filter surface. DDM shows desirable dewatering efficiency, relatively high flux, and high ultimate biomass concentration at a low TMP [22] for a 24 h filtration. The unique structure of DDM protects the membrane against both reversible and irreversible fouling [16]. Nonetheless, such membrane is only effective at a low aeration rate to prevent turbulence from destroying the dynamic cake layer (the construction of the membrane itself). The excellent behavior of DDM

TABLE 7.1
Application and Recent Developments of Ceramic-Based Pressure-Driven Membranes.

Membrane Types	Performance Summary	References
Dynamic membrane	Diatomite dynamic membrane shows high filtration flux, high biomass concentration, and low operation TMP.	[22]
Ceramic membrane	Membranes coated with functional polymer during initiated chemical vapor deposition process have better permeate flux, breakthrough pressure, and separation yield.	[23]
Supported ceramic membrane	A support mixture with 87 wt% kaolin, 10 wt% starch, and 3 wt% sand for porous ceramic membrane produces the best materials for microalgae filtration.	[24]

is due to the high shear stress on the direct surface of the membrane, which results in slower fouling rate.

The traditional ceramic membranes are not desirable for the concentration of microalgae due to their high shrinkage, small pore size, low porosity, high thickness, and poor flexibility [24]. To overcome this, starch was added to kaolin powder in a novel ceramic membrane. The starch addition helps to increase membrane porosity and water permeability, but in expense of slight decreases in the mechanical strength. Mechanical strength can also be improved to such membrane by addition of small amount of sand to the mixture of kaolin powder and starch. A study found that a mixture of 87 wt% kaolin, 10 wt% starch, and 3 wt% sand produced the best ceramic membrane for microalgae concentration [24].

Other membrane development includes the fabrication of electromembrane through modification of carbon electrodes. Such modification enhances harvesting performance in terms of flux, concentration factor, and water removal by 150%. Electromembrane exhibits antifouling properties from the electronic repulsion between the membrane and microalgae cells [25,26]. Researchers have also developed a new type of membrane in which a steel-use-stainless (SUS) is coated with functional polymer via an initiated chemical vapor deposition process. SUS shows better permeate flux, breakthrough pressure (5.96 kPa), and separation yield (95.5 wt%) [23].

Dynamic Filtration System

In addition to the conventional filtration system, recent technology implemented dynamic process in forms of vibration and rotation. Magnetically induced membrane vibrating system (MMV) allows microalgae to

be harvested via microfiltration [27]. MMV consists of a vibration driver, electric wire, vibration engine, and the membrane [28]. The membrane sheet is fixed onto a vibration frame, which is directly connected to an engine. In this system, vibration is generated to enhance shear rates at the liquid-membrane interface. This advancement is economically attractive besides being able to efficiently control fouling and allows the filtration to occur at high fluxes.

Another study shows that a vibratory microfiltration system achieves high permeate fluxes at extended period to reach biomass concentration of as high as 100 g/L. The shear imparted on the system does not have to be at its maximum to enhance the flux [29]. The beauty of this system is on the ability to exert shear on the membrane surface where it is really wanted [28].

Dynamic filtration based on the rotation of perforated disk using UF or MF membranes also increases the shear stress, especially at higher rotating speed. Therefore, it allows membranes to obtain higher permeate flux. The permeate flux reaches a plateau, around 9× to 11× higher than the ones without rotation [30]. A combination of rotational perforated disk with UF membrane for dynamic filtration of *Scenedesmus obliquus* reduces cake layer and pore adsorption, hence lowering the total fouling rate [30]. However, it is important to be noted that the dynamic filtration or mechanically induced filtration requires a more complex filtration system at a higher scale and might increase energy consumption if the cycles period is not properly calculated [31].

OSMOTICALLY DRIVEN MEMBRANES

Osmotically driven membranes utilize osmotic pressure difference between draw and feed solution to drive the

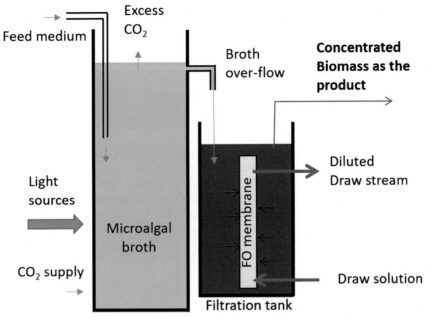

FIG. 7.3 Illustration of microalgae harvesting using forward osmosis process.

transport of water across the membrane. Forward osmosis (FO) is an osmotically driven process that has recently been applied for microalgae harvesting and gains a more attention. Illustration of FO system for microalgae concentration/harvesting is shown in Fig. 7.3. FO transports water through a selective semipermeable membrane from a solution with a lower (microalgae broth) to a higher solute concentration (a concentrated draw solution, or simply seawater) by an osmotic pressure gradient [32]. As illustrated in Fig. 7.3, the water is drawn from the growth medium, resulting in more concentrated biomass. However, when compared with RO, FO still has lower water flux, especially when low draw solution (such as seawater) is used. A reverse draw solute flux presents more prominently in the FO than in the RO. Biofouling potentials occur in both sides of the FO membrane and only the feed side of the RO membranes [33]. As a low-fouling process, FO has been extensively explored for application in microalgae harvesting as summarized in Table 7.2.

TABLE 7.2		
Applications and Recent Developments of Osmotically Driven Membranes.		
Dewatering Process	**Remarks**	**References**
Draw solution versus microalgae species	*Chlorella vulgaris* is the most suitable species to be dewatered via FO with high recoveries of >81% and negligible flux decline regardless of draw solution type.	[37]
Alternative draw solution	Crude glycerol can achieve 2.5-fold concentration factor.	[39]
Membrane orientation	Active layer facing draw solution orientation is more prone from fouling than the active layer facing feed solution.	[35]
Alternative FO membrane	UF/MF-like membranes can be used as osmotically driven process by using poly(sodium 4-styrene-sulfonate) as a draw solution.	[38]

FO, Forward osmosis; *UF,* Ultrafiltration; *MF,* Microfiltration.

Operating Conditions

Important operating conditions in FO include draw solution concentration, cross-flow velocity, membrane orientation, and temperature [32]. There are two membrane orientations: the active layer facing draw solution (AL-DS) and active layer facing feed solution (AL-FS) [34]. The latter is more efficient in resisting membrane fouling and is preferred for microalgae harvesting [35]. When using FO system employing the real sweater as draw solution, it is observed that dewatering rates oscillate throughout the experiments. Long exposure in the ocean (for 45 days) leads to severe biofouling, and at 52 days, the membrane suffers from leakage [36]. The carbohydrate composition of algae cell wall significantly affects the dewatering performance of FO membranes [37].

Recent studies have enabled FO to be done using porous membranes, which have characteristics like UF or MF membranes. It shows a high FO water flux of 7.6 $L/m^2.h$ even at low osmotic driving force of merely 0.11 bar by using 0.1% poly(sodium 4-styrene-sulfonate) as the draw solution [38]. This application is very attractive because of the availability of UF membrane with tunable properties that can be readily implemented. However, the leak of the draw solute into the feed side in the form of reverse salt flux may interrupt the microalgal medium and alter the cultivation condition.

Draw Solutes

Selection of draw solute and microalgae species significantly affects the FO performance. It seems that the draw solute must be optimized for each species, and there is no universal solute that is effective for all species. Application of Ca^{2+} as draw solution for harvesting *S. obliquus* and *Chlamydomonas reinhardtii* leads to severe Ca^{2+} back diffusion causing the species to produce more extracellular carbohydrates. Excretion of extracellular carbohydrate then results in microalgae cells aggregation and eventually decreases the water flux. Microalgae biomass rich in galactose in its cell wall exhibits similar behavior but at a lower extent. The same study also proves that *Chlorella vulgaris* without fructose is the best species to be dewatered via FO. It has negligible flux decline regardless of the type of draw solutes [37]. When crude glycerol was used as draw solution, the microalgal broth could be concentrated into 2.5× in an hour and thus allows substantial reduction of broth volume [39].

MEMBRANE FOULING CONTROL

Membrane fouling is the greatest obstacle during microalgae harvesting using membrane technology. It is a loss of permeability overtime due to the accumulation of deposits on the membrane surface, either externally or internally. It is inevitable in microalgae harvesting due to the high cell density in the feed [40] and due to the high fouling prone nature of the microalgae cell. A low membrane permeability leads to an increase in energy consumption, which is commercially unfavorable [41]. A summary on the membrane fouling control in microalgae harvesting is presented in Table 7.3.

Types of Fouling

Depending on the cleanability, there are two types of fouling: reversible and irreversible fouling [15]. Reversible foulant loosely attaches to the membrane surface and thus effectively be removed via physical cleaning. Foulant tightly bounded to the membrane, which can only be removed via chemical cleaning, is categorized as irreversible [42]. Reversible fouling reduces productivity and increases operational cost, while irreversible fouling increases operational complexity as well as reduces membrane life span [43].

Fouling can also be classified into internal fouling, pore blocking, and cake layer formation based on its location on the membrane material. Internal fouling occurs due to foulant penetration into the membrane pores, while cake layer is formed from the accumulation of foulants on the membrane surface [7]. The earliest stage of fouling is referred to as concentration polarization, in which a gradient of excluded products forms near the membrane surface [44].

Fouling can also be grouped based on the type of foulants: biofouling, organic, and inorganic. Biofouling comprises living cells, while organic fouling and inorganic fouling are due to the deposition of organic and inorganic materials, respectively [7]. Common foulants include proteins, polysaccharides, bacteria, and microalgae cells [45]. Hydrophobic organic foulants are removed easily via chemical cleaning, but hydrophilic organic foulants still stick strongly onto the membrane surface [46].

Membrane Fouling Factors in Pressure-Driven Processes

Internal fouling in MF and UF is caused by different factors. Fouling in MF is predominantly a result of monolayer adsorption of biopolymers, while fouling in UF is mostly due to the monolayer adsorption of humic substances [47]. Different algogenic organic matters (AOMs) also have different fouling propensities as shown from combinations of fluorescence excitation-emission matrix. Membrane fouling is also affected by

TABLE 7.3
Types, Factors, and Methods of Membrane Fouling Mitigating.

Factors	Remarks	References
Type	Depending on cleanability: Reversible: Loosely attached to the membrane and can be removed via physical cleaning. Irreversible: Tightly bounded to the membrane that can only be removed via chemical cleaning.	[42]
	Depending on foulant location: internal fouling, pore blocking and cake layer.	[7]
	Depending on materials: Biofouling is caused by the deposition of living cells, organic fouling is due to the deposition of organics, and inorganic fouling is due to the deposition of inorganics.	[7]
Fouling in FO	Fouling in an FO is divided into three stages. Stage 1: adhesion between foulants and a clean membrane. Stage 2: formation of foulant layer on the membrane surface. Stage 3: formation of irreversible fouling.	[56]
	Organic fouling is greater in FO membranes, but the high cross-flow velocity in FO system can restore the initial flux better than RO.	[45]
Factors affecting fouling	Membrane pore size: more irreversible fouling to membranes with looser and opener pores.	[47]
	Pressure: high transmembrane pressure causes a rapid formation of foulant layer on the external surface of the membrane, while low TMP increases the degree of internal fouling.	[51]
	Cross-flow velocity: high value decreases microalgae cells depositing on membrane surface.	[20]
	Temperature: 35°C is optimum for filtration but less so for cultivation.	[61]
Mitigating membrane fouling	Important parameters: temperature, pressure, time, pH, and surface area of the membrane.	[59]
	Surface modification: implantation of polar organic functional groups via plasma treatment, surface coating, surface blending, and surface grafting.	[54]
	Turbulence promoter: spacers can be placed in between cross-flow filtration membranes.	[72]
	Vibration: axial vibration helps to impede the deposition of microalgae cells.	[78,79]
	Physical cleaning: relaxation, backflushing, and ultrasonication.	[62]
	Chemical cleaning: sodium hypochlorite, hydrochloric acid, nitric acid, and sodium hydroxide.	

FO, Forward osmosis; *RO*, Reverse osmosis; *TMP*, Transmembrane pressure.

the sizes of the AOMs and/or membrane material surface chemistry. In a recent study, the strongest flux decline is observed in cyanobacteria of *Aphanizomenon flosaquae* followed by *Anabaena flosaquae* and *Microcystis aeruginosa* [48]. Natural organic matters (NOMs) are also significant contributors to membrane fouling such as humic acid, which is notable one.

When comparing MF and UF, the former exhibits a faster fouling rate because of its larger pores. Large pore sizes of MFs are more vulnerable from blocking and clogging by the microalgae cells [20]. Higher rate of irreversible fouling is generally expected in

membranes with more open pores, as more foulant penetrates into the pores [47]. However, this trend is not universal. A report on dewatering of *Chlorella pyrenoidosa* at low fluxes reveals that for membrane with higher pore size, less microalgae biomass is adsorbed on the membrane surface due to the low permeate flow drag force. As a result, less fouling occurs [49]. For membranes with similar pore size, membranes with greater pore density experienced lesser total resistance and smaller fraction of irreversible fouling because of their lower local flux and hence lower convective drag force [50].

Studies show that high TMP causes a rapid formation of external foulant layer, while low TMP increases the degree of internal fouling due to the penetration of smaller foulants into the membrane pores [51]. An increase in cross-flow velocity in the feed side decreases microalgae cells deposition and hence increases the flux [20]. At constant TMP, membranes show similar fouling behavior regardless of the flux. In addition, at a flux above the threshold flux, fouling becomes severe [52].

The important role of flow hydrodynamics is the basis for development of corrugated membrane. In a prism-patterned surface, fouling is predominant on the valley of the pattern and less on the hills. Formation of vortex on the valleys causes solid aggregation, which favors fouling. Local wall shear stress, which is higher on the hills part due to the vortex, eases detachment and prohibits attachment of foulant materials [53].

Electrostatic interactions have been found to be important fouling factor. Based on the DLVO theory, particle and surface with opposite charges exhibits attraction force, while particle and surface with the same charge forms repulsive force, which is desirable to minimize fouling [54]. Increasing the particle surface charges in the feed also enhances the specific resistance and compressibility of the cake. This lowers the rate of cake layer formation, and vice versa [55].

Fouling Factors in Forward Osmosis Process

Fouling in an FO process constitutes of three consecutive stages. The first stage involves the adhesion between foulants and a clean membrane, which significantly lowers the permeability. In the second stage, foulant layer is formed by depositing foulants. This is promoted by van der Waals interaction between foulants and the rough membrane surface and drag force due to water transport toward the membrane surface. The last stage is basically the consequences of the first two, which is continuous permeability drops [56].

FO experiences lesser biofouling than RO, and when membrane fouling exists, it is more reversible. FO membrane has a greater vulnerability from organic fouling. However, high cross-flow velocity in FO system is able to restore the initial flux better than RO. Physical cleaning is effective to remove organic and inorganic fouling in FO by a simple osmotic backwashing, while chemical cleaning using chlorine is proven to be effective to control biofouling [45].

Studies have shown that membrane fouling is less affected by osmotic pressure in FO but is highly influenced by the feed salinity. High feed salinity leads to excessive secretion of extracellular polymeric substances (EPS) under high salt stress and the change in hydrophobicity of the feed [57]. Increasing salt concentration promotes pore blocking, which eventually increases membrane fouling [58].

Membrane fouling in FO depends on the type of salts in draw and feed solutions. Draw solution containing Mg^{2+} ions causes more severe and irreversible fouling due to the reverse diffusion of the ions into the feed solution [59], while microalgae with larger concentration of EOM have higher degrees of fouling [41].

Temperature also plays part in membrane fouling of a FO process. At higher temperature, the increase in osmotic pressure leads to greater net driving force, and this improves the water flux [60]. A temperature of 35°C leads to the best filtration performance with least fouling as shown by a study using PVDF MF membrane for harvesting of C. pyrenoidosa. However, high temperature lowers the rate of microalgae growth. The optimum solution is then to maintain the temperature at a range of 25–30°C for microalgae cultivation but carries out filtration at 35°C for best results [61].

Mitigation of Membrane Fouling

Controlling membrane fouling can be done by feed pretreatments, optimization of operational conditions, modifying membrane, and conducting maintenance and intensive membrane cleaning [62]. Pretreatment of feed deals with removal of NOMs. Oxidants such as sodium hypochlorite (NaOCl) change the molecular characteristics of NOM during feed peroxidation and have been tried to mitigate membrane fouling [62–64]. Feed can also be perchlorinated to influence the molecular weight distribution of NOM by breaking down the molecules into substances with lower molecular weight due to oxidation [65]. These compounds hence do not attach to the membrane because of their reduced chemical affinity to the membrane surface [19,56].

Operational conditions that affect membrane fouling include temperature, pressure, time, and pH. Adjustment of pH to 5.5 inhibits scaling, but fouling still persists [66]. Extreme pH in the feed is not desirable as it damages membrane material and reduces membrane life span and permeability [54]. High pH causes additives in the membrane to swell because of high repulsive force. This then alters membrane structure and lowers the pore size [67]. An increase in ionic strength in the feed solution does not affect the fouling propensity of membrane [68]. On the other hand, hydraulic pressure only slightly affects the fouling rate but greatly affects the foulant layer reversibility. Hydraulic pressure compacts the foulant cake layer due to permeate drag force combined with hydraulic pressure. This eventually leads to an increase in the fouling irreversibility [69].

Membrane orientation in FO also affects membrane fouling. The AL-DS mode exhibits rapid flux decline and lower flux recovery due to internal membrane fouling, which includes internal pore clogging, adsorption, and internal concentration polarization in the support layer [70]. The AL-FS mode has greater flux recovery when cleaned physically and chemically because no internal clogging occurs as foulants accumulate as cake on the surface of the tight active layer.

Minimizing cake formation can be done by increasing the tangential flow velocity in a pressure-driven membrane, but this requires higher pump energy, which results in an increase of production costs [71]. For biofouling control, feed spacers can be introduced in the flow channel. An optimum feed spacer allows good shear enhancement on the membrane surface and drag over the spacer [72]. In a study on the effect of feed spacers thicknesses (28, 31, and 46 mm), the 46 mm spacer shows the best filtration performance, proving that a thicker spacer reduces the effect of biofouling on membranes. This can be explained by mixing effect induced by the spacer and the formation of a less compact biofilm when a thick spacer is used, hence resulting in a lower flux decline by reducing intracranial pressure [33].

Another most common approach to tackle membrane fouling—related issue is through innovative developments of membrane materials through exploiting preparation parameters and through performing surface modifications. Surface modification can be done by implantation of polar organic functional groups on the membrane surface through plasma treatment, surface coating, surface blending, and surface grafting [62]. Through surface modification, the hydrophilicity of membranes tremendously increases, hence improving their performance. Ultraviolet (UV)-initiated graft polymerization is used to develop membranes with enhanced resistance toward fouling. It helps to alter colloidal interactions and membrane surface chemistry. With the incorporation of TiO_2 nanoparticles along with UV radiation, membranes obtain photocatalytic properties to decompose foulant deposited on the membrane surfaces [73]. Coating porous FO membranes with silver (Ag) and TiO_2 nanoparticles shows significant antibacterial effect by enhancing surface hydrophilicity, which is almost 11 times higher than uncoated membranes. Coated membranes exhibit 67%–72% of initial flux recovery, twice of the reference. This proves that the coated membrane is easier to clean after fouling due to the alteration of interfacial interactions between the membrane and fouling biofilms, leading to improved fouling resistance of the FO membranes [74]. Surface charges also found promoted different biofouling adherence rate on membrane surface. Depending on the charge of microorganisms attached to the membrane surface, surface charge modification of the membrane could prevent a more attractive biofouling control [75]. Another method of mitigating fouling in FO membrane is via the incorporation of $CaCO_3$ nanoparticles as sacrificial medium during the fabrication of PSf substrate. This technique results in membrane with great porosity and smaller structural parameter, allowing internal concentration polarization to be reduced through etching [76]. Application of new type of filter in the form of nanofiber has also shown some promise as a fouling-resistant material for microalgae harvesting [77].

Membrane modification can also be done by integrating membrane filtration with electropolarization using an electroconductive nanocarbon. It shows greater permeate flux compared with unpolarized membrane. After filtration, permeate flux of the cathodic polarization, the anodic polarization, and the alternating polarization are approximately 2.9×, 6.7×, and 8.1× higher than ones without electropolarization, respectively. These results can be explained by the electrochemical sterilization process during the electropolarization. Without living bacteria, suspended cells cannot be adhered to the membrane surface, potentially reducing membrane fouling [25].

The axial vibration membrane (AVM) is better than the traditional submerged aerated membrane in mitigating membrane fouling in the pressure-driven membrane process. Vibration is efficient to impede the deposition of microalgae cells [78]. AVM increases shear rate, which is determined by frequency and amplitude of vibration. The generated shear rate decreases membrane fouling and increases critical flux. At high shear rate, a long-range repulsive force prevents deposition of microalgae cells. While at low shear rate, a long-range attractive force draws microalgae to the membrane. At small amplitudes, the AVM system produces enough shear force for membrane fouling control without affecting the microalgae. However, this system is still prone from adhesion of EOM and be deposited into the pores to promote irreversible fouling. It happens because vibration is unable to remove the deposited microalgae on the membrane as inertial lift force is lesser than the attractive force [79].

Physical and chemical cleanings are the most common approach for maintaining performance of a membrane process. The former include aeration, membrane relaxation, backflushing, and ultrasonication. The latter is a more effective way to clean irreversible foulant [54,62]. When a bubble generator plate is placed underneath the membrane sheet, great hydrodynamic power efficiently reduces membrane fouling [80]. Aeration and microbubbles are possible cleaning techniques, but microbubbles are proven to exhibit greater impact for MF especially when combined with panel design in the form of tilted panel system [81]. They are effective specifically for cleaning EPS and soluble microbial products [82]. For feed with high biomass concentration (>0.6 g/L), aeration is inefficient for fouling control due to the quick formation of cake layer and high degree of concentration polarization [83]. Ultrasound can clean membranes by producing acoustic and microstreaming, microstreamers, microjets, and shock waves. These waves help to control fouling by both preventing the deposition of microalgae on the membrane and removing the layer of microalgae adhered onto the membrane. In ultrasonic cleaning, low frequency (35 kHz) was found better than a high frequency (130 kHz) in slowing down the rate of fouling. Low-frequency sonication generates bigger sized cavitation bubbles at long rarefaction cycles of sonic waves. Higher frequency (130 kHz) was found effective at removing NOMs [84].

Membrane relaxation is less efficient in removing the cake layer foulant. Cake layer removal can be achieved by a combination of backwash and high cross-flow shear. Nonetheless, a thin layer remains on the membrane surface [55]. The optimum relaxation time is normally feed-to-membrane specific. As an example, 0.2−4 min relaxation time was found effective in a recent study [85]. In addition to scouring-off foulant, excessive backwashing can alter the intrinsic membrane properties by enlarging the pore size.

Chemicals such as NaOCl, hydrochloric acid, nitric acid, and sodium hydroxide are commonly used as membrane cleaning agents [54]. Among sodium hydroxide, sodium hypochlorite, sulfuric acid, nitric acid, citric acid, and oxalic acid, the latest exhibits the best cleaning efficacy. Typically, until a certain extent, the longer the cleaning duration, the greater the flux recovery [46]. Alkaline agents are better at removing foulant layer than acidic agents with NaOCl showing the best cleaning effect followed by NaOH [86]. However, NaOCl can swell and solubilize protein besides — most importantly — breaking up the adhesion between foulant and membrane. Excessive use of NaOCl also deteriorates membrane stability [87] and causes bacteria lysis [88]. Acidic agents have relatively poor efficacy with only 68% of flux recovery [86].

Chemical-enhanced backwashing (i.e., using NaOH) is effective in reducing irreversible and total fouling compared with pure water backwashing. A study using NaOH-assisted backwashing reveals its effectiveness in enhancing the detachment of organic components from the fouled membranes due to hydraulic actions and chemical alterations [89].

FUTURE PROSPECT OF MEMBRANE TECHNOLOGY

Recent advancements have brought significant improvements to the membrane technology in microalgae dewatering. The most common pressure-driven membranes are MF and UF. In addition to the phase-inverted membrane, unconventional membranes in forms of dynamic and ceramic materials have also been explored and show promising results.

FO has extensively been explored recently for microalgae harvesting. Up to date, the research still focuses on proof-of-concept and the feasibility of FO. Exploration of operational condition such as membrane orientation, hydrodynamics, selection of draw solution, on other parameters has been reported. A research must be focused in the context of osmotic concentration to avoid draw solution recovery.

Research on application of pressure and osmotically driven membrane processes for microalgae harvesting is mostly in lab scale. Despite showing promise as energy-efficient technology for microalgae harvesting, only limited study reports on its scale-up performance. Good lab-scale performance does not guarantee similar performance in larger scale. A recent report found encouraging finding on implementation of pressure-driven membrane process for microalgae harvesting [90]. However, the pilot data were only used as a basis for large-scale projection. Therefore, study in a larger scale membrane-based microalgae harvesting is required and thus should become the focus for the near future. The ultimate good is to have a solid proof that demonstrates a net energy balance. Else, focus must be shifted toward other application that can offer economic sustainability, such as for aquaculture feed [91].

REFERENCES

[1] D.P. Ho, H.H. Ngo, W. Guo, A mini review on renewable sources for biofuel, Bioresour. Technol. 169 (2014) 742−749.

[2] P.C. Hallenbeck, M. Grogger, M. Mraz, D. Veverka, Solar biofuels production with microalgae, Appl. Energy 179 (2016) 136−145.

[3] I. Rawat, R. Ranjith Kumar, T. Mutanda, F. Bux, Biodiesel from microalgae: a critical evaluation from laboratory to large scale production, Appl. Energy 103 (2013) 444−467.

[4] T.M. Mata, A.A. Martins, N.S. Caetano, Microalgae for biodiesel production and other applications: a review, Renew. Sustain. Energy Rev. 14 (2010) 217−232.

[5] M.A. Borowitzka, N.R. Moheimani, Sustainable biofuels from algae, Mitig. Adapt. Strateg. Glob. Change. 18 (2013) 13−25.

[6] L. Brennan, P. Owende, Biofuels from microalgae—a review of technologies for production, processing, and

extractions of biofuels and co-products, Renew. Sustain. Energy Rev. 14 (2010) 557–577.

[7] M.R. Bilad, H.A. Arafat, I.F.J. Vankelecom, Membrane technology in microalgae cultivation and harvesting: a review, Biotechnol. Adv. 32 (2014) 1283–1300.

[8] Y. Chisti, Biodiesel from microalgae, Biotechnol. Adv. 25 (2007) 294–306.

[9] N. Uduman, Y. Qi, M.K. Danquah, G.M. Forde, A. Hoadley, Dewatering of microalgal cultures: a major bottleneck to algae-based fuels, J. Renew. Sustain. Energy 2 (2010) 012701.

[10] H.C. Greenwell, L.M.L. Laurens, R.J. Shields, R.W. Lovitt, K.J. Flynn, Placing microalgae on the biofuels priority list: a review of the technological challenges, J. R. Soc. Interface 7 (2010) 703–726.

[11] R. Harun, M. Singh, G.M. Forde, M.K. Danquah, Bioprocess engineering of microalgae to produce a variety of consumer products, Renew. Sustain. Energy Rev. 14 (2010) 1037–1047.

[12] M.R. Tredici, Photobiology of microalgae mass cultures: understanding the tools for the next green revolution, Biofuels 1 (2010) 143–162.

[13] I.L.C. Drexler, D.H. Yeh, Membrane applications for microalgae cultivation and harvesting: a review, Rev. Environ. Sci. Biotechnol. 13 (2014) 487–504.

[14] G. Keir, V. Jegatheesan, A review of computational fluid dynamics applications in pressure-driven membrane filtration, Rev. Environ. Sci. Biotechnol. 13 (2014) 183–201.

[15] M.R. Bilad, Membrane bioreactor for domestic wastewater treatment: principles, challenges and future research directions, Indones. J. Sci. Technol. 2 (2017) 97.

[16] Y. Zhang, Y. Zhao, H. Chu, X. Zhou, B. Dong, Dewatering of *Chlorella pyrenoidosa* using diatomite dynamic membrane: filtration performance, membrane fouling and cake behavior, Colloids Surf. B Biointerfaces. 113 (2014) 458–466.

[17] M.R. Bilad, V. Discart, D. Vandamme, I. Foubert, K. Muylaert, I.F.J. Vankelecom, Coupled cultivation and pre-harvesting of microalgae in a membrane photobioreactor (MPBR), Bioresour. Technol. 155 (2014) 410–417.

[18] F. Gao, Z.-H. Yang, C. Li, G.-M. Zeng, D.-H. Ma, L. Zhou, A novel algal biofilm membrane photobioreactor for attached microalgae growth and nutrients removal from secondary effluent, Bioresour. Technol. 179 (2015) 8–12.

[19] N. Javadi, F. Zokaee Ashtiani, A. Fouladitajar, A. Moosavi Zenooz, Experimental studies and statistical analysis of membrane fouling behavior and performance in microfiltration of microalgae by a gas sparging assisted process, Bioresour. Technol. 162 (2014) 350–357.

[20] H. Elcik, M. Cakmakci, B. Ozkaya, The fouling effects of microalgal cells on crossflow membrane filtration, J. Membr. Sci. 499 (2016) 116–125.

[21] X. Sun, C. Wang, Y. Tong, W. Wang, J. Wei, Microalgae filtration by UF membranes: influence of three

membrane materials, Desalin. Water Treat. 52 (2014) 5229–5236.

[22] H. Chu, Y. Zhao, Y. Zhang, L. Yang, Dewatering of Chlorella pyrenoidosa using a diatomite dynamic membrane: characteristics of a long-term operation, J. Membr. Sci. 492 (2015) 340–347.

[23] M.J. Kwak, Y. Yoo, H.S. Lee, J. Kim, J.-W. Yang, J.-I. Han, S.G. Im, J.-H. Kwon, A simple, cost-efficient method to separate microalgal lipids from wet biomass using surface energy-modified membranes, ACS Appl. Mater. Interfaces 8 (2016) 600–608.

[24] M. Issaoui, L. Limousy, B. Lebeau, J. Bouaziz, M. Fourati, Manufacture and optimization of low-cost tubular ceramic supports for membrane filtration: application to algal solution concentration, Environ. Sci. Pollut. Res. 24 (2017) 9914–9926.

[25] X. Fan, H. Zhao, X. Quan, Y. Liu, S. Chen, Nanocarbon-based membrane filtration integrated with electric field driving for effective membrane fouling mitigation, Water Res. 88 (2016) 285–292.

[26] D.-Y. Kim, T. Hwang, Y.-K. Oh, J.-I. Han, Harvesting *Chlorella* sp. KR-1 using cross-flow electro-filtration, Algal Res. 6 (2014) 170–174.

[27] M.R. Bilad, V. Discart, D. Vandamme, I. Foubert, K. Muylaert, I.F.J. Vankelecom, Harvesting microalgal biomass using a magnetically induced membrane vibration (MMV) system: filtration performance and energy consumption, Bioresour. Technol. 138 (2013) 329–338.

[28] M.R. Bilad, G. Mezohegyi, P. Declerck, I.F.J. Vankelecom, Novel magnetically induced membrane vibration (MMV) for fouling control in membrane bioreactors, Water Res. 46 (2012) 63–72.

[29] C.S. Slater, M.J. Savelski, P. Kostetskyy, M. Johnson, Shear-enhanced microfiltration of microalgae in a vibrating membrane module, Clean Technol. Environ. Policy 17 (2015) 1743–1755.

[30] K. Kim, J.-Y. Jung, H. Shin, S.-A. Choi, D. Kim, S.C. Bai, Y.K. Chang, J.-I. Han, Harvesting of Scenedesmus obliquus using dynamic filtration with a perforated disk, J. Membr. Sci. 517 (2016) 14–20.

[31] M.Y. Jaffrin, Dynamic shear-enhanced membrane filtration: a review of rotating disks, rotating membranes and vibrating systems, J. Membr. Sci. 324 (2008) 7–25.

[32] M. Shibuya, M. Yasukawa, T. Takahashi, T. Miyoshi, M. Higa, H. Matsuyama, Effects of operating conditions and membrane structures on the performance of hollow fiber forward osmosis membranes in pressure assisted osmosis, Desalination 365 (2015) 381–388.

[33] R. Valladares Linares, S.S. Bucs, Z. Li, M. AbuGhdeeb, G. Amy, J.S. Vrouwenvelder, Impact of spacer thickness on biofouling in forward osmosis, Water Res. 57 (2014) 223–233.

[34] M.R. Bilad, L. Qing, A.G. Fane, Non-linear least-square fitting method for characterization of forward osmosis membrane, J. Water Process Eng. 25 (2018) 70–80.

[35] M. Larronde-Larretche, X. Jin, Microalgae (*Scenedesmus obliquus*) dewatering using forward osmosis membrane:

influence of draw solution chemistry, Algal Res. 15 (2016) 1–8.

[36] P. Buckwalter, T. Embaye, S. Gormly, J.D. Trent, Dewatering microalgae by forward osmosis, Desalination 312 (2013) 19–22.

[37] M. Larronde-Larretche, X. Jin, Microalgal biomass dewatering using forward osmosis membrane: influence of microalgae species and carbohydrates composition, Algal Res. 23 (2017) 12–19.

[38] S. Qi, Y. Li, Y. Zhao, W. Li, C.Y. Tang, Highly efficient forward osmosis based on porous membranes—applications and implications, Environ. Sci. Technol. 49 (2015) 4690–4695.

[39] T. Mazzuca Sobczuk, M.J. Ibáñez González, E. Molina Grima, Y. Chisti, Forward osmosis with waste glycerol for concentrating microalgae slurries, Algal Res. 8 (2015) 168–173.

[40] T. Hwang, S.-J. Park, Y.-K. Oh, N. Rashid, J.-I. Han, Harvesting of *Chlorella* sp. KR-1 using a cross-flow membrane filtration system equipped with an antifouling membrane, Bioresour. Technol. 139 (2013) 379–382.

[41] F. Qu, H. Liang, J. Zhou, J. Nan, S. Shao, J. Zhang, G. Li, Ultrafiltration membrane fouling caused by extracellular organic matter (EOM) from Microcystis aeruginosa: effects of membrane pore size and surface hydrophobicity, J. Membr. Sci. 449 (2014) 58–66.

[42] K. Kimura, K. Tanaka, Y. Watanabe, Microfiltration of different surface waters with/without coagulation: clear correlations between membrane fouling and hydrophilic biopolymers, Water Res. 49 (2014) 434–443.

[43] J. Tian, M. Ernst, F. Cui, M. Jekel, Correlations of relevant membrane foulants with UF membrane fouling in different waters, Water Res. 47 (2013) 1218–1228.

[44] H. Norafifah, M.Y. Noordin, K.Y. Wong, S. Izman, A.A. Ahmad, A study of operational factors for reducing the fouling of hollow fiber membranes during wastewater filtration, Procedia CIRP 26 (2015) 781–785.

[45] H. Yoon, Y. Baek, J. Yu, J. Yoon, Biofouling occurrence process and its control in the forward osmosis, Desalination 325 (2013) 30–36.

[46] Y.C. Woo, J.J. Lee, L.D. Tijing, H.K. Shon, M. Yao, H.-S. Kim, Characteristics of membrane fouling by consecutive chemical cleaning in pressurized ultrafiltration as pre-treatment of seawater desalination, Desalination 369 (2015) 51–61.

[47] R. Shang, F. Vuong, J. Hu, S. Li, A.J.B. Kemperman, K. Nijmeijer, E.R. Cornelissen, S.G.J. Heijman, L.C. Rietveld, Hydraulically irreversible fouling on ceramic MF/UF membranes: comparison of fouling indices, foulant composition and irreversible pore narrowing, Sep. Purif. Technol. 147 (2015) 303–310.

[48] W. Huang, H. Chu, B. Dong, Fouling and cake behavior of algal organic foulants on microfiltration membranes in various growth phases, Clean. Soil Air Water 44 (2016) 1661–1671.

[49] F. Zhao, H. Chu, Z. Yu, S. Jiang, X. Zhao, X. Zhou, Y. Zhang, The filtration and fouling performance of membranes with different pore sizes in algae harvesting, Sci. Total Environ. 587–588 (2017) 87–93.

[50] E. Kanchanatip, B.-R. Su, S. Tulaphol, W. Den, N. Grisdanurak, C.-C. Kuo, Fouling characterization and control for harvesting microalgae *Arthrospira* (Spirulina) maxima using a submerged, disc-type ultrafiltration membrane, Bioresour. Technol. 209 (2016) 23–30.

[51] I. Rosas, S. Collado, A. Gutiérrez, M. Díaz, Fouling mechanisms of *Pseudomonas putida* on PES microfiltration membranes, J. Membr. Sci. 465 (2014) 27–33.

[52] D.J. Miller, S. Kasemset, D.R. Paul, B.D. Freeman, Comparison of membrane fouling at constant flux and constant transmembrane pressure conditions, J. Membr. Sci. 454 (2014) 505–515.

[53] Y.K. Lee, Y.-J. Won, J.H. Yoo, K.H. Ahn, C.-H. Lee, Flow analysis and fouling on the patterned membrane surface, J. Membr. Sci. 427 (2013) 320–325.

[54] T. Lin, Z. Lu, W. Chen, Interaction mechanisms and predictions on membrane fouling in an ultrafiltration system, using the XDLVO approach, J. Membr. Sci. 461 (2014) 49–58.

[55] S. Lorenzen, Y. Ye, V. Chen, M.L. Christensen, Direct observation of fouling phenomena during cross-flow filtration: influence of particle surface charge, J. Membr. Sci. 510 (2016) 546–558.

[56] L. Li, X. Liu, H. Li, A review of forward osmosis membrane fouling: types, research methods and future prospects, Environ. Technol. Rev. 6 (2017) 26–46, https://doi.org/10.1080/21622515.2016.1278277.

[57] G. Qiu, Y.-P. Ting, Direct phosphorus recovery from municipal wastewater via osmotic membrane bioreactor (OMBR) for wastewater treatment, Bioresour. Technol. 170 (2014) 221–229.

[58] D. Jang, Y. Hwang, H. Shin, W. Lee, Effects of salinity on the characteristics of biomass and membrane fouling in membrane bioreactors, Bioresour. Technol. 141 (2013) 50–56.

[59] S. Zou, Y.-N. Wang, F. Wicaksana, T. Aung, P.C.Y. Wong, A.G. Fane, C.Y. Tang, Direct microscopic observation of forward osmosis membrane fouling by microalgae: critical flux and the role of operational conditions, J. Membr. Sci. 436 (2013) 174–185.

[60] J. Heo, K.H. Chu, N. Her, J. Im, Y.-G. Park, J. Cho, S. Sarp, A. Jang, M. Jang, Y. Yoon, Organic fouling and reverse solute selectivity in forward osmosis: role of working temperature and inorganic draw solutions, Desalination 389 (2016) 162–170.

[61] H. Chu, F. Zhao, X. Tan, L. Yang, X. Zhou, J. Zhao, Y. Zhang, The impact of temperature on membrane fouling in algae harvesting, Algal Res. 16 (2016) 458–464.

[62] H. Lin, W. Peng, M. Zhang, J. Chen, H. Hong, Y. Zhang, A review on anaerobic membrane bioreactors: applications, membrane fouling and future perspectives, Desalination 314 (2013) 169–188.

[63] V. Discart, M.R. Bilad, L. Marbelia, I.F.J. Vankelecom, Impact of changes in broth composition on *Chlorella vulgaris* cultivation in a membrane photobioreactor (MPBR) with permeate recycle, Bioresour. Technol. 152 (2014) 321–328.

[64] V. Discart, M.R. Bilad, D. Vandamme, I. Foubert, K. Muylaert, I.F.J. Vankelecom, Role of transparent exopolymeric particles in membrane fouling: Chlorella vulgaris broth filtration, Bioresour. Technol. 129 (2013) 18–25.

[65] V. Discart, M.R. Bilad, I.F.J. Vankelecom, Critical evaluation of the determination methods for transparent exopolymer particles, agents of membrane fouling, Crit. Rev. Environ. Sci. Technol. 45 (2015) 167–192.

[66] S.C. Chen, G.L. Amy, T.-S. Chung, Membrane fouling and anti-fouling strategies using RO retentate from a municipal water recycling plant as the feed for osmotic power generation, Water Res. 88 (2016) 144–155.

[67] D. Zhang, K.Y. Fung, K.M. Ng, Novel filtration photobioreactor for efficient biomass production, Ind. Eng. Chem. Res. 53 (2014) 12927–12934.

[68] M.M. Motsa, B.B. Mamba, A. D'Haese, E.M.V. Hoek, A.R.D. Verliefde, Organic fouling in forward osmosis membranes: the role of feed solution chemistry and membrane structural properties, J. Membr. Sci. 460 (2014) 99–109.

[69] M. Xie, J. Lee, L.D. Nghiem, M. Elimelech, Role of pressure in organic fouling in forward osmosis and reverse osmosis, J. Membr. Sci. 493 (2015) 748–754.

[70] R. Honda, W. Rukapan, H. Komura, Y. Teraoka, M. Noguchi, E.M.V. Hoek, Effects of membrane orientation on fouling characteristics of forward osmosis membrane in concentration of microalgae culture, Bioresour. Technol. 197 (2015) 429–433.

[71] M.L. Gerardo, D.L. Oatley-Radcliffe, R.W. Lovitt, Integration of membrane technology in microalgae biorefineries, J. Membr. Sci. 464 (2014) 86–99.

[72] X. Liu, W. Li, T.H. Chong, A.G. Fane, Effects of spacer orientations on the cake formation during membrane fouling: quantitative analysis based on 3D OCT imaging, Water Res. 110 (2017) 1–14.

[73] V. Kochkodan, D.J. Johnson, N. Hilal, Polymeric membranes: surface modification for minimizing (bio) colloidal fouling, Adv. Colloid Interface Sci. 206 (2014) 116–140.

[74] A. Nguyen, L. Zou, C. Priest, Evaluating the antifouling effects of silver nanoparticles regenerated by TiO2 on forward osmosis membrane, J. Membr. Sci. 454 (2014) 264–271.

[75] Y. Wibisono, W. Yandi, M. Golabi, R. Nugraha, E.R. Cornelissen, A.J.B. Kemperman, T. Ederth, K. Nijmeijer, Hydrogel-coated feed spacers in two-phase flow cleaning in spiral wound membrane elements: a novel platform for eco-friendly biofouling mitigation, Water Res. 71 (2015) 171–186.

[76] W. Kuang, Z. Liu, G. Kang, D. Liu, M. Zhou, Y. Cao, Thin film composite forward osmosis membranes with poly(2-hydroxyethyl methacrylate) grafted nano-TiO2 as additive in substrate, J. Appl. Polym. Sci. 133 (2016).

[77] M.R. Bilad, A.S. Azizo, M.D.H. Wirzal, L. Jia Jia, Z.A. Putra, N.A.H.M. Nordin, M.O. Mavukkandy, M.J.F. Jasni, A.R.M. Yusoff, Tackling membrane fouling in microalgae filtration using nylon 6,6 nanofiber membrane, J. Environ. Manage. 223 (2018) 23–28.

[78] F. Zhao, H. Chu, Y. Zhang, S. Jiang, Z. Yu, X. Zhou, J. Zhao, Increasing the vibration frequency to mitigate reversible and irreversible membrane fouling using an axial vibration membrane in microalgae harvesting, J. Membr. Sci. 529 (2017) 215–223.

[79] F. Zhao, H. Chu, X. Tan, Y. Zhang, L. Yang, X. Zhou, J. Zhao, Comparison of axial vibration membrane and submerged aeration membrane in microalgae harvesting, Bioresour. Technol. 208 (2016) 178–183.

[80] T. Hwang, Y.-K. Oh, B. Kim, J.-I. Han, Dramatic improvement of membrane performance for microalgae harvesting with a simple bubble-generator plate, Bioresour. Technol. 186 (2015) 343–347.

[81] A. Eliseus, M.R. Bilad, N.A.H.M. Nordin, Z.A. Putra, M.D.H. Wirzal, Tilted membrane panel: a new module concept to maximize the impact of air bubbles for membrane fouling control in microalgae harvesting, Bioresour. Technol. 241 (2017) 661–668.

[82] E.-J. Lee, Y.-H. Kim, H.-S. Kim, A. Jang, Influence of microbubble in physical cleaning of MF membrane process for wastewater reuse, Environ. Sci. Pollut. Res. 22 (2015) 8451–8459.

[83] A. Alipourzadeh, M.R. Mehrnia, A.H. Sani, A. Babaei, Application of response surface methodology for investigation of membrane fouling behaviours in microalgal membrane bioreactor: the effect of aeration rate and biomass concentration, RSC Adv. 6 (2016) 111182–111189.

[84] V. Naddeo, L. Borea, V. Belgiorno, Sonochemical control of fouling formation in membrane ultrafiltration of wastewater: effect of ultrasonic frequency, J. Water Process Eng. 8 (2015) e92–e97.

[85] M.L. Christensen, T.V. Bugge, B.H. Hede, M. Nierychlo, P. Larsen, M.K. Jørgensen, Effects of relaxation time on fouling propensity in membrane bioreactors, J. Membr. Sci. 504 (2016) 176–184.

[86] A.L. Ahmad, N.H. Mat Yasin, C.J.C. Derek, J.K. Lim, Chemical cleaning of a cross-flow microfiltration membrane fouled by microalgal biomass, J. Taiwan Inst. Chem. Eng. 45 (2014) 233–241.

[87] M.F. Rabuni, N.M. Nik Sulaiman, M.K. Aroua, C. Yern Chee, N. Awanis Hashim, Impact of in situ physical and chemical cleaning on PVDF membrane properties and performances, Chem. Eng. Sci. 122 (2015) 426–435.

[88] W. Cai, Y. Liu, Enhanced membrane biofouling potential by on-line chemical cleaning in membrane bioreactor, J. Membr. Sci. 511 (2016) 84–91.

[89] Z. Zhou, F. Meng, H. Lu, Y. Li, X. Jia, X. He, Simultaneous alkali supplementation and fouling mitigation in

membrane bioreactors by on-line NaOH backwashing, J. Membr. Sci. 457 (2014) 120–127.

[90] L. Wang, B. Pan, Y. Gao, C. Li, J. Ye, L. Yang, Y. Chen, Q. Hu, X. Zhang, Efficient membrane microalgal harvesting: pilot-scale performance and techno-economic analysis, J. Clean. Prod. 218 (2019) 83–95.

[91] A.L.K. Sheng, M.R. Bilad, N.B. Osman, N. Arahman, Sequencing batch membrane photobioreactor for real secondary effluent polishing using native microalgae: process performance and full-scale projection, J. Clean. Prod. 168 (2017) 708–715.

CHAPTER 8

Processing of Microalgae to Biofuels

EMMANUEL MANIRAFASHA • KAILIN JIAO • XIANHAI ZENG • YUANCHAO XU •
XING TANG • YONG SUN • LU LIN • THEOPHILE MURWANASHYAKA •
THEONESTE NDIKUBWIMANA • KEJU JING • YINGHUA LU

INTRODUCTION

Microalgae resources are being believed as a prominent and sustainable feedstock for biofuels and other bioproducts production. The prominence and sustainability of microalgae are associated with their high productivity within a short period, leading to higher biomass yield, and the ability of cultivation and biomass harvesting round all the year by integrated cultivation modes, as compared with other energy crops. The most important advantage associated with microalgae utilization is that microalgal biomass production has no competition with food production. Microalgae exploitation is also considered as a solution of environmental protection where domestic, industrial wastewater is used as low-price sources of nutrients to replace artificial culture media for microalgal biomass production coupled with phycoremediation, which is considered as an environmentally friendly, innovative solution and sustainable bioenergy generation [1].

The most considered potential associated with microalgae as prominent feedstock is that the microalgae can accumulate a broad range of components as well as resilience and autoprotection to the harsh environmental conditions [2]. Microalgal biomass is an essential resource whose exploitation can lead to many sorts of products such as biofuels, as well as other bioproducts with potential uses as plastics, nutraceuticals, pharmaceuticals, food, feed, fertilizers, and personal care products [3] (see Fig. 8.1A). Regardless of the advantages and potentials mentioned above of microalgae as a sustainable industrial feedstock, microalgal biomass exploitation needs advanced knowledge because there is no one-step process and conventional technique/strategy for processing microalgal biomass to those microalgae-derived products. The production of bioenergy products, as well as other bioproducts, requires multiple processes, including cell wall disruption, extraction, and conversion. Each process has different categories depending on the targeted end product (see Fig. 8.1B).

This chapter attempts to highlight and describe the current technologies published in recent studies, aimed to cut down on the cost of microalgae downstream process, which thus in return can lead to the price reduction of microalgae-derived products including bioenergy products. Downstream processes of microalgal biomass into bioproducts are subdivided into three parts: (1) harvesting and dewatering; (2) extraction; and (3) fractionation and conversion [4]. An overview of the microalgal products including biodiesel, biogas, hydrogen, pigments, and fatty acids, as end products of those downstream processes, is also highlighted in this chapter.

HARVESTING AND DEWATERING

Microalgae harvesting is the intermediate process that comes after the cultivation process and precedes the extraction process. It aims to separate culture broth and microalgal cells that are referred to as biomass. Microalgae are typically cultured in a dilute medium, which makes microalgal biomass harvesting cost about a quarter of the total production process. There are various harvesting techniques/strategies including filtration, centrifugation, flocculation, flotation, and sedimentation (see Fig. 8.2).

The selection of the most appropriate microalgal biomass harvesting strategy considers several factors such as cell size, strain phenotype, biomass concentration, pH of culture medium, ionic strength, the final quality of harvested microalgal biomass, and recycling of spent medium [5]. Those harvesting techniques can be used in a single step or combined mode depending on microalgae strains, biomass concentrations, and available facilities. Microalgae size is among key factors since low-cost harvesting procedures are concerned. In

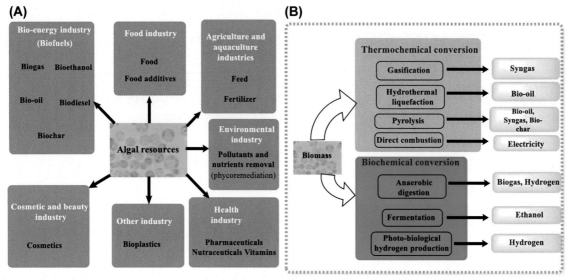

FIG. 8.1 **(A)** Some applications of microalgae and **(B)** conversion techniques for microalgal biofuels.

recent researches, the flocculation, under its cost-efficient type of bioflocculation, has to be coupled with sedimentation, centrifugation, or filtration techniques [6]; in those combined techniques, the flocculation helps to aggregate the microalgae cells for effective particle size, which facilitates the further filtration or sedimentation processes [6,7]. Each technique has associated advantages and drawbacks (see Table 8.1). For instance, sedimentation is such a harvesting technique that is cheap, energy efficient, cost-effective assembling, maintenance, and control, but it takes long time and is inefficient for small size microalgae species [8].

There is no conclusive decision to a given harvesting technique that is effective, and it can independently be utilized for all microalgae species harvesting. Therefore, all these microalgae harvesting techniques have to be adequately selected and utilized to build an environmentally friendly and economically harvesting process and reducing energy consumption.

Filtration

The process of filtration is made up of a porous medium that retains solid particles and allows the escape of liquids [11]. Filtration process can be categorized into different types of filtration (including magnetic, pressure, membrane, tangential flow, vacuum, and cross-flow) that is applicable for microalgae harvesting [12]. Soomro et al. [13] highlighted yield and input in terms of the energy of various filtration types applied for microalgae dewatering from various studies. Fundamentally, all filtration processes focus on the removal

of solids from solid-liquid mixtures by retaining those particles with sizes larger than the pore size of the permeable membrane with the defined differential pressure drop across the membrane bed [14]. Filtration as a microalgae dewatering technology has many advantages such as eco-friendly (low cost in terms of energy input, setup, and operation) and environmentally friendly (water recycle and reuse) [13]. It also has drawbacks such as long processing time, inefficient scalability due to frequent clogging, and the high tendency of fouling to filtration membranes with biological cells because organic materials with different physical and chemical characteristics (sizes, shapes, and compressibility) may mix in the biological feedstock [14].

Also, the polarization phenomenon becomes significant because of the surface charge of cells, affecting the nature of the cell, exogenous matter, and the membrane surface [14]. Microalgae release extracellular organic matter and pollute filtration membranes, the result being that the cake resistance increases regardless of membrane material. It exponentially increases with microalgae deposition rate but is negligible under low feed concentrations. The microalgal cake deposit is squeezable, and membrane fouling by microalgae is proportional to the number of organic polymers released, as extracellular organic matter, which is indicated by trans-membrane pressure on cake resistance [13,15].

Centrifugation

Centrifugation is a separation process with a principle based on the application of centrifugal force.

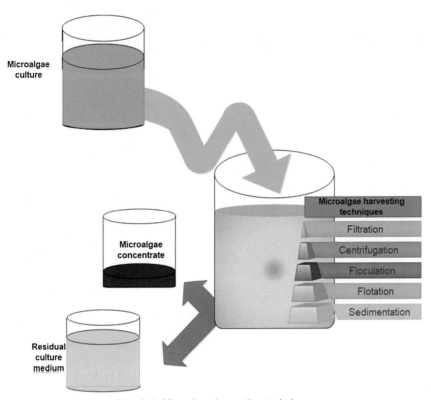

FIG. 8.2 Microalgae harvesting techniques.

Centrifugation is applied for microalgal biomass separation from the culture medium according to rotor speed, cell size, the viscosity, and density of the medium [11,16]. This technique is advantageous as it is time effective, easy [17], continuous, highly recovery, and efficient for large-scale processing [4]. However, structural damage and unwanted cell disruption would be the drawback associated with high gravitational and shear forces applied to the microalgae cells during centrifugation process [18]. Also, the process of centrifugation is economically unattractive because the processing at large scale takes a long time and leads to large energy consumption [16]. In other words, the centrifugation technique is in high capital cost. Soomro et al. [13] summarized the output in terms of performance and energy input of different centrifugation systems reported for microalgae dewatering.

Flocculation

Flocculation is a harvesting technique that promotes the aggregation of the microalgal cells to form flocs and assists in their settling process, which facilitates further dewatering steps such as sedimentation and filtration processes [7,11,13,19]. There are some advantages and drawbacks associated with flocculation depending on flocculants used during the process. Flocculation process is efficient in microalgal biomass harvest, has high volumetric capacity and ability, and is easy to scale up [20]. However, it might be expensive on account of the high cost of some flocculants; toxic materials and high dosage of flocculants may affect the water recycling ability and the quality of final extracted products. The microalgae harvesting by flocculation can be categorized into three main categories depending on the flocculation agent used in that process: chemical flocculation, autoflocculation, and bioflocculation.

Chemical flocculation

Chemicals (inorganic or organic coagulants) have been extensively used to induce flocculation, and the process has been exhibited adequate and plausible results in aggregating algae with low concentration from the culture media [18,21−24], and this flocculation process is also referred as "chemical coagulation." Soomro et al. [13] summarized the output in terms of the performance and input in terms of the energy for chemical

TABLE 8.1
Advantages, Disadvantages, and Performance of Different Harvesting Methods Applied to Microalgal Biomass [9,10].

Harvesting Method	Advantages	Disadvantages	Process Performance
Filtration	High recovery efficiency. Low cost in energy input, set up and operation. Water recycle and reuse.	Long processing time. The frequent clogging leads to inefficient scalability of the technology. High fouling tendency of filtration membranes.	1%–98%
Centrifugation	Fast method. High recovery efficiency. Easy large-scale processing.	Expensive method. High energy consumption. The possibility of cell damage due to high gravitational and shear forces.	0.4%–95%
Chemical flocculation	Simple and fast method. No energy requirements.	Chemical flocculants may be expensive and toxic and may also affect the quality of algal extracts. Recycling of culture medium is limited.	15%–99%
Autoflocculation	Inexpensive method. Allows culture medium recycling. Nontoxic to microalgal biomass.	Changes in cellular composition.	40%–98%
Bioflocculation	Inexpensive method. High recovery efficiency. Biodegradable flocculants. Allows culture medium recycling.	The possibility of microbiological contamination is subsequently interfering with food and feed applications of algal biomass.	83%–100%
Flotation	Feasible for large-scale applications. Low-cost method. Low space requirements. Short operation times.	Generally, it requires the use of chemical flocculants. Unfeasible for marine microalgae harvesting.	76.6%–99%
Gravity sedimentation	Simple and inexpensive method. Allows culture medium recycling.	Land-intensive and time-consuming. Species are dependent. The possibility of biomass deterioration. Low concentration of the algal cake.	Depends on pretreatment such as flocculation

flocculation systems of microalgae dewatering. Negative surface charge electrostatically stabilizes microalgal cells in solutions at most pH levels [21]. For a given flocculant dosage, the flocculation efficiency can be decreased, and it is considered as a challenge linked to the presence of secreted extracellular organic matter in solution [23].

In such a case, the superior flocculation efficiency can be achieved by increasing the quantity and quality of flocculant to be used. The rate of formation of microalgae flocs (AB) during flocculation of x units of microalgal cells (A) using y units of flocculant (B) can be expressed by the following equation [25]:

$$\frac{d[AB]}{dt} = k[A]^x[B]^y \qquad (8.1)$$

where k represents the flocculation rate constant, which is expressed as per time function. At any point during the flocculation process, the concentration of microalgae in suspension [A] is equal to the difference between the initial microalgae concentration [Ao] and the flocculated microalgae concentration [AB]. Also, the concentration of the flocculant [B] can be expressed as the difference between the initial flocculant concentration [Bo] and the concentration used in the formation of flocs [AB]. Thus, Eq. (8.1) becomes:

$$\frac{d[AB]}{dt} = k([A_0] - [AB])^x([B_0] - [AB])^y \qquad (8.2)$$

By specifying the initial concentration of microalgae and the concentration of flocculant, the above equation can be solved to obtain the mass concentration rate of flocculated microalgae under varying rate constants.

Different theories describe the flocculation of microalgae in different ways [26]. The principles of the flocculation process are based on the theories that rely on the nature of alkalinity of flocculants, which is the reason why in most case the flocculation process is in conjunction with pH adjustment. Under these theories, the flocculants play the role of neutralization of the repelling surface charge of microalgal cells, and then they facilitate microalgal cells aggregation into flocs followed by settling. The same theories can be explained through the assumption that the increase of flocculant dosage favors the increase of electrostatic flocculation in a linear stoichiometric way. Even though the adsorption of flocculation and formation of flocs are two stages that describe the flocculation mechanisms, the flocculation mechanisms are generally classified into four processes: interparticle collision, reduction of electrical charge, synthetic bridging flocculants, and natural bridging polymers. Those flocculation processes can lead to the high flocculation efficiency depending on the control of factors that may affect the flocculation process including the pH, polymer dosage, shear on the flocs, particle size, the pulp density, molecular weight, and temperature. It is imperative to maintain optimum agitation to avoid its excess that breaks up the formed flocs, while the reverse process of reforming the flocs is practically ineffective even though the agitation is removed. The size of microalgal cells is a critical function of flocculant dosage and nature, which implicates the surface area of the solid; for instance, if the microalgal cells that are considered as particles are too small, it will involve the increase of the amount of flocculant to

be used for the optimum flocculant/solid ratio. The nature of flocculation is considered because the charge of flocculant should be opposite to the charge of the cell surface, which means that the neutralization and stability proportional of the algal cells take place with the increase in the amount of adsorbed flocculant.

Moreover, in the case of polymeric flocculants, such as chitosan, the high microalgal concentration favors the formation of flocs, and the flocculation process is done by "bridging" (cross-linking) between cells. The self-same surface charges of microalgal cells that cause the repulsion force among them are one of the principles of flocculants, and the adherence of microalgal cells in the presence of flocculants can prove this principle where flocculants prevent this repel by binding to each other at low pH [26]. The research conducted by Chen et al. [27] is one behind this principle; they successfully achieved high *Scenedesmus* sp. cells flocculation efficiency by adjusting the pH of the culture medium in the absence of any aggregation agents as flocculants or coagulants [27].

Different flocculant types and dosages are critical factors that determine the rate of flocculation and further settling in conjunction with the optimum agitation. In other words, the optimal dosage of flocculants and sedimentation time (i.e., settling time) are two critical parameters in conjunction with initial microalgal biomass concentration that should be undertaken to evaluate the yield of flocculation process regarding the ratio of the output to the input of the process [26]. That evaluation is very crucial for life cycle assessment of microalgal harvesting process. Pretreatments aimed to concentrate the microalgal biomass can help to reduce the required amount of flocculants because the surface charge of the microalgal cells requires higher flocculants dosages, as the increase of biomass concentration influences the increase in the number of formed flocs, also known as aggregates [28].

Autoflocculation

Flocculation often occurs spontaneously in microalgal suspensions due to changes in culture pH [29–31]. Autoflocculation is defined as a naturally induced microalgae aggregation by merely evaluating and controlling the optimum pH of the culture medium [32]. The pH changes associated with photosynthetic CO_2 depletion result in the culture medium superior dissolution state; thus, the simultaneous microalgae settling is induced by calcium and phosphate [32–34].

The negatively charged algal cells adsorb to the calcium and phosphate precipitates, and this results in heavy flocs that can settle down easily [35]. Usually, the flocculation process takes place through the charge neutralization and/or sweep flocculation induced by calcium and phosphate precipitates that carry positive charges, depending on the conditions [32].

Sukenik and Shelef [36] conducted autoflocculation studies using both laboratory and outdoor experiments with cultures of *Scenedesmus dimorphus* and *Chlorella vulgaris*. A removal efficiency up to 96% was achieved with a comparable decrease in alkalinity, calcium, and orthophosphate concentrations. However, the authors observed that a relatively low removal percentage of about 40% occurred at pH 3, and no flocculation was observed at pH 5 to 7.5, whereas an algae removal of 98% could be achieved at pH above 8.5. Furthermore, the authors observed calcium-induced flocculation above pH 8.5, and no flocculation was observed with the addition of calcium in the absence of phosphate. The interaction between the negative surface charge of microalgal cells and the positive charge of calcium phosphate precipitates is the base of autoflocculation [37]; inevitably, calcium and phosphate ions have to be present together for the autoflocculation to occur at a specific pH range [32,36].

Also, some experimental studies at the laboratory level have suggested that the addition of NaOH to the microalgae cultures to reach specific pH values can initiate autoflocculation [33]. Moreover, $Mg(OH)_2$ also precipitates at high pH [33,36]. Up to pH value of 12, these precipitates are still positively charged. Therefore, they can interact with microalgae cells once present in the culture medium, resulting in flocculation [30,36,38]. The flocculation of freshwater microalgae (including *Chlorella vulgaris*, *Scenedesmus* sp., and *Chlorococcum* sp.) and marine microalgae (*Nannochloropsis oculata* and *Phaeodactylum tricornutum*) was evaluated and reported by Wu et al. [30]. Their results revealed that the adjustment of pH of microalgal culture media could stimulate the microalgae flocculation process. Freshwater microalgae were successfully harvested with relatively high flocculation efficiencies (90%) at pH above 9. The authors suggest that the mechanism behind this flocculation is a result of $Mg(OH)_2$ precipitate formed as the Mg^{2+} present in the culture medium hydrolyzed, and the formed cations stimulate the coagulation of microalgae by charge neutralization and sweeping flocculation. However, the algal biomass concentration and polysaccharides secreted by microalgal cell could induce the flocculation process and involve in the achievement of flocculation efficiency [30]. Recently, Liu et al. [31] have proved that the flocculation can be induced by the action of decreasing the pH of freshwater microalgal culture media, and the

efficiency of the flocculation could be reached above 90% for various microalgae species such as *Scenedesmus* sp., at acidic pH of 4. Based on the obtained results, the interpretation of results described the neutralization of negative charges of microalgal cells occurred through the electrostatic interaction between positively charged carboxylate ions of secreted organic matter and negative charges of microalgal cells by the action of decreasing the pH of the culture medium [30]. Therefore, the dispersing stability of algal cells is disrupted, and consequently, cells are flocculated.

Autoflocculation by calcium and phosphate precipitation requires relatively high concentrations of phosphate, and the increasing cost of phosphate due to the declining phosphates reserves; therefore, this process is not suitable except in the case where microalgae are used for wastewater treatment and excess phosphate needs to be further separated [32]. Furthermore, flocculation induced by high pH is due to ionic precipitates formation coupled with the pH adjustment at an optimum value. It is imperative to avoid the addition of the high concentrations of minerals during the harvesting process because they may cause toxicity in the harvested biomass. Moreover, they can interfere with subsequent downstream processes [12,32], as they should inevitably be removed from the biomass.

Bioflocculation

Bioflocculation is known as one of the microbial-based harvesting processes by the action of certain bacteria and/or microalgae. Bioflocculation is a promising microalgae dewatering process that can significantly contribute to the process cost reduction because no costs associated with the use of chemical flocculants are incurred with little to no energy consumption [39]. Soomro et al. [13] depicted the yield of the process by evaluating the output in terms of performances through harvesting efficiency of some reported microalgae bioflocculation methodologies. Multiple microalgae species are effectively and naturally being harvested through the coculture of bacterial microorganisms that secrete the natural flocculants into microalgae culture or by adding bioflocculants that are also produced by bacterial microorganisms cultivated/fermented in different bioreactors to one of the microalgae culture. This novel flocculation technique prominently exhibits to enhance bioflocculation efficiency in terms of yield and process cost reduction [39]. For example, some fungi possess hyphae with positive charges that can induce the flocculation by interacting with the negative surface charges of microalgal cells [40,41]. During the coculture of bacteria or fungi with microalgae (cultivation in the same reactor), optimum conditions such as optimum carbon source and optimal aeration/agitation should be supplied to facilitate the formation of mixed microalgal-bacterial flocs/granules with big size compared with single microalgal cells, which leads to autosettling and in return facilitate the further harvesting process [42]. Even though the coculturing process of bacteria and microalgae is advantageous for microalgal biomass harvesting, it could probably interfere with the microalgal biomass applications, as food and feedstock have interfered due to microbiological contamination by bacteria or fungi applied as flocculating agents. Despite any advantage associated to the utilization of microorganisms in the production of bioflocculants, the harvesting efficiency and biodegradability advantages of bioflocculants have made them the suitable assets to be used in harvesting microalgal biomass with aquaculture and biodiesel production [43,44].

Flotation

Flotation is a separation technique rooted in the mineral industry. In various industries including biotechnology, it has great application potential in high valued end products such as proteins [45]. The principle of flotation process is based on the solid-liquid suspension, where bubbles are formed by air or gas, and then particles (i.e., particles to be separated) adhere to the formed bubbles. These bubbles carry particles to the lipid surface, where they are usually harvested by skimming [37,46].

The flocculation process can be classified into six flotation units/classes based on the methods of producing bubbles. Those classes are (1) suspended air flotation, (2) dispersed air flotation, (3) dissolved air flotation, (4) Jameson cell or jet flotation, (5) dispersed ozone flotation, and (6) electrolytic flotation [37,45]. Dispersed air flotation consists of the usage of a surface-active chemical to create foam by continuously pumping the air into a flotation cell. Suspended air flotation consists of creating small bubbles without the usage of compressor and saturator. Dissolved air flotation consists of injection of air-supersaturated water in the flotation cell under pressure [37,45]. Jet flotation, also referred to Jameson cell technology, involves the generation of small bubbles through high mixing, leading to faster recovery. Compared with other flotation methods, dispersed ozone flotation is a little bit different because the air commonly used to generate bubbles is replaced by ozone. Electrolytic flotation is a simple flotation process. It produces microbubbles by water electrolysis, hydrogen gas bubbles at anode, and

oxygen gas bubbles at the cathode. After the gas bubbles are generated, the bubbles are attached to the colloidal molecules and then move upward together to the surface where the particles may be harvested [45]. It has also been reported that most of flotation units could be expensive due to the involvement of expensive surfactants and/or collectors used during the flotation process to improve the performance of flotation performance.

For most processes, there are advantages and drawbacks. The flotation units also have some advantages and disadvantages linked to their applications. For instance, electrolytic flotation has shown numerous advantages, such as the high quality of formed gas bubbles and easy manipulation. However, there are also some drawbacks associated with the regular replacement of electrodes used during the electroflotation process; otherwise, they could affect the features of formed gas bubbles [45].

Apart from general advantages and disadvantages, Ndikubwimana et al. [45] also highlighted the advantages and potentials of jet flotation associated with its application in the various fields, especially microalgal biomass harvesting with separation efficiency that can reach 98% above. The dissolved air flotation has been shown as the best flotation method, usually providing higher flotation efficiency, but it is also more expensive than other flotation categories due to the high energy cost required to produce supersaturated water under air pressure [37].

Under various studies, the flotation process is evaluated, described, and applied [45,47]. It is regarded to be a suitable method for microalgal biomass harvesting and has potential applications, especially in the production of microalgae-derived biofuels [39,48–50]. According to the procedure adopted in the flotation process, many authors have different opinions on the application of flotation process for microalgal biomass harvesting [13]. One of the assumptions is that flotation may be regarded as a prominent microalgae harvesting technique compared with natural settling [11], where the microalgae are harvested/separated on the liquid surface instead of under the receptor as natural settling does. Under such assumption, flotation is preferable due to the high overflow rate requisite by microalgae mass cultivation [46]. Even though dispersed air flotation is not widely applied in microalgal biomass harvesting, it has been exhibited to be useful for concentrating microalgal biomass as the primary method [37].

Some parameters need to be considered before selecting flotation methods/units for microalgal biomass harvesting. These parameters are availability, affordability, sustainability, easy to maintain, and quality of materials, among others. Innovation and technology are regarded as a promising solution to the challenges of various industries. Regarding the microalgal resources exploitation, especially in microalgae-derived biofuels, it is essential to develop novel technologies that can solve the main impeder of the microalgae harvesting, which is important among the necessities. Therefore, it is imperative to develop flotation technology in term of large-scale and commercial optimization, including continuous systems, as well as the establishment of life cycle assessment for evaluation of the sustainability of the whole production system [45].

The mechanism of the flotation process is presumed to be a first-order reaction, and linked kinetics for the removal rate of particles can be described by Eq. (8.3) under batch cultivation mode [45]:

$$\frac{dX}{dt} = -kX \tag{8.3}$$

where X, k, and t represent the number of particles inside the flotation unit, the flotation rate constant, and the run time, respectively. Presuming X_0 as an initial number of particles at initial time $t = 0$, the number of particles X in the flotation cell at a given time can be determined by the integration of Eq. (8.4) [45]:

$$X = X_0 e^{-t} \tag{8.4}$$

The number of particles to be floated is represented by the number of particles inside the flotation unit at $t = 0$. Ndikubwimana et al. [45] depicted and compared the output in terms of performances and input in term of energy consumption of some reported microalgae biomass harvesting and dewatering by flotation methods/units. Apart from generalized mechanisms and kinetics of flotation, the DLVO (Derjaguin, Landau, Verwey, and Overbeek) modeling can be applied to interpret both simple flotation process and flotation coupled with other harvesting techniques for microalgal biomass harvesting [47].

Sedimentation

Sedimentation process, also known as gravity settling, consists of settling the biomass on the bottom of the container from the remainder of the culture medium, as well as produced extracellular compounds. Generally, the sedimentation is coupled with other harvesting and dewatering strategies, such as flocculation and filtration; the flocculation is intended to improve the size of microalgal cells by forming flocs [6]. Sedimentation coupled with flocculation is a prevalent practice for microalgae dewatering, where after flocculation, the formed flocs tend to fall out of solution under the influence of gravity [6]. The sedimentation time (i.e., settling time) depends on the size of formed flocs, where larger flocs settle faster than smaller flocs. When the settling time is shorter, the quality of particle removal is better as the formed flocs have a larger size than the simple cells [19,51]. The settling efficiency can be approximately 80%—98%, depending on compaction provided by the flocculant and dosage. The advantages associated with sedimentation are that it is eased in setup and maintenance, inexpensive, and energy efficient. However, it is somehow time-consuming and inefficient for small microalgae species. Hence, it is commonly in conjunction with flocculation, which is intended to aggregate the microalgae cells.

Based on recently published researches, it is apparent that the coupling of membrane bioreactor (MBR) with microalgae cultivation is promising. This coupling process will economically and environmentally contribute to the resolution of current microalgal biomass harvesting (which is mainly linked to the small size of microalgae species and dilute microalgal biomass concentration in culture broth/water), as well as the wastewater treatment and reuse [52]. Integration of MBR in microalgal cultivation will also be ease to establish reasonable procedures for further microalgal biomass harvesting and dewatering processes. The granulation such as activated algae granules is regarded as a new technology, which can contribute to the reduction of high cost of microalgae biomass recovery processes [53], thereby reducing the production cost of microalgae-derived biofuels. The automation of existing methods could also facilitate in cost reduction of microalgal biomass harvesting and dewatering processes.

Recent researches have revealed some upstream processes that facilitate and reduce the tasks in downstream processes [54]. Microbial consortium is one of those cultivation processes that has associated harvesting benefits. The advantages of microalgae-bacteria symbiosis are not only beneficial to the harvesting process but also enhance biomass productivity and quality as well as environmental remediation [54]. Therefore, there is a necessity to link the microalgae cultivation process and harvesting process to enhance productivity and reduce the production cost (see Fig. 8.3). The coculture of microalgae and other microorganisms, especially bacteria, is regarded as a promising solution to some harvesting challenges. The symbiosis cultivation helps increase the size of particles and biomass concentration, so as to improve the biomass harvesting.

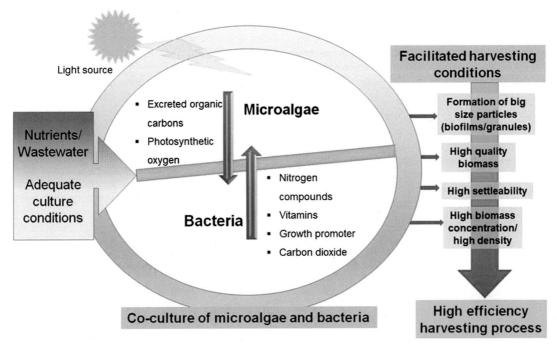

FIG. 8.3 Microbial consortia cultivation with facilitated microalgae harvesting conditions.

EXTRACTION

Microalgae store energy in cells through photosynthesis where CO_2 and nutrients, with the assistance of environmental conditions, are converted into microalgal biomass and release oxygen. The microalgae biomass contains plentiful valuable compounds, such as lipid, protein, and carbohydrate [55]. Microalgal biomass contains significant quantities of lipids, which are more suitable for biofuel production than other oil sources [56]. Similar to any other oleaginous biomass, for energy generation purpose, algal biomass can be utilized directly or processed to yield biofuels by chemical, biochemical, or thermochemical processes [49] (Fig. 8.1B). Direct use consists of combustion of dried biomass. Chemical processes are to convert the algae-derived lipids into biodiesel. Both fermentation and anaerobic digestion, which are classified as biochemical conversion, are applied to convert the carbohydrate fraction and total biomass into bioethanol and biogas, respectively. Thermochemical processes comprising the hydrothermal liquefaction, pyrolysis, and gasification of algal biomass decompose microalgal biomass organic compounds into different forms of biofuels, including liquid and gaseous.

These energy sources can be extracted as raw materials for synthetic biofuels and bioactive and chemical products. The main energy sources are lipids. The microalgal lipid content varies from 2% to 75% [57]. Lipid extraction is a crucial process for the synthesis of biofuels from microalgal resources. As the lipid is an intracellular metabolite, the cell wall disruption is required to release lipid from the cell before the lipid extraction process. There are several extraction methods including physical, chemical, solvent, and enzymatic methods. Recent researches have reported the lipid extraction protocols from wet microalgae biomass without the need for harvesting and drying steps [58]. The reduction of some steps contributes to reduce the production cost of microalgae-derived biofuels.

Physical

Physical method is among effective extraction methods. It is suitable for most microalgae species for extracting microalgal lipids. The risk of contamination of lipid products is low, and sometimes, no solvent is needed during the extraction process. However, physical extraction methods require higher energy and cooling system, which can make them more expensive when applied at large scale. Ultrasonication, bead-beating, hydrothermal liquefaction, bead mills, expeller press, and microwave are among the physical extraction methods that are commonly used in microalgal lipids. Solvent-free extraction is another type of the physical extraction methods, and it exhibits numerous advantages, including the following: it is simple and easy, has high purity of the final product, has lipid extraction efficiency with the nontoxic process, and can be applied for both wet and dry microalgae biomass.

Chemical

Chemicals (such as alkalis, acids, and surfactants) are applied for chemical linkages on microalgal cell walls through the process known as chemical treatments. As results of chemical linkages, the permeability of microalgal cell walls is increased and thus leading to the disruption of cell walls [59]. Chemical methods are considered as the pretreatment methods in microalgae extraction processes. The chemical method is usually coupled with solved methods for the extraction of microalgal lipids [60,61]. The chemicals to be used have to be carefully selected in avoidance of adverse effects such as degradation or degeneration of targeted compounds [61].

Solvent

Microalgae-derived energy sources can be extracted using solvent methods, where various organic solvents or a different combination of solvents is utilized for extracting lipids from microalgal biomass. Chloroform is an extracting solvent that is commonly used for effective extraction. The Folch, and the Blish and Dyer methods are well known and widely used as solvent methods for extracting lipids from homogenized cell suspension using extracting solvents. The 2:1 v/v ratio of coupled chloroform-methanol is used as an extraction solvent in the Folch method. The later method is similar to the Floch method except the ratio of solvents, which is interchanged into 1:2 v/v chloroform-methanol instead of 2:1 v/v. Both methods have a typical advantage of recovering a high amount of lipids at pilot-scale and large-scale extraction processes. Various modified methods based on both previous methods are being used for recovery of microalgae-derided lipids [60]. The commercial and industrial application of solvent methods for microalgae-derived lipid recovery is impeded by environmental and health risks [62]. Apart from chloroform and methanol, other less-toxic solvents, such as hexane, ethanol, isopropanol, butanol, acetic acid esters, and methyl-tert-butyl ether, are employed singly or by various combinations, for extracting lipids from microalgae cells.

Mechanisms of solvent extraction methods are based on the following principles: (1) In the case of single solvent extraction method, selection of the solvent depends on the similarity to the lipids, and it facilitates to increase the rate of mass transfer, solvent accessibility, and reduction of the immiscible solvent dielectric constant. The extraction process takes place at high temperatures and high pressure. (2) In case of a combination of different solvents, for example, chloroform/methanol, one of them should be more polar cosolvent that facilitates the microalgal cell wall disruption, and the second less polar cosolvent facilitates lipid extraction, which is the reason why it should have similar polarity with the lipids to be extracted. For example, a combination of chloroform/methanol results in the extraction interaction of water/methanol greater than that of methanol/chloroform and greater than that of chloroform/lipid.

Enzymatic

The microalgal lipids require the cell wall disruption for their extraction. The enzymatic extraction method is one of the promising ways to extract microalgal lipids. This method is similar to solvent extraction; the only difference is to add the enzyme rather than the solvent. In other words, this method consists of adding enzymes that ease cell disruption, and then the intracellular lipids are easily released into extraction broth. It does not require additional physical methods for microalgal cell wall disruption.

CONVERSION

Microalgae biomass conversion technologies, also known as the third-generation options, commonly consist of transesterification, hydroprocessing, fermentation, gasification, anaerobic digestion, and pyrolysis. The thermochemical conversion and biochemical conversion are two groups of all conversion technologies. Selection of the conversion technology depends on the raw materials/feedstock and targeted end product. Some conversion process, such as anaerobic digestion, can combine both microalgae cell wall disruption and extraction processes.

Transesterification

Transesterification is a catalytic reaction that converts lipids into fatty ester methyl esters and glycerol as coproduct through four steps. The first step is preparing the catalyst and mixing the catalyst with alcohol. The second step is the transesterification reaction that takes place through the reaction between the mixture of catalyst/alcohol and with the fatty acid (triglycerides). The optimum temperature range is about 40–60°C for transesterification reaction, and the rate of reaction increases with the increase of temperature. The third step is to separate glycerol from biodiesel, and the fourth step concerns the addition of water to both the biodiesel and glycerol for purification, especially the glycerol that may remain in the biodiesel. The acid should be added to glycerol for the production of neutralized glycerol.

Transesterification process facilitates the recovery of the high amount of mixture of fatty esters, but it requires a high amount of alcohol to prevent the production of soap, which is an unwanted reaction. Enzymes (i.e., enzymatic reaction) can be used instead of catalysts, but the lipases that are mostly used in the enzymatic reaction are expensive; thus the high cost makes the enzymatic reaction less effective than chemical reactions, and the reaction rate is slow and the productivity is low.

Hydroprocessing

The wet microalgal biomass contains about 75%—98% moisture that can be transformed into biocrude oil by hydrothermal liquefaction process. The process takes place at catalytic and high conditions (temperatures [200—500°C] and pressures [>20 bar]) [9,63]. The mixture of gases (CO_2, H_2, CH_4, C_2H_4, C_2H_6, N_2), the residual solids, and the aqueous phase are the by-products of the process. The high content of nutrients in those by-products can be recycled back to grow microalgae [64].

Hydrothermal liquefaction process is efficiently high, but its development and application are still in the embryonic stage [56]. The associated equipment is complex and expensive, which may hamper its broad application. Many extensive types of research are needed for improvement and commercial exploitation. Moreover, the high nitrogen and oxygen contents in algal biocrude oil make the oil quality a concern and unstable if it is not upgraded. The content of nitrogen increases with the conversion temperature, which can reach 5% wt of the oil; the combustion of the oil causes the conversion of that nitrogen into nitrogen oxides (NOx), which can react to form smog and acid rain [63].

Fermentation

Bioethanol is one of the bioenergy products obtained through a fermentation process of the carbohydrate, which is a component of microalgal biomass. After the fermentation process, the bioethanol is concentrated through the distillation process, which is followed by the bioethanol recovery from the fermentation broth. The gasification process can also be applied to bioethanol production from algal resources [9]. The high content of carbohydrates (greater than 50% of biomass components) can be accumulated in many microalgae species, such as *Chlorella*, *Dunaliella*, *Chlamydomonas*, and *Scenedesmus*, and this high carbohydrate content makes them a potential feedstock for producing algal bioethanol. After fermentation, the resulting solid residue can be used as food for animals or subjected to other processes to further the energy recovery from it. Moreover, though bioethanol fermentation generates a significant amount of by-product "CO_2," this by-product can be fully recovered to grow microalgae.

Both the simplicity in processing and operation conditions and the low energy requirements of fermentation make bioethanol production more advantageous. However, compared with biodiesel production, as a biofuel production technology coupled with wastewater treatment, it gained little attention.

Recent researches have proposed the perspective of other bioproducts such as biobutanol from the fermentation of microalgae [65]. Biobutanol has higher energy density and is molecularly similar to gasoline, and thus it is more effective than bioethanol. This kind of new biofuels would increase the value of microalgae as feedstocks for the fermentation process.

Gasification

Gasification is a thermochemical conversion method that is being used to convert microalgal biomass into biofuels. Gasification process produces gases such as hydrogen, methane, and carbon monoxide. Till now, it is obvious that the gasification of microalgae to biofuels is somehow expensive because the microalgal biomass contains the high moisture content. The catalytic gasification can rapidly convert microalgal biomass into gases [66]. For example, the supercritical water gasification is regarded as a promising technology that is suitable for gasifying microalgal biomass because it does not require a dry feedstock.

Anaerobic Digestion

Anaerobic digestion is classified as a biological conversion process that breaks feedstocks into CH_4 and CO_2 under anaerobic conditions (absence of oxygen). The anaerobic digestion is subdivided into hydrolysis, acidogenesis, acetogenesis, and methanogenesis. During hydrolysis, the addition of water facilitates the degradation of macromolecules into monomers and release of hydrogen, where the reaction can be chemically or enzymatically catalyzed. The formed monomers are converted into different organic compounds such as volatile acids and alcohol; this second step is referred to acidogenesis. During the third step, acetogenic bacteria (including *Syntrophobacter* sp., *Clostridium* sp., *Lactobacillus*, and *Actinomyces*) convert the acidogenesis intermediates into acetic acid, carbon dioxide, and hydrogen. The intermediate products in the previous step are mainly converted into biomethane during the methanogenesis step.

Anaerobic digestion technology has shown potential as a key process for managing large quantities of biomass residues, including microalgal residue after extraction of high-value compounds and/or other metabolites. The microalgal biomass and microalgal biomass residues after extracting high-value compounds such as oil are sustainable raw materials for anaerobic digestion, which can be fed to the anaerobic digester. The microalgae-derived feedstocks are promising and sustainable feedstocks for biogas production through anaerobic digestion due to their high nutrients, carbon, and moisture content and absence of lignin [67]. The microalgal biomass is also being exploited as a source of biogas and biofertilizer under anaerobic digestion process [68], and this will be sustainable when the renewable energy production is coupled with wastewater treatment.

Pyrolysis

Pyrolysis gives three kinds of products, including biooil, gases, and biochar [9,63,69]. Algal biomass can be converted to biooil by pyrolysis. In other words, pyrolysis occurs at a high temperature of about 400−1000°C under anaerobic conditions, and the temperature increases proportionally with an increase of heating rate and a decrease of residence time. The original small size with no fibrous tissue of algal biomass makes it a more ideal feedstock than other kinds of biomass sources, such as lignocellulose, for biooil production. The characteristics of primary end products of pyrolysis process are affected by the conditions such as pressure, temperature, and heating rate. The advantages associated with the pyrolysis process are evaluated in terms of environmentally friendly and eco-friendly potentials, including self-sustaining energy and production of chemicals from bioresources. However, its practice is limited by the high costs of the required energy [9]. Also, the poor thermal stability and corrosiveness of biooil are the challenges of pyrolysis-based biooil [70].

In some cases, the downstream (including harvesting, dewatering, extraction, and conversion) techniques can be combined in a single step for the process cost reduction. For example, the culture broth can be transferred in the anaerobic digester or fermenter without harvesting and dewatering (see Fig. 8.4). The optimization of microalgal biofuels has to be considered from the cultivation up to the final products.

PRODUCTS
Fatty Acids

The microalgae fatty acids, like any other seed plant fatty acids, are produced in the plastid and then stored in triglycerides through the biochemical process between the plastid envelope and the endoplasmic reticulum [71]. Some fatty acids can then be converted into biodiesel by transesterification conversion processes. Microalgae are especially rich in long-chain polyunsaturated fatty acids (LC-PUFAs), such as arachidonic acid (ARA, C20:4ω6), eicosapentaenoic acid (20:5ω3), and docosahexaenoic acid (22:6ω3). These LC-PUFAs are of great nutritional importance and are of increasing demand in fields of healthy food, medicine, and cosmetics [72−74].

Pigments and Proteins

Microalgal proteins can be used as food, feeds, fertilizers, industrial enzymes, bioplastics, and surfactants [75,76]. Some algae species including *Spirulina*, *Anabaena*, *Chlorella*, *Dunaliella*, and *Euglena* are known to have high protein content. Many health benefits linked to algal proteins make such species to be used as a supplement in both human and animal nutrition [77]. Microalgae are also a valuable source of nonstandard amino acids such as ornithine [78] and almost all essential vitamins [75].

Microalgae also contain a multitude of pigments alongside chlorophyll, phycobiliproteins, and carotenoids being the most relevant. Carotenoids include astaxanthin, β-carotene, bixin, fucoxanthin, lutein, lycopene, zeaxanthin, etc. These compounds are used in food industry as natural food colorants and as an animal feed additive and in cosmetics. They also have many positive health effects including antiinflammatory properties [77]. The medicinal use of the ketocarotenoid astaxanthin is well recognized in the prevention and treatment of a diversity of health conditions [75]. Phycobiliproteins, including phycocyanin and phycoerythrin, are often used in food and cosmetics area. They also have several health-promoting properties and a broad range of pharmaceutical applications and are very powerful and highly sensitive fluorescent reagents. Phycocyanin is known to stimulate the immune defense system and has antioxidant, antiinflammatory, antiviral, anticancer, and cholesterol-lowering properties [77,79].

Most algal products other than biofuels are considered high-value products. When bringing the production

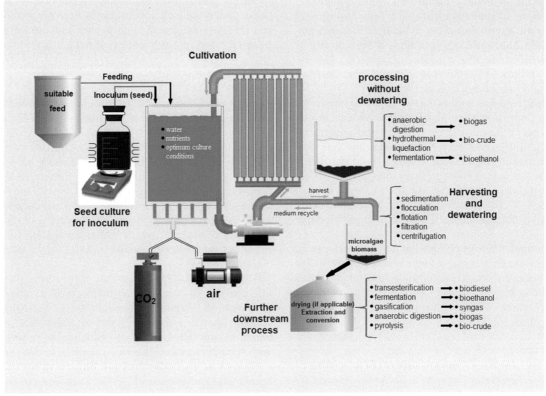

FIG. 8.4 Processing of microalgae to the biofuels flow sheet.

of high-value products to the matter, microalgae are the ideal source for the production of fine chemicals, human nutrition, pharmaceuticals, and cosmetics. However, for safety reasons as stated by Ref. [80], wastewater-grown biomass is rarely suitable for the production of food or even high-value chemicals due to high-quality requirements and public acceptance. Fertilization also, as it ends up affecting life, should only be conducted if algal biomass has no heavy metals or recalcitrant compounds (which are often found in industrial effluents). Consequently, biofuels production seems the remaining option to use wastewater-grown algal biomass though biofuel is a low-value algal product. There are different conversion processes of microalgae biomass to biofuels, and the selection of the most appropriate conversion process is a function of the biomass biochemical composition and water content, the desired type of biofuels, economic and technical considerations, and environmental standards [81]. In this respect, to find efficient, low-cost processes that require minimal energy inputs is a challenging issue [9]. The combination

of various conversion pathways would play a pivotal role to exploit the biomass optimally. Furthermore, with a mind to utilize all of the raw material components while improving resource flow and reducing resource losses, whenever safe and feasible, the generation of high-value products from biomass and/or by-products should be of priority as long as it is safe and feasible, since the cost of algal biofuel alone may not cover the production cost in most cases.

Biodiesel

Biodiesel is an alternative to the conventional diesel fuel. It is a mixture of monoalkyl esters of oils (long-chain fatty acids) derived from a renewable lipid feedstock by transesterification [82] with alcohol (chiefly methanol or ethanol), which is a catalytic reaction by alkali, acid, or enzyme and yields glycerol as a by-product [9,49,83]. The use of alkali catalyst is the most common [83]. The enzyme-catalyzed process does not need neutralization and requires less alcohol. It can convert feedstocks with a high free fatty acid content, but it is still quite expensive [49]. Both production efficiency and quality of biodiesel are affected by the lipid profile. Due to the increased content of free fatty acid in algal lipids, the transesterification to biodiesel is to be catalyzed by acids, such as hydrochloric or sulfuric acid or boron trifluoride, to avoid their saponification. However, the process here requires heating and longer reaction time and uses more excess alcohol [49].

Algal biodiesel production requires the release of lipids from their intracellular location, which should be done in the most energy-efficient and economical ways [75]. However, on the top of steps like the dewatering and drying of algal biomass, lipid extraction, and subsequent solvent recovery that are all energy intensive, there comes the low lipid content of algal biomass ($\leq 30\%$ mostly), which is another challenge of an economical production of microalgal biodiesel [63].

Glycerol, which is obtained as a by-product, can be utilized as a carbon source and converted into valuable metabolic products. This utilization constitutes another avenue to add value to the waste glycerol, including the optimization of high content lipid accumulation in microalgae cells [75,84].

Hydrogen

Direct photolysis of water can achieve microalgae-based hydrogen generation, also referred to biohydrogen, by microalgae or by a dark anaerobic fermentation that converts microalgae carbohydrates to hydrogen. The later process is assisted by either photosynthetic bacteria or a combination of both photosynthetic and anaerobic bacteria. Some algal strains, including *Spirulina*, *Anabaena*, *Chlorella*, *Synechococcus*, *Chlamydomonas*, and *Scenedesmus*, were reported as good candidates for hydrogen production [9,84,85].

Biogas

Biogas production is quite essential to promote the expansion and optimization of the entire biofuels production process at a low cost [84]. The process requires high moisture content and organic waste for anaerobic digestion, which is a bacteria-assisted conversion of organic material into biogas [63]. Both harvested biomass and residual biomass (after lipid extraction) are suitable feedstocks for biogas production. The produced biogas is a mixture consisting chiefly of methane (55%–75%) and carbon dioxide CO_2 (25%–45%), with a just detectable amount of other gases, such as hydrogen sulfide (below the standard limit). Since microalgae cell wall is composed of lipids and proteins together with little cellulose and almost no lignin content, Mussgnug et al. [86] state that, by anaerobic digestion, microalgae biomass has the potential for superior quality methane production. With its potential to recover energy from algal biomass after lipid extraction, biogas production has recently received much attention [84].

Syngas

Microalgae biomass can by gasification be converted to both syngas (a combustible gas mixture composed of H_2, CH_4, CO_2, and C_2H_4) and a solid combustible termed "biochar" [64]. Due to the low heating value, syngas is suitable to be either burned directly, used as a gas engine fuel, or used as a feedstock for the production of chemicals such as methanol [87]. Besides the energy used for biomass conversion, additional energy is used in the required pretreatment (drying and/or grinding) of biomass, which makes gasification technique not economically profitable [63].

Other Bioproducts

On the top of algal carbohydrates such as starch, glucose, and cellulose/hemicellulose, microalgae contain also a variety of polysaccharides that are currently considered as a class of high-value products with wide applications in food, cosmetics, textiles, stabilizers, emulsifiers, flocculant, lubricants, thickening agents, and clinical drugs [75,88]. Microalgal polysaccharides in general and sulfated polysaccharides such as spirulan [89], sulfate-containing exo- or extracellular polysaccharides, in particular, have been reported to exhibit a wide range of pharmacological activities. Microalgal polysaccharides, being a source of various biologically active molecules, have gained particular attention as natural therapeutic agents, cosmetic additives, and functional food ingredients [88,90]. A polysaccharide β-1,3-glucan found in *Chlorella* species is an active immunostimulator, a free radical scavenger, and a blood lipid reducer. It has been reported that β-1,3-glucan can also work effectively against gastric ulcers, wound, and constipation, prevent atherosclerosis and hypercholesterolemia, and act as antitumor [77,91].

CONCLUSIONS

Although microalgal biofuels are of great potential in the future, processing is still one of the bottlenecks for microalgal biofuels at the moment. The downstream route of microalgal industry should be economically feasible, energy cost controllable, and environmentally friendly. The reduction of energy cost during harvesting, extraction, and conversion would be a potential means for enhancing the economy of microalgal industry, while the utilization of by-products and upgrading the high-valued biochemicals in microalgae cells are the efficient strategies for increasing revenue.

ACKNOWLEDGMENTS

The following sponsors are acknowledged for supporting this work: the special fund for Fujian Ocean High-Tech Industry Development (No. FJHJF-L-2018-1), China; the National Natural Science Foundation of China (Grant Nos.21506177; 21736009); the Fundamental Research Funds for the Central Universities (20720160077); the Natural Science Foundation of Fujian Province of China (Grant Nos. 2018J01017); the Energy Development Foundation of the College of Energy, Xiamen University (Grant No. 2017NYFZ02); and the Disciplinary Cluster of Energy Science and Engineering, World-Top Class Universities and Disciplines Development Program, Xiamen University, China.

REFERENCES

[1] J. Hwang, J. Church, S. Lee, J. Park, W.H. Lee, Use of microalgae for advanced wastewater treatment and sustainable bioenergy generation, Environ Eng Sci. 33 (2016) 882−897.

[2] E. Manirafasha, T. Ndikubwimana, X. Zeng, Y. Lu, K. Jing, Phycobiliprotein: Potential microalgae derived pharmaceutical and biological reagent, Biochem. Eng. J. 109 (2016) 282−296.

[3] M. Kesaano, R.C. Sims, Algal biofilm based technology for wastewater treatment, Algal Res. 5 (2014) 231−240.

[4] F. Fasaei, J.H. Bitter, P.M. Slegers, A.J.B. van Boxtel, Techno-economic evaluation of microalgae harvesting and dewatering systems, Algal Res. 31 (2018) 347−362.

[5] J. Fan, L. Zheng, Y. Bai, S. Saroussi, A.R. Grossman, Flocculation of Chlamydomonas reinhardtii with different phenotypic traits by metal cations and high pH, Front. Plant Sci. 8 (2017) 1−10.

[6] L. Zhu, Z. Li, E. Hiltunen, Microalgae *Chlorella vulgaris* biomass harvesting by natural flocculant: effects on biomass sedimentation, spent medium recycling and lipid extraction, Biotechnol. Biofuels 183 (2018) 1−10.

[7] M. Mondal, S. Goswami, A. Ghosh, G. Oinam, O.N. Tiwari, P. Das, K. Gayen, M.K. Madal, G.N. Halder, Production of biodiesel from microalgae through biological carbon capture: a review, 3 Biotech. 99 (2017) 1−21.

[8] L. Zhu, Y.K. Nugroho, S.R. Shakeel, Z. Li, B. Martinkauppi, E. Hiltunen, Using microalgae to produce liquid transportation biodiesel: what is next? Renew. Sustain. Energy Rev. 78 (2017) 391−400.

[9] A.E.M. Abdelaziz, G.B. Leite, P.C. Hallenbeck, Addressing the challenges for sustainable production of algal biofuels: II. Harvesting and conversion to biofuels, Environ. Technol. 34 (2013) 1807−1836.

[10] A.I. Barros, A.L. Gonçalves, M. Simões, J.C.M. Pires, Harvesting techniques applied to microalgae: a review, Renew. Sustain. Energy Rev. 41 (2015) 1489−1500.

[11] N. Uduman, Y. Qi, M.K. Danquah, G.M. Forde, A. Hoadley, Dewatering of microalgal cultures: a major bottleneck to algae-based fuels, J. Renew. Sustain. Energy 2 (2010), 012701-15.

[12] K.Y. Show, D.J. Lee, J.S. Chang, Algal biomass dehydration, Bioresour. Technol. 135 (2013) 720−729.

[13] R.R. Soomro, T. Ndikubwimana, X. Zeng, Y. Lu, L. Lin, M.K. Danquah, Development of a two-stage microalgae dewatering process − a life cycle assessment approach, Front. Plant Sci. 7 (2016) 113.

[14] S.D. Ríos, J. Salvadó, X. Farriol, C. Torras, Antifouling microfiltration strategies to harvest microalgae for biofuel, Bioresour. Technol. 119 (2012) 406−418.

[15] S. Babel, S. Takizawa, Microfiltration membrane fouling and cake behavior during algal filtration, Desalination 261 (2010) 46−51.

[16] I. Rawat, R.R. Kumar, T. Mutanda, F. Bux, Biodiesel from microalgae: a critical evaluation from laboratory to large scale production, Appl. Energy 103 (2013) 444−467.

[17] E.M. Grima, E.H. Belarbi, F.G.A. Fernández, A.R. Medina, Y. Chisti, Recovery of microalgal biomass and metabolites: process options and economics, Biotechnol. Adv. 20 (2003) 491−515.

[18] R.M. Knuckey, M.R. Brown, R. Robert, D.M.F. Frampton, Production of microalgal concentrates by flocculation and their assessment as aquaculture feeds, Aquacult. Eng. 35 (2006) 300−313.

[19] T. Ndikubwimana, X. Zeng, T. Murwanashyaka, E. Manirafasha, N. He, W. Shao, Harvesting of freshwater microalgae with microbial bioflocculant: a pilot-scale study, Biotechnol. Biofuels 9 (2016) 47−58.

[20] T. Ndikubwimana, X. Zeng, N. He, Z. Xiao, Y. Xie, J.S. Chang, Microalgae biomass harvesting by bioflocculation-interpretation by classical DLVO theory, Biochem. Eng. J. 101 (2015) 160−167.

[21] N.B. Wyatt, L.M. Gloe, P.V. Brady, J.C. Hewson, A.M. Grillet, M.G. Hankins, Critical conditions for ferric chloride-induced flocculation of freshwater algae, Biotechnol. Bioeng. 109 (2012) 493−501.

[22] J.J. Chen, H.H. Yeh, The mechanisms of potassium permanganate on algae removal, Water Res. 39 (2005) 4420−4428.

[23] R. Henderson, S.A. Parsons, B. Jefferson, The impact of algal properties and pre-oxidation on solid-liquid separation of algae, Water Res. 42 (2008) 1827−1845.

[24] D. Vandamme, I. Foubert, B. Meesschaert, K. Muylaert, Flocculation of microalgae using cationic starch, J. Appl. Phycol. 22 (2010) 525−530.

[25] N. Uduman, Dewatering of Microalgae Using Flocculaiton and Electrocoagulation, Department of Chemical Engineering, Monash University, Clayton Victoria Australia, 2012.

[26] A. Schlesinger, D. Eisenstadt, A. Bar-Gil, H. Carmely, S. Einbinder, J. Gressel, Inexpensive non-toxic flocculation of microalgae contradicts theories; overcoming a major hurdle to bulk algal production, Biotechnol. Adv. 30 (2012) 1023–1030.

[27] L. Chen, C. Wang, W. Wang, J. Wei, Optimal conditions of different flocculation methods for harvesting *Scenedesmus* sp. cultivated in an open-pond system, Bioresour. Technol. 133 (2013) 9–15.

[28] D.G. Kim, H.J. La, C.Y. Ahn, Y.H. Park, H.M. Oh, Harvest of *Scenedesmus* sp. with bioflocculant and reuse of culture medium for subsequent high-density cultures, Bioresour. Technol. 102 (2011) 3163–3168.

[29] K. Spilling, J. Seppälä, T. Tamminen, Inducing autoflocculation in the diatom *Phaeodactylum tricornutum* through CO_2 regulation, J. Appl. Phycol. 23 (2011) 959–966.

[30] Z. Wu, Y. Zhu, W. Huang, C. Zhang, T. Li, Y. Zhang, Evaluation of flocculation induced by pH increase for harvesting microalgae and reuse of flocculated medium, Bioresour. Technol. 110 (2012) 496–502.

[31] J. Liu, Y. Zhu, Y. Tao, Y. Zhang, A. Li, T. Li, Freshwater microalgae harvested via flocculation induced by pH decrease, Biotechnol. Biofuels 6 (2013) 1–11.

[32] D. Vandamme, I. Foubert, K. Muylaert, Flocculation as a low-cost method for harvesting microalgae for bulk biomass production, Trends Biotechnol. 31 (2013) 233–239.

[33] C.Y. Chen, K.L. Yeh, R. Aisyah, D.J. Lee, J.S. Chang, Cultivation, photobioreactor design and harvesting of microalgae for biodiesel production: a critical review, Bioresour. Technol. 102 (2011) 71–81.

[34] J. Benemann, B. Koopman, J. Weissman, D. Eisenberg, R. Goebel, Development of Microalgae Harvesting and High-Rate Pond Technologies in California, Algae biomass: production and use, Munich, Germany, 1980.

[35] A. Lavoie, J. De la Noüe, Harvesting of *Scenedesmus obliquus* in wastewaters: auto-or bioflocculation? Biotechnol. Bioeng. 30 (1987) 852–859.

[36] A. Sukenik, G. Shelef, Algal autoflocculation-verification and proposed mechanism, Biotechnol. Bioeng. 26 (1984) 142–147.

[37] D. Vandamme, Flocculation Based Harvesting Processes for Micralgae Biomass Production, KU LEUVEN, Belgium, 2013.

[38] D. Vandamme, I. Foubert, I. Fraeye, B. Meesschaert, K. Muylaert, Flocculation of *Chlorella vulgaris* induced by high pH: role of magnesium and calcium and practical implications, Bioresour. Technol. 105 (2012) 114–119.

[39] L. Christenson, R. Sims, Production and harvesting of microalgae for wastewater treatment, biofuels, and bioproducts, Biotechnol. Adv. 29 (2011) 686–702.

[40] W. Zhou, Y. Cheng, Y. Li, Y. Wan, Y. Liu, X. Lin, Novel fungal pelletization-assisted technology for algae harvesting and wastewater treatment, Appl. Biochem. Biotechnol. 167 (2012) 214–228.

[41] J. Zhang, B. Hu, A novel method to harvest microalgae via co-culture of filamentous fungi to form cell pellets, Bioresour. Technol. 114 (2012) 529–535.

[42] J. Han, L. Zhang, S. Wang, G. Yang, L. Zhao, K. Pan, Co-culturing bacteria and microalgae in organic carbon containing medium, J. Biol. Res. 23 (2016) 8.

[43] D. Manheim, Y. Nelson, Settling and bioflocculation of two species of algae used in wastewater treatment and algae biomass production, Environ. Prog. Sustain. Energy 32 (2013) 946–954.

[44] H.M. Oh, S.J. Lee, M.H. Park, H.S. Kim, H.C. Kim, J.H. Yoon, Harvesting of *Chlorella vulgaris* using a bioflocculant from *Paenibacillus* sp. AM49, Biotechnol. Lett. 23 (2001) 1229–1234.

[45] T. Ndikubwimana, J. Chang, Z. Xiao, W. Shao, X. Zeng, I.-S. Ng, Y.L. Flotation, A promising microalgae harvesting and dewatering technology for biofuels production, Biotechnol. J. 11 (2016) 315–326.

[46] N. Pragya, K.K. Pandey, P. Sahoo, A review on harvesting, oil extraction and biofuels production technologies from microalgae, Renew. Sustain. Energy Rev. 24 (2013) 159–171.

[47] L. Xia, Y. Li, R. Huang, S. Song, Effective harvesting of microalgae by coagulation-flotation, R. Soc. Open Sci. 4 (2017) 170867.

[48] P.M. Schenk, H.S.R. Thomas, E. Stephens, U.C. Marx, J.H. Mussgnug, C. Posten, O. Kruse, B. Hankamer, Second generation biofuels: high-efficiency microalgae for biodiesel production, Bioenergy Res. 1 (2008) 20–43.

[49] L. Brennan, P. Owende, Biofuels from microalgae-a review of technologies for production, processing, and extractions of biofuels and co-products, Renew. Sustain. Energy Rev. 14 (2010) 557–577.

[50] T. Coward, J.G.M. Lee, G.S. Caldwell, Harvesting microalgae by CTAB-aided foam flotation increases lipid recovery and improves fatty acid methyl ester characteristics, Biomass Bioenergy 67 (2014) 354–362.

[51] T. Chatsungnoen, Y. Chisti, Harvesting microalgae by flocculation–sedimentation, Algal Res. 13 (2016) 271–283.

[52] Z. Chen, D. Wang, M. Sun, N.H. Hao, W. Guo, G. Wu, W. Jia, L. Shi, Q. Wu, F. Guo, H. Hu, Sustainability evaluation and implication of a large scale membrane bioreactor plant, Bioresour. Technol. 269 (2018) 246–254.

[53] O. Tiron, C. Bumbac, E. Manea, M. Stefanescu, L.M. Nita, Overcoming microalgae harvesting barrier by activated algae granules, Sci. Rep. 7 (2017) 4646.

[54] J.L. Fuentes, I. Garbayo, M. Cuaresma, Z. Montero, M. González-Del-Valle, C. Vílchez, Impact of microalgae-bacteria interactions on the production of algal biomass and associated compounds, Mar. Drugs 14 (2016) 1–16.

[55] M. Axelsson, F. Gentili, A single-step method for rapid extraction of total lipids from green microalgae, PLoS One 9 (2014) e89643.

[56] D. Chiaramonti, M. Prussi, M. Buffi, D. Casini, A.M. Rizzo, Thermochemical conversion of microalgae: challenges and opportunities, Energy Procedia 75 (2015) 819–826.

[57] P. Vo Hoang Nhat, H.H. Ngo, W.S. Guo, S.W. Chang, D.D. Nguyen, P.D. Nguyen, X.T. Bui, X.B. Zhang, J.B. Guo, Can algae-based technologies be an affordable green process for biofuel production and wastewater remediation? Bioresour. Technol. 256 (2018) 491−501.

[58] Y. Du, B. Schuur, S.R.A. Kersten, D.W.F. Brilman, Multistage wet lipid extraction from fresh water stressed Neochloris oleoabundans slurry-Experiments and modelling, Algal Res. 31 (2018) 21−30.

[59] Z. Chen, L. Wang, S. Qiu, S. Ge, Determination of microalgal lipid content and fatty acid for biofuel production, BioMed Res. Int. 2018 (2018) 1503126.

[60] Y. Li, F. Ghasemi Naghdi, S. Garg, T.C. Adarme-Vega, K.J. Thurecht, W.A. Ghafor, S. Tannock, P.M. Schenk, A comparative study: the impact of different lipid extraction methods on current microalgal lipid research, Microb. Cell Fact. 13 (2014) 14.

[61] F. Ghasemi Naghdi, L.M. González González, W. Chan, P.M. Schenk, Progress on lipid extraction from wet algal biomass for biodiesel production, Microbial Biotechnol. 9 (2016) 718−726.

[62] R.R. Kumar, R.P. Hanumantha, M. Arumugam, Lipid extraction methods from microalgae: a comprehensive review, Frontiers Energy Res. 2 (2015), 00061.

[63] A. Mehrabadi, R. Craggs, M.M. Farid, Wastewater treatment high rate algal ponds (WWT HRAP) for low-cost biofuel production, Bioresour. Technol. 184 (2015) 202−214.

[64] D. López Barreiro, C. Zamalloa, N. Boon, W. Vyverman, F. Ronsse, W. Brilman, W. Prins, Influence of strain-specific parameters on hydrothermal liquefaction of microalgae, Bioresour. Technol. 146 (2013) 463−471.

[65] T.K. Yeong, K. Jiao, X. Zeng, L. Lin, S. Pan, M.K. Danquah, Microalgae for biobutanol production − technology evaluation and value proposition, Algal Res 31 (2018) 367−376.

[66] Q. Guan, C. Wei, P. Ning, S. Tian, J. Gu, Catalytic gasification of algae Nannochloropsis sp. in sub/supercritical water, Procedia Environ. Sci. 18 (2013) 844−848.

[67] R. Paul, L. Melville, M. Sulu, Anaerobic digestion of micro and macro algae, pre-treatment and Co-Digestion-Biomass-A review for a better practice, Int. J. Environ. Sustain. Dev. 7 (2016) 646−650.

[68] E. Doğan-Subaşı, G.N. Demirer, Anaerobic digestion of microalgal (Chlorella vulgaris) biomass as a source of biogas and biofertilizer, Environ. Prog. Sustain. Energy 35 (2016) 936−941.

[69] Y. Chen, Y. Wu, D. Hua, C. Li, M.P. Harold, J. Wang, M. Wang, Thermochemical conversion of low-lipid microalgae for the production of liquid fuels: challenges and opportunities, RSC Adv. 5 (2015) 18673−18701.

[70] I.V. Babich, M. van der Hulst, L. Lefferts, J.A. Moulijn, P. O'Connor, K. Seshan, Catalytic pyrolysis of microalgae to high-quality liquid bio-fuels, Biomass Bioenergy 35 (2011) 3199−3207.

[71] U. Lionel, B. Vincent, M. Virginie, X.A. Mats, S. Benoit, C. Benoit, Microalgal fatty acids and their implication in health and disease, Mini Rev. Med. Chem. 17 (2017) 1112−1123.

[72] X. Guo, X. Su, G. Li, J. Chang, X. Zeng, Y. Sun, L. Lin, Light intensity and N/P nutrient affect the accumulation of lipid and unsaturated fatty acids by Chlorella sp, Bioresour. Technol. 191 (2015) 385−390.

[73] K. Jiao, J. Chang, X. Zeng, I.-S. Ng, Z. Xiao, Y. Sun, L. Lin, 5-Aminolevulinic acid promotes arachidonic acid biosynthesis in the red microalga Porphyridium purpureum, Biotechnol. Biofuels 10 (2017) 168.

[74] G. Su, K. Jiao, Z. Li, X. Guo, J. Chang, T. Ndikubwimana, L. Lin, Phosphate limitation promotes unsaturated fatty acids and arachidonic acid biosynthesis by microalgae Porphyridium purpureum, Bioproc. Biosyst. Eng. 39 (2016) 1129−1136.

[75] J. Trivedi, M. Aila, D.P. Bangwal, S. Kaul, M.O. Garg, Algae based biorefinery-How to make sense? Renew. Sustain. Energy Rev. 47 (2015) 295−307.

[76] X. Zeng, M.K. Danquah, S. Zhang, X. Zhang, M. Wu, X.D. Chen, Y. Lu, Autotrophic cultivation of Spirulina platensis for CO_2 fixation and phycocyanin production, Chem. Eng. J. 183 (2012) 192−197.

[77] P. Spolaore, C. Joannis-Cassan, E. Duran, A. Isambert, Commercial applications of microalgae, J. Biosci. Bioeng. 101 (2006) 87−96.

[78] P.M. Foley, E.S. Beach, J.B. Zimmerman, Algae as a source of renewable chemicals: opportunities and challenges, Green Chem. 13 (2011) 1399−1405.

[79] N.T. Eriksen, Production of phycocyanin−a pigment with applications in biology, biotechnology, foods and medicine, Appl. Microbiol. Biotechnol. 80 (2008) 1−14.

[80] R. Muñoz, B. Guieysse, Algal-bacterial processes for the treatment of hazardous contaminants: a review, Water Res. 40 (2006) 2799−2815.

[81] J.K. Pittman, A.P. Dean, O. Osundeko, The potential of sustainable algal biofuel production using wastewater resources, Bioresour. Technol. 102 (2011) 17−25.

[82] E.A. Mahmoud, L.A. Farahat, Z.K. Abdel Aziz, N.A. Fatthallah, R.A. Salah El Din, Evaluation of the potential for some isolated microalgae to produce biodiesel, Egyptian J. Petroleum 24 (2015) 97−101.

[83] S.A. Razzak, M.M. Hossain, R.A. Lucky, A.S. Bassi, H. de Lasa, Integrated CO_2 capture, wastewater treatment and biofuel production by microalgae culturing-A review, Renew. Sustain. Energy Rev. 27 (2013) 622−653.

[84] I. Rawat, V. Bhola, R.R. Kumar, F. Bux, Improving the feasibility of producing biofuels from microalgae using wastewater, Environ. Technol. 34 (2013) 1765−1775.

[85] D. Nagarajan, D.J. Lee, A. Kondo, J.S. Chang, Recent insights into biohydrogen production by microalgae − from biophotolysis to dark fermentation, Bioresour. Technol. 227 (2017) 373−387.

[86] J.H. Mussgnug, V. Klassen, A. Schluter, O. Kruse, Microalgae as substrates for fermentative biogas production in a combined biorefinery concept, J. Biotechnol. 150 (2010) 51−56.

[87] I. Rawat, R. Ranjith Kumar, T. Mutanda, F. Bux, Dual role of microalgae: phycoremediation of domestic wastewater

and biomass production for sustainable biofuels production, Appl. Energy 88 (2011) 3411−3424.

[88] H.W. Yen, I.C. Hu, C.Y. Chen, S.H. Ho, D.J. Lee, J.S. Chang, Microalgae-based biorefinery - from biofuels to natural products, Bioresour. Technol. 135 (2013) 166−174.

[89] H.C. Jun, K. Seung, J. Sung, Spirulan from blue-green algae inhibits fibrin and blood clots: its potent antithrombotic effects, J. Biochem. Mol. Toxicol. 29 (2015) 240−248.

[90] M.F. de Jesus Raposo, R.M.S.C. de Morais, A.M.M.B. de Morais, Bioactivity and applications of sulphated polysaccharides from marine microalgae, Mar. Drugs 11 (2013) 233−252.

[91] K. Hudek, L.C. Davis, J. Ibbini, L. Erickson, Commercial products from algae, in: R. Bajpai, A. Prokop, M. Zappi (Eds.), Algal Biorefineries: Volume 1: Cultivation of Cells and Products, Springer Netherlands, Dordrecht, 2014, pp. 275−295.

CHAPTER 9

Microalgal Cell Disruption and Lipid Extraction Techniques for Potential Biofuel Production

AHASANUL KARIM • M. AMIRUL ISLAM • ZAIED BIN KHALID •
CHE KU MOHAMMAD FAIZAL • MD. MAKSUDUR RAHMAN KHAN • ABU YOUSUF

INTRODUCTION

The global energy crisis is mounting day by day due to the exponentially growing population and extensive industrial development. The conventional fossil fuels such as petrol, diesel, coal, and natural gas are considered as the basic sources to meet this energy demand [1]. However, the progressive depletion of these petroleum-based fuels is recognized as a future challenge. In this context, the concern regarding alternative sources of energy to replace the fossil fuels is increasing tremendously [1,2]. The increasing demand for bioenergy sources and bioactive compounds has intensified research into biofuels as a viable renewable source to fulfill these needs. The biofuels (e.g., biochar, biogas, biohydrogen, biodiesel, bioethanol, etc.) produced from biomass (e.g., wood and wood residues, plants, animal matters, waste energy feedstocks, algae or algae-derived biomass, etc.) have been considered as sustainable renewable sources to meet the future energy demand [3,4]. Among them, biodiesel is regarded as a promising alternative to the petroleum-based fuels for the transportation sector. Recently, microalgae-derived biodiesel has gained widespread interest as one of the promising substitutes to the nonrenewable fossil fuels.

Microalgae and macroalgae are a diverse cluster of aquatic organisms, usually found in the freshwater and the marine environments, which possess the ability to fix 1.83 kg of carbon dioxide (CO_2) while producing 1 kg of algal biomass [5]. Apart from that, microalgae also produce a wide range of valuable nutrients useful in various industries, such as proteins and carbohydrates, along with the lipids used to produce biofuels [6]. The microalgal lipid yield was estimated to be 20,000−80,000 L/acre/year, which is about 30 times more than that obtained from seed crops fuels [7]. In addition, microalgae are considered as effective source of biodiesel due to the rapid growth rate; ability to grow in various complex environments including wastewaters; high biomass productivity; low land use; nontoxic; biodegradable; less harmful gas emissions [8]; diverse biochemical composition [9]; limited competition with the edible crops; etc. [3]. There are several steps and techniques for processing of microalgae to different valuable products (Fig. 9.1). However, large quantities of chemicals or high energy inputs are required for the extraction of intracellular compounds from the cell compartment due to the recalcitrance, complexity, and diversity of microalgal cell walls [10]. Consequently, the use of microalgae as a feedstock for biofuel production is hindered by the process economics and sustainability [2]. For instance, about 25%−75% of the algal biomass comprises stored lipids; however, extraction of lipids is the most challenging and energy-intensive procedure due to the tiny algal cells, rigid cell walls, and limited contact between the solvents during lipid extraction [3].

Generally, the cell walls of microalgae are structurally robust, complex, and chemically diverse, and therefore, the disruption of the microalgal cell wall is the most crucial step to extract different valuable biomolecules (Fig. 9.2) from the cells. Moreover, the lipid extraction process is often influenced by the water content of biomass, selection of suitable solvents, the blocking effects from insoluble biomass residues, the limited lipid accessibility, the formation of stable emulsions, etc. Therefore, several alternative approaches (e.g., use of green solvents, direct transesterification, pretreatment for disrupting microalgal cells, etc.) for lipid extraction have been proposed to mitigate the

Microalgae Cultivation for Biofuels Production. https://doi.org/10.1016/B978-0-12-817536-1.00009-6

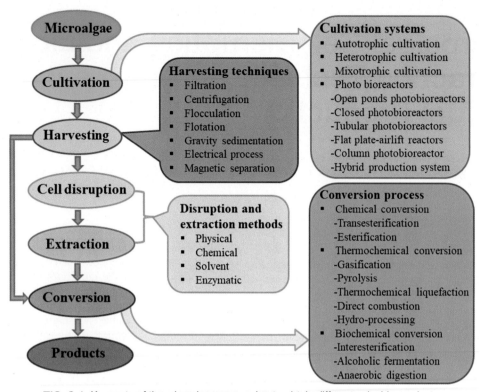

FIG. 9.1 Key route of the microalgae processing to obtain different valuable products.

problems associated with conventional lipid extraction process. Among them, the pretreatment approach in conjunction with solvent-based extraction can be an emerging approach to address the problems during the extraction of microalgal lipids. Recently, the extraction processes are often conducted with solvents using untreated, chemically treated, or mechanically treated cells. These methods of treating microalgal cell to enhance intracellular lipid extraction are often called pretreatment methods or cell disruption techniques [2]. Various cell disruption techniques (e.g., physical, chemical, and biological methods, etc.) have been reported to extract the desired compounds from microalgal biomass [3,5,11]. The physical techniques include bead-beating, high-pressure homogenization, ball milling, microwave heating [12], ultrasonication, hydrodynamic cavitation, thermolysis, electrocoagulation, osmotic shocks, electroporation, and laser treatments [13], whereas the chemical methods are based on selective interaction of the cell walls with certain chemicals, such as chloroform, methanol, hexane, isopropanol, acetone, dichloromethane, and so on. Some other methods include autoclaving and lyophilization [14],

supercritical carbon dioxide extraction [15], etc. However, there is no agreed conclusion of the most suitable pretreatment method for different algal species. Therefore, in this chapter, the challenges of lipid extraction and the proposed alternatives to overcome the challenges for enhanced product recovery are discussed. The importance of a suitable cell disruption technique for enhanced product recovery is also focused in this regard.

MICROALGAL LIPID COMPOSITION AND DISTRIBUTION

Microalgae are gaining increased interest compared with other sources [16,17], since they can accumulate diverse kinds of value-added components (e.g., proteins, lipids, carbohydrates, and pigments) (Fig. 9.2). Generally, microalgae contain a wide range of lipid classes (Fig. 9.3), such as free fatty acids (FFAs), glycolipids, phospholipids, acylglycerides, lipoprotein, hydrocarbons, sterols, and pigments. The lipid classes have different physical and chemical properties, such as solubility, polarity, and viscosity [7]. The lipids can be

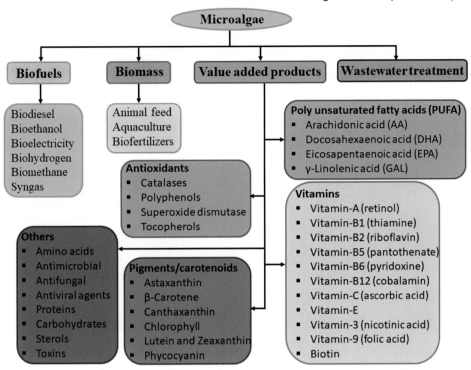

FIG. 9.2 The potential applications of microalgae and different valuable products.

classified as polar and nonpolar, based on their polarity and chemical structure of molecular head group. The polar lipids of microalgae are used to form the cell membranes and include glycolipids and phospholipids. On the other hand, the nonpolar lipids (i.e., neutral lipids) are usually used as the energy source and comprise FFAs and acylglycerols (mono, di, and tri) [18]. The key ingredients of both polar and nonpolar lipid molecules are the long-chain (comprising 12−22 carbon in length) fatty acids, which can be unsaturated (at least one double bond) or saturated (no double bonds). It is worth noting that the quality and production of biodiesel is directly affected by the composition of fatty acids [13]. However, triacylglycerols (TAGs) are the main target to produce biodiesel among the different lipid groups, because of their lower degree of unsaturation compared with other lipid fractions.

Generally, lipids are stored in different locations in the cell and play a significant role since they have specific cellular functions. The lipid bodies (containing mostly of sterol esters and TAGs) are surrounded by a phospholipid monolayer and exist in the cytoplasm as a form of energy storage. These lipid-rich cell compartments are found in all eukaryotic organisms, and also in some prokaryotic genera, e.g., *Streptomyces* and *Rhodococcus* [11]. However, the rigidity of microbial cell wall impedes the efficient extraction of lipids for biodiesel production. Eventually, the yield of biodiesel depends on the lipid contents of individual cells, which varies significantly for different microalgae species. Biodiesel production from microalgae includes several upstream and downstream operations, including microalgae cultivation, biomass harvesting, lipid extraction, and transesterification of the lipids (Fig. 9.4). Among them, the extraction of lipids is the most challenging step, and it represents a significant bottleneck for industrial-scale biodiesel production [19]. Since lipids are generally synthesized in the cellular compartments of the microalgal cells in the form of lipid droplets, it is essential to break up the cell walls to improve the extraction yield of the intracellular lipids from biomass. Various chemical, mechanical, biological, and physiochemical pretreatment methods could be applied to disintegrate the microalgal cellular membranes [20]. Therefore, a suitable cell disruption technique can be considered as a key factor to improve the lipid extraction efficiency in microalgae-based biodiesel production [10].

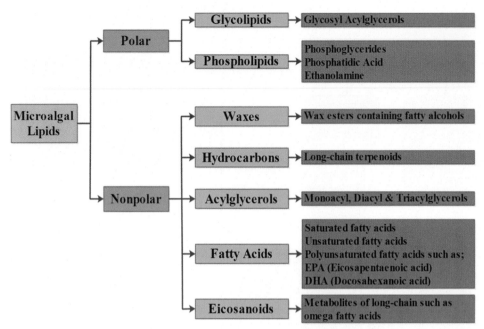

FIG. 9.3 Different classes of microalgal lipids with examples.

MICROALGAL LIPID EXTRACTION MECHANISMS

The lipid extraction from microalgae is generally carried out from dried biomass (dry route) or wet biomass concentrate (wet route) [13]. In most cases, lipids are extracted using solvent-based techniques; hence, solvents play a vital role in both routes. In some cases, solvents are subjected directly into cell pellets, while in other cases, solvents are employed during cell disruptions [2,3]. Generally, lipids are dissolved into a number of solvents, such as methanol, ethanol, butanol, isopropanol, chloroform, n-hexane, acetone, benzene, and cyclohexane; however, hexane, chloroform, and methanol are regarded as the most potential solvents to extract microalgal lipids [5,11]. The solvents should have high specificity for intracellular lipids, insoluble in water, greater penetration ability through the cell matrix, low boiling point, volatile, inexpensive, and nontoxic characteristics to be efficient for product recovery [13].

In solvent extraction, generally, a mixture of polar and nonpolar solvent, with a particular mixing ratio, has been used for complete recovery of intracellular lipids [5]. The solvent molecules penetrate the cell walls, subsequently enter to the cytoplasm, and form the complex structures with neutral lipids (which exist in the cytoplasm in the form of globules), while microalgal biomass is subjected to a nonpolar solvent (Fig. 9.4). These lipid-solvent complexes diffuse out of the cell walls into the bulk solvent due to variation of concentration gradient [11]. However, a certain fraction of neutral lipids is still remained inside the cell as a complex with polar lipids (which are attached to the cell membrane protein via hydrogen bond). Hence, the use of polar solvent is emergent to disperse these lipid fractions. The polar solvents separate these neutral lipids from the complex via formation of hydrogen bonds with the polar lipids [13]. This process inevitably brings out the polar lipids into the bulk solvent. Finally, they are recovered by the distillation or evaporation of the solvents.

CHALLENGES IN MICROALGAL LIPID EXTRACTION PROCESS

The rigid cell wall of microalgae may prevent direct contact between the solvent and the cell membrane and, ultimately, hinders the lipid extraction. Besides, the physiological properties, e.g., the location where the lipid is stored and the process by which lipid contents accumulated in the cell, can also influence the efficiency of the solvent [5]. Apart from these, lipid recovery from wet biomass in a practical extraction process can be also affected by other factors, such as water content in cell biomass, selection of the solvents, mass transfer mechanism, lipid accessibility, and formation of emulsions. [11,21]. However, these important factors are overlooked since the cell disruption was mostly focused in

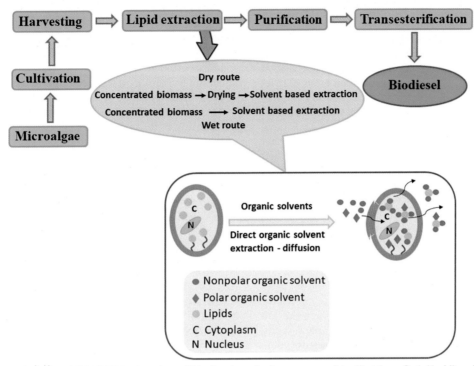

FIG. 9.4 Key steps of microalgae-based biodiesel production process. (Modified from Sati, H., Mitra, M., Mishra, S., & Baredar, P. (2019). Microalgal lipid extraction strategies for biodiesel production: a review. Algal Res. 38, 101413.)

all studies of wet lipid extraction, even though these factors are crucial to develop an economic and sustainable process.

The Role of Cell Wall Structure and Composition on Product Yield

Microalgae form robust cell walls having a tensile strength of ∼9.5 MPa, which is about three times stronger than the plant cells [3]. The robust cell wall structure of various species acts as a barrier to the commercial utilization of microalgae. It is important to note that some species of microalgae, such as Chlorophyceae and Trebouxiophyceae (e.g., *Botryococcus* sp.), have a nonhydrolyzable algaenan structure, cellulose, or fibrillar cell wall, and the algaenan has been widely observed to contribute to resistance of cell wall and act as a barrier to effective recovery of bioproducts [2,3]. Generally, cell wall comprises cellulose, glycoproteins, protein, uronic acid, xylan, mannose, layers of algaenan, and minerals such as calcium or silicate [3]. For example, the complex sugars forming microalgal cell walls (e.g., *Tetraselmis striata*, *Tetraselmis suecica*, etc.) are mainly made up of galactose, mannose, xylose, rhamnose,

and arabinose. The high content of glucose and mannose contributes to the rigidity of cell wall [2]. Consequently, the cell wall structure and composition of microalgae have a significant effect in lipid extraction efficiency and performance as well as in cell disruption techniques. Therefore, a deep insight into the role of cell wall structure, composition, and cell size needs to be considered for selecting an appropriate pretreatment technique to increase product recovery.

Effect of Biomass Water Content on Lipid Extraction

Drying of the biomass is an emergent step for lipid extraction and transesterification from microbial cells. However, the drying of microalgal biomass significantly increases the time and cost of the biodiesel production through transesterification [5]. Moreover, the drying temperature may affect the composition and amounts of lipids during the extraction from algal biomass. For instance, while drying the biomass at 60−70°C, the TAG in lipids remains unchanged, and the total lipid content may decrease slightly; however, both the concentration of TAG and lipid yield may decrease with a higher temperature than that of normal range [5,21]. Therefore, the wet biomass has been preferred to minimize the production cost and process difficulties; however, the extraction efficiency of lipids would be decreased if microalgal biomass contains high water content [21]. In a recent study, Ehimen et al. [22] reported that the oil to fatty acid methyl ester (FAME) conversion was practically hindered with microalgae biomass having a water content more than 32%; however, 82% FAME conversion was achieved when water content was reduced to 0.7%. In these cases, a higher proportion of alcohol to total lipid content is required, which may not be eco-friendly. Nevertheless, the extraction of lipids from wet microalgal biomass experienced several challenges such as reduced mass transfer, limited accessibility of lipids, and formation of stable emulsions. [11]. Therefore, lipid extraction from wet microalgal biomass using solvents, is difficult without a suitable pretreatment [5].

The Solvent Extraction

The solvent extraction is usually employed in extraction of lipids from wet biomass in most of the transesterification studies, particularly in direct transesterification. However, heating for reflux needs higher energy consumption and suffers from many difficulties in solvent extraction; hence, the scale-up process is still limited [21,23]. Cheirsilp and Louhasakul [24] observed that the direct transesterification without addition of nonpolar solvent required longer reaction time (~6 h) with a methanol/biomass ration of 125:1. However, the reaction time decreased to 1 h with an increase in methanol/biomass ratio to 209:1. This is due to the higher water content of reaction medium. In another study, Wahlen et al. [25] reported that wet microalgae cells having a water content of greater than 50% required a higher amount of methanol to obtain a FAME content of more than 70%. However, the uses of an excess amount of solvent would not be environmentally friendly; therefore, it is important to overwhelm this limitation. It is presumed that a suitable pretreatment method for cell disruption prior to the extraction could be effective to solve this problem.

Selection of Solvent for Lipid Extraction

In the solvent extraction, a suitable partition coefficient of extraction solvent is imposing the solute to migrate into the solvent phase from the aqueous phase. Therefore, a solvent should be selected so that the lipids have higher partition coefficient in it. However, neutral lipids (i.e., hydrophobic lipids) would be favorably partitioned into the nonpolar solvent phase based on the "like dissolve like" principle, whereas polar lipids could not be extracted so readily with nonpolar solvents because of their bindings with biomass matrix [11,13]. Thus, the cosolvents are typically used in lab-scale extraction to disrupt the linkage between polar lipids and biomass matrix and to improve the solubility of polar lipids as well.

The efficiency of the solvent extraction is reliant on the selection of solvents and the ratio of the solvents in which these are used. Although lots of individual solvents and their combinations have been employed so far, the chloroform:methanol mixture has been reported as a quick, effective, and quantitative combination for extraction of microalgal lipid in a number of comparative studies [11,13,26,27]. Besides, the mostly used mixture ratios of the chloroform:methanol for lab-scale studies are 1:2 (v/v) (widely known as Bligh-Dyer method) [26] and 2:1 (v/v) (commonly known as Folch method) [27], having to the high effectiveness. Bligh-Dyer and Folch methods employed methanol and chloroform to increase solubility and accessibility of polar lipids that eventually improve the total yield of lipids [11]. However, the extraction of lipids in lab-scale studies is mostly conducted in batch mode and limits the lipid extraction once the solvent becomes saturated with the lipid. As a consequent, a continuous organic solvent method (i.e., Soxhlet extraction) is commonly used to overcome this problem. In this lipid extraction method, the solvent evaporation and condensation cycles were used, continuously replacing the biomass with fresh solvents [13]. Among various solvents, hexane or a mixture of hexane with alcohol has been reported as a feasible less toxic substitute to the chloroform:methanol mixture and employed in several studies of Soxhlet extraction technique [11,13]. The high volume of solvents and the high energy requirement for their evaporation in Soxhlet extraction have a negative impact for the overall efficiency of this technique. To address this issue, a solar-driven extraction was employed in a recent study by Bhattacharya et al. [28] to extract the nonpolar lipids from *Chlorella variabilis*, in which the energy requirement was 11 times less than the conventional Soxhlet method. Although these methods of lipid extraction can be simply implemented, they need higher implementing time. Apart from the solar-driven extraction, a modified solvent-based extraction technique, namely, accelerated solvent extraction, has been studied with organic solvents at high temperature (50−200°C) and pressure (500-300 psi) for short time periods (5−10 min) to reduce the extraction time.

Lipid Extraction by Using Nonpolar Solvents

Generally, nonpolar lipids (e.g., TAGs and FFAs) exist as the form of small oil droplets in an aqueous environment, since those have reduced solubility in water. These lipids seem to float on the surface of the aqueous phase because of their lighter density. However, these lipid bodies may adhere to or be encapsulated by insoluble cellular debris. Therefore, the mechanism of nonpolar lipid (by using nonpolar solvents) from aqueous environment is to dissolve the tiny lipid droplets into the bulk solvent phase (Fig. 9.5A); thereafter, it can be separated from the aqueous biomass residue by phase separation [11].

In contrast, the mechanism of polar lipid extraction (by using nonpolar solvents) is more critical compared with the nonpolar lipids. The polar lipids are generally ingredients of cell membranes or closely connected with other cellular components and might not be extracted so easily with nonpolar solvents. Even though it is assumed that the polar lipids are released from the ruptured cellular debris and exist as free molecules, a hydration shell may exist around the polar lipid because of the electrostatic attraction of the water molecules [11,29]. The polar lipid extraction from an aqueous phase would be imagined by releasing an ion (or hydrophilic moiety) from its linked water molecule

followed by transfer into the solvent. However, the breaking up of the liquid-water interaction needs additional energy to liberate the polar lipids [29,30]. Consequently, the extraction of these ionic (or amphiphilic) lipid molecules needs more energy input. In addition, the polar lipids that adhere to the solvent-water interface may behave as surfactants and easily lead to emulsion formation through the reduction of interfacial tension (Fig. 9.5A). On the other hand, some of the polar lipids (such as glycolipids and phospholipids) partially dissolve in water to form micelles if the critical micelle concentration is reached. The micelles are capable of encapsulating nonpolar lipids and solvent, which in turn reduces the total lipid yield and leads to solvent loss. The amphiphilic lipids could be extracted by using solvents; however, they may form reverse micelles in a nonpolar environment [29]. The reverse micelles have a polar internal environment and a nonpolar external surface, which can hold proteins, water, and even entire cells. The reverse micelles are also recognized to make the organic solvent phase turbid or even can form gels, both of the facts make the lipid and solvent recovery much more difficult [11,30].

In conclusion, the cosolvents could be employed to increase the polar lipid recovery as illustrated above; however, they are unlikely to be practical for an

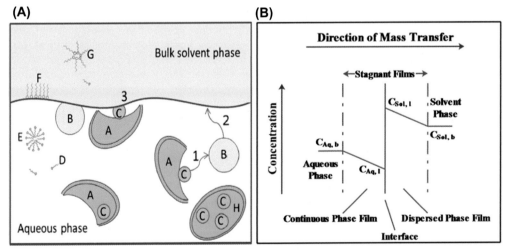

FIG. 9.5 **(A)** The mechanism of lipid extraction from wet biomass (A: ruptured cells; B: solvent droplet dispersed in aqueous phase; C: neutral lipids droplet; D: polar lipids; E: polar lipid micelle; F: polar lipids accumulated on the interfacial surface; G: reverse micelle; H: intact microbial cell). **(B)** The concentration profile for solute transfer from the bulk (*b*) aqueous (*aq*) across the interface (*i*) to the bulk solvent (*sol*) phase based on two film theory. (**(A)** Courtesy Dong, T., Knoshaug, E. P., Pienkos, P. T., & Laurens, L. M. (2016). *Lipid recovery from wet oleaginous microbial biomass for biofuel production: a critical review. Appl. Energy, 177,* 879–895; **(B)** Based on Liddell, J. (1994). *Solvent Extraction Processes for Biological Products Engineering Processes for Bioseparations.* Butterworth-Heinemann Ltd, Oxford.)

industrial-scale application due to the complex downstream processing. Alternatively, these lipids could be efficiently released by the biomass pretreatments, such as changing pH or chemical/enzymatic hydrolysis. On the contrary, extraction of the neutral lipids (e.g., TAGs and FFAs) is favored compared with the polar lipids because of their higher partitioning in nonpolar solvents, better solvent recovery, improved mass transfer, and reduced emulsion formation. Therefore, converting the polar lipids into the FFAs by a biomass pretreatment prior to the extraction would be preferred from an engineering perspective.

The Mechanism of Mass Transfer in Lipid Extraction From Wet Biomass

The mass transfer in extraction process is presumed to be occurred through the two films (two-film theory) with concentration gradients in both films and equilibrium at the interface [29]. However, in a practical lipid extraction process, one phase is usually dispersed as droplets in the other phase so that mass transfer occurs (Fig. 9.5B). A higher concentration difference and larger interfacial area are favored to generate higher rate of solute transfer. Generally, the algal biomass is composed of carbohydrates, proteins, and polar lipid compounds, which have many surface activities forming particles. Hence, the biomass components and surfactants have a great impact on the mass transfer mechanism in a solvent extraction of biologically derived products. The surface activity forming particles of various molecules typically displays varying degree of surface polarity, triggering them to concentrate at the phase interface, which in turn affects the mass transfer process [31]. If biomass residue absorbs to the interface, it can create resistance to solute mass transfer because of the physical obstruction by the absorbed layer. In addition to the physical barrier effects, these absorbed solid materials also reduce mass transfer by dampening of the interfacial turbulence. It was reported in a study by Pursell et al. [31]; the surface tension and overall mass transfer coefficient were decreased with the increment of biomass concentration in an extraction process. However, there is only a little resistance to mass transfer if the interface is clean [11]. Since the presence of biomass residues can hinder the lipid extraction efficiency by reducing mass transfer or physical encapsulation, a significant reduction of nonlipid insoluble biomass debris through a physical, chemical, or enzymatic treatment prior to lipid extraction might be an effective approach to improve the efficiency of lipid extraction.

Formation of Stable Emulsion

The formation of stable emulsion is always considered as a problem in lipid extraction process because it can limit the lipid yield and may cause losses of the product or the solvent [11]. Generally, fatty acid–based biosurfactants, such as monoacylglycerol and diacylglycerol, act to stabilize the oil-in-water emulsion, while several phospholipids demonstrate excellent surfactant properties. Microbial glycolipids, lipopeptides, and sophorolipids have also been reported as dominant biosurfactants [32]. In the lipid extraction process, biosurfactants lower the surface tension (interfacial tension) between water and oil phases to a great extent depending on the concentration. Although the lower level of interfacial tension allows a lesser energy input to force oil droplet contact with solvent and enhance the extraction efficiency [11], the lower surface tension can make the phase separation more problematic as mentioned above. Hence, the interfacial tension should not be less or more; it must be appropriate for better solvent-water contact [11,32]. On the other hand, the emulsifiers are generally responsible for film forming properties around oil droplets, thus preventing the dispersion of oil droplets to the solvent phase. The soluble surface-active compounds also act as emulsifiers, reducing the interfacial tension in a biomass slurry. Moreover, the solid residues that present in wet biomass can likely to decrease the surface tension [11]. Consequently, the settling of dispersion in wet extraction process might take longer time for phase separation due to the presence of solid biomass residues and emulsifiers (e.g., polar lipids, proteins, polysaccharides, etc.) [32]. However, the emulsifiers could be degraded to smaller units or fractions, which have no emulsification capability. Therefore, an approach to remove emulsion (i.e., by breaking the emulsifiers) should be developed to ensure an efficient and complete lipid recovery.

LIMITATIONS OF CONVENTIONAL SOLVENT-BASED EXTRACTION AND ALTERNATIVES

The solvent extraction has widely been employed to extract lipids from microalgae. However, the practical application of conventional solvent extraction is limited due to the huge amount of solvents requirement. In addition, the solvent-based lipid extraction methods are time-consuming, are not eco-friendly, and possess low efficiency (Table 9.1) [11,13]. Therefore, intensive research efforts need to be addressed to minimize the problems associated with solvent extraction. Currently,

TABLE 9.1
The Advantages and Limitations of Microalgal Lipid Extraction Methods.

Methods of Extraction	Advantages	Limitations	References
Conventional solvent extraction	- Using solvents is generally inexpensive - Solvents can be recycled - No setup cost - Ease of extraction	- Toxic - Adverse effect on environment	[13,14].
Folch method	- A standard method of extraction - Widely reported to extract lipid	- Laborious method - Adverse effect of chloroform on the environment	[7]
Bligh and Dyer method	- A simple and standard method of extraction - Total lipids can be determined - Samples can be analyzed directly without predrying	- Laborious method - Adverse effect of chloroform on the environment	[7]
Soxhlet extraction	- Continuous process - Less toxic solvents - Solvents can be simply evaporated and highly selective for neutral lipids	- Higher amount of solvent - Higher energy requirement for evaporation of solvents	[13]
Accelerated solvent extraction	- Short time period - Relatively higher extraction yield	- High temperature and pressure - High organic solvents	[13]
Bio-based solvents	- Derived from bio-based feedstocks—eco-friendly - Efficient extraction - Cost-effective	- Continuous feedstocks supply may not possible - Yet to be standardized and studied at large-scale	[13]
Ionic liquid extraction	- Synthetically flexible - Thermally stable - It is nonflammable in a wide temperature range - High conductivity - Broad miscibility range - Can be recycled - Reduced energy consumption - Enables solvent extraction	- Possible pathway into the environment through wastewater - Solvents synthesis is not eco-friendly - Higher energy requirement for distillation of solvents	[13,33]
Supercritical fluid extraction	- Quick separation - Gives highly purified extracts - Low toxicity - Highly selective since flexibility of temperature and pressure variations - No separation step is required	- High residual water content within microalgal biomass is not accepted because it results in flow impedance and restrictor plugging - High equipment and operational cost (e.g., pressure vessel is expensive)	[13,34]
Expeller press	- No solvent is required - Easy to use - Choking problems - Heat generation	- End product may have traces of other cellular contents along with extracted lipids	[35]
Ultrasound-assisted extraction	- Solvent consumption is reduced - Greater penetration of solvents into cell compartment - Lower extraction time - Higher yield of lipids	- Energy intensive - Possess scale-up difficulties	[2,14]

Continued

TABLE 9.1
The Advantages and Limitations of Microalgal Lipid Extraction Methods.—cont'd

Methods of Extraction	Advantages	Limitations	References
Microwave-assisted extraction	- Simple, easy, highly effective - Can be easily scaled up - Higher yield of lipids with superior quality - Short duration of extraction time	- Energy intensive	[3,14]
Switchable solvents	- Easily recyclable - Green approach	- Technical viability of the process is yet to be studied - Synthesis of solvent can be environmentally damaging	[13]
Direct transesterification	- Economical process - Less energy - Reduce time and solvent uses	- Paired with organic solvents; hence, it is toxic - Less efficient with high water content; thus typically requires dry biomass - Technical viability of the process is yet to be studied	[13]

the efforts have been focused in three parallel directions such as finding green solvents (e.g., bio-based solvents, ionic liquids, supercritical fluids, switchable solvents, etc.); direct transesterification (i.e., concurrent extraction of lipids and conversion of lipids to biodiesel); and exploring pretreatment approaches of algal biomass (e.g., expeller press, microwave, ultrasound, chemical treatments, enzymatic methods, etc.) [13].

Green Solvents Approach

The uses of green solvents in microalgal lipid extraction are of great interest nowadays, with an intention to address the problems raised by organic solvents. As reported in several studies [13,36,37], different kinds of green solvents such as bio-based solvents, ionic liquids, and supercritical fluids have been used to replace the organic solvents. Dejoye Tanzi [36] reported that the bio-based solvents, derived from agricultural sources, such as terpenes (e.g., gum terpene, D-limonene, and p-cymene, etc. produced from citrus species, pine trees) and tree leaf oils, could be successfully implemented to extract lipid from *Chlorella vulgaris*. The solvents, derived from citrus, corn, and soybeans, e.g., ethyl-lactate, methyl-soyate, ethyl-acetate, 2-methyl tetrahydrofuran, and cyclopentyl-methyl ether, have also been regarded as a potential replacement of the organic solvents [37]. These solvents were demonstrated to improve the biodiesel quality of the extracted lipids because they resulted in a comparatively low level of

polyunsaturated fatty acids (PUFAs) in lipid extraction [38]. Although the bio-based solvents are biodegradable, nontoxic, and able to replace many hazards, the consistent feedstock supply is of a great concern [13]. Furthermore, the supercritical fluids have been appeared to be an efficient substitute to organic solvents due to safety, health, and environmental concerns. In this technique, the microalgal biomass is subjected to the supercritical fluid under controlled conditions of temperature and pressure; then the lipid contents of microalgal cells desorbed in the fluid stream are finally recovered by condensation. CO_2 is widely used in supercritical fluid extraction (SFE) due to its low flammability, lack of reactivity, low toxicity, and recoverable characteristics [39]. In a recent study, Tai and Kim [40] obtained maximum yield of lipids ($\sim 6.2\%$) by employing SFE compared with organic solvent-based lipids yield ($<5\%$). Other than that of bio-based solvents and supercritical fluids, the ionic liquids (i.e., nonaqueous salt solutions composed of an organic cation and a polyatomic inorganic anion) and the switchable solvents (a subclass of ionic liquids) have also been considered as green solvents [13]. In a recent study, Kim et al. [41] obtained 1.6-fold higher lipid yield by using ionic liquid namely, $[Bmim][MeSO_4]$, to extract lipids from *C. vulgaris* with the assistance of ultrasonic pretreatment. Therefore, it can be supposed that the direct use of green solvents is desirable but still an active area of research and development. As mentioned above, green solvents have been employed in various studies to replace the organic solvents; however, they used them in combination with a pretreatment method to enhance the efficiency of lipid extraction.

Direct Transesterification

Direct transesterification of microalgal lipids has been reported as an easy, simple, and rapid approach for quantifying fatty acids by integrating the extraction and transesterification into a single step (Fig. 9.6), also termed as "*in situ* transesterification" [21]. This technique entails a solvent-mediated extraction of microalgal lipids followed by solvent evaporation and, thereafter, production of FAME [13]. In this technique, wet or dry biomass is treated with a mixture of methanol and inorganic acid or base catalyst in a single reactor, resulting in the reactive extraction of lipids as FAAEs (i.e., fatty acid alkyl esters), typically, FAME. Methanol acts both as an extraction solvent and as an esterification reagent [23]. The process of simultaneous lipid extraction and transesterification of lipids to FAME not only saves the time but also reduces the addition of organic solvents in large amounts. Moreover, this process can decrease the cost of instrument installation and maintenance, and the energy consumption [13,21]. In addition to that, there are enough reports where direct transesterification process has been found to be advantageous due to the higher FAME yield compared with two-step processes. For instance, Vicente et al. [42] achieved a higher yield of FAME (>99% of

FAME), whereas the conventional processes produced 91.4%–98% of FAME. This is attributed to the involvement of FFAs, phospholipids, and glycerides that resulted in higher FAME yield in the *in situ* transesterification process [23]. Furthermore, this technique has been proved to be efficient in forming biodiesel from both the mono- and mixed cultures of oleaginous microbes [23,42].

Although the direct transesterification process offers a shorter processing time and lower cost of production, it requires further study to improve the factors that affect the efficacy of biodiesel production from microalgal biomass. The major hurdle that must be faced in direct transesterification is the disruption of microalgal cell wall to increase the release of intracellular lipids [21,43]. Another major challenge in this process is selection of the catalysts. Usually, homogeneous base catalysts have been used in transesterification reaction due to their moderate reaction conditions and faster reaction rate than the acid catalysts (generally, acid-catalyzed reactions need higher temperature and longer reaction time) [21]. Nevertheless, acid-catalyzed esterification reactions of FFAs result in the water formation, which limits the completion of the reaction. Base-catalyzed reactions, on the other hand, cause soaps formation from cellular FFAs [13]. Furthermore, the water level of the biomass significantly effects the production cost of biodiesel in direct transesterification process. The lipid extraction efficiency is decreased, if biomass contains high amounts of water. In addition, this technique has been developed mainly for laboratory-scale

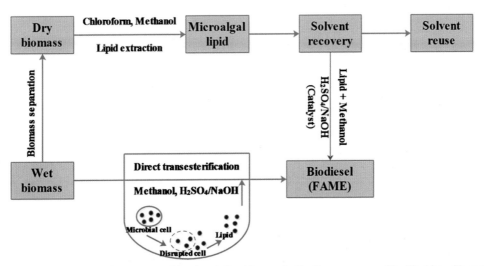

FIG. 9.6 The difference between traditional and direct transesterification processes. (Modified from Yousuf, A., Khan, M. R., Islam, M. A., Ab Wahid, Z., & Pirozzi, D. (2017). Technical difficulties and solutions of direct transesterification process of microbial oil for biodiesel synthesis. Biotechnol. Lett., 39(1), 13–23.)

reactions and is used for analytical quantification of lipids. Generally, most of the laboratory methods use excess amounts of solvents and prolonged reaction time to ensure complete recovery [21]. Therefore, the method needs to be further modified and optimized for industrial-scale applications.

Pretreatment Methods for Microalgal Cell Disruption

The pretreatment approach of microalgal biomass aims at disruption of algal cells by using chemical, mechanical, or biological techniques to obtain better product yield. Disruption of microalgal cells is often required to improve intracellular product release from cell compartments because the rigid cell walls and membranes can reduce the extraction efficiency as they are reducing biodegradability of the cells [2,3]. Therefore, in most cases, cell disruption prior to lipid extraction is considered as an essential step to eradicate or weaken the protective cell walls of algal cells to make the intracellular lipids more accessible in solvent extraction to enhance the lipid extraction yield.

Effects of biomass pretreatment on lipid extraction

A suitable biomass pretreatment method or cell disruption technique could mitigate the problems associated with the lipid extraction process to increase extraction efficiency (Fig. 9.7). In algae-based biofuel production process, the cell disruption prior to lipid extraction is not only necessary to break up the cell walls exposing lipids [19,44] but would also help to liberate combined lipids for better extraction, reduce insoluble solid residues to enhance mass transfer, increase lipid accessibility, and decrease the formation of stable emulsions to improve solvent recovery (i.e., ease phase separation) [32,45]. In this way, the amphiphilic polar lipids could be converted into the hydrophobic FFA for a better extraction and utilized as preferred biofuel precursors with reduced toxicity in a downstream catalytic upgrading [11]. Therefore, future research for biomass pretreatment should be comprehensively assessed regarding the lipid extraction efficiency, scalability, energy consumption, and compacivity with downstream processing.

Different pretreatment techniques

Various physical, chemical, and enzymatic methods have been reported to disrupt the microalgal cell walls (Fig. 9.8) [46]. There are two common terms—"cell wall disruption" and "cell disintegration." The cell wall disruption implies to disrupt outer cell wall structure rather inner cellular organizations. However, the cell disintegration entails the rupturing of the entire cell, where the cells are no longer recognized as intact cells under microscope. In general, both methods can be used to release the bioactive compounds embedded within the cells. It is demonstrated that the cell wall disruption and cell disintegration could be performed by either mechanical or nonmechanical treatments [47,48]. The mechanical methods include different

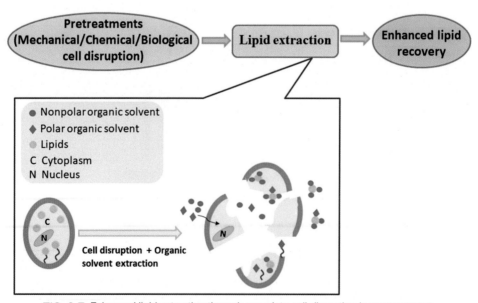

FIG. 9.7 Enhanced lipid extraction through complete cell disruption by pretreatment.

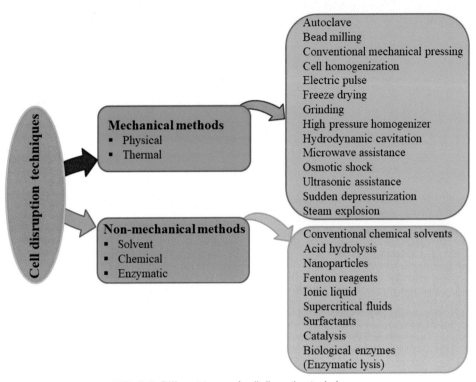

FIG. 9.8 Different types of cell disruption techniques.

physical (e.g., ultrasonication, osmotic shock, microwave, etc.), mechanical (e.g., ball milling, bead beating, high-pressure homogenizer, etc.) or thermal methods (e.g., autoclave, thermolysis, steam explosion, etc.), whereas the chemical methods are based on selective interaction of the cell walls with certain chemicals such as chloroform, methanol, hexane, isopropanol, acetone, and dichloromethane [3,49].

The most commonly used mechanical methods are oil expeller, bead milling, high-pressure homogenizer, high-speed homogenizer, ultrasonication, microwave, autoclaving, electroporation (i.e., pulsed electric field), electrocoagulation, hydrodynamic cavitation, osmotic shocks, etc., and nonmechanical methods include enzymatic treatments and chemical treatments such as Fenton's reagents, ionic liquids, supercritical fluids, and surfactants [20]. However, different pretreatment methods have their own mechanism to treat the algal cells, and the efficiency of a method mostly depends on the several operating parameters (Table 9.2). For instance, microwave shatters the cells using shock of high-frequency waves [50], while ultrasonication cracks the cell wall and membrane due to a cavitation effect [51]. On the other hand, bead-beating causes direct

mechanical damage to the cells based on high-speed spinning with fine beads [51], whereas expeller press crushes and breaks the cells by high mechanical pressure [7]. In Fenton's method, the hydroxyl radicals ($\cdot$OH) are generated via a reaction between H_2O_2 and Fe^{2+} ions $(Fe^{2+} + H_2O_2 \rightarrow Fe^{3+} + OH^- + \cdot OH)$, and they can attack specific zones of the algal cell wall composed of organic compounds [44]. In electroporation, on the other hand, the higher treatment intensity of electrical field can induce irreversible permeabilization of the cell wall leading to its disruption by triggering fore formation [19,44].

Advantages and limitations of different pretreatment methods

The major concern during the selection of cell disruption techniques is the maintaining of nutritive quality of products (i.e., lipids and other bioactive compounds) within the algae cells. It is believed that some of the nonmechanical methods (i.e., chemical and enzymatic treatments) can modify the nutritive quality of the intracellular constituents [14]. Moreover, these methods are often limited to the small scale and low process efficiency [44,47]. Unfortunately, the

TABLE 9.2

Summary of Process Parameters and Mode of Action for Different Cell Wall Disruption Techniques.

Pretreatment Methods	Mode of Action	Control Parameters	Energy Consumption	References
Expeller/oil press	- Mechanical compaction and shear forces	- Pressure - Configurations of various press (e.g., screw, expeller, piston, etc.)	High	[20,49]
High-speed homogenization	- Hydrodynamic cavitation and shear forces	- Stirring rates - Number of passes used	High/medium	[20,49]
High-pressure homogenization	- Mechanical stress - The effect of cavitation and shear forces	- Number of passes used - Operating temperature - Pressure	High/medium	[20,49]
Bead milling	- Mechanical compaction and shear forces	- Suspension feed rate for continuous operation - Process time for batch operation - Agitator design and speed - Design of milling chamber - Bead diameter - Bead filling - Bead density - Biomass concentration	High/medium	[20,49]
Ultrasonication	- The effect of cavitation - The acoustic streaming and liquid shear stress	- Frequency/cycle number - Power - Time - Temperature	Medium/low	[20,49]
Microwave	- Temperature and molecular energy increase	- Agitation - Time - Power - Frequency	High/medium	[20,49]
Chemical treatments	- Degradation mechanism, i.e., the protein, cellulose or/and pectin of microalgal cell are degraded by different chemical reactions	- Concentration of chemicals, such as NaOH and KOH	Low	[20,49]
Enzymatic treatments	- Enzymes hydrolyze the chemical bonds by binding to specific molecules in the cell wall	- Agitation - Enzymatic type	Low	[20,49]
Autoclave	- High thermal stress	- Temperature - Pressure - Time	High/medium	[20,49]
Steam explosion	- Sudden pressure drop, e.g., pressure wave and rapidly expanding steam can cause severe water hammer	- Temperature - Pressure - Retention time - Microalgae species	High/medium	[20,49]
Pulsed electric field	- Electroporation phenomena, including the electrochemical compression and electric field induced tension - Cell membrane permeabilization and pore formation	- Current intensity - Voltage - Electrode distance - Conductivity - Pulse duration	High/medium	[20,49]

TABLE 9.3
The Advantages and Limitations of Different Pretreatment Techniques.

Pretreatment Methods	Advantages	Limitations	References
Autoclave	- Extraction is efficient, and increased lipid yield - Decreases the degradation of the desired product	- High energy consumption - Time-consuming - Not scaleable	[14,49].
Freeze drying	- Does not damage cellular components	- Expensive - Dehydration may occur	[52]
Grinding	- No external energy is required - Easy and fast	- The heat generated during grinding can degrade certain compounds (e.g., cellular components) - Not feasible for large-scale production	[53]
Bead milling	- Can be scaled up to a few m^3 - No solvent uses - Suitable for samples with high moisture content	- May not extract majority of lipid - Low efficiency with rigid cells - Susceptible to lipid degradation - Further process is required to remove undesirable products as well as beads	[2,54]
Fenton reagents	- Fast - Less energy consumption	- Expensive chemicals - Use of toxic chemicals and regents - Chemical contamination during lipid extraction - High possibility to form inhibitors	[49,55]
Acid hydrolysis	- Easy and simple method	- Costly and toxic - Biomass drying is required - Risk of chemical contamination - High possibility to form inhibitors	[13]
Oxidation	- Higher efficiency	- Still new, large-scale application is yet to be studied - The catalysts cost needs to be addressed - High-energy requirements	[13]
Detergents	- Causes damage to cellular contents - Easy and fast	- Causes environmental pollution - Expensive	[3]
Enzymatic	- Mild operating temperature - Low time and energy requirements - No harmful solvent uses - Can devoid of harsh physical conditions	- Expensive - Long reaction time - Since it is depending on the cell wall characteristics; hence, less selective	[2,13]
Osmotic shock	- Simple, efficient	- Longer treatment time - May not scale up	[14]
Mechanical pressing	- No solvent is required - Simple and easy to use	- Requires dry biomass, and thus energy intensive - Costly and longer duration	[3]

Continued

TABLE 9.3
The Advantages and Limitations of Different Pretreatment Techniques.—cont'd

Pretreatment Methods	Advantages	Limitations	References
Microwave	- Simple and rapid - Can be scaled up - The energy requirement can range from 90 to 540 MJ kg^{-1} - Reduced extraction time - Environmentally friendly - Reduced solvent usage - Improved extraction yield	- High temperatures may result in lipid oxidation - Unpredictable efficiency specially, for nonpolar and volatile solvents - Easy scale-up process; yet to be standardized - Undesirable solid residue formation and filtration/centrifugation needed to remove them - Higher capital investment and operational cost	[2,20,47,49]
Pulsed electric field	- Higher yield - Short time requirement - No chemical usage - No cell debris formation - Relatively simple - Eco-friendly	- New method, has not been investigated with more algal species - Further research needed - Dependence on medium composition - Sensitive to medium conductivity	[3,13,20]
Surfactant treatment	- Less capital and operation costs - Easy to scale up - Surfactants (e.g., CTAB) are used in food industries - Biodegradable - Nontoxic	- High concentrations of the surfactants are harmful to aquatic life - Long duration of time - Selection and recovery are challenging - Efficiency depends on structure of the surfactants	[2,13]
Enzymatic disruption	- Enzymatic cell disruption is useful because it is specific - Less energy requirements - Used for higher yields of extractions - Nontoxic	- Higher capital cost - Limited application due the high cost of the enzymes - Longer time requirement - Sterile conditions - Proper selection of enzymes and their combination is crucial	[3,49]
Hydrogen peroxide	- Simple and low cost - Can be applied to large-scale production	- Less selective - Can damage sensitive substances (e.g., if product is enzyme)	[3]
Ultrasonic pretreatment	- Lower extraction time - Less solvent requirements - Higher penetration of solvent into cellular compartment - Greater release of lipids into bulk medium - Environmentally friendly	- Energy intensive - Difficulties in scale-up process - Operational cost may be prohibitive	[20,49]
Steam explosion	- Accessibility of lipid recovery without release of hazardous wastes - Cost is relatively low	- Efficiency depends on the species, and hence less selectivity	[49]

TABLE 9.3
The Advantages and Limitations of Different Pretreatment Techniques.—cont'd

Pretreatment Methods	Advantages	Limitations	References
Oil/expeller press	- Easy process - No solvent	- Slow process, and hence time-consuming - Large amount of sample requirement - Not suitable for samples with high moisture content	[20]
High-speed homogenization	- Easy and simple method - Efficient extraction - Short duration of contact time - Solvent free	- Energy intensive - High heat generation during operation	[20]
High-pressure homogenization	- Solvent free - Easy and simple method - Efficient extraction - Short duration of contact time	- Maintenance cost is high - Less efficient with filamentous microbes	[20]
Hydrodynamic cavitation	- High extraction efficiency - Lower extraction time - It can be employed for wet biomass slurries	- Energy intensive - The capital cost is high	[13]

practice of using chemicals and enzymes can lead to the greater complexity in minimizing environmental impacts. Consequently, these methods have found to be limited commercial applications to date [19,44] and are less favorable compared with some mechanical methods, such as bead milling [47,48], ultrasonication [47], and homogenizer [14]. However, the mechanical methods also have varying degree of effects on microalgae cells. For instance, bead milling could be used to completely break up the cell wall, leading to full disintegration of cells, but lipids are susceptible to degradation [47,48]. Although the mechanical methods, e.g., microwave, sonication, osmotic shocks, and high-pressure homogenizer, are preferred due to the high intracellular product release and better process efficiency, they are associated with high energy input, greater heat generation, and longer time requirement [19,47]. In addition, some of these methods can also destroy the biomolecules of interest, such as denature of the proteins [19]. The main advantages and limitations of different cell disruption techniques have been concised in Table 9.3.

CONCLUSION

Microalgae are regarded as one of the most promising feedstocks for producing biofuels and diverse kinds of value-added products. However, the extraction of intracellular lipids is the most crucial step, which may affect the product quality, quantity, and cost efficiency of the operation in algae-based biodiesel production. Lipid extraction efficiency is mainly affected by the rigid structure and composition of microalgal cell walls, water content of the biomass, limited accessibility of lipids, reduced mass transfer, and the formation of stable emulsions, etc. Several modifications, such as uses of green solvents, direct transesterification, and pretreatment of biomass, to the conventional solvent-based lipid extraction process have been proposed to minimize the problems so far. Direct transesterification and uses of green solvent can simplify the transesterification process to some extent, but lipid extraction yield is still hampered due to the rigid microalgal cell wall, which ultimately affects the process efficiency. A suitable biomass pretreatment for cell disruption prior to wet extraction can mitigate these problems to increase lipid extraction efficiency. However, the pretreatment method should be energy efficient, environmentally friendly, and scalable for industrial application to develop an economical and sustainable extraction process. Therefore, the research initiatives need to be emphasized on developing an understanding of the relationship between pretreatment mechanisms and challenges (e.g., lipid accessibility, mass transfer mechanisms, cell wall composition, etc.) of lipid extraction to develop a practical and scalable process for extracting lipids from wet biomass. More research and review are still required for establishing an efficient, practical, and scalable process to extract microalgal lipids for biodiesel production.

ACKNOWLEDGMENT

The financial support from the Ministry of Higher Education Malaysia (MOHE) through Universiti Malaysia Pahang (UMP), Grant # PGRS RDU 160150 is acknowledged.

REFERENCES

[1] A. Yousuf, S. Sultana, M.U. Monir, A. Karim, S.R.B. Rahmaddulla, Social business models for empowering the biogas technology, Energy Sources B Energy Econ. Plan. Policy 12 (2) (2017) 99−109.

[2] M. Alhattab, A. Kermanshahi-Pour, M.S.-L. Brooks, Microalgae disruption techniques for product recovery: influence of cell wall composition, J. Appl. Phycol. 31 (1) (2019) 61−88.

[3] S. Bharte, K. Desai, Techniques for harvesting, cell disruption and lipid extraction of microalgae for biofuel production, Biofuels (2018) 1−21.

[4] A. Karim, M.A. Islam, C.K. Mohammad Faizal, A. Yousuf, M. Howarth, B.N. Dubey, et al., Enhanced biohydrogen production from citrus wastewater using anaerobic sludge pretreated by an electroporation technique, Ind. Eng. Chem. Res. 58 (2) (2018) 573−580.

[5] L. Brennan, P. Owende, Biofuels from microalgae—a review of technologies for production, processing, and extractions of biofuels and co-products, Renew. Sustain. Energy Rev. 14 (2) (2010) 557−577, https://doi.org/10.1016/j.rser.2009.10.009.

[6] K.W. Chew, J.Y. Yap, P.L. Show, N.H. Suan, J.C. Juan, T.C. Ling, J.-S. Chang, Microalgae biorefinery: high value products perspectives, Bioresour. Technol. 229 (2017) 53−62.

[7] M.K. Enamala, S. Enamala, M. Chavali, J. Donepudi, R. Yadavalli, B. Kolapalli, et al., Production of biofuels from microalgae-A review on cultivation, harvesting, lipid extraction, and numerous applications of microalgae, Renew. Sustain. Energy Rev. 94 (2018) 49−68.

[8] E. Arenas, M. Rodriguez Palacio, A. Juantorena, S. Fernando, P. Sebastian, Microalgae as a potential source for biodiesel production: techniques, methods, and other challenges, Int. J. Energy Res. 41 (6) (2017) 761−789.

[9] A. Papazi, E. Andronis, N.E. Ioannidis, N. Chaniotakis, K. Kotzabasis, High yields of hydrogen production induced by meta-substituted dichlorophenols biodegradation from the green alga Scenedesmus obliquus, PLoS One 7 (11) (2012) e49037.

[10] H.G. Gerken, B. Donohoe, E.P. Knoshaug, Enzymatic cell wall degradation of Chlorella vulgaris and other microalgae for biofuels production, Planta 237 (1) (2013) 239−253.

[11] T. Dong, E.P. Knoshaug, P.T. Pienkos, L.M. Laurens, Lipid recovery from wet oleaginous microbial biomass for biofuel production: a critical review, Appl. Energy 177 (2016) 879−895.

[12] J. Iqbal, C. Theegala, Microwave assisted lipid extraction from microalgae using biodiesel as co-solvent, Algal Res. 2 (1) (2013) 34−42.

[13] H. Sati, M. Mitra, S. Mishra, P. Baredar, Microalgal lipid extraction strategies for biodiesel production: a review, Algal Res. 38 (2019) 101413.

[14] J.-Y. Lee, C. Yoo, S.-Y. Jun, C.-Y. Ahn, H.-M. Oh, Comparison of several methods for effective lipid extraction from microalgae, Bioresour. Technol. 101 (1) (2010) S75−S77.

[15] C. Crampon, A. Mouahid, S.-A.A. Toudji, O. Lépine, E. Badens, Influence of pretreatment on supercritical CO_2 extraction from Nannochloropsis oculata, J. Supercrit. Fluids 79 (2013) 337−344.

[16] M.A. Borowitzka, High-value products from microalgae—their development and commercialisation, J. Appl. Phycol. 25 (3) (2013) 743−756.

[17] P. Mehta, D. Singh, R. Saxena, R. Rani, R.P. Gupta, S.K. Puri, A.S. Mathur, High-value Coproducts from Algae—An Innovational Way to Deal with Advance Algal Industry Waste to Wealth, Springer, 2018, pp. 343−363.

[18] G.V. Subhash, M. Rajvanshi, B.N. Kumar, S. Govindachary, V. Prasad, S. Dasgupta, Carbon streaming in microalgae: extraction and analysis methods for high value compounds, Bioresour. Technol. 244 (2017) 1304−1316.

[19] A. Karim, A. Yousuf, M.A. Islam, Y.H. Naif, C.K.M. Faizal, M.Z. Alam, D. Pirozzi, Microbial lipid extraction from Lipomyces starkeyi using irreversible electroporation, Biotechnol. Prog. 34 (4) (2018) 838−845.

[20] A. Patel, F. Mikes, L. Matsakas, An overview of current pretreatment methods used to improve lipid extraction from oleaginous microorganisms, Molecules 23 (7) (2018) 1562.

[21] A. Yousuf, M.R. Khan, M.A. Islam, Z. Ab Wahid, D. Pirozzi, Technical difficulties and solutions of direct transesterification process of microbial oil for biodiesel synthesis, Biotechnol. Lett. 39 (1) (2017) 13−23.

[22] E. Ehimen, Z. Sun, C. Carrington, Variables affecting the in situ transesterification of microalgae lipids, Fuel 89 (3) (2010) 677−684.

[23] J.-Y. Park, M.S. Park, Y.-C. Lee, J.-W. Yang, Advances in direct transesterification of algal oils from wet biomass, Bioresour. Technol. 184 (2015) 267−275.

[24] B. Cheirsilp, Y. Louhasakul, Industrial wastes as a promising renewable source for production of microbial lipid and direct transesterification of the lipid into biodiesel, Bioresour. Technol. 142 (2013) 329−337.

[25] B.D. Wahlen, R.M. Willis, L.C. Seefeldt, Biodiesel production by simultaneous extraction and conversion of total lipids from microalgae, cyanobacteria, and wild mixed-cultures, Bioresour. Technol. 102 (3) (2011) 2724−2730.

[26] E.G. Bligh, W.J. Dyer, A rapid method of total lipid extraction and purification, Can. J. Biochem. Physiol. 37 (8) (1959) 911–917.

[27] J. Folch, M. Lees, G. Sloane Stanley, A simple method for the isolation and purification of total lipides from animal tissues, J. Biol. Chem. 226 (1) (1957) 497–509.

[28] S. Bhattacharya, R. Maurya, S.K. Mishra, T. Ghosh, S.K. Patidar, C. Paliwal, S. Mishra, Solar driven mass cultivation and the extraction of lipids from *Chlorella variabilis*: a case study, Algal Research 14 (2016) 137–142.

[29] J. Liddell, Solvent extraction processes for biological products, in: Engineering Processes for Bioseparations, Butterworth-Heinemann Ltd, Oxford, 1994.

[30] P. Walde, A.M. Giuliani, C.A. Boicelli, P.L. Luisi, Phospholipid-based reverse micelles, Chem. Phys. Lipids 53 (4) (1990) 265–288.

[31] M.R. Pursell, M.A. Mendes-Tatsis, D.C. Stuckey, Effect of fermentation broth and biosurfactants on mass transfer during liquid–liquid extraction, Biotechnol. Bioeng. 85 (2) (2004) 155–165.

[32] A. Schwenzfeier, A. Helbig, P.A. Wierenga, H. Gruppen, Emulsion properties of algae soluble protein isolate from *Tetraselmis* sp, Food Hydrocolloids 30 (1) (2013) 258–263.

[33] S.-A. Choi, J.-S. Lee, Y.-K. Oh, M.-J. Jeong, S.W. Kim, J.-Y. Park, Lipid extraction from *Chlorella vulgaris* by molten-salt/ionic-liquid mixtures, Algal Research 3 (2014) 44–48.

[34] B.-C. Liau, C.-T. Shen, F.-P. Liang, S.-E. Hong, S.-L. Hsu, T.-T. Jong, C.-M.J. Chang, Supercritical fluids extraction and anti-solvent purification of carotenoids from microalgae and associated bioactivity, J. Supercrit. Fluids 55 (1) (2010) 169–175.

[35] N.S. Topare, S.J. Raut, V. Renge, S.V. Khedkar, Y. Chavanand, S. Bhagat, Extraction of oil from algae by solvent extraction and oil expeller method, Int. J. Chem. Sci. 9 (4) (2011) 1746–1750.

[36] C. Dejoye Tanzi, M. Abert Vian, C. Ginies, M. Elmaataoui, F. Chemat, Terpenes as green solvents for extraction of oil from microalgae, Molecules 17 (7) (2012) 8196–8205.

[37] S.J. Kumar, G.V. Kumar, A. Dash, P. Scholz, R. Banerjee, Sustainable green solvents and techniques for lipid extraction from microalgae: a review, Algal Research 21 (2017) 138–147.

[38] W.M.A.W. Mahmood, C. Theodoropoulos, M. Gonzalez-Miquel, Enhanced microalgal lipid extraction using bio-based solvents for sustainable biofuel production, Green Chem. 19 (23) (2017) 5723–5733.

[39] F. Sahena, I. Zaidul, S. Jinap, A. Karim, K. Abbas, N. Norulaini, A. Omar, Application of supercritical CO_2 in lipid extraction—A review, J. Food Eng. 95 (2) (2009) 240–253.

[40] H.P. Tai, K.P.T. Kim, Supercritical carbon dioxide extraction of Gac oil, J. Supercrit. Fluids 95 (2014) 567–571.

[41] Y.-H. Kim, S. Park, M.H. Kim, Y.-K. Choi, Y.-H. Yang, H.J. Kim, S.H. Lee, Ultrasound-assisted extraction of lipids from *Chlorella vulgaris* using [Bmim][MeSO4], Biomass Bioenergy 56 (2013) 99–103.

[42] G. Vicente, L.F. Bautista, R. Rodríguez, F.J. Gutiérrez, I. Sádaba, R.M. Ruiz-Vázquez, et al., Biodiesel production from biomass of an oleaginous fungus, Biochem. Eng. J. 48 (1) (2009) 22–27.

[43] H. Kakkad, M. Khot, S. Zinjarde, A. RaviKumar, Biodiesel production by direct in situ transesterification of an oleaginous tropical mangrove fungus grown on untreated agro-residues and evaluation of its fuel properties, Bio-Energy Res. 8 (4) (2015) 1788–1799.

[44] M.A. Islam, A. Yousuf, A. Karim, D. Pirozzi, M.R. Khan, Z. Ab Wahid, Bioremediation of palm oil mill effluent and lipid production by *Lipomyces starkeyi*: a combined approach, J. Clean. Prod. 172 (2018) 1779–1787.

[45] R. Halim, T.W. Rupasinghe, D.L. Tull, P.A. Webley, Modelling the kinetics of lipid extraction from wet microalgal concentrate: a novel perspective on a classical process, Chem. Eng. J. 242 (2014) 234–253.

[46] S.Y. Lee, J.M. Cho, Y.K. Chang, Y.-K. Oh, Cell disruption and lipid extraction for microalgal biorefineries: a review, Bioresour. Technol. 244 (2017) 1317–1328.

[47] E. Günerken, E. d'Hondt, M. Eppink, L. Garcia-Gonzalez, K. Elst, R. Wijffels, Cell disruption for microalgae biorefineries, Biotechnol. Adv. 33 (2) (2015) 243–260.

[48] A.K. Lee, D.M. Lewis, P.J. Ashman, Disruption of microalgal cells for the extraction of lipids for biofuels: processes and specific energy requirements, Biomass Bioenergy 46 (2012) 89–101.

[49] C. Onumaegbu, J. Mooney, A. Alaswad, A. Olabi, Pre-treatment methods for production of biofuel from microalgae biomass, Renew. Sustain. Energy Rev. 93 (2018) 16–26.

[50] G. Cravotto, L. Boffa, S. Mantegna, P. Perego, M. Avogadro, P. Cintas, Improved extraction of vegetable oils under high-intensity ultrasound and/or microwaves, Ultrason. Sonochem. 15 (5) (2008) 898–902.

[51] S.J. Lee, B.-D. Yoon, H.-M. Oh, Rapid method for the determination of lipid from the green alga *Botryococcus braunii*, Biotechnol. Tech. 12 (7) (1998) 553–556.

[52] K.H. Wiltshire, M. Boersma, A. Möller, H. Buhtz, Extraction of pigments and fatty acids from the green alga *Scenedesmus obliquus* (Chlorophyceae), Aquat. Ecol. 34 (2) (2000) 119–126.

[53] E.M. Grima, E.-H. Belarbi, F.A. Fernández, A.R. Medina, Y. Chisti, Recovery of microalgal biomass and metabolites: process options and economics, Biotechnol. Adv. 20 (7–8) (2003) 491–515.

[54] A.R. Byreddy, C.J. Barrow, M. Puri, Bead milling for lipid recovery from *thraustochytrid* cells and selective hydrolysis of *Schizochytrium DT3* oil using lipase, Bioresour. Technol. 200 (2016) 464–469.

[55] A. Concas, M. Pisu, G. Cao, Disruption of microalgal cells for lipid extraction through Fenton reaction: modeling of experiments and remarks on its effect on lipids composition, Chem. Eng. J. 263 (2015) 392–401.

CHAPTER 10

Conversion of Microalgae Biomass to Biofuels

MIN-YEE CHOO • LEE ENG OI • TAU CHUAN LING • ENG-POH NG • HWEI VOON LEE • JOON CHING JUAN

INTRODUCTION TO MICROALGAE-BASED BIOFUEL

The combustion of fossil fuel is the primary contributing factor to the emissions of greenhouse gases (GHG), and subsequently resulted in the climate change and global warning. The world energy demands continue to grow progressively annually due to the growth in world population and the advancement of technology. Today, fossil fuel supplies about 90% of world energy demand, while only 10% is supplied by renewable energy [1]. The United State Environmental Protection Agency (USEPA) has stated that transportation has consumed 40% of primary energy in 2010 but accounted for 71% of GHG emission [2]. The major GHG, carbon dioxide (CO_2), could retain in the atmosphere for up to two centuries, trapping the heat and resulted in global warming [3]. Therefore, the scientific community is sourcing for alternatives to the fossil fuels by complementary use of renewable energy such as biofuel, solar energy, wind energy, and hydropower. Among these potential renewable energy sources, biofuels have perceived as ideal candidate to reduce the reliance of fossil fuel, at the same time providing carbon offset due to the modern carbon cycle [4].

The advantages of biofuel include sustainability, nontoxicity, biodegradability, and renewability. As shown in Fig. 10.1, the feedstock for biofuel production evolved from the first generation that uses edible feedstock, to second generation that uses lignocellulosic biomass and agricultural waste, to third generation that uses microalgae, and to the current fourth generation that utilizes engineered microalgae as the feedstock [5,6]. Microalgae have been recognized as the promising feedstock for biofuel production, as microalgae possess (1) high biomass and lipid productivity, (2) no competition for food production, (3) high tolerance toward environmental stress, (4) high CO_2 sequestration ability, and (5) biorefineries capability [3,5]. Microalgae could double its mass daily, which is hundred-fold faster than terrestrial plant [7]. Besides, the microalgae oil yield could be up to 5000 gallons per acre, whereas the yield of typical oil crops such as palm, corn, and coconut could be less than 1000 gallons per acre [8].

The composition of microalgae differs with species and the cultivation conditions. Generally, microalgae contain 20%–40% lipids, 0%–20% carbohydrates, 30%–50% proteins, and 0%–5% nucleic acids [9]. Some microalgae species could accumulate cellular lipids up to 85% [4]. The absence of lignin and hemicellulose components in microalgae composition showed greater effectiveness in biofuel production. Microalgae species such as *Scenedesmus obliquus*, *Chlorella vulgaris*, *Chlamydomonas reinhardtii*, and *Nannochloropsis oculata* are recognized as promising species for biofuel production [10–14]. Different types of biofuel can be produced from the microalgae such as fermentative biofuels (bioethanol and biobutanol), solid biofuel (biochar), liquid biofuel (biodiesel and green hydrocarbon), and gaseous biofuel (hydrogen, methane). Apart from microalgae-based biofuels, microalgae-based protein and bioactive compounds such as antioxidants could be used also for valuable health supplements and pharmaceutical purposes [15]. High biomass and lipid productivity of microalgae can be achieved by proper selection or genetic modification of microalgae strain, optimizing the cultivation condition (carbon source, nutrients, light intensity, etc.), which has been discussed in the previous chapter. However, the biofuel production from this highly potential feedstock needs a powerful downstream processing technology. Therefore, in this chapter, the latest research and development on the conversion route to produce high-quality microalgae-based biofuel will be discussed in detail.

Microalgae Cultivation for Biofuels Production. https://doi.org/10.1016/B978-0-12-817536-1.00010-2

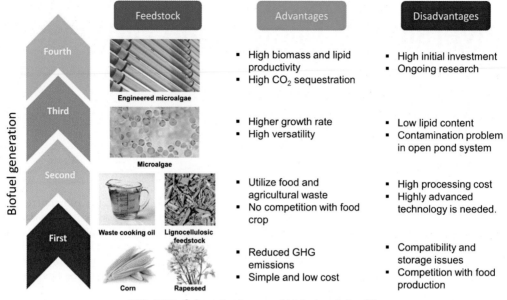

FIG. 10.1 Schematic diagram of biofuel evolution [5].

CONVERSION OF MICROALGAE BIOMASS INTO BIOFUEL

For the production of biofuel, the microalgae biomass can be transformed via biochemical and thermochemical process. Fig. 10.2 shows the transformation route of microalgae biomass into biofuel and their products. The biochemical process such as fermentation and anaerobic digestion converts biomass into methane, biohydrogen, syngas, and methanol. Although biochemical processes are more environmentally friendly and less energy-consuming, they are not suitable for large-scale production because of their low conversion efficiency, tedious reaction steps, and non-cost-effective nature [16]. Thermochemical processes include gasification, liquefaction,

and pyrolysis, which convert biomass into biofuel such as biochar, biogas, syngas, and hydrocarbon fuel.

Transesterification

Biodiesel in the form of fatty acid methyl esters (FAMEs) are produced from transesterification. Triacylglycerol or triglyceride is converted into nontoxic, low-molecular-weight, less viscous, and biodegradable FAME (biodiesel) to be directly applied for engine use [17–21]. The reaction is reversible; thus, surplus amount of methanol is supplied together with the catalyst, to shift the reaction equilibrium toward FAME production and catalyze the reaction rate. Fig. 10.3 illustrates that the transesterification reaction in the

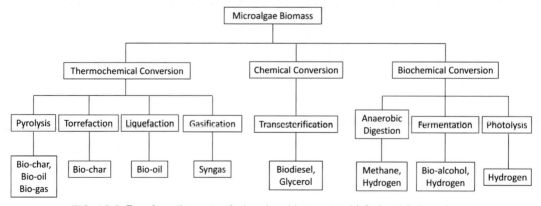

FIG. 10.2 Transformation route of microalgae biomass into biofuel and their products.

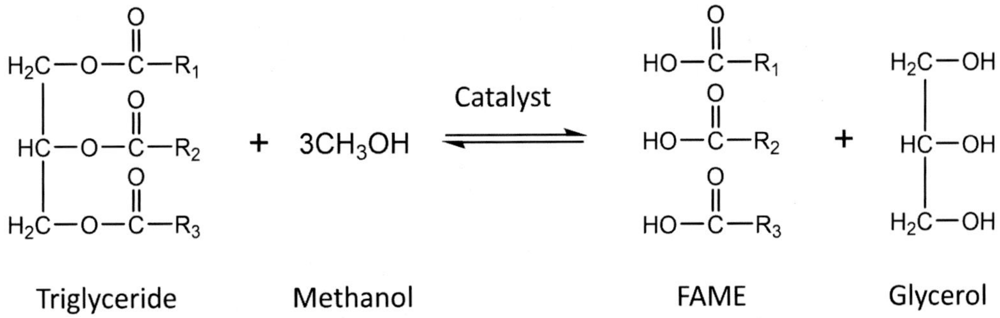

FIG. 10.3 The triglycerides transesterification reaction producing fatty acid methyl esters.

presence of methanol and catalyst produces FAMEs and glycerol. The glycerol, by-product from this reaction, could be used for pharmaceutical and cosmetic purposes [22]. The reaction rate is governed by the reaction conditions (temperature, duration, alcohol quantity, and catalyst amount). Homogeneous acid and alkaline catalysts such as potassium hydroxide or sodium hydroxide in methanol, hydrochloric, or sulfuric acid in methanol are used [23]. The catalytic activity using acid catalyst is low and thus requires a high-temperature operation, in the first step of the transesterification reaction. For instance, the reaction rate of alkaline catalyst is 4000 times faster than acid-catalyzed reaction; thus it is commonly used in the larger-scale operation [9,21].

Enzymatic catalysis is also applied, which is carried out at 35–45°C, provided enzyme itself has to be tolerant of methanol inactivation. Lipase-producing microorganisms include *Pseudomonas cepacia*, *Mucor miehei*, *Candida antarctica*, and *Rhizopus oryzae* [18]. However, high enzyme cost is the main constraint to its application for industrial-scale operation. The quality of fuel produced via enzymatic reaction may not meet biodiesel ASTM standards [21].

Direct transesterification or *in situ* transesterification is a rapid method that combines lipid extraction and transesterification in a single-step process. This single-step process produces higher FAME yield than the conventional two-step method, by shifting the reaction equilibrium toward more FAME yields and enhancing extraction efficiency [18]. Chemical solvent works as a solvent for lipid extraction out of microalgae cell, as well as the reactant for transesterification reaction [23]. The yield of more than 98% FAME was obtained through *in situ* transesterification, with n-hexane and chloroform as solvents, methanol as reactant, and sulfuric acid as catalyst, respectively [24]. Biodiesel yield was improved by 20% through direct transesterification, as compared with conventional method [17]. *Chlorella pyrenoidosa* with the lipid content of 56.2% obtained 95%

biodiesel yield via direct transesterification by using hexane as cosolvent and 0.5 M sulfuric acid at 90°C for 2 h [23]. Nevertheless, challenges confronted with large-scale direct transesterification would affect the reaction efficiency. These include high chemicals consumption, energy-intensive, the high water content of biomass cells, the requirement for temperature and stirring speed control, and difficulty in recovering glycerol [17,19,21]. Moisture content of 37.1% would completely inhibit the *in situ* transesterification reaction, by producing fatty acid and diglyceride rather than fatty acid esters [23].

Biophotolysis

Hydrogen gas is a highly versatile, efficient fuel energy carrier due to the fact that it is renewable, does not produce CO_2 during burning, has higher heating value (HHV) per unit mass (141.65 MJ/kg), and is convertible into electricity by fuel cells [25]. Currently, hydrogen is produced from reforming process of fossil fuels, which generated 9 kg of CO_2 per kg of H_2 produced and released trace amounts of nitrogen dioxide and sulfur dioxide [26]. These air pollutants will result in the formation of acid rain when it enters the atmosphere without proper treatment. Biohydrogen production is a method to produce hydrogen gas from microorganisms. Photosynthesis involves biological electrolysis, in which the water molecules are being split into a proton (H^+) and oxygen (O_2). In typical photosynthesis pathway, carbohydrate and oxygen are formed when hydrogen is reacted with carbon dioxide. However, under an anaerobic condition or specific microalgae species, instead of converting into carbohydrate during photosynthesis, hydrogen is being produced. A number of microalgae species that belong to *Chlamydomonas*, *Chlorella*, *Scenedesmus*, *Tetraspora*, etc. may utilize hydrogenase enzyme to produce hydrogen.

The production of biohydrogen can occur via three pathways, namely, (1) direct biophotolysis where hydrogen is generated from the electrons and proton from water splitting at photosystem II (PSII), (2) indirect biophotolysis where electrons and proton are generated by the degradation of intracellular carbon compounds, and (3) fermentation where electrons derived from oxidation of cellular endogeneous substate react with H^+ to form H_2 as end product [22]. In direct biophotolysis, the water is being split into molecular oxygen (O_2) and proton (H^+), and the electrons originated from the water splitting at PSII are transferred to PSI by the photosynthetic transport chain [4]. At PSI, upon utilization of sunlight, the elevated potential energy of these electrons reduces the ferredoxin (Fd) to Fd_{red} as stated in Eq. (10.1).

$$2H_2O + light + Fd_{ox} \rightarrow O_2 + 4H^+ + Fd_{red}(4e^-) \quad (10.1)$$

In indirect biophotolysis, instead of water splitting, the electrons and protons can be produced from the glycolysis of endogenous substrate (carbon source) [27]. In fermentation pathway, the starch reserves that build up during the photosynthesis are biochemically degraded into pyruvate and then oxidized by pyruvate-ferredoxin oxidoreductase (PFOR) enzyme to form acetyl-CoA and CO_2, together with the production of Fd_{red} [27]. Other products such as ethanol and acetate may be formed from the acetyl-CoA as well [28,29]. The Fd_{red} combines with the proton to produce molecular H_2 in a reversible reaction catalyzed by [FeFe]-hydrogenase, as shown in Eq. (10.2) [30].

$$4H^+ + Fd_{red}(4e^-) \xleftrightarrow{[FeFe]hydrogenase} 2H_2 + Fd_{ox} \quad (10.2)$$

The [FeFe]-hydrogenase found in the chloroplast of microalgae is 10- to 100-fold more reactive than [NiFe]-hydrogenase found in cyanobacteria. However, the hydrogen production stops after a few minutes as [FeFe]-hydrogenase is very extremely sensitive to the O_2 produced during the PSII activity [30]. The photosynthetic rate is usually four to seven times higher than respiration rate under normal condition; sulfur deprivation can be used to reduce the oxygenic PSII activity [25,31]. As a result, the culture medium becomes anaerobic as the photosynthetic oxygen generation rate is slower than the oxygen uptake rate by respiration and maintaining the activity of the [FeFe]-hydrogenase. The duration of photosynthetic hydrogen production has been extended from 2 min to about 5 days with this approach [32]. However, this process requires alternation of the stage between photosynthesis and sulfur deprivation as the cells will eventually run out of starch. Other than sulfur deprivation, phosphorus, nitrogen, and potassium deprivation have been reported to be effective in hydrogen photoproduction [27]. Some microalgae strain have been developed to increase their

hydrogen production [16,33]. For example, *C. reinhardtii* D1 mutant strain shows a higher rate of hydrogen production under sulfur deprivation due to the low chlorophyll content and improved respiration rate [33]. Immobilization of microorganism has several advantages such as high cell density, ease control of culture parameter, and cell protection from undesirable conditions (change in pH, salinity, heavy metal, etc.) [34]. The common immobilization matrices used in photohydrogen production include alginate film, agar, porous glass, and carrageenan gels [27]. For example, *C. reinhardtii* under sulfur/phosphorus deprivation immobilized in alginate films exhibited higher cell density (2000 µg Chl mL), higher hydrogen evolution (12.5 mmol/mg Chl h), and >1% quantum yield and resisted to [Fe-Fe] hydrogenase deactivation [35]. The improved production of hydrogen is due to the presence of acetate, which maintains higher respiration rates and effective O_2 removal by the alginate film. In the study by Stojkovic et al. [36], encapsulation of *C. reinhardtii* in the TiO_2 under sulfur deprivation has achieved a twofold higher field of H_2 production than their nonencapsulated counterpart. Besides, Das et al. [10] reported that *C. reinhardtii* immobilized in "artificial leaf device," which is made up of fabric and alginate, has improved the hydrogen yield up to 20-fold as compared with that obtained from the batch reactor.

Fermentation

Fermentation is a metabolic process that converts organic substrates such as sucrose, bagasse, cellulose, or starch to ethanol via microbial activities [37,38]. Basically, fermentation is classified into two types: (1) aerobic and (2) anaerobic depending on the requirement of oxygen in the process. Microalgae with relatively high carbohydrates concentration are suitable for fermentation to produce bioethanol [38]. However, most of the microalgae species have very low carbohydrates; the biomass composition of the algae can be modified by controlling the cultivation conditions such as nutrient level and light source [4]. Sugar source from carbohydrates component of algae biomass such as laminarin, alginate, agar, sulfated polysaccharides, and mannitol is broken down into ethanol by yeast [39]. The bioethanol yield produced from microalgae is two to five times higher than ethanol produced from sugarcane and corn, respectively [40]. Eq. (10.3) shows breakdown of the sugar stored in microalgae to ethanol by yeast:

$$C_6H_{12}O_6 \xrightarrow{yeast} 2C_2H_5OH + 2CO_2 \quad (10.3)$$

C. vulgaris is an ideal microalgae species for bioethanol production; this species possesses high starch content (ca. 37%), which can achieve up to 65% of ethanol conversion [41]. *Chlamydomonas, Dunaliella, Scenedesmus,* and *Spirulina* with more than 50% starch are also considered as potential species for bioethanol production [42]. Fermentation of microalgae biomass or starch requires pretreatment to convert algae biomass or starch into sugars by milling, and then it is mixed with *Saccharomyces cerevisiae* yeast and water in a fermenter for fermentation [43]. The diluted alcohol produced after fermentation is pumped to a holding tank for distillation. Commonly, the diluted alcohol product contains 10%–15% of ethanol [37]. The distillation process can purify the product and produced concentrated ethanol with 95% purity, which is suitable as a supplement or substitute for transport fuel [44]. This fermentation process requires less energy consumption; the CO_2 formed during the process can be channeled for microalgae cultivation, while the residue can be used as animal feed [37] or recycled as medium for microalgae growth. Red, green, and brown microalgae are all studied for fermentation to produce ethanol. Among the microalgae types, brown algae are recommended as the main feedstock due to the high carbohydrate concentration and ease of mass cultivation [45]. Fig. 10.4 shows the overall fermentation process.

In the past decade, ethanol is widely used as the substitution for gasoline, due to the potential of CO_2 and some hazardous gases (CO, NO) reduction during combustion [46]. Bioethanol is market available and readily used as transportation fuel; statistics shows that in year 2008, 86% of cars sold in Brazil are using gasoline with blending of ethanol [47]. However, bioethanol has some drawbacks such as corrosive, low energy value, toxic to ecosystem, and low vapor pressure. Bioethanol is currently ready to integrate with existing transportation fuel, but, to achieve long-term benefits and sustainability, more efforts are needed to upgrade the bioethanol to higher quality fuel that can be stand alone and compatible with fossil fuel.

Anaerobic Digestion for Biogas Production

Anaerobic digestion (AD) is the process of converting organic matters into a biogas, primarily methane and carbon dioxide [22] in the absence of oxygen or with the aid of hydrogen (refer to Fig. 10.5). The produced biogas has an energy content of 20%–40% of the lower heating value of the feedstock [37]. AD is commonly used for organic waste with high moisture content (80%–90%). Hence, it is applicable for conversion of wet microalgae biomass, which is an environmentally feasible way to create renewable energy source for domestic and industrial consumption.

The AD conversion of organic matters releases methane gas by extracting the carbon and nitrogen content from lipids of microalgae biomass. The carbohydrates, lipids, and proteins content of the microalgae will directly affect the AD process. Lipids of microalgae biomass can produce more biogas as compared with carbohydrate or protein [45]. The microalgae such as *Ulva lactuca* can produce 271 m^3 of methane per ton of organic matter [48]. The pathways involved in AD process are shown in Fig. 10.6. Basically, AD proceeds through four sequential stages: (1) hydrolysis, (2) acidogenesis, (3) acetogenesis, and (4) methanogenesis.

Hydrolysis, the first stage in AD process, is the common process that is applied in pretreatment of wastewater. The organic substrate of microalgae cell lipids is degraded into simple monomers (soluble sugars and amino acids, respectively) [37,41,49]. The second stage involved in AD process is acidogenesis, where volatile fatty acid (VFA), carbon dioxide, and hydrogen are being produced by acidogenic bacteria [50]. In the third stage of AD, acetogenesis, the VFA is oxidized to acetate substrates that are easy for methanogenesis [49]. The acetogenesis can be aided by adding hydrogen pressure for the oxidation of acids produced from the acidogenesis process [50]. Methanogenesis is the last stage of AD, which converts acetate substrates to produce methane (60%–70%) and carbon dioxide (30%–40%) [51]. The extracted microalgae lipids can be used as a fresh medium for microalgae cultivation [49,52]. The methane gas produced from AD has 650–750 Btu/ft^3 of heating value [53]. Combustion of methane fuel releases less carbon dioxide as compared with combustion of hydrocarbon fuel, which contributes to

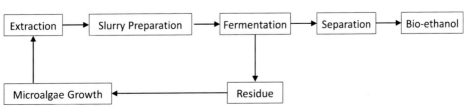

FIG. 10.4 Fermentation process of microalgae feedstock.

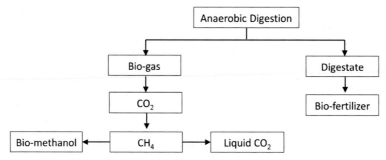

FIG. 10.5 Anaerobic digestion of microalgae biomass.

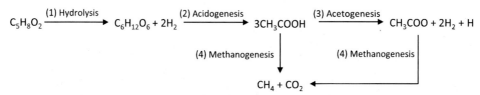

FIG. 10.6 Pathways involved in the anaerobic digestion of microalgae biomass.

pollution reduction [4]. Report shows that methane has lower combustion heat (891 kJ/mol); however, the heat per unit mass of methane (16 g/mol) is higher than other complex hydrocarbon fuel [4]. Therefore, methane fuel that is cleaner and environmentally friendly can be an alternative energy source.

The microbes involved in the AD for methane production are sensitive to the algae biomass chemical composition. Algae with high protein content such as *C. vulgaris* (58%), *Dunaliella salina* (57%), *Euglena gracilis* (61%), and *Spirulina maxima* (71%) [22,54] may possess lower carbon to nitrogen mass ratio (ca. 10). High protein content in the system tends to reduce the AD performance. Hence, Suganya et al. [22] proposed to increase the carbon to nitrogen mass of algae by adding codigester such as waste paper. A similar issue was highlighted by Yen and Brune [55], where double the methane yield (1.17 mL/L per day compared with pure microalgae biomass production of 0.57 mL/L) is achieved by blending 50% of waste paper to the algae biomass. Basically, brown algae produce higher methane yield than green algae [56]. Green algae have higher sulfate content, which leads to formation of H_2S [57]. The unfavorable H_2S can inhibit the production of methane gas and produce foul odor and sulfur dioxide emissions, which results in corrosive environment [45,57,58]. However, the emission of H_2S is controllable by using air scrubber treatment or adding metal ions [57,58]. Due to the high energy cost, the AD of algae biomass, which is suitable for biogas and biofertilizer production, is gaining more attention.

Pyrolysis

Pyrolysis is a thermal treatment method carried out without oxygen or air, at a temperature of 400−600°C and atmospheric pressure, transforming dry biomass into solid fuel (biochar), liquid fuel (biooil), and gaseous product [59]. The pyrolysis process is aimed to produce biofuel with a medium-low calorific value [22]. Depending on the operating condition (temperature, heating rate, and duration), pyrolysis technique can be classified into different modes as shown in Table 10.1.

Slow heating rate and long heating duration produce biochar, which can be used not only for fuel purpose but also for generation of power and sequestration of carbon and as an adsorbent and

TABLE 10.1
Different Modes of Pyrolysis.

Pyrolysis Mode	Temperature (°C)	Heating Rate (°C/min)	Residence Time
Slow	400	0,1−1.0	>10 s
Fast	580−1000	10−200	0.5−10 s
Flash	700−1000	>1000	< 0.5 s

Adapted from T. Suganya, M. Varman, H.H. Masjuki, S. Renganathan, Macroalgae and microalgae as a potential source for commercial applications along with biofuels production: a biorefinery approach, Renew. Sustain. Energy Rev., 55 (2016) 909−941.

biofertilizer [60,61]. With the high carbon content (>50%), this biochar can help in reducing the GHG effect and increase the water retention in the soil due to its highly porous structure [61]. Besides, biochar contains functional groups and inorganic minerals, which make it a good bioadsorbent in water treatment to remediate organic and inorganic contaminants [62,63]. While faster heating rate and shorter duration in fast pyrolysis tend to produce biooil, greater liquid and gas yield (around 70%−80%) is attained with fast pyrolysis compared with that in slow pyrolysis (15%−65%) [64]. Francavilla et al. [65] reported 45.13 wt.% of biooil yield is produced from the fast pyrolysis of the residue of *Dunaliella tertiolecta* at 600°C. Lack of phenolic compounds in microalgae biomass makes its fast pyrolysis more simple and efficient than other lignocellulosic biomass [41]. However, the biooil produced from pyrolysis suffered from drawbacks such as high acidic content, high oxygen content, and high viscosity. Besides, high nitrogenous content resulted from decomposition of chlorophyll and protein present in the microalgae will increase the NOx emission in the transportation fuel [66,67]. Therefore, these pyrolysis oils require future upgrading process such as hydrogenation and deoxygenation to produce biofuel that has same properties as fossil fuels [68,69]. These drawbacks can be overcome by addition of a catalyst in the pyrolysis. For example, catalytic pyrolysis with HZSM-5 has reduced the oxygen content in biooil from 30.09% by direct pyrolysis to 19.53% and therefore increased the HHV from 24.6 to 32.7 MJ/kg, respectively [12]. In the study of Gao et al. [70], catalytic pyrolysis of cyanobacteria with MgAl-LDO not only improved the liquid yield from 22.0% (direct pyrolysis) to 41.1%, but also reduced the nitrogenous compounds from 50.91% to 54.314%. Upon incorporation of Ni onto zeolite Y in the pyrolysis of *C. vulgaris* at 500°C, the nitrogenous and oxygenous compounds were reduced from 33.33% to 29.09% and from 31.75% to 16.36%, respectively [11]. In flash pyrolysis, a very high temperature with very fast heating rate and short heating duration produces syngas, which has half the energy density of natural gas and can be used as a feedstock for methanol synthesis [22]. Besides the temperature effect, the carbohydrate and protein content in microalgae biomass will affect the pyrolysis product. In the study of Hong et al. [71], higher biooil yield (13 wt %) is produced from protein-rich biomass (spirulina) due to their low PAHs content, while higher gas yield is produced from carbohydrate-rich biomass (porphyra). It is worth to mention that the gaseous product from pyrolysis of porphyra contains high HHV (13−19

MJ/Nm^3) and comparable $H_2 + CO$ content with the conventional gasification process [71].

Torrefaction

Torrefaction is the thermochemical conversion method to produce coal fuel (biochar) from biomass. The main product of the torrefaction is solid coal fuel, while liquids and gases are produced as by-products [72]. Biochar is a carbon-rich material from a biomass, which is produced by thermal composition of organic feedstock in the oxygen-free condition [73]. As mentioned in the previous section, biochar can be obtained via slow pyrolysis of biomass. The biochar produced by pyrolysis has high surface area, which is more suitable in the soil fertility improvement and waste treatment. However, as compared with the torrefaction, the increase in the calorific value is minimal. Therefore, high-quality solid fuel can be obtained via torrefaction as compared with pyrolysis [74].

Torrefaction can be divided into two groups, namely, (1) dry torrefaction and (2) wet torrefaction. Dry torrefaction is operated at temperature range between 200 and 300°C, under atmospheric pressure and inert nitrogen gas atmosphere with absence of oxygen. Dry torrefaction is also known as mild pyrolysis or low temperature pyrolysis, as lower reaction temperature is used as compared with conventional pyrolysis (>400°C) [74]. In this torrefaction, nitrogen gas is commonly used as a torrefaction gas to prevent the oxidation of the biochar during the reactions. Recently, there are some studies that used noninert gas (combustion gas or CO_2) to reduce the energy and the cost [75,76]. However, noninert promotes the oxidative reactions. As a result, CO and CO_2 gas are present in the gas phase, while water, phenol, and acetic acid are present in the liquid phase of end product. Besides, the solid yield using noninert gas is lower compared with that of inert nitrogen gas, which may be due to the Boudouard reaction [77]. Torrefaction temperature is a critical factor in the torrefaction process. For example, in the torrefaction of *Chlamydomonas* sp. JSC4, the solid yield has decreased from 93.9% to 51.3% when the torrefaction temperature increased from 200 to 300°C [78]. Similar trend was observed in another study where solid yield decreased from 86.37% to 63.23% when temperature increased from 200 to 300°C in torrefaction of *S. obliquus* CNW-N [79]. The main challenge of the dry torrefaction is that the predrying step is required to reduce the moisture content to less than 10 wt% [80]. Large amount of energy (3−5 MJ) is required to reduce the moisture the content from 50−60 wt% to 10−15 wt%, which in turn largely increased the operating cost and reduced the overall performance of dry torrefaction [81].

Wet torrefaction is an attractive technique to produce coal fuel from biomass without predrying step. Similar to dry torrefaction, the process is operated in the inert condition but with a temperature of 180–260°C, which is relatively lower than that of dry torrefaction [80]. In wet torrefaction, subcritical water is used a reaction medium. Its low dielectric constant enhances the ionic reaction and thus effectively solubilizes the biomass elements [82]. In terms of reaction mechanism, presence of subcritical water hydrolyzed the C-O bond in the ether and ester bonds between monomeric sugar, which reduce the activation energy; thus, degradation of hemicellulose in the wet torrefaction is more efficient compared with dry torrefaction [83]. As discussed by Bach and Skreiberg [80], wet torrefaction is more efficient than dry torrefaction as it requires relatively low temperature and holding due to the highly reactive reaction media (hydrothermal treatment). Similar solid yield (~52%) can be achieved via wet torrefaction at 180°C for 10 min or 170°C for 30 min as compared with dry torrefaction (300°C for 60 min) [76,84].

Liquefaction

Liquefaction process is carried out in a low temperature typically at 250–350°C and high H_2 pressure in the range of 5–20 MPa, and a catalyst is used to assist the process to produce biooil [59]. This technique is very suitable for moisture feedstock like microalgae as it has very high water content (>80%) and no drying process is required [66]. Due to the small size of microalgae, it is a good feedstock for liquefaction as it enhances the thermal transfer during the process [85]. However, this process is expensive as compared with pyrolysis and gasification, owing to the complex reactors setup and fuel-feed system [37] that decomposes the biomass into smaller and shorter molecular materials that possess high energy density by using high subcritical water activity [86]. Through liquefaction, microalgae can be decomposed into four types of products: biooil, biochar, aqueous phase, and gaseous product [87]. One advantage of the liquefaction process is that aqueous phase that contains C, N, and P nutrients can be recycled and used in the cultivation of microalgae [88,89]. Usually, 10% –15% higher biooil yield than the lipid content is produced from the liquefaction as the carbohydrates and proteins fraction can also be converted into biooil through hydrolysis and repolymerization in hot compressed water condition [87,90,91]. Therefore, this process could apply to the carbohydrate- and

protein-rich residue biomass from biodiesel processes and lipid-extracted biomass residues (LMBRs) [4]. Otherwise, these materials will be discarded as waste, and their disposal cost will surge the biodiesel production cost. The biooil produced from the liquefaction process contains lower oxygen and moisture content, which resulted in the more stable product than that from the pyrolysis process [92]. Addition of homogeneous catalysts such as KOH [93], Na_2CO_3 [94], and acetic acid [95] has increased the biooil yield. As an example, maximum oil yield has increased to 22.67% with KOH-catalyzed liquefaction of *Cyanidioschyzon* at 300°C as compared with the noncatalyzed system (16.98%) [93]. Besides, coliquefaction with other species [96,97] or feedstock like swine manure [98] can also improve the yield of biooil. Coliquefaction of *Cyanidioschyzon merolae* and *Galdieria sulphuraria* in 80:20 mass ratio at 300°C increased the biooil yield to 25.5% from individual liquefaction of *C. merolae* and *G. sulphuraria* (18.9% and 14.0%, respectively) [97]. However, similar to pyrolysis, the product from liquefaction also contains oxygen and nitrogen, which invite undesirable properties such as high viscosity and therefore require pretreatment, further upgrading or introducing catalyst during the liquefaction process [87,99,100]. By applying pretreatment step (mild treatment < 200°C), the nitrogen content has reduced by 55% compared with single-stage hydrothermal liquefaction (250–350°C) [101]. Li et al. [100] have studied the catalytic upgrading of biooil from liquefaction of *Chlorella* over series of modified SBA-15 catalysts. Compared with thermal upgrading, the amount of heavy oil (>400°C) has reduced from 9.57 wt% to 1.89 wt%, respectively, while 65.7 wt% hydrocarbon yield was achieved by CuO/Al-SBA-15 as compared with thermal upgrading (25.1 wt%) [100]. Due to the simplicity and low energy consumption, direct introduction of catalyst (one-pot method) is preferred over the catalytic upgrading process (two-step method) [102]. At 2011, Duan and Savage [103] reported the liquefaction of *Nannochloropsis* at 350°C over noble metals (Pd, Pt, and Ru) supported on carbon and transition metals (Ni, Co, and Mo) supported on Al_2O_3, SiO_2, or zeolite. They discovered that biooil from Pt/C, Ni/SiO_2-Al_2O_3, and Co-Mo/Al_2O_3 showed a lower O/C ratio than catalysis-free reaction, while zeolite produced most N_2 in the gas product. Zeolites catalysts are active in removing the nitrogenous compounds and acid compounds in the liquefaction process. For instance, in the liquefaction of *Euglena* sp. over zeolite H-beta at 280°C, the amount of nitrogenous compounds and that of acid compounds have been reduced from

30.87% and 16.44% to 16.68% and 9.50%, respectively [104]. The inclusion of metal (Ni, Fe, Ce) on zeolites could further improve biooil yield [102,105,106]. In the liquefaction of *Nannochloropsis* at 365°C, highest biooil yield (38.1%) was achieved by Fe/HZSM-5, which is 25% improvement over catalysis-free counterpart [102].

Gasification

Gasification is a partial oxidation of microalgae biomass in the air atmosphere, oxygen, or steam at high temperature of 800–1000°C [107]. Through gasification, carbonaceous materials in microalgae are transformed into synthesis gas (syngas). The syngas is a mixture of gases such as hydrogen, carbon monoxide, methane, and ethylene and has a low calorific content (4–6 MJ/m^3) [108]. Both hydrogen and methane are considered as a clean energy source in the future global energy portfolio [109]. The burning of syngas can produce heat for electricity generation or be used as feedstock for methanol production and hydrogen as a transportation fuel [64]. The gasification processes begin with pyrolysis of biomass for biochar production; then the produced biochar is gasified in the presence of gasifying agents such as oxygen and water to produce syngas. There are two branches of gasification processes: (1) conventional gasification and (2) supercritical water gasification (SCWG). In conventional gasification, higher temperature (800–1000°C) and pressure (1–10 bar) are used, where dry microalgae react with the oxidizer such as air, oxygen, or steam [107]. Khoo et al. [107] studied the gasification of *Nannochloropsis* sp. with a fixed-bed reactor at temperature of 850°C; the products comprise 58.18 wt.% char, 13.74 wt.% biooil, and 28.08 wt.% gas, with HHVs of 17.5, 34.1, and 32.9 MJ/kg, respectively. However, in conventional gasification, drying process of microalgae is very energy intensive to remove moisture from wet microalgae biomass. In SCWG, supercritical water (374°C and 22.1 MPa) is used as a reaction medium, and microalgae are directly converted to a gas product without drying process [110,111]. The SCWG is carried out at the low reaction temperature (300–600°C) [111]. For instance, 50% less energy is needed to heat the water from 25 to 300°C rather than vaporize it [112]. Unlike liquefaction, supercritical water in SCWG breaks the C-C bonds for production of burnable gases such as CH$_4$ and H$_2$, while subcritical water produces the liquid hydrocarbon in gasoline and diesel range [111,112]. According to literature, 10.37 mol/kg hydrogen yield was obtained from the SCWG of *Posidonia oceanica* at 600°C [111]. In SCWG, there are three main reactions,

namely, steam reforming (Eq. 10.4), water-gas shift (Eq. 10.5), and methanation (Eq. 10.6) [113].

$$\text{Steam reforming: } C_xH_yO_z + (2x-z)H_2O \rightarrow \left(2x-z+\frac{y}{2}\right)H_2 + xCO_2 \tag{10.4}$$

$$\text{Water gas shift (WGS): } CO + H_2O \rightarrow CO_2 + H_2 \tag{10.5}$$

$$\text{Methanation: } CO + 3H_2 \rightarrow CH_4 + H_2O \tag{10.6}$$

However, low reaction rate in WGS reaction leads to the high CO content in the noncatalyzed SCWG, thus requiring higher reaction temperature (~ 600°C) to achieve better gasification efficiency [113]. A catalyst such as noble metal catalyst is commonly used in the gasification of microalgae to enhance the gasification of tar and char and produce more H$_2$ and CH$_4$ and therefore can be carried out at lower temperature [114]. The yield of H$_2$ and CH$_4$ can increase up to three and nine times, respectively, as compared with noncatalyzed SCWG process [113]. Jiao et al. [113] have studied the SCWG of *C. pyrenoidosa* at 430°C over series of noble metals supported on activated carbon. They claimed that Ru/C, Pt/C, and Ir/C favored WGS and steam-reforming process, while methanation is preferred over Pd/C and Rh/C. Particularly, 5.97 mmol/g H$_2$ yield and 7.19 mmol/g methane yield were attained with Ru/C and Rh/C, respectively. Recently, same research group has reported that the synergistic effect between Ru/C and Rh/C in 1:1 mass ratio has increased the methane yield to 19.85 mmol/g, as compared with individual catalyzed system (Ru/C: 16.58 mmol/g and Rh/C: 18.20 mmol/g) [115].

CONCLUSION

The growing concern over the utilization of fossil fuels, GHG emissions, and the competition of biofuel feedstocks with agricultural resources has motivated the research interest in the biofuel production from microalgae. However, the existing upstream and downstream technologies are not effective enough to unleash all the potential of microalgae. Recently, a number of achievements in genetic engineering and cultivation method have successfully boosted the biomass growth and lipid productivity of microalgae. The successive growth of microalgae needs to be supported by highly efficient and cost-effective downstream processing to achieve high-quality biofuel production to reduce the dependence of fossil fuels. Downstream processing technology involves harvesting, lipid-extracting mechanisms, and transforming biomass into solid, liquid, and gaseous biofuel. The main challenge faced in the harvesting of microalgae and extraction is related to the energy input, time, and cost in concentrating and dewatering of diluted microalgae culture. Harvesting and lipid-extracting methods such as flocculation and mechanical extraction are still preferred for large-scale application. Thermochemical conversion methods that do not require predrying process of wet microalgae biomass such as liquefaction and SCWG are preferred over pyrolysis and conventional gasification. Transesterification of microalgae resulted in the production of biodiesel, which can be blended with petroleum diesel. Biochemical conversion involving carbohydrates proceeds through fermentation, anaerobic digestion, and biophotolysis to produce bioethanol, methane gas, and hydrogen gas, respectively. Continuous research and development of the downstream processing technology is crucial to accelerate the production of high-quality biofuel from microalgae.

REFERENCES

[1] H.-W. Yen, I.C. Hu, C.-Y. Chen, S.-H. Ho, D.-J. Lee, J.-S. Chang, Microalgae-based biorefinery — from biofuels to natural products, Bioresour. Technol. 135 (2013) 166–174.

[2] T.-H. Pham, B.-K. Lee, J. Kim, Novel improvement of CO_2 adsorption capacity and selectivity by ethylenediamine-modified nano zeolite, J. Taiwan Inst. Chem. Eng. 66 (2016) 239–248.

[3] W.Y. Cheah, T.C. Ling, J.C. Juan, D.-J. Lee, J.-S. Chang, P.L. Show, Biorefineries of carbon dioxide: from carbon capture and storage (CCS) to bioenergies production, Bioresour. Technol. 215 (2016) 346–356.

[4] E.S. Shuba, D. Kifle, Microalgae to biofuels: 'Promising' alternative and renewable energy, review, Renew. Sustain. Energy Rev. 81 (2018) 743–755.

[5] M.Y. Choo, L.E. Oi, P.L. Show, J.S. Chang, T.C. Ling, E.P. Ng, S.M. Phang, J.C. Juan, Recent progress in catalytic conversion of microalgae oil to green hydrocarbon: a review, J. Taiwan Inst. Chem. Eng. 79 (2017) 116–124.

[6] J. Popp, Z. Lakner, M. Harangi-Rákos, M. Fári, The effect of bioenergy expansion: food, energy, and environment, Renew. Sustain. Energy Rev. 32 (2014) 559–578.

[7] M.K. Lam, K.T. Lee, A.R. Mohamed, Current status and challenges on microalgae-based carbon capture, Int. J. Greenh. Gas Con. 10 (2012) 456–469.

[8] A. Galadima, O. Muraza, Biodiesel production from algae by using heterogeneous catalysts: a critical review, Energy 78 (2014) 72–83.

[9] C. Zhao, T. Brück, J.A. Lercher, Catalytic deoxygenation of microalgae oil to green hydrocarbons, Green Chem. 15 (2013) 1720.

[10] A.A.K. Das, M.M.N. Esfahani, O.D. Velev, N. Pamme, V.N. Paunov, Artificial leaf device for hydrogen generation from immobilised C. reinhardtii microalgae, J. Mater. Chem. 3 (2015) 20698–20707.

[11] N.H. Zainan, S.C. Srivatsa, F. Li, S. Bhattacharya, Quality of bio-oil from catalytic pyrolysis of microalgae *Chlorella vulgaris*, Fuel 223 (2018) 12–19.

[12] P. Pan, C. Hu, W. Yang, Y. Li, L. Dong, L. Zhu, D. Tong, R. Qing, Y. Fan, The direct pyrolysis and catalytic pyrolysis of *Nannochloropsis* sp. residue for renewable bio-oils, Bioresour. Technol. 101 (2010) 4593–4599.

[13] Y. Zhang, C. Xiong, A new way to enhance the porosity and Y-faujasite percentage of in situ crystallized FCC catalyst, Catal. Sci. Technol. 2 (2012) 606–612.

[14] Y. Zhang, X. Kong, Z. Wang, Y. Sun, S. Zhu, L. Li, P. Lv, Optimization of enzymatic hydrolysis for effective lipid extraction from microalgae *Scenedesmus* sp, Renew. Energy 125 (2018) 1049–1057.

[15] M.N. García-Casal, J. Ramírez, I. Leets, A.C. Pereira, M.F. Quiroga, Antioxidant capacity, polyphenol content and iron bioavailability from algae (*Ulva* sp., *Sargassum* sp. and *Porphyra* sp.) in human subjects, Br. J. Nutr. 101 (2008) 79–85.

[16] M. Nyberg, T. Heidorn, P. Lindblad, Hydrogen production by the engineered cyanobacterial strain Nostoc PCC 7120 ΔhupW examined in a flat panel photobioreactor system, J. Biotechnol. 215 (2015) 35–43.

[17] I. Rawat, R. Ranjith Kumar, T. Mutanda, F. Bux, Biodiesel from microalgae: a critical evaluation from laboratory to large scale production, Appl. Energy 103 (2013) 444–467.

[18] N. Pragya, K.K. Pandey, P.K. Sahoo, A review on harvesting, oil extraction and biofuels production technologies from microalgae, Renew. Sustain. Energy Rev. 24 (2013) 159–171.

[19] I. Rawat, R. Ranjith Kumar, T. Mutanda, F. Bux, Dual role of microalgae: phycoremediation of domestic wastewater and biomass production for sustainable biofuels production, Appl. Energy 88 (2011) 3411–3424.

[20] S.A. Razzak, M.M. Hossain, R.A. Lucky, A.S. Bassi, H. de Lasa, Integrated CO_2 capture, wastewater treatment and biofuel production by microalgae culturing—a review, Renew. Sustain. Energy Rev. 27 (2013) 622–653.

[21] B. Kiran, R. Kumar, D. Deshmukh, Perspectives of microalgal biofuels as a renewable source of energy, Energy Convers. Manag. 88 (2014) 1228–1244.

[22] T. Suganya, M. Varman, H.H. Masjuki, S. Renganathan, Macroalgae and microalgae as a potential source for commercial applications along with biofuels production: a biorefinery approach, Renew. Sustain. Energy Rev. 55 (2016) 909–941.

[23] M.K. Lam, K.T. Lee, Microalgae biofuels: a critical review of issues, problems and the way forward, Biotechnol. Adv. 30 (2012) 673–690.

[24] W.Y. Cheah, T.C. Ling, P.L. Show, J.C. Juan, J.-S. Chang, D.-J. Lee, Cultivation in wastewaters for energy: a microalgae platform, Appl. Energy 179 (2016) 609–625.

[25] W. Khetkorn, R.P. Rastogi, A. Incharoensakdi, P. Lindblad, D. Madamwar, A. Pandey, C. Larroche, Microalgal hydrogen production — a review, Bioresour. Technol. 243 (2017) 1194–1206.

[26] R.G. Lemus, J.M. Martínez Duart, Updated hydrogen production costs and parities for conventional and

renewable technologies, Int. J. Hydrogen Energy 35 (2010) 3929–3936.

[27] E. Eroglu, A. Melis, Microalgal hydrogen production research, Int. J. Hydrogen Energy 41 (2016) 12772–12798.

[28] S.J. Burgess, H. Taha, J.A. Yeoman, O. Iamshanova, K.X. Chan, M. Boehm, V. Behrends, J.G. Bundy, W. Bialek, J.W. Murray, P.J. Nixon, Identification of the elusive pyruvate reductase of *Chlamydomonas reinhardtii* chloroplasts, Plant Cell Physiol. 57 (2016) 82–94.

[29] W. Yang, C. Catalanotti, S. D'Adamo, T.M. Wittkopp, C.J. Ingram-Smith, L. Mackinder, T.E. Miller, A.L. Heuberger, G. Peers, K.S. Smith, M.C. Jonikas, A.R. Grossman, M.C. Posewitz, Alternative acetate production pathways in Chlamydomonas reinhardtiiduring dark anoxia and the dominant role of chloroplasts in fermentative acetate production, Plant Cell 26 (2014) 4499.

[30] M.L. Ghirardi, Implementation of photobiological H2 production: the O2 sensitivity of hydrogenases, Photosynth. Res. 125 (2015) 383–393.

[31] A. Melis, Green alga hydrogen production: progress, challenges and prospects, Int. J. Hydrogen Energy 27 (2002) 1217–1228.

[32] A. Melis, L. Zhang, M. Forestier, M.L. Ghirardi, M. Seibert, Sustained photobiological hydrogen gas production upon reversible inactivation of oxygen evolution in the green alga Chlamydomonas reinhardtii, Plant Physiol. 122 (2000) 127

[33] G. Torzillo, A. Scoma, C. Faraloni, A. Ena, U. Johanningmeier, Increased hydrogen photoproduction by means of a sulfur-deprived Chlamydomonas reinhardtii D1 protein mutant, Int. J. Hydrogen Energy 34 (2009) 4529–4536.

[34] Y. Liu, M.H. Rafailovich, R. Malal, D. Cohn, D. Chidambaram, Engineering of bio-hybrid materials by electrospinning polymer-microbe fibers, Proc. Natl. Acad. Sci. USA 106 (2009) 14201.

[35] S.N. Kosourov, M. Seibert, Hydrogen photoproduction by nutrient-deprived *Chlamydomonas reinhardtii* cells immobilized within thin alginate films under aerobic and anaerobic conditions, Biotechnol. Bioeng. 102 (2008) 50–58.

[36] D. Stojkovic, G. Torzillo, C. Faraloni, M. Valant, Hydrogen production by sulfur-deprived TiO2-encapsulated *Chlamydomonas reinhardtii* cells, Int. J. Hydrogen Energy 40 (2015) 3201–3206.

[37] P. McKendry, Energy production from biomass (part 2): conversion technologies, Bioresour. Technol. 83 (2002) 47–54.

[38] O.K. Lee, E.Y. Lee, Sustainable production of bioethanol from renewable brown algae biomass, Biomass Bioenergy 92 (2016) 70–75.

[39] S.-M. Lee, J.-H. Lee, Ethanol fermentation for main sugar components of brown-algae using various yeasts, J. Ind. Eng. Chem. 18 (2012) 16–18.

[40] M. Veillette, A. Giroir-Fendler, N. Faucheux, M. Heitz, Biodiesel from microalgae lipids: from inorganic carbon to energy production, Biofuels 9 (2018) 175–202.

[41] O.M. Adeniyi, U. Azimov, A. Burluka, Algae biofuel: current status and future applications, Renew. Sustain. Energy Rev. 90 (2018) 316–335.

[42] J. Hu, F. Yu, Y. Lu, Application of fischer–tropsch synthesis in biomass to liquid conversion, Catalysts 2 (2012) 303.

[43] A. Demirbaş, Biomass resource facilities and biomass conversion processing for fuels and chemicals, Energy Convers. Manag. 42 (2001) 1357–1378.

[44] P. McKendry, Energy production from biomass (part 1): overview of biomass, Bioresour. Technol. 83 (2002) 37–46.

[45] J. Milledge, B. Smith, P. Dyer, P. Harvey, Macroalgae-derived biofuel: a review of methods of energy extraction from seaweed biomass, Energies 7 (2014) 7194.

[46] A. Demirbas, Biofuels sources, biofuel policy, biofuel economy and global biofuel projections, Energy Convers. Manag. 49 (2008) 2106–2116.

[47] D.A. Walker, Biofuels – for better or worse? Ann. Appl. Biol. 156 (2010) 319–327.

[48] M. Alvarado-Morales, A. Boldrin, D.B. Karakashev, S.L. Holdt, I. Angelidaki, T. Astrup, Life cycle assessment of biofuel production from brown seaweed in Nordic conditions, Bioresour. Technol. 129 (2013) 92–99.

[49] E.A. Ehimen, J.B. Holm-Nielsen, M. Poulsen, J.E. Boelsmand, Influence of different pre-treatment routes on the anaerobic digestion of a filamentous algae, Renew. Energy 50 (2013) 476–480.

[50] J. Trivedi, M. Aila, D.P. Bangwal, S. Kaul, M.O. Garg, Algae based biorefinery—how to make sense? Renew. Sustain. Energy Rev. 47 (2015) 295–307.

[51] K.B. Cantrell, T. Ducey, K.S. Ro, P.G. Hunt, Livestock waste-to-bioenergy generation opportunities, Bioresour. Technol. 99 (2008) 7941–7953.

[52] S.M. Phang, M.S. Miah, B.G. Yeoh, M.A. Hashim, Spirulina cultivation in digested sago starch factory wastewater, J. Appl. Phycol. 12 (2000) 395–400.

[53] S.N. Naik, V.V. Goud, P.K. Rout, A.K. Dalai, Production of first and second generation biofuels: a comprehensive review, Renew. Sustain. Energy Rev. 14 (2010) 578–597.

[54] N.A.R. Kaliaperumal, J.R. Ramalingam, S. Kalimuthu, R. Ezhilvalavan, Seasonal changes in growth, biochemical constituents and phycocolloid of some marine algae of Mandapam coast, Seaweed Res. Util. (2002) 73–77.

[55] H.-W. Yen, D.E. Brune, Anaerobic co-digestion of algal sludge and waste paper to produce methane, Bioresour. Technol. 98 (2007) 130–134.

[56] A.D. Sutherland, J.C. Varela, Comparison of various microbial inocula for the efficient anaerobic digestion of *Laminaria hyperborea*, BMC Biotechnol. 14 (2014) 7.

[57] M.G. Hilton, D.B. Archer, Anaerobic digestion of a sulfate-rich molasses wastewater: inhibition of hydrogen sulfide production, Biotechnol. Bioeng. 31 (1988) 885–888.

[58] F. Murphy, G. Devlin, R. Deverell, K. McDonnell, Biofuel production in Ireland—an approach to 2020 targets with a focus on algal biomass, Energies 6 (2013) 6391.

[59] W.-H. Chen, B.-J. Lin, M.-Y. Huang, J.-S. Chang, Thermochemical conversion of microalgal biomass into biofuels: a review, Bioresour. Technol. 184 (2015) 314–327.

[60] F.R. Amin, Y. Huang, Y. He, R. Zhang, G. Liu, C. Chen, Biochar applications and modern techniques for characterization, Clean Technol. Envir. 18 (2016) 1457–1473.

[61] K. Chaiwong, T. Kiatsiriroat, N. Vorayos, C. Thararax, Study of bio-oil and bio-char production from algae by slow pyrolysis, Biomass Bioenergy 56 (2013) 600–606.

[62] H. Zheng, W. Guo, S. Li, Y. Chen, Q. Wu, X. Feng, R. Yin, S.-H. Ho, N. Ren, J.-S. Chang, Adsorption of p-nitrophenols (PNP) on microalgal biochar: analysis of high adsorption capacity and mechanism, Bioresour. Technol. 244 (2017) 1456–1464.

[63] M.I. Inyang, B. Gao, Y. Yao, Y. Xue, A. Zimmerman, A. Mosa, P. Pullammanappallil, Y.S. Ok, X. Cao, A review of biochar as a low-cost adsorbent for aqueous heavy metal removal, Crit. Rev. Environ. Sci. Technol. 46 (2016) 406–433.

[64] J.J. Milledge, B. Smith, W.P. Dyer, P. Harvey, Macroalgae-derived biofuel: a review of methods of energy extraction from seaweed biomass, Energies 7 (2014).

[65] M. Francavilla, P. Kamaterou, S. Intini, M. Monteleone, A. Zabaniotou, Cascading microalgae biorefinery: fast pyrolysis of *Dunaliella tertiolecta* lipid extracted-residue, Algal Res. 11 (2015) 184–193.

[66] U. Jena, K.C. Das, Comparative evaluation of thermochemical liquefaction and pyrolysis for bio-oil production from microalgae, Energy Fuel. 25 (2011) 5472–5482.

[67] Z. Hu, Y. Zheng, F. Yan, B. Xiao, S. Liu, Bio-oil production through pyrolysis of blue-green algae blooms (BGAB): product distribution and bio-oil characterization, Energy 52 (2013) 119–125.

[68] W.-c. Zhong, Q.-j. Guo, X.-y. Wang, L. Zhang, Catalytic hydroprocessing of fast pyrolysis bio-oil from Chlorella, J. Fuel Chem. Technol. 41 (2013) 571–578.

[69] J. Wang, P. Bi, Y. Zhang, H. Xue, P. Jiang, X. Wu, J. Liu, T. Wang, Q. Li, Preparation of jet fuel range hydrocarbons by catalytic transformation of bio-oil derived from fast pyrolysis of straw stalk, Energy 86 (2015) 488–499.

[70] L. Gao, J. Sun, W. Xu, G. Xiao, Catalytic pyrolysis of natural algae over Mg-Al layered double oxides/ZSM-5 (MgAl-LDO/ZSM-5) for producing bio-oil with low nitrogen content, Bioresour. Technol. 225 (2017) 293–298.

[71] Y. Hong, W. Chen, X. Luo, C. Pang, E. Lester, T. Wu, Microwave-enhanced pyrolysis of macroalgae and microalgae for syngas production, Bioresour. Technol. 237 (2017) 47–56.

[72] S.-X. Li, C.-Z. Chen, M.-F. Li, X. Xiao, Torrefaction of corncob to produce charcoal under nitrogen and carbon dioxide atmospheres, Bioresour. Technol. 249 (2018) 348–353.

[73] Y.-M. Chang, W.-T. Tsai, M.-H. Li, Chemical characterization of char derived from slow pyrolysis of microalgal residue, J. Anal. Appl. Pyrolysis 111 (2015) 88–93.

[74] M.A. Sukiran, F. Abnisa, W.M.A. Wan Daud, N. Abu Bakar, S.K. Loh, A review of torrefaction of oil palm solid wastes for biofuel production, Energy Convers. Manag. 149 (2017) 101–120.

[75] Y. Uemura, V. Sellappah, T.H. Trinh, S. Hassan, K.-i. Tanoue, Torrefaction of empty fruit bunches under biomass combustion gas atmosphere, Bioresour. Technol. 243 (2017) 107–117.

[76] W.-H. Chen, M.-Y. Huang, J.-S. Chang, C.-Y. Chen, Torrefaction operation and optimization of microalga residue for energy densification and utilization, Appl. Energy 154 (2015) 622–630.

[77] D. Eseltine, S.S. Thanapal, K. Annamalai, D. Ranjan, Torrefaction of woody biomass (Juniper and Mesquite) using inert and non-inert gases, Fuel 113 (2013) 379–388.

[78] Y.-C. Chen, W.-H. Chen, B.-J. Lin, J.-S. Chang, H.C. Ong, Impact of torrefaction on the composition, structure and reactivity of a microalga residue, Appl. Energy 181 (2016) 110–119.

[79] W.-H. Chen, Z.-Y. Wu, J.-S. Chang, Isothermal and non-isothermal torrefaction characteristics and kinetics of microalga Scenedesmus obliquus CNW-N, Bioresour. Technol. 155 (2014) 245–251.

[80] Q.-V. Bach, Ø. Skreiberg, Upgrading biomass fuels via wet torrefaction: a review and comparison with dry torrefaction, Renew. Sustain. Energy Rev. 54 (2016) 665–677.

[81] L. Fagernäs, J. Brammer, C. Wilén, M. Lauer, F. Verhoeff, Drying of biomass for second generation synfuel production, Biomass Bioenergy 34 (2010) 1267–1277.

[82] Y. Yu, X. Lou, H. Wu, Some recent advances in hydrolysis of biomass in hot-compressed water and its comparisons with other hydrolysis methods, Energy Fuel. 22 (2008) 46–60.

[83] J.A. Libra, K.S. Ro, C. Kammann, A. Funke, N.D. Berge, Y. Neubauer, M.-M. Titirici, C. Fühner, O. Bens, J. Kern, K.-H. Emmerich, Hydrothermal carbonization of biomass residuals: a comparative review of the chemistry, processes and applications of wet and dry pyrolysis, Biofuels 2 (2011) 71–106.

[84] Q.-V. Bach, W.-H. Chen, S.-C. Lin, H.-K. Sheen, J.-S. Chang, Wet torrefaction of microalga *Chlorella vulgaris* ESP-31 with microwave-assisted heating, Energy Convers. Manag. 141 (2017) 163–170.

[85] M.K. Lam, K.T. Lee, Microalgae biofuels: a critical review of issues, problems and the way forward, Biotechnol. Adv. 30 (2012) 673–690.

[86] L. Brennan, P. Owende, Biofuels from microalgae—a review of technologies for production, processing, and extractions of biofuels and co-products, Renew. Sustain. Energy Rev. 14 (2010) 557–577.

[87] P.E. Savage, Algae under pressure and in hot water, Science 338 (2012) 1039.

[88] T. Selvaratnam, A.K. Pegallapati, H. Reddy, N. Kanapathipillai, N. Nirmalakhandan, S. Deng,

P.J. Lammers, Algal biofuels from urban wastewaters: maximizing biomass yield using nutrients recycled from hydrothermal processing of biomass, Bioresour. Technol. 182 (2015) 232–238.

[89] G. Yu, Y. Zhang, B. Guo, T. Funk, L. Schideman, Nutrient flows and quality of bio-crude oil produced via catalytic hydrothermal liquefaction of low-lipid microalgae, Bio-Energ. Res. 7 (2014) 1317–1328.

[90] P. Biller, A.B. Ross, Potential yields and properties of oil from the hydrothermal liquefaction of microalgae with different biochemical content, Bioresour. Technol. 102 (2011) 215–225.

[91] D. López Barreiro, W. Prins, F. Ronsse, W. Brilman, Hydrothermal liquefaction (HTL) of microalgae for biofuel production: state of the art review and future prospects, Biomass Bioenergy 53 (2013) 113–127.

[92] N. Neveux, A.K.L. Yuen, C. Jazrawi, M. Magnusson, B.S. Haynes, A.F. Masters, A. Montoya, N.A. Paul, T. Maschmeyer, R. de Nys, Biocrude yield and productivity from the hydrothermal liquefaction of marine and freshwater green macroalgae, Bioresour. Technol. 155 (2014) 334–341.

[93] T. Muppaneni, H.K. Reddy, T. Selvaratnam, K.P.R. Dandamudi, B. Dungan, N. Nirmalakhandan, T. Schaub, F. Omar Holguin, W. Voorhies, P. Lammers, S. Deng, Hydrothermal liquefaction of *Cyanidioschyzon merolae* and the influence of catalysts on products, Bioresour. Technol. 223 (2017) 91–97.

[94] Y.F. Yang, C.P. Feng, Y. Inamori, T. Maekawa, Analysis of energy conversion characteristics in liquefaction of algae, Resour. Conserv. Recycl. 43 (2004) 21–33.

[95] A.B. Ross, P. Biller, M.L. Kubacki, H. Li, A. Lea-Langton, J.M. Jones, Hydrothermal processing of microalgae using alkali and organic acids, Fuel 89 (2010) 2234–2243.

[96] B. Jin, P. Duan, Y. Xu, F. Wang, Y. Fan, Co-liquefaction of micro- and macroalgae in subcritical water, Bioresour. Technol. 149 (2013) 103–110.

[97] K.P.R. Dandamudi, T. Muppaneni, N. Sudasinghe, T. Schaub, F.O. Holguin, P.J. Lammers, S. Deng, Co-liquefaction of mixed culture microalgal strains under sub-critical water conditions, Bioresour. Technol. 236 (2017) 129–137.

[98] W.-T. Chen, Y. Zhang, J. Zhang, L. Schideman, G. Yu, P. Zhang, M. Minarick, Co-liquefaction of swine manure and mixed-culture algal biomass from a wastewater treatment system to produce bio-crude oil, Appl. Energy 128 (2014) 209–216.

[99] A.J. Ramirez, J.R. Brown, J.T. Rainey, A review of hydrothermal liquefaction bio-crude properties and prospects for upgrading to transportation fuels, Energies 8 (2015).

[100] J. Li, X. Fang, J. Bian, Y. Guo, C. Li, Microalgae hydrothermal liquefaction and derived biocrude upgrading with modified SBA-15 catalysts, Bioresour. Technol. 266 (2018) 541–547.

[101] C. Jazrawi, P. Biller, Y. He, A. Montoya, A.B. Ross, T. Maschmeyer, B.S. Haynes, Two-stage hydrothermal liquefaction of a high-protein microalga, Algal Res 8 (2015) 15–22.

[102] Z. Liu, H. Li, J. Zeng, M. Liu, Y. Zhang, Z. Lius, Influence of Fe/HZSM-5 catalyst on elemental distribution and product properties during hydrothermal liquefaction of *Nannochloropsis* sp, Algal Res. 35 (2018) 1–9.

[103] P. Duan, P.E. Savage, Hydrothermal liquefaction of a microalga with heterogeneous catalysts, Ind. Eng. Chem. Res. 50 (2011) 52–61.

[104] B. Zhang, Q. Lin, Q. Zhang, K. Wu, W. Pu, M. Yang, Y. Wu, Catalytic hydrothermal liquefaction of Euglena sp. microalgae over zeolite catalysts for the production of bio-oil, RSC Adv. 7 (2017) 8944–8951.

[105] J. Li, G. Wang, C. Gao, X. Lv, Z. Wang, H. Liu, Deoxy-liquefaction of Laminaria japonica to high-quality liquid oil over metal modified ZSM-5 catalysts, Energy Fuel. 27 (2013) 5207–5214.

[106] Y. Xu, X. Zheng, H. Yu, X. Hu, Hydrothermal liquefaction of Chlorella pyrenoidosa for bio-oil production over Ce/HZSM-5, Bioresour. Technol. 156 (2014) 1–5.

[107] H.H. Khoo, C.Y. Koh, M.S. Shaik, P.N. Sharratt, Bioenergy co-products derived from microalgae biomass via thermochemical conversion – Life cycle energy balances and CO₂ emissions, Bioresour. Technol. 143 (2013) 298–307.

[108] R. Saidur, E.A. Abdelaziz, A. Demirbas, M.S. Hossain, S. Mekhilef, A review on biomass as a fuel for boilers, Renew. Sustain. Energy Rev. 15 (2011) 2262–2289.

[109] J.A. Onwudili, P.T. Williams, Hydrogen and methane selectivity during alkaline supercritical water gasification of biomass with ruthenium-alumina catalyst, Appl. Catal. B Environ. 132–133 (2013) 70–79.

[110] B. Patel, M. Guo, C. Chong, S.H.M. Sarudin, K. Hellgardt, Hydrothermal upgrading of algae paste: inorganics and recycling potential in the aqueous phase, Sci. Total Environ. 568 (2016) 489–497.

[111] I. Deniz, F. Vardar-Sukan, M. Yüksel, M. Saglam, L. Ballice, O. Yesil-Celiktas, Hydrogen production from marine biomass by hydrothermal gasification, Energy Convers. Manag. 96 (2015) 124–130.

[112] A. Kruse, Supercritical water gasification, Biofuel Bioprod. Bioref. 2 (2008) 415–437.

[113] J.L. Jiao, F. Wang, P.G. Duan, Y.P. Xu, W.H. Yan, Catalytic hydrothermal gasification of microalgae for producing hydrogen and methane-rich gas, Energy Sources Part A 39 (2017) 851–860.

[114] M.R. Díaz-Rey, M. Cortés-Reyes, C. Herrera, M.A. Larrubia, N. Amadeo, M. Laborde, L.J. Alemany, Hydrogen-rich gas production from algae-biomass by low temperature catalytic gasification, Catal. Today 257 (2015) 177–184.

[115] P.-G. Duan, S.-C. Li, J.-L. Jiao, F. Wang, Y.-P. Xu, Supercritical water gasification of microalgae over a two-component catalyst mixture, Sci. Total Environ. 630 (2018) 243–253.

CHAPTER 11

Microalgal Biorefinery

ELENI KOUTRA • PANAGIOTA TSAFRAKIDOU • MYRSINI SAKARIKA •
MICHAEL KORNAROS

INTRODUCTION

Microalgae, as an indispensable part in food chain and aqueous ecosystems, represent extraordinary photosynthetic microorganisms, with numerous advantages compared with terrestrial plants, and great potential, in terms of bio-based products and derived applications [1]. Microalgal biomass is mainly composed of proteins, lipids, and carbohydrates, while plenty of bioactive substances and molecules, including pigments, macro- and micronutrients, vitamins, phenols, and sterols, are also produced [2]. Depending on the species used, the culture conditions and therefore the characteristics of the produced biomass and profuse valorization options become available, as presented in Fig. 11.1. Both biofuels production and extraction/formation of added-value products from microalgae can meet several challenges in developed and developing countries, including fossil fuel depletion, CO_2 emissions, climate change, water, food and feed scarcity, and renewable energy demand. However, large-scale energy uses of microalgae remain unfeasible up to date, and the available microalgal products represent only a small share on markets, resulting in high product values [3]. Furthermore, current industrial production has focused either on the whole microalgal biomass for neutraceuticals and feed uses or on one single fraction, leaving up to 70% of the produced biomass totally unexploited [4]. To this end, feasibility and sustainability of microalgal technology necessitates the complete and optimal use of all the available biomass compounds, in terms of an integrated biorefinery approach [5].

Similar to a petroleum refinery, or more suitably to a dairy plant processing raw materials rich in proteins, lipids, and carbohydrates, microalgal biorefinery can result in multiple products, given that the applied techniques are mild and well determined, retaining the functionality of the available fractions [6]. Biomass processing largely depends on the desired end use and quality of the target product [7] and can include various harvesting, drying, cell disruption, extraction, fractionation, and purification steps. In case intact cells are targeted, extraction is not needed, while for extracellular substances such as polysaccharides, cell disruption is unnecessary [8]. In total, however, several processing steps hinder technology scale-up and make microalgal biorefinery too costly compared with common industrial downstream processes [5,9]. Technical breakthroughs, recycling options, and concurrent production of low-value biofuels and more profitable products, along with enhanced biomass production, are anticipated to make the overall approach feasible [6,8,10,11]. The objective of this chapter is to highlight the valorization options of microalgal biomass. In this regard, production of biofuels, including biodiesel, biogas, biohydrogen, and bioethanol, and added-value products, such as fertilizer, forage, and bioactive compounds, are fully described.

MICROALGAE TO BIODIESEL

Biodiesel is the product of transesterification reaction, during which neutral microalgal lipids in the form of triacylglycerols (TAGs) are converted to fatty acid alkyl esters, in the presence of an alcohol, commonly methanol and an acid/base or enzyme catalyst, while glycerol is also coproduced [12]. It is one of the most common biofuels, with enhanced fuel properties such as high energy density, limited generation of air pollutants during combustion, and compatibility with existing engines [13]. Besides neutral lipids serving as energy reserves, the lipid fraction of microalgal biomass is also composed of structural lipids, mainly polar lipids (phospholipids and glycolipids) and sterols, as well as waxes and hydrocarbons [14]. Concerning lipid composition, chain length of microalgal fatty acids varies between C10 and C24; however, C16−C18 are considered the most suitable for high-quality biodiesel, as determined by the International Standards ASTM D6751 in the United States and EN 14214 in Europe.

Microalgae Cultivation for Biofuels Production. https://doi.org/10.1016/B978-0-12-817536-1.00011-4

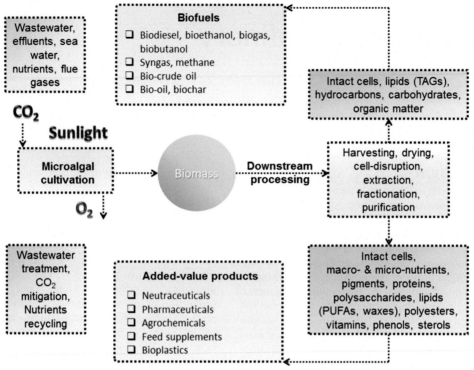

FIG. 11.1 Microalgal biorefinery. Biomass can be produced through photosynthesis, and/or effluents and flue gases recycling, and it can be further processed toward production of biofuels and added-value products.

According to the latter, C18:3 must be lower than 12%, and polyunsaturated fatty acids (PUFAs) must not surpass 1% (w/w). Furthermore, the presence of free fatty acids (FFAs) can hinder biodiesel production due to saponification under alkaline transesterification, resulting in decreased yield and difficult product separation. Therefore, acid esterification of FFA, which is primarily affected by methanol:FFA ratio, is performed as a pretreatment process at industrial scale, followed by alkali transesterification [15].

Due to the high viscosity of microalgal lipids, transesterification is a necessary process that produces a compatible fuel to existing diesel engines. Concerning the reactants used, methanol is most usually applied due to its low cost, while available catalysts have both merits and disadvantages, which are extensively reviewed in the literature [12,16]. Between acid- and alkali-based reactions, alkali-catalysts, such as NaOH, result in lower biodiesel yield and soap formation; therefore, acids, including HCl, H_2SO_4, and CH_3COCl, are commonly used for algal oil conversion to biodiesel. In contrast, transesterification catalyzed by enzymes is more efficient, lower reaction temperatures

are needed, and downstream processing is less demanding, while the catalyst cost is high. An emerging option for biodiesel production from microalgae includes direct conversion to biodiesel, skipping the lipid extraction step. The outcome of this in situ transesterification, in terms of fuel yield, is comparable with the conventional process, and since the latter is more energy demanding, in situ transesterification is preferred [17]. To this end, wet biomass of a high-lipid strain of *Chlorella vulgaris* was ultrasonicated and then subjected to in situ transesterification catalyzed by an immobilized lipase, resulting in high biodiesel conversion [18].

According to the existing standards, biodiesel must abide by several criteria to be used as a vehicle fuel or for heating purposes [19]. Therefore, microalgal biodiesel should be assessed in terms of its cetane number, indicative of its ignition capacity and combustion quality, iodine value that results from total unsaturation degree, heating value showing the released energy upon combustion, as well as its viscosity, behavior under low temperatures, and oxidation stability [16]. Since lipid composition of microalgae determines the values of biodiesel properties and dictates the appropriate use of algal oil, fatty acids (FAs) profiling is of utmost importance. The oil produced by *Neochloris oleoabundans* was characterized by higher concentration of C16:0 than soybean oil that is commonly used as a biodiesel feedstock, resulting in fuel with higher cetane number and stability upon storage, though inferior cold properties [20].

The percentage of lipids in microalgal biomass is species specific and can greatly vary, reaching up to 85% of dry weight, while environmental factors and especially harsh conditions usually favor lipid accumulation. Nutrient deficiency, high pH values, increasing light intensity, and salinity are between the most common factors inducing neutral lipids accumulation over membrane lipid synthesis [21]. For effective biodiesel production, however, high lipid content is not the only prerequisite since lipid productivity, determined simultaneously by biomass production, should be at least 40 mg/L d [22]. Cultivation of two microalgal strains, *Chlorella* SDEC-18 and *Scenedesmus* SDEC-8 in digested kitchen waste, resulted in high lipid productivity of 21.31 and 20.27 mg/L d, respectively, and enhanced biodiesel properties, in terms of cetane number, iodine value, and cloud point [23]. Significantly higher lipid productivity of 238 mg/L d was observed in case of *Nannochloropsis* sp. grown in a nitrogen-depleted synthetic medium, under 190 $\mu mol/m^2$ s illumination and 5% v/v CO_2 [24]. In lab-scale conditions and under nutrient limitation, lipid productivity of *C. vulgaris* reached up to 1.425 g/L d, while sufficient productivity of 0.33 g/L d was also observed in large scale, with high content of unsaturated fatty acids [25].

Salinity stress, as another significant factor affecting lipid accumulation, resulted in increased lipid content of *Scenedesmus* sp., 33% compared with 19% under control conditions [26]. Addition of 3% NaCl also triggered higher lipid accumulation than N-stress in *Coelastrella* sp., 37% over 24%; however, the produced biodiesel would not be suitable for cold climates due to its high percentage of saturated and polyunsaturated fatty acids [27]. The addition of 200 mM NaCl resulted in 33% lipids in *Acutodesmus dimorphus* cells, while an increase in polar lipids over the neutral lipid fraction was also observed, a response that seems to be species specific [28]. In total, accumulation of lipids has been proposed as a protective cellular strategy toward osmotic stress that can be caused by different salts, such as in case of 40% and 44% lipids in *Chlorella sorokiniana* and *Desmodesmus* sp. induced by $CaCl_2$ [29], leading to improved biodiesel potential of microalgae. Furthermore, high light intensity increased lipid percentage in *Scenedesmus*

abundans, from 21% under 3000 lux, to 33% under 6000 lux [30]. Also, heterotrophic metabolism has been proposed as an effective way to improve biodiesel production from microalgae [31]. In case of microalgal consortia heterotrophically cultivated in industrial wastewaters, extremely high lipid content values, varying between 33% and 85%, were recorded in potato starch effluent; however, the highest percentages were accompanied by low growth rates [32]. In other cases, glucose addition and nutrient deficiency not only increased lipid productivity of mixed microalgal cultures but mixotrophy also enhanced biodiesel properties through increased saturated fatty acids [33].

To surpass bottlenecks associated with low yields, both biomass and lipid productivity can substantially increase through genetic engineering and molecular tools aiming at recombination, silencing, or overexpression of key genes involved in microalgal lipid metabolism. Currently, several strategies have been proposed, including improved performance under high irradiation levels or enhanced CO_2 fixation, thioesterase, and acyl-CoA synthetase manipulation for FFA release and reduction of downstream processing cost, as well as improved fuel properties through oleic acid and tocopherol synthesis [34]. Recently, increased TAG accumulation and unsaturated fatty acid synthesis were achieved through knockout of the key enzyme PLA_2 in the model alga *Chlamydomonas reinhardtii* [35] improving the potential of biodiesel synthesis.

MICROALGAE TO BIOGAS

Microalgal biomass can be exploited as a feedstock for anaerobic digestion (AD) toward biogas production. AD is a process that has been widely used for the treatment of various organic wastes, as it is considered an efficient technology that serves waste management as well as biofuel production [36]. AD comprises four individual stages, namely, hydrolysis, acidification, acetogenesis, and methanogenesis, where organic materials are decomposed to biogas under anaerobic conditions (Fig. 11.2). Biogas is a mixture of CH_4 (60%–70%), CO_2 (20%–40%), and trace amounts of NH_3, H_2S, and N_2 [37].

To succeed an economically feasible process, parameters such as type of inoculum, type of substrate, bioreactor type, organic loading rate, hydraulic retention time (HRT) of the bioreactors, operational pH, and temperature must be taken into consideration. In addition to the above, when microalgae are used as feedstock for the AD, their species, potential pretreatment methods, and cultivation methods are the key points that affect the viability of the process [38].

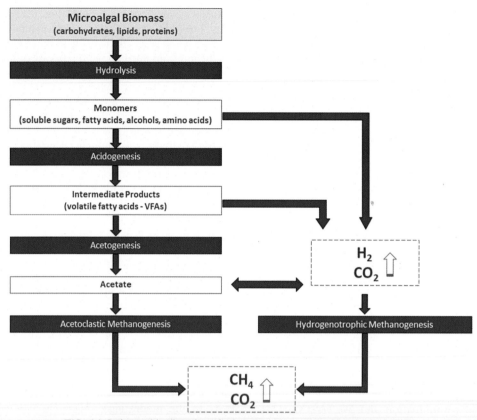

FIG. 11.2 Anaerobic digestion of microalgal biomass for biogas production.

AD of algal biomass is a process that is naturally occurring when cells are deposited at zones without oxygen and light at the bottom of water bodies, with estimated CH_4 emissions of $525-715 \times 10^6$ t per year and an annual rising rate of 1% [39]. The first research attempt of the biochemical methane potential of microalgae biomass was published over 60 years ago, where AD of green microalgae resulted in comparable CH_4 yields with raw sewage [40]. Recent studies showed that CH_4 content of the biogas obtained from AD of maize silage, a substrate that is widely used in biogas industry, is 7%–13% lower than that obtained from microalgae [41].

The composition of algal biomass in lipids, carbohydrates, and proteins is related to the theoretical methane yield of microalgae. Among those substrates, lipids present the highest value at 1.390 L_{CH4}/g VS, whereas proteins and carbohydrates follow with 0.800 L_{CH4}/g VS and 0.746 L_{CH4}/g VS, respectively [42]. According to literature, microalgal biomass composition depends on species, growth, and harvest conditions,

and the values may vary significantly [37,39]. Jankowska et al. in a recent review reported that lipid content ranges from 7% to 23%, carbohydrate content from 5% to 64%, and protein content from 6% to 71% [43]. Actual biogas yields obtained are lower than theoretical calculated values, ranging from 0.47 to 0.79 L_{CH4}/g VS, mainly due to difficulties on the hydrolysis of the microalgal cell, low C/N ratio, and other operational conditions [44].

Successful hydrolysis for further exploitation of the microalgal biomass depends on the physicochemical characteristics of their cell wall. Resistance of the cell wall to hydrolytic enzymes is correlated with the complexity of the molecules that form it. Most microalgal species' cell walls are composed of polysaccharides such as cellulose, hemicellulose, and pectin or glycoproteins [45]. Schwede et al. reported also the existence of an outer cell wall, in some species, made of algaenan, a polymer that cannot be hydrolyzed with chemical or enzymatic methods [46]. Although *Scenedesmus* sp. and *Chlorella* sp. are the most studied species, they have a hard cell wall that in most cases demands pretreatment to augment the biomass biodegradability. Species with thin or no cell walls are expected to be easier to disrupt and be converted to biogas or other biofuels more efficiently. These species include *Dunaliella* sp. that has no cell wall and *Chlamydomonas* sp. that lack cellulose [37]. Physical, chemical, and enzymatic pretreatments or combinations of them are used to break down cell walls and are evaluated regarding their enhancement on biodegradability of specific microalgal species or their potential application in all types of microalgal biomass. Selection of the appropriate pretreatment method is of great importance for the cost-effectiveness of the whole process [37,43].

Low temperatures (75–95°C) were used as thermal pretreatment for microalgal biomass grown in a pilot wastewater treatment plant. Methane yield noted a 70% increase compared with the nonpretreated biomass while energy gain after thermal pretreatment was at about 2.7 GJ/d. Microscopy in digestate samples showed only species with rigid cell walls, such as diatoms, indicating that thermal pretreatment was effective and microalgal cells were disrupted [47]. Likewise, Lavrič et al. reported the improvement of CH_4 yield after thermal pretreatment of a mixed microalgal biomass by 62%, while bioaugmentation with *Clostridium thermocellum* at 55°C showed an increment of 12% [48]. Enzymatic hydrolysis of *C. vulgaris* and *Scenedesmus* sp. prior to AD enhanced methane yields 1.72-fold and 1.53-fold, respectively. Interestingly, tested enzyme that resulted in the highest methane yield was protease, indicating that proteins were the inhibiting factor of AD and not macromolecules of carbohydrates as it is widely believed [49].

At this point, it is worth mentioning that AD as a process and biogas as a biofuel present significant assets compared with biodiesel or bioethanol production, namely, (1) all the macromolecules of the biomass are valorized, (2) the digestate of the process that is rich in nitrogen and phosphorus could be further utilized as cultivation medium of the algae, (3) there is no need for drying or oil extraction, (4) intact cells or remaining biomass from other processes can be used as feedstock, and most importantly (5) the biofuel (methane) is captured in the gaseous phase [37,43].

Codigestion of microalgal biomass with other substrates to balance the high nitrogen levels and fix the C/N ratio to an optimum value for AD is proposed as an alternative method to enhance the obtained biogas and also treat another carbon-rich biomass [50]. Substrates such as glycerol, waste-activated sludge, food waste, wheat straw, barley straw, pig manure, and olive mill solid waste have been codigested with various microalgal species and resulted in higher methane yields compared with the monodigested algal biomass [51–55].

Bohutskyi et al. studied the codigestion of algal and bacterial biomass that were grown in wastewater with cellulose, a substrate that lacks nitrogen, to determine the optimum C/N ratio. The optimum values were 35%:65% algal-bacterial biomass: cellulose (20: 80 based on VS). Methane production rate increased by 35%, while lag phase decreased by 50%. The researchers evaluated also four kinetic models, in an attempt to estimate total energy output and net energy ratio in a scale-up anaerobic system. Gompertz kinetic model presented better accuracy regarding prediction of methane production and had the best fitting to experimental data among the tested models [56]. Residual coffee husks, a novel waste biomass that has not been widely valorized for energy production up until now, have been codigested with microalgal biomass after thermal pretreatment at 120°C for 60 min. The pretreated codigested coffee husks gave 196 mL_{CH4}/g VS and codigestion increased by 17% the theoretical methane yield value if both substrates were treated with AD separately [47].

Besides codigestion, the development of two stage anaerobic systems enhances the energy return due to simultaneous hydrogen production and provides better stability to the process. The steps of hydrolysis and acidification are performed separately from methanation in the two-stage setup. Different bioreactors are used and operate at different pH conditions. The anaerobic fermentative bacteria of the acidification stage manage to hydrolyze the rigid algal cell in short time periods (small HRTs), and macromolecules are fermented to produce H_2, CO_2, and low-molecular-weight by-products such as volatile fatty acids, alcohols, and lactic acid. The liquid effluent from this process is used as feedstock for methanogenic bacteria that produce biogas in the second bioreactor. With this process, methane yields also increase compared with one-stage systems [57]. Microalgae *C. vulgaris* was used as a feedstock in a two-stage anaerobic system and resulted in 14.46 and 1.46 kJ/g VS in terms of energy production from methane and H_2, respectively, whereas direct methane production from one-stage systems is reported to be 14.86 kJ/g VS [58]. In another study, macroalgae *Laminaria digitata* and microalgae *Arthrospira platensis* were the substrates of the two-stage continuous anaerobic system. The highest yields reached 55.3 mL/g VS for H_2 and 245.0 mL/g VS for CH_4. The overall energy yield

(9.4 kJ/g VS) was equal to the 77.7% of that obtained in an optimum batch system [57].

An innovative holistic approach regarding the methods of biogas production is the use of AD digestate as a source of nutrients for the cultivation of microalgae and the upgrading of the produced biogas through microalgal biomass to remove CO_2. This closed-loop scenario is a promising zero-waste process that takes advantage of all the participating streams in AD and could be the key solution to an effective and feasible microalgae-based biorefinery [59].

MICROALGAE TO BIOHYDROGEN

Hydrogen (H_2) is considered to be a very promising biofuel, since its energy yield (122 kJ/g) per unit weight is threefold higher compared with hydrocarbon fuels. In addition, it is characterized by high cleanness, due to lack of CO_2 and only water vapor production from its combustion [60]. Up to date, H_2 production methods evolve fossil fuels consumption and are defined as high greenhouse gas footprint processes representing a contradiction to the whole clean fuel concept [61]. An alternative approach to the former processes relies on the use of biological methods for the production of the so-called biohydrogen. Microalgae, as presented in Fig. 11.3, can be considered as a sustainable platform

for biohydrogen production, either directly from their cells via biophotolysis or indirectly through dark fermentation of their biomass [62].

Biophotolysis

The biological electrolysis of water through photosynthetic organisms is the breakdown of water to oxygen and hydrogen. The fate of this hydrogen is to react with CO_2 and form carbohydrate, while initial O_2 is released in the atmosphere. Gaffron and Rubin in 1942 published a research on the ability of green algae, such as *Scenedesmus obliquus* and *C. reinhardtii*, to release H_2 under certain conditions [63,64]. Since then, many research studies have been conducted to enlighten the pathways of biophotolysis phenomenon that green algae follow. *C. reinhardtii* is considered a model microorganism for research on this topic and a much promising species for H_2 production with this method. Attempts of genetic modification of this species have been reported, with elevated productivities compared with wild types. A H_2 productivity of more than 300 mL/L was observed in some mutants [62].

Breakdown of water in biophotolysis is catalyzed from the photosystem II (PSII) and the [FeFe]-hydrogenase. Lack of sulfur minimizes O_2 production from PSII and creates anoxic conditions in the culture, thus leading to hydrogenase synthesis and activity. In

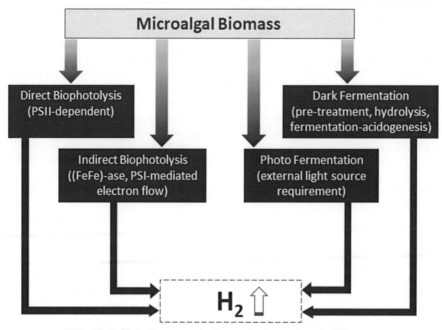

FIG. 11.3 Biohydrogen production routes from microalgal biomass.

other words, by regulating sulfur availability to the algal cell, it is possible to interfere with the internal oxygen flow and result in H_2 production [62,64].

Although biophotolysis is an eco-friendly method of biofuel production, its efficiency is rather low (under 1%) and the operational cost is relatively high, with the current technologies and knowledge. Vargas et al. studied two strains of *Chlamydomonas* sp. that were cultivated under sulfur deprivation in two stages: aerobic and anaerobic. They reported a maximum production of H_2 from *C. reinhardtii* at 5.95 ± 0.88 µmol/mg with a H_2 productivity at 17.02 ± 3.83 µmol/L h [65].

An innovative approach to overcome the inhibitory effect of O_2, which plays a crucial role on the hydrogenase activity, is the use of immobilization carriers or encapsulation materials to protect algal cells [66,67]. A thin alginate film was used as immobilization carrier of *C. reinhardtii*, at a study of the pathways followed by this alga. It was observed that the entrapment of the cells elevated PSII-independent pathway of H_2 production and sustained the high productivity rates [68].

With the continuous research and technological evolvement, the optimum cultivation methods and conditions of specific strains as well as development of efficient photobioreactors may enable a cost-effective H_2 production through biophotolysis for commercial application [69].

Bio-H_2 via dark fermentation

Dark fermentation (DF) is a simple method for H_2 production through the valorization of various organic wastes that can be used as feedstock. Long chain polymers such as carbohydrates, proteins, and lipids are hydrolyzed by mixed or pure cultures that produce hydrogen through two main pathways: (1) acetate pathway that has a theoretical yield of 4 mol H_2 from 1 mol glucose and (2) butyrate pathway with a theoretical yield of 2 mol H_2 from 1 mol glucose [70]. The microorganisms that are involved in DF are facultative or obligate anaerobic bacteria, such as *Escherichia coli*, *Enterobacter aerogenes*, *Citrobacter intermedius*, *Enterobacter cloacae*, *Ruminococcus albus*, *Clostridium beijerinckii*, and *Clostridium paraputrificum* [71,72].

Microalgal biomass is considered an alluring feedstock for DF, because of its high carbohydrates content, since the latter has been correlated with elevated H_2 productivities [38]. DF of microalgae presents a higher production rate compared with photofermentation, biophotolysis, and microbial electrolysis [73]. Research efforts of microalgae use in DF are summarized in three main categories: (1) untreated biomass, (2) pretreated biomass, and (2) lipid-extracted biomass. The type of biomass has a direct effect on the H_2 production, since it may influence severely the biodegradability through the presence of chemical barriers [74]. Pretreatments, such as cell disruption with enzymes, acid hydrolysis along with thermal treatment, and even ultrasonication, may increase the final H_2 yield and reach 50%–70% of the theoretical value [73].

The most studied microalgal species for H_2 production via DF are *Chlorella* sp., *Scenedesmus* sp., and *Saccharina* sp., and depending on the pretreatment of biomass, hydrogen yields range from 0.37 mL H_2/g VS for untreated microalgae to 338 mL H_2/g VS for biomass that was thermally pretreated and even 958 mL H_2/g VS for *Chlorella* sp. after combining acid hydrolysis and heat treatment [75]. A comparison between the two H_2 production processes, biophotolysis and dark fermentation, is presented in Table 11.1.

MICROALGAE TO BIOETHANOL

Bioethanol is considered as the most prominent alternative to fossil fuels, with the United States and Brazil, as shown in Table 11.2, being the countries with the largest volumes produced up to date [81].

Microalgae biomass constitutes a great substitute for the first- and second-generation bioethanol feedstock, such as edible crops and lignocellulosic materials, and is representing a promising substrate for third generation bioethanol, as it faces the challenges of the previous generations' feedstock [82]. The process steps of bioethanol production from microalgal biomass are presented in Fig. 11.4.

Research on the topic is still in its infancy, although microalgae present promising potentials regarding renewable energy production [83]. The carbohydrate content of marine organisms that could be valorized can reach up to 50% of the dry cell weight [84].

Various parameters affect the total intracellular content in carbohydrates, such as nutrient limitations, cultivation temperature, O_2 and CO_2 levels, light intensity, pH, salinity as well as the presence of compounds with toxic effects [83]. In general, limitations on N and P seem to increase carbohydrates accumulation [59]. Species of microalgae used for bioethanol production play the most important role on the efficiency of the process, as the carbohydrate content varies significantly among them. According to literature, *Scenedesmus*, *Chlorella*, and *Chlamydomonas* present the higher accumulation of carbohydrates under specific cultivation methods and conditions [82,83,85].

Different kinds of carbohydrates are present in the microalgal cell. Agar, alginate, pectin, cellulose, and hemicellulose can be found as structural elements of the cell wall, while starch and glycogen are the storage

TABLE 11.1
Comparison of Photo- and Dark Fermentation for H_2 Production.

Parameter	Photofermentation	Dark Fermentation
Light demand	+	−
O_2 demand	−	−
Formula	$CH_3COOH + 2H_2O + light \rightarrow 4H_2 + 2CO_2$	$C_6H_{12}O_6 + 6H_2O \rightarrow 12H_2 + 6CO_2$
Related enzymes	PSII, [Fe]-hydrogenase	CoA, acetyl-CoA
Fermentation microorganisms	Algal species and photosynthetic bacteria (e.g., *Anabaena variabilis, Rhodobacter sphaeroides*, etc.)	Anaerobic bacteria (e.g., *Escherichia coli, Clostridium, Sporolactobacillus,* etc.)
Advantages	Wide range of feedstock, wide spectrum of light	Cost-effective, high yields, wide range of feedstock
Drawbacks	Expensive photobioreactors, light dependency, low conversion efficiency of solar energy	Need for effluent treatment, additional gaseous by-products (H_2S, CO_2, CO, CH_4)

Adapted from K. Bolatkhan, B.D. Kossalbayev, B.K. Zayadan, T. Tomo, T.N. Veziroglu, S.I. Allakhverdiev, Hydrogen production from phototrophic microorganisms: reality and perspectives, Int. J. Hydrog. Energy 44 (2019) 5799–5811; A. Sharma, S.K. Arya, Hydrogen from algal biomass: a review of production process, Biotechnol. Reports 15 (2017) 63–69.

products in microalgae and cyanobacteria, respectively [86]. The absence of lignin in the microalgal biomass makes the latter more susceptible to its conversion to monosaccharides, in comparison with other materials,

TABLE 11.2
Bioethanol Fuel Producing Countries, 2018.

Country (Region)	Production (Million Liters)	Feedstock
United States	60,945	Corn, wheat
Brazil	30,094	Sugarcane
European Union	5,413	Corn, wheat, sugar-based (beets and cane), other cereals, lignocellulosics
China	4,467	Corn, cassava, rice
Canada	1,817	Corn, wheat
Thailand	1,476	Sugarcane, molasses, cassava
India	1,249	Sugarcane, molasses
Argentina	1,098	Sugarcane
Rest of world	2,082	Various

Data adapted from https://www.epure.org/media/1763/180905-def-data-epure-statistics-2017-designed-version.pdf; https://gain.fas.usda.gov/Recent%20GAIN%20Publications/Forms/AllItems.aspx. https://www.statista.com/

such as plants or crops, of lignocellulosic nature [87]. Besides total carbohydrates content, the effective extraction method of this content plays a crucial role on the final yield of sugars that will be fermented to produce ethanol and should be chosen thoroughly to avoid further degradation of products or operational cost increase. There are various pretreatment methods that aim at the lysis of the algal cell and hydrolysis of the carbohydrates polymers that fall into three main categories, namely, mechanical (physical), chemical, and enzymatic pretreatment methods or combinations of them [85].

Mechanical methods

These methods are based on the use of physical forces to disrupt the cell wall. They are considered to be less invasive methods than chemical or enzymatic approaches; thus, the functionality of the algal materials is preserved, and any contamination of the biomass with other substances is avoided. Mechanical methods include ultrasonication, bead-beating, and high-pressure homogenization. Thermal adjustment at high temperatures (above 50°C) and milling are usually reported as pretreatment methods for lipid extraction and the augmentation of biogas in AD processes [88]. Kim et al. reported the comparison of thermal (autoclave at 121°C and 120 kPa for 20 min), milling (bead-beating) at 15 Hz for 30 s, and sonication (24 kHz on 40% amplitude for 15 min) pretreatments in combination with enzymatic hydrolysis for

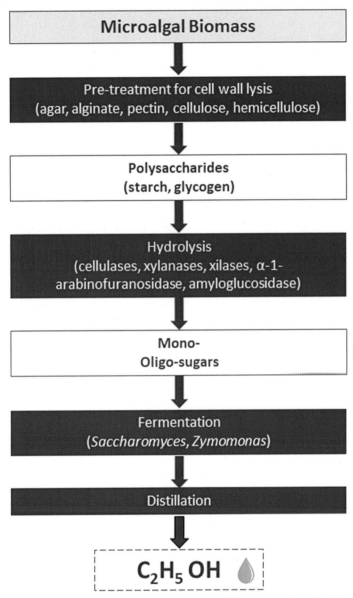

FIG. 11.4 Bioethanol production from microalgal biomass used as feedstock.

bioethanol production from *C. vulgaris* using immobilized *Saccharomyces cerevisiae*. The produced reducing sugars differed significantly, with the autoclaved and sonicated biomass presenting zero differences between the treated and nontreated cells, in contrast with the samples that were milled. The latter pretreatment showed 25% increase of the sugar conversion rate [84]. Likewise, a pretreatment of 45 min with bead-milling using zirconia beads resulted in 68% release of sugars from *N. oleoabundans* cells [89]. Sonication

of *Chlamydomonas mexicana* under optimum conditions (at 40 kHz, 2.2 Kw, 50°C for 15 min) quadrupled total reducing sugars [90].

It should also be mentioned that the decrease of temperature at cooling (4°C) and freezing (−20°C) points and freeze drying have been studied and evaluated as storage methods of *Acutodesmus obliquus* biomass in combination with thermal pretreatment (100°C for 60 min), milling, ultrasonication, enzymes, and combinations of them [91].

As drawbacks of mechanical methods, researchers mention high energy requirements that lead to production cost increase and lack of targeted specification among the different fractions of the algal cells [86].

Chemical methods

Chemical reagents used in these methods are either acids that hydrolyze glycosidic bonds or alkalis that saponify lipids of the cell membrane and thus disrupt the cell wall. Besides the type of reagent used, its concentration, the duration of the treatment, concentration of biomass as well as species of the algae and temperature are the key parameters that dictate the final yield of the produced reducing sugars [92]. Chemical methods are considered rather easy to perform and cost-effective and require small amount of time.

Among acids, H_2SO_4 is most commonly used for acid hydrolysis of the algal biomass [83]. *C. reinhardtii* was pretreated with 3% sulfuric acid at 110°C for 15−20 min, and glucose release up to 58% was achieved [93]. Various combinations of H_2SO_4 concentrations (0%, 0.5%, 1%, 2%, and 4% v/v), temperatures (60, 90, or 120°C), duration (0, 15, 30, or 60 min), and biomass load (2.5%, 5%, or 10% w/v) were tested to optimize the pretreatment of native South American microalgae regarding sugars' extraction [87]. Other acids such as HNO_3, HCl, and H_3PO_4 have also been reported as potential pretreatment reagents in the literature [94]. Alkali treatments rely mainly on the use of NaOH. Hamouda et al. studied the hydrolysis of *C. vulgaris* with chemical and enzymatic methods. Different concentrations of H_2SO_4 or NaOH and α-amylase were compared regarding the saccharification efficiency. The best conditions determined for *C. vulgaris* hydrolysis were one round of autoclave (20 min) and 5% H_2SO_4 where 382 mg/g dry weight sugars were obtained [95]. NaOH was also used for the pretreatment of *Nannochloropsis oculata* and *Tetraselmis suecica* that were cultivated in treated municipal wastewater and evaluated for bioethanol production [96].

The bottleneck of these methods is the potential corrosion of equipment (depending on the reagent concentration), need for pH modification prior to fermentation, and most importantly formation of toxic compounds for yeasts, such as furfural or 5-hydroxymethylfurfural, which could have an inhibitory effect on the subsequent fermentation process [86]. The aforementioned substances are the result of the extended degradation of sugars due to high temperatures and pressure, along with the concentrations of acid or alkali used. Therefore, it is of great importance to choose the optimum values of reaction reagent concentration, temperature, and pressure to achieve operational efficiency [92].

Enzymatic methods

Enzymes have the advantage of selectivity on the bond that they hydrolyze and require milder conditions, compared with chemical methods. By-products formation is minimized this way, while high yields of reducing sugars are obtained. From a technoeconomic point of view, enzymatic methods are considered more expensive than chemical hydrolysis because enzymes cannot be effectively recovered, leading this way to a nonfeasible process in large scale, although there is a significantly lower energy consumption [92].

Due to the presence of various types of polysaccharides in the algal cell, different enzymes should be applied for their conversion to simple monomers. Cellulases (endoglucanases, exoglucanases, and β-glucosidase), xylanases, xilases, α-1-arabinofuranosidase, and amyloglucosidase are the main enzymes needed for the hydrolysis of the carbohydrate macromolecules found in the algal cell [82,86].

In a recently published study, commercial cellulase from *Trichoderma reesei* was used to develop a kinetic model describing the enzymatic hydrolysis of cellulose from a mixed microalgal biomass for bioethanol production. Temperature, biomass concentration, and pH were the parameters under investigation [97]. Hamouda et al. reported that α-amylase from *Bacillus subtilis* SH04 was more efficient in terms of final sugar content obtained from *C. vulgaris*, than the α-amylase from (B) *cereus* SH06. The use of the former resulted in 220.46 ± 2.57 mg/g biomass, while the latter gave 107.34 ± 1.83 mg/g biomass [95]. *C. mexicana* was enzymatically pretreated with cellulase, to obtain 445.5 mg/g biomass total reducing sugar yield and an ethanol concentration of 10.5 g/L [90].

The step of hydrolysis is followed by fermentation toward bioethanol production. The conversion of sugar monomers to ethanol by yeasts such as *Saccharomyces* and *Zymomonas* can take place either in the same vessel (fermentor) where hydrolysis is performed (simultaneous hydrolysis and fermentation [SSF]) or in different vessels (separated hydrolysis and fermentation [SHF]) [92]. Both strategies present advantages and disadvantages. With the SHF, it is easier to control parameters of both hydrolysis and fermentation that affect the final product, such as time, temperature, and pH, and thus optimize separately each process. It is also feasible to run a continuous mode of fermentation with cell recycling. Demerits of SHF include inhibition of enzyme activity from the product during hydrolysis; higher enzyme concentrations and lower biomass concentrations are needed in order to achieve adequate bioethanol yield, which leads to operational cost increase; and finally, separate hydrolysis requires longer time (up to 4 days), and the risk of microbial contamination is relatively high [98]. On the other hand, SSF method simplifies the whole process with lower required capital cost, since only one reactor is needed for hydrolysis (saccharification) and fermentation and less operational time than SHF is needed. As stated before though, this approach makes it more difficult to optimize the conditions, and the recovery of both enzymes and yeast cells is challenging, so operation of a scale-up system is hard from an economic and technical view [86].

Stoichiometry of ethanol production from glucose shows that 0.511 $g_{ethanol}/g_{glucose}$ is the maximum yield that can be achieved:

$$C_6H_{12}O_6 \rightarrow 2CH_3CH_2OH + 2CO_2$$

$$1g \rightarrow 0.511g + 0.489g$$

Singh et al. published recently that 0.116 $g_{ethanol}/g_{algal\ biomass}$ (0.305 $g_{ethanol}/g_{glucose}$) was produced through SSF process using *S. cerevisiae* to ferment carbohydrates obtained from *Chlorella* sp. after cycloheximide treatment to increase its starch content [99]. Fermentation of concentrated sugar medium prepared after the hydrolysis of mixed microalgal biomass gave an ethanol yield 0.46 g $_{ethanol}/g_{glucose}$ [97]. An 87.4% of the maximum theoretical value was obtained with two novel strains, *Desmodesmus* sp. FG and a green microalga, which is yet unidentified (strain SP2-3). Ethanol reached 24 $g_{ethanol}/L_{hydrolysate}$ in SHF using *S. cerevisiae* [87].

MICROALGAE TO FERTILIZER

Microalgae can be used as microbial fertilizers, also known as biofertilizers. They can act as slow-release fertilizers [100], liberating nitrogen (N), phosphorus (P), and potassium (K) in a gradual manner. This nutrient release is comparatively aligned to the nutrient demand pattern of the plant, and therefore, nutrient losses can be prevented, especially when compared with chemical fertilizers [101]. Furthermore, microalgae contain plant growth—promoting substances and antifungal compounds that can further benefit the plants [102].

Even though microalgal fertilizers present comparatively low N, P, and K values, they can be highly efficient due to their higher micronutrient content (Table 11.3). Wuang et al. [103] reported that the micronutrient content (calcium (Ca), iron (Fe), manganese (Mn), zinc

(Zn), and selenium (Se)) of *A.* (*Spirulina*) *platensis* was higher than in commercial fertilizers. These micronutrients play an important role in the plant's development [103]: calcium is involved in the production of new root tips and growing points; iron is a structural enzyme component; manganese plays a pivotal role in the synthesis of chlorophyll and activates enzymes related to nitrogen assimilation; zinc contributes to the synthesis of enzymes; and finally, selenium, even though it is not important for the plant itself, plays an important role in the metabolism of humans and herbivorous animals.

Mineralization tests on a mixed microalgal population grown on dairy and pig manure effluents showed that c.a. 5% of the microalgal N is present in a phytoavailable form, whereas this value increased at up to 29% and 41% after 21 and 63 days [101]. In the same study, fertilization experiments using corn (*Zea mays* L.) seedlings presented similar performance to commercial fertilizers. Apart from the good mineralization pattern, microbial fertilizers aid in the reduction of nutrient losses. In this respect, Das et al. [104] used *Tetraselmis* sp. and *Nannochloropsis* sp. grown in municipal wastewater, to test their effects on the fertilization of wheat. The microbial fertilizer presented better performance compared with the commercial fertilizer used, possibly due to the washout of nutrients due to the irrigation, on an equal amount of nitrogen basis. Nevertheless, the authors noted that attention should be paid when marine microalgae are used, as recurring application can increase soil salinity.

The marine microalga *N. oculata* as well as microalgal-bacterial flocs (MaB flocs) have been tested as slow-release organic fertilizers on tomato plants (*Solanum lycopersicum* cv. Maxifort and *Solanum lycopersicon* cv. Merlice) [100]. The nitrogen content of the leaves was higher than the commercial organic fertilizer demonstrating the good N-fertilizing properties of microalgal fertilizers. On the other hand, the provision of MaB flocs resulted in lower potassium and magnesium content. The latter though rarely results in tomato yield reduction. Finally, the *N. oculata* treatment induced calcium deficiency as a result of the water stress due to increased salinity. Besides the increased plant performance, microalgae have been reported to reduce the germination time. Specifically, when Garcia-Gonzalez and Sommerfeld [105] supplied the culture and 50% aqueous extract (individual treatments) of *A. dimorphus* on Roma tomato seeds (*Solanum lycopersicum* var. Roma), the germination was 2 days faster than the control. At the same time, the plant's performance increased as foliar spray of 50% aqueous *A. dimorphus* extract resulted in better growth performance (increased

TABLE 11.3
Nutrient Content of the Biomass of Several Microalgae Used as Microbial Fertilizers. The Composition of a Chemical Fertilizer is Presented for Comparison.

	Arthrospira (Spirulina) platensis	Tetraselmis sp.	Nannochloropsis sp.	Chlorella sp.	Scenedesmus sp.	Nannochloropsis oculata	Microalgal-bacterial Flocs	Chemical Fertilizer (Triple Pro 15–15–15)
N (%)	7.80	6.85	7.43	4.82	5.08	8.07	2.44	12.4
P (%)	0.80	0.82	0.75	0.28	0.36	1.29	0.59	6.6
K (%)	1.60	1.33	1.17	0.73	0.76	1.36	0.18	12.5
N:P:K	9.8:1:2	8.3:1:1.6	9.2:1:1.6	18:1:2.6	14:1:2.1	6.3:1:1.1	4.1:1:0.3	1.9:1:1.9
Ca (%)	0.40	2.54	1.49	0.68	0.66	0.2	20.4	0.1
Fe (ppm)	1057	1830	1370	2540	3190	328	143	455
Mn (ppm)	42	176	154	570	230	42.5	113	26.1
Zn (ppm)	155	63	41	1290	380	143	67	n/d
Se (ppm)	0.018	—	—	—	—	—	—	n/d
Reference	[103]	[104]	[104]	[104]	[104]	[100]	[100]	[103]

"—", indicates not available values; n/d, not detected.

plant height as well as numbers of flowers and branches per plant). Finally, the application of dry biomass 22 days prior to the transplantation of seedlings resulted in enhanced growth performance.

Wuang et al. [103] noted that the efficacy of the microalgal fertilizers is species dependent. More specifically, the effect of *Spirulina*-based fertilizers was tested on Arugula (*Eruca sativa*), Bayam Red (*Ameranthus gangeticus*), Pak Choy (*Brassica rapa* ssp. chinensis), Chinese Cabbage (*Brassica rapa* ssp. chinensis), and Kai Lan (*Brassica oleracea alboglabra*). Overall, this study demonstrated the significant enhancement of growth of Bayam Red, Arugula, and Pak Choy, which presented a performance similar to the plants cultivated with the chemical fertilizer. Meanwhile, the germination of Chinese Cabbage and Kai Lan notably increased (in terms of plant dry weight). The growth performance of Bayam Red was enhanced as indicated by the increase (58% −156%) of growth parameters like plant height as well as fresh and dry weight, compared with the control. At the same time, the overall growth performance was comparable with the commercially available fertilizer tested (Triple Pro 15−15−15), with only the fresh and dry weights of *Spirulina*-grown plants being lower. Arugula cultivated using *Spirulina* presented very promising results with all the growth parameters being comparable with the plants cultivated using the commercial fertilizer. *Spirulina*-cultivated Pak Choy presented longer roots than the control being comparable with plants cultivated with the commercial fertilizer. The seed germination, which is an important indicator for the plant's field performance, of all the plants tested except White Crown, was significantly increased through the use of the *Spirulina* fertilizer.

There are several studies reporting the increase in crop yield and quality through the use of microalgal fertilizers. For instance, two microalgal consortia, one (MC1) composed of unicellular microalgae (*Chlorella*, *Scenedesmus*, *Chloroc occum*, and *Chroococcus* sp.) and the second (MC2) composed of filamentous strains (*Phormidium*, *Anabaena*, *Westiellopsis*, *Fischerella*, and *Spirogyra* sp.), were tested as a microbial fertilizer on wheat crop (*Triticum aestivum* L.) [106]. Both microalgal fertilizers resulted in notable increase in N, P, and K content of the roots, shoots, and grains, while they increased by 7.4%−33% the plant dry weight. Most importantly, the 1000-grain weight increased by up to 8.4% compared with the provision of full dosage of fertilizer. This was further supported by Dineshkumar et al. [107] who observed 7%−21% increase in rice yield when supplying *C. vulgaris* and *Spirulina platensis* microbial fertilizer. The authors of this study found

that the addition of dry microalgal biomass presents better results than the soil drench application. In the study of Coppens et al. [100], the microalgal fertilizers presented comparable results with the commercial organic fertilizer, while improving the quality of tomato fruit, through increasing sugar and carotenoid content (up to 18% and 44%, respectively). No significant differences were observed in the number of fruits; nevertheless, a lower yield was observed through the algal treatments. This could be attributed to the higher salinity, which confines the fruit's water influx, or to the supply of ammonium as a nitrogen source (instead of nitrate that is the preferred nitrogen source). When *Spirulina* was tested in mung bean plants (*Vigna radiata* L.), the results showed high crop growth rate, dry matter, as well as harvest index and yield [108]. Furthermore, Faheed et al. [109] tested the use of *C. vulgaris* on lettuce (*Lactuca sativa*). This study revealed that the provision of this microalga can significantly increase the growth of lettuce (fresh and dry weight), as well as the pigment content. Safinaz and Ragaa [110], tested the effect of the use of red marine algae on corn plants (*Z. mays* L.). Specifically, the microalgae tested were *Laurencia obtusa*, *Corallina elongata*, and *Jania rubens*. All treatments presented promising results, with the application of *L. obtusa* and *C. elongata* increasing the plant fresh weight by 90.9%. The plant's phosphorus content increased up to 74.0%, while the application in *J. rubens* resulted in 129% nitrogen increase.

Finally, large amounts of lipid-extracted biomass are expected to be produced though the large-scale microalgal biodiesel production [111]. A sustainable approach requires further valorization of this biomass. In that context, Maurya et al. [111] tested the use of lipid-extracted *Chlorella variabilis* and *Lyngbya majuscula* biomass as a fertilizer for corn (*Z. mays* L.). Both treatments increased the grain yield, while *L. majuscula* resulted in the highest grain P and K content. On the other hand, *C. variabilis* treatment resulted in the highest carbohydrate and crude fat content.

Even though microalgal fertilizers present promising results, further research should be performed to elucidate their effect on plant growth and crop productivity, as well as establish the optimal fertilizer composition. In this respect, Coppens et al. [100] recommended that the optimal fertilizer mixture to achieve good yields of high-quality fruit would be composed of a mixture of mainly inorganic (containing nitrate) and smaller percentage of microalgal fertilizer. Nevertheless, the ingredient compositions are yet to be identified. Furthermore, research should focus on establishing the effect of different dosages as well as forms of microbial fertilizer supplied (e.g., viable vs. nonviable cells). Finally, field trials are required to assess the long-term effects of the microbial fertilizer on field and soil properties.

MICROALGAE TO ANIMAL FEED

Microalgae have gained recognition as a food supplement due to their high nutritional value and completeness, as a food source. They present the potential to substitute commonly used ingredients such as soybean meal and can therefore alleviate the competition between food/feed/biofuel [112]. They are highly nutritious [113,114] because of the following reasons:

a. They contain high amounts of protein. This protein is often referred to as single-cell protein or microbial protein, and the content of microalgae can range between 8% and 71% (Table 11.4).

b. The amino acid profile is well balanced with the exception of the sulfur-containing amino acids, methionine, and cysteine. Specifically, amino acids can be classified into nutritionally essential amino acids (EAAs) and nutritionally non−essential amino acids (NEAAs). EAAs cannot be synthesized de novo, and for animals, they include cysteine (Cys), histine (His), isoleucine (Ile), leucine (Leu), lysine (Lys), methionine (Met), phenylalanine (Phe), threonine (Thr), tyrosine (Tyr), and valine (Val),

whereas NEEAs include arginine (Arg), glutamic acid (Glu), glutamine (Gln), glycine (Gly), and proline (Pro). NEAAs play an important role in growth, development, reproduction, health as well as survival and should therefore be included in the animal feed [117]. All these amino acids are produced by microalgae [114].

c. They contain lipids at a percentage varying between 16% and 75% [110], most importantly, PUFAs, specifically docosahexaenoic acid (DHA), eicosapentaenoic acid (EPA), arachidonic acid (AA), and γ-linolenic acid.

d. They contain 8%−57% (Table 11.4) carbohydrates, such as starch, sugars, cellulose, and polysaccharides, characterized by excellent digestibility. These carbohydrates contain a significant amount of dietary fiber, which is beneficial to animals' digestive system [112]. Given that carbohydrates comprise the biggest part of livestock feed and energy source, this characteristic renders them a noteworthy animal feed ingredient.

e. They enclose pigments in their cells, i.e., chlorophylls, carotenes, xanthophylls, and phycobiliproteins that act as antioxidants.

f. They are an important source of vitamins (A, B group, C, D, and E), including thiamine, riboflavin, and folic acid while they also contain B_{12} (cyanocobalamin).

g. Finally, research has shown that due to other bioactive compounds contained within the cells, they can have antiviral, antitumor, antiinflammatory, antiallergenic, antidiabetic, and antibacterial properties [118] as well as hypocholesterolemic and hypotriglyceridemic effects [119].

Nevertheless, there are some limitations regarding the use of microalgae as feed [114,119]: (1) the cell composition variates depending on the cultivation conditions and medium composition; thus research needs to focus on establishing the effect of these process parameters to achieve comparative results; (2) the sulfur-containing amino acids methionine and cysteine are commonly contained in limiting concentrations and therefore need to be externally supplied; (3) the rigid cell wall renders microalgae indigestible for some animals—especially monogastrics [112]. This requires the implementation of pretreatment techniques, such as mechanical or heat treatment, so that the cells are disrupted, and the enclosed nutrients become bioavailable for potential consumers. Madeira et al. [112] noted that carbohydrate-active enzymes (CAZymes) are expected to increase the efficiency of microalgae in monogastric

TABLE 11.4
Protein and Carbohydrate Content of Microalgae Commonly Used as Feed/Food Source, Presented in Percentage (%) of Dry Weight Basis.

	Protein Content (%)	Carbohydrate Content (%)
Arthrospira (Spirulina) maxima	46−71	13−16
Arthrospira (Spirulina) platensis	46−70	8−14
Chlorella vulgaris	38−58	12−17
Chlorella sp.	40−58	−
Haematococcus pluvialis	45−52	−
Nannochloropsis oculata	22−49	−
Porphyridium cruentum	8−58	40−57
References	[114−116]	[115]

"−", indicates not available values.

animal diets, via inducing the breaking of the cell well. These enzymes are already well accepted as feed additives for poultry and pigs. Finally, (4) the high content of nucleic acids in comparison with other food sources can lead to the formation of gout (kidney stones), due to the accumulation of uric acid. Specifically, the nucleic acid content can range between 2.9% and 5% for eukaryotic algae and 1%—17% for cyanobacteria. The latter requires downstream processing steps, for instance heat-shock treatment, to reduce the nucleic acid content [120].

Microalgae have proven to be nutritious when included in the feed of aquaculture, pigs, poultry, sheep, cows, as well as pets [112]. Some specific requirements for microalgae to be used as feed or feed ingredients are [113] (1) acceptance from the animal; (2) nutritionality and digestibility of the product; (3) absence of toxins; (4) effect on animals' weight gain; (5) smell and taste of the product. Nevertheless, each animal presents different responses to microalgae-containing feed, and therefore, the discussion will focus on different animal categories.

Aquaculture and shrimp feed

Given the projection that microalgal biomass will be highly demanded for aquaculture applications in the near future [121], a strong focus has been put in this application of microalgae. Strains commonly used in aquaculture feed belong to the genera *Haematococcus*, *Isochrysis*, *Nannochloropsis*, *Phaeodactylum*, and *Tetraselmis* [122], as well as the diatoms *Skeletonema* and *Thalassiosira* [123]. Microalgae can be used as feed for herbivore [113], carnivore [124] and juvenile omnivore [125] fish. Apart from fish nutrition, microalgae are also used as feed for mollusks (clams, oysters, and scallops) [123] and shrimps [126]. *Hypnea cervicornis* and *Cryptonemia crenulata* were successfully used in shrimp feeding trials, resulting in significant increase of the shrimp's growth rate [126]. They can be supplied as paste, powder, or pellets [113], live or dried [123]. Key parameters for the success of microalgae as aquaculture feed are the PUFA and pigment content, as both comprise ingredients that are essential for fish and shrimp nutrition [127]. Some marine fish are not able to convert C18 fatty acids to C20 and C22, and therefore, microalgae show the potential of replacing the essential PUFA, EPA, and DHA, of fish meal [128]. *Pavlova viridis* and *Nannochloropsis* sp. presented promising results as n—3 feed source on European sea bass (*Dicentrarchus labrax* L.), substituting fish oil [129]. Concerning pigments, astaxanthin presents beneficial effects in fish; specifically, apart from the increased pigmentation, it

increases survival and improves growth performance, reproductivity, and disease resistance [130]. A microalga known for its high astaxanthin content is *Haematococcus pluvialis* [130]. Importantly, *H. pluvialis* has been approved by the US Food and Drug Administration (FDA) as feed ingredients that enhance the color of salmonoids [131]. Finally, lipid-extracted algae meal derived from *Navicula* sp., *Chlorella* sp., and *Nannochloropsis salina* showed promising results in feeding trials with juvenile red drum (*Sciaenops ocellatus*) [124]. Patterson and Galtin [124] commented that the substitution of 10% of the soymeal or fishmeal crude protein is possible, noting though that the whole algal biomass is more nutritious and therefore preferred.

Monogastric livestock

Microalgae in the genera *Chlorella*, *Dunaliella*, *Desmodesmus*, *Nannochloropsis*, *Oocystis*, *Porphyridium*, *Scenedesmus*, *Arhtrospira* (*Spirulina*), and *Staurosira* have shown promising results in monogastric animals [113]. Nevertheless, there are limited available data about the effect of microalgae inclusions in fodder.

Poultry is the most targeted livestock for microalgal feed utilization as they present particularly favorable commercial uses; for instance, DHA-enriched eggs due to feeding with microalgae can already be found in the market [112]. The characteristics that are most affected by microalgae incorporation in feed are the poultry skin as well as yolk color [127]. Microalgae provided at a content of up to 10% in poultry feed have presented promising results, including increased disease resistance, decreased cholesterol, and better (darker) yolk color [113]. The latter is mainly attributed to β-carotene as well as other carotenoids produced by microalgae, especially *Dunaliella* sp. Furthermore, *A. platensis* supplementation in feed has been reported to increase the EPA and DHA content in thigh meat [132], whereas other studies report n—3 PUFA-enriched breast meat through feeding with *Schizochytrium* sp. [112]. Finally, poultry fed with *Porphyridium* sp. resulted in egg yolk with reduced cholesterol and improved color [133].

Existing literature regarding the inclusion of microalgae in pig feed present promising results. Feedings trials with *L. digitata* revealed that the inclusion of microalgae in the fodder can increase the daily pig weight up to 10% [134]. Moreover, lipid-extracted *Desmodesmus* sp. containing protease and non—starch polysaccharide-degrading enzymes (NSPase) resulted in comparable performance to control feeds in pigs [135]. When *Schizochytrium* sp. was used as a feed additive, the DHA content of pork meat increased [136], while *A. platensis* inclusion in pig feed resulted in up to 26% increase of the average daily weight gain [137].

Ruminants

The digestive system of ruminants renders them as the most eligible candidates to digest microalgae [113]. Their complex digestive system is suitable for the digestion of the recalcitrant cell walls that can contain cellulose [113]. Nevertheless, the literature about the effect of microalgal diet on ruminants is quite limited. In vitro digestibility of lipid-extracted *Nannochloropsis* sp. or *Chlorella* sp. presented similar behavior to soybean meal [138]. Moreover, feeding trials with *Arthrospira* sp. in cattle resulted in increased weight gain and final weight, improved feed intake and fertility, enhanced the immune system, and increased milk production and milk protein content [113]. Research shows that *A. platensis* shows promising results when supplied as feed ingredient for growth promotion in cattle [112,127]. The results also showed that the incorporation of n−3 PUFA in their diet results in higher EPA and DHA in their meat [112]. Finally, the provision of *Schizochytrium* sp. in dairy cows resulted in reduced milk fat as well as increased conjugated linolenic acid [139].

Overall, the most prominent effect of microalgal supplementation in fodder concerns the increase of PUFAs, e.g., EPA and DHA, in the meat, milk, and eggs. However, further research is required to determine the effect of each microalgal nutritional element on animal diet.

MICROALGAE TO BIOACTIVE COMPOUNDS

Microalgae have been widely recognized as outstanding producers of a great variety of natural products, applicable in several sectors, including energy, nutrition, pharmaceutics, and cosmetics. Activity of microalgal ingredients extracted or released from microalgal cells has been long tested and confirmed, as extensively reviewed in several articles available in the literature [140,141]. In the framework of an integrated biorefinery approach, bioactive compounds can greatly enhance economics of microalgal technology, due to their high market values [2].

Between the most prevalent biomass components, though a smaller fraction compared with lipids, carbohydrates, and proteins, pigments have received huge biotechnological and industrial interest due to their nutritional and pharmaceutic, mainly neuroprotective, antibiotic, and antiinflammatory properties [142]. Microalgal pigments, including chlorophylls, carotenoids, and phycobilins, are of vital importance for photosynthesis, light harnessing, and excess energy dissipation maintaining the proper function and integrity of algal cells. Due to their high antioxidant activity, their use as natural colorants, nutritional supplements, and cosmetics' ingredients, the most well-known pigments include chlorophylls, β-carotene, astaxanthin, violaxanthin, lutein, and fucoxanthin. Accumulation of the most abundant photosynthetic pigments and chlorophylls *a* and *b* depends on several factors, including the species used and several environmental conditions [143]. An increase in chlorophyll concentration is usually observed under low light intensity, as a result of shade adaptation. In contrast, chlorophyll reduction can be observed under nutrient, mainly nitrogen and phosphorus depletion, a trend also observed during micronutrient deficiency.

Accumulation of accessory photosynthetic pigments, carotenoids, is triggered by unfavorable environmental conditions including high light intensity and increased temperatures, as well as osmotic stress [94]. Carotenoid extracts from *H. pluvialis*, as the main source of natural astaxanthin, as well as from *Dunaliella salina*, a well-known alga for β-carotene production, were evaluated in terms of antitumor activity suggesting their high health protective role [144]. Moreover, carotenoids extracted from *N. oleoabundans*, mainly composed of lutein, violaxanthin, and monoesters, were effectively tested against colon cancer cells, with the antitumor activity mainly correlated with carotenoid monoesters [145]. Another target pigment is fucoxanthin, which is mainly produced by brown algae. Fucoxanthin is well known for its high antioxidant activity, contributing to the prevention of many diseases correlated with oxidative stress, while antidiabetic, antitumor activity, and chemoprevention have been attributed to fucoxanthin consumption [146]. By appropriately regulating culture conditions, fucoxanthin productivity from *Phaeodactylum tricornutum* reached 2.3 mg/L d [147].

PUFAs, including EPA, DHA, γ-linolenic, and arachidonic acid, represent valuable target products from microalgal biomass, as promising alternatives to fish oil [148]. Cultivation of microalgae for PUFA production must be well determined, since long-chain fatty acid content is significantly affected by the species used, environmental conditions such as light, temperature, and carbon supply, as well as by the growth phase of the culture [149]. PUFAs consumption offers numerous health benefits against cardiovascular diseases, cognitive decline, and mental disorders such as depression [150]. Under optimized photoautotrophic conditions, EPA productivity of the marine diatom *Fistulifera solaris* reached up to 136 mg/L d, suggesting its potential use for large-scale EPA production [151]. Also, EPA and arachidonic acid produced by *Koliella antarctica*, grown at low temperatures, accounted for 4.9% and 3.5% of total FAs, respectively [152].

Polymers excreted usually to induce cell adhesion and maintain cell vitality under stress conditions are also common microalgal products, with several potential uses. Extracellular polymeric substances, EPS, commonly composed of polysaccharides, proteins, lipids, and other organic and inorganic compounds, are involved in multiple biological processes and both medical and industrial applications [153]. EPS production by immobilized *Netrium digitus* was increased under high light intensity and nitrogen concentration, as well as acidic conditions [154]. Also, optimum light and nitrogen conditions resulted in 2 mg/mL EPS production in *Botryococcus braunii* cultures [155]. EPS extracts from *Gloeocapsa* sp. have showed high antimicrobial activity against different gram$^+$ and gram$^-$ bacteria and fungi, with inhibitory concentrations of extracts varying from 0.125 to 1 mg/ mL [156]. Depending on the solvent used to produce microalgal extracts (acetone, methanol, ethanol, chloroform, water, hexane) different compounds, including phenols, flavonoids, tannins, fatty acids, and pigments, are delivered and can be assessed concerning their potential antimicrobial activity. Aqueous extract of *C. vulgaris* failed to inhibit growth of *E. coli*, while ethanol-delivered substances, at a concentration of 3.12 mg/mL, inhibited *E. coli* proliferation [157]. Lastly, plenty of vitamins (including A, B, C), minerals (such as Ca, Mg, K), and chemical compounds, including phenols and sterols, provide multiple valorization options for biomass produced during cultivation of microalgae. In case of sterols, which are associated with prevention of cardiovascular diseases, microalgae have high potential as phytosterol producers reaching up to 5.1% of DW [158]. Also, antioxidant activity of microalgae can be attributed to several phenolic compounds (gallic, cinnamic, salicylic, caffeic, ferulic, and p-coumaric acids) that have been identified in different microalgal species [159].

CONCLUDING REMARKS

In conclusion, microalgae have attracted increasing attention over the past decades, as potential feedstock for a wide variety of products, characterized by low up to extremely high prices. Between the alternative uses of biomass, biofuels are a highly promising option with numerous advantages in terms of renewable energy production and CO_2 mitigation. However, major technical advancements as well as enhanced biomass productivity are needed prior to industrial application of algal technology and commercialization of microalgal biofuels. Recently, it has been made clear that fuels production alone is not sustainable; thus a biorefinery approach is highly recommended to make full use of microalgal biomass, exploit valuable fractions that are available, and enhance economics toward a prevalent technology in years to come.

REFERENCES

[1] J. Milano, H.C. Ong, H.H. Masjuki, W.T. Chong, M.K. Lam, P.K. Loh, V. Vellayan, Microalgae biofuels as an alternative to fossil fuel for power generation, Renew. Sustain. Energy Rev. 58 (2016) 180−197.

[2] M. Koller, A. Muhr, G. Braunegg, Microalgae as versatile cellular factories for valued products, Algal Res. 6 (2014) 52−63.

[3] J. Ruiz, et al., Towards industrial products from microalgae, Energy Environ. Sci. 9 (2016) 3036−3043.

[4] R. Sankaran, P.L. Show, D. Nagarajan, J.-S. Chang, Chapter 19 − Exploitation and Biorefinery of Microalgae, Waste Biorefinery, Potential and Perspectives, 2018, pp. 571−601.

[5] M. Vanthoor-Koopmans, R.H. Wijffels, M.J. Barbosa, M.H.M. Eppink, Biorefinery of microalgae for food and fuel, Bioresour. Technol. 135 (2013) 142−149.

[6] M.H.M. Eppink, G. Olivieri, H. Reith, C. van den Berg, M.J. Barbosa, R.H. Wijffels, From current algae products to future biorefinery practices: a review, in: Advances in Biochemical Engineering/Biotechnology, Springer, Berlin, Heidelberg, 2017, pp. 1−25.

[7] M.L. Gerardo, S. Van Den Hende, H. Vervaeren, T. Cowarda, S.C. Skill, Harvesting of microalgae within a biorefinery approach: a review of the developments and case studies from pilot-plants, Algal Res. 11 (2015) 248−262.

[8] I. Hariskos, C. Posten, Biorefinery of microalgae − opportunities and constraints for different production scenarios, Biotechnol. J. 9 (2014) 739−752.

[9] G.P. 'tLam, M.H. Vermuë, M.H.M. Eppink, R.H. Wijffels, C. van den Berg, Multi-product microalgae biorefineries: from concept towards reality, Trends Biotechnol. 36 (2018) 216−227.

[10] L. Zhu, Biorefinery as a promising approach to promote microalgae industry: an innovative framework, Renew. Sustain. Energy Rev. 41 (2015) 1376−1384.

[11] Y. Chen, S.-H. Ho, D. Nagarajan, N. Ren, J.-S. Chang, Waste biorefineries − integrating anaerobic digestion and microalgae cultivation for bioenergy production, Curr. Opin. Biotechnol. 50 (2018) 101−110.

[12] S. Rangabhashiyam, B. Behera, N. Aly, P. Balasubramanian, Biodiesel from microalgae as a promising strategy for renewable bioenergy production − a review, J. Environ. Biotechnol. Res. 6 (2017) 260−269.

[13] N. Rashid, M.S.U. Rehman, M. Sadiq, T. Mahmood, J.-I. Han, Current status, issues and developments in microalgae derived biodiesel production, Renew. Sustain. Energy Rev. 40 (2014) 760–778.

[14] K.K. Sharma, H. Schuhmann, P.M. Schenk, High lipid induction in microalgae for biodiesel production, Energies 5 (2012) 1532–1553.

[15] M. Chai, Q. Tu, M. Lu, Y.J. Yang, Esterification pretreatment of free fatty acid in biodiesel production, from laboratory to industry, Fuel Process. Technol. 125 (2014) 106–113.

[16] M. Veillette, A. Giroir-Fendler, N. Faucheux, M. Heitz, Biodiesel from microalgae lipids: from inorganic carbon to energy production, Biofuels 9 (2018) 175–202.

[17] G. Uctug, D.N. Modi, F. Mavituna, Life cycle assessment of biodiesel production from microalgae: a mass and energy balance approach in order to compare conventional with in situ transesterification, Int. J. Chem. Eng. Appl. 8 (2017) 355–360.

[18] D.-T. Tran, K.-L. Yeh, C.-L. Chen, J.-S. Chang, Enzymatic transesterification of microalgal oil from *Chlorella vulgaris* ESP-31 for biodiesel synthesis using immobilized *Burkholderia* lipase, Bioresour. Technol. 108 (2012) 119–127.

[19] Y. Chisti, Biodiesel from microalgae, Biotechnol. Adv. 25 (2007) 294–306.

[20] R.B. Levine, M.S. Costanza-Robinson, G.A. Spatafora, Neochloris oleoabundans grown on anaerobically digested dairy manure for concomitant nutrient removal and biodiesel feedstock production, Biomass Bioenergy 35 (2011) 40–49.

[21] Q. Hu, M. Sommerfeld, E. Jarvis, M. Ghirardi, M. Posewitz, M. Seibert, A. Darzins, Microalgal triacylglycerols as feedstocks for biofuel production: perspectives and advances, Plant J. 54 (2008) 621–639.

[22] Y. Gong, M. Jiang, Biodiesel production with microalgae as feedstock: from strains to biodiesel, Biotechnol. Lett. 33 (2011) 1269–1284.

[23] Z. Yu, M. Song, H. Pei, F. Han, L. Jiang, Q. Hou, The growth characteristics and biodiesel production of ten algae strains cultivated in anaerobically digested effluent from kitchen waste, Algal Res. 24 (2017) 265–275.

[24] G. Benvenuti, R. Bosma, M. Cuaresma, M. Janssen, M.J. Barbosa, R.H. Wijffels, Selecting microalgae with high lipid productivity and photosynthetic activity under nitrogen starvation, J. Appl. Phycol. 27 (2015) 1425–1431.

[25] P. Přibyl, V. Cepák, V. Zachleder, Production of lipids in 10 strains of *Chlorella* and *Parachlorella*, and enhanced lipid productivity in *Chlorella vulgaris*, Appl. Microbiol. Biotechnol. 94 (2012) 549–561.

[26] I. Pancha, K. Chokshi, R. Maurya, K. Trivedi, S.K. Patidar, A. Ghosh, S. Mishra, Salinity induced oxidative stress enhanced biofuel production potential of microalgae *Scenedesmus* sp. CCNM 1077, Bioresour. Technol. 189 (2015) 341–348.

[27] R. Karpagam, K.J. Raj, B. Ashokkumar, P. Varalakshmi, Characterization and fatty acid profiling in two fresh water microalgae for biodiesel production: lipid enhancement methods and media optimization using response surface methodology, Bioresour. Technol. 188 (2015) 177–184.

[28] K. Chokshi, I. Pancha, A. Ghosh, S. Mishra, Salinity induced oxidative stress alters the physiological responses and improves the biofuel potential of green microalgae *Acutodesmus dimorphus*, Bioresour. Technol. 244 (2017) 1376–1383.

[29] G. Srivastava, Nishchal, V.V. Goud, Salinity induced lipid production in microalgae and cluster analysis (ICCB 16-BR_047), Bioresour. Technol. 242 (2017) 244–252.

[30] S.K. Mandotra, P. Kumar, M.R. Suseela, S. Nayaka, P.W. Ramteke, Evaluation of fatty acid profile and biodiesel properties of microalga *Scenedesmus abundans* under the influence of phosphorus, pH and light intensities, Bioresour. Technol. 201 (2016) 222–229.

[31] O. Perez-Garcia, F.M.E. Escalante, L.E. de-Bashan, Y. Bashan, Heterotrophic cultures of microalgae: metabolism and potential products, Water Res. 45 (2011) 11–36.

[32] E. Jordaan, M.P. Roux-van der Merwe, J. Badenhorst, G. Knothe, B.M. Botha, Evaluating the usability of 19 effluents for heterotrophic cultivation of microalgal consortia as biodiesel feedstock, J. Appl. Phycol. 30 (2018) 1533–1547.

[33] R. Chandra, M.V. Rohit, Y.V. Swamy, S.V. Mohan, Regulatory function of organic carbon supplementation on biodiesel production during growth and nutrient stress phases of mixotrophic microalgae cultivation, Bioresour. Technol. 165 (2014) 279–287.

[34] Y.-S. Chung, J.-W. Lee, C.-H. Chung, Molecular challenges in microalgae towards cost-effective production of quality biodiesel, Renew. Sustain. Energy Rev. 74 (2017) 139–144.

[35] Y.S. Shin, J. Jeong, T.H.T. Nguyen, J.Y.H. Kim, E. Jin, S.J. Sim, Targeted knockout of phospholipase A2 to increase lipid productivity in *Chlamydomonas reinhardtii* for biodiesel production, Bioresour. Technol. 271 (2019) 368–374.

[36] L.M. González-González, L. Zhou, S. Astals, S.R. Thomas-Hall, E. Eltanahy, S. Pratt, P.D. Jensen, P.M. Schenk, Biogas production coupled to repeat microalgae cultivation using a closed nutrient loop, Bioresour. Technol. 263 (2018) 625–630.

[37] E. Kendir, A. Ugurlu, A comprehensive review on pretreatment of microalgae for biogas production, I, J. Energy Res. 42 (2018) 3711–3731.

[38] A. Ghimire, G. Kumar, P. Sivagurunathan, S. Shobana, G.D. Saratale, H.W. Kim, V. Luongo, G. Esposito, R. Munoz, Bio-hythane production from microalgae biomass: key challenges and potential opportunities for algal bio-refineries, Bioresour. Technol. 241 (2017) 525–536.

[39] E. Kwietniewska, J. Tys, Process characteristics, inhibition factors and methane yields of anaerobic digestion process, with particular focus on microalgal biomass

fermentation, Renew. Sustain. Energy Rev. 34 (2014) 491–500.

[40] C.G. Golueke, W.J. Oswald, H.B. Gotaas, Anaerobic digestion of algae, Appl. Microbiol. 5 (1957) 47–55.

[41] J.H. Mussgnug, V. Klassen, A. Schlüter, O. Kruse, Microalgae as substrates for fermentative biogas production in a combined biorefinery concept, J. Biotechnol. 150 (2010) 51–56.

[42] J.D. Murphy, B. Drosg, E. Allen, J. Jerney, A. Xia, C. Herrmann, A Perspective on Algal Biomass, IEA Bioenergy, 2015.

[43] E. Jankowska, A.K. Ahu, P. Oleskowicz, Biogas production from microalgae: review on microalgae's cultivation, harvesting and pretreatment for anaerobic digestion, Renew. Sustain. Energy Rev. 75 (2017) 692–709.

[44] M.E. Montingelli, S. Tedesco, A.G. Olabi, Biogas production from algal biomass: a review, Renew. Sustain. Energy Rev. 43 (2015) 961–972.

[45] H.G. Gerken, B. Donohoe, E.P. Knoshaug, Enzymatic cell wall degradation of *Chlorella vulgaris* and other microalgae for biofuels production, Planta 237 (2013) 239–253.

[46] S. Schwede, A. Kowalczyk, M. Gerber, R. Span, Influence on Different Cell Disruption Techniques on Mono Digestion of Algal Biomass, World Renewable Energy Congress, Sweden, 2011 (Linköping).

[17] F. Passos, I. Ferrer, Microalgae conversion to biogas: thermal pretreatment contribution on net energy production, Environ. Sci. Technol. 48 (2014) 7171–7178.

[48] L. Lavrič, A. Cerar, L. Fanedl, B. Lazar, M. Žitnik, R.M. Logar, Thermal pretreatment and bioaugmentation improve methane yield of microalgal mix produced in thermophilic anaerobic digestate, Anaerobe 46 (2017) 162–169.

[49] A. Mahdy, L. Mendez, E. Tomás-Pejó, M. del Mar Morales, M. Ballesteros, C. González-Fernández, Influence of enzymatic hydrolysis on the biochemical methane potential of *Chlorella vulgaris* and *Scenedesmus* sp, J. Chem. Technol. Biotechnol. 91 (2016) 1299–1305.

[50] B. Rincón, M.J. Fernández-Rodríguez, D. de la Lama-Calvente, R. Borja, The Influence of Microalgae Addition as Co-substrate in Anaerobic Digestion Processes, Microalgal Biotechnology, Eduardo Jacob-Lopes, Leila Queiroz Zepka and Maria Isabel Queiroz, Intech Open, June 27, 2018.

[51] E.A. Ehimen, S. Connaugthon, Z. Sun, G.C. Carrigton, Energy recovery from lipids extracted, transesterified and glycerol co-digested microalgae biomass, Glob. Change Biol. Bioenergy 1 (2009) 371–381.

[52] M. Wang, A.K. Sahu, B. Rusten, C. Park, Anaerobic co-digestion of microalgae *Chlorella* sp. and waste activated sludge, Bioresour. Technol. 142 (2013) 585–590.

[53] G. Zhen, X. Lu, T. Kobayashi, G. Kumar, K. Xu, Anaerobic co-digestion on improving methane production from mixed microalgae (*Scenedesmus* sp., *Chlorella* sp.) and food waste: kinetic modeling and synergistic evaluation, Chem. Eng. J. 299 (2016) 332–341.

[54] C. Herrmann, N. Kalita, D. Wall, A. Xia, J.D. Murphy, Optimized biogas production from microalgae through co-digestion with carbon-rich co-substrates, Bioresour. Technol. 214 (2016) 328–337.

[55] M.J. Fernández-Rodríguez, B. Rincón, F.G. Fermoso, A.M. Jimenez, R. Borja, Assessment of two-phase olive mill solid waste and microalgae co-digestion to improve methane pro- duction and process kinetics, Bioresour. Technol. 157 (2014) 263–269.

[56] P. Bohutskyi, D. Phan, A.M. Kopachevsky, S. Chow, E.J. Bouwer, M.J. Betenbaugh, Synergistic co-digestion of wastewater grown algae-bacteria polyculture biomass and cellulose to optimize carbon-to-nitrogen ratio and application of kinetic models to predict anaerobic digestion energy balance, Bioresour. Technol. 269 (2018) 210–220.

[57] L. Ding, E.C. Gutierrez, J. Cheng, A. Xia, R. O'Shea, A.J. Guneratnam, J.D. Murphy, Assessment of continuous fermentative hydrogen and methane co-production using macro- and micro-algae with increasing organic loading rate, Energy 151 (2018) 760–770.

[58] N. Wieczorek, M.A. Kucuker, K. Kuchta, Fermentative hydrogen and methane production from microalgal biomass (*Chlorella vulgaris*) in a two-stage combined process, App. Energy 132 (2014) 108–117.

[59] E. Koutra, C.N. Economou, P. Tsafrakidou, M. Kornaros, Bio-based products from microalgae cultivated in digestates, Trends Biotechnol. 36 (2018) 819–833.

[60] N. Qi, X. Hu, X. Zhao, L. Li, J. Yang, Y. Zhao, X. Li, Fermentative hydrogen production with peanut shell as supplementary substrate: effects of initial substrate, pH and inoculation proportion, Renew. Energy 127 (2018) 559–564.

[61] A. Ghimire, L. Frunzo, F. Pirozzi, E. Trably, R. Escudie, P.N.L. Lens, G. Esposito, A review on dark fermentative biohydrogen production from organic biomass: process parameters and use of by-products, Appl. Energy 144 (2015) 73–95.

[62] G. Buitron, J. Carrillo-Reyes, M. Morales, C. Faraloni, G. Torzillo, Biohydrogen production from microalgae, in: Microalgae-Based Biofuels and Bioproducts from Feedstock Cultivation to End-Products, Woodhead Publishing Series in Energy, 2017, pp. 209–234.

[63] H. Gaffron, J. Rubin, Fermentation and photochemical products of hydrogen in algae, J. Gen. Physiol. 26 (1942) 219–240.

[64] E.S. Shuba, D. Kifle, Microalgae to biofuels: 'Promising' alternative and renewable energy, review, Renew. Sust. Energy Rev. 81 (2018) 743–755.

[65] S.R. Vargas, P.V. dos Santos, L.A. Giraldi, M. Zaiat, M. do Carmo Calijuri, Anaerobic phototrophic processes of hydrogen production by different strains of microalgae *Chlamydomonas* sp, FEMS Microbiol. Lett. 365 (2018) fny073.

[66] T.K. Antal, T.E. Krendeleva, E. Tyystjarvi, Multiple regulatory mechanisms in the chloroplast of green algae: relation to hydrogen production, Photosynth. Res. 125 (2015) 357−381.

[67] D. Stojkovic, G. Torzillo, C. Faraloni, M. Valant, Hydrogen production by sulfur-deprived TiO_2-encapsulated *Chlamydomonas reinhardtii* cells, Int. J. Hydrog. Energy 40 (2015) 3201−3206.

[68] T.K. Antal, D.N. Matorin, G.P. Kukarskikh, M.D. Lambreva, E. Tyystjarvi, T.E. Krendeleva, A.A. Tsygankov, A.B. Rubin, Pathways of hydrogen photoproduction by immobilized *Chlamydomonas reinhardtii* cells deprived of sulfur, Int. J. Hydrog. Energy 39 (2014) 18194−18203.

[69] J.V.C. Vargas, V. Kava, W. Balmant, A.B. Mariano, J.C. Ordonez, Modeling microalgae derived hydrogen production enhancement via genetic modification, Int. J. Hydrog. Energy 41 (2016) 8101−8110.

[70] C. Sambusiti, M. Bellucci, A. Zabaniotou, L. Beneduce, F. Monlau, Algae as promising feedstocks for fermentative biohydrogen production according to a biorefinery approach: a comprehensive review, Renew. Sust. Energy Rev. 44 (2015) 20−36.

[71] B. Bharathiraja, M. Chakravarthy, R. Ranjith Kumar, D. Yogendran, D. Yuvaraj D, J. Jayamuthunagai, Aquatic biomass (algae) as a future feedstock for bioerefineries: a review on cultivation, processing and products, Renew. Sust. Energy Rev. 47 (2015) 634−653.

[72] T. Mutanda, D. Ramesh, S. Karthikeyan, S. Kumari, A. Anandraj, F. Bux, Bioprospecting for hyperelipid producing microalgal strains for sustainable biofuel production, Bioresour. Technol. 102 (2011) 57−70.

[73] S. Shobana, G.D. Saratale, A. Pugazhendhi, S. Arvindnarayan, S. Periyasamy, G. Kumar, S.-H. Kim, Fermentative hydrogen production from mixed and pure microalgae biomass: key challenges and possible opportunities, Int. J. Hydrogen Hydrog. Energy 42 (2017) 26440−26453.

[74] V.T.D.C. Neves, E. Andrade, L.W. Perelo, Influence of lipid extraction methods as pre-treatment of microalgal biomass for biogas production, Renew. Sustain. Energy Rev. 59 (2016) 160−165.

[75] J. Wang, Y. Yin, Fermentative hydrogen production using pretreated microalgal biomass as feedstock, Microb. Cell Fact. 17 (2018) 22.

[76] K. Bolatkhan, B.D. Kossalbayev, B.K. Zayadan, T. Tomo, T.N. Veziroglu, S.I. Allakhverdiev, Hydrogen production from phototrophic microorganisms: reality and perspectives, Int. J. Hydrog. Energy 44 (2019) 5799−5811.

[77] A. Sharma, S.K. Arya, Hydrogen from algal biomass: a review of production process, Biotechnol. Reports 15 (2017) 63−69.

[78] https://www.epure.org/media/1763/180905-def-data-epure-statistics-2017-designed-version.pdf.

[79] https://gain.fas.usda.gov/Recent%20GAIN%20Publications/Forms/AllItems.aspx.

[80] https://www.statista.com/.

[81] Q.C. Doan, N.R. Moheimani, A.J. Mastrangelo, D.M. Lewis, Microalgal biomass for bioethanol fermentation: implications for hypersaline systems with an industrial focus, Biomass Bioenergy 46 (2012) 79−88.

[82] M.M. El-Dalatony, E.-S. Salama, M.B. Kurade, S.H.A. Hassan, S.-E. Oh, S. Kim, B.-H. Jeon, Utilization of microalgal biofractions for bioethanol, higher alcohols, and biodiesel production: a review, Energies 10 (2017) 2110.

[83] S.A. Jambo, R. Abdulla, S.H.M. Azhar, H. Marbawi, J.A. Gansau, P. Ravindra, A review on third generation bioethanol feedstock, Renew. Sustain. Energy Rev. 65 (2016) 756−769.

[84] H.M. Kim, C.H. Oh, H.-J. Bae, Comparison of red microalgae (*Porphyridium cruentum*) culture conditions for bioethanol production, Bioresour. Technol. 233 (2017) 44−50.

[85] C.K. Phwan, H.C. Ong, W.-H. Chen, T.C. Ling, E.P. Ng, P.L. Show, Overview: comparison of pretreatment technologies and fermentation processes of bioethanol from microalgae, Energy Convers. Manag. 173 (2018) 81−94.

[86] J. Martin-Juarez, G. Markou, K. Muylaert, A. Lorenzo-Hernando, S. Bolado, Breakthroughs in bioalcohol production from microalgae: solving the hurdles, in: Microalgae-Based Biofuels and Bioproducts From Feedstock Cultivation to End-Products, Woodhead Publishing Series in Energy, 2017, pp. 183−207.

[87] L. Sanchez Rizza, M.E. Sanz Smachetti, M. Do Nascimento, G.L. Salerno, L. Curatti, Bioprospecting for native microalgae as an alternative source of sugars for the production of bioethanol, Algal Res. 22 (2017) 140−147.

[88] F. Passos, E. Uggetti, H. Carrere, I. Ferrer, Algal biomass: physical pretreatments,", in: A. Pandey, S. Negi, P. Binod, C. Larroche (Eds.), Pretreatments of Biomass, Elsevier, Amsterdam, 2015, pp. 195−226.

[89] E. Gunerken, E. D'Hondt, M. Eppink, K. Elst, R. Wijffels, Influence of nitrogen depletion in the growth of *N. oleoabundans* on the release of cellular components after bead milling, Bioresour. Technol. 214 (2016) 89−95.

[90] M.M. Eldalatony, A.N. Kabra, J.H. Hwang, S.P. Govindwar, K.H. Kim, H. Kim, B.H. Jeon, Pretreatment of microalgal biomass for enhanced recovery/extraction of reducing sugars and proteins, Bioproc. Biosys. Eng. 39pp (2016) 95−103.

[91] M.R. Gruber-brunhumer, J. Jerney, E. Zohar, M. Nussbaumer, C. Hieger, G. Bochmann, *Acutodesmus obliquus* as a benchmark strain for evaluating methane production from microalgae: influence of different storage and pretreatment methods on biogas yield, Algal Res. 12 (2015) 230−238.

[92] M.K. Lam, K.T. Lee, Bioethanol production from microalgae, in: S.K. Kim (Ed.), Handbook Marine Microalgae: Biotechnology Advances, Janice Audet, USA, 2015, pp. 197−208.

[93] C.M. Nguyen, T.N. Nguyen, G.J. Choi, Y.H. Choi, K.S. Jang, Y.J. Park, J.C. Kim, Acid hydrolysis of Curcuma longa residue for ethanol and lactic acid fermentation, Bioresour. Technol. 151 (2014) 227−235.

[94] G. Markou, E. Nerantzis, Microalgae for high value compounds and biofuels production: a review with focus on cultivation under stress conditions, Biotechnol. Adv. 31 (2013) 1532−1542.

[95] R.A. Hamouda, S.A. Sherif, M.M. Ghareeb, Bioethanol production by various hydrolysis and fermentation processes with micro and macro green algae, Waste Biomass Valor. 9 (2018) 1495−1501.

[96] Z. Reyimu, D. Ozçimen, Batch cultivation of marine microalgae *Nannochloropsis oculata* and *Tetraselmis suecica* in treated municipal wastewater toward bioethanol production, J. Clean. Prod. 150 (2017) 40−46.

[97] H. Shokrkar, S. Ebrahimi, M. Zamani, Enzymatic hydrolysis of microalgal cellulose for bioethanol production, modeling and sensitivity analysis, Fuel 228 (2018) 30−38.

[98] A.R. Sirajunnisa, D. Surendhiran, Algae quintessential and positive resource of bioethanol production: a comprehensive review, Renew. Sustain. Energy Rev. 66 (2016) 248−267.

[99] S. Singh, I. Chakravarty, K.D. Pandey, S. Kundu, Development of a Process Model for Simultaneous Saccharification and Fermentation (SSF) of Algal Starch to Third Generation Bioethanol, Biofuels, 2018.

[100] J. Coppens, et al., The use of microalgae as a high−value organic slow−release fertilizer results in tomatoes with increased carotenoid and sugar levels, J. Appl. Phycol. 28 (2016) 2367−2377.

[101] W. Mulbry, S. Kondrad, C. Pizarro, Biofertilizers from algal treatment of dairy and swine manure effluents, J. Veg. Sci. 12 (2007) 107−125.

[102] P. Spolaore, C. Joannis−Cassan, E. Duran, A. Isambert, Commercial applications of microalgae, J. Biosci. Bioeng. 101 (2006) 87−96.

[103] S.C. Wuang, M.C. Khin, P.Q.D. Chua, Y.D. Luo, Use of *Spirulina* biomass produced from treatment of aquaculture wastewater as agricultural fertilizers, Algal Res. 15 (2016) 59−64.

[104] P. Das, M.A. Quadir, M.I. Thaher, G.S.H.S. Alghasal, H.M.S.J. Aljabri, Microalgal nutrients recycling from the primary effluent of municipal wastewater and use of the produced biomass as bio−fertilizer, Int. J. Environ. Sci. Technol. (2018) 1−10.

[105] J. Garcia−Gonzalez, M. Sommerfeld, Biofertilizer and biostimulant properties of the microalga *Acutodesmus dimorphus*, J. Appl. Phycol. 28 (2016) 1051−1061.

[106] N. Renuka, et al., Exploring the efficacy of wastewater−grown microalgal biomass as a biofertilizer for wheat, Environ. Sci. Pollut. Res. 23 (2016) 6608−6620.

[107] R. Dineshkumar, R. Kumaravel, J. Gopalsamy, M.N.A. Sikder, P. Sampathkumar, Microalgae as bio−fertilizers for rice growth and seed yield productivity, Waste Biomass Valor. 9 (2017) 793−800.

[108] K.L.N. Aung, Effect of *Spirulina* biofertilizer suspension on growth and yield of *Vigna radiata* (L.) Wilczek, Univ. Res. J. 4 (2011) 9−14.

[109] F.A. Faheed, Effect of *Chlorella vulgaris* as bio−fertilizer on growth parameters and metabolic aspects of lettuce plant, ISSN OnlineAWB, J. Agri. Soc. Sci. 4 (2008) 1813−2235.

[110] A.F. Safinaz, A.H. Ragaa, Effect of some red marine algae as biofertilizers on growth of maize (*Zea mays* l.) plants, Int. Food Res. J. 20 (2013) 1629−1632.

[111] R. Maurya, et al., Lipid extracted microalgal biomass residue as a fertilizer substitute for *Zea mays* L. Front. Plant Sci. 6 (2016) 1266.

[112] M.S. Madeira, et al., Microalgae as feed ingredients for livestock production and meat quality: a review, Livest. Sci. 205 (2017) 111−121.

[113] M. Hayes, et al., Microalgal Proteins for Feed, Food and Health, Microalgae−Based Biofuels Bioprod. From Feed. Cultiv. to End−Products, 2017, pp. 347−368.

[114] M. García−Garibay, L. Gómez−Ruiz, A.E. Cruz−Guerrero, E. Bárzana, Single cell protein: the algae, Encycl. Food Microbiol. vol. 3 (2014) 425−430.

[115] E. Becker, Microalgae − Biotechnology and Microbiology 183, Cambridge University Press, 1994.

[116] C. Safi, et al., Release of hydro−soluble microalgal proteins using mechanical and chemical treatments, Algal Res. 3 (2014) 55−60.

[117] G. Wu, Dietary requirements of synthesizable amino acids by animals: a paradigm shift in protein nutrition, J. Anim. Sci. Biotechnol. 5 (2014) 34.

[118] M.F. de Jesus Raposo, R.M.S.C. de Morais, A.M.M.B. de Morais, Health applications of bioactive compounds from marine microalgae, Life Sci. 93 (2013) 479−486.

[119] F. Navarro, et al., Microalgae as a safe food source for animals: nutritional characteristics of the acidophilic microalga *Coccomyxa onubensis*, Food Nutr. Res. 60 (2016) 30472.

[120] W.E. Trevelyan, Processing yeast to reduce its nucleic acid content. Induction of intracellular RNase action by a simple heat−shock procedure, and an efficient chemical method based on extraction of RNA by salt solutions at low pH, J. Sci. Food Agric. 29 (1978) 141−147.

[121] I.G. Anemaet, M. Bekker, K.J. Hellingwerf, Algal photosynthesis as the primary driver for a sustainable development in energy, feed, and food production, Mar. Biotechnol. 12 (2010) 619−629.

[122] M.S. Chauton, K.I. Reitan, N.H. Norsker, R. Tveterås, H.T. Kleivdal, A techno−economic analysis of industrial production of marine microalgae as a source of EPA and DHA−rich raw material for aquafeed: research challenges and possibilities, Aquaculture 436 (2015) 95−103.

[123] S. Ghosh, B. Xavier, L. Edward, B. Dash, Live Feed for Marine Finfish and Shellfish Culture, 2016. http://eprints.cmfri.org.in/10857/.

[124] D. Patterson, D.M. Gatlin, Evaluation of whole and lipid−extracted algae meals in the diets of juvenile red drum (*Sciaenops ocellatus*), Aquacult 416−417 (2013) 92−98.

[125] R.S.C. Barone, D.Y. Sonoda, E.K. Lorenz, J.E.P. Cyrino, Digestibility and pricing of *Chlorella sorokiniana* meal for use in tilapia feeds, Sci. Agric. 75 (2018) 184−190.

[126] R. Harun, M. Singh, G.M. Forde, M.K. Danquah, Bioprocess engineering of microalgae to produce a variety of consumer products, Renew. Sustain. Energy Rev. 14 (2010) 1037−1047.

[127] Z. Yaakob, E. Ali, A. Zainal, M. Mohamad, M.S. Takriff, An overview: biomolecules from microalgae for animal feed and aquaculture, J. Biol. Res. 21 (2014).

[128] D.R. Tocher, Omega−3 long−chain polyunsaturated fatty acids and aquaculture in perspective, Aquacult 449 (2015) 94−107.

[129] S. Haas, et al., Marine microalgae *Pavlova viridis* and *Nannochloropsis* sp. as n−3 PUFA source in diets for juvenile European sea bass (*Dicentrarchus labrax* L.), J. Appl. Phycol. 28 (2016) 1011−1021.

[130] K.C. Lim, F.M. Yusoff, M. Shariff, M.S. Kamarudin, Astaxanthin as feed supplement in aquatic animals, Rev. Aquacult 10 (2017) 738−773.

[131] FDA, CFR − Code of Federal Regulations Title 21, Code of Federal Regulations, 2018 [Online]. Available: https://www.accessdata.fda.gov/scripts/cdrh/cfdocs/cfcfr/CFRSearch.cfm?fr=73.185.

[132] E. Bonos, E. Kasapidou, A. Kargopoulos, A. Karampampas, E. Christaki, *Spirulina* as a functional ingredient in broiler chicken diets, S. Afr. J. Anim. Sci. 46 (2016) 94−102.

[133] A. Ginzberg, M. Cohen, U. Sod−moriah, S. Shany, A. Rosenshtrauch, S.M. Arad, Chickens fed with biomass of the red microalga *Porphyridium* sp. have reduced blood cholesterol level and modified fatty acid composition in egg yolk, J. Appl. Phycol. 12 (2000) 325−330.

[134] M.L. He, W. Hollwich, W.A. Rambeck, Supplementation of algae to the diet of pigs: a new possibility to improve the iodine content in the meat, J. Anim. Physiol. Anim. Nutr. 86 (2002) 97−104.

[135] R. Ekmay, S. Gatrell, K. Lum, J. Kim, X.G. Lei, Nutritional and metabolic impacts of a defatted green marine microalgal (*Desmodesmus* sp.) biomass in diets for weanling pigs and broiler chickens, J. Agric. Food Chem. 62 (2014) 9783−9791.

[136] E. Vossen, K. Raes, D. Van Mullem, S. De Smet, Production of docosahexaenoic acid (DHA) enriched loin and dry cured ham from pigs fed algae: nutritional and sensory quality, Eur. J. Lipid Sci. Technol. 119 (2017) 1600144.

[137] A. Simkus, A. Simkiené, J. Cemauskiené, N. Kvietkuté, A. Cemauskas, The effect of blue algae *Spirulina platensis* on pig growth; Melsvadumblio *Spirulina platensis*, Vet. ir Zootech. 61 (2013) 70−74.

[138] S.L. Lodge−Ivey, L.N. Tracey, A. Salazar, Ruminant nutrition symposium: the utility of lipid extracted algae as a protein source in forage or starch−based ruminant diets, J. Anim. Sci. 92 (2014) 1331−1342.

[139] C. Boeckaert, et al., Effect of dietary starch or micro algae supplementation on rumen fermentation and milk fatty acid composition of dairy cows, J. Dairy Sci. 91 (2008) 4714−4727.

[140] M.P. Caporgno, A. Mathys, Trends in microalgae incorporation into innovative food products with potential health benefits, Front. Nutr. 5 (2018) 1−10.

[141] M.H. Bule, I. Ahmed, F. Maqbool, M. Bilal, H.M.N. Iqbal, Microalgae as a source of high-value bioactive compounds, Front. Biosci. 10 (2018) 197−216.

[142] H. Begum, F.M.D. Yusoff, S. Banerjee, H. Khatoon, M. Shariff, Availability and utilization of pigments from microalgae, Crit. Rev. Food Sci. Nutr. 56 (2016) 2209−2222.

[143] V. da Silva Ferreira, C. Sant'Anna, Impact of culture conditions on the chlorophyll content of microalgae for biotechnological applications, World J. Microbiol. Biotechnol. 33 (2017) 20.

[144] F.K. El-Baz, R.A. Hussein, K. Mahmoud, S.M. Abdo, Cytotoxic activity of carotenoid rich fractions from *Haematococcus pluvialis* and *Dunaliella salina* microalgae and the identification of the phytoconstituents using LC-DAD/ESI-MS, Phytother Res. 32 (2018) 298−304.

[145] M. Castro-Puyana, A. Pérez-Sánchez, A. Valdés, O.H.M. Ibrahim, S. Suarez-Álvarez, J.A. Ferragut, V. Micol, A. Cifuentes, E. Ibáñez, V. García-Cañas, Pressurized liquid extraction of *Neochloris oleoabundans* for the recovery of bioactive carotenoids with antiproliferative activity against human colon cancer cells, Food Res. Int. 99 (2017) 1048−1055.

[146] K. Mikami, M. Hosokawa, Biosynthetic pathway and health benefits of fucoxanthin, an algae-specific xanthophyll in Brown seaweeds, Int. J. Mol. Sci. 14 (2013) 13763−13781.

[147] D.D. McClure, A. Luiz, B. Gerber, G.W. Barton, J.M. Kavanagh, An investigation into the effect of culture conditions on fucoxanthin production using the marine microalgae *Phaeodactylum tricornutum*, Algal Res. 29 (2018) 41−48.

[148] M.A. Borowitzka, High-value products from microalgae—their development and commercialisation, J. Appl. Phycol. 25 (2013) 743−756.

[149] P. Boelen, A. van Mastrigt, H.H. van de Bovenkamp, H.J. Heeres, A.G.J. Buma, Growth phase significantly decreases the DHA-to-EPA ratio in marine microalgae, Aquacult. Int. 25 (2017) 577−587.

[150] F. Shahidi, P. Ambigaipalan, Omega-3 polyunsaturated fatty acids and their health benefits, Annu. Rev. Food Sci. Technol. 9 (2018) 345−381.

[151] T. Tanakaa, Y. Yabuuchia, Y. Maedaa, D. Nojimaa, M. Matsumotob, T. Yoshino, Production of eicosapentaenoic acid by high cell density cultivation of the

marine oleaginous diatom *Fistulifera solaris*, Bioresour. Technol. 245 (2017) 567–572.

[152] H. Suzuki, C.J. Hulatt, R.H. Wijffels, V. Kiron, Growth and LC-PUFA production of the cold-adapted microalga *Koliella antarctica* in photobioreactors, J. Appl. Phycol. (2018) 1–17.

[153] R. Xiao, Y. Zheng, Overview of microalgal extracellular polymeric substances (EPS) and their applications, Biotechnol. Adv. 34 (2016) 1225–1244.

[154] A. Ekelhof, M. Melkonian, Enhanced extracellular polysaccharide production and growth by microalga *Netrium digitus* in a porous substrate bioreactor, Algal Res. 28 (2017) 184–191.

[155] V. Cepák, P. Přibyl, Light intensity and nitrogen effectively control exopolysaccharide production by the green microalga *Botryococcus braunii* (Trebouxiophyceae), Genet. Plant Physiol. 8 (2018) 24–37.

[156] H.M. Najdenski, L.G. Gigova, I.I. Iliev, P.S. Pilarski, J. Lukavský, I.V. Tsvetkova, M.S. Ninova, V.K. Kussovski, Antibacterial and antifungal activities of selected microalgae and cyanobacteria, Int. J. Food Sci. Technol. 48 (2013) 1533–1540.

[157] J. Annamalai, J. Shanmugam, T. Nallamuthu, Phytochemical screening and antimicrobial activity of *Chlorella vulgaris* BEIJERINCK, Int. J. Curr. Res. Rev. 4 (2012) 33–38.

[158] X. Luo, P. Su, W. Zhang, Advances in microalgae-derived phytosterols for functional food and pharmaceutical applications, Mar. Drugs 13 (2015) 4231–4254.

[159] H. Safafar, J. van Wagenen, P. Møller, C. Jacobsen, Carotenoids, phenolic compounds and tocopherols contribute to the antioxidative properties of some microalgae species grown on industrial wastewater, Mar. Drugs 13 (2015) 7339–7356.

Microalgal Biorefineries for Industrial Products

SABEELA BEEVI UMMALYMA • DINABANDHU SAHOO • ASHOK PANDEY

INTRODUCTION

The shortage of conventional energy along with global warming effect throughout the world leads to a search for renewable energy sources as an alternative for fossil-based fuels. Microalgae are reported as green microscopic unicellular or multicellular photosynthetic plants and used as renewable fuels. Most of the published studies on microalgae are their exploitation for bioenergy such as biodiesel and bioethanol due to high photosynthetic efficiency, growth, year-round availability, noncompetitive with food, and possibility of mass cultivation in degraded land. Commercial exploitation of microalgae is still challenging due to the high cost associated with biomass processing. Another opportunity of algal biomass is exploitation of high-value product. High-value metabolite along with lipids and carbohydrates for biofuels can be economically viable. Several industries cultivate microalgae for food supplements and antioxidants [1].

Any of novel developed bioprocess technology is only sustainable if the process can able to answer the questions such as i) whether the technology is feasible ii) if it is economically profitable process: can the novel technology be produced at lower cost than its market values iii) is the technology environmentally sustainable, and iv) does the novel technology have an acceptable environmental impact? [2].

If any process can able to answer the abovementioned question, then the technology and product are commercially viable. Mass cultivation of microalgae can be conducted either in raceway reactors or photobioreactors under phototrophic and mixotrophic growth conditions. Heterotrophic cultivation is performed in photobioreactors in a controlled environment. Biorefineries are utilizing complete exploitation of raw materials into a marketable product. The generation of high-value products in biorefinery can lead to reducing the environmental impact along with an increase in revenues [3]. However, the process technology is not yet commercialized in a full-scale operation due to challenges associated with the energy-intensive and costly process. Algal biomass harvesting from its broth itself is 30% of the cost of the whole process. Hence the objective of the chapter is to address the challenges of microalgal cultivation and exploitation of different algal metabolites for various industrial products in biorefinery concept to solve the issues associated with microalgal biomass and its future perspectives.

CHALLENGES OF ALGAL CULTIVATION

Microalgae can be cultivated in wastewater and seawater for viable technologies. Freshwater algae are commonly cultivated in freshwater medium, which is not practically viable process due to escalating demands for freshwater. However, microalgae adapted to grow both waste streams and seawater are an economically sustainable alternative. Another challenge is a suitable organism able to adapt in any harsh conditions and resistance to attack by other microorganisms and high biomass with high yields of metabolite productions suitable for industries. Very few microalgae are well suited for an industrial operation, which include *Chlorella* sp., *Chlorococcum, and Scenedesmus* sp. [4]. Limitation of light penetrations in phototrophic cultivations associated with low biomass production is another bottleneck in mass cultivation of algae and availability of suitable low-cost carbon source for mixotrophic and heterotrophic cultivations. Contaminations are other issues in heterotrophic cultivation of microalgae. Industrial-scale production of by-products from microalgae is not yet cost-effective, which is primarily due to the high energy associated with algal biomass harvesting [5]. Algal biomass generation consists of growing of algae in an environment that favors the accumulation of target product and recovery of biomass for

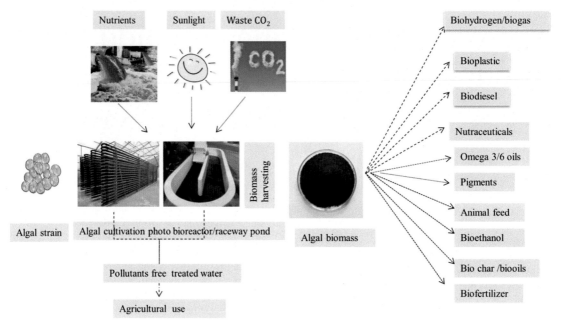

FIG. 12.1 Microalgae-based biorefinery.

downstream processing. Major hurdles in many of the industries dealing with algal biomass are harvesting the biomass from the growth medium. Most challenging research areas in microalgal biofuel are targeting cost-effective harvesting methods [6]. The reports showed that 20%–30% of the total production cost is involved in the biomass harvesting [7,8]. Other researchers showed that the cost of the recovery process in their investigation contributed about 50% to the final cost of oil production [6,9]. Several types of research on microalgal biofuel production have been focused on the yield of lipids and composition of biomass rather than harvesting process. Therefore, it is necessary to develop effective and economic technologies for harvesting the biomass from the suspended water along with complete utilization of biomass in a zero waste biorefinery process. Moreover, choice of algal cultivation location is an impact on the economic and environmental sustainability of algal biorefinery.

ALGAL BIOREFINERY CONCEPTS

Microalgal biomass has great potential to produce various metabolites suitable for bioenergy, food additives, pigments, bioplastics, polymers, and chemicals. The algae-based biorefinery is the integration of mass production of algal biomass for energy along with

different industrially important molecules for balancing the sustainability of microalgal industries. Complete utilization of algal biomass is feasible with biorefineries. Algae-based refineries are represented in Fig. 12.1.

Oleaginous microalgae can be feedstock for biodiesel application due to their high oil content approximately $5000-100,000$ L/ha year [10]. The ethanol can be produced from hydrolysis and fermentation process of carbohydrates from deoiled biomass; biomethane and hydrogen can be generated from anaerobic digestion (AD). Protein-rich biomass from algae can be exploited for the production of bioplastics and thermoplastics [11]. Microalgae biomass enriched with polyunsaturated fatty acids (PUFAs) such as docosahexaenoic acids (DHA) and eicosapentaenoic acid (EPA) along with pigments can be applied in the food and pharmaceutical industries [12]. However, low-cost process for algae technologies can be linked with exploitation of wastewaters, and waste CO_2 along with complete utilization of biomass will be viable for sustainability and environmental safe, which might be a successful biorefinery approach.

Microalgae to Biodiesel

Microalgae have high potential to produce biomass with a high content of lipids. It has been reported that some microalgae have lipids content of up to 50%

−80% of their dry cell weight. Microalgal oil rich in triglycerides is considered as suitable feedstock for biodiesel production. Microalgae biodiesel is third-generation carbon-neutral biofuel; it can able to assimilate maximum CO_2 during algal growth as it is produced up on combustion of fuel, and hence, it is an effective, sustainable solution for climate change [13,14]. Fatty acid alkyl esters (biodiesel) are produced by many processes such as microemulsification, catalytic cracking, and transesterification. The most common method to produce biodiesel from oil is transesterification, while others are costly and produce low-quality diesel [15]. Transesterification process converts raw algal lipids (triglycerides) to low-molecular-weight fatty acid alkyl esters with the help of catalysts such as acids, alkalis, and enzymes [16,17]. Commonly used alkali catalysts include sodium hydroxides, potassium hydroxides, and sodium methoxide. Acids such as hydrochloric acids, sulfuric acids, phosphoric acids, and sulfonic acids have been utilized. Lipase is the preferred enzymatic catalyst; inorganic heterogenous catalyst is also preferred for esterification. Transesterification reaction is influenced by the nature of alcohol, molar ratio of alcohol to oil, type and amount of catalyst, reaction time, and temperature [18].

The enzymatic route is recently getting more attraction for transesterification reaction. Lipase-based reaction can be performed with both extracellular and intracellular lipases. Lipases are given more preferences due to reusability and simple downstream processing for purification of biodiesel. The bottleneck of enzyme-based reaction is the cost of the enzyme and inactivation mainly because of methanol and glycerol [19,20]. Practical use of algal biodiesel for vehicle applications should meet the international biodiesel standards (EN14214). Microalgae selection for biodiesel production should be based on the biodiesel physico-chemical properties of algal oils along with their emission characteristics and engine performance. Glycerol is a by-product produced during transesterification process, which can be used as a carbon source for heterotrophic and mixotrophic cultivation of algae for recycling back this glycerol for oil production for biodiesel [21,22].

Microalgae to Biogas

Deoiled algal biomass and biomass itself can be a raw material for biogas production via AD. AD is a biochemical process where specific anaerobic microorganisms act on complex organic substrates to produce methane (55%−75%) and CO_2 (25%−45%) [23,24]. In biogas production process, many steps are involved, such as hydrolysis, fermentation, acetogenesis, and methanogenesis [25,26]. Microalgae-based AD was first showed by Golueke et al. [27]. Many investigations on microalgae as raw material for AD have shown significant methane yield compared with normal substrates such as manures and sewage sludge [28,29]. Biogas production studies with microalgae of *Isochrysis* sp. *and Scenedesmus* sp. showed methane production yield of 400 mL CH4 g^{-1} VS [30,31]. Biogas yield may be further increased with the appropriate pretreatment process applied. The energy content of AD methane is 16,200−30,600 kJ/m^3, which further depends on the type of raw material [9]. Many microorganisms are involved in the AD process. Hydrolytic bacteria are involved in the initial stage for breakdown of substrate molecules. Later the action of acidogenic bacteria along with hydrogenated and sulfate reducers played a major role. As a result of oxidized organic matter, gases and organic acids are produced from this process. Methane, CO_2, and reduced organics are produced as a result of methanogenic microorganisms [32,33]. Biogas produced from AD can be used as a fuel or electricity [34]. The residues obtained after AD can be used as a fertilizer, which could encourage sustainable agriculture practices with better efficiencies along with balancing algal production cost. Lack of lignin and less cellulosic content in microalgal biomass showed better conversion efficiencies for AD [35].

Microalgae to biohydrogen

Biohydrogen is considered as one of the cleanest fuels because water is the only by-product and there is no CO_2 emission to the atmosphere during its combustion [36]. It has higher specific energy content (142 MJ/kg) than other fuel such as methane (56), natural gas (54), and gasoline (47) [37]. These qualities make hydrogen a good choice for future fuel. Microalgae-based biohydrogen production has been an alternate solution for minimizing the production cost and environmental impact probably due to its production use; sunlight is the driving force to split water into H_2 and O_2 [36,38]. Microalgae-based biohydrogen production was initiated by Gaffron and Rubin [39]. Algae have the potential of changing their metabolism based on the growth condition to produce hydrogen. Biohydrogen from microalgae can be produced from many mechanisms such as direct and indirect photolysis, photofermentation, and dark fermentation [40]. Hydrogen production via the abovementioned process from algae can be further improved through extensive investigation to make biological hydrogen production economically viable [41]. Many researchers have reviewed recently different perspectives of the bio-hydrogen production process from different microalgae [42,43].

Microalgae to Sugarcane Ethanol Industry

Bioethanol is an alternative to gasoline due to its identical physical and chemical properties [44,45]. First- and second-generation biofuels from food crops and ligno-cellulosic biomass have food versus fuel dispute and involve additional pretreatment processes to break down the recalcitrance of lignin and hemicellulose complex and exposure of cellulose for further hydrolysis and fermentations [46]. Due to these challenges, industries are looking for appropriate feedstock for ethanol production that could not be competitive with food and should be available throughout the year. However, algae are found to be potential candidates as raw material for third-generation biofuel. Microalgal bioethanol is known as third-generation ethanol due to its potential to produce biomass from sunlight. Carbohydrates-enriched microalgae are selected as feedstock for anaerobic fermentation to produce ethanol [47,48]. Many of the microalgal species contain high lipids and low carbohydrates content. Hence, algal biomasses are applied for biodiesel conversion [49]. Macroalgae are attracted more as ethanol feedstock due to the presence of polysaccharides such as glucan and mannitol that can be easily transformed into ethanol by industrial yeast [50,51].

Carbohydrates accumulation in microalgae is dependent on cultivation condition especially with CO_2 concentration, light intensity, and photosynthetic efficiency of organisms. It has been reported that the highest carbohydrates (70%)-accumulating microalga is *Chlorococcum littorale* when cultivated in photobioreactor with optimum light and CO_2 [52]. Fed-batch cultivation of *Chlamydomonas reinhardtii* with acetic acid as a carbon source enhanced the carbohydrates content up to 57% and has been utilized for ethanol production [53]. Other carbohydrates-rich microalgae are *Chlorella* sp. and *Scenedesmus* sp., and their carbohydrates contents are increased under nutrients-limited stress conditions [54].

Biomasses from microalgae are processed for ethanol production via hydrolysis and fermentation. Sometimes, pretreatment of biomass is also required for better efficiency. Polysaccharides found in microalgae are starch, cellulose, and glycogen. Under stress condition, these carbohydrates can be reached up to 70%. *Saccharomyces* and *Zymomonas* are the preferred microorganisms for the fermentation process. Pretreatment approaches used for biomass processing are acidic, alkali, and enzymatic hydrolyses and finally deoiling with solvent extractions. These processes assist with getting fermentable sugars and algal oil [55]. Physical pretreatments such as bead-beating sonications are effective in algal biomass. It has been reported that bead-beating with pectinase enzyme treatment of algal biomass of *Chlorella vulgaris* cells produced fermentable sugar release that enhanced from 45% to 70% with 90% of fermentation yield [56]. However, potent algal strains having high starch accumulation and robustness are required to support industrial mass production of biomass for biorefinery products. Fig. 12.2 represents bioethanol production process from algal biomass.

Microalgae to Fertilizer

Biofertilizers consist of live microorganisms, or their powder form originated from algae, bacteria, fungi, and their metabolites that are nourishing the soil for crop growth and its productivity. Biofertilizers can enhance soil quality and its fertility to nutrients transfers, stabilizes aggregates of soils, and increase the population of beneficial microorganisms. The accessibility of micro/macronutrients is a crucial factor for achieving higher crop yields. Biomass of microalgae contains high nitrogen content, which serves as fertilizers for different crops. Algal biofertilizer along with associated microbes can able to colonize soil, and rhizosphere, which further improves nutrients level and growth of plants

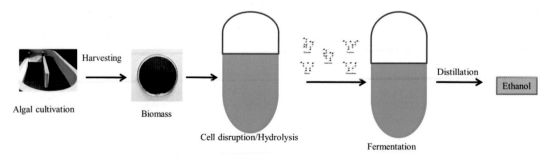

FIG. 12.2 Bioethanol production process from algal biomass.

[57,58]. Biofertilizers are available in different forms of carrier-based, pellets, or liquid formulations [59]. Microalgal biofertilizers, especially from blue green algae of *Nostoc* sp., *Anabaena* sp., *Tolypothrix* sp., *Aulosira* sp., etc., are used in the paddy field for nitrogen fixation. Deoiled algal biomass is rich in good sources of NPK, and other minerals suitable for plant growth can be used as biofertilizers. Leftover solids of AD of algal biomass can be exploited as fertilizers. For cultivation of microalgae as biofertilizers, wastewater is the best option to produce biomass [22,60]. Renuka et al. [60] showed that the formulation of biofertilizers made from microalgae and filamentous cyanobacteria used for wheat growth under controlled conditions improved the plant growth and yield. Algae can absorb maximum nutrient from wastewater and return to the soil as fertilizers.

Microalgae to Animal Feed

Microalgal biomass can be used alone or mixed with other material as healthy feed and food. It has been reported that ~75% of annual biomass production is utilized for the manufacturing of algal powder, tablets, and capsules for animal and human food [61]. Algal lipids are enriched with EPAs and DHAs, which are important for the health of human. Algal biomasses are exploited as aquaculture feed. Algal biomasses are blended with other feed material, utilized as pellets or as paste or powder form. Microalgae of *Tetraselmis*, *Nannochloropsis*, *Phaeodactylum*, *Haematococcus, and Isochrysis* are commonly used as aquaculture feed [62]. Important nutrient contents of aquaculture feed are proteins and PUFAs, which are abundantly available in algae. Microalgal biomass can be used in poultry feed pellets, and approximately 10% of algal biomass has been successfully utilized in poultry industry [63,64]. Comparative studies on poultry feed of algae and traditional food showed enhanced resistance to disease, improved color of yolk, and low cholesterol obtained from algal feed [63,65,66]. Utilization of defatted *Desmodesmus* biomass with polysaccharide degrading enzymes along with protease helps pig growth compared to control [64]. Recently, Raja et al. [67] updated a review on utilization of different microalgal species as feed or food. It has been evident that algae as a protein source had great potential as nutritional supplement resource as an animal feed. Table 12.1 represents the application of potent microalgal species for feed and food application.

Microalgae to Bioactive Compounds

Microalgae are enriched with different types of active compounds such as phycobiliproteins, fatty acids, vitamins, fatty acids, antioxidants, and pigments, which can be a potential application in many pharmaceutical industries. The bioactive compounds found in algae are research target for many diseases and have more feasibilities for mass production at an industrial level shortly. Bioactive compounds from microalgae act as antiviral, antibacterial, antimalarial, antioxidant, antifungal, antitumor, and antiinflammatory components. In comparison with areal plants, the natural bioactive compound from algae is underexplored [68,69].

Chlorophyll, carotenoids, and phycobiliproteins and their derivatives are natural pigments and have been evaluated for their effectiveness against cancer cells as chemopreventive agents [70]. These pigments have

TABLE 12.1
Utilization of Potent Algal Biomass as Food and Feed Applications.

Potent Microalgae	Product	Application Food/Feed
Spirulina platensis *Spirulina pacifica* *Chlorella vulgaris* *Schizochytrium limacinum*	Single-cell protein	Protein supplements
Schizochytrium sp. *Crypthecodinium cohnii* *Nannochloropsis* *Nitzschia* sp. *Phaeodactylum tricornutum*	Fatty acids/ polyunsaturated fatty acids Docosahexaenoic acid Eicosapentaenoic acid	Food supplements
Dunaliella salina *Dunaliella bardawil* *Haematococcus pluvialis* *S. platensis*	Pigments β-carotene Astaxanthin Phycocyanin	Food colorants/ feed additives Food supplements/ feed additives Food colorants

antitumor and antimutagenic effects, which help in suppressing cancer formation. Cytotoxic assay of the pigment for anticancer activity showed that higher activity obtained under nutrients-stressed conditions [71]. Cytotoxic activity of marine microalgae against human tumor cells showed anticancer properties of different extracts of algae [72]. These reports showed that algal pigments could be exploited as potential anticancer compounds. Microalgal PUFAs and pigments can act as potential antiinflammatory components and can be used in a dietary supplement to reduce inflammatory diseases [73]. Reported review of Deng and Chow [74] represented clinical studies based on algal antiinflammatory effect. Many researchers have studied antiinflammatory and antioxidant activities of different marine and freshwater microalgae [73–75]. This evidence proved potential of microalgae as antiinflammatory components.

Microalgae to Plastics

Bioplastics from natural raw material present a biodegradable alternative to conventional petrochemical-based plastic and are environmentally safe and reducing dependency on fossil reserves. Polymers of biomass such as cellulose and starch are used as a starting material for the conversion of polylactic acids (PLAs), thermoplastic starch, and cellulose acetate (CA) [76]. These molecules are produced from food crop and not viable alternatives. When compared with conventional plastic from food staples, microalgae can be exploited as an excellent source of bioplastic production due to a high percentage of carbohydrate polymers and protein [11]. It has been reported that *Spirulina* has 46%–63%

protein content and *Chlorella* has 51%–58% dry weight. It has been reported that bioplastic and thermostable plastic blends are prepared from *Chlorella* and *Spirulina* biomass [11]. Another report showed that leftover biomass after biodiesel production is chemically treated to produce bioplastic mainly polyhydroxy butyrate (PHB), which is produced at 27% after 14 days of cultivation from *Chlorella pyrenoidosa* [77]. Fig. 12.3 shows algae to bioplastic production. For sustainable algal biomass for biorefinery products including bioplastic, wastewater-based cultivation could be encouraged to remediate water for future use and exploitation of resulting biomass for value-added products. Bioplastic from algae is a by-product in algal refinery; it will help for generating revenue along with biofuels. Microalgae-based plastic is in its infancy stage, and it can play a vital role as an environmentally friendly future product.

FUTURE PERSPECTIVES

Algal biomass has many capabilities to produce different metabolites for an extensive range of application. Freshwater-based algal biorefinery is not a viable process due to a projected demands for freshwater resources in the future. Algal biomass cultivation should be based on wastewater for biomass production of low-value products such as bioethanol, biodiesel, and biogas, and treated water can be used for other purposes of the refinery. For high-value products such as PUFAs, pigments, and animal feeds, algal cultivation can be routed to the exploitation of seawater resources, and hence it will address freshwater shortages and fertilizer

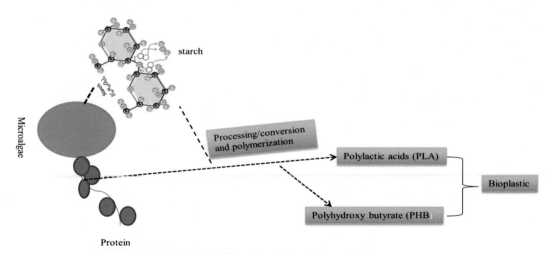

FIG. 12.3 Microalgae to bioplastic production.

cost for algae. More research needs to be focused on the harvesting of biomass, which is a costly process; self-flocculating algae and robust strain to adapt to any fluctuating environment should be chosen for biorefinery application. Algal biorefineries should be coupled with wastewater and seawater utilization along with CO_2 from industries that help to balance the sustainability with complete exploitation of algal biomass assisting the overall sustainability and viability of the technology.

CONCLUSIONS

Microalgae-based biorefineries are exploiting complete valorization of biomass for minimizing the production cost and generating extrarevenue for the sustainable processing of algae-based biomass to biofuels. The further economy of the process can be improved by adapting alternative cost-reducing activities such as utilization of wastewater or seawater for mass production of algae with nutrients recycling and exploitation of waste CO_2 from industries as a carbon source.

Microalgal biorefinery will be a sustainable process to run the industry in a profitable way in future by producing various products from algal biomass.

ACKNOWLEDGMENT

Sabeela Beevi Ummalyma is grateful to the Institute of Bioresources and Sustainable Development (IBSD), a national institute under Department of Biotechnology, Government of India, for providing necessary help and support to this work.

REFERENCES

[1] J.A.V. Costa, B.C.B. De Freitas, C.R. Lisboa, T.D. Santos, L.R.de F. Brusch, M.G. de Morais, Microalgal biorefinery from CO_2 and the effect under blue economy, Renew. Sustain. Energy Rev. 99 (2019) 58−65.

[2] G. Thomassen, M.V. Dael, S.V. Passel, Potential of microalgae biorefineries in Belgium and India; An environmental techno-economic assessment, Bioresour. Technol. 267 (2018) 271−280. Bioresour Technol 267.

[3] K.W. Chew, J.Y. Yap, P.L. Show, N.H. Suan, J.C. Juan, T.C. Ling, D.J. Lee, J.S. Chang, Microalgae biorefinery: high value products perspectives, Bioresour. Technol. 229 (2017) 53−62.

[4] R.H. Wijffels, M.J. Barbosa, An outlook on microalgal biofuels, Science (Washington, D.C.) 329 (5993) (2010) 796−799.

[5] M.A.S. Alkarawi, G.S. Caldwell, J.G.M. Lee, Continuous harvesting of microalgae biomass using foam flotation, Algal Res. 36 (2018) 125−138.

[6] H.C. Greenwell, L.M.L. Laurens, R.J. Shields, R.W. Lovitt, K.J. Flynn, Placing Microalgae on the Biofuels Priority List: A Review of the Technological Challenges, J. Royal Society Interface, 2009 rsif20090322.

[7] R. Gutierrez, I. Ferrer, A. Gonzalez-Molina, H. Salvado, J. Garcia, E. Uggetti, Microalgae recycling improves biomass recovery from waste water treatment high rate algal ponds, Water Res. 106 (2016) 539−549.

[8] E.M. Grima, E.H. Belarbi, F.A. Fernández, A.R. Medina, Y. Chisti, Recovery of microalgal biomass and metabolites: process options and economics, Biotechnol. Adv. 20 (7) (2003) 491−515.

[9] Y. Chisti, Biodiesel from microalgae beats bioethanol, Trends Biotechnol. 26 (3) (2008) 126−131.

[10] P.J. McGinn, K.E. Dickinson, S. Bhatti, J.C. Frigon, S.R. Guiot, S.J.B. O'Leary, Integration of microalgae cultivation with industrial waste remediation for biofuel and bioenergy production: opportunities and limitations, Photosynth. Res. 109 (1−3) (2011) 231−241.

[11] K. Wang, A. Mandal, E. Ayton, R. Hunt, M.A. Zeller, S. Sharma, Modification of protein rich algal biomass to form bioplastics and odor removal, Protein Byproduct (2016) 107−117.

[12] S.H. Ho, S.W. Huang, C.Y. Chen, T. Hasunuma, A. Kondo, J.S. Chang, Bioethanol production using carbohydrate-rich microalgae biomass as feedstock, Bioresour. Technol. 135 (2013) 191−198.

[13] P.D. Patil, H. Reddy, T. Muppaneni, S. Deng, Biodiesel fuel production from algal lipids using supercritical methyl acetate (glycerine free) technology, Fuel 195 (2017) 201−207.

[14] J. R Ziolkowska, L. Simon, Recent development and prospects for algae based fuels in US, Renew. Sustain. Energy Rev. 29 (2014) 847−853.

[15] M. Faried, M. Samer, E. Abdelsalam, R.S. Yousef, Y.A. Attia, A.S. Ali, Biodiesel production from microalgae:Processes, technologies and recent advancement, Renew. Sustain. Energy Rev. 79 (2017) 893−913.

[16] I. Rawat, R.R. Kumar, T. Mutanda, F. Bux, Dual role of microalgae: phycoremediation of domestic wastewater and biomass production for sustainable biofuel production, Appl. Energy 88 (2011) 3411−3424.

[17] J.S. Lemoes, R.C.M. Alves Sobrinho, S. P Farias, R.R. de Moura, E.G. Primel, P.C. Abreu, A.F. Martins, M.G. Montes D'Oca, Sustainable production of biodiesel from microalgae by direct transesterification, Sust. Chem. Pharm. 3 (2016) 33−38.

[18] J.Y. Park, M.S. Park, Y.C. Lee, J.W. Yang, Advances in direct transesterification of algal oil from Wet Biomass, Bioresour. Technol. 184 (2015) 267−275.

[19] N. Pragya, K.K. Pandey, P.K. Sahoo, A review on harvesting, oil extraction and biofuel production technologies from microalgae, Renew. Sustain. Energy Rev. 24 (2013) 159−171.

[20] A. Arumugam, V. Ponnusami, Production of biodiesel by enzymatic transestrification of waste sardine oil and evaluation of its engine performance, Heliyon 3 (12) (2017) e00486.

[21] S.B. Ummalyma, R.K. Sukumaran, Cultivation of the fresh water micro alga *Chlorococcum* sp. RAP13 in sea water for producing oil suitable for biodiesel, J. Appl. Phycol. 27 (1) (2015) 141–147.

[22] S.B. Ummalyma, R.K. Sukumaran, Cultivation of microalgae in dairy effluent for oil production and removal of organic pollution load, Bioresour. Technol. 165 (2014) 295–301.

[23] E. Kwietniewska, J. Tys, Process characteristic, inhibition factors and methane yield of anaerobic digestion process with particular focus on microalgal biomass fermentation, Renew. Sustain. Energy Rev. 34 (2014) 491–500.

[24] E. Jankowska, A.K. Sahu, P. Oleskowicz, Biogas from microalgae: review on microalgae cultivation, harvesting and pretreatment for anaerobic digestion, Renew. Sustain. Energy Rev. 75 (2017) 692–709.

[25] R. Diltz, P. Pullammanappallil, Biofuels from algae liquid, Gaseous Solid Biofuels Convers. Tech. 14 (2013) 432–449.

[26] M. Parsaee, M. Kiani Deh Kiani, K. Karimi, A review of biogas production from sugarcane vinasse, Biomass Bioenergy 122 (2019) 117–122.

[27] C.G. Golueke, W.J. Oswald, H.B. Gotaas, Anaerobic digestion of algae, Adv. Appl. Microbiol. 5 (1) (1957) 47–55.

[28] V. Klassen, O. Blifernez-Klassen, L. Wobbe, A. Schluter, O. Cruse, J.H. Mussgnug, Efficiency and biotechnological aspects of biogas production from microalgal substrates, Biotechnol. 234 (2016) 7–26.

[29] D. Fozer, B. Kiss, L. Lorincz, E. Szekely, P. Mizsey, A. Nemeth, Improvement of microalgae biomass productivity and subsequent biogas yield of hydrothermal gasification via optimization of illumination, Renew. Energy (2019), https://doi.org/10.1016/j.renene.2018.12.122.

[30] L.M. González-González, D.F. Correa, S. Ryan, P.D. Jensen, S. Pratt, P.M. Schenk, Integrated biodiesel and biogas production from microalgae: towards a sustainable close loop through nutrient recycling, Renew. Sust. Eneg. Rev. 82 (2018) 1127–1148.

[31] B. Tartakovsky, F. Matteau-Lebrun, P. J Mcginn, S.J. O'leary, S.R. Guiot, Methane production from microalgae *Scenedesmus sp.*, AMDD in continuous anaerobic reactor, Algal Res. 2 (2013) 394–400.

[32] A. Omer, Y. Fadalla, Biogas energy technology in Sudan, Renew. Energy 28 (3) (2003) 499–507.

[33] M. Cooney, N. Maynard, C. Cannizzaro, C.J. Benemann, Two-phase anaerobic digestion for production of hydrogen–methane mixtures, Bioresour. Technol. 98 (14) (2007) 2641–2651.

[34] M. Shukla, S. Kumar, Algal growth in photosynthetic algal microbial fuel cell and its subsequent utilization for biofuel, Renew. Sustain. Energy Rev. 82 (2018) 402–414.

[35] A. Vergara-Fernández, G. Vargas, N. Alarcón, A. Velasco, Evaluation of marine algae as a source of biogas in a two-stage anaerobic reactor system, Biomass Bioenergy 32 (4) (2008) 338–344.

[36] L.B. Brentner, J. Peccia, J.B. Zimmerman, Challenges in developing biohydrogen as a sustainable energy source: implications for a research agenda, Environ. Sci. Technol. 44 (7) (2010) 2243–2254.

[37] M.K. Lam, K. T Lee, Renewable and sustainable biopenergies production from palm oil mill effluent (POME): win win strategies towards better environmental protection, Biotechnol. Adv. 29 (1) (2011) 124–141.

[38] E. Eroglu, A. Melis, Photobiological hydrogen production: recent advances and state of the art, Bioresour. Technol. 102 (18) (2011) 8403–8413.

[39] H. Gaffron, J.J. Rubin, Fermentative and photochemical production of hydrogen in algae, J. Gene. Physiol. 26 (2) (1942) 219–240.

[40] K.Y. Show, Y. Yan, M. Ling, G. Ye, T. Li, D.J. Lee, Hydrogen production from algal biomass- Advances, challenges and prospects, Bioresour. Technol. 257 (2018) 290–300.

[41] J. Mathews, G. Wang, Metabolic pathway engineering for enhanced biohydrogen production, Int. J. Hydrogen Energy 34 (17) (2009) 7404–7416.

[42] E. Eroglu, A. Melis, Microalgal hydrogen production research, J. Hydro Energ. 41 (30) (2016) 12772–12798.

[43] W. Khelkorn, R.P. Rastogi, A. Incharoensakdi, P. Lindblad, D. Madamwar, A. Pandey, C. Lorroche, Microalgal biohydrogen production, Bioresou. Technol. 243 (2017) 1194–1206.

[44] A.R. Sirajunnisa, D. Surendhiran, Algae —A quintessential and positive resource bioethanol production: a comprehensive review, Renew. Sustain. Energy Rev. 66 (2016) 248–267.

[45] S.R. Chia, H.C. Ong, K.W. Chew, P.L. Show, S.M. Phang, T.C. Ling, D. Nagarajan, D.J. Lee, J.S. Chang, Sustainable approaches for algae utilization in bioenergy production, Renew. Energy 129 (2018) 838–852.

[46] R. Bibi, Z. Ahmad, M. Imran, S. Hussain, A. Ditta, S. Mahmood, A. Khalid, Algal bioethanol production technology: a trend towards sustainable development, Renew. Sustain. Energy Rev. 71 (2017) 976–985.

[47] C.K. Phwan, H.C. Ong, W.H. Chen, T.C. Ling, E.P. Ng, P.L. Show, Overview: comparison of pretreatment technologies and fermentation processes of bioethanol from microalgae, Energy Convers. Manag. 173 (2018) 81–94.

[48] R. Paul, L. Melville, M. Sulu, Anaerobic digestion of micro and macro algae, pretreatment and Co- Digestion biomass- A review of better practice, Int. J. Environ. Sustain. Dev. 7 (2016) 646–650.

[49] L. Brennan, P. Owende, Biofuel from microalgae-a review of technologies for production, processing and extractions of biofuels and co-products, Renew. Sustain. Energy Rev. 14 (2010) 557–577.

[50] G. Roesijadi, A.E. Copping, M.H. Huesemann, J. Forster, J.R. Benemann, Techno Economic Feasibility Analysis of

Offshore Seaweed Farming for Bioenergy and Biobased Products, in: Battelle Pacific Northwest Division Report Number PNWD-3931, 2008, pp. 1–115.

[51] J.J. Milledge, B. Smith, P.W. Dyer, P. Harvey, Microalgae derived biofuel: a review of methods of energy extraction from seaweed biomass, Energies 7 (2014) 7194–7222.

[52] Q. Hu, N. Kurano, M. Kawachi, I. Iwasaki, S. Miyachi, Ultra high cell density culture of a marine green algae *Chlorococcum littorale* in a flate-plate photobioreactor, Appl. Micrbiol. Biotechnol. 49 (1998) 655–662.

[53] S.P. Choi, M.T. Nguyen, S.J. Sim, Enzymatic pretreatment of *Chlamydomonas reinhardtii* biomass for ethanol production, Bioresour. Technol. 101 (2010) 5330–5336.

[54] S.R. Chia, H.C. Ong, K.W. Chew, P.L. Phang, S.M. Ling, J.S. Chang, Sustainable approaches for algae utilization in bioenergy production, Renew. Energy 129 (2018) 838–852.

[55] E.E. de Farias Silva, A. Bertucco, Bioethanol from microalgae and cyanobacteria: a review and technological outlook, Process Biochem. 51 (2016) 1833–1842.

[56] K.H. Kim, I.S. Choi, H.M. Kim, S.G. Wi, H. Bae, Bio-ethanol production from nutrient stress –induced microalgae *Chlorella vulgaris* by enzymatic hydrolysis and immobilized yeast fermentation, Bioresour. Technol. 153 (2014) 47–54.

[57] R. Wang, B. Peng, K. Huang, The research progress of CO_2 sequestration by algal biofertilizer in China, J. CO_2 Util. 11 (2015) 67–70

[58] R. Prasanna, M. Joshi, A. Rana, Y.S. Shivay, L. Nain, Influence of bacteria cyanobacteria on crop yield and C-N Sequestration in soil under rice crop, World J. Microbiol. Biotechnol. 28 (2012) 1223–1235.

[59] Y. Yan, H. Hou, T. Ren, Y. Xu, Q. Wang, W. Xu, Utilization of environmental waste cyanobacteria as pesticide carrier: studies on controlled release and photostability of avermectin, Colloids Surf. B. 102 (2013) 341–347.

[60] N. Renuka, R. Prasanna, A. Sood, A.S. Ahluwalia, R. Bansal, S. Babu, L. Nain, Exploring efficacy of wastewater grown microalgal biomass as a biofertilizer for wheat, Environ Pollu Res 23 (7) (2015) 6608–6620.

[61] T.L. Chacon-LeeChacón-Lee, G.E. Gonzalez-Marino, Microalgae for healthy foods possibilities and challenges : comprehensive review, Food Sci. Food Safety 9 (6) (2010) 655–675.

[62] M.S. Chauton, K.I. Reitana, N.H. Norsker, R. Tveterås, H.T. Kleivdal, A technoeconomic analysis of industrial production of marine microalgae as a source of EPA and DHA- rich raw material for aquafeed: research challenges and possibilities, Aquaculture 436 (2015) 95–103.

[63] M.S. Madeira, C. Cardoso, P.A. Lopes, D. Coelho, C. Afonso, N.M. Bandarra, J.A.M. Prates, Microalgae as feed ingredients for livestock production and meat quality: a review, Livest. Sci. 205 (2017) 111–121.

[64] R. Ekmay, S. Gatrell, K. Lum, J. Kim, X.G. Lei, Nutritional and metabolic impacts of a defatted green marine microalgal (*Desmodesmus* sp.) biomass in diets for weanling pigs and broiler chickens, J. Agric. Food Chem. 62 (40) (2014).

[65] S. Fredriksson, K. Elwinger, J. Pickova, Fatty acid and carotenoid composition of egg yolk as an effect of microalgae addition to feed formula for laying hens, Food Chem. 99 (3) (2006) 530–537.

[66] R. Gonzalez-Esquerra, S. Leeson, Effects of feeding hens regular or deodorized menhaden oil on production parameters, yolk fatty acid profile, and sensory quality of eggs, Poult. Sci. 79 (2000) 1597–1602.

[67] [67a] A. Saeid, K. Chojnacka, M. Korczyński, D. Korniewicz, Z. Dobrzański, Effect on supplementation of *Spirulina maxima* enriched with Cu on production performance, metabolical and physiological parameters in fattening pigs, J. Appl. Phycol. 25 (2013) 1607–1617; [67b] R. Raja, A. Coelho, S. Hemaiswarya, P. Kumar, I.S. Carvalho, A. Alagarsamy, Applications of microalgal paste and powder as food and feed: an update text mining tool, BJBAS 7 (4) (2018) 740–747.

[68] E. Jacob-Lopes, M.M. Maroneze, M. C Depra, R.B. Sartori, R.R. Dias, L.Q. Zepka, Bioactive compounds founds from microalgae: an Innovative framework on industrial biorefineries, Curr. Opin. Food Sci. (2018), https://doi.org/10.1016/j.cofs.2018.12.003.

[69] M.C. Pina-Perez, A. Rivas, A. Martinez, D. Rodrigo, Antimicrobial potential of macro and microalgae against pathogenic spoilage microorganisms in food, Food Chem. 235 (2017) 34–44.

[70] M.E. Abd El-Hack, S. Abdelnour, M. Alagawany, M. Abdo, M.A. Sakr, A.F. Khafaga, S.A. Mahgoub, S.S. Elnesr, M.G. Gebriel, Microalgae in modern cancer therapy: current knowledge, Biomed. Pharmacother. 111 (2019) 42–50.

[71] I.C. Dewi, C. Falaise, C. Hellio, N. Bourgougnon, J.L. Mouget, Anticancer, antiviral, antibacterial, and antifungal properties in microalgae, Microalgae Health Dis. Preven. (2018) 235–261.

[72] E.A.C. Guedes, T.G. da Silva, J.S. Aguiar, L.D. de Barros, L.M. Pinotti, A.E.G. Sant'Ana, Cytotoxic activity of marine algae against cancerous cells, Revista Brasileira de Farmacognosia 23 (4) (2013) 668–673.

[73] R. Robertson, F. Guihéneuf, B. Bahar, M. Schmid, D. Stengel, G. Fitzgerald, R.P. Ross, C. Stanton, The anti-inflamatory effect of algae derived lipid extract on Lipopolysaccharides (LPS) Stimulated Human THP-1 Macrophages, Mar. Drugs 13 (8) (2015) 5402–5424.

[74] R. Deng, T.J. Chow, Hypolipidemic, antioxidant, and anti-inflammatory activities of microalgae *Spirulina*, Cardiovasc. Ther. 28 (2010) 33–45.

[75] C. Lauritano, J.H. Andersen, E. Hansen, M. Albrigtsen, L. Escalera, F. Esposito, A. Ianora, Bioactivity screening of microalgae for antioxidant, anti-inflammatory, anticancer, antidiabetics, and antimalarial activities, Front Marine Sci. 3 (2016).

[76] H. Karan, C. Funk, M. Grabert, M. Oey, B. Hankamer, Green bioplastic as a part of a circular bioeconomy, Trends Plant Sci. (2019) 1–13.

[77] K.A. Das, A. Sathish, J. Stanley, Production of biofuel and bioplastic from *Chlorella pyrenoidosa*, Proce. Material Today 5 (8) (2018) 16774–16781.

Biorefinery of Microalgae for Nonfuel Products

AMBREEN ASLAM* • TAHIR FAZAL*,** • QAMAR UZ ZAMAN • ALI SHAN •
FAHAD REHMAN • JAVED IQBAL • NAIM RASHID • MUHAMMAD SAIF UR REHMAN

INTRODUCTION

Microalgae are unique creature having more than 50,000 species on earth, which can thrive in a diverse environment. Microalgae are classified as unicellular and multicellular photosynthetic microorganisms. Microalgae can be eukaryotic as well as prokaryotic. Based on their habitat, they are categorized as freshwater and marine algae. Marine algae, also known as seaweed, are multicellular photosynthetic microorganisms, which can grow in coastal areas. Microalgae fix CO_2 during photosynthesis and accumulate macromolecules during their growth. They can use organic carbon as a food source. Thus, the organic carbon present in wastewater can be sourced for microalgae growth [1]. Microalgae can grow in seawater and freshwater too. The growth conditions of microalgae impact the biomass composition of microalgae. Microalgae growth depends on several abiotic and biotic factors including temperature, pH, light source, carbon source, medium composition, and the pattern of nutrients supply. Alteration in growth condition impacts the quality and quantity of macromolecules including carbohydrate, protein, and lipid. The final use of microalgae biomass largely depends on its composition. For example, microalgae biomass rich in lipids yield is preferred for biooil production, while protein-rich biomass is used as a food supplement. In past years, the use of microalgae for biooil or biofuels production received great momentum to qualify them as alternative to conventional fuel resources. The research momentum took off recently, realizing that microalgae-based biofuels production is not economically sustainable [2]. To this end, interest is growing to exploit the use of microalgae biomass for other biorefinery application or integrate with other technologies to offset the cost of microalgae bioprocessing. The ambitious goal of sustainable microalgae biorefinery can be accomplished by exploring its use as a coproduct. During microalgae growth, along with macromolecules, some other metabolites are also produced in trace amount, called *coproducts*.

Among many coproducts, the most notable are polysaccharides, chitin, fucoidans, agar, carrageenan, alginate, terpenoids, tocopherols, phlorotannins, beta-carotenoids, phycobiliproteins, and polyunsaturated fatty acids [3]. These coproducts are regarded as valuable feedstock for medicine, pharmaceutical, cosmetic, and biochemical industries [4]. Another preferentially studied aspect of microalgae biomass is its use as a food and feed source. Several studies have reported that microalgae biomass can be used as a feed for animals, livestock, aquaculture, and poultry. The use of microalgae offers interesting benefits to the human health too. Unfortunately, little research focus has been dedicated to such uses of microalgae biomass. It raises the need for discussing the possibility of redirecting the microalgae research focus, documenting its contemporary uses, and providing perspective on its future outlook. This chapter outlines the recent uses of microalgae, their economic sustainability, and future research trends to realize the importance of microalgae biorefinery.

BIOACTIVE COMPOUND FROM MICROALGAE

The earth's surface is covered with 70% of the oceans, which represents over 95% of the biosphere. Thus, the marine world is rich of valuable products [5]. Among many creatures growing in the marine environment, marine algae present high biorefinery interest. Marine algae are categorized as microalgae and macroalgae. Marine

---

*Contributed equally.
**Co-first author.

Microalgae Cultivation for Biofuels Production. https://doi.org/10.1016/B978-0-12-817536-1.00013-8

algae, based on its inherited characteristic, are red, brown, and green algae. Marine algae produce metabolites that are structurally diverse and possess different bioactivities [6]. The organic molecules isolated from marine natural sources are termed as bioactive compounds. So far, almost 18,000 novel compounds have been extracted from oceanic resources, especially the marine algae. Most of them have not been described nor extracted until today [7]. Since two decades, the search for marine-derived novel compounds has been extended across different countries. Microalgae and marine algae, in particular, have been identified as a source of various bioactive compounds, which have received intense research interest in recent year. Some of the bioactive compounds are presented in Fig. 13.1 and discussed in the following section.

Polysaccharides

Polysaccharides are extracted from marine algal resources. They offer cutting-edge advantages to human health. Polysaccharides are used as a source of carrageenans, alginates, and agar. Polysaccharides are classified as insoluble and soluble fibers, based upon their solubility. The insoluble fibers, commonly known as sulfated galactans (carrageenans, agars, floridean, xylans, and starch) are obtained from the red algae (Rhodophyta). The soluble fibers (fucans, laminarans, and alginates) are extracted from the brown marine algae (Phaeophyta) [8]. Xylans, ionic, mannans, and starch polysaccharides having sulfate groups are obtained from the marine green algae [9]. Bioactive compounds such as xylose, arabinose, galactose, rhamnose, and uronic acids are also extracted from the marine algae [10].

Chitin

Chitin is the second-most abundant natural polymer in the universe. The cell wall of plants, bacteria, and fungi contains chitin. Microalgal species also have a cell wall containing chitin [11]. Chitin is a valuable bioproduct and can be used for various purposes, including wastewater treatment and biosequestration of greenhouse gases, and as a feed for microalgae. The chitin-grown green microalgae produce a high amount of pigments including carotenoids [12].

Fucoidans

The cell walls of brown marine algae contain sulfated polysaccharides fucoidans. These bioactive compounds have a number of biological and physiological features. They can be used as antioxidant, antitumor, antiviral, anticoagulant, and antithrombotic activities. Fucoidan also has a major role as a dietary complement and as an antioxidant [13].

Agar

Agar is an end product of polysaccharides. It contains agro pectin and agarose. It has resemblance with the carrageenan on the basis of their structure and function. It is generally sulfated polysaccharides and extracted from the Phaeophyceae family, microalgae (*Gelidium* sp. and *Gracilaria* sp.) [14]. This bioactive compound is largely used in gelling and blending to change the viscosity properties of food items at commercial scale and scientific levels [15]. Lower-quality agar is used in different food items such as frozen food, fruit juice, desert gels, meringues, bakery, ice creams, and candies. Medium-quality agar is commonly used as the gel

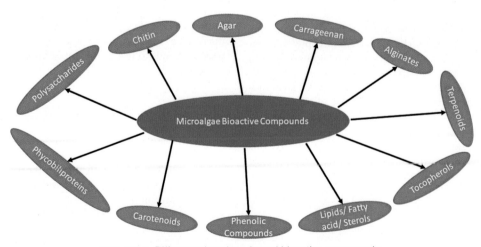

FIG. 13.1 Different microalgae-based bioactive compounds.

substrate in biological cultures media. The fine quality of agar (agarose) is used in intermolecular biology for different separation techniques such as in immunodiffusion, gel chromatography, and electrophoresis. Agar is also used for the absorption of ultraviolet rays. It can decrease the level of glucose in blood and employ effects such as antiaggregation on erythrocytes [16].

Carrageenan

Carrageenan belongs to the family of natural sulfated galactans, which are water soluble and made up of alternative units of α (1-4)-3, 6-anhydro-D-galactose and β (1-3)-D-galactose [17]. It is extensively used as stabilizer and emulsifiers in various food items especially formed from the milk products rather than agar. Carrageenan is prominently used in milk products such as deserts gels, chocolate, jams, evaporated milk, pudding, jellies, ice creams, etc. due to its suspension and thicking property. For the potential pharmaceutical opinion, it has anticoagulant, antitumor, immune modulation, and antiviral activities [18]. It can also be used as an anticoagulant in eating items and for the management of bowel issues, i.e., dysentery constipation and diarrhea. The *Chondrus crispus*−extracted carrageenan gel may hunk the diffusion of HIV (human immunodeficiency virus) and STD (sexually transmitted infection) viruses, such as genital warts, herpes simplex viruses, and gonorrhea [19].

Alginates

Its first production on the industrial scale started in 1929, California. It has a collective name for ancestors of linear polysaccharide. The cell wall of brown algae (Phaeophyceae) contains alginites, having α-L-guluronic acid and 1,4-linked β-D-mannuronic acid. They are chain-forming heteropolysaccharides and arranged in nonregular block-wise order [20]. They are available in saltish form and acidic form. The cell wall of brown algal species has saltish alginate and contains 40% −47% dry weight of algal biomass, while the acidic form of alginate is a linear polyuronic acid, known as alginic acid [5]. Alginates also have the ability to reduce the cholesterol concentration and can avert the assimilation of lethal biochemical substance [21]. They also have the ability to absorb cholesterol and eliminated it from the intestine, which results in hypolipidemic and hypocholesterolemic responses [22].

Terpenoids

The terpenoids (isoprenoids) are extensively dispersed group of secondary metabolites, present largely in aquatic algal resources. Many novel terpenoids including bromo- and chloro-substituents are present in microalgae species. Terpenoids play an active role for the prevention and therapy of different cancers. It is helpful in premature of the skin aging and treatment of skin cancer. It inhibits the carcinomas, which harm erythema development and progression of photodermatoses. Novel marine algal terpenoids are great resources of novel antioxidant agents. They are used in the formulations of cosmetics because of their little systemic toxicity, lower levels of irritation, and penetration-enhancing abilities [23].

Tocopherols

Fat-soluble compounds are the main sources of vitamin E in the body. A group of eight compounds includes four key tocopherols, and four key tocotrienols refer to vitamin E [24]. They are normally obtained in the form of α-tocopherols (90%) and predominant with the α-tocopherol stereoisomer (D-α-tocopherol). They have strong antioxidant action [25]. Different findings showed that tocopherols are obtained from marine algal resources. α-Tocopherols can act as a shielding agent to decrease the material look of aging due to the exposure of sunlight. They are used as an excipient in association with the sun-blocking mediators [26]. The biotechnology yield of microalgae with various concentrations and compositions are considered as tocopherols. Presence of tocopherols has been reported in different oceanic microalgae species such as *Nannochloropsis oculata* and *Tetraselmis suecica* (Kylin) Butcher [13].

Lipids, Fatty Acids, and Sterols

Some well-known ω-3 and ω-6 fatty acids, known as essential fatty acids, are required for human needs. The essential fatty acids are significant for the reliability of tissues (soft and fleshy). Arachidonic acids are attained from linoleic acid. It is a crucial fatty acid due to the incapability of alteration or insufficiency in linoleic acid [27]. Some microalgae species have been used to promote the growth of adolescents, particularly oysters for their connection in sterols. Different marine algae species contain various sterols. Green marine algae contain 28 of uco-cholesterol, cholesterol, 24 methylene cholesterol, and β-sitosterol, while brown algae hold fucosterol, cholesterol, and brassicasterol. Red marine algae contain various types of sterols like desmosterol, cholesterol, sitosterol, fucosterol, and chalinasterol [28].

Phenolic Compounds

Various antioxidant phenolic compounds are found in marine algae. Clusters of phenolic compounds (phlorotannins) provide the basis for polymers of phloroglucinol. These compounds are extracted from the famous constituents of various brown algal groups such as *Sargassaceae*, *Fucaceae*, and *Alariaceae*. Terpenoids, phlorotannins, phenolic pigments, and bromophenols are the major classes of phenols found in marine algal resources. Phlorotannins are the most important group and usually used in the cosmetics industry. These bioactive compounds are extracted from the brown algae like *Ecklonia cava* Kjellman, *Fucus vesiculosus*, and *Ascophyllum nodosum* [29]. The main function of the phlorotannin phenolic compound is to provide shield against photooxidative stress induced by UV β radiation and show the inhibitory effects on melanogenesis [30].

Carotenoids

Algae and plants are photosynthetic organisms. Photosynthesis carotenoids act as basic pigments in light reaping during the light phase of photosynthesis. They are also helpful to protect the photosynthetic apparatus from excess light by hunting reactive oxygen types with individual oxygen and supplementary free radicals. Almost all carotenoids are elaborate in satisfying singlet oxygen and catching peroxyl activists [31]. Human health is affected due to the bioactivities of astaxanthin. They have antiinflammatory action and UV light protection because of their stronger antioxidant activity [32]. However, some green microalgae such as *Haematococcus* sp. can collect xanthophylls carotenoid in the cytoplasm [33]. Manufacturing of xanthophylls in algal cells can be affected by nitrogen limitation, temperature, corrosion, the intensity of light, metal(loids), and salts concentration [34].

Phycobiliproteins

Phycobiliproteins are water-soluble pigments present in the family of red algae (*rhodophytes, cryptomonads, glaucocystophytes*). Cyanobacteria show important fluorescent properties. Some phycobiliproteins consist of phycoerythrin (PE), phycocyanin (PC), and allophycocyanin (APC) constituents [35]. These pigments are heterooligomers and organized into different complexes called "phycobilisomes." They have alliance of subunits inside the producing cells (Cyanophyta) or chloroplasts (Rhodophyta) [36]. They have been used as natural dyes and as nutraceutical in various applications of biotechnology [37]. Phycobilins (phycoerythrin and phycocyanin) are commonly present in stroma of chloroplasts of red algae (rhodophyta), cyanobacteria, some cryptomonads, and glaucophyta. All chlorophyta including green algae do not produce chromophores naturally. Chemically, they organize open-chain tetrapyrroles

basically associated with the pigments of the mammalian bile. Blue pigment such as phycocyanin is primarily present in cyanobacteria, whereas phycoerythrin is characteristically responsible for red coloration in red algae. Along all known pigments of photosynthetic stains, phycobilins are unique as they bonded with phycobiliproteins (water soluble-proteins). Its basic function is to forward the gathered light energy to chlorophylls for photosynthesis. Hence, carotenoids also function as secondary light-harvesting pigments [38]. Phycobilin fluorescence is often used as chemical tags at a particular wavelength. In immunofluorescence techniques, phycobilin fluorescence is used in the binding of phycobiliproteins to antibodies. Such phycobiliproteins have the far highest market values as they are the algal-derived products. Besides, using as chemical tags, phycobilins are also utilized in cosmetics and as food colorants because of their high coloration effects [39]. Marine algae also contains different kind of other bioactive compounds, such as fucan, phycocyanin, allophycocyanin and diacylglycerol etc. as listed in (Table 13.1).

DIVERSE USE OF MICROALGAE BIOMASS

Microalgae received significant attention in the past few years due to their unique ability to provide environmental and ecological services. Microalgae can fix atmospheric CO_2 and can use a wide variety of bioactive compounds as a food source. They accumulate various biochemicals during their growth, which can be used as feedstock for diverse biorefinery applications. The use of microalgae biomass for biofuels production has been extensively studied; however, their use as a food and feed source has been preferentially studied [2]. Microalgae can be used as nutritional supplements to human feeding [51]. *Spirulina* (cyanobacterial) and *Chlorella* (green algae) microalgae products are successfully used at industrial scale [52]. Microalgae biomass can be used as a source of, e.g., β-carotene, ω-3-fatty acids, astaxanthin biopolymers, bioplastics, nutraceuticals, and drugs (e.g., lutein, zeaxanthin). Microalgae are also used as a source of pigments. Green-colored pigment known as chlorophyll is present in cyanobacteria algae and plants as it plays a vital role in photosynthesis. In the electromagnetic spectrum, chlorophyll captures light mostly in the blue region and slightly in the red region. Conversely, the green and near-green portions are absorbed by chlorophyll. This causes a well-known and typical green pattern of chlorophyta (green algae), which harbor chlorophylls as major pigments. *Chlorella* sp. (green microalgae) is a famous

TABLE 13.1
Bioactive Compounds in Marine Algae.

Species Name	Bioactive Compound	References
Fucus vesiculosus *Cystoseira nodicaulis* *Cystoseira tamariscifolia* *Cystoseira usneoides* *Ecklonia cava*	Fucophloroethol, fucodiphloroethol Fucotriphloroethol Phlorofucofuroeckol bieckol/dieckol	[40]
Porphyridium sp. *Costaria costata* *Ulva lactuca* *Cylindrotheca closterium*	Fucoidans, fucans Highly sulfated galactan	[41]
Isochrysis galbana	b-type heteropolysaccharide	[42]
Gracilaria gracilis *Spirulina platensis* *Porphyridium* sp.	R-Phycoerythrin Phycocyanin Allophycocyanin	[43]
Dunaliella salina *Haematococcus* sp.	ß-Carotene	[44]
D. salina *Chlorella sorokiniana*	Lutein	[44]
Synechocystis sp. *Chlorella saccharophila*	Zeaxanthin	[45]
Porphyra sp.	Zeaxanthin, α-β-carotene Lutein, anteraxanthin	[46]
D. salina *U. lactuca*	β-Carotene, zeaxanthin, neoxanthin, anteraxanthin, violaxanthins, siphonein siphonaxanthin	[40]
Haematococcus pluvialis *Chlorella zofingiensis* *Chlorococcum* sp.	Astaxanthin	[47]
Sargassum siliquastrum *Chaetoceros* sp. *Odontella aurita*	Fucoxanthin	[48]
Tetraselmis sp. *Nannochloropsis* sp. *Porphyridium* sp. *Cyanobacteria* *Spirulina platensis*	Eicosapentaeonic, docosahexaeonic, eicosatetraenoic, polyunsaturated ω- 3 fatty acids	[43]
Stephanodiscus sp.	Monogalactosyldiacylglyceride, digalactosyldiacylglyceride, sulfoquinovosyl diacylglycerol	[49]
Porphyra sp. *Catenella repens* *Chlamydomonas hedleyi* *Padina crassa* *Desmarestia aculeatia*	Aminocyclohexenone type Aminocyclohexene imine type	[50]

source for the production of chlorophyll as it has high content of natural chlorophyll [53].

Chlamydomonas has been extensively studied to produce natural pigments. *Haematococcus* is also used to produce a variety of unique pigments, especially astaxanthin [54]. Due to the high antioxidant capacity of astaxanthin, it offers high health benefits. Production of astaxanthin from *Haematococcus* genus is highly valuable [55]. Mostly, algal species are used in astaxanthin production, which belong to *Haematococcus* genus and *Chlorella zofingiensis* (*Chlorella* species) [56]. *Chlamydomonas nivalis* is also able to produce astaxanthin, which takes part to UV light protection and oxidative defense in their natural territories [57]. Furthermore, microalgal pigments are also utilized in industries to form anti-aging creams, regenerating or refreshing maintenance goods for repairing and healing of spoiled skin with diets [58]. For instance, *Haematococcus pluvialis* microalgae have a natural basis to produce keto-carotenoid astaxanthin. Red pigment astaxanthin is the ancestor particle of vitamin A, which plays an important part in cell reproduction, development of the embryo at aquaculture, and poultry firms [59]. Astaxanthin has antioxidant properties as compared with α-carotene, β-carotene, lycopene, canthaxanthin, and lutein. Meanwhile, vitamin E can be used as an appropriate supplement for human feed [32].

Lutein is an essential carotenoid mainly found in human serum and foods. It is present in lens and macula lutea in eye retina. It is not only useful as a colorant in food items and makeups [60] but also responsible for coloring in poultry and fish. Various epidemiological studies have evaluated that lutein is now considered as an effective agent to cure various human diseases. High amount of lutein is found in *Scenedesmus almeriensis* and *Muriellopsis* sp. under optimal temperature conditions [61]. The amount of lutein is also affected due to the variation of pH values in *Auxenochlorella* (*Chlorella*) *prototothecoides*, *Dunaliella salina*, and *Chlamydomonas zofingiensis* [60]. Microalgae have not been used for commercial production of lutein. Although *Muriellopsis* sp. and *Scenedesmus* are cultivated outdoor for pilot-scale production of lutein [61]. *Dunaliella* sp., *Scenedesmus* sp., and *Chlorella* sp., and cyanobacterial strains (*Nostoc* sp. and *Spirulina* sp.) are consumed as sources of compounds and nutrient food supplements [62]. For vitamin A biosynthesis, β-carotene is measured as ancestor molecule in the human body, and its basic role is to provide egg yolk orange color. β-Carotene production from halotolerant *D. salina* is preferred among all other species because of highest carotenoids content (~10% of dry biomass) [63]. *Chlorella* sp. has chlorophyll-gratified biomass as 7% that is five times higher than of *Spirulina* chlorophyll content.

Previous studies on red algae revealed that phycoerythrobilin is the pigment moiety of phycoerythrin as it exhibited antiinflammatory activity, which inhibited mast cell degranulation [64]. However, the present studies evaluated that dulse water extract inhibited the secretion of proinflammatory mediators by phycobilin fraction in lipopolysaccharide-stimulated macrophages. It follows that phycobiliprotein-rich algae produce phycobilin that contributes to antiinflammatory activity [65].

Polyunsaturated long-chain fatty acids (PUFAs) produced by microalgae are of great interest. PUFAs production from microalgae could be used in advanced microalgae technology as a sustainable source because of large biomass production and high oil content [66]. However, PUFAs with nutritional supplements have displayed health benefits [67], whereas traditionally PUFAs production through the fish oils comprised of unmanageable fishing techniques that not only cause the hazard of metal(loid) contamination but also deplete global fisheries [68]. *Nitzschia, Monodus, Crypthecodinium, Isochrysis, Phaeodactylum*, and *Nannochloropsis* are among industrially important species regarding PUFAs production as health supplements [69]. Moreover, *Chlamydomonas debaryana* and *Nannochloropsis gaditana* were used to produce moderately oxidized n-3 PUFAs, e.g., oxylipins, which also show antiinflammatory effects [70]. PUFAs are extensively assisted in the inhibition of numerous cardiac syndromes and recognized as essential nutritional components. Due to the rapid depletion of marine resources, the retrieval of essential complexes from marine fish has been searched [47]. *Arthrospira* is a cyanobacterium (relevant prokaryote) and can be used for livestock feed. In another case, *Chlorella* sp. is widely sold out as an important nutrient source of livestock feed [71]. A group of promising microalgae (*Isochrysis galbana*) is used in food industry due to their value-added source of n-3 long-chain fatty acids (>C18, PUFA [n-3 LCPUFA], mostly docosahexaenoic acid [DHA, 22:6n-3] and eicosapentaenoic acid [EPA, 20:5n-3]) and high lipids content (over 20%). Some microalgae species are used in industries for the β-carotene production from *D. salina* and astaxanthin production through *H. pluvialis* [72]. The details of the biochemical compounds originated from microalgae are given in Table 13.2.

TABLE 13.2
Examples of Essential Fatty Acids and Carotenoid From Microalgal Species.

Substance	Microalgal Species	Dry Weight (%)	References
Carotenoids			
Lutein	*Dunaliella salina*	0.4%–0.8%	[73]
β-Carotene	*D. salina*	10%	[63]
Astaxanthin	Especially *Haematococcus pluvialis* and *Haematococcus* sp.	1%–8%	[74]
Essential fatty acids			
Eicosapentaenoic acid	*Porphyridium cruentum, Monodus subterraneus, Phaeodactylum tricornutum, Amphora* sp.	Total lipids (0.7%–6.1%)	[75]
Docosahexaenoic acid	*Isochrysis, Thalassiosira stellaris, Rhizosolenia setigera, Spirulina platensis, Crypthecodinium cohnii*	Total lipids (17.5%–30.2%)	[75]

Microalgae as Livestock Feed

Microalgae have can be used as a livestock feed. Different algal strains have been studied for their organic compositions (molecular or structural) to evaluate their suitability for livestock feeds [76]. *Spirulina* contains blend comprising the nutrients such as proteins, minerals, B-complex vitamins, super antioxidants (i.e., vitamin E, β-carotene, trace elements), γ-linolenic acid, and various unexplored bioactive compounds. *Spirulina* sp. has advantages over other microorganisms as it can be cultivated in high alkaline conditions and saline water [77]. *Spirulina* as a supplement can increase live weight, body, and growth conformation responses of genetically divergent Australian sheep [78].

The research revealed that dietary supplementation with ω-3 PUFAs is used as fish meal and in early-lactating dairy cows for enhancing milk production with no change in milk structure [20]. Furthermore, brown algae (e.g., *Hizikia fusiforme, Laminaria* sp., and *Undaria* sp.) and red algae (sea weeds) (i.e., *Kappaphycus* sp., *Porphyra* sp., and *Gracilaria* sp.) are utilized in human foods, which display their potential to be used as feed for livestock. *Grateloupia* sp., *Kappaphycus* sp., *Gracilaria* sp., and *Porphyra* sp. can also be used to enhance the yield of particular products in foodstuffs [79].

Microalgae and cyanobacteria can be used as feed in the poultry industry. The *Arthrospira* sp. is very rich in dietary ordinary functional constituent (proteins) and shows an impact on the development of broiler chickens [80]. DHAs are also extracted from *Schizochytrium* sp. microalga. They have ω-3 fatty acid compounds and utilized as a food supplement in poultry feed. This extraction compound might have a positive effect on broiler productive performance in terms of final enhancement in the weight of the body. *Schizochytrium* sp. also has a negative effect due to the increase of average daily feed intake [81]. In another study, *Arthrospira platensis* provides valuable nutrients such as PUFAs (docosapentaenoic acid and EPA.

Microalgae as Aquaculture Feed

Microalgae supplementation to aquaculture might have a direct or indirect impact on the nutritional values of different aquatic organisms such as fish and rotifers. Some gastropods larvae and fish species directly use microalgae for food [82]. Microalgae can be used indirectly as feed on *Zooplankton*. Some rotifers consume particular organic microalgae as essential nutrients and contribute to nutrient recycling for some fish larvae. *Chlorella* sp. has been reviewed as versatile specie and plays a vital role in antimicrobial activities. It is used as an antimicrobial agent such as antifungal, antibacterial, and antiviral activities against aquaculture-related diseases and decreasing the risk of transmission to fish larvae [83].

β-Glucans (BGs) have long been used in aquaculture as bioactive compounds either as vaccine adjuvants, as feed additives, or in prophylactic baths. They exist naturally as the components of the cell wall in microalgae, bacteria, molds, plants, and yeast [84]. When used in fish diets as a supplement, such molecules are indigestible because of the β-glycosidic linkages among glucose molecules. Under heterotrophic situations, *Euglena* accumulate up to 90% of the cell mass as Paramylon (linear β-1,3 glucan), and it is a favorable nutraceutical applicant [85]. Paramylon used to feed multicolored trout (*Oncorhynchus mykiss*) as diet supplement enhanced their adaptive and innate immune reactions after tested with *Yersinia ruckeri* [86]. For organic response modifiers, β-glucans have been used as fodder flavors. The plentiful brewery by-product (Brewers' yeast) is recognized as the source of BG and feed additive that grasps the extensive research attention in aquafeeds [87]. Various studies have been focusing on trying brewery remains from *Saccharomyces cerevisiae*, whereas Algamune (profitable stabilizer) manufactured from *Euglena gracilis*, and the β-glucan purified by *Euglena* (Paramylon) may increase immune responses of red barrel both in vivo and ex-vivo [88].

Microalgae can be used as a fish feed because of their rich nutritional value (vital amino acids, lipids, reserves, vitamins, and crucial fatty acids). The feeding trials on aquatic animals show that different microalgae are used to promote growth, survival rate, and feed utilization [89]. The integrated culture of both shrimp and microalgae had a significant impact on water quality and shrimp growth performance. Three marine feed microalgae (*Chlorella vulgaris*, *Chaetoceros muelleri*, and *Platymonas helgolandica*) are cocultured with *Litopenaeus vannamei* (white shrimp), as they have a great influence on water quality and the productive performance of shrimp [90]. Recently, research has been focused on the use of microalgae as a source of metabolite for antimicrobial activities. *Dunaliella* sp. is identified for the production of β-carotene, which prevents infection from secondary and opportunistic pathogens to improve immature immune system in shrimp. For instance, coculture of shrimps with 3% microalgae in diet indicates the highest survival rates of any infected group [91].

Microalgae are used as a feed on shrimp to alive organisms in aquaculture. The optimal level of microalgae nutrients (lipids, carbohydrates, and protein) is quite effective in reducing water pollution and feed cost. The growth of aquaculture organisms is dependent on the microalgae nutrients level and their species. The feeding of different microalgal species has an impact on the biochemical profile of *Litopenaeus vannamei* and *Penaeus monodon* shrimp culture. In aquaculture hatcheries, local microalgal species are continuously used as feed to shrimp hatcheries. Local isolated microalgae are explored to meet the nutritional demands in shrimp hatcheries. It can ensure hatchery protocol to the rearing of shrimp larvae and postlarvae. The results show that microalgae can be used as a healthy, robust, and cost-effective feedstock for commercial use [92].

Preserved Microalgae as Feed

For the immediate use, on-site microalgae production as live feeds is necessary in most hatcheries because of logistics, high nutritional value, and their cost for the critical larval feeding stages. However, significant developments have been observed in off-the-shelf alternatives during the past few decades. In few cases, their use is partially increased in total replacement of fresh cultured microalgae [93]. Many studies demonstrated that concentrated and preserved microalgae can act as a significant substitute for fresh microalgae in greenwater applications, for shrimp larvae feed, juvenile, larval, and the culture of rotifers [82,94]. Furthermore, survival and growth of some larvae can even improve while feeding concentrated microalgae [94]. Specific harvesting techniques are used to develop concentrates from microalgae culture followed by storage protocol or processing to preserve the preparation. Chemical flocculation and centrifugation are quite appropriate harvesting techniques to prepare concentrates for aquaculture feed, based on harvesting efficiency and cell density [93]. Coupling of storage techniques with centrifugation is effective for the concentrates production of *N. oculata*, *Tetraselmis* spp., and *Chaetoceros calcitrans* with shelf lives beyond 1−2 months [93,94]. However, this whole process may cause damage to larger diatoms (*C. muelleri*) and fragile flagellates (*Isochrysis* sp. (T.ISO), *Pavlova lutheri*) and reduce their shelf life [94]. Flocculation, using nonionic polymer and pH adjustment, is an effective harvesting technique for diatoms (*T. suecica* and *P. lutheri*) having shelf lives of 1−3 weeks [160]. Shelf life of concentrates can extend through various processes such as storage at low temperature, the addition of antioxidants, cryoprotectants, food acids, airbubbling, etc. [94].

Specialized algal mass culture is produced economically for on-selling to hatcheries. In the United States, Reed Mariculture Inc. have been the largest operator and distributor of concentrates over 70 countries (Table 13.3). The concentrates of *Thalassiosira weissflogii*, *Pavlova* sp., *Nannochloropsis* sp., *Tetraselmis* sp., and *Isochrysis* sp. have been stored at 1−4°C to maintain their shelf lives for at least 4 months. However, dried preparation of microalgae can extend their shelf life, especially for the aquaculture strains because the drying

TABLE 13.3
Commercially Provided dried and Concentrated Microalgae [98].

Provider and Website	Product Description	Target Species
Reed Mariculture Inc. (www.reed.mariculture.com): United States	Concentrates of microalgae including *Tetraselmis* sp., *Nannochloropsis oculata*, *Pavlova* sp. *Isochrysis* sp., and *Thalassiosira weissflogii*	Shrimp larviculture, bivalve and postlarvae; live feed enrichments for fish
Innovative Aquaculture Products Ltd (www.innovativeaqua.com): (Canada)	Concentrates of microalgae comprising *Isochrysis* sp. (T.ISO) and *Isochrysis galbana*, *Nannochloropsis oculata*, *Phaeodactylum tricornutum*, *Pavlova lutheri*, and *Chaetoceros* sp.	Bivalve postlarvae; shrimp larviculture, live feed enrichments for fish
Necton SA (www.phytobloom.com): (Portugal)	Concentrates of microalgae includes *Nannochloropsis* sp.	Shrimp larviculture, live feed enrichments for fish; greenwater
Fitoplancton Marino SL (www.easyalgae.com): (Spain	Concentrates of microalgae comprising *Chaetoceros* sp., *Tetraselmis* sp., *Nannochloropsis* sp., *Isochrysis* sp., *Phaeodactylum* sp. *Skeletonema* sp., and *Rhodomonas* sp.; freeze-dried *Tetraselmis* sp. and *Nannochloropsis* sp.	Bivalve postlarvae; shrimp larviculture, live feed enrichments for fish
Pacific Trading Co Ltd (www.pacific-trading.co.jp): (Japan)	Freshwater microalgae includes *Chlorella* sp., with or without enrichment with docosahexaeonic and	Live feed enrichments for fish, shrimp larviculture; greenwater

Continued

TABLE 13.3
Commercially Provided dried and Concentrated Microalgae [98].—cont'd

Provider and Website	Product Description	Target Species
	eicosapentaenoic acids	
Aquafauna Bio-Marine Inc.: (www.aquafauna.com)	Dried preparations of DHA-rich *Schizochytrium* sp. and astaxanthin-rich *Haematococcus* sp	Enriched nutrient of live feeds for shrimp larviculture and fish

process significantly damages the cells. Dried *Tetraselmis chuii* also contains a moderate nutritional value for the clam larvae [95]. The integration of dried *Chlorella* into formulated feed successfully enhances the growth of abalone (*Haliotis diversicolor supertexta* postlarvae) [96]. Heterotrophically grown microalgae organisms (thraustochytrid *Schizochytrium* and dinoflagellate *Crypthecodinium*) are commercially used as spray-dried preparation in Aquafauna Biomarine Inc., USA. These organisms have higher content of DHA (more than 40% of fatty acids) and 2 years shelf life as compared with phototrophically grown microalgae [97].

FUTURE TRENDS

Microalgae have been extensively studied to promote carbon-neutral and resource-efficient bioeconomy. Microalgae have many attributes, but it was mainly studied in the perspective of biofuels production. However, commercial entities of microalgae technology show that it is not economically supported. It emerges the need for finding alternative uses of microalgae and integrating it with other closely related technologies. It constitutes a new research niche that should be focused more onto nonconventional uses of microalgae. The basic notion of this research should be to exploit the use of microalgae biomass as coproducts. The use of microalgae as a food and feed source has an appeal and can offer great market pull. Comprehensive technoeconomic analyses can assess its technology-readiness level that can dictate the choice of using microalgae biomass for alternative bioproducts. Pivoting the research focus from fuel to nonfuel products requires many forward-looking steps to realize its large-scale potential. Bioprospecting can be a powerful abstraction to rationalize microalgae research potential and tailor it according to market demand.

REFERENCES

[1] A. Souliès, J. Pruvost, C. Castelain, T. Burghelea, Microscopic flows of suspensions of the green non-motile Chlorella micro-alga at various volume fractions: applications to intensified photobioreactors, J. Non-newtonian Fluid Mech. 231 (2016) 91–101.

[2] A. Ghosh, S. Khanra, M. Mondal, G. Halder, O.N. Tiwari, S. Saini, K. Gayen, Progress toward isolation of strains and genetically engineered strains of microalgae for production of biofuel and other value added chemicals: a review, Energy Convers. Manag. 113 (2016) 104–118.

[3] M. Hultberg, A.S. Carlsson, S. Gustafsson, Treatment of drainage solution from hydroponic greenhouse production with microalgae, Bioresour. Technol. 136 (2013) 401–406.

[4] I. Priyadarshani, B. Rath, Commercial and industrial applications of micro algae–A review, J. Algal Biomass. 3 (4) (2012) 89–100.

[5] R.S. Rasmussen, M.T. Morrissey, Marine biotechnology for production of food ingredients, Adv. Food Nutr. Res. 52 (2007) 237–292.

[6] S. Cox, N. Abu-Ghannam, S. Gupta, An assessment of the antioxidant and antimicrobial activity of six species of edible Irish seaweeds, Int. Food Res. J. 17 (2012) 205–220.

[7] S. Lordan, R.P. Ross, C. Stanton, Marine bioactive as functional food ingredients: potential to reduce the incidence of chronic diseases, Mar. Drugs 9 (2011) 1056–1060.

[8] S. Kraan, Carbohydrates Comprehensive Studies on Glycobiology and Glycotechnology, InTech, Rijeka, Croatia, 2012, pp. 489–524. Algal Polysaccharides, Novel Applications and Outlook.

[9] T.S. Zaporozhets, N.N. Besednova, T.A. Kuznetsova, T.N. Zvyagintseva, I.D. Makarenkova, S.P. Kryzhanovsky, V.G. Melnikov, The prebiotic potential of polysaccharides and extracts of seaweeds, Russ. J. Mar. Biol. 40 (2014) 1–9.

[10] A. Bocanegra, S. Bastida, J. Benedí, S. Rodenas, F.J. Sánchez-Muniz, Characteristics and nutritional and cardiovascular-health properties of seaweeds, J. Med. Food 12 (2009) 236–258.

[11] B.A. Cheba, Chitin and chitosan: marine biopolymers with unique properties and versatile applications, Glob. J. Biotechnol. Biochem. 6 (2011) 149–153.

[12] P.K. Dutta, J. Dutta, V.S. Tripathi, Chitin and chitosan: chemistry properties and applications, J. Sci. Ind. Res. (India) 63 (2004) 20–31.

[13] A.M. O'Sullivan, Y.C. O'Callaghan, M.N. O'Grady, B. Quegui-neur, D. Hanniffy, D.J. Troy, J.P. Kerry, N.M. O'Brien, In vitro and cellular antioxidant activities of seaweed extracts prepared from five brown seaweeds harvested in spring from the west coast of Ireland, Food Chem. 126 (2011) 1064–1070.

[14] FAO, FAO Statistics, 2008. http://wwwfaoorg/figis/servlet/static?dom=root&xml=aquaculture/indexxml.

[15] J.A. Hemmingson, R.H. Furneaux, V.H. Murray-Brown, Biosynthesis of Agar Polysaccharides in Gracilaria chilensis Bird, McLachlan et Oliveira, Carbohydr. Res. 287 (1996) 101–115.

[16] M. Murata, J. Nakazoe, Production and use of marine algae in Japan, Jpn. Agric. Res. Q. 35 (2001) 281–290.

[17] A.G. Goncalves, D.R. Ducatti, M.E. Duarte, M.D. Noseada, Sulfated and pyruvylated disaccharide alditols obtained from a red seaweed galactan: ESIMS and NMR approaches, Carbohydr. Res. 337 (2002) 2443–2453.

[18] A. Antonopoulos, P. Favetta, W. Helbert, M. Lafosse, On-line liquid chromatography electrospray ionization mass spectrometry for the characterization of κ- and ι-carrageenans. Application to the hybrid ι-/ν-carrageenans, Anal. Chem. 77 (2005) 4l–4136.

[19] M.J. Carlucci, C.A. Pujol, M. Ciancia, M.D. Noseda, M.C. Matulewicz, E.B. Damonte, A.S. Cerezo, Antiherpetic and anticoagulant properties of carrageenans from the red seaweed Gigartina skottsbergii and their cyclized derivatives: correlation between structure and biological activity, Int. J. Biol. Macromol. 20 (1997) 97–105.

[20] L.R. Andrade, L.T. Salgado, M. Farina, M.S. Pereira, P.A. Mourão, G.M. Amado Filho, Ultrastructure of acidic polysaccharides from the cell walls of Brown algae, J. Struct. Biol. 145 (2004) 216–225.

[21] I.H. Kim, J.H. Lee, Antimicrobial activities against methicillin-resistant Staphylococcus aureus from macroalgae, J. Ind. Eng. Chem. 14 (2008) 568–572.

[22] L.N. Panlasigui, O.Q. Baello, J.M. Dimatangal, B.D. Dumelod, Blood cholesterol and lipid-lowering effects of carrageenan on human volunteers, Asia Pac. J. Clin. Nutr. 12 (2003) 209–214.

[23] R. Paduch, M. Kandefer-szerszen, M. Trytek, J. Fiedurek, Terpenes, Substances useful in human healthcare, Arch. Immunol. Ther. Exp. 55 (2007) 315–327.

[24] J.M. Zingg, Vitamin E: an overview of major research directions, Mol. Asp. Med. 28 (2007) 400–422.

[25] J.K. Lodge, Vitamin E bioavailability in humans, J. Plant Physiol. 162 (2005) 790–796.

[26] P. Kullavanijaya, H. Lim, J. Photoprotection, American Acad. Dermatol. 52 (2005) 937–958.

[27] A.C. Guedes, H.M. Amaro, F.X. Malcata, Microalgae as sources of carotenoids, Mar. Drugs 9 (2011) 625–644.

[28] D.I. Sanchez-Machado, J. Lopez-Hernandez, P. Paseiro-Losada, J. Lopez-Cervantes, An HPLC method for the quantification of sterols in edible seaweeds, Biomed. Chromatogr. 18 (2004) 183–190.

[29] S. Holdt, S. Kraan, Bioactive compounds in seaweed: functional food applications and legislation, J. Appl. Phycol. 23 (2011) 543–597.

[30] N.V. Thomas, S.K. Kim, Beneficial effects of marine algal compounds in cosmeceuticals, Mar. Drugs 11 (1) (2013) 146–164.

[31] L.H. Skibsted, Carotenoids in antioxidant networks. Colorants or radical scavengers, J. Agric. Food Chem. 60 (2012) 2409–2417.

[32] M. Guerin, M.E. Huntley, M. Olaizola, *Haematococcus astaxanthin*: applications for human health and nutrition, Trends Biotechnol. 21 (2003) 210−216.

[33] K. Grünewald, H. Hirschberg, C. Hagen, Ketocarotenoid biosynthesis outside of plastids in the unicellular green alga *Haematococcus pluvialis*, J. Biol. Chem. 276 (2001) 6023−6029.

[34] R. Pallela, Y. Na-Young, S.K. Kim, Anti-photo aging and photo protective compounds derived from marine organisms, Mar. Drugs 8 (2010) 1189−1202.

[35] T. Prices, Anandans, The TrpA protein of *Trichodesmiumery threaum* I MS101 is a non-fibril −forming collagen and a component of the outer sheath, An. Microbiol. 160 (2014) 2148−2156.

[36] J. Wiedenmann, Marine proteins, in: P.J. Walsh, S.L. Smith, L.E. Fleming, H.M. Solo-Gabriele, W.H. Gerwick (Eds.), Oceans and Human Health. Risks and Remedies from the Sea, Academic Press, St. Louis, MO, 2008, pp. 469−495.

[37] M.T. Calejo, A.J. Almeida, A.I. Fernandes, Exploring a new jellyfish collagen in the production of microparticles for protein delivery, J. Microencapsul. 29 (2012) 520−531.

[38] A. Parmar, N.K. Singh, A. Kaushal, S. Sonawala, D. Madamwar, Purification, characterization and comparison of phycoerythrins from three different marine cyanobacterial cultures, Bioresour. Technol. 102 (2) (2011) 1795−1002.

[39] S.M. Arad, A. Yaron, Natural pigments from red microalgae for use in foods and cosmetics, Trends Food Sci. Technol. 3 (1992) 92−97.

[40] E. Talero, S. García-Maurino, J. Avila-Roman, A. Rodriguez-Luna, A. Alcaide, V. Motilva, Bioactive compounds isolated from microalgae in chronic inflammation and cancer, Mar. Drugs 13 (2015) 6152−6209.

[41] M.F. Raposo, R.M. de Morais, A.M. Bernardo de Morais, Bioactivity and applications of sulphated polysaccharides from marine microalgae, Mar. Drugs 11 (2013) 233−252.

[42] Y. Sun, H. Wang, G. Guo, Y. Pu, B. Yan, The isolation and antioxidant activity of polysaccharides from the marine microalgae *Isochrysis galbana*, Carbohydr. Polym. 113 (2014) 22−31.

[43] M. Francavilla, M. Franchi, M. Monteleone, C. Caroppo, The red seaweed *Gracilaria gracilis* as a multi products source, Mar. Drugs 11 (2013) 3754−3776.

[44] A. Martins, H. Vieira, H. Gaspar, S. Santos, Marketed marine natural products in the pharmaceutical and cosmeceutical industries: tips for success, Mar. Drugs 12 (2014) 1066−1101.

[45] K.N. Chidambara, Murthy, A. Vanitha, J. Rajesha, M. Mahadeva Swamy, P.R. Sowmya, G.A. Ravishankar, In vivo antioxidant activity of carotenoids from *Dunaliella salina* − a green microalga, Life Sci. 76 (2005) 1381−1390.

[46] N. Schubert, E. García-Mendoza, I. Pacheco-Ruiz, Carotenoid composition of marine red algae, J. Phycol. 42 (2006) 1208−1216.

[47] T. Wang, R. Jonsdottir, H. Liu, L. Gu, H.G. Kristinsson, S. Raghavan, Antioxidant capacities of phlorotannins extracted from the brown algae *Fucus vesiculosus*, J. Agric. Food Chem. 60 (2012) 5874−5883.

[48] J. Liu, F. Chen, Biology and industrial applications of Chlorella: advances and prospects, in: Microalgae Biotechnology, Springer, Cham, 2014, pp. 1−35.

[49] J. Peng, J.P. Yuan, C.F. Wu, J.H. Wang, Fucoxanthin, A marine carotenoid present in brown seaweeds and diatoms: metabolism and bioactivities relevant to human health, Mar. Drugs 9 (2011) 1806−1828.

[50] Z. Su, R. Kang, S. Shi, W. Cong, Z. Cai, An economical device for carbon supplement in large-scale micro-algae production, Bioproc. Biosyst. Eng. 31 (6) (2008) 641−645.

[51] P. Spolaore, C. Joannis-Cassan, E. Duran, A. Isambert, Commercial applications of microalgae, J. Biosci. Bioeng. 101 (2) (2006) 87−96.

[52] Y. Chisti, Large-scale production of algal biomass: raceway ponds, in: Algae Biotechnology, Springer, Cham, 2016, pp. 21−40.

[53] D. Bewicke, B.A. Potter, Chlorella: The Emerald Food, Ronin Publishing, 2009.

[54] D.Y. Kim, D. Vijayan, R. Praveenkumar, J.I. Han, K. Lee, J.Y. Park, Y.K. Oh, Cell-wall disruption and lipid/astaxanthin extraction from microalgae: Chlorella and Haematococcus, Bioresour. Technol. 199 (2016) 300−310.

[55] M.E. Hong, Y.Y. Choi, S.J. Sim, Effect of red cyst cell inoculation and iron (II) supplementation on autotrophic astaxanthin production by Haematococcus pluvialis under outdoor summer conditions, J. Biotechnol. 218 (2016) 25−33.

[56] R.R. Ambati, S.M. Phang, S. Ravi, Aswathanarayana, R.G. Astaxanthin, sources, extraction, stability, biological activities and its commercial applications—a review, Mar. Drugs 12 (1) (2014) 128−152.

[57] D. Remias, U. Lütz-Meindl, C. Lütz, Photosynthesis, pigments and ultra structure of the alpine snow alga *Chlamydomonas nivalis*, Eur. J. Phycol. 40 (3) (2005) 259−268.

[58] L. Zhu, Biorefinery as a promising approach to promote microalgae industry: an innovative framework, Renew. Sustain. Energy Rev. 41 (2015) 1376−1384.

[59] M. Goto, H. Kanda, S. Machmudah, Extraction of carotenoids and lipids from algae by supercritical CO_2 and subcritical dimethyl ether, J. Supercrit. Fluids 96 (2015) 245−251.

[60] E.S. Jin, J.E. Polle, H.K. Lee, S.M. Hyun, M. Chang, Xanthophylls in microalgae: from biosynthesis to biotechnological mass production and application, J. Microbiol. Biotechnol. 13 (2) (2003) 165−174.

[61] J.A. Del Campo, M. García-González, M.G. Guerrero, Outdoor cultivation of microalgae for carotenoid production: current state and perspectives, Appl. Microbiol. Biotechnol. 74 (6) (2007) 1163−1174.

[62] Y. Seo, W. Choi, J. Park, K.H. Jung, H. Lee, Stable isolation of phycocyanin from Spirulina platensis associated with high-pressure extraction process, Int. J. Mol. Sci. 14 (1) (2013) 1778−1787.

[63] A. Prieto, J.P. Canavate, M. García-González, Assessment of carotenoid production by *Dunaliella salina* in different culture systems and operation regimes, J. Biotechnol. 151 (2) (2011) 180−185.

[64] S. Sakai, Y. Komura, Y. Nishimura, T. Sugawara, T. Hirata, Inhibition of mast cell degranulation by phycoerythrin and its pigment moiety phycoerythrobilin, prepared from Porphyra yezoensis, Food Sci. Technol. Res. 17 (2) (2011) 171−177.

[65] D. Lee, M. Nishizawa, Y. Shimizu, H. Saeki, Anti-inflammatory effects of dulse (*Palmaria palmata*) resulting from the simultaneous water-extraction of phycobiliproteins and chlorophyll a, Food Res. Int. 100 (2017) 514−521.

[66] K.W. Chew, J.Y. Yap, P.L. Show, N.H. Suan, J.C. Juan, T.C. Ling, J.S. Chang, Microalgae biorefinery: high value products perspectives, Bioresour. Technol. 229 (2017) 53−62.

[67] J. Wang, X.D. Wang, X.Y. Zhao, X. Liu, T. Dong, F.A. Wu, From microalgae oil to produce novel structured triacylglycerols enriched with unsaturated fatty acids, Bioresour. Technol. 184 (2015) 405−414.

[68] S.M. Kitessa, M. Abeywardena, C. Wijesundera, P.D. Nichols, DHA-containing oilseed: a timely solution for the sustainability issues surrounding fish oil sources of the health-benefitting long-chain omega-3 oils, Nutrients 6 (5) (2014) 2035−2058.

[69] T. Mutanda, D. Ramesh, S. Karthikeyan, S. Kumari, A. Anandraj, F. Bux, Bioprospecting for hyper-lipid producing microalgal strains for sustainable biofuel production, Bioresour. Technol. 102 (1) (2011) 57−70.

[70] C. de los Reyes, J. Ávila-Román, M.J. Ortega, A. de la Jara, S. García-Mauriño, V. Motilva, E. Zubía, Oxylipins from the microalgae Chlamydomonas debaryana and Nannochloropsis gaditana and their activity as TNF-α inhibitors, Photochemistry 102 (2014) 152−161.

[71] W.M. Bishop, H.M. Zubeck, Evaluation of microalgae for use as nutraceuticals and nutritional supplements, J. Nutr. Food Sci. 2 (5) (2012) 1−6.

[72] K. Pirwitz, L. Rihko-Struckmann, K. Sundmacher, Valorization of the aqueous phase obtained from hydrothermally treated Dunaliella salina remnant biomass, Bioresour. Technol. 219 (2016) 64−71.

[73] K. Wichuk, S. Brynjólfsson, W. Fu, Biotechnological production of value-added carotenoids from microalgae: emerging technology and prospects, Bioengineered 5 (3) (2014) 204−208.

[74] L. Brennan, P. Owende, Biofuels from microalgae—a review of technologies for production, processing, and extractions of biofuels and co-products, Renew. Sustain. Energy Rev. 14 (2) (2010) 557−577.

[75] J. Xue, Y.F. Niu, T. Huang, W.D. Yang, J.S. Liu, H.Y. Li, Genetic improvement of the microalga Phaeodactylum tricornutum for boosting neutral lipid accumulation, Metab. Eng. 27 (2015) 1−9.

[76] J. Singh, S. Gu, Commercialization potential of microalgae for biofuels production, Renew. Sustain. Energy Rev. 14 (9) (2010) 2596−2610.

[77] A. Kulshreshtha, U. Jarouliya, P. Bhadauriya, G.B.K.S. Prasad, P.S. Bisen, Spirulina in health care management, Curr. Pharmaceut. Biotechnol. 9 (5) (2008) 400−405.

[78] B.W.B. Holman, A. Kashani, A.E.O. Malau-Aduli, Growth and body conformation responses of genetically divergent Australian sheep to Spirulina (*Arthrospira platensis*) supplementation, Am. J. Exp. Agric. 2 (2012) 160−173.

[79] S. Qin, P. Jiang, C. Tseng, Transforming kelp into a marine bioreactor, Trends Biotechnol. 23 (5) (2005) 264−268.

[80] E. Bonos, E. Kasapidou, A. Kargopoulos, A. Karampampas, E. Christaki, P. Florou-Paneri, I. Nikolakakis, Spirulina as a functional ingredient in broiler chicken diets, S. Afr. J. Anim. Sci. 46 (1) (2016) 94−102.

[81] T. Ribeiro, M.M. Lordelo, P. Costa, S.P. Alves, W.S. Benevides, R.J.B. Bessa, J.A.M. Prates, Effect of reduced dietary protein and supplementation with a docosahexaenoic acid product on broiler performance and meat quality, Br. Poult. Sci. 55 (6) (2014) 752−765.

[82] M. Brown, R. Robert, Preparation and assessment of microalgal concentrates as feeds for larval and juvenile Pacific oyster (*Crassostrea gigas*), Aquaculture 207 (3−4) (2002) 289−309.

[83] C. Falaise, C. François, M.A. Travers, B. Morga, J. Haure, R. Tremblay, V. Leignel, Antimicrobial compounds from eukaryotic microalgae against human pathogens and diseases in aquaculture, Mar. Drugs 14 (9) (2016) 159.

[84] J. Raa, Immune modulation by non-digestible and non-absorbable beta-1,3/1, 6-glucan, Microb. Ecol. Health Dis. 26 (1) (2015) 27824.

[85] J. Krajčovič, M. Vesteg, S.D. Schwartzbach, Euglenoid flagellates: a multifaceted biotechnology platform, J. Biotechnol. 202 (2015) 135−145.

[86] J. Skov, P.W. Kania, L. Holten-Andersen, B. Fouz, K. Buchmann, Immunomodulatory effects of dietary β-1,3-glucan from Euglena gracilis in rainbow trout (*Oncorhynchus mykiss*) immersion vaccinated against Yersinia ruckeri, Fish Shellfish Immunol. 33 (1) (2012) 111−120.

[87] R.A. Dalmo, J. Bøgwald, ß-glucans as conductors of immune symphonies, Fish Shellfish Immunol. 25 (4) (2008) 384−396.

[88] F.Y. Yamamoto, F. Yin, W. Rossi Jr., M. Hume, D.M. Gatlin III, β-1, 3 glucan derived from *Euglena gracilis* and Algamune™ enhances innate immune responses of red drum (*Sciaenops ocellatus* L.), Fish Shellfish Immunol. 77 (2018) 273−279.

[89] D. Medina-Félix, J.A. López-Elías, L.R. Martínez-Córdova, M.A. López-Torres, J. Hernández-López, M.F. Rivas-Vega, F. Mendoza-Cano, Evaluation of the productive and physiological responses of *Litopenaeus vannamei* infected with WSSV and fed diets enriched with *Dunaliella* sp, J. Invertebr. Pathol. 117 (2014) 9−12.

[90] H. Ge, J. Li, Z. Chang, P. Chen, M. Shen, F. Zhao, Effect of microalgae with semicontinuous harvesting on water quality and zootechnical performance of white shrimp

reared in the zero water exchange system, Aquacult. Eng. 72 (2016) 70–76.

[91] D.M. Félix, J.A.L. Elías, Á.I.C. Córdova, L.R.M. Córdova, A.L. González, E.C. Jacinto, M.G.B. Zazueta, Survival of Litopenaeus vannamei shrimp fed on diets supplemented with *Dunaliella* sp. is improved after challenges by Vibrio parahaemolyticus, J. Invertebr. Pathol. 148 (2017) 118–123.

[92] R.K. Karthik, Thamizharasan, D. Sankari, C.K.R. Ram, A. Ashwitha, Biochemical profile of shrimp larvae fed with five different micro algae and enriched *Artemia salina* under laboratory conditions, Int. J. Fisheries Aquatic Stud. 4 (4) (2016) 376–379.

[93] M.R. Tredici, N. Biondi, E. Ponis, L. Rodolfi, G.C. Zittelli, Advances in microalgal culture for aquaculture feed and other uses, in: New Technologies in Aquaculture, Woodhead Publishing, 2009, pp. 610–676.

[94] M. Heasman, J. Diemar, W. O'connor, T. Sushames, L. Foulkes, Development of extended shelf-life microalgae concentrate diets harvested by centrifugation for bivalve molluscs—a summary, Aquacult. Res. 31 (8-9) (2000) 637–659.

[95] I. Laing, A.R. Child, A. Janke, Nutritional value of dried algae diets for larvae of Manila clam (*Tapes philippinarum*), J. Marine Biol. Assoc. United Kingdom 70 (1) (1990) 1–12.

[96] W.R. Chao, C.Y. Huang, S.S. Sheen, Development of formulated diet for post-larval abalone, Haliotis diversicolor supertexta, Aquaculture 307 (1–2) (2010) 89–94.

[97] M. Harel, W. Koven, I. Lein, Y. Bar, P. Behrens, J. Stubblefield, A.R. Place, D.H.A. Advanced, EPA and ArA enrichment materials for marine aquaculture using single cell heterotrophs, Aquaculture 213 (1–4) (2002) 347–362.

[98] M.R. Brown, S.I. Blackburn, Live microalgae as feeds in aquaculture hatcheries, in: Advances in Aquaculture Hatchery Technology, Wood head Publishing, 2013, pp. 117–158.

CHAPTER 14

Recent Trends in Strain Improvement for Production of Biofuels From Microalgae

S.V. VAMSI BHARADWAJ • SHRISTI RAM • IMRAN PANCHA • SANDHYA MISHRA

INTRODUCTION

Most climatologists concur that anthropogenic activity is driving our planet toward disaster. The 2°C rise in temperature above preindustrialization levels is accepted as the safe limit by the world's governments; however, this limit is just a beginning toward preventing global catastrophe [1–4]. In December 2015, the Paris Agreement was adopted under the United Nations Framework Convention on Climate Change (UNFCCC). The countries involved in this agreement accepted to limit the global temperature rise by 1.5°C by 2030–50, by using their legal systems [4]. The most straightforward solution to the global warming problem is to reduce our energy consumption derived from fossil fuels and shift to cleaner, sustainable fuels, ultimately achieving zero net-carbon emission. Biofuels are touted as carbon neutral fuels since the carbon in them is derived from recently captured atmospheric carbon dioxide, using sunlight via photosynthesis. Biofuels are defined as fuels obtained from biomass; these are a renewable source of energy and can contribute toward the sustainable development of humankind. One of the earliest recorded uses of biofuels, after industrialization, was during 1931–60 in China, and during the World War II in Europe when gasoline was scarce, many automobiles were powered by wood gas, which is a mixture of gases, mostly containing carbon monoxide and hydrogen generated by partially burning wood in a gasifier [5]. In the 1970s following oil embargo by Organization of the Petroleum Exporting Countries (OPEC), countries such as the United States and Brazil focused on biofuels among other sources of renewable energy. In recent times, depleting fossil fuel resources and global warming have been the major driving forces for our renewed interest in biofuels. Biofuels are classified into four generations, based on the feedstock used

for their production. First-generation biofuels are derived from oil seeds and carbohydrate-rich plants such as sugarcane, beet, and corn. Biofuel obtained from first-generation sources is not favorable since they compete with food production. This gave way to second-generation biofuels, which utilize cellulosic biomass. Cellulose is one of the most abundant organic polymers on earth; this makes cellulosic biomass a good source of feedstock for biofuel production. However, cellulosic biomass is most commonly found in nature as lignocellulose, which is recalcitrant to enzymatic reactions. Currently, there is much research devoted to finding the optimum pretreatment methods for obtaining sugars from cellulosic biomass since; the industrial effectiveness of most cellulases limits biofuel production from cellulosic biomass. Microalgae are gaining prominence over time as a feedstock for biofuel production, due to their rapid growth rate. The fuels derived from microalgae are termed as third-generation biofuels overcoming the disadvantages of previous generations of biofuels; they do not compete with food crops for land and can be grown in facilities that are based on barren land or at sea. The fourth or the latest generation of biofuels are derived from engineered algae and other microbes, which are tailored to withstand stresses and produce more biofuel compared with naturally occurring, unmodified algae [6,7]. Fourth-generation biofuels are advantageous for industrial-scale production of biofuels and carbon sequestration since they can be metabolically engineered to have a higher growth rate and productivity, along with modifications for easier downstream processing. To realize the full potential of microalgae, some improvements must be made in our understanding of microalgae, using various integrative approaches.

Microalgal biomass is considered as a promising feedstock for biofuel production due to its various

Microalgae Cultivation for Biofuels Production. https://doi.org/10.1016/B978-0-12-817536-1.00014-X

advantages such as high photosynthetic efficiency, high accumulation of energy reserve molecules, and ability to grow in wastewater and saline water; over and above, recently, various molecular and genetic techniques are available to improve the production of energy reserve molecules in various microalgae. Genetic modification of algae is easier and quicker in comparison with terrestrial energy crop plants [8,9]. The primary objective of this chapter is to discuss recent findings in the metabolic engineering approach to enhance biofuel and chemical production using various microalgae. We have also briefly discussed the utilization of various tools and latest techniques for genome engineering in model microalgae. Fig. 14.1 is a graphical representation of strain improvement of microalgae for various biofuels.

SCOPE OF STRAIN IMPROVEMENT

Microalgal strain improvement for production of the desired molecule is essential for the commercial viability of microalgal-based biofuels. Industrial production of biofuels requires strains that ideally produce high titers of the product while being genetically stable over several rounds of reinoculation and storage while posing no threat to human health or environment. Microalgae for industrial production must also be able to proliferate quickly on cheap medium that is available throughout the year with consistency in its composition. It is also advisable that the strain should be robust and tolerate various stresses such as heavy metals, salinity, and temperature. On the downstream processing aspect, the biofuel producing strains must be economical to harvest, and the desired end product should be easy to recover from the culture broth. In addition to the above requirements, an ideal strain must be able to grow in an environment that does not allow the growth of contaminants, thereby reducing the cost of sterilization. Any microorganism has two primary objectives: reproduction and survival; production of biofuels actively diverts the available resources away from biomass generation, which counters the objective of the microorganism to reproduce. Some biofuels such as ethanol are toxic to the producing organism when they are present in high titers. These factors are inherent to all biological systems and prevent efficient production of biofuels from substrates. Advances in metabolic engineering and systems biology can help us in developing strains that are resistant to high titers of biofuels and allow us to control on the gene expression to temporally regulate the biofuel production. Fig. 14.2 represents overview of the various strategies to produce biofuel and valuable chemicals using microalgae as a host organism.

Methods for Strain Improvement

The biochemical engineering approach refers to the strategy of enhancing lipid production of microalgae by changing the nutritional or cultivation conditions (e.g., temperature, pH, salinity, and media ingredients)

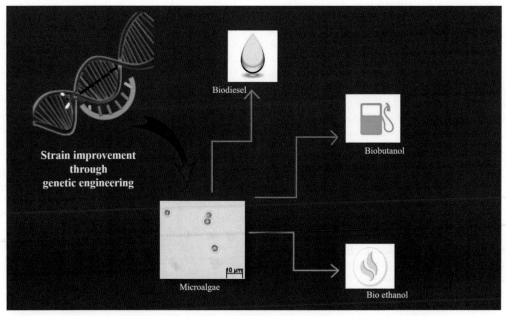

FIG. 14.1 Strain improvement of microalgae for the production of various biofuels.

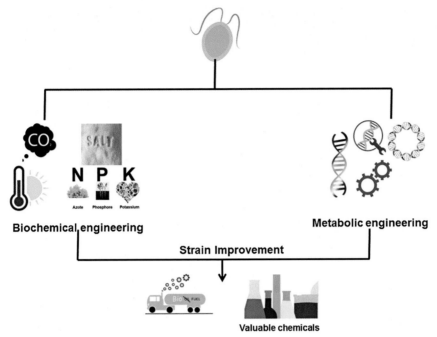

FIG. 14.2 Biochemical and metabolic engineering in microalgae for production of biofuel and valuable chemicals.

to channel metabolic flux generated in photosynthesis to lipid biosynthesis [10]. With the help of biochemical engineering, the ability of microalgae to adapt their metabolism to varying culture conditions provides opportunities to modify, control, and thereby maximize the formation of targeted compounds with nonrecombinant microalgae. Under normal growth conditions, higher biomass and a lower amount of TAGs are produced; however, under stress conditions, many microalgae alter their lipid production pathway and produce a high amount of TAGs (more than 50% of CDW) to withstand stress or adverse environmental conditions [11]. Table 14.1 shows various strain improvement methods in microalgae for the production of biofuels and valuable chemicals.

Classical strain improvement techniques are based on obtaining a large number of mutants, followed by screening for desired phenotypes. Such methods are highly laborious and time-consuming. In classical mutagenesis-based strain improvement, tangible improvements may be obtained at any point during the subsequent mutagenesis-screening cycles. Due to the nature of random mutagenesis, and the lack of understanding of the pathway organization, this process can take up a few weeks to months, depending on the

organism and the phenotype being selected. This process is highly species dependent and is without any guarantee for immediate improvements; however, the chances of success improve with the number of mutants screened. One advantage of developing strains in this manner is that they are currently considered non-genetically modified organisms, thereby improving their acceptance by both regulatory agencies and consumers. Genome shuffling is an improvement over classical methods; it utilizes the power of whole-genome recombination, to select for the favorable phenotype(s). Genome shuffling is similar to classical strain improvement techniques in that it does not require any genomic information and thus can be easily applied to any strain with a quantifiable target phenotype. A library of parental strains is created using classical techniques or chosen from closely related species to improve a strain using genome shuffling. Protoplasts are generated and are allowed to fuse with other protoplasts in the pool. The fused protoplasts are allowed to recover and are subjected to screening and further rounds of strain improvement [28]. Genome shuffling can be applied to improve the quantity of biofuel produced by selecting variants that are resistant to higher titers of metabolites in the fermentation broth; this would help in

TABLE 14.1
Strain Improvement Methods in Microalgae for the Production of Biofuels and Valuable Chemicals.

S. No.	Organisms	Strategy Used	Product Yield Improvement	References
1	*Phaeodactylum tricornutum*	Overexpression of DGAT2 gene	Lipid content increased by 35%	[12]
2	*Synechococcus elongatus* sp. *PCC7942*	Exogenous pathway	2.4 g/L 2,3-butanediol production	[13]
3	*S. elongatus* sp. *PCC7942*	Modification of glycolytic pathway and Calvin Benson cycle	12.6 g/L 2,3-butanediol production	[14]
4	*S. elongatus* sp. *PCC7942*	Synthetic pathway using glycerol dehydratase	3.79 mM 1,3-propanediol	[15]
5	*S. elongatus* sp. *PCC7942*	Use of metabolic flux modeling	100-fold increase in limonene production	[16]
6	*Synechocystis* sp. *PCC 6803*	Reconstruction of PDCzm-slr1192 pathway	2.3 g/L ethanol production	[17]
7	*Synechocystis* sp. *PCC 6803*	Heterologous expression of acyl-Co-A	2.9 g/g of DCW fatty alcohol	[18]
8	*Synechocystis* sp. *PCC6803*	By the addition of BiCA transporter	Produce 2-fold higher biomass	[19]
9	*Phaeodactylum tricornutum*	Over expression of PtME	57.8% of neutral lipid	[20]
10	*Chlamydomonas reinhardtii*	Disruption of ADP-glucose pyrophosphorylase or isoamylase genes	10-fold higher TAG accumulation	[21]
11	*Schizochytrium* sp.	Overexpression of acetyl-CoA synthetase	11.3% higher fatty acid	[22]
12	*Chlorella ellipsoidea*	Expression of GmDof4 TFs	52% increase in total lipids	[23]
13	*C. reinhardtii*	Expression of citrate synthase	169% increase in TAG level	[24]
14	*Cyanidioschyzon merolae*	Overexpression of ER-localized CmGPAT	56.1-fold increase in TAG productivity	[25]
15	*C. merolae*	Overexpression of CmGLG1	4.7-fold increase in starch content	[26]
16	*C. merolae*	Overexpression of CmFAX1	2.4-fold increase in TAG content	[27]

eliminating wild-type strains and strains not having beneficial mutations from the pool. Such strategies do not require sophisticated facilities and are simple to implement. Rational modification of microalgal metabolism is dependent extensively on the existing knowledge base. Knockout of competing pathways, overexpression of cofactors, and helper proteins are considered part of rational metabolic engineering. Such modifications are planned using in silico methods and confirmed by biochemical analyses such as ^{13}C metabolic flux analysis. Rational modification of

cyanobacteria is severely hampered by the unavailability of a repertoire of genetic tools like those available for *Escherichia coli*.

State of Cyanobacterial Genetic Engineering
Transformation

The ability to transform cyanobacteria is an essential requirement for the development of molecular biological techniques and subsequent biotechnological applications [29]. Many researchers have developed various transformation techniques for a different group of cyanobacteria [29−33]. Cyanobacteria can be transformed directly using exogenous DNA fragments (natural transformation) [32] or can be transformed using plasmid delivered by E. coli via conjugation [34]. The ability to undergo natural transformation made Synechocystis and Synechococcus become model organisms for the study of photosynthesis and circadian rhythm, respectively, since genetic engineering such as knockout and genetic complementation played an essential role in understanding several molecular mechanisms. Various factors, such as the cyanobacterial strain, length, and concentration of DNA, affect the overall transformation ability of cyanobacteria. It has been reported that the transformation efficiency is $10^{-3}-10^{-8}$ [35,36].

Apart from the natural transformation of cyanobacteria, as described by Shestakov et al. [30], it can also be transformed using techniques such as electroporation and ultrasonic treatment. Transformation through electroporation has been used for transformation of Anabaena sp., Synechococcus sp., Synechocystis sp., Fremyella diplosiphon, and Plectonema boryanum, while ultrasonic treatment has been used to transform Synechocystis sp. [30,36].

Synechocystis sp. PCC 6803 can be transformed through well-established double homologous recombination [35,36]. Despite thorough investigations on cyanobacterial transformation, very little information is available on the mechanisms by which exogenous DNA binds to the cell and its subsequent processing and transport across the cell membrane [35,37,38]. However, it is hypothesized that type IV pili are involved in the exogenous DNA uptake process [31]. The development of tools for genetic modification of cyanobacteria will help in accelerating basic cyanobacterial research, which in turn will ultimately improve our understanding and its ease of use for industrial biotechnological applications as is currently the case of E. coli. Standardization of genetic tools for cyanobacteria will go a long way in cyanobacterial research by making it easy to develop tools not only for model organisms but to quickly adapt those tools for other microalgae that show promising characteristics for industrial utilization. Traits such as fast growth rate, resistance to high temperatures, or alkaline or saline conditions are desirable since they help in preventing contamination by other microbes. The green alga Chlamydomonas reinhardtii is a model organism that has been used for the study of photosynthesis, motility, cell−cell interactions, etc. It has been a good photosynthetic model organism since various advanced molecular genetic techniques have been developed and sequenced

nuclear mitochondrial and chloroplast genomes are available [39−41]. However, the process of transformation in eukaryotic microalgae is more complex when compared with cyanobacterial transformation.

Selectable markers

Selectable markers are used to select for successful transformants, from untransformed cells, they provide a survival advantage to the cells containing exogenous DNA. It is necessary to select for the survival of transformants since the number of transformed cells is very less in the total population of treated cells [32]. Systematic and efficient methods for identification and isolation of successful transformants are well established. Survival advantage to transformed host cells is usually conferred by the use of antibiotic-resistant genes [42]. When cultured in the presence of the antibiotic, only those cells to which the antibiotic-resistant gene containing construct has been transferred would grow. Since cyanobacteria are gram-negative bacteria, antibiotics that affect gram-negative bacteria can also be used for selection of cyanobacterial transformants [43]. Antibiotics used as selectable markers in cyanobacteria are neomycin and kanamycin, and the corresponding antibiotic resistance is conferred by neomycin phosphotransferase (npt)−encoding genes from transposons Tn5 and Tn903; and streptomycin and spectinomycin antibiotic resistance is conferred by aminoglycoside adenyltransferase (aadA)−encoding genes from Tn7 [44] and the omega interposon [45]. Chloramphenicol resistance conferred by chloramphenicol acetyltransferase (cat) can be used for a section of unicellular cyanobacteria and Anabaena sp. strain 90 [46,47]. Other antibiotics, such as tetracycline and rifampicin, are light sensitive and are of little use for cyanobacterial selection [42,48]. Multiple genes can be manipulated sequentially in a single cyanobacterial strain by using a combination of antibiotic resistance cassettes [49−51].

Plasmid expression vectors

Plasmid vectors used for cyanobacterial transformation can be divided into replicative plasmids and integrative DNA fragments [42]. Integrative DNA fragments are derived from fast-growing bacteria such as E. coli or can be constructed using PCR-based approach; these DNA fragments contain flanking homologous regions along with selection marker and gene of interest, and these DNA fragments cannot replicate in the host cell independently [52]. Replicative plasmids include shuttle vectors and plasmids that are native to the cyanobacteria and having the ability to replicate in both cyanobacteria and E. coli [42,52,53] since they contain the origin of replication from both *E. coli* and the cyanobacterial host [48]. Both integrative and replicative plasmids have been developed for cyanobacteria and have specific experimental uses, for example a plasmid vector for transferring foreign DNA must contain homologous sequences for integration into the genome to be stably maintained are the integrative vectors [31,54]. Integrative vectors are ideal for targeted mutagenesis in which the gene in the chromosome is replaced by the selection marker, which is delivered by the integrative vector. The cyanobacterial host needs to be selected for the recombination event by using a selection marker. It has been observed that common cloning vectors from *E. coli* seem to be unable to replicate in cyanobacteria [32] since they do not have the origins of replication such as colE1, p15A, and PSC101 and are not recognized by the DNA-replicating machinery of cyanobacteria [31]. Replicative plasmids can be divided into two subcategories: replicative plasmids that contain replicons from broad-host-range plasmids [55] and the replicative plasmids that are derived from endogenous plasmids of closely related cyanobacterial strains [53,56]. Various replicative plasmids having a variety of selection markers have been reported for cyanobacteria such as the plasmid pFCLV7 [46] and plasmids derived from RSF1010 (e.g., pSL1211, pPMQAK1, and pFC1) [52,57,58]. Replicative plasmids that can replicate in multiple hosts (either broad or narrow host range) offer a convenient method to study the role of a gene in various hosts. However, till date, most shuttle vectors for cyanobacteria have not been well characterized for their copy number, which limits the control over heterologous gene expression. For example, the copy number of plasmids derived from RSF1010 is 10 per chromosome in *E. coli* cells and approximately 10−30 per cell in *Synechocystis* sp. PCC 6803 (1−3 per chromosome) [52]. It has been reported that the transformation of *Synechocystis* sp. PCC 6803 with plasmids obtained from already transformed *Synechocystis* sp. PCC 6803 cultures results in 10−20 times higher transformation rates rather than transformations done using the same plasmid obtained from bacteria; this observation might be due to a compatible DNA methylation pattern. Nevertheless, plasmids extracted from *E. coli* are generally used as they are much more convenient to prepare due to faster growth of *E. coli* [59].

SYSTEMS BIOLOGY: A NOVEL PARADIGM

Integration of various omics technologies (genomics, metabolomics, transcriptomics, and proteomics) is a goal of systems biology. Recently, several reports suggest altering biochemical parameters for obtaining higher lipid inducing nitrogen starvation, phosphate starvation, and high salinity starvation in microalgae. Although biochemical starvation results in lipid accumulation, it diminishes the growth rate and overall biofuel productivity. Genetic engineering has led to many advances in technologies for biofuel production from microalgae, such as flux balance analysis (FBA), constraint-based flux analysis, and dynamic FBA. Omics (genomics, transcriptomics, proteomics, and metabolomics) knowledge has enabled us to predict the microalgal metabolism better, thus reducing wet-lab work. A common goal of these technologies is to develop a robust microalgal strain, which could help us in commercializing biofuels. System biology aims at understanding the metabolic pathways, thereby generating leads for improvement in the organism, which may result in the development of efficient algae by modification of the pathways, leading to higher lipid content and quality. Overexpression of selected genes has been shown to increase the production of target compounds; these approaches do not take into consideration the efficiency of production. Systems-oriented studies of microalgae would help us understand the bottlenecks in strain engineering and would accelerate the creation of more rationally designed synthetic pathways, which would put lower pressure on the fitness of the cell, thus improving the productivity of the process. Systems biology research is dependent on the data from high-throughput omics technologies; these data sets would become a source for the development of an in silico model, which would help us in predicting cellular behavior under certain conditions or might help us in the identification of genetic circuits that are activated under stress. Systems biology provides us with powerful tools for engineering microalgae.

Role of Genomics in Improving Alcoholic Biofuel Production

Whole-genome sequencing of microalgae has provided us with information that we could use to devise metabolic engineering strategies. The metabolism of a microalgal cell can be reconstructed using various systems biology approaches. One such approach is constraint-based modeling, notably, FBA. One of the main advantages of using FBA is that it does not require kinetic characterization of all the enzymes in a pathway. Hence, this technique of metabolic pathway reconstruction can be applied to large-scale metabolic networks. FBA is used to predict the optimal steady-state fluxes that maximize a given objective function, such as the production of biomass for growth. To apply FBA, one must first reconstruct the metabolic pathway from the genome; this draft metabolic pathway is then refined by identifying gaps and inconsistencies within the draft network. Finally, the set of reactions are converted into a mathematical model using markup languages such as SBML (Systems Biology Markup Language). Metabolic networks are determined by the genome and its interaction with its environment. Based on the networks, i.e., biochemical reactions in the metabolic reconstruction, the matrix, $S(m \times n)$, is generated, which represents all the chemical reactions that take place in the target microalgae. Therefore, in the matrix S, S_{ij} represents the stoichiometric coefficient of the ith metabolite in the jth reaction. This stoichiometric reaction matrix can be solved for the steady-state condition by FBA based on linear programming, since the overall flux of metabolites through is zero at steady state, which is represented by the equation $S.v = 0$ where v is the matrix representing the flux through all the reactions in the metabolic reconstruction. The model is then evaluated using constraint-based analysis and other methods of computational systems biology. This is followed by validation of predictions by using the phenotypic data, either available already or generated after prediction by the model using targeted experiments. The errors in the model are rectified, and the above steps are repeated until a reasonably accurate model of the strain's metabolism has been constructed. The reconstruction of microalgal metabolism can be used to predict cell growth, nutrient uptake, and product output under various defined conditions. The phenotype can also be predicted accurately by single- or multiple-gene deletion simulations if the model is accurate. FBA models allow us to couple regulatory genes with proteins, which allow us to predict knockouts effectively. These predictions act as a guide for genetic engineering and also to predict the robustness of the metabolic network. Another area where metabolic reconstructions help us is in assessing the availability of cofactors such as ATP, NADPH, and NADH in a metabolic network; it is particularly useful since many studies indicate the availability of these cofactors as the limiting factor for product synthesis. Phenotypic phase plane analysis is another effective method for identifying the steady-state solution space of the metabolic network matrix. This could help us understand how the biomass objective function could change by the addition of metabolites to the network and assess the limiting factors for improving the growth of microalgae. Methods such as MOMA, OptKnock, OptGene, GDLS, and OptFlux help us in devising gene deletion strategies to improve the microorganisms. However, these methods have not been applied to microalgal

models because metabolic reconstruction of microalgae is at the early stage of metabolic reconstruction and is limited to two strains (*C. reinhardtii* and *Synechocystis* sp. PCC6803). This limitation can be overcome with time and efforts devoted to the development of high-quality metabolic reconstructions of other microalgae using high-throughput omics data sets [60]. Knoop et al. in 2010 [61] have successfully reconstructed a steady-state metabolic model of *Synechocystis* sp. PCC 6803; this model simulates the steady-state metabolism and is optimized for biomass objective function (biomass production as the highest priority). This model has been validated using various experiments, and it has been shown to accurately predict various mutations required to optimize a pathway and also lethal mutations. Since the whole-genome sequence of the prokaryotic microalga *Synechocystis* sp. PCC6803 was published in 1996, various in silico models of the *Synechocystis* central metabolism or a whole-genome metabolic network have been built and used to predict cell growth under various trophic conditions or to optimize H_2 production by FBA. Recently, optimization of ethanol production for a *Synechocystis* model by MOMA and ROOM was published. Two mutants were identified as candidates to increase ethanol productivity [62]. Compared with cyanobacterial in silico models, metabolic reconstruction of eukaryotic microalgae is lagging due to their large genome size. The in silico model of the central metabolism in *C. reinhardtii* was built in 2009 and was simulated under three trophic conditions by FBA [63]. Two years later, lipid metabolism and biofuel production were included in the whole-genome metabolic reconstruction [64]. This in silico model was validated using cellular growth under trophic conditions and FBA-based gene deletion analysis comparisons with gene knockouts. Additionally, photosynthetic efficiency was evaluated for this constraint model using different light sources. The in silico model of *Chlorella* sp. FC2 IITG was also simulated by FBA and dynamic FBA to maximize biomass and neutral lipid production [65] and to explore cellular physiology and changes in intracellular flux during the transition from nutrient sufficiency to nutrient deficiency. *C. reinhardtii* is a metabolically versatile organism that can perform photosynthetic carbon dioxide fixation, aerobic respiration, and anaerobic fermentation metabolism. Many of the pathways and specific enzymes associated with fermentation metabolisms in this organism are just being defined, and there is little known about the mechanisms by which these pathways are regulated [66]. The full genome has been sequenced [40], and there are well-established molecular tools, including transformation, which has been devised for the species [67]. The reconstructed metabolic network of *C. reinhardtii* consists of 458 metabolites and 484 metabolic reactions. Almost half of the metabolites included in the network are present in the chloroplast, which is a result of a large number of reactions localized to the chloroplast (212 out of 484) [60]. Understanding biological systems on a systems level, which is the goal of systems biology, can give rise to improvements in the rational design of biological systems and strategies for modification of biological systems. Also, understanding mechanisms that systematically control the state of the cell can be modulated to minimize malfunction of biological systems in large-scale settings. While many engineering efforts are being attempted to design optimal production hosts, the inherent complexity of microbial systems presents a formidable challenge to cellular engineering.

Systems biology provides the opportunity to study the cell from a global perspective to gain a snapshot of the systems that are being affected during production and give clues about where pathway bottlenecks lie. It is important to note that existing studies with model systems such as *E. coli* and *Saccharomyces cerevisiae* provide a wealth of information that is highly pertinent to many of the conditions necessary in industrial processes. As a result of integrating multiple omics data, it is expected to discover genes that contribute to microalgal tolerance to biofuel molecules. It is a well-known fact that alcoholic biofuels are toxic to the host organism when present in high titers.

One approach to overcome this problem is to identify genes that are upregulated under stress conditions and engineer microorganisms accordingly to improve their tolerance to high titers of biofuel. In a recent study, Zhu et al. [68] have utilized an integrated omics method, in which transcriptomics was combined with metabolomics to identify gene targets to improve butanol tolerance in *Synechocystis* sp. PCC 6803. Such integrative and model-driven characterization of microalgae is needed to make the production of biofuels from microalgae economically feasible.

Role of Transcriptomics in Alcoholic Biofuel Production

Transcriptomics is the study of global gene expression of a cell. It is one of the important tools for measuring gene expression levels and for the study of differential expression of genes under various conditions. Also, the study of the transcriptome has important implications in the eukaryotic organisms where small RNA molecules play an important role in gene expression

and regulation. The information obtained can be used to construct synthetic regulatory networks using small RNA molecules that allow organisms to be more tolerant of biofuel molecules and produce high titers of biofuel. There is a large amount of data from microarray that reveals the genes that are downregulated or upregulated under certain conditions. To properly utilize the microarray data, systems biology approaches need to be undertaken to streamline the information and interpret the data with the help of a model; then, strategies can be devised to engineer organisms to meet our requirements. With the advent of next-generation sequencing technologies, whole-transcriptome shotgun sequencing, the presence and the quantity of RNA can now be estimated simultaneously.

Role of Proteomics in Alcoholic Biofuel Production

Proteomics deals with the qualitative and quantitative determination of all the proteins in a cell; two-dimensional polyacrylamide gel electrophoresis followed by various technologies such as mass spectrometry, iTRAQ, and ICAT has allowed accurate and precise determination of the proteins in a proteome. Proteomics provides us with insights such as protein stability, which is dependent on the degradation tags; such information cannot be provided by other omics technologies such as transcriptomics and genomics. Proteomics can also help us devise strategies for sustainable production of biofuels rather than subjecting the microalgae to stress for increasing biofuel production.

SYNTHETIC BIOLOGY APPROACHES FOR STRAIN IMPROVEMENT

Synthetic biology is an interdisciplinary approach that combines biotechnology, molecular biology, systems biology, evolutionary biology, and biophysics. Synthetic biology may be easily confused with genetic engineering; however, synthetic biology differs from the latter in terms of scale, scope, and techniques of manipulation and application. Synthetic biology is defined as the design and engineering of biologically based parts, novel devices, and systems as well as the redesign of existing, natural biological systems [69]. Synthetic biology aims to bring few, important reforms in the conventional strain improvement approaches. Primarily, it deals with the application of engineering principles of abstraction and modularization to biological entities and considers "chassis" as a system that enables the functionality of a device, unlike the usual concept of cellular chassis. One such implementation of this ideology is the BioBricks standard [70]. Maintaining a part registry is integral to the synthetic biology regime. A part registry contains a list of standardized biological parts that are available to a community of users, governed by various agreements depending on the registry management. These have the advantage of reliability through characterized biological parts and easy availability of modules that can be used in a strain improvement project rather than constructing a device from scratch and characterizing it. The emphasis on the standardization of biological parts is to allow the ability to assimilate into a larger structure without any complications. Synthetic biology could be used to improve enzymes and metabolic pathways for existing industrial processes and to create cellular chassis that is optimized for production with industry in focus, unlike natural systems that are entities whose primary goal is to "survive and reproduce" [71]. In the postgenomic era, we have information about most of the genes in an organism; this information can be used to analyze knockouts and determine the best suitable phenotype using FBA and other methods described above. The real challenge is to implement the predictions of the tools. The discovery of CRISPR-Cas systems, MAGE (multiplex-automated genomic engineering) using recombineering, genome-editing tools such as TALEN's help us to make large-scale modifications to a genome and are expected to become widely used tools in synthetic biology.

Novel Pathway for Biofuels Generation Using a Synthetic Biology Approach

Advances in protein engineering and systems biology in the synthetic biology context make possible the construction of novel pathways that enable us to engineer microorganisms that produce novel fuel molecules for specific purposes, rather than adapting vehicles for fuel molecules provided by nature. Yim et al. [72] in their seminal study have engineered *E. coli* to produce 1,4-butanediol, whose biosynthetic pathway is not yet found in any microbial isolates. 1,4-Butanediol is an important commodity chemical that is used in the manufacture of polymers and plastics. The 1,4-butanediol pathway was constructed by assembling genes from various microbes; the computational approach to the selection of genes from an existing knowledge base represents the synthetic biology approach for engineering microbes for the production of next-generation biofuels. Xin et al. [73] have modulated the expression of type 2 DGATs to engineer the high lipid-producing microalga *Nannochloropsis oceanica*

that has various degrees of unsaturation, depending on the expression/repression of type 2 DGATs which make the strains fit for a specific application ranging from the food industry to the biofuel industry. Isoprenoids are naturally occurring organic compounds derived from isoprene units. Approximately 80,000 isoprenoid structures have been described [74]. Various enzymes that perform reactions such as benzoylation, glycosylation, acetylation, ring closure, and hydroxylation contribute toward the diversity of isoprenoids. This diversity of isoprenoids makes them a perfect starting point to search for future biofuels that are easy to harvest in an industrial setting, have low toxicity to the host cell, and are energetically dense. Various isoprenoids have been explored for their biofuel potential; isopentenol is considered as a gasoline replacement fuel/fuel additive, and it has been expressed in *E. coli*, but metabolic engineering of microalgae for isopentenol production is still underway [75]. Pinene and α-pinene production in cyanobacteria was reported by Kallas et al. [76] in *Synechococcus* sp. PCC 7002; pinene dimers have been shown to contain high volumetric energy similar to that found in the tactical fuel JP-10, which is a high-energy-dense fuel [77]. Sabinene and terpinene are considered as potential next-generation jet fuel components. Various methods for novel bioproduct discovery/synthesis have been summarized in Fig. 14.3.

Pathway Scaffolding for Improved Tolerance and Efficiency

The main idea behind pathway scaffolding is the enzymes in a pathway that are clustered together in such a way that the reaction intermediates do not have to diffuse long distances for subsequent reactions; often, enzyme clustering reduces the reaction times, reduces the toxicity of intermediates, and prevents the reaction intermediates from being used up by other competing enzymes. Pathway scaffolding has been observed in a wide range of organisms [78]. Natural systems exhibit metabolite channeling, where metabolites pass through a tunnel in the enzyme to the active site of another enzyme. Metabolite channeling has been observed in various natural pathways such as the synthesis of tryptophan, which is mediated by the polyaromatic pathway that consists of 13 total enzymatic reactions [79,80]. These structures illustrate that physical tunnels can protect reactive intermediates from competing reactions and prevent diffusion into the bulk solution. Cellulosomes are another example of natural systems that utilize pathway enzyme crowding to optimize pathways. These multienzyme complexes are produced by anaerobic cellulolytic bacteria for the degradation of

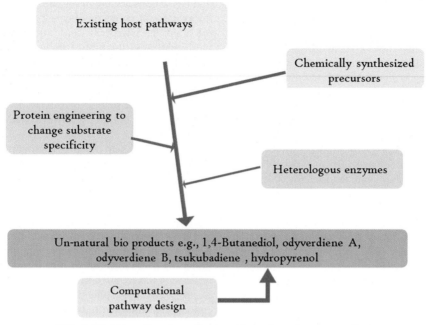

FIG. 14.3 Generation of unnatural products via various approaches.

lignocellulosic biomass. They consist of a complex of scaffolding, which is the structural subunit, and various enzymatic subunits. The interactions in these multienzyme complexes are facilitated by cohesin and dockerin modules [81]. Liu et al. [82] have constructed a multienzyme methanol oxidation pathway based on cohesin-dockerin interactions. The NADH production rate was five times higher than the unassembled enzyme mixture as a result of increased reaction rates facilitated by substrate channeling effects. Blatti et al. [83] demonstrate that protein-protein interactions between the fatty acid acyl carrier protein and thioesterase govern fatty acid hydrolysis within the *Chlamydomonas* chloroplast. Protein-based scaffolds have been largely successful, but compatibility across host systems is a challenge that needs to be addressed. Nucleic acids—based scaffolds have also been explored and are relatively host independent since their structure is largely dependent on temperature and ion concentration. Nucleic acid scaffolds self-assemble in accordance with well-studied interactions; also, new structures can be readily designed computationally with high predictability. This approach allows for the construction of complex nanostructures that can impose detailed spatial organization on interacting or bound proteins [84]. These factors make nucleic acid scaffolds favorable in a synthetic biology approach. Use of nucleic acids as molecular scaffolds has gained a significant foothold with the

advent of nucleic acid—based nanotechnology. Rajendran et al. [85] review the use of nucleic acids as molecular scaffolds. However, the use of pathway scaffolding in algae has been hampered, partly due to limited methods for genetic engineering in algae compared with *E. coli*. Compartmentalization of proteins in organelles provides a unique system of enzyme clustering that has the advantage of selective control over the substrate and product fluxes in and out of the compartment, and creation of a segregated environment that is more favorable for certain reactions [86]. Utilization of prokaryotic microcompartments such as carboxysomes for 1,2-propanediol perfectly represents the case of compartmentalized enzymes that improve the rate of reactions. Such systems consist of porous proteinaceous shells that encapsulate enzymes/enzyme complexes. The pores on these microcompartments control the diffusion of metabolites such as CO_2. Diffusion from the carboxysomal shell is prevented by the walls of the microcompartment, which improves the CO_2 fixation efficiency [87]. In one representative example of pathway localization in mitochondria, Avalos et al. [88] have increased the production of isobutanol in *S. cerevisiae* by the consolidation of the pathway's enzymes into the mitochondria. The production of isobutanol in native *S. cerevisiae* was conducted in two stages: an upstream pathway, located in the mitochondria, which converts pyruvate to α-ketoisovalerate, and a downstream pathway in the cytosol, which converts α-ketoisovalerate to isobutyraldehyde and then to isobutanol. Overexpression of the upstream pathway improved isobutanol production, while overexpression of the downstream pathway with localization at native sites did not improve isobutanol production.

Transcriptional Factor Engineering

To enhance the biofuel production capabilities of microalgae, the molecular mechanism behind the accumulation of such reserved molecules as well as genetic control points for such synthesis should be understood well. In conventional metabolic engineering approach generally, one or more gene(s) are either overexpressed or knock out to generate the desired microalgal strain with high capability for the production of biofuel. In this approach, sometimes overexpression or knockout does not give desired results, or sometimes other characteristics of the microalgae may be hampered. Transcription factors are the proteins that regulate various genes by binding to specific DNA sequence and regulating the activity of RNA polymerase. Modifying the transcription factor will provide a way to enhance the production of desired molecules in microalgae through regulating multiple genes of the whole pathway in the organisms [89]. In the case of plants, it has been reported that changing the transcription factor enhances the production of pigments, secondary metabolites, etc. [90,91]. However, very less information is available in case of microalgal transcription factor engineering.

Due to the advancement of various genetic and omics platforms, different transcription factors have been identified in various microalgae, and different studies also indicate the role of various transcription factors in the cellular processes in microalgae [89]. Among studied transcription factor, basic-region/leucine zipper (bZIP) is an important transcription factor and plays a central role in the various processes such as stress tolerance and cell signaling. [92]. Recently, Li et al. [93] indicated that bZIP-type transcription factor NobZIP1 in microalga *N. oceanica* regulates lipid biosynthesis and secretion. They indicated that NobZIP1 positively regulates key genes such as ACBP, KAS, LC-FACS, and LPAAT involved in lipid biosynthesis while negatively regulates CPS and UGDH involved in cell wall synthesis and carbohydrate metabolism. Another important transcription factor that regulates TAG biosynthesis in model microalga *C. reinhardtii* was PSR1 [94]. Overexpression of PSR1 resulted in upregulation of key genes involved in lipid and starch accumulation in microalgae. The results also indicate that overexpression of PSR1 does not

have any negative effect on the growth of microalgae, which is a promising way to enhance both TAG and biomass production in microalgae and overreduce the biofuel production cost. Identification of transcription factor(s) affected under stress conditions is a promising way to identify key factors that regulate TAG accumulation under various stress conditions such as nitrogen starvation [9]. A study by Ajjawi et al. [95] has identified 20 transcription factors under nitrogen starvation conditions and also constructed insertional knockout strain for 18 transcription factors. Among all the strain transcription factors, ZnCys is a key transcription factor that regulates TAG biosynthesis in *N. gaditana* via carbon portioning from protein. They identified that the ZnCys KO mutant resulted in a 200% increase in carbon partitioning into lipid compared with the wild-type strain.

Genome Editing

Genome editing is an important method for the development of strains with desired properties; unlike genetic engineering, genome editing attempts to insert multiple genes and modify the genome of the host in multiple sites; this was made possible by the development of TALENs, zinc finger nucleases, which can be tailored to cut the genome at any desired site. CRISPR/CAS9 (clustered regularly interspaced short palindromic repeats/CRISPR-associated protein) has been a revolution in genome editing, due to its efficiency, ease of use, and specificity. The CRISPR/CAS9 system is a handy genome editing tool, which can be used to edit various sites in a genome simultaneously. CRISPR/Cas9 is involved in the adaptive immunity of bacteria where it protects the bacteria from invading phages and mobile genetic elements. Three types of CRISPR/Cas systems have been identified, among which type II has been well studied. The CRISPR/Cas9 system of *Streptococcus pyogenes* is the most widely applied system for genome editing. The RNA is transcribed from the CRISPR locus and then processed to generate small RNAs known as crRNA (CRISPR RNA); these crRNAs guide effector nucleases such as Cas9 to bind to the complementary DNA [96]. The Cas protein recognizes the target sequence with the help of guide RNA. The Cas protein then causes double-stranded breaks in the invading DNA, which allows the exonucleases to digest the foreign DNA. Many laboratories all over the world have rapidly adopted the CRISPR/Cas9 system. A mutated variant of Cas9, known as dCas9, is used for silencing/activation of gene expression; this variant has mutations in the active site, which makes the protein incapable of causing double-stranded breaks. Another variant of the Cas9 nuclease, known as Cas9n (Cas9 nickase), causes single-strand breaks; this nuclease helps avoid the occurrence of indels, generally associated with nonhomologous end joining (NHEJ)—based repair of the double-stranded breaks in DNA.

CONCLUSIONS

Currently, there is a dire need to devise methods to integrate data from various sources to fully utilize the information available for metabolic engineering of microalgal hosts. Systems biology aims to devise a quantitative model of cellular metabolism that could be used as a source for the hypothesis for future experiments. These mathematical models of cellular expression take into consideration the dynamic nature of a cell's metabolism; thus, a good model of cellular metabolism must be able to predict the effect of environmental stresses reliably; such a model would allow us to study in silico the effect of gene knockouts or overexpression of few genes or any other genetic manipulation. Model-driven hypothesis generation would help us by reducing a lot of wet lab work, but a mathematical model can be reliable only when large data sets from various sources are integrated into the model. The omics-based approach to metabolic engineering coupled with systems biology is expected to make biofuels from microalgae economically feasible and simpler to process.

ACKNOWLEDGMENTS

This manuscript has been assigned PRIS number CSIR-CSMCRI—168/2016. Vamsi Bharadwaj S.V. would like to acknowledge CSIR, New Delhi, for awarding senior research fellowship. Shristi Ram would like to acknowledge CSIR for financial support. Vamsi Bharadwaj S.V. and Shristi Ram wish to acknowledge AcSIR-CSMCRI for PhD enrollment.

Imran Pancha would like to acknowledge CSIR for financial support.

REFERENCES

[1] M.E. Mann, Defining dangerous anthropogenic interference, Proc. Natl. Acad. Sci. USA. 106 (2009) 4065—4066.

[2] A. Macintosh, Keeping warming within the 2°C limit after Copenhagen, Energy Policy 38 (2010) 2964—2975.

[3] R. Knutti, J. Rogelj, J. Sedláček, E.M. Fischer, A scientific critique of the two-degree climate change target, Nat. Geosci. 9 (2016) 13—18.

[4] J. Rogelj, M. den Elzen, N. Höhne, T. Fransen, H. Fekete, H. Winkler, R. Schaeffer, F. Sha, K. Riahi,

M. Meinshausen, Paris agreement climate proposals need a boost to keep warming well below 2°C, Nature 534 (2016) 631—639.

[5] K.D. Decker, Wood Gas Vehicles: Firewood in the Fuel Tank, Barcelona Low-Tech Mag., 2010.

[6] K. Dutta, A. Daverey, J.-G. Lin, Evolution retrospective for alternative fuels: first to fourth generation, Renew. Energy 69 (2014) 114—122.

[7] W. Klinthong, Y.-H. Yang, C.-H. Huang, C.-S. Tan, A Review: microalgae and their applications in CO_2 capture and renewable energy, Aerosol Air Qual. Res. 15 (2015) 712—742.

[8] P.M. Schenk, S.R. Thomas-Hall, E. Stephens, U.C. Marx, J.H. Mussgnug, C. Posten, O. Kruse, B. Hankamer, Second generation biofuels: high-efficiency microalgae for biodiesel production, BioEnergy Res. 1 (2008) 20—43.

[9] I. Pancha, K. Chokshi, B. George, T. Ghosh, C. Paliwal, R. Maurya, S. Mishra, Nitrogen stress triggered biochemical and morphological changes in the microalgae Scenedesmus sp. CCNM 1077, Bioresour. Technol. 156 (2014) 146—154.

[10] N.M.D. Courchesne, A. Parisien, B. Wang, C.Q. Lan, Enhancement of lipid production using biochemical, genetic and transcription factor engineering approaches, J. Biotechnol. 141 (2009) 31—41.

[11] K.K. Sharma, H. Schuhmann, P.M. Schenk, K.K. Sharma, H. Schuhmann, P.M. Schenk, High lipid induction in microalgae for biodiesel production, Energies 5 (2012) 1532—1553.

[12] Y.-F. Niu, M.-H. Zhang, D.-W. Li, W.-D. Yang, J.-S. Liu, W.-B. Bai, H.-Y. Li, Improvement of neutral lipid and polyunsaturated fatty acid biosynthesis by overexpressing a type 2 diacylglycerol acyltransferase in marine diatom Phaeodactylum tricornutum, Mar. Drugs 11 (2013) 4558—4569.

[13] J.W.K. Oliver, I.M.P. Machado, H. Yoneda, S. Atsumi, Cyanobacterial conversion of carbon dioxide to 2,3-butanediol, Proc. Natl. Acad. Sci. USA 110 (2013) 1249—1254.

[14] M. Kanno, A.L. Carroll, S. Atsumi, Global metabolic rewiring for improved CO_2 fixation and chemical production in cyanobacteria, Nat. Commun. 8 (2017) 14724.

[15] Y. Hirokawa, Y. Maki, T. Tatsuke, T. Hanai, Cyanobacterial production of 1,3-propanediol directly from carbon dioxide using a synthetic metabolic pathway, Metab. Eng. 34 (2016) 97—103.

[16] X. Wang, W. Liu, C. Xin, Y. Zheng, Y. Cheng, S. Sun, R. Li, X.-G. Zhu, S.Y. Dai, P.M. Rentzepis, J.S. Yuan, Enhanced limonene production in cyanobacteria reveals photosynthesis limitations, Proc. Natl. Acad. Sci. USA 113 (2016) 14225—14230.

[17] G. Luan, Y. Qi, M. Wang, Z. Li, Y. Duan, X. Tan, X. Lu, Combinatory strategy for characterizing and understanding the ethanol synthesis pathway in cyanobacteria cell factories, Biotechnol. Biofuels 8 (2015) 184.

[18] L. Yao, F. Qi, X. Tan, X. Lu, Improved production of fatty alcohols in cyanobacteria by metabolic engineering, Biotechnol. Biofuels 7 (2014) 94.

[19] N.A. Kamennaya, S. Ahn, H. Park, R. Bartal, K.A. Sasaki, H.-Y. Holman, C. Jansson, Installing extra bicarbonate transporters in the cyanobacterium *Synechocystis* sp. PCC6803 enhances biomass production, Metab. Eng. 29 (2015) 76–85.

[20] J. Xue, Y.-F. Niu, T. Huang, W.-D. Yang, J.-S. Liu, H.-Y. Li, Genetic improvement of the microalga *Phaeodactylum tricornutum* for boosting neutral lipid accumulation, Metab. Eng. 27 (2015) 1–9.

[21] V.H. Work, R. Radakovits, R.E. Jinkerson, J.E. Meuser, L.G. Elliott, D.J. Vinyard, L.M.L. Laurens, G.C. Dismukes, M.C. Posewitz, Increased lipid accumulation in the *Chlamydomonas reinhardtii* sta7-10 starchless isoamylase mutant and increased carbohydrate synthesis in complemented strains, Eukaryot. Cell 9 (2010) 1251.

[22] J. Yan, R. Cheng, X. Lin, S. You, K. Li, H. Rong, Y. Ma, Overexpression of acetyl-CoA synthetase increased the biomass and fatty acid proportion in microalga *Schizochytrium*, Appl. Microbiol. Biotechnol. 97 (2013) 1933–1939.

[23] J. Zhang, Q. Hao, L. Bai, J. Xu, W. Yin, L. Song, L. Xu, X. Guo, C. Fan, Y. Chen, J. Ruan, S. Hao, Y. Li, R.R.-C. Wang, Z. Hu, Overexpression of the soybean transcription factor GmDof4 significantly enhances the lipid content of *Chlorella ellipsoidea*, Biotechnol. Biofuels 7 (2014) 128.

[24] X. Deng, J. Cai, X. Fei, Effect of the expression and knockdown of citrate synthase gene on carbon flux during triacylglycerol biosynthesis by green algae *Chlamydomonas reinhardtii*, BMC Biochem. 14 (2013) 38.

[25] S. Fukuda, E. Hirasawa, T. Takemura, S. Takahashi, K. Chokshi, I. Pancha, K. Tanaka, S. Imamura, Accelerated triacylglycerol production without growth inhibition by overexpression of a glycerol-3-phosphate acyltransferase in the unicellular red alga *Cyanidioschyzon merolae*, Sci. Rep. 8 (2018) 12410.

[26] I. Pancha, H. Shima, N. Higashitani, K. Igarashi, A. Higashitani, K. Tanaka, S. Imamura, Target of rapamycin-signaling modulates starch accumulation *via* glycogenin phosphorylation status in the unicellular red alga *Cyanidioschyzon merolae*, Plant J. 97 (2019) 485–499.

[27] T. Takemura, S. Imamura, K. Tanaka, Identification of a chloroplast fatty acid exporter protein, CmFAX1, and triacylglycerol accumulation by its overexpression in the unicellular red alga *Cyanidioschyzon merolae*, Algal Res. 38 (2019) 101396.

[28] D. Biot-Pelletier, V.J.J. Martin, Evolutionary engineering by genome shuffling, Appl. Microbiol. Biotechnol. 98 (2014) 3877–3887.

[29] A. Vioque, Transformation of cyanobacteria, In: Transgenic Microalgae as Green Cell Factories, Springer, New York, New York, NY, 2007, pp. 12–22.

[30] S.V. Shestakov, N.T. Khyen, Evidence for genetic transformation in blue-green alga *Anacystis nidulans*, MGG Mol. Gen. Genet. 107 (1970) 372–375.

[31] J.J. Eaton-Rye, The construction of gene knockouts in the cyanobacterium *Synechocystis sp.* PCC 6803., In:

Photosynth. Res. Protoc, Humana Press, New Jersey, 2004, pp. 309–324.

[32] R.D. Porter, Transformation in cyanobacteria, CRC Crit. Rev. Microbiol. 13 (1986) 111–132.

[33] S.S. Golden, L.A. Sherman, Optimal conditions for genetic transformation of the cyanobacterium *Anacystis nidulans* R2, J. Bacteriol. 158 (1984) 36–42.

[34] J. Elhai, C.P. Wolk, [83] Conjugal transfer of DNA to cyanobacteria, Methods Enzymol. 167 (1988) 747–754.

[35] G.I. Kufryk, M. Sachet, G. Schmetterer, W.F. Vermaas, Transformation of the cyanobacterium *Synechocystis* sp. PCC 6803 as a tool for genetic mapping: optimization of efficiency, FEMS Microbiol. Lett. 206 (2002) 215–219.

[36] X. Zang, B. Liu, S. Liu, K.K.I.U. Arunakumara, X. Zhang, Optimum conditions for transformation of *Synechocystis* sp. PCC 6803, J. Microbiol. 45 (2007) 241–245.

[37] W. Vermaas, Molecular genetics of the cyanobacterium *Synechocystis* sp. PCC 6803: principles and possible biotechnology applications, J. Appl. Phycol. 8 (1996) 263–273.

[38] R. Barten, H. Lill, DNA-uptake in the naturally competent cyanobacterium, *Synechocystis* sp. PCC 6803, FEMS Microbiol. Lett. 129 (1995) 83–88.

[39] J.E. Maul, J.W. Lilly, L. Cui, C.W. dePamphilis, W. Miller, E.H. Harris, D.B. Stern, The *Chlamydomonas reinhardtii* plastid chromosome, Plant Cell 14 (2002) 2659–2679.

[40] S.S. Merchant, S.E. Prochnik, O. Vallon, E.H. Harris, S.J. Karpowicz, G.B. Witman, A. Terry, A. Salamov, L.K. Fritz-Laylin, L. Maréchal-Drouard, W.F. Marshall, L.-H. Qu, D.R. Nelson, A.A. Sanderfoot, M.H. Spalding, V. V Kapitonov, Q. Ren, P. Ferris, E. Lindquist, H. Shapiro, S.M. Lucas, J. Grimwood, J. Schmutz, P. Cardol, H. Cerutti, G. Chanfreau, C.-L. Chen, V. Cognat, M.T. Croft, R. Dent, S. Dutcher, E. Fernández, H. Fukuzawa, D. González-Ballester, D. González-Halphen, A. Hallmann, M. Hanikenne, M. Hippler, W. Inwood, K. Jabbari, M. Kalanon, R. Kuras, P.A. Lefebvre, S.D. Lemaire, A. V Lobanov, M. Lohr, A. Manuell, I. Meier, L. Mets, M. Mittag, T. Mittelmeier, J. V Moroney, J. Moseley, C. Napoli, A.M. Nedelcu, K. Niyogi, S. V Novoselov, I.T. Paulsen, G. Pazour, S. Purton, J.-P. Ral, D.M. Riaño-Pachón, W. Riekhof, L. Rymarquis, M. Schroda, D. Stern, J. Umen, R. Willows, N. Wilson, S.L. Zimmer, J. Allmer, J. Balk, K. Bisova, C.-J. Chen, M. Elias, K. Gendler, C. Hauser, M.R. Lamb, H. Ledford, J.C. Long, J. Minagawa, M.D. Page, J. Pan, W. Pootakham, S. Roje, A. Rose, E. Stahlberg, A.M. Terauchi, P. Yang, S. Ball, C. Bowler, C.L. Dieckmann, V.N. Gladyshev, P. Green, R. Jorgensen, S. Mayfield, B. Mueller-Roeber, S. Rajamani, R.T. Sayre, P. Brokstein, I. Dubchak, D. Goodstein, L. Hornick, Y.W. Huang, J. Jhaveri, Y. Luo, D. Martínez, W.C.A. Ngau, B. Otillar, A. Poliakov, A. Porter, L. Szajkowski, G. Werner, K. Zhou, I. V Grigoriev, D.S. Rokhsar, A.R. Grossman, A.R. Grossman, The *Chlamydomonas* genome reveals the

evolution of key animal and plant functions, Science 318 (2007) 245–250.

[41] C.E. Popescu, R.W. Lee, Mitochondrial genome sequence evolution in *Chlamydomonas*, Genetics 175 (2007) 819–826.

[42] O. Koksharova, C. Wolk, Genetic tools for cyanobacteria, Appl. Microbiol. Biotechnol. 58 (2002) 123–137.

[43] W.F.J. Vermaas, Targeted genetic modification of cyanobacteria: new biotechnological applications, In: Handb. Microalgal Cult, Blackwell Publishing Ltd, Oxford, UK, 2004, pp. 455–470.

[44] C.P. Wolk, J. Elhai, T. Kuritz, D. Holland, Amplified expression of a transcriptional pattern formed during development of *Anabaena*, Mol. Microbiol. 7 (1993) 441–445.

[45] P. Prentki, A. Binda, A. Epstein, Plasmid vectors for selecting IS1-promoted deletions in cloned DNA: sequence analysis of the omega interposon, Gene 103 (1991) 17–23.

[46] F. Chauvat, L. De Vries, A. Van der Ende, G.A. Van Arkel, A host-vector system for gene cloning in the cyanobacterium *Synechocystis* PCC 6803, Mol. Gen. Genet. MGG 204 (1986) 185–191.

[47] L. Rouhiainen, L. Paulin, S. Suomalainen, H. Hyytiainen, W. Buikema, R. Haselkorn, K. Sivonen, Genes encoding synthetases of cyclic depsipeptides, anabaenopeptilides, in *Anabaena* strain 90, Mol. Microbiol. 37 (2000) 156–167.

[48] A.M. Ruffing, Engineered cyanobacteria: teaching an old bug new tricks, Bioeng. Bugs. 2 (2011) 136–149.

[49] C. Jansson, R.J. Debus, H.D. Osiewacz, M. Gurevitz, L. McIntosh, Construction of an obligate photoheterotrophic mutant of the cyanobacterium *Synechocystis* 6803 : inactivation of the psbA gene family, Plant Physiol 85 (1987) 1021–1025.

[50] H.-A. Chu, A.P. Nguyen, R.J. Debus, Site-directed photosystem II mutants with perturbed oxygen-evolving properties. 1. Instability or inefficient assembly of the manganese cluster in vivo, Biochemistry 33 (1994) 6137–6149.

[51] T.R. Morgan, J.A. Shand, S.M. Clarke, J.J. Eaton-Rye, Specific Requirements for cytochrome *c* -550 and the manganese-stabilizing protein in photoautotrophic strains of *Synechocystis* sp. PCC 6803 with mutations in the domain Gly-351 to Thr-436 of the chlorophyll-binding protein CP47, Biochemistry 37 (1998) 14437–14449.

[52] B. Wang, J. Wang, W. Zhang, D.R. Meldrum, Application of synthetic biology in cyanobacteria and algae, Front. Microbiol. 3 (2012) 344.

[53] M.D. Deng, J.R. Coleman, Ethanol synthesis by genetic engineering in cyanobacteria, Appl. Environ. Microbiol. 65 (1999) 523–528.

[54] S.S. Golden, J. Brusslan, R. Haselkorn, [12] Genetic engineering of the cyanobacterial chromosome, Methods Enzymol. 153 (1987) 215–231.

[55] H.-H. Huang, D. Camsund, P. Lindblad, T. Heidorn, Design and characterization of molecular tools for a Synthetic Biology approach towards developing cyanobacterial biotechnology, Nucleic Acids Res. 38 (2010) 2577–2593.

[56] C.P. Wolk, A. Vonshak, P. Kehoe, J. Elhai, Construction of shuttle vectors capable of conjugative transfer from *Escherichia coli* to nitrogen-fixing filamentous cyanobacteria, Proc. Natl. Acad. Sci. USA 81 (1984) 1561–1565.

[57] W.-O. Ng, R. Zentella, Y. Wang, J.-S.A. Taylor, H.B. Pakrasi, phrA, the major photoreactivating factor in the cyanobacterium *Synechocystis* sp. strain PCC 6803 codes for a cyclobutane-pyrimidine-dimer-specific DNA photolyase, Arch. Microbiol. 173 (2000) 412–417.

[58] B.M. Berla, R. Saha, C.M. Immethun, C.D. Maranas, T.S. Moon, H.B. Pakrasi, Synthetic biology of cyanobacteria: unique challenges and opportunities, Front. Microbiol. 4 (2013) 246.

[59] J. Labarre, F. Chauvat, P. Thuriaux, Insertional mutagenesis by random cloning of antibiotic resistance genes into the genome of the cyanobacterium *Synechocystis* strain PCC 6803, J. Bacteriol. 171 (1989) 3449–3457.

[60] I. Thiele, B.Ø. Palsson, A protocol for generating a high-quality genome-scale metabolic reconstruction, Nat. Protoc. 5 (2010) 93–121.

[61] H. Knoop, Y. Zilliges, W. Lockau, R. Steuer, The metabolic network of *Synechocystis* sp. PCC 6803: systemic properties of autotrophic growth, Plant Physiol 154 (2010) 410–422.

[62] T. Sengupta, M. Bhushan, P.P. Wangikar, Metabolic modeling for multi-objective optimization of ethanol production in a *Synechocystis* mutant, Photosynth. Res. 118 (2013) 155–165.

[63] N.R. Boyle, J.A. Morgan, Flux balance analysis of primary metabolism in *Chlamydomonas reinhardtii*, BMC Syst. Biol. 3 (2009) 4.

[64] R.L. Chang, L. Ghamsari, A. Manichaikul, E.F.Y. Hom, S. Balaji, W. Fu, Y. Shen, T. Hao, B.Ø. Palsson, K. Salehi-Ashtiani, J.A. Papin, Metabolic network reconstruction of *Chlamydomonas* offers insight into light-driven algal metabolism, Mol. Syst. Biol. 7 (2011) 518.

[65] M. Muthuraj, B. Palabhanvi, S. Misra, V. Kumar, K. Sivalingavasu, D. Das, Flux balance analysis of *Chlorella* sp. FC2 IITG under photoautotrophic and heterotrophic growth conditions, Photosynth. Res. 118 (2013) 167–179.

[66] A.R. Grossman, E.E. Harris, C. Hauser, P.A. Lefebvre, D. Martinez, D. Rokhsar, J. Shrager, C.D. Silflow, D. Stern, O. Vallon, Z. Zhang, *Chlamydomonas reinhardtii* at the crossroads of genomics, Eukaryot. Cell 2 (2003) 1137–1150.

[67] T.L. Walker, C. Collet, S. Purton, Algal transgenics in the genomic era, J. Phycol. 41 (2005) 1077–1093.

[68] H. Zhu, X. Ren, J. Wang, Z. Song, M. Shi, J. Qiao, X. Tian, J. Liu, L. Chen, W. Zhang, Integrated OMICS guided engineering of biofuel butanol-tolerance in photosynthetic *Synechocystis* sp. PCC 6803, Biotechnol. Biofuels 6 (2013) 106.

[69] L.J. Clarke, R.I. Kitney, Synthetic biology in the UK – an outline of plans and progress, Synth. Syst. Biotechnol. 1 (2016) 243–257.

[70] T. Knight, Idempotent Vector Design for Standard Assembly of Biobricks, 2003.

[71] G.-Q. Chen, New challenges and opportunities for industrial biotechnology, Microb. Cell Fact. 11 (2012) 111.

[72] H. Yim, R. Haselbeck, W. Niu, C. Pujol-Baxley, A. Burgard, J. Boldt, J. Khandurina, J.D. Trawick, R.E. Osterhout, R. Stephen, J. Estadilla, S. Teisan, H.B. Schreyer, S. Andrae, T.H. Yang, S.Y. Lee, M.J. Burk, S. Van Dien, Metabolic engineering of *Escherichia coli* for direct production of 1,4-butanediol, Nat. Chem. Biol. 7 (2011) 445–452.

[73] Y. Xin, Y. Lu, Y.-Y. Lee, L. Wei, J. Jia, Q. Wang, D. Wang, F. Bai, H. Hu, Q. Hu, J. Liu, Y. Li, J. Xu, Producing designer oils in industrial microalgae by rational modulation of Co-evolving Type-2 Diacylglycerol Acyltransferases, Mol. Plant 10 (2017) 1523–1539.

[74] T.A. Pemberton, M. Chen, G.G. Harris, W.K.W. Chou, L. Duan, M. Köksal, A.S. Genshaft, D.E. Cane, D.W. Christianson, Exploring the influence of domain architecture on the catalytic function of diterpene synthases, Biochemistry 56 (2017) 2010–2023.

[75] K.W. George, M.G. Thompson, A. Kang, E. Baidoo, G. Wang, L.J.G. Chan, P.D. Adams, C.J. Petzold, J.D. Keasling, T. Soon Lee, Metabolic engineering for the high-yield production of isoprenoid-based C5 alcohols in *E. coli*, Sci. Rep. 5 (2015) 11128.

[76] T. Kallas, M. Nelson, E. Singsaas, Methods for isoprene and pinene production in cyanobacteria, WiSys Technology Foundation Inc, 2016, pp. 382–554. U.S. Patent 9.

[77] B.G. Harvey, M.E. Wright, R.L. Quintana, High-density renewable fuels based on the selective dimerization of pinenes, Energy Fuels 24 (2010) 267–273.

[78] R.J. Conrado, J.D. Varner, M.P. DeLisa, Engineering the spatial organization of metabolic enzymes: mimicking nature's synergy, Curr. Opin. Biotechnol. 19 (2008) 492–499.

[79] C.C. Hyde, S.A. Ahmed, E.A. Padlan, E.W. Miles, D.R. Davies, Three-dimensional structure of the tryptophan synthase alpha 2 beta 2 multienzyme complex from *Salmonella typhimurium*, J. Biol. Chem. 263 (1988) 17857–17871.

[80] E.W. Miles, S. Rhee, D.R. Davies, The molecular basis of substrate channeling, J. Biol. Chem. 274 (1999) 12193–12196.

[81] E.A. Bayer, J.-P. Belaich, Y. Shoham, R. Lamed, The cellulosomes: multienzyme machines for degradation of plant cell wall polysaccharides, Annu. Rev. Microbiol. 58 (2004) 521–554.

[82] F. Liu, S. Banta, W. Chen, Functional assembly of a multienzyme methanol oxidation cascade on a surface-displayed trifunctional scaffold for enhanced NADH production, Chem. Commun. 49 (2013) 3766.

[83] J.L. Blatti, J. Beld, C.A. Behnke, M. Mendez, S.P. Mayfield, M.D. Burkart, Manipulating fatty acid biosynthesis in microalgae for biofuel through protein-protein interactions, PLoS One 7 (2012) e42949.

[84] V. Linko, H. Dietz, The enabled state of DNA nanotechnology, Curr. Opin. Biotechnol. 24 (2013) 555–561.

[85] A. Rajendran, E. Nakata, S. Nakano, T. Morii, Nucleic-acid-templated enzyme cascades, Chembiochem 18 (2017) 696–716.

[86] A.H. Chen, P.A. Silver, Designing biological compartmentalization, Trends Cell Biol. 22 (2012) 662–670.

[87] S. Cheng, Y. Liu, C.S. Crowley, T.O. Yeates, T.A. Bobik, Bacterial microcompartments: their properties and paradoxes, Bioessays 30 (2008) 1084–1095.

[88] J.L. Avalos, G.R. Fink, G. Stephanopoulos, Compartmentalization of metabolic pathways in yeast mitochondria improves the production of branched-chain alcohols, Nat. Biotechnol. 31 (2013) 335–341.

[89] A.K. Bajhaiya, J. Ziehe Moreira, J.K. Pittman, Transcriptional engineering of microalgae: prospects for high-value chemicals, Trends Biotechnol. 35 (2017) 95–99.

[90] E. Butelli, L. Titta, M. Giorgio, H.-P. Mock, A. Matros, S. Peterek, E.G.W.M. Schijlen, R.D. Hall, A.G. Bovy, J. Luo, C. Martin, Enrichment of tomato fruit with health-promoting anthocyanins by expression of select transcription factors, Nat. Biotechnol. 26 (2008) 1301–1308.

[91] C. Schluttenhofer, L. Yuan, Regulation of specialized metabolism by WRKY transcription factors, Plant Physiol. 167 (2015) 295–306.

[92] W. Dröge-Laser, C. Weiste, The C/S1 bZIP Network: a regulatory hub orchestrating plant energy homeostasis, Trends Plant Sci. 23 (2018) 422–433.

[93] D.-W. Li, S. Balamurugan, Y.-F. Yang, J.-W. Zheng, D. Huang, L.-G. Zou, W.-D. Yang, J.-S. Liu, Y. Guan, H.-Y. Li, Transcriptional regulation of microalgae for concurrent lipid overproduction and secretion, Sci. Adv. 5 (2019) eaau3795.

[94] A.K. Bajhaiya, A.P. Dean, L.A.H. Zeef, R.E. Webster, J.K. Pittman, PSR1 is a global transcriptional regulator of phosphorus deficiency responses and carbon storage metabolism in *Chlamydomonas reinhardtii*, Plant Physiol. 170 (2016) 1216–1234.

[95] I. Ajjawi, J. Verruto, M. Aqui, L.B. Soriaga, J. Coppersmith, K. Kwok, L. Peach, E. Orchard, R. Kalb, W. Xu, T.J. Carlson, K. Francis, K. Konigsfeld, J. Bartalis, A. Schultz, W. Lambert, A.S. Schwartz, R. Brown, E.R. Moellering, Lipid production in *Nannochloropsis gaditana* is doubled by decreasing expression of a single transcriptional regulator, Nat. Biotechnol. 35 (2017) 647–652.

[96] M. Jinek, K. Chylinski, I. Fonfara, M. Hauer, J.A. Doudna, E. Charpentier, A programmable dual-RNA-guided DNA endonuclease in adaptive bacterial immunity, Science 337 (2012) 816–821.

Microalgae to Biogas: Microbiological Communities Involved

OLIVIA CÓRDOVA • ROLANDO CHAMY

INTRODUCTION

At global level, numerous studies are carried out about the generation of renewable energies, such as solar and wind energy, biomass energy and others, to meet the high demands. For processes where energy production is based on the degradation of biomass, different organic substrates are biomethanized through anaerobic digestion, generating biogas, which can be used as a biofuel [1–3]. Thus, the study of the anaerobic digestion process, from its research to the applicability at industrial scale, is of great interest since this process is efficient in the treatment of wastewater and solid waste [4,5]. In addition, its final product, biogas, is of great interest as a source of renewable energy [6,7].

In general terms, anaerobic digestion is a process by which organic matter is transformed into biogas through a series of biochemical phases, which in turn are carried out by different microorganisms [4,8,9]. Anaerobic digestion is presented as an alternative with low operational costs in relation to other processes for biofuel production [1,10,11]. The organic matter used can come from different sources, which are characterized and evaluated in relation to their biomethanization potential in terms of biogas production values. Most used substrates are municipal waste, waste and wastewater from the agriculture industry, industrial waste [2], some higher plants crops [12], and some microalgae species, which, in recent years, have been of great interest [3,13,14].

Anaerobic digestion stability process is fully dependent on the syntrophic activity of the microorganisms that belong to the different functional guilds that carry out the different phases of the entire process [9,15]. Thus, biogas production process involves a series of successive metabolic reactions and requires the combined activity of several groups of microorganisms, which differ in their growth requirements and metabolic

capacities, which are associated to each of the phases of anaerobic digestion. For biogas production process efficiency, each one of the biochemical phases for the conversion and the microorganisms involved must operate in a synchronized manner [15–18].

Even though anaerobic digestion of a microalgal biomass is well established as a biological process, certain drawbacks are identified that still do not allow it to be at industrial scale. Therefore, the challenges to be addressed, such as determining the interaction between microalgae and bacteria: what are the forms of interaction and whether they are positive or not for biogas production process? Advancement of molecular tools allows an approach that covers whole microbiological community, not only the function of few microorganisms. Thus, new molecular technologies will answer these questions to implement techniques that improve biogas production from microalgal cultures.

In this chapter, we will introduce the background related to microbiological communities associated with anaerobic sludge for the biomethanization of a microalgal biomass, where we will discuss various aspects that should be considered for the optimal development of anaerobic digestion and, therefore, the production of biogas, using strategies obtained from the information about the microbiological community involved, with a specific focus on operations management and microbiology.

MICROALGAE TO BIOGAS

Currently, microalgae are used as an alternative source of organic matter for biogas generation mainly due to their favorable biological characteristics, for example, their CO_2 requirement for their growth, and biochemical and cultural requirements [14,19,20].

Microalgae Cultivation for Biofuels Production. https://doi.org/10.1016/B978-0-12-817536-1.00015-1

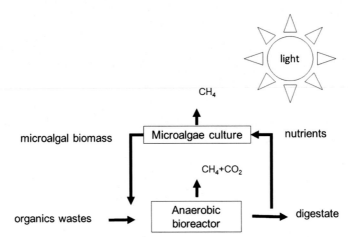

FIG. 15.1 Biorefinery concept based on microalgal cultures. Anaerobic digestion process coupled to microalgae cultures for biogas production in a closed system.

If we consider a sustainable process for the generation of clean energy (Fig. 15.1), we can consider that the combination of anaerobic digestion and microalgae growth can be used for the development of closed systems, where carbon (released CO_2) and nutrients (released during the anaerobic conversion of the microalgae) are recycled, which could be used for microalgae growth [21]. This microalgal biomass produced would feed an anaerobic bioreactor, converting biomass into biogas.

The biogas produced by the process of organic matter anaerobic digestion consists mainly of two components: (1) methane (CH_4), with a percentage that varies between 55% and 70% of its total volume and (2) carbon dioxide (CO_2), with a percentage that varies between 30% and 40% of its total volume [4,22]. Depending on the sources of the substrate used to obtain the biogas, other compounds such as hydrogen, nitrogen, oxygen (<1%), hydrogen sulfide (0−50 ppm), sulfur compounds, aromatics, halogenated volatile organic compounds, and siloxanes could be present [23].

Several methane yields have been reported from different microalgal biomasses through anaerobic digestion processes, which vary between 50 and 600 mL CH_4/g of volatile solids of the biomass, with percentages of methane that vary between 40% and 72% of the biogas from different operational parameters, making their comparison complex (Table 15.1).

The advantages of using microalgae as a substrate for the production of biogas began to be known since the mid-20th century. However, it is not yet a scalable system to an industrially viable scale [11,25,26]. It is for

this reason that there are large investments in new technologies that allow the exploration of industrial production mechanisms. This is mainly due to the fact that, despite presenting a series of advantages (Table 15.2), for its industrial scaling, it is necessary to solve some disadvantages (Table 15.3) that the process shows.

A series of biochemical processes are involved in the process of anaerobic digestion (Table 15.4).

The hydrolysis of organic matter occurs in the first phase. This phase becomes a constraint in case the particles of organic matter to be digested by the bacteria are too complex to degrade, which can make the anaerobic digestion become a more extensive process than usual, requiring more time, and therefore the productivity of biogas decreases [4,20]. In this sense, the microalgal biomass has a series of characteristics (mainly its composition) that could cause the hydrolysis (first phase of anaerobic digestion) to become a limiting step throughout the anaerobic digestion process, consequently resulting in a deficient and unprofitable biogas production [43,44]. This is due to the fact that the microalgae cell wall is mainly characterized by being composed of biopolymers that are difficult to degrade, such as the sporopollenin and algaenan macromolecules (as recalcitrant compounds in photosynthetic organisms) [45−47]. Extracellular enzymes released by hydrolytic bacteria are not able to completely biodegrade the cell wall [48,49]. Thus, the concentration of organic matter available to continue the anaerobic digestion process is limited, and consequently, the production of biogas is deficient.

TABLE 15.1
Main Advantages in Biogas Production From Microalgae Culture [13,20,24].

MAIN ADVANTAGES	
• High productivity	Algae have a high efficiency of light energy conversion to biomass ($\geq$5−10%) compared with higher plants ($\geq$0.5% −3%).
• Water quality for cultivation	Depending on the specific requirements, wastewater, brackish water, seawater, and freshwater can be used for microalgae culture.
• Arable land for cultivation	Culture microalgae can be carried out both in indoor growing system and in noncultivable land or in artificial lagoons. In addition, it is possible to carry out these cultures in arid zones.
• Carbon dioxide fixation	Microalgae converted carbon dioxide to biomass, and media culture can be enriched with different sources of carbon dioxide.
• Biomass uses	In anaerobic digestion process of microalgae, all organic material fractions, including lipids, proteins, carbohydrates, and nucleoids acids, can be converted to different biofuels.
• Nutrients reuses	In anaerobic digestion process, the microalgae culture surplus can be released. Thus, the liquid supernatant can be used as a fertilizer for a new microalgae cultivation, since it is rich in nitrogen and phosphorus. The solid part can also be used for a new culture or as a biofertilizer for the agriculture industry.
• Environmentally sustainable process	Microalgal biomass can be used from another environmental process, such as wastewater treatment. On the other hand, the digestate can be used as a fertilizer. In this way, the process is environmentally sustainable.

TABLE 15.2
Main Disadvantages in Biogas Production From Microalgae [24,27,28].

MAIN DISADVANTAGES	
• High cost to microalgae cultures	Some elements of the culture medium are expensive, if complementary sources of nutrients are not available.
• Possible deficiencies in photosynthetic process	Photoinhibition and low-carbon assimilation processes are possible under stress conditions.
• Incomplete digestion of microalgal cells	Some microalgae species contains recalcitrant matter, which cannot be hydrolyzed by a conventional anaerobic digestion process. Requires additional processes such as pretreatments, which significantly increase production costs.
• Slow conversion rate	In general, algal biomass residence time in the anaerobic digester is 10 −30 days.

Microalgae Cell Wall—Structure and Composition

Cell wall exerts a fundamental biomechanical role regarding its environmental interactions [50]. Thus, cell wall structure and composition will depend on multiple factors (growth stage, cellular form, environmental factors), in addition to the phylogenetic base factor that incorporates functional genes, which will be responsible for the cell wall synthesis and formation, and its subsequent modifications [50,51].

The strength of the microalgae cell wall when subjected to a hydrolysis process is attributed to the polymers that form it, since they are difficult to biodegrade. Sporopollenin and algaenan polymers are generally associated with the strength of the microalgae hydrolysis process. Sporopollenin is a fairly stable compound, which can be presented as an oxygenated aromatic element or as an aliphatic biopolymer depending on the lipids oxidative polymerization [46,52]. Research reports that cell walls that present algaenan have an efficient barrier to extracellular enzymes involved in the hydrolysis process. Species

TABLE 15.3
Conditions and Results of Anaerobic Digestion for Biogas Production From Different Species of Microalgae.

Microalgae	Reactor Type	Temp. (°C)	HRT (days)	CH₄ Production (mL CH₄/g. VS)	Biogas, % CH₄	Reference
Chlorella vulgaris	Batch	28–31	64	320–380	–	[29]
Spirulina maxima	RTA	35	16	150	65,6	[30]
Scenedesmus obliquus	Batch	38	35	178	62	[31]
Chlorella vulgaris	Batch	–	–	150	–	[32]
Arthrospira máxima	BMP	37	–	185	–	[33]
Scenedesmus obliquus	BMP	37	30	210	–	[34]
Scenedesmus sp.	Batch	–	–	162	–	[35]
Chlorella kessleri	Batch	38	35	218	65	[31]
Nannochloropsis gaditana	BMP	–	–	228	–	[36]
Chlorella sp.	Batch	–	–	415	–	[37]
S. obliquus	BMP	37	–	265	60	[33]
Acutodesmus obliquus	BMP	–	–	289	–	[38]
Chroococcus sp.	BMP	37	–	317	–	[39]
Dunaliella salina	RTA	38	32	320	64	[31]
Botryococcus braunii	BMP	37	–	326	–	[40]
Stigeoclonium sp., Monoraphidium sp., Nitzschia sp., and Navicula sp.	Batch	–	–	181	–	[41]
Phaeodactylum tricornutum	BMP	37	30	350	–	[34]
Chlorella sp.	BMP	37	38	470	–	[42]

HRT, Hydraulic retention time.

such as *Chlorella* sp. present this trilaminar membrane-resistant structure of the nonhydrolyzable biopolymer (algaenan). The main function of this biopolymer is to protect against parasites and dissection problems [47,53−55]. The presence of any of the aforementioned biopolymers adversely affects the efficiency of the entire anaerobic digestion process, since hydrolysis becomes a limiting step.

Regarding the differences between species, studies that address the microalgae cell wall structure, formation, and composition by means of phylogenetic analysis distinguish these differences, mentioning that they have been acquired from different phylogenetic lines associated with various cellular functions such as cellular form, environmental signal perception, intercellular connection, and cell motility, among others [50]. Cell wall formation in microalgae and differences in cell wall formation have been described between species of the same genus. Thus, within *Chlorella* genus

microalgae, widely used in biofuels production, two different forms have been defined for their formation, which would be associated, in turn, with the characteristics of the cell wall composition [56]. On the other hand, cell wall structure is usually continuous and rigid, enveloping the cell. Main components are described in Fig. 15.2. Cell wall microalgae can be extracellular or intracellular. In some species, cell wall can be associated with the internal surface of the plasma cells membrane, as it happens with microalgae belonging to Cryptophyceae, Dinophyceae, and Euglenophyceae classes [57]. However, a shared feature is usually the presence of the sporopollenin and algaenan macromolecules, which have been described in some microalgae species used for biofuels production (*Chlorella* sp., *Chlorella vulgaris*, *Nannochloropsis* sp., *Pediastrum boryanum*, *Scenedesmus obliquus*) [45]. It is also relevant to consider the cell wall composition, as this will release organic matter to be digested by bacteria.

TABLE 15.4
Description of Anaerobic Digestion Process.

Phase	Process	Products	Microorganisms
Hydrolysis	Enzymes excreted by hydrolytic bacteria (exoenzymes) are released to converting complex polymers into compounds that can traverse cell walls and membranes of bacteria	Simple sugar Amino acids Fatty acids	*Cellulomonas* *Bacillus* *Mycobacterium*
Acidogenesis (fermentation)	Simple compounds are incorporated by endoenzymes that are then excreted in volatile fatty acids, alcohols, lactic acid, carbon dioxide, hydrogen, ammonia, and hydrogen sulfide as well as new cellular material	Amino acid and sugar anaerobic oxidation—fatty acid and alcohol anaerobic oxidation	*Nitrobacter* sp. *Nitrosomonas* sp.
Acetogenesis	Products of fermentation are converted into acetate, hydrogen, and carbon dioxide	Acetic acid and hydrogen	*Desulfovibrio desulfuricans*
Methanogenesis	Acetate, hydrogen, and carbon dioxide are converted in methane and carbon dioxide	Methane from acetic acid Methane from hydrogen and carbon dioxide	*Methanobacterium,* *Methanosarcina* *Methanomicrobium,* *Methanospirillum*

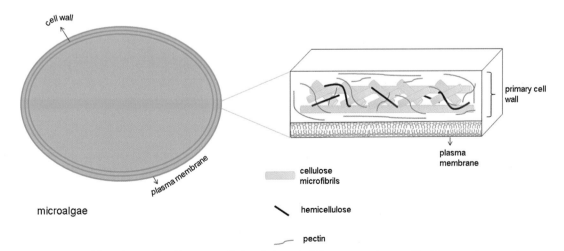

FIG. 15.2 Microalgae cell wall structure. Cell wall characteristics are species specific. However, in general, a microalgae cell wall contains cellulose microfibrils, hemicellulose, and pectin, which provide the cell tensile strength and protection.

Characteristics of their composition are that they are specific and can vary from glycoproteins, glucosamine, pectins, and microcellulose fibrils as main components, to other components such as polysaccharides, silica, or calcium carbonates. Thus, Takeda [54,58] made an analysis of the carbohydrates present in nine microalgae species (*Chlorella* genus) and separated them into two clusters according to sugar predominant in cell wall:

(1) *Chlorella fusca, Chlorella luteoviridis, Chlorella minutissima, Chlorella protothecoides, Chlorella saccharophila,* and *Chlorella zofingiensis* presented a cell wall mainly composed of glucose-mannose, and (2) *Chlorella kessleri, Chlorella sorokiniana,* and *Chlorella vulgaris* presented a cell wall mainly composed of amino sugar and glucosamine. Abo-Shady et al. analyzed and compared the monosaccharides composition and constituent amino acids of the microalgae cell wall, determining that the main monosaccharides present in the cell wall are glucose, rhamnose, arabinose, and mannose [59].

BIOCHEMICAL PROCESS FROM THE SEWAGE MICROBIOLOGY COMMUNITY

As mentioned above, the efficiency and stability of the anaerobic digestion process depend on the synchronized activity and effective interaction between the microorganisms of the different functional guilds that carry out the hydrolysis, fermentation, acetogenesis, and methanogenesis phases (Fig. 15.3) [4,60]. Thus, hydrolytic microorganisms produce enzymes that hydrolyze complex particles, in addition to fermenting microorganisms

that convert simple sugars and proteins into organic acids, alcohol, hydrogen, and carbon dioxide. Acetogenic microorganisms produce acetate, which is converted by aceticlastic methanogenic archaea to methane, while carbon dioxide and hydrogen are converted to methane by hydrogenotrophic methanogenic archaea. Homoacetogenic bacteria participate in the conversion of the monocarbonated compounds to acetic acid. Sulfur-reducing bacteria are responsible for the reduction of sulfur compounds to hydrogen sulfides [61].

Hydrolysis

Hydrolysis → *Degradation of particles or colloidal biopolymeric forms to monomers.*

Hydrolysis is the splitting (lysis) of complex carbohydrates, proteins, and lipids into simple sugars, amino acids, and fatty acids, respectively.

To carry out this process, various groups of hydrolytic bacteria are fundamental for the degradation of complex polymeric compounds, which will be converted to oligomers or monomers, and then they are digested [9,62,63].

Enzymes are protein molecules that catalyze biochemical reactions. There are two types of enzymes

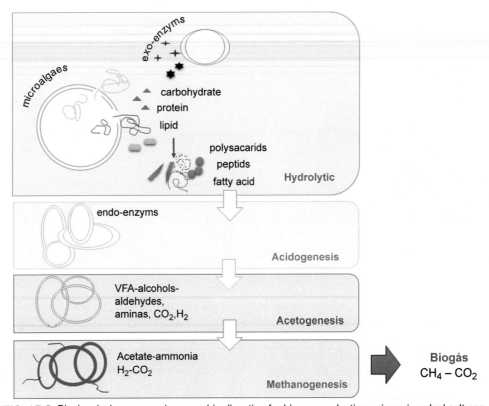

FIG. 15.3 Biochemical processes in anaerobic digestion for biogas production using microalgal cultures.

involved in substrate degradation: endoenzymes and exoenzymes. The former are produced inside the cells and degrade soluble substrate inside the cell. The latter ones are also produced in the cell, but they are released through a mucus that contains the cell, which adheres to the insoluble substrate, which then solubilizes the colloidal particles and substrates. Once solubilized, the substrate enters the cell, and it is degraded by the endoenzymes. The exoenzymes production, the particle solubilization, and the colloidal substrate usually occur in a few hours [9].

All bacteria produce endoenzymes, but not all bacteria produce exoenzymes. Each exo- and endoenzyme degrades only one specific substrate (or group); therefore, there is a large and diverse community of bacteria that are necessary to ensure that the appropriate exo- and endoenzymes are available for the degradation of the different substrates [9,64]. However, this situation is rarely achieved, especially when using microalgae with complex cell wall composition. Depending on the species used, hydrolysis can be a limiting step due to the cell wall structure and composition (Microalgae cell wall — structure and composition section). During hydrolysis, different groups of fermentative bacteria are able to excrete extracellular enzymes.

A strategy commonly used for microalgal cultures is the application of pretreatments, which is an alternative to conditioning the substrate [37,41]. The main objective of the application of a specific pretreatment, physical or enzymatic, is to achieve the damage and/or rupture of the microalgal cell wall and thereby increase the amount of organic matter available for assimilation by the fermenting microorganisms [65].

Acidogenesis

Acidogenesis → *Fermentation of amino acids and acetate to sugars, hydrogen, and some intermediates such as propionate, butyrate, lactate, and ethanol. The β-oxidation of long chains of fatty acids and the alcoholic fermentation to volatile fatty acids (VFAs) and hydrogen.*

The monomers produced as a result of the above-mentioned hydrolytic process are taken up by different facultative and obligatory anaerobic bacteria for their degradation into different short-chain organic acids such as butyric acids, propanoic acids, acetic acids, some alcohols, hydrogen, and carbon dioxide [27].

In this second phase, the concentrations of the products generated must be balanced to maintain a stable and efficient process. Gerardi mentions that the hydrogen concentration as an intermediate product during acidogenesis impacts the type of final product formed [9]. Hence, if hydrogen partial pressure is too

significantly higher, it could decrease the number of reduced compounds.

Acetogenesis

Acetogenesis → *Anaerobic oxidation of intermediates such as VFAs to acetate, carbon dioxide, and hydrogen.*

In the third phase of the anaerobic digestion biochemical process, acidogenic products are consumed as substrates by different microorganisms. Thus, acids such as propionic and butyric acid and some alcohols are degraded by acetogenic bacteria in carbon dioxide and hydrogen and acetic acid [9,66].

This phase turns out to be quite more complex than the previous ones. Consider the following: the hydrogen produced in this phase plays a role of relevant intermediate product, since the reaction only occurs if the partial hydrogen pressure is low enough to thermodynamically allow the conversion of all the acids. The decrease in partial pressure is carried out by hydrogen-depleting bacteria.

Acetogenesis products cannot be directly converted to methane. Methanogenic bacteria are converted into methanogenic substrates. Alcohols and VFAs are oxidized into methanogenic substrates: hydrogen, acetate, and carbon dioxide. VFAs with carbon chains longer than one unit are oxidized into hydrogen and acetate.

The syntrophic activity between the microorganisms is fundamental in this phase, since they perform the anaerobic oxidation reactions with methane-forming microorganisms; this syntrophic activity be subject to hydrogen partial pressure in the system [67]. In anaerobic oxidation, protons are used as the final electron acceptors, which lead to hydrogen production [66,68]. Nevertheless, oxidation reactions can only occur with a low hydrogen partial pressure, and that is why this interaction (syntrophic activity) is fundamental because they will consume hydrogen to produce methane.

Methanogenesis

Methanogenesis → *Transformation of molecular hydrogen and carbon dioxide into methane acetate into methane by acetoclastic methanogenic microorganisms.*

Anaerobic digestion final phase is methanogenesis, where methane is produced by methanogenic archaea. Methane is produced as a metabolic by-product. In the global metabolism of carbon, it is possible to distinguish three types of routes for the synthesis of methane that are distinguished by the substrate that can be (1) CO_2, (2) methanol, or (3) acetate [69].

Methanogenesis is a critical step in the entire anaerobic digestion process as it is the slowest biochemical reaction of the process. This is due to the fact that the microorganisms are relatively sensitive to environmental factors and their growth is relatively slow compared with fermenting or acetogenic bacteria [70].

The anaerobic digestion methanogenesis is performed by archaeas from two different groups: (1) acetoclastic methanogens, a group that uses acetate as a substrate for methane and carbon dioxide, and (2) hydrogenophilic methanogens, a group that uses hydrogen as an electron donor and carbon dioxide as an acceptor for the final production of methane. The microorganisms that participate in these processes are strict anaerobes, and *Methanobacterium, Methanobacillus,* and *Methanosarcina* are the most representative and described microorganisms, and (3) methylotrophic methanogens grow on substrates that contain the methyl group ($-CH_3$) [43,68].

Of both methanogenic archaea, the one that plays a dominant role in the production of methane corresponds to the acetoclastic methanogenic archaea, being responsible for 70% of the methane produced in the digesters and where the acetate-methane equivalence is one to one. The remaining 30% of methane corresponds to the transformation of hydrogen to methane by the hydrogenotrophic methanogenic archaea, with a hydrogen-methane equivalence of four to one. This species plays a key role for the whole process of anaerobic digestion, since it keeps the partial pressure of H_2 low (<10 Pa), which is necessary for an efficient functioning of the trophic intermediate groups and syntrophic bacteria responsible for the conversion of the intermediate products of direct methane precursors (organic acids and alcohols) [9,70,71].

Microbial Community

In recent years, the amount of research related to determining the composition and functionality of the microbiological communities of the anaerobic sludge used for biomethanization of organic matter (Fig. 15.4) has increased significantly [18,72–74]. However, for research on the anaerobic digestion of microalgal biomasses, there are only a few studies reported in the literature [48,75–77].

Although the initial research is exclusively related to identify with the highest possible taxonomic resolution what bacteria and archaea participate in this process, we are currently seeking to identify their functional roles, which depends on a highly complex interaction between microorganisms.

To optimize biogas production, it is important to have knowledge and an understanding of how the microbial community works: metabolic capacities, mechanisms for interspecies interactions, and level of functional redundancy, among others. However, as with other complex environments, most microorganisms in the anaerobic digestion process have not yet been identified. As from amplicon sequencing studies targeting 16S rDNA and metagenome, an increasing

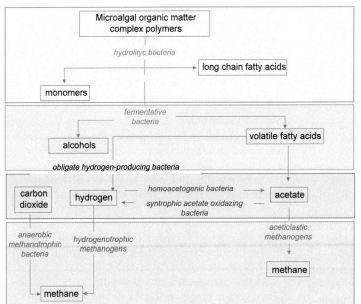

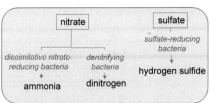

FIG. 15.4 Community microbiological structure for anaerobic digestion process, highlighting each stage of the process. The blue box describes a community associated with inhibitory process.

number of microorganisms have been found, but not all have been identified.

Bacterial Communities

Many studies have correlated microbial composition with digester performance to identify the factors that determine the community conformation since it is known that microbial community and their interactions are strongly related to biogas production operating parameters, such as organic loading rate, retention time, substrate composition, and temperature. However, as mentioned above, most species present in anaerobic digestion process are not identified. Higher taxonomic levels are used to identify functional groups within the community. In Table 15.5, we provide a list with the main bacterial phyla described in different researches, when a microalgae biomass was used as a substrate.

In general, Bacteriodetes, Proteobacteria, Firmicutes, Chloriflexi, Actinobacteria, Spirochaetes, Thermotogae, and Synergistetes are commonly found in anaerobic reactor, which are reported at different levels of abundances (dominance or rarely) depending on the operating conditions [18,75,76,78]. Also, core populations and cooccurrence patterns in digesters are suggested with variation associated to substrate and process conditions.

In the first phase of anaerobic digestion, bacteria degrade organic matter using two kinds of enzymes: endoenzymes and exoenzymes. Members of the orders Clostridiales, Bacteroidales, and Proteobacteria were identified as key microorganisms in this anaerobic digestion process [68,76,79], specifically in the process of organic matter degradation (remember that this process has been described as limiting for several microalgae species). The relative abundance of bacteria within an anaerobic digester often is greater than 10^{16} cells per milliliter, divided into saccharolytic bacteria ($\sim 10^8$ cells/mL), proteolytic bacteria ($\sim 10^6$ cells/mL), lipolytic bacteria ($\sim 10^5$ cells/mL), and methane-forming bacteria ($\sim 10^8$ cells/mL) [9].

In acetogenesis phase, acetate-forming bacteria grow in a symbiotic relationship with methane-forming bacteria. Acetate serves as a substrate for methane-forming bacteria. Acetogenic bacteria reproduce very slowly, with a generation time usually greater than 3 days. *Acetobacterium*, *Clostridium*, and *Thermoacetogenium* are organisms included as acetogenic bacteria [76,78,80,81].

On the other hand, sulfate is reduced to sulfide by bacteria to use sulfate as the principal sulfur nutrient, and during sulfide fermentation or desulfurication, sulfate is reduced to sulfide as organic compounds are oxidized. These processes are carried out by sulfate-reducing bacteria—incomplete oxidizers and complete oxidizers [9].

At this point of anaerobic digestion process, it is important to consider in the study of the microbiological community the syntrophic activity that occurs between sulfate-reducing bacteria and archaea methanogens [68,70,75] since both need hydrogen. Sulfate-reducing bacteria such as *Desulfovibrio*, *Desulfobacter*, *Desulfococcus*, and *Desulfonema* are commonly mentioned for anaerobic digestor [43,71,75].

Numerous studies report diversity and microbiological abundance. Nevertheless, a new direction of the next research on the label and the possible functions in the microorganisms is fundamental. The above is possible to the enormous advance of technology in molecular biology.

Archaeal Communities

Methanogenic is a small percentage of whole microbiological community in digesters but still performing a fundamental role. All methanogens archaea produce methane, are obligate anaerobes, and grow as microbial consortia. Generation times for methanogens archaea range from 3 days at 35°C to 50 days at 10°C. Most methanogens are mesophiles or thermophiles, with some bacteria growing at temperatures above 100°C. These microorganisms obtain energy by reducing compounds or substrates such as acetate and carbon dioxide. The activity of these organisms is usually determined by measuring changes in volatile acid concentration or methane production [9].

Methane produced in digester comes mainly from hydrogen and acetate as a substrate by methanogens, which requires the presence of not only a large number of methanogens archaea but also a large diversity. About 50 species of archaea methanogens are classified in three orders: (1) Methanobacteriales, family Methanobacteriaceses; (2) Methanococcales, family Methanococcaceae; and (3) Methanomicrobials, family Methanomicrobiaceas Methanosarcinaceae.

In Table 15.6, we provide a list with the main archaea methanogens that composed the microbial communities described when a microalgae biomass was used as a substrate.

As we mentioned above, there are three groups of methanogens archaea: (1) hydrogenotrophic

TABLE 15.5

Main Bacteria and Their Function Described in Anaerobic Digestion Process Using Microalgae as a Substrate[a] [77,78].

Bacterial Microorganism	Function
HYDROLYSIS	
Bacteroides	Exoenzymes to organic matter degradation—carbohydrates and proteins
Clostridium	Exoenzymes to organic matter degradation—carbohydrates, proteins, and lipids
Micrococcus	Exoenzymes to organic matter degradation—lipids
Bacillus	Exoenzymes to organic matter degradation—proteins
Peptococcus	Exoenzymes to organic matter degradation—proteins
Enterobacter	Alga-lytic activity, exoenzymes to organic matter degradation
Streptococcus	Exoenzymes to organic matter degradation
Mycobacterium	Exoenzymes to organic matter degradation
FERMENTATION AND ACETOGENESIS	
Clostridium	Hydrogen producing, acetogenic bacteria
Escherichia	Endoenzymes to organic matter fermentation
Lactobacillus	Endoenzymes to organic matter fermentation
Pseudomonas	Endoenzymes to organic matter fermentation
Desulfococcus	Sulfate-reducing bacteria
Sporanaerobacter	Acetogenic bacteria
Coprothermobacter	Secondary fermenting
Azoarcus	Fixing N_2
Paludibacter	Secondary fermenting
Thermoacetogenium	Syntrophic acetate-oxidizing bacteria
Desulfobacter	Acetogenic bacteria
Desulfuromonas	Sulfate-reducing bacteria
Desulfovibrio	Sulfate-reducing bacteria
Desulfomicrobium	Sulfate-reducing bacteria
Desulfohalobium	Sulfate-reducing bacteria

TABLE 15.5

Main Bacteria and Their Function Described in Anaerobic Digestion Process Using Microalgae as a Substrate[a] [77,78].—cont'd

Bacterial Microorganism	Function
Geobacter	Sulfate metal-reducing bacteria
Thermotogae	Fermentation and hydrogen production
Syntrophomonas	Hydrogen producing
Tepidanaerobacter	Acetate-oxidant bacteria
Syntrophobacter	Homoacetogenic bacteria

[a] *Chlorella sorokiniana, Chlorela vulgaris, Scenedesmus* sp., *Chlamydomonas reinhardtii, Haematococcus pluvialis.*

TABLE 15.6

Main Archaea Methanogens and Their Function Described in Anaerobic Digestion Process Using Microalgae as a Substrate[a].

Microorganism	Function
METHANOGENESIS	
Methanosarcina	Acetoclastic methanogenic
Methanospirillum	Acetoclastic methanogenic
Methanosaeta	Acetoclastic methanogenic
Methanobacterium	Hydrogenotrophic methanogenic
Methanoplanus	Hydrogenotrophic methanogenic
Methanobervibacterium	Hydrogenotrophic methanogenic

[a] *Chlorella sorokiniana, Chlorela vulgaris, Scenedesmus* sp., *Chlamydomonas reinhardtii, Haematococcus pluvialis.*

methanogens; (2) acetotrophic methanogens; and (3) methylotrophic methanogens. Hydrogenotrophic and acetotrophic methanogens produce methane from carbon dioxide and hydrogen. Methylotrophic methanogens produce methane directly from methyl groups and not from carbon dioxide [82].

Methanobacteriales, Methanomicrobiales, Methanococcales, and Methanosarcinales are commonly found in anaerobic reactor [68,82—85].

Most of the studies that describe methanogenic communities use diversity indices. Specific studies on the methanogen community are carried out to describe the community in relation to its diversity and abundances, in addition to evaluating the activated metabolic pathways for the process of methanogenesis in relation to the different substrates used [76,82]. This last point is relevant since it is possible to hypothesize about the association between degraded substrates (microalgal biomass, municipal wastes, among others) with the specific metabolic pathways activation in methane production.

INTERACTION MECHANISMS BETWEEN MICROALGAE—BACTERIA IN THE ANAEROBIC DIGESTION PROCESS

The type of interaction between bacteria-bacteria and/or bacteria-archaea are processes that have been studied in greater depth in recent years, where, for example, several syntrophic acetate oxidant bacteria, *Clostridium ultunense* [86], have been reported. These studies have led to the conclusion that the interactions between microorganisms are fundamental for the stability of anaerobic digestion. However, in studies and subsequent analyses of these biochemical processes in which a microalgal biomass is used for anaerobic digestion, the possible interactions between microalgae and bacteria have been poorly studied; in fact, they are considered independently.

In natural environments, microalgae-bacteria interactions cover the entire range of conceivable forms of ecological interaction, ranging from mutualism to parasitism. Microalgae, bacteria, and archaea have been described as the main producers and decomposers, respectively, which makes them structuring organisms of different environments [87]. However, most types of interactions between microalgae and bacteria are poorly studied in their natural environments, especially in productive systems such as anaerobic digesters. This is due to the time-consuming task of separating these microorganisms, which are naturally linked to each other. Therefore, it is relevant that the analyses are not considered or carried out independently.

Microalgae and bacteria are affected synergistically both in their physiology and in their metabolism [88,89]. However, often bacteria are not considered. An example of the above in productive systems, such as microalgal farming systems, is that bacteria are even considered a mere contamination. This vision has changed in recent years, as the interactions between microalgae and bacteria are being considered as promising tools in the area of biotechnology, as recent studies have shown a positive effect of the microalgae-bacteria interaction in biotechnological processes [87].

Some biotechnological processes where these interactions are used as strategies in favor of productive processes are the growth of microalgal crops and their subsequent harvesting, which are well established. However, processes specifically associated with the production of renewable energies from this microalgal biomass, as for example, during anaerobic digestion, have been scarcely reported.

Currently, we do not know what type of microalgae-bacteria interaction generated in productive systems such as anaerobic digesters. The information on how this interaction is, at what levels it is produced, by whom it is produced, and the way in which it could be used as a biotechnological tool for the production of renewable energies is scarce. For example, it has been reported that an anaerobic environment suitable for the production of microalgal hydrogen can be conferred by bacteria that can consume the O_2 generated photosynthetically by microalgae, without damaging the photosynthetic apparatus [90]. However, the possible benefits of the microalgae-bacteria interaction, whatever they may be, have not been described for the production of biogas.

In a metatranscriptomic expression analysis during the production of biogas from *C. sorokiniana*, Cordova et al. report the activation of the quorum sensing (QS) mechanism, which is a regulatory system that allows bacteria to share information about cell density and adjust the gene expression. Processes controlled by QS include antibiotic production, competition, conjugation, motility, virulence, and biofilm formation as bacterial defense mechanisms [91]. These defense mechanisms consist of a cell-to-cell communication that causes the expression and release of bioactive substances in the surrounding environment and influences the behavior of other microorganisms present in the environment [91]. Córdova et al. report that the activation of QS comes from the *Defluviitoga tunisiensis* (Petrotogales) and *Thermosipho africanus* (Thermotogales) species during the anaerobic digestion of *C. sorokiniana* [78]. However, the authors could not determine the causes of the QS activation, that is, if it is due to a microalgae-bacteria or bacteria-bacteria defense mechanism. In the case of the microalgae-bacteria mechanism, the initial step of the microalgae-bacteria interaction is the bacterial detection of microalgal cells. The detection signal is precise and modulated according to the microalgae, their stage of growth, and the density of their biomass [92]. However, the expression of "algicidal" substances or enzymes does not depend on the detection of the presence of microalgal cells, but it is regulated by bacterial cell density. This bacterial cell-to-cell communication is given through the

QS that detects the biomass [93] and involves the secretion of bacteria from signaling compounds called autoinducers. These are detectable by the same population of bacteria by means of a receptor molecule that can specifically process the signaling molecule (inducer). When the inducer binds to the receptor, it activates the transcription of certain genes, important for growth, survival, and pathogenicity, including the synthesis of inducers. In addition, QS acts as a bacteria inducer to produce and secrete "algicidal" substances in the surrounding medium [94]. On the other hand, microalgae are able to secrete compounds that mimic the QS detection signals of many gram-negative bacteria, which result in stimulant or inhibitory effects. Thus, for example, *Chlorella vulgaris* on growing with different strains of *Clostridium botulinum*, incubated under anaerobic conditions, determining that *C. vulgaris* prevented the growth of *C. botulinum*, in all the strains studied [95].

The evidence provided by the different studies aims to hypothesize about certain types of biochemical mechanisms that would be activated by the microalgae-bacteria interaction.

MOLECULAR TECHNIQUES FOR THE ANAEROBIC SLUDGE MICROBIOLOGY ANALYSIS—CONTRIBUTION FOR BIOGAS PRODUCTION

Currently, the process of anaerobic digestion is well established and understood in literature [15,62,96]. However, the microbial community responsible for each of the biochemical phases associated with the conversions only recently began to be of greater interest for its in-depth study. Due to the large number of microorganisms involved, this community was often considered a black box for a long time [97,98].

The above mentioned not only is an underestimation of the biological and ecological components of the community but also responds to the limitations of the molecular biology methods available for an adequate study of the microorganisms. Techniques and analysis such as isolation, identification, and physiological characterization required a large amount of laboratory resources and time. The concern to deepen the knowledge of microbiological communities has allowed the gap between microbiologists and engineers in the field of anaerobic digestion to decrease more and more, in such a way that studies have increased.

Considering that, the data obtained from the application of molecular techniques have allowed creating strategies for the modeling and optimization of the operation of various anaerobic digesters. The detailed knowledge about the dynamics of the microbial community (diversity, structure, and functional roles) is fundamental not only for basic scientific research but also to understand the link that exists with the production of renewable energies [80,99]. In this way, the data provided by the different molecular techniques allow validating engineering models, preventing failures, and optimizing the biogas production. Moreover, it is expected that the new knowledge obtained from the molecular analysis of the community is even more suitable as a control parameter in the operation of an anaerobic digester, in replacement of other indicators such as volatile suspended solids such as an indicator of active biomass and/or concentration of VFAs as an inhibitions indicator [100].

Thanks to molecular techniques, it is possible to determine significant changes in the ecology of microbiological communities, that is, in their diversity and abundance dynamics, when different substrates are used as biomass to produce biogas under the same laboratory operational conditions [69,101].

On the other hand, molecular techniques allow us to determine possible changes in microbiological communities when a certain organic matter is found under different conditions. Thus, Bareither et al. characterized the microbial diversity (bacteria and archaea) during the biodegradation of urban solid waste under two conditions, such as solids and leachates, to later correlate it with methane generation [102]. This study led to conclude that the microbiological communities were not similar between each condition, where solid waste contained a larger bacterial abundance, with organisms capable of degrading cellulose (for example, Firmicutes). In addition, they concluded that the concentration of methanogenic microorganisms also differed, finding an association with methane production, which can be attributed to the initial cellulose concentration in the digested waste.

Molecular Techniques Fundamentals

The microorganisms that participate in anaerobic digestion began to be studied in depth in recent decades by more advanced molecular techniques that allow knowing not only what type of microorganisms participate (potentiality) but also what functions they fulfill in this biochemical process (expression) [103].

If we compare the current existing techniques for the analysis of microorganisms and their functions, we can mention the application of culture-dependent methods that have allowed evaluating, for example, key populations capable of carrying out specific metabolic processes in the anaerobic digestion. The problem with the use of these techniques is that most of the microorganisms that participate in this process have not been cultured [16]. In addition, even if they were cultured, they could not be studied in their syntrophic activity nor the influence of environmental factors that also condition the gene expression and their functionality within the entire process.

On the other hand, in the scientific community, it is mostly accepted that the proportion of known microorganisms, both bacteria and archaea, is quite low compared with the total diversity of existing microorganisms, so that it is far from having a full overview of the microbial diversity of an ecosystem if only conventional molecular methodologies are used.

7000 bacteria species have been described up to now, but according to the molecular and ecological estimation, the actual number varies in orders of magnitude higher (Amann et al., 1995), not reflecting the real composition and diversity of a microbial community.

The techniques are based on the small subunit of ribosomal RNA, the 16S rRNA, or its corresponding genes, considering them as a *molecular clock*. Ribosomal RNA was chosen for its universality and abundance in all living beings [85,104,105]. It is also a highly conserved molecule during evolution, although it has certain regions that are highly variable. The above characteristics allow comparing between organisms within the same domain, as well as differentiating strains of the same species [106].

On the other hand, the gene sequence is long enough to generate adequate statistical data and can easily be sequenced with the current technology. This allows studying the biodiversity of a natural habitat in a relatively simple but complete manner [107,108].

The possibility of identifying specific populations of microorganisms without the need to isolate them has revolutionized the microbial ecology and has given rise to a wide range of new applications in different fields of research. Therefore, molecular techniques are an excellent method for a fast and efficient analysis of microorganisms in their communities [109].

Molecular Techniques

Traditional microbiological analysis methods include techniques such as count-based cultures, microscopic enumeration, biochemical analysis, and immunostaining. These techniques do not provide detailed information about the structure, function, dynamics, and diversity of microbial communities. Molecular techniques, which can measure the DNA (deoxyribonucleic acid) and RNA (ribonucleic acid) of cells, are more direct and robust in relation to the ecology of the communities. DNA can be monitored at any point in the existence of a specific organism. These techniques can characterize genes and proteins (enzymes), in addition to metabolic products. Then, the molecular analysis of microbial cells, therefore, can provide useful information about the structure (who they are), function (what they do), and dynamics (how they change through space and time) [104,110].

A critical point in molecular techniques is the nucleic acid extraction and DNA and RNA extraction, to obtain information at a later stage. The extraction procedure must be adapted and optimized not only necessarily depending on the type of sample to be evaluated but also in addition to our research hypothesis. This is where we must consider that environmental samples are highly susceptible to the degradation of their nucleic acids due to the presence of endogenous nucleases. Incorrect handling of the samples will not deliver the actual information of our samples. Therefore, it is essential to stabilize the nucleic acids, particularly RNA, for which it is recommended that the samples should be immediately frozen by immersion in liquid nitrogen and stored at $-80°C$ [111].

Subsequently, to obtain the DNA and RNA molecules, an enzymatic digestion or mechanical rupture of the cells is required to obtain the molecules. The extracted nucleic acids must be purified to eliminate possible inhibitors. In the literature, there are numerous protocols for extracting DNA from environmental samples [98,112], but there are also many commercial kits.

As mentioned above, to determine the molecular technique to be used, it is essential to know what we want to respond in relation to our samples. Next, we describe the molecular techniques (Fig. 15.5), which we have divided into two groups: (1) what microorganism is present associated with the autoecology of microorganisms and (2) what is its function within the community, associated with ecology population and community.

Polymerase Chain Reaction and amplification

Polymerase chain reaction (PCR) is a technique used in molecular laboratories to study genetic information in an efficient manner [104,113,114]. Amplifications of almost all nucleotide sequences are possible using specific primers.

Primers with the sequence of their functional genes are used for the amplification of different microorganisms, but for a specific molecular identification, it is

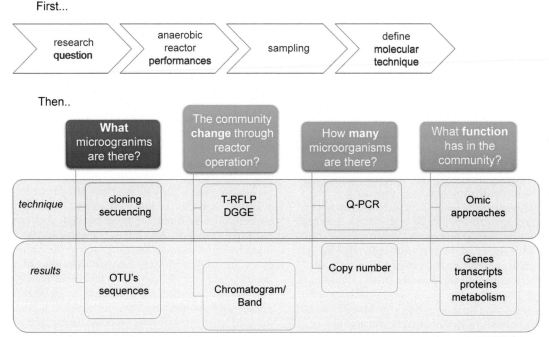

FIG. 15.5 Experimental design for application of molecular techniques for the study of the microbiology associated with anaerobic digestion for biogas production.

more common to use primers of the 16S rRNA gene [113].

16S rRNA gene of about 1500 nucleotides length contains enough reliable information for phylogenetic analysis [115]. One of the advantages of the 16S rRNA sequences is the existence of a large database with currently around 500,000 (>300 bp) ssRNA sequences that can be used for the design of primers [116].

Detection and amplification at different levels of taxonomic can be achieved since the 16S rRNA gene has fixed regions, with the same sequence in many microorganisms, and variable regions, which are the sequences of each species. In addition, the use of fixed 16S rRNA sequences also allows the detection of previously unknown and noncultured microorganisms [117,118].

Some drawbacks reported include the biases caused by the amplification, such as preferentially amplifying certain sequences [105], or during the amplification, there can be discrepancies between the ratio of the 16S rRNA gene fragments and the original mixture, caused by differences in the activity of Taq polymerase or other effects that may arise during the amplification, especially in complex microbial communities with several similar phylotypes.

Cloning and sequencing

For a correct phylogenetic characterization of the microorganisms, the sequencing of the gene of interest is necessary. The sequence comparison of the 16S rRNA gene has become one of the most widely used measures for the analysis of microbial diversity [85,111,119]. Currently, the ideal and most used method is the sequencing of the complete genome. All the genes of a community (metagenome) can be subjected to cloning and sequencing [120].

Pyrosequencing is a large-scale DNA sequence determination technology, which is applicable to whole genomes through luminescence [121,122]. This technology has been used for the estimation of microbiological diversities in different environments, through the operational taxonomic units [123].

Cloning refers to the process of isolating a DNA sequence of interest and inserting it into a plasmid to obtain multiple copies of the sequence in an organism vector. Usually, the *Escherichia coli* bacteria are used for this function. The cloning process consists of a fragmentation, where the DNA of interest is fragmented to obtain the desired segment of DNA to be cloned. A ligation continues, which is used when the amplified fragment is inserted into a vector. That vector is generally circular, and it is converted into a linear sequence using restriction enzymes. Subsequently, the transformation occurs, where the vector with the gene of interest is transfected into the cells. Finally, the success of the cloning is evaluated through a selection of the transfected cells by culturing them, where the colonies of cells that have been successfully transfected with the vector containing the desired gene will be identified. The 16S rRNA sequences obtained after cloning can be analyzed by other molecular techniques such as denaturing gradient gel electrophoresis (DGGE) or their sequencing.

The identification of cloned fragments through the analysis of 16S rRNA gene sequences is a standard procedure, since it is impossible to adequately describe microbial communities without these data [124]. For the comparison of nucleic acid sequences, there are several alternatives on Internet: http://www.ncbi.nlm.nih.gov/ BLAST [125](McGinnis et al., 2004); http://www.rdp. cme.msu.edu [106].

Restriction enzymes

Specific molecular techniques use enzymes to make modifications in nucleic acid fragments of a given microorganism. Restriction enzymes can recognize a specific nucleotides sequence within a DNA molecule and cut the DNA at that specific point of interest.

Amplified ribosomal DNA restriction Analysis (ARDRA) is a technique in which the ribosomal RNA gene is amplified by PCR and digested in fragments [126,127]. After an incubation with the restriction enzyme, the fragments are separated by high-resolution gel electrophoresis, obtaining patterns of different specific sequences. This technique has been used when the objective is to make comparisons between rRNA genes [107,128]. Typically, this technique is carried out on agarose gels for a restriction digestion for strains or clones; however, the analysis of a community that potentially has a large number of fragments is done through the use of polyacrylamide gels to produce specific patterns for the community [127].

Restriction terminal fragment length polymorphism (T-RFLP) is a molecular technique that uses restriction enzymes and only detects terminal restriction fragments (T-RF) [84]. It is used for both qualitative and quantitative analyses [128]. T-RFLP uses the PCR technique, in which one of the two primers is labeled by fluorescent probe (5′end). After the amplification, the PCR products are fragmented with a restriction endonuclease at specific sites, obtaining genetic fingerprints from a microbial community or a specific product from a single microorganism [129]. The terminal restriction fragments are measured using an automatic DNA sequencer. This technique allows evaluating the diversity of complex microbial communities and a quick comparison of the structure and diversity of different ecosystems [129].

One of the disadvantages of the T-RFLP technique is that microbial populations that are not quantitatively dominant will not be represented, because the DNA of these populations represents a small fraction of the total community's DNA and, consequently, it will result in an underestimation of the diversity of species of the microbial community [128]. In addition, the determination of similarities in complex microbial communities using the profiles generated by T-RFLP with a single restriction enzyme can lead to erroneous conclusions. For an adequate analysis and correct conclusions about the samples, it is recommended to use multiple restriction enzymes separately, generating a large amount of data [126].

Single-strand conformation polymorphism (SSCP) is a technique that is based on the separation by capillary electrophoresis of different amplicons of DNA previously denatured with a chemical agent such as formamide, where the sample is renatured, allowing the formation of secondary structures in the strands of single-stranded DNA. The electrophoresis is carried out at low temperatures and under nondenaturing conditions to maintain the intrachain connections [129].

Denaturing gradient gel electrophoresis

The DGGE technique is based on the double-stranded DNA fragments fusion behavior. The amplified DNA fragments may have the same length, but with different base pairs in the sequence, revealing the diversity of the microbial community [117,126,130]. This technique was widely used in environmental microbiology research to study the diversity and relative changes in abundance of the richness of microbial taxonomic groups [130] in productive systems such as anaerobic digesters [117,126,131]. The taxonomic specificity of the primers used in the PCR amplification process determines which groups of microorganisms will be analyzed. The amplified DNA fragments are separated in DGGE according to the differences in the DNA sequences. The principle is based on the dissociation of the double DNA strand under the influence of a denaturing chemical, such as formamide and urea at a constant temperature. During the electrophoresis, the PCR products progress through the gel migrating to the positive pole. But as they move through the gel, because of the changing composition of the gel, the two strands of the DNA molecule denaturalize by separating at a specific point. The separation of the double DNA strand depends on the hydrogen bonds formed between the complementary base pairs and the attraction between neighboring bases of the same strand [132]. This separation always starts from one end, since one of the primers used has a GC clamp. A GC clamp is a 40-nucleotide sequence that is very rich in cytosine and guanine, which is associated with one of the starters [133,134]. Thus, all PCR products have at their end a sequence of 40 nucleotides that are very rich in GC content, and therefore, it is very difficult to carry out the separation to avoid the total dissociation of the PCR products. The percentage of separation of the DNA fragment depends on the sequence of the PCR products. The greater the amount of G and C present in the fragment, the greater the percentage of denaturing that is needed for the separation of the DNA strands. When the PCR product is converted into a single strand with a branched structure, the mobility capacity of the molecules through the gel is reduced, resulting in a partial dissociation of the fragment. Accordingly, the final location of the molecules in the gel will depend on the nucleotide sequences of the fragments. Even in the case of the substitution of a single nucleotide, the migration of the fragment is different [130,134].

This molecular technique has been used as a method to track changes in microbial communities over time. In addition, it allows establishing the presence and/or relative frequency of a sequence of complex communities when they are affected by disturbances in their environment [133].

It has been reported that DGGE is sensitive enough to detect the organisms that constitute up to 1% of the total microbial community [130]. Hence, when the primers are specific for a domain, both the dominant bacteria of the community, as well as the "strange" ones may be represented in the DGGE profiles. Through the use of statistical software, it is possible to calculate and estimate similarity indexes and analysis of community profiles through, for example, the Shannon-Wiener index, analysis of hierarchical grouping expressed in a dendrograms and the individual bands, and their intensities comparing through logistic regression analysis [135].

"Omics" approaches

Due to technological advances in DNA sequencing, nearly 1000 microbial genomes have been sequenced. The first annotations of DNA sequences and comparative analyses have provided new information about functional genes, metabolic pathways, regulatory pathways, etc., increasing knowledge and understanding about how microbiological communities operate [136,137].

Cells are complete systems of different functional molecules, which finally determine the phenotype of the cells. Such molecules include mRNA (messenger RNA) transcribed from DNA, proteins translated from mRNA, and various metabolites of small molecular mass generated by various enzymatic activities. The mere analysis of the microbial genomes DNA sequences is not enough to obtain crucial information regarding the functionality of these molecules and the regulatory mechanisms involved in their generation [138]. Therefore, to elucidate it in its entirety, it is necessary to include a gene quantitative and functional characterization, mRNA, proteins, as well as their interaction. In relation to this type of analysis, new "omics" techniques have risen: metagenomics, metatranscriptomics, proteomics, and metabolomics. The omics approaches are characterized by (1) being holistic, high performance, and data-based methodologies; (2) their analysis is based on the understanding of an integrated system; and (3) they generate a large amount of information, which requires a huge analysis effort [103,137].

The "omics" techniques have allowed evaluating community compositions and functional populations [136], particularly for methanogenic communities. This type of technique has also allowed us to understand how the configuration of a reactor and the operational conditions influence the structure and dynamics and how this, in turn, connects the efficiency of the performance with the stability of the entire process [103,139]. On the other hand, these types of techniques have revealed a great phylogenetic and metabolic diversity, showing that the vast majority of anaerobic digesters are dominated by microorganisms that have not yet been identified [140].

Understanding that the information that reveals this type of molecular techniques allows us to analyze the metabolic capacities of microorganisms, functional levels of the community, and fundamental mechanisms of interaction between microorganisms, this type of information will help us to have more knowledge to optimize the anaerobic digestion process and the formation of compounds of interest [139,141,142].

CASE STUDY: MICROBIOLOGICAL COMMUNITIES INVOLVED IN BIOGAS PRODUCTION FROM A MICROALGAE CULTURE

The study of metatranscriptomics implicates the sequencing of the mRNA from the microbial community, which allows measuring the expression of genes in situ, focusing exclusively on the microorganisms that are active, unlike the metagenomics that includes all microorganisms. Thus, metatranscriptomics can provide information about the real active functions of a microbial community [76,119,143].

The "reads" or readings obtained from this type of analysis must be mapped with reference genomes or metagenomes of similar environments; thus, the levels of expression of differential genes can be estimated. This type of technology uses very sensitive and highly specific sequencing platforms in the measurement of gene expression, which allows the identification of transcripts without any a priori knowledge of their nucleotide sequence [144].

One of the advantages of the metatranscriptomics study is that it allows determining population functions, including the contribution of low abundance microorganisms to the stability and efficiency of the entire process. This is especially interesting when trying to achieve approximations in phylogenies whose functionality varies strongly with changes in environmental conditions [119,145]. Also, it can be used to measure immediate regulatory responses in anaerobic digesters caused by changes and/or disturbances in metabolic profiles and in the functional guilds that characterize a community [103].

Although the metatranscriptomics analysis to study anaerobic digesters has been limited to describing only active populations, this approach has the potential to discover highly expressed metabolic pathways involved in the conversion of organic biogas raw materials [137]. Thus, the study of the metatranscriptomics expression of the microbiological community can provide information related to active metabolic pathways, delivering new knowledge, which can be used to direct the microbial community toward a certain path to achieve a certain objective such as higher biogas production [15].

Cordova et al. evaluated the changes in the expression of the microbial community metatranscriptomics for anaerobic sludge in the biomethanization of *C. sorokiniana* without pretreatment and pretreated with the cellulase enzyme to achieve increases in soluble microalgal organic matter [78]. The authors determined significant differences in relation to the diversity and expression of microbial metabolic pathways. While microalgae samples without pretreatment showed the presence of live microalgal cells, a bacterial community dominated by δ-proteobacteria and γ-proteobacteria, and a community of archaea dominated by *Methanospirillum hungatei*, associated to low biogas productivity, the samples subjected to an enzymatic pretreatment were characterized by the absence of living microalgal cells, a dominance of bacteria of the Clostridia order, associated with significant increases in biogas production. The differences in the microbial diversity between the two different conditions of the microalgal biomass were associated to differences in the transcription of enzymes and, consequently, to the activation of associated metabolic pathways/routes from the degradation of organic matter to their methanization [78].

The change in the structure of the microbiological communities was observed in four levels: changes in the source of bacteria energy generation, ecological interactions, dominant taxa, and metabolic pathways for methanogenesis. In the case of the activation of microalgal cells defense mechanisms, the authors hypothesize that this type of mechanism would lead to new interactions between microorganisms [78]. This could favor certain conditions for the development of a community bacterial phenotype that adapts to this new configuration of the microalgal biomass.

Finally, we can conclude that the molecular biology tools used allowed to deepen the knowledge on how microbiological communities are modulated in relation to the characteristics of biomethanized microalgal biomass, allowing to relate the active microbial metabolic pathways identified with the active diversity of prokaryotes under two conditions (with or without pretreatment).

CONCLUDING REMARKS

Although the study of biogas production from biomass microalgae has increased significantly since the past decade, many aspects of the process are still quite unknown.

Currently, it is not only relevant to know which microorganisms are part of the anaerobic digestion, but rather what kind of function/role they fulfill and at what levels of the process. Activated metabolic pathways, interactions between microorganisms are areas to approach more deeply in future studies related to biogas production from microalgal biomass. This is especially relevant for the development of strategies as a biotechnological tool that allows producing improvements in the process of generating this type of renewable energy.

In addition, it is necessary to deepen the knowledge and control of the mechanisms involved in the microalgae-bacteria interaction, since, according to the few studies reported, the identification of these interaction mechanisms could help to improve the production processes.

Finally, molecular techniques are essential to obtain information on ecological patterns, such as diversity and functionality, to model the proper functioning of anaerobic digesters predicting possible stability problems in digesters, to determine changes produced by external disturbances, and to improve biogas production, among other purposes.

REFERENCES

[1] F. Cherubini, The biorefinery concept: using biomass instead of oil for producing energy and chemicals, Energy Convers. Manag. 51 (2010) 1412–1421.

[2] J.B. Holm-Nielsen, T. Al Seadi, P. Oleskowicz-Popiel, The future of anaerobic digestion and biogas utilization, Bioresour. Technol. 100 (2009) 5478–5484.

[3] L. Zhu, Biorefinery as a promising approach to promote microalgae industry: an innovative framework, Renew. Sustain. Energy Rev. 41 (2015) 1376–1384.

[4] I. Angelidaki, D. Batstone, Anaerobic digestion, solid waste, Technol. Manag 1–2 (2010) 583–600.

[5] C. Mao, Y. Feng, X. Wang, G. Ren, Review on research achievements of biogas from anaerobic digestion, Renew. Sustain. Energy Rev. 45 (2015) 540–555.

[6] N. Abas, A. Kalair, N. Khan, Review of fossil fuels and future energy technologies, Futures 69 (2015) 31–49.

[7] B. Demirel, P. Scherer, O. Yenigun, T.T. Onay, Production of Methane and Hydrogen from Biomass through Conventional and High-Rate Anaerobic Digestion Processes, 2010.

[8] J.A. Fiestas Ros de Ursinos, R. Borja-Padilla, Biomethanization, Int. Biodeterior. Biodegradation 38 (1996) 145–153.

[9] M. Gerardi, The Microbiology of Anaerobic Digesters, 2003.

[10] F.K. Kazi, J.A. Fortman, R.P. Anex, D.D. Hsu, A. Aden, A. Dutta, G. Kothandaraman, Techno-economic comparison of process technologies for biochemical ethanol production from corn stover, Fuel 89 (2010) S20–S28.

[11] C. Zamalloa, E. Vulsteke, J. Albrecht, W. Verstraete, The techno-economic potential of renewable energy through the anaerobic digestion of microalgae, Bioresour. Technol. 102 (2011) 1149–1158.

[12] E.M. Rubin, Genomics of cellulosic biofuels, Nature 454 (2008) 841–845.

[13] M.M. Gruber, Anaerobic digestion of microalgal biomass, Conf. CEBC Graz. 34 (2014) 2098–2103.

[14] M. Rizwan, J.H. Lee, R. Gani, Optimal design of microalgae-based biorefinery: economics, opportunities and challenges, Appl. Energy 150 (2015) 69–79.

[15] K. Venkiteshwaran, B. Bocher, J. Maki, D. Zitomer, Relating anaerobic digestion microbial community and process function, Microbiol Insights 8 (2015) 37–44.

[16] T. Amani, M. Nosrati, T.R. Sreekrishnan, Anaerobic digestion from the viewpoint of microbiological, chemical, and operational aspects — a review, Environ. Rev. 18 (2010) 255–278.

[17] M. Gerardi, The Microbiology of Anaerobic Digesters, John Wiley & Sons, Inc., Hoboken, New Jersey, 2003.

[18] J. De Vrieze, W. Verstraete, Perspectives for microbial community composition in anaerobic digestion: from abundance and activity to connectivity, Environ. Microbiol. 18 (2016) 2797–2809.

[19] F.G. Acién, E. Molina, J.M. Fernández-Sevilla, M. Barbosa, L. Gouveia, C. Sepúlveda, J. Bazaes, Z. Arbib, Economics of microalgae production, Microalgae-Based Biofuels Bioprod. (2017) 485–503.

[20] C. González-Fernández, B. Sialve, N. Bernet, J.P. Steyer, Impact of microalgae characteristics on their conversion to biofuel. Part I: focus on cultivation and biofuel production, Biofuels Bioprod. Biorefining. 6 (2012) 246–256.

[21] W.A.V. Stiles, D. Styles, S.P. Chapman, S. Esteves, A. Bywater, L. Melville, A. Silkina, I. Lupatsch, C. Fuentes Grünewald, R. Lovitt, T. Chaloner, A. Bull, C. Morris, C.A. Llewellyn, Using microalgae in the circular economy to valorise anaerobic digestate: challenges and opportunities, Bioresour. Technol. 267 (2018) 732–742.

[22] V.N. Gunaseelan, Anaerobic digestion of biomass for methane production: a review, Biomass Bioenergy 13 (1997) 83–114.

[23] L. Brennan, P. Owende, Biofuels from Microalgae: Towards Meeting Advanced Fuel Standards, 2013.

[24] P. Bohutskyi, E. Bouwer, Biogas production from algae and cyanobacteria through anaerobic digestion: a review, analysis, and research needs, in: W.J. Lee (Ed.), Adv. Biofuels Bioprod., Springer New York, New York, NY, 2013, pp. 873–975.

[25] L. Moreno-Garcia, K. Adjall??, S. Barnab??, G.S. V Raghavan, Microalgae biomass production for a biorefinery system: recent advances and the way towards sustainability, Renew. Sustain. Energy Rev. 76 (2017) 493–506.

[26] D.U. Santos-Ballardo, S. Rossi, C. Reyes-Moreno, A. Valdez-Ortiz, Microalgae potential as a biogas source: current status, restraints and future trends, Rev. Environ. Sci. Biotechnol. (2016) 1–22.

[27] C. González-Fernández, B. Sialve, N. Bernet, J.P. Steyer, Impact of microalgae characteristics on their conversion to biofuel. Part II: focus on biomethane production, Biofuels Bioprod. Biorefining. 6 (2012) 205–218.

[28] J. Milano, H.C. Ong, H.H. Masjuki, W.T. Chong, M.K. Lam, P.K. Loh, V. Vellayan, Microalgae biofuels as an alternative to fossil fuel for power generation, Renew. Sustain. Energy Rev. 58 (2016) 180–197.

[29] E.P. Sanchez Hernandez, L. Travieso Cordoba, Anaerobic digestion of *Chlorella vulgaris* for energy production, Resour. Conserv. Recycl. 9 (1993) 127–132.

[30] V.H. Varel, H. Chen, A.G. Hashimoto, T.H. Chen, Thermophilic and mesophilic methane production from anaerobic degradation, Resour. Conserv. Recycl. 1 (1988) 19–26.

[31] J.H. Mussgnug, V. Klassen, A. Schlüter, O. Kruse, Microalgae as substrates for fermentative biogas production in a combined biorefinery concept, J. Biotechnol. 150 (2010) 51–56.

[32] A. Mahdy, M. Ballesteros, C. González-Fernández, Enzymatic pretreatment of *Chlorella vulgaris* for biogas

production: influence of urban wastewater as a sole nutrient source on macromolecular profile and biocatalyst efficiency, Bioresour. Technol. 199 (2016) 319–325.

[33] F. Ometto, G. Quiroga, P. Psenicka, R. Whitton, B. Jefferson, R. Villa, Impacts of microalgae pretreatments for improved anaerobic digestion: thermal treatment, thermal hydrolysis, ultrasound and enzymatic hydrolysis, Water Res. 65 (2014) 350–361.

[34] C. Zamalloa, N. Boon, W. Verstraete, Anaerobic digestibility of Scenedesmus obliquus and Phaeodactylum tricornutum under mesophilic and thermophilic conditions, Appl. Energy 92 (2012) 733–738.

[35] A. Mahdy, L. Mendez, M. Ballesteros, C. González-Fernández, Autohydrolysis and alkaline pretreatment effect on Chlorella vulgaris and Scenedesmus sp. methane production, Energy 78 (2014) 48–52.

[36] J.-C. Frigon, F. Matteau-Lebrun, R. Hamani Abdou, P.J. McGinn, S.J.B. O'Leary, S.R. Guiot, Screening microalgae strains for their productivity in methane following anaerobic digestion, Appl. Energy 108 (2013) 100–107.

[37] S. He, X. Fan, N.R. Katukuri, X. Yuan, F. Wang, R.B. Guo, Enhanced methane production from microalgal biomass by anaerobic bio-pretreatment, Bioresour. Technol. 204 (2016) 145–151.

[38] M.R. Gruber-Brunhumer, J. Jerney, E. Zohar, M. Nussbaumer, C. Hieger, G. Bochmann, M. Schagerl, J.P. Obbard, W. Fuchs, B. Drosg, Acutodesmus obliquus as a benchmark strain for evaluating methane production from microalgae: influence of different storage and pretreatment methods on biogas yield, Algal Res. 12 (2015) 230–238.

[39] S.K. Prajapati, P. Kumar, A. Malik, V.K. Vijay, Bioconversion of algae to methane and subsequent utilization of digestate for algae cultivation: a closed loop bioenergy generation process, Bioresour. Technol. 158 (2014) 174–180.

[40] G. Ciudad, O. Rubilar, L. Azócar, C. Toro, M. Cea, Á. Torres, A. Ribera, R. Navia, Performance of an enzymatic extract in Botrycoccus braunii cell wall disruption, J. Biosci. Bioeng. 117 (2014) 75–80.

[41] F. Passos, S. Astals, I. Ferrer, Anaerobic digestion of microalgal biomass after ultrasound pretreatment, Waste Manag. 34 (2014) 2098–2103.

[42] E.A. Ehimen, Z.F. Sun, C.G. Carrington, E.J. Birch, J.J. Eaton-Rye, Anaerobic digestion of microalgae residues resulting from the biodiesel production process, Appl. Energy 88 (2011) 3454–3463.

[43] C. Gonzalez-Fernández, B. Sialve, B. Molinuevo-Salces, Anaerobic digestion of microalgal biomass: challenges, opportunities and research needs, Bioresour. Technol. 198 (2015) 896–906.

[44] B. Sialve, N. Bernet, O. Bernard, Anaerobic digestion of microalgae as a necessary step to make microalgal biodiesel sustainable, Biotechnol. Adv. 27 (2009) 409–416.

[45] P. Biller, A.B. Ross, S.C. Skill, Investigation of the presence of an aliphatic biopolymer in cyanobacteria: implications for kerogen formation, Org. Geochem. 81 (2015) 64–69.

[46] J. Burczyk, J. Dworzanski, Comparison of sporopollenin-like algal resistant polymer from cell wall of Botryococcus, scenedesmus and Lycopodium clavatum by GC-pyrolysis, Phytochemistry 27 (1988) 2151–2153.

[47] F. Gelin, J.K. Volkman, C. Largeau, S. Derenne, J.S. Sinninghe Damst??, J.W. De Leeuw, Distribution of aliphatic, nonhydrolyzable biopolymers in marine microalgae, Org. Geochem. 30 (1999) 147–159.

[48] O. Córdova, J. Santis, G. Ruiz-Fillipi, M.E. Zuñiga, F.G. Fermoso, R. Chamy, Microalgae digestive pretreatment for increasing biogas production, Renew. Sustain. Energy Rev. 82 (2018).

[49] J. Jiang, Q. Zhao, L. Wei, K. Wang, D.J. Lee, Degradation and characteristic changes of organic matter in sewage sludge using microbial fuel cell with ultrasound pretreatment, Bioresour. Technol. 102 (2011) 272–277.

[50] Z.A. Popper, G. Michel, C. Herve, D.S. Domozych, W.G. Willats, M.G. Tuohy, B. Kloareg, D.B. Stengel, Evolution and diversity of plant cell walls: from algae to flowering plants, Annu. Rev. Plant Biol. 62 (2011) 567–590.

[51] Z.A. Popper, M.G. Tuohy, Beyond the green: understanding the evolutionary puzzle of plant and algal cell walls, Plant Physiol. 153 (2010) 373–383.

[52] A.W. Atkinson, B.E.S. Gunning, P.C.L. John, Sporopollenin in the cell wall of Chlorella and other algae: ultrastructure, chemistry, and incorporation of 14C-acetate, studied in synchronous cultures, Planta 107 (1972) 1–32.

[53] Y. Nemcova, T. Kalina, Cell wall development, microfibril and pyrenoid structure in type strains of Chlorella vulgaris, C. kessleri, C. sorokiniana compared with C. luteoviridis (Trebouxiophyceae, Chlorophyta), Arch. Hydrobiol. Suppl. Algol. Stud. 100 (2000) 95–105.

[54] H. Takeda, Sugar composition of the cell wall and the taxonomy of Chlorella (Chlorophyceae)1, J. Phycol. 27 (1991) 224–232.

[55] H. Takeda, Cell wall sugars of some Scenedesmus species, Phytochemistry 42 (1996) 673–675.

[56] M. Yamamoto, I. Kurihara, S. Kawano, Late type of daughter cell wall synthesis in one of the Chlorellaceae, Parachlorella kessleri (Chlorophyta, Trebouxiophyceae), Planta 221 (2005) 766–775.

[57] T. Cavalier-Smith, Symbiogenesis: mechanisms, evolutionary consequenses, and systematic implications, Annu. Rev. Ecol. Evol. Syst. 44 (2013) 145–172.

[58] H. Takeda, Classification of Chlorella strains by cell wall sugar composition, Phytochemistry 27 (1988) 3823–3826.

[59] A.M. Abo-Shady, Y.A. Mohamed, T. Lasheen, Chemical composition of the cell wall in some green algae species, Biol. Plant. 35 (1993) 629–632.

[60] A.L. Singh, P.K. Singh, M.P. Singh, Biomethanization of coal to obtain clean coal energy: a review, Energy Explor. Exploit. 30 (2012) 837–852.

[61] G. Muyzer, A.J.M. Stams, The ecology and biotechnology of sulphate-reducing bacteria, Nat. Rev. Microbiol. 6 (2008) 441–454.

[62] I. Angelidaki, D.J. Batstone, Anaerobic digestion: process, in: T. Christensen (Ed.), Solid Waste Technol. Manag., Vol. 1 2, John Wiley & Sons, Ltd., Chichester, UK., 2010, pp. 583–600.

[63] Y. Vazana, Y. Barak, T. Unger, Y. Peleg, M. Shamshoum, T. Ben-Yehezkel, Y. Mazor, E. Shapiro, R. Lamed, E.A. Bayer, A synthetic biology approach for evaluating the functional contribution of designer cellulosome components to deconstruction of cellulosic substrates, Biotechnol. Biofuels 6 (2013) 182.

[64] G. Guerriero, J.-F. Hausman, J. Strauss, H. Ertan, K.S. Siddiqui, Lignocellulosic biomass: biosynthesis, degradation, and industrial utilization, Eng. Life Sci. 16 (2016) 1–16.

[65] O. Córdova, F. Passos, R. Chamy, Physical pretreatment methods for improving microalgae anaerobic biodegradability, Appl. Biochem. Biotechnol. (2017).

[66] J. Ma, Q.-B. Zhao, L.L.M. Laurens, E.E. Jarvis, N.J. Nagle, S. Chen, C.S. Frear, Mechanism, kinetics and microbiology of inhibition caused by long-chain fatty acids in anaerobic digestion of algal biomass, Biotechnol. Biofuels 8 (2015) 141.

[67] C.B. Walker, A.M. Redding-Johanson, E.E. Baidoo, L. Rajeev, Z. He, E.L. Hendrickson, M.P. Joachimiak, S. Stolyar, A.P. Arkin, J.A. Leigh, J. Zhou, J.D. Keasling, A. Mukhopadhyay, D.A. Stahl, Functional responses of methanogenic archaea to syntrophic growth, ISME J. 6 (2012) 2045–2055.

[68] H.M. Jang, J.H. Kim, J.H. Ha, J.M. Park, Bacterial and methanogenic archaeal communities during the single-stage anaerobic digestion of high-strength food wastewater, Bioresour. Technol. 165 (2014) 174–182.

[69] C. Lee, J. Kim, K. Hwang, V. O'Flaherty, S. Hwang, Quantitative analysis of methanogenic community dynamics in three anaerobic batch digesters treating different wastewaters, Water Res. 43 (2009) 157–165.

[70] S. Kato, K. Watanabe, Ecological and evolutionary interactions in syntrophic methanogenic consortia, Microb. Environ. 25 (2010) 145–151.

[71] D. Ozuolmez, H. Na, M.A. Lever, K.U. Kjeldsen, Methanogenic archaea and sulfate reducing bacteria co-cultured on acetate: teamwork or coexistence? Front Microbiol. 6 (2015) 1–12.

[72] A. Briones, L. Raskin, Diversity and dynamics of microbial communities in engineered environments and their implications for process stability, Curr. Opin. Biotechnol. 14 (2003) 270–276.

[73] L. Chistoserdova, Is metagenomics resolving identification of functions in microbial communities? Microb. Biotechnol. 7 (2014) 1–4.

[74] R. Heyer, F. Kohrs, U. Reichl, D. Benndorf, Metaproteomics of complex microbial communities in biogas plants, Microb. Biotechnol. 8 (2015) 749–763.

[75] S. Greses, J.C. Gaby, D. Aguado, J. Ferrer, A. Seco, S.J. Horn, Microbial community characterization during anaerobic digestion of *Scenedesmus* spp. under mesophilic and thermophilic conditions, Algal Res. 27 (2017) 121–130.

[76] V. Nolla-Ardevol, M. Strous, H.E. Tegetmeyer, Anaerobic digestion of the microalga Spirulina at extreme alkaline conditions: biogas production, metagenome and metatranscriptome, Front. Microbiol. 6 (2015) 1–21.

[77] J.L. Sanz, P. Rojas, A. Morato, L. Mendez, M. Ballesteros, C. González-Fernández, Microbial communities of biomethanization digesters fed with raw and heat pretreated microalgae biomasses, Chemosphere 168 (2017) 1013–1021.

[78] O. Córdova, R. Chamy, L. Guerrero, A. Sánchez-rodríguez, D.E. Marco, Assessing the effect of pretreatments on the structure and functionality of microbial communities for the bioconversion of microalgae to biogas, Front Microbiol. 9 (2018) 1–11.

[79] X. Goux, M. Calusinska, S. Lemaigre, M. Marynowska, M. Klocke, T. Udelhoven, E. Benizri, P. Delfosse, Microbial community dynamics in replicate anaerobic digesters exposed sequentially to increasing organic loading rate, acidosis, and process recovery, Biotechnol. Biofuels 8 (2015) 1–18.

[80] S. Aydin, Enhancement of microbial diversity and methane yield by bacterial bioaugmentation through the anaerobic digestion of *Haematococcus pluvialis*, Appl. Microbiol. Biotechnol. 100 (2016) 5631–5637.

[81] R. Wirth, G. Lakatos, T. Böjti, G. Maróti, Z. Bagi, M. Kis, A. Kovács, N. Ács, G. Rákhely, K.L. Kovács, Metagenome changes in the mesophilic biogas-producing community during fermentation of the green alga *Scenedesmus obliquus*, J. Biotechnol. 215 (2015) 52–61.

[82] S. Campanaro, L. Treu, P.G. Kougias, D. De Francisci, G. Valle, I. Angelidaki, Metagenomic analysis and functional characterization of the biogas microbiome using high throughput shotgun sequencing and a novel binning strategy, Biotechnol. Biofuels 9 (2016) 1–17.

[83] G. Borrel, P.W. O'Toole, H.M.B. Harris, P. Peyret, J.F. Brugère, S. Gribaldo, Phylogenomic data support a seventh order of methylotrophic methanogens and provide insights into the evolution of methanogenesis, Genome Biol. Evol. 5 (2013) 1769–1780.

[84] F. Bühligen, R. Lucas, M. Nikolausz, S. Kleinsteuber, A T-RFLP database for the rapid profiling of methanogenic communities in anaerobic digesters, Anaerobe 39 (2016) 114–116.

[85] Y. Stolze, M. Zakrzewski, I. Maus, F. Eikmeyer, S. Jaenicke, N. Rottmann, C. Siebner, A. Pühler, A. Schlüter, Comparative metagenomics of biogas-producing microbial communities from production-scale biogas plants operating under wet or dry fermentation conditions, Biotechnol. Biofuels 8 (2015) 14.

[86] S. Hattori, Syntrophic acetate-oxidizing microbes in methanogenic environments, Microb. Environ. 23 (2008) 118–127.

[87] R. Ramanan, B. Kim, D. Cho, H. Oh, H. Kim, Algae – bacteria interactions: evolution, ecology and emerging applications, Biotechnol. Adv 34 (2016) 14–29.

[88] juan luis Fuentes, I. Garbayo, M. Cuaresma, M. Zaida, Impact of Microalgae-Bacteria Interactions on the Production of Algal Biomass and, 2016.

[89] S.R. Subashchandrabose, B. Ramakrishnan, M. Megharaj, Consortia of cyanobacteria/microalgae and bacteria : biotechnological potential, Biotechnol. Adv. 29 (2011) 896–907.

[90] Z. He, J. Kan, F. Mansfeld, L.T. Angenent, K.H. Nealson, Self-sustained phototrophic microbial fuel cells based on the synergistic cooperation between photosynthetic microorganisms and heterotrophic bacteria, Environ. Sci. Technol. 43 (2009) 1648–1654.

[91] C.M. Waters, B.L. Bassler, Quorum sensing : communication in bacteria, Annu. Rev. Cell Dev. Biol. 21 (2005) 319–346.

[92] A. Mitsutani, I.Y. I, H. Kitaguchi, J. Kat, S. Ven, Y. Ishida, Analysis of algicidal proteins of a diatom-lytic marine bacterium *Pseudoalteromonas* sp . strain A25 by two-dimensional electrophoresis, Phycologia 40 (2001) 286–291.

[93] P. Williams, SGM Special Lecture Quorum Sensing, Communication and Cross-Kingdom Signalling in the Bacterial World, 2017, pp. 3923–3938.

[94] M. Demuez, C. González-Fernández, M. Ballesteros, Algicidal microorganisms and secreted algicides: new tools to induce microalgal cell disruption, Biotechnol. Adv. 33 (2015) 1615–1625.

[95] A. Shehata, W. Schrödi, J. Neuhaus, M. Krüger, Antagonistic effect of different bacteria on Clostridium botulinum types A, B, D and E in vitro, Vet. Rec. 172 (2013) 47.

[96] J. Jimenez, E. Latrille, J. Harmand, A. Robles, J. Ferrer, D. Gaida, C. Wolf, F. Mairet, O. Bernard, V. Alcaraz-Gonzalez, H. Mendez-Acosta, D. Zitomer, D. Totzke, H. Spanjers, F. Jacobi, A. Guwy, R. Dinsdale, G. Premier, S. Mazhegrane, G. Ruiz-Filippi, A. Seco, T. Ribeiro, A. Pauss, J.P. Steyer, Instrumentation and control of anaerobic digestion processes: a review and some research challenges, Rev. Environ. Sci. Biotechnol. 14 (2015) 615–648.

[97] P. Dabert, J. Delgen, Contribution of Molecular Microbiology to the Study in Water Pollution Removal of Microbial Community Dynamics Contribution of Molecular Microbiology to the Study in Water Pollution Removal of Microbial Community Dynamics, 2002.

[98] L. Raskin, D. Zheng, M.E. Griffin, P.G. Stroot, P. Misra, Characterization of Microbial Communities in Anaerobic Bioreactors Using Molecular Probes, 1995, pp. 297–308.

[99] H.P.J. Buermans, J.T. den Dunnen, Next generation sequencing technology: advances and applications, Biochim. Biophys. Acta (BBA) — Mol. Basis Dis. 1842 (2014) 1932–1941.

[100] I.M. Ahring, B.K. Rintala, J. Nozhevnikova, A.N. Mathrani, Metabolism of acetate in thermophilic (55°C) and extreme thermophilic (70°C) UASB granules, Proc. Int. Meet. Anaer. Proc. Bioenergy Env. (1995) 1–130.

[101] K. Kampmann, S. Ratering, I. Kramer, M. Schmidt, W. Zerr, S. Schnell, Unexpected stability of Bacteroidetes and Firmicutes communities in laboratory biogas reactors fed with different defined substrates, Appl. Environ. Microbiol. 78 (2012) 2106–2119.

[102] C.A. Bareither, G.L. Wolfe, K.D. McMahon, C.H. Benson, Microbial diversity and dynamics during methane production from municipal solid waste, Waste Manag. 33 (2013) 1982–1992.

[103] I. Vanwonterghem, P.D. Jensen, D.P. Ho, D.J. Batstone, G.W. Tyson, Linking microbial community structure, interactions and function in anaerobic digesters using new molecular techniques, Curr. Opin. Biotechnol. 27 (2014) 55–64.

[104] J.E. Edwards, R.J. Forster, T.M. Callaghan, V. Dollhofer, S.S. Dagar, Y. Cheng, J. Chang, S. Kittelmann, K. Fliegerova, A.K. Puniya, J.K. Henske, S.P. Gilmore, M.A. O'Malley, G.W. Griffith, H. Smidt, PCR and omics based techniques to study the diversity, ecology and biology of anaerobic fungi: insights, challenges and opportunities, Front. Microbiol. 8 (2017).

[105] I. Head, J. Saunder, R. Pickup, Microbial Evolution, Diversity, and Ecology: A Decade of Ribosomal RNA Analysis of Uncultivated Microorganisms, 1998, pp. 1–21.

[106] J.R. Cole, B. Chai, R.J. Farris, Q. Wang, S.A. Kulam, D.M. Mcgarrell, G.M. Garrity, J.M. Tiedje, The Ribosomal Database Project (RDP-II): sequences and tools for high-throughput rRNA analysis, Nucleic Acids Res. 33 (2005) 294–296.

[107] G.M. Guebitz, A. Bauer, G. Bochmann, A. Gronauer, S. Weiss, Biogas Science and Technology, 2015.

[108] B.E. Rittmann, G. Lov, S. Ok, B. Oerther, M. Hausner, F. Loffler, N.G. Love, G. Muyzer, S. Okabe, D.B. Oerther, J. Peccia, L. Raskin, M. Wagner, B.E. Rittm, A vista for microbial ecology and environmental biotechnology, Environ. Sci. Technol. 40 (2006) 1096–1103.

[109] A. Cabezas, J.C. de Araujo, C. Callejas, A. Galès, J. Hamelin, A. Marone, D.Z. Sousa, E. Trably, C. Etchebehere, How to use molecular biology tools for the study of the anaerobic digestion process? Rev. Environ. Sci. Biotechnol. 14 (2015) 555–593.

[110] R. Kumar, S. Singh, O.V. Singh, Bioconversion of lignocellulosic biomass: biochemical and molecular perspectives, J. Ind. Microbiol. Biotechnol. 35 (2008) 377–391.

[111] L. Sanz, T. Ko, Molecular biology techniques used in wastewater treatment : an overview, Process Biochem. 42 (2007) 119–133.

[112] M.E. Griffin, K.D. Mcmahon, R.I. Mackie, L. Raskin, Methanogenic Population Dynamics during Start-Up of Anaerobic Digesters Treating Municipal Solid Waste and Biosolids, 1998.

[113] C. Gawad, W. Koh, R. Quake, Single-cell genome sequencing : current state of the science, Nat. Publ. Gr. 17 (2016) 175–188.

[114] S.K. Prajapati, A. Bhattacharya, A. Malik, V.K. Vijay, Pretreatment of algal biomass using fungal crude enzymes, Algal Res. 8 (2015) 8–14.

[115] R.I. Amann, W. Ludwig, K. Schleifer, Phylogenetic identification and in situ detection of individual microbial

cells without cultivation, Microbiol Rev. 59 (1995) 143–169.

[116] W. Ludwig, O. Strunk, R. Westram, L. Richter, H. Meier, A. Buchner, T. Lai, S. Steppi, G. Jobb, W. Fo, I. Brettske, S. Gerber, A.W. Ginhart, O. Gross, S. Grumann, T. Liss, R. Lu, S. Hermann, R. Jost, A. Ko, È. Nonhoff, B. Reichel, R. Strehlow, A. Stamatakis, M. May, N. Stuckmann, A. Vilbig, M. Lenke, T. Ludwig, A. Bode, ARB: a software environment for sequence data, Nucleic Acids Res. 32 (2004) 1363–1371.

[117] S. Aydin, A. Shahi, E.G. Ozbayram, B. Ince, O. Ince, Use of PCR-DGGE based molecular methods to assessment of microbial diversity during anaerobic treatment of antibiotic combinations, Bioresour. Technol. 192 (2015) 735–740.

[118] T. Urich, A. Lanzén, J. Qi, D.H. Huson, C. Schleper, S.C. Schuster, Simultaneous assessment of soil microbial community structure and function through analysis of the meta-transcriptome, PLoS One 3 (2008).

[119] M. Zakrzewski, A. Goesmann, S. Jaenicke, S. Jünemann, F. Eikmeyer, R. Szczepanowski, W.A. Al-Soud, S. Sørensen, A. Pühler, A. Schlüter, Profiling of the metabolically active community from a production-scale biogas plant by means of high-throughput metatranscriptome sequencing, J. Biotechnol. 158 (2012) 248–258.

[120] S.G. Tringe, C. Von Mering, A. Kobayashi, A.A. Salamov, K. Chen, H.W. Chang, M. Podar, J.M. Short, E.J. Mathur, J.C. Detter, P. Bork, P. Hugenholtz, E.M. Rubin, Communities comparative metagenomics of microbial communities, Science 554 (2005) 554–558.

[121] J.M. Heather, B. Chain, The sequence of sequencers: the history of sequencing DNA, Genomics 107 (2016) 1–8.

[122] M. Ronaghi, S. Karamohamed, B. Pettersson, M. Uhle, Real-time DNA sequencing using detection of pyrophosphate release, Anal. Biochem. 89 (1996) 84–89.

[123] N. Nguyen, T. Warnow, M. Pop, B. White, A perspective on 16S rRNA operational taxonomic unit clustering using sequence similarity, Pharm. J. 272 (2016) 5008–5016.

[124] T.D. Leser, J.Z. Amenuvor, T.K. Jensen, R.H. Lindecrona, M. Boye, K. Møller, Culture-independent analysis of gut Bacteria: the pig gastrointestinal tract microbiota revisited, Appl. Environ. Microbiol. 68 (2002) 673–690.

[125] S. Mcginnis, T.L. Madden, BLAST: at the core of a powerful and diverse set of sequence analysis tools, Nucleic Acids Res. 32 (2004) 20–25.

[126] A. Cabezas, J. Calabria, D.A. Cecilia, E. Trably, C. Etchebehere, How to Use Molecular Biology Tools for the Study of the Anaerobic Digestion Process?, 2015.

[127] A. Martinez-Murcia, S. Acinas, F. Rodriguez-Valera, Evaluation of prokaryotic diversity by restrictase digestion of 16s rDNA directly amplified from hypersaline environments, FEMS Microbiol. Energy 17 (1995) 247–255.

[128] W. Liu, T.L. Marsh, H. Cheng, L.J. Forney, Characterization of microbial diversity by determining terminal restriction fragment length polymorphisms of genes encoding 16S rRNA, Appl. Environ. Microbiol. 63 (1997) 4516–4522.

[129] N.J. Fredriksson, M. Hermansson, B.M. Wilén, Impact of T-RFLP data analysis choices on assessments of microbial community structure and dynamics, BMC Bioinf. 15 (2014) 1–18.

[130] G. Muyzer, DGGE/TGGE a method natural ecosystems for identifying genes from, Curr. Opin. Microbiol. (1999) 317–322.

[131] G. Collins, S. Connaughton, V. OFlaherty, Development of microbial community structure and actvity in a high-rate anaerobic bioreactor at 18°C, Water Res. (2006).

[132] E. Adil, Corrective measures of denaturing gradient gel electrophoresis limitations, J. Environ. Sci. Technol. 8 (2015) 1–12.

[133] N. Fromin, J. Hamelin, S. Tarnawski, D. Roesti, N. Forestier, F. Gillet, M. Aragno, P. Rossi, Minireview Statistical Analysis of Denaturing Gel Electrophoresis (DGE) Fingerprinting Patterns, 2007, pp. 634–643.

[134] K. Roest, H.H.G.J. Heilig, H. Smidt, W.M. de Vos, A.J.M. Stams, A.D.L. Akkermans, Community analysis of a full- scale anaerobic bioreactor treating paper mill wastewater, Syst. Appl. Microbiol. (2005) 175–185.

[135] J. Birtel, J.C. Walser, S. Pichon, H. Bürgmann, B. Matthews, Estimating bacterial diversity for ecological studies: methods, metrics, and assumptions, PLoS One 10 (2015) 1–23.

[136] J.K. Jansson, J.D. Neufeld, Omics for Understanding Microbial Functional Dynamics, 2013.

[137] W. Zhang, F. Li, L. Nie, Integrating multiple "omics" analysis for microbial biology: application and methodologies, Microbiology 156 (2010) 287–301.

[138] W.C. Nierman, J.A. Eisen, R.D. Fleischmann, C.M. Fraser, Genome data: what do we learn? Curr. Opin. Struct. Biol. (2000) 343–348, n.d.

[139] D.J. Beale, A. V Karpe, J.D. McLeod, S. V Gondalia, T.H. Muster, M.Z. Othman, E.A. Palombo, D. Joshi, An 'omics' approach towards the characterisation of laboratory scale anaerobic digesters treating municipal sewage sludge, Water Res. 88 (2016) 346–357.

[140] M.C. Nelson, M. Morrison, Z. Yu, A meta-analysis of the microbial diversity observed in anaerobic digesters, Bioresour. Technol. 102 (2011) 3730–3739.

[141] A. Cabezas, J. Calabria, D.A. Cecilia, E. Trably, C. Etchebehere, How to Use Molecular Biology Tools for the Study of the Anaerobic Digestion How to Use Molecular Biology Tools for the Study of the Anaerobic Digestion Process? Anaerobic Digestion, 2015.

[142] S.C.J. De Keersmaecker, I.M. V Thijs, J. Vanderleyden, K. Marchal, Integration of omics data: how well does it work for bacteria? Mol. Microbiol. 62 (2006) 1239–1250.

[143] M.A. Moran, Metatranscriptomics: eavesdropping on complex microbial communities, Microbe 4 (2009) 329–335.

[144] K.-O. Mutz, A. Heilkenbrinker, M. Lönne, J.-G. Walter, F. Stahl, Transcriptome analysis using next-generation sequencing, Curr. Opin. Biotechnol. 24 (2013) 22–30.

[145] G. Carrier, M. Garnier, L. Le Cunff, G. Bougaran, I. Probert, C. De Vargas, E. Corre, J.P. Cadoret, B. Saint-Jean, Comparative transcriptome of wild type and selected strains of the microalgae Tisochrysis lutea provides insights into the genetic basis, lipid metabolism and the life cycle, PLoS One 9 (2014).

CHAPTER 16

Microalgae-Based Biofuel Production Using Low-Cost Nanobiocatalysts

S. VIJAYALAKSHMI • M. ANAND • J. RANJITHA

INTRODUCTION

Energy plays a significant role either directly (or) indirectly in the socioeconomic development and human welfare of any nation [1]. The energy need has become an exploding issue, as there is a decline in nonrenewable energy resources that resulted in fuel crisis. Similarly, the depletion of fossil fuel resources and the growing global population have created an energy demand in the worldwide. As a result, the fuel prices have gone high in the present scenario, threatening the world's economic security [2]. Recent study has found that as the fossil fuel risk is at the zenith, the situation will become grave by 2020 [3]. Consequently, the concentration of greenhouse gas (GHG) in our planet is increasing extremely due to the development of industrialization and other human activities [4]. Therefore, there is rising worldwide attention toward sustainable biofuel production to reduce the GHG emission and diminish dependence on the fossil fuel usage. Biofuel classification is based on the technological construction: first-, second-, third-, and fourth-generation biofuels [5−7]. First-generation biofuels have obtained from food crops such as rapeseed, soybean, sugarcane, palm oil, and also animal fats using conventional technology [8]. These fuels are considerably restricted, due to the presence of freshwater. The significant usage of first-generation biofuel feedstock on farmland competes with food crops grown for global consumption; as a result of this, the price of food oil is increased. Due to this, food security problem arises, and this will naturally end up with the increase in the price of biofuel. Generally, second-generation biofuels are produced from nonedible plant and woody materials, which can reduce the food security issue, but the production is limited. It includes bioethanol production from cellulosic materials, hemicelluloses, sugar, starch, etc. Third-generation biofuels are produced from nonagricultural crops such as microalgae, which can grow in fresh- or saline water and on nonarable land [9,10]. The

fourth-generation biofuel production is on the transformation of vegetable oil into biodiesel and biogasoline using progressive technologies [11]. Recently, scientists are more focused toward microorganisms-based biofuel production. Microorganisms such as microalgae, filamentous fungi, yeast, and bacteria are investigated for producing biodiesel from their lipids [12]. Researchers are mainly focusing on microalgae-based biodiesel production; because of their photosynthesis ability, microalgae capture atmospheric carbon dioxide and convert that into lipids using sunlight, and also their growth rate is very fast compared with other conventional energy crops. The objective of the chapter is to explore the microalgae-based biofuel production using various low-cost nanomaterials.

Microalgae as a Suitable Source for Biofuel Production

Microalgae are unicellular (or) simple multicellular, photoautotrophic microorganisms that capture CO_2 in the presence of solar energy 50 times greater than the plants during photosynthesis [6,13,14]. The size of microalgae is mainly focused on the species and can vary from 1 to 100 μm [15,16]. A lot of attention has been gained by the oleaginous microalgae feedstock as a new source for biodiesel production to find out the alternative energy sources. Compared with oil crops, microalgae have important features, such as high biomass yield, higher photosynthetic efficiency, and faster growth, and synthesize and accumulate larger quantities of oil contents [5,17]. The microalgal biomass generally consists of three compounds: carbohydrates, proteins, and lipids/natural oil. The main composition of lipids present in the microalgal species exists in the form of triacylglycerides (TAGs), which are responsible for the production of biodiesel [18]. Microalgae grow very quickly; when compared with other crops, growth rate doubles every 36 h [19]. The percentage of oil content present in the different types of microalgal species typically ranges between 20% and 50% (dry weight). Spolaore et al. [20] reported that some microalgal strains can reach up to 80% lipid content. The algal cells consist of fatty acids adhered to the TAGs within both short- and long-chain hydrocarbons. In the production of biodiesel, the short-chain hydrocarbons are ideal, and other beneficial uses can be obtained from longer-chain hydrocarbons (Fig. 16.1). The main merits of microalgae-based biofuel production [13,20,21] are mentioned below:

- Microalgae can utilize carbon dioxide from the atmosphere by photosynthesis.
- Most of the microalgae species have high growth rate.
- Microalgae can be used for biofuel production along with wastewater treatment by decreasing the carbon, nitrogen, and phosphorus content in industrial, municipality, and agricultural wastewater.
- Various other value-added products can also be extracted from microalgae, which can be used in different industries.

- The residual algal biomass after extraction process can be utilized as fertilizer, biomethane and bioethanol production.
- They have the capability to flourish in harsh state; they can be grown in any season and can use wastewater as a culture medium.
- Deposition of huge amounts of lipids due to their higher photosynthesis efficiency.
- Swift growth rates result in microalgae biomass, doubling up within 24 h.
- Lack of competition with food crops on land.

Recently, scientists and engineers are continuously working on the improvement of the several components of the biofuel industry such as feedstock pretreatment, product quality and yields, process optimization, process capital costs, public acceptance, and market availability for various biofuels [22].

Importance of Nanomaterials in Biofuel Production

Recently, in biofuel industry, nanomaterials (NMs) are used to enhance the biofuel product quality and yield

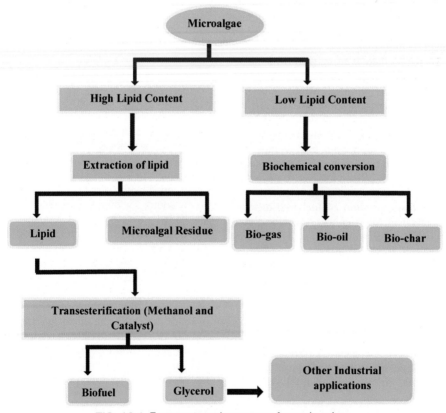

FIG. 16.1 Energy conversion process from microalgae.

efficiency [23,24]. The key principle of introducing nanotechnology in the biofuel industry is to validate both the scientific (green and catalytic chemistry) and engineering solutions in an eco-friendly energy sources [25]. The main merits of NMs in various biofuel systems such as high surface areas and unique characteristics such as high degree of crystallinity, catalytic activity, stability, adsorption capacity, durability, and efficient storage, which could enhance the efficiency of biofuel production. Similarly, the NMs have high potential for recovery, reusability, and recycling [26]. The successful usage of nanotechnology in biofuel technologies includes transesterification, anaerobic digestion, pyrolysis, gasification, and hydrogenation for the production of fatty esters, biogas, renewable hydrocarbons, etc. The combinations of these technologies with nanotechnology have been proven to be efficient, economical, and mature at laboratory and pilot scale and are expected to be promising way forward for replacing conventional systems, when developed at commercial scale [27]. The field nanotechnology has attracted a significant transaction of interest for the optimization of biodiesel production through NM-based catalysts. The use of NMs as a catalyst has led to the development of more efficient, economic, durable, and stable. In transesterification process, different types of nanocatalysts such as titanium dioxide, calcium oxide, magnesium oxide, and strontium oxide have been utilized to increase the catalytic performance in biodiesel production.

CATALYST

The catalyst is a mediator or compound that is supplementary to the process to speed up the chemical reaction. The catalyst is able to enhance the reaction rate of condensation and leads to the secondary reactions. The catalyst plays an important role in the transesterification process of biodiesel production. Generally, different types of catalysts are used in the biofuel production and include homogeneous catalysts (acid, base, enzyme) and heterogeneous catalysts. The disadvantages of using the chemical acid/base catalysts [28,29] in transesterification process include the following:

- Additional steps for neutralization and product purification are required.
- Energy requirement is high.
- The consumption of water during downstream processing was more, which leads to the generation in manifold of wastewater, which is disposed in the environment causing pollution.
- Homogeneous chemical catalyst cannot be reused.
- Formation of unwanted soap during the transesterification process affects the biodiesel yield.

Acid Catalysts

The acid catalysts such as sulfuric acid, sulfonic acid, hydrochloric acid, organic sulfonic acid, and ferric sulfate are most commonly used acids in the transesterification process of biodiesel production. The acid-catalyzed transesterification is carried out in the absence of water, to avoid the by-product formation of carboxylic acids, which reduces the yield of alkyl esters. Generally, the waste oil containing free fatty acids cannot be converted into biodiesel using an alkaline catalyst [30–32]. These free fatty acids can produce soap that will interact in the separation of fatty acid methyl ester (FAME). Hence, liquid acid–catalyzed transesterification is proposed to overcome the soap formation caused by using base catalysts. The reaction mechanism involved in the acid-catalyzed transesterification is the initial protonation of the acid to give an oxonium ion and further it initiates the alcohol to undergo exchange reaction to form the main product as ester. The main factor influencing the chemical reaction is alcohol to oil molar ratio. Freedman et al. [33] described an effective optimized reaction condition for the conversion of fatty acid methyl ester and achieved significant FAME conversion using 1 mol% of sulfuric acid with a molar ration of 30:1 at 65°C. The reaction rates in acid-catalyzed processes can be increased using large quantity of catalyst. Several studies reported that the catalyst concentrations in the reaction mixture can be varied from 1 to 5 wt% using sulfuric acid. The experimental studies revealed that the rate enhancement is observed with the increased quantity of catalyst, which enhanced the percentage of ester yield. The dependence of the rate of the reaction is mainly focused on the catalyst concentration. The main demerit of acid catalyst is usage at very high concentration in the reaction result in neutralization process and occur before product formation. The addition of calcium oxide neutralizes the high acid concentration in the transesterification process. Hence, the high acid concentration leads to increased calcium oxide cost and higher production cost. The liquid acid–catalyzed transesterification process does not suit for commercial industrial applications. The experimental studies revealed that the homogeneous acid-catalyzed reaction is about 4000 times slower than the homogeneous base-catalyzed reaction.

Alkali Catalysts

The common homogeneous base catalysts used in the transesterification process are sodium hydroxide, sodium methoxide, and potassium hydroxide. All the homogeneous base catalysis processes are done either with sodium hydroxide (NaOH) or potassium hydroxide

(KOH) in the presence of methanol (or) ethanol. The soluble metal oxide (CH3ONa) is the most active catalyst as it can yield high conversion rates (498%) within shorter period of time (30 min) even at minimum concentration (0.5 mol%). In alkali-catalyzed process, the effect of moisture has negative impact on soap formation that consumes the catalyst as well as decreases the efficiency, causing an increase in viscosity [34]. To achieve the maximum conversion efficiency, the desired molar ratio of 6:1 is considered to be favorable ratio. The most common solvent used in this reaction is ethanol and methanol. The reaction with alkali catalysts is found to be significant than the acid-catalyzed reaction. Generally, the mechanism of the base-catalyzed transesterification of biooils involves four steps. The first step is the reaction of the base with the alcohol, producing an alkoxide and the protonated catalyst. The second step is the nucleophilic attack of the alkoxide at the carbonyl group of the triglyceride, generating a tetrahedral intermediate. The third step involves the formation of the alkyl ester and the corresponding anion of diglyceride [35]. The final step involves deprotonating the catalyst, thus regenerating the active species, which is now able to react with a second molecule of the alcohol, starting another catalytic cycle. Diglycerides and monoglycerides are converted by the same mechanism to a mixture of alkyl esters and glycerol. The main demerits of alkali-catalyzed process include postreaction treatment to remove the catalyst from the product biodiesel, interferences occasioned by the presence of free fatty acid and water during the reaction, difficulty in the recovery of glycerol after the reaction, and finally the postreaction treatment of the alkaline wastewater to avoid the environmental effects of its disposal.

Homogeneous Catalysts

In the production of biodiesel, homogeneous catalysts such as Brønsted and Lewis acids, metals, organometallic groups, and organocatalysts are used to enhance the efficiency of the FAME. In the homogeneous reaction, the reaction mixture and catalyst both are present in same phase, and the catalyst and reactants are used to determine the high homogeneity, which results in the high reactivity and selectivity of the reaction under controlled reaction conditions [36]. The main contribution of homogeneous catalytic process is in chemical industry, and the usage of heterogeneous catalytic process is significantly lesser, which is only about 17%–20%. The importance of homogeneous catalysis is increasing significantly in the field of pharmaceutical and polymer industry. However, homogeneous processes are also associated with some of the major demerits, which result in the limited use of these processes [37,38]. The homogeneous catalysts are stable only in relatively mild conditions, which limit their applicability. Since the catalysts are molecularly dispersed in the phase as the reactants, products, and solvents, the separation at the end of the process is difficult and expensive. In most of the cases, it is not possible to recover the catalyst.

Heterogeneous Catalysts

In the heterogeneous catalysis reaction, the catalysts can be separated from final product by filtration and can be reused. The most commonly used heterogeneous catalyst are La_2O_3, SrO, BaO KNO_3, KF, CaO, $CaCO_3$, etc. To enhance the catalytic performance, catalysts such as alumina, silica, zirconia, and zinc oxide are used to improve the mass transfer limitation of the three-phase reaction. The mass transfer limitations in heterogeneous catalyst are due to the two-phase zone interaction (solid-Liquid). It requires well mixing efficiency to reduce the external limitations. The usage of heterogeneous catalysts is to be operated at high temperature and high alcohol to oil mole ratio [39]. The well-known examples for the utilization of heterogeneous catalysts in the chemical reactions are Haber-Bosch and Fischer-Tropsch processes for the synthesis of hydrocarbon derivatives. Recently, in biodiesel production, acidic heterogeneous catalysts are used for their insensitivity for water content and free fatty acid content. Heterogeneous catalyst can be used in both batch and continuous reactor systems. The main merits of the heterogeneous catalysts are noncorrosive and high selectivity toward the reactants. These catalysts can be used in the form of fine particles, powders, and granules. These catalysts might be stored on the strong support material (bolstered catalysts) or utilized as sole material (unsupported catalysts). As heterogeneous catalysis is a surface phenomenon, performance of catalysts is mainly based on the uncovered surface zone. The uncovered surface region is increased by weakening the molecule portion; however, the slighter elements in general result in the deactivation of the catalyst [40]. The binding capacity toward the active site on solid support retains the group of the reactant particles, thus enhancing the reactant performance. For industrial applications, solid supports are considered to have high chemical, mechanical, and thermal stability.

Biocatalysts

Enzymes are protein molecules in cells that act as a catalyst in the chemical reaction. The enzymes are produced from tissues, plants, and organisms (yeast, microbes, or fungi). It has salient features such as high selectivity, high proficiency, and ecofriendly. The major applications of enzymes are biofuel production (e.g., lipase for the generation of biodiesel from vegetable oil), dairy industry (e.g., protease, lipase for lactose evacuation, renin for cheddar planning), cooking industry (e.g., amylase for bread delicateness and volume, glucose oxidase for mixture reinforcing), cleanser production (e.g., proteinase, lipase, amylase used to expel stains of proteins, fats, starch, individually), leather industry (e.g., protease for unhairing and bating), paper industry, material industry (e.g., amylase for expelling starch from woven textures), etc. In this chapter, we are mainly focusing on biofuel production from microalgal species using lipase-immobilized nanoparticles (NPs) as a biocatalyst. The enzymes can be directly used as a biocatalyst in the transesterification reaction, but immobilized lipase enhanced the biodiesel efficiency and reusability of biocatalysts. Lipase enzyme has several advantages that include high enzymatic activity toward the reaction under moderate conditions, and it has the capability to perform the catalytic activity on oils with high free fatty acid and water and reduces saponification and emulsions [41]. The lipase enzyme is classified into three types depending on the enzyme specificity to region or position, specificity toward fatty acid types, and specificity for a particular type of acylglycerols (mono-, di-, or triglycerides). Enzyme catalyst is too costly compared with other catalysts, and to overcome this disadvantage, the enzymes are used by immobilizing it on supporting materials so that it can be reused and its stability also improved [42]. The immobilization of enzyme is carried out using different methods such as cross-linked enzyme totality and microwave-assisted immobilization. The concentration of the enzyme immobilized by the NP was significantly high, when compared with that provided by investigational procedures based on the immobilization on planar 2-D surfaces because of their higher surface to quantity ratio. When compared with native enzymes, the enzyme immobilized on NP has good thermal stability at a controlled pH and temperature range. The following are the advantages of immobilization lipase [43] on NPs:

- It is easy to incorporate in high solid contented.
- Surface-active agents and harmful reagents are not needed.
- Thick enzyme shell can be obtained with a homogeneous and clear-cut core-shell NPs.
- Particle can be resized according to the need.

- Coimmobilization of multienzymes may attain when NPs are used as supporting materials.

Lipase and Its Immobilization

Lipases (triacylglycerol acyl hydrolases) are defined as a class of enzymes able to hydrolyze insoluble triacylglycerols. Generally, lipases had shown lots of interest in different biocatalytic processes as a result of their capability to catalyze a wide range of chemical reactions. If the water concentration is low in the chemical reaction system like esterification, alcoholysis and acidolysis, etc., can result in reversed reaction condition. The lipases play a key role in commercial utilities, such as textile, detergent, food processing, pulp and paper, fats hydrolysis and modification, oils and oleochemical industry, pharmaceutical processing, resolution of racemic mixtures, chemical analyses as well as biochemistry research. Lipase is completely soluble in aqueous solution and very difficult to separate the lipase from the reaction system. The usage of free lipase in the chemical reaction suffers from the difficulty of reusability, which hinders the easier and more expansive commercial utilities of lipase. To overcome this problem of reuse and stability, recently, immobilization technique has been explored. In immobilization technique, different classes of enzymes have been immobilized on different supports, such as synthetic organic polymers, biopolymers, hydrogels, and inorganic supports. As a result of this, it has been found that the immobilized enzymes exhibit higher activity over free ones. The main merits of using lipase in immobilized forms are more convenient handling, facile separation, more efficient recovery and reuse, capability of continuous fixed-bed operation, and often enhanced stability. Recently, NMs have been used as a novel and promising support in lipase immobilization due to their larger specific surface area and less diffusion limitation. In lipase immobilization, NPs, nanotubes, and nanofibrous membranes are the most commonly used NMs. Immobilized enzymes are the enzymes physically restricted or localized in specific limited area of the space, which has catalytic activity, and they were used repeatedly in continuous process [44]. In current years, researchers are using immobilization technique to develop the strength of lipase for the biodiesel production. Lipases of microbial origin, mainly bacterial and fungal, are most widely used in biotechnological applications and organic chemistry. Several methods are used for lipase immobilization technique; they are covalent bonding, adsorption, entrapment, cross-linking, and encapsulation.

Cross-linking immobilization

It is an irreversible method performed by the formation of intermolecular cross-linkages between the enzyme molecules by covalent bonds. The cross-linking immobilization technique is carried out in the presence of multifunctional reagent, which acts as likers to connect the enzyme molecules into three-dimensional cross-linked aggregates. In cross-linking immobilization, usually two approaches are used, one is cross-linking enzyme aggregate and the other one is cross-linking enzyme crystal [45]. Both the methods require the cross-linking reagent such as glutaraldehyde to cross-link enzyme molecules through the reactions with the free amino groups of lysine residues on the reactive site of neighboring enzyme molecules. In the second cross-link enzyme crystal (CLEC)based approach, after crystallization, glutaraldehyde is added to CLECs. The cross-linking enzyme aggregates is very easy method and it includes precipitation of enzyme on aqueous solutions by adding organic solvents, salts, or nonionic polymers. The enzyme lipases from algae are precipitated in enzyme cross-linking with glutaraldehyde in the presence of sodium dodecyl sulfate surfactant with ammonium sulfate. The cross-linking enzyme aggregates of hydrolytic activity and lipases are enhanced threefold and twofold, respectively, in the excess of free enzymes.

Adsorption

In enzyme immobilization technique, adsorption is one of the most significant methods [46]. This method mainly depends on the binding process of Van der Waals force and hydrophobic, hydrogen bonding, or dipole-dipole mechanism. Initially, enzyme is dissolved in the solution and reacted with the solid support for a fixed period of time under suitable conditions. The unreacted enzyme molecules are removed from the surface by washing with buffer. The performance of physical binding is carried out under ambient condition and resulted in high loading enzyme. The main demerit of adsorption immobilization process is very poor stability, due to the weak binding of enzyme and support may cause loss of enzyme molecules at the time of operation and washing.

Covalent immobilization

In enzyme immobilization technique, covalent binding is one of the most widely used methods. Covalent immobilization provides strong binding between enzymes and support matrix to form a stable complex [47]. The enzyme functional groups used in covalent coupling are amino group, carboxylic group, phenolic group, sulfhydryl group, thiol group, imidazole group, indole group, hydroxyl group, etc. Generally, the binding procedure of enzyme to the solid support involves two stages: (1) activation of the surface using linker molecules such as glutaraldehyde or carbodiimide and (2) enzyme covalent coupling to the activated support. Consequently, linker molecules act as a multifunctional reagent between the surface and enzyme via covalent bonding. The attachment of enzyme with NP through covalent bonding of enzymes and supports is known as covalent immobilization of biocatalysts. When the enzyme and support matrix are strong, the covalent bond protects the enzyme leakage from the surface matrix and improves the thermal stability. In this method, the deactivation of enzyme occurs due to conformational restriction by covalent binding of the enzyme. This technique can be used for lipase immobilization on different types of NPs, and it has various applications in the field of biofuel production, bioremediation, biosensors, and bioconversion of hydrocarbon derivatives.

Entrapment immobilization

The enzyme entraps the fibers or porous gel, known as entrapment immobilization technique. In this method, enzyme is not directly attached to the support surface but entrapped within a polymeric network. It allows only the traverse of substrate and products and retains the enzyme to avoid enzyme diffusion [48]. This process is conducted through two steps: (1) mixing enzyme into a monomer solution and (2) polymerization of monomer solution by a chemical reaction or changing experimental conditions. The enzyme is physically limited within a polymer lattice network, and the enzyme does not chemically interact with the entrapping polymer. The method can improve enzyme stability and minimize enzyme leaching and denaturation. The NPs used for entrapment immobilization usually depend on the sol-gel technique or reverse-micelle method (Fig. 16 2).

Biodiesel Production Using Lipase-Immobilized Nanocatalyst

The biodiesel production from microalgae appeared as a potential alternative to first- and second-generation biodiesel. Microalgae can grow rapidly compared with oil crops and can double their biomass within 24 h, and their oil content can exceed 80% by weight of dry biomass. Algae can frequently live in extreme environments (light, salinity, and temperature) and are easy to cultivate at industries for large-scale production [49]. Recently, genetic engineering plays an important role in increasing the lipid content in microalgae. The

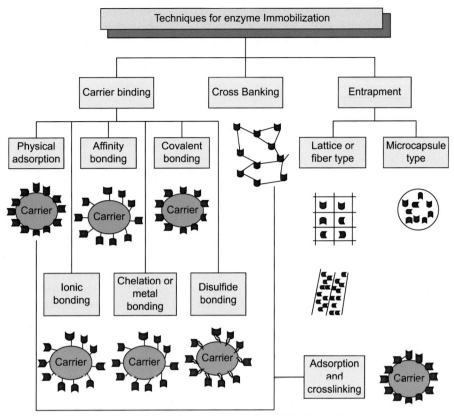

FIG. 16 2   Different types of enzyme immobilization techniques.

average yield of biodiesel produced from microalgae is 10–20 times higher, and the land area required is 49–132 times less than the other oleaginous seeds [50]. It was reported that microalgae have the highest oil yield among various plant oils. It was found to produce up to 100,000 oil per ha year, whereas palm, coconut, castor, and sunflower produce up to 5950, 2689, 1413, and 952l per ha year, respectively. Xu et al. [14] obtained high-quality biodiesel from microalgae *Chlorella protothecoides* grown in heterotrophic medium through acidic transesterification process, and the heterotrophic *C. protothecoides* contained the crude lipid content of 55.2%. In this work, an integrated method for the production of biodiesel from heterotrophic *C. protothecoides* was studied, and the results proved that the new process was a low-cost, feasible, and effective method for the production of high-quality biodiesel from microalgal species. Effects of concentration of CO_2 aeration on the biomass production and lipid accumulation of *Nannochloropsis oculata* in a semicontinuous culture were investigated, and it was found that the optimized condition for long-term

biomass and lipid yield from *N. oculata* NCTU-3 was to grow in the semicontinuous system aerated with 2% CO_2 and operated by 1-day replacement [4].

The microalgal species *C. vulgaris* was cultivated semicontinuously in artificial wastewater and achieved highest lipid content (42%) and lipid productivity (147 mg/L/d), and the cost analyses showed that the algal biomass can be competitive with petroleum at US$ 63.97 per barrel with the potential credit for wastewater treatment [51]. The effects of temperature and nitrogen concentration on the lipid content of *N. oculata* and *C. vulgaris* were studied, and it was found that an increase in temperature from 20 to 25°C practically doubled the lipid content of *N. oculata* (from 7.90% to 14.92%), and a 75% decrease of the nitrogen concentration in the medium, with respect to the optimal values for growth, increased the lipid fractions of *N. oculata* from 7.90% to 15.31% and of *C. vulgaris* from 5.90% to 16.41%, respectively [52]. Effects of iron on growth and lipid accumulation in marine microalgae *C. vulgaris* were investigated, and it was reported that both the relative fluorescence intensity of neutral lipids and the extractable lipid content in cultures supplemented with $1.2 \times 10^{-5} mol^{-1} Fe^{3+}$ were approximately thrice to sevenfold that in other cultures [53]. Hsieh and Wu [54] cultivated *Chlorella* sp. in batch, fed-batch, and semi continuous culture modes to assess biomass and lipid productivity with the respect to urea concentration and found that the maximum lipid productivity of 0.139 g d L in the semicontinuous culture was highest in comparison with those in the batch and fed-batch cultivations. To choose microalgae with a high biomass and lipid productivity, *Botryococcus braunii, C. vulgaris, and Scenedesmus* sp. were cultivated with ambient air containing 10% CO_2 and flue gas, and the results suggested that *Scenedesmus* sp. is appropriate for mitigating CO_2, due to its high biomass productivity and C-fixation ability, whereas *B. braunii* is appropriate for producing biodiesel, due to its high lipid content and oleic acid proportion [55]. Tamilarasan and Sahadevan [56] employed two steps for biodiesel production from *Caulerpa peltata* under ultrasonic irradiation. The microalgal species are photosynthetic organisms that use solar energy, consume more carbon dioxide than the conventional food crops, and produce huge amount of biomass per area, and they grow in large amount in freshwater, slightly brackish water, and the marine water around the world [57]. The oil from the algal species such as *Fucus spiralis* and *Pelvetia canaliculata* was transesterified into biodiesel, and its properties were analyzed that showed that it is suitable for manufacturing fuel pellets [58]. Tamilarasan et al. [59] developed a two-step process of biodiesel production from microalgal species. They also executed kinetic study to acquire reaction rate constant for the transesterification reaction and found that transesterification follows first-order reaction kinetics with the maximum biodiesel yield of 90.6% when base catalyst was used. Compared with microalgae, the microalgal biomass has lower lipid content (0.3% −6%), and therefore, macroalgae would not appear to be a suitable feedstock for the production of biodiesel via transesterification [60].

Recently, NMs play an important role in the enzyme immobilization technique (Table 16 1), due to their stability and reusability, and act as a biocatalyst in various chemical reaction. In the biodiesel production process, the lipase-immobilized NP was used to the increase the biodiesel efficiency. The lipid extracted from the *Scenedesmus* species was transesterified into biodiesel using lipase biocatalyst. The produced biodiesel yield using lipase-immobilized magnetic NP catalyst was compared with other catalysts such as NaOH and KOH, and it was found that nanobiocatalysts showed

a high yield of 95.1% [61]. Wang et al. [62] successfully developed a packed-bed reactor system for biodiesel production using lipase-immobilized magnetic NPs, and the biodiesel conversion was found to be 90%. It showed higher conversion and stability and proved to have a great potential in biodiesel production for large-scale nanobiocatalytic systems. Xie et al. [63] investigated that the lipase-immobilized magnetic Fe_3O_4 NPs can be used as a nanobiocatalyst in the production of biodiesel. The enzyme lipase was covalently bonded with the amino-functionalized NPs using glutaraldehyde. The activity healing of the enzyme binding was found to be 70%−84%, respectively. When 60% of enzyme has been used to catalyze the alcoholysis of biooil, it showed a highest conversion of fatty acids into FAME up to 90%. It was reported that immobilized lipase can be utilized as a biocatalyst for about four times without precision of the enzyme action. The methanol with biooil was transesterified into biodiesel, and the reaction was carried out in a batch operation process using enzyme catalyst (11 wt %) and the percentage yield was found to be 90%. After completing the transesterification process, the catalyst can be reused for about 10 cycles [64].

Tran et al. [65] studied the biodiesel production from the microalgal species *C. vulgaris* ESP-31 using immobilized *Burkholderia* lipase and produced high FAME conversion efficiency of 97.25%. Similarly, Lai et al. [66] firstly studied the enzymatic conversion of microalgal oils to biodiesel in ionic liquids. Initially, lipids of four microlagal strains of *B. braunii*(BB763 and BB764), *C. vulgaris*, and *Chlorella pyrenoidosa* were extracted and converted into biodiesel using two immobilized lipases extracted from *Penicillium expansum* lipase and *Candida antarctica* lipase B (Novozym 435), in two solvent systems: an ionic liquid (1-butyl-3-methylimidazolium hexafluorophosphate [BMIm][PF6]) and an organic solvent (tert-butanol). The lipase extracted from *P. expansum* lipase was found more efficient for this application, and the ionic liquid [BMIm][PF6] showed a greater conversion yield (90.7% and 86.2%). The green marine microalgae (*Tetraselmis* sp.) produce lipids, which were converted into biodiesel. In exponential phase, the microalgae were harvested using the solvent chloroform or methanol, and then, the lipid was extracted and confirmed by Nile-Red staining method [67]. Similarly, lipase produced by *Burkholderia* sp. C20 was immobilized in magnetic NP used as a biocatalyst in enzymatic transesterification for conversion of algal oil to biodiesel [68]. The extracted microalgal oil (method I) is converted into biodiesel via transesterification, and the microalgal biomass (method II) was disrupted directly via

TABLE 16.1
Different Nanoparticles Used for Microalgal Oil Transesterification.

Nanoparticles (NPs)	Algal Species	Synthesis	Size	Yield %
Gold (magnetic NP)	*Chlorella vulgaris*	Intracellularly	40–60	8.50 ± 1.74
Cadmium sulfide	*Scenedesmus*	Intracellularly	–	11.80 ± 1.52
Silver (magnetic NP)	*Parachlorella kessleri*	Extracellularly	9, 14, 18	5.00 ± 024
Palladium (immobilized NP)	*C. vulgaris*	Microalga culture	7	14–22
Gold (magnetic NP)	*Eolimna minima*	Intracellularly	5–100	11.80 ± 0.74
Silver (magnetic NP)	*Botryococcus braunii*	Extracellularly	15.67	24
Nickel (immobilized NP)	*Chlorella sorokiniana*	Microalga culture	–	5–25
Gold (magnetic NP)	*E. minima*	Intracellularly	5–100	4.00 ± 0.23
Silver (magnetic NP).	*Scenedesmus*	Intracellularly	15–20	12–14
Palladium (immobilized NP)	*Plectonema boryanum*	Extracellularly	<30	5–12
Palladium (immobilized NP)	*C. vulgaris*	Intracellularly	5–12	8.50 ± 1.74
Gold (magnetic NP)	*Tetraselmis suecica*	Extracellularly	79	3
Nickel (immobilized NP)	*Nostoc muscorum*	Intracellularly	5–24	7.40 ± 0.74
Gold (magnetic NP)	*Anabaena flousaquae*	Intracellularly	40	5.00 ± 0.61
Crystallized silver (magnetic NP)	*Spirulina platensis*	Extracellularly	12–16	5.00 ± 0.51
Gold (magnetic NP)	*Anabaena oryzae*	Extracellularly	<22	4.70 ± 0.12
Silver (magnetic NP)	*Nostoc humifusum*	Intracellularly	15–67	11.80 ± 1.52
Silver (magnetic NP)	*Wollea* sp.	Extracellularly	24	4.00 ± 0.22
Gold (magnetic NP)	*Phormidium* sp	Intracellularly	5–24	8.40 ± 0.65
Silica (immobilized NP)	*Chlamydomonas reinhardtii*	Intracellularly	13	4.8 ± 0.12
Silver (magnetic NP)r	*Gelidium latifolium*	Extracellularly	4–14	1.3 ± 0.0
Silver (magnetic NP)	*Oscillatoria willei*	Intracellularly	3	4.30 ± 0.32
Silver (magnetic NP)	*S. platensis*	Intracellularly	12–34	7.5 ± 0.30
Silver (magnetic NP)	*Jania rubens*	Extracellularly	11	0.25 ± 0.01
Gold (magnetic NP)	*Schizochytrium* sp.	Intracellularly	12–16	50–77

transesterification. The M-II (97.3 wt% oil) shows that the biodiesel conversion is higher than that in the M-I (72.1 wt% oil). When wet microalgae biomass is used as an oil substrate, the immobilized lipase biocatalyst metabolic activity works very well. The high molar ratio of methanol and oil is tolerated by immobilized lipase enzyme. The M-II method is repeated for six cycles or 288 h with no major defeat on the original activity of the lipase-immobilized nanobiocatalyst. The microalgae *Scenedesmus quadricauda* was cultivated in a photobioreactor supplying carbon dioxide and urea as a nitrogen source (0.075 g/L). The extracted oil was converted into biodiesel using enzymatic transesterification process, and the biodiesel yield is 96.4% [69]. An indigenous

microalga *C. vulgaris* was grown in a photobioreactor with CO_2 aeration that obtained a high oil content [70]. The microalgal oil was then converted to biodiesel by enzymatic transesterification using an immobilized lipase originating from *Burkholderia* sp. A research work was performed using water sampled from Lake Malaren in Sweden. Major focus of this research is to enhance indigenous algae production rather than inoculate new species into the system [71]. In a research work [53], *Burkholderia* lipase was immobilized onto hydrophobic magnetic particles for biodiesel production. A one-step extraction/transesterification process was developed to directly convert wet oil-bearing microalgal biomass of *C. vulgaris* into biodiesel using immobilized *Burkholderia* lipase as the catalyst [65]. The microalgal biomass (water content of 86%−91%; oil content of 14%−63%) was pretreated by sonication to disrupt the cell walls and then directly mixed with methanol and solvent to carry out the enzymatic transesterification. A research focuses on the production of biodiesel from wet, lipid-rich algal biomass using a two-step process involving hydrothermal carbonization and supercritical in situ transesterification [53].

Raita et al. [17] studied the biocatalytic activity of vegetable oil by immobilizing *Thermomyces lanuginosus* lipase on Fe_3O_4 magnetic NP using carrier modified by 3-aminopropyl triethoxysilane and covalently linked by 1-ethyl-3-(3-dimethylaminopropyl)carbodiimide and N-hydroxysuccinimide. This showed highest yield when transesterified palm oil. Using central composite design, reaction variables were optimized and identified 23.2% w/w enzyme of loading and 4.7:1 methanol to free fatty acids molar ratio with 3.4% water content in the presence of 1:1 (v/v) tert-butanol to oil as optimal conditions, leading to 97.2% FAME yield after incubation at 50°C for 24 h. The catalyst was separated by magnet, and it was reused for five cycles without any change in stability and activity. Enzymes immobilized on NPs will increase enzyme loading capacity and significant mass transfer efficiency when compared with enzyme immobilized on micrometric supports. Likewise, nanosized particles of magnetite Fe_3O_4 are called as magnetosomes, which have a wide range of applications, and this is biomineralized by a phylogenetically diverse group of prokaryotes, the magnetotactic bacteria [72]. The immobilized lipase was identified as an effective biocatalyst in the production of biodiesel from biooil, and percentage yield was found to be 90%. The Lipozyme TL IM immobilized on hydrotalcite gave a yield of 92.8% in the transesterification reaction for biodiesel production. Watanabe et al. [73] reported that three-step transesterification of biodiesel production from biooil to FAMEs uses immobilized lipase, and the lipase could be reused for 25 cycles and the percentage yield obtained was found to be 93.8%. Using immobilized *C. antarctica* lipase, conversion of biooil into corresponding methyl esters could be achieved up to 98%. All of these results represent that NMs can be used as a good carrier for lipase immobilization and that nanoimmobilized lipase has potential for commercial application in biodiesel production.

CONCLUSION

The nanobiocatalysts increase the efficiency of the biofuel production from the oleaginous microalgal species. The various types of algal species are found and gained several advantages toward biofuel production. The biofuel production from algae is not new, but it needs improvements in the process, product separation, and materials used in the biofuel production. Recently, various NMs have attracted much attention as carriers in the enzyme immobilization techniques. The advantages of NMs such as their large surface areas compared with other materials used in enzyme immobilization process. Many scientists have effectively used immobilized lipases on functionalized NMs in the production of biodiesel, and their applications appear promising in the industrial aspects. The enzymatic transesterification process enhances higher yield of biodiesel with less contamination, low cost, and simple operation. The immobilization of enzyme is an important parameter that influences the efficiency of lipase used to maintain stability. In particular, the application of nanoimmobilized lipase in bioreactors resulted in high enzyme loading, multiple reuses, and effective protection from enzyme denaturation in biodiesel production, showing the potential of nanoimmobilization technology in the biofuel industry. Further investigation, especially for the scale-up of the biodiesel production process, using nanoimmobilized lipase is necessary to implement these technologies on an industrial level. The nanoimmobilization technique will play a major role in cost-effective biodiesel production.

REFERENCES

[1] J. Singh, S. GU, Biomass conversion to energy in India-A Critique, Renew. Sustain. Energy Rev. 14 (2010) 1367−1378.

[2] A.I.A.R. Shanab, B. Jeon, H. Song, Y. Kim, J. Hwang, Algae-biofuel: potential use as sustainable alternative green energy, Online J. Power Energy Eng. 1 (1) (2010) 4−6.

[3] L.M.W. Leggett, D.A. Ball, The implication for climate change and peak fossil fuel of the continuation of the current trend in wind and solar energy production, Energy Policy 41 (2012) 610−617.

[4] S.Y. Chu, C.Y. Kao, M.T. Sai, S.C. Ong, C.H. Chen, C.S. Lin, Lipid accumulation and CO_2 utilization of *Nannochloropsis oculata* in response to CO_2 aeration, Bioresour. Technol. 100 (2009) 833−838.

[5] A. Demirba, M.F. Demirbas, Importance of algae oil as a source of biodiesel, Energy Convers. Manag. 52 (2011) 163−170.

[6] M.F. Demirbas, Bio refineries for biofuel upgrading: a critical review, Appl. Energy 86 (2009) 151−161.

[7] M.F. Demirbas, Biofuels from algae for sustainable development, Appl. Energy 88 (2011) 3473−3480.

[8] L. Brennan, P. Owende, Biofuels from microalgae − a review of technologies for production, processing, and extractions of biofuels and co-products, Renew. Sustain. Energy Rev. 14 (2010) 557−577.

[9] K. Heiman, Novel approaches to microalgal and cyanobacterial cultivation for bioenergy and biofuel production, Curr. Opin. Biotechnol. 38 (2016) 183—189.

[10] T. Hasunuma, F. Okazaki, N. Okai, K.Y. Hara, J. Ishii, A. Kondo, A review of enzymes and microbes for lignocellulosic biorefinery and the possibility of their application to consolidated bioprocessing technology, Bioresour. Technol. 135 (2013) 513—522.

[11] M.A. Mujeeb, A.B. Vedamurthy, C.T. Shivasharana, Current strategies and prospects of biodiesel production: a review, Adv. Appl. Sci. Res. 7 (1) (2016) 120—133.

[12] P. Vinuselvi, J.M. Park, J.M. Lee, K. Oh, C.-M. Ghim, S.K. Lee, Engineering microorganisms for biofuel production, J. Biofuels 2 (2) (2011) 153—166.

[13] S. Amin, Review on biofuel oil and gas production processes from microalgae, Energy Convers. Manag. 50 (2009) 1834—1840.

[14] H. Xu, X. Miao, Q. Wu, High quality biodiesel production from a microalga *Chlorella protothecoides* by heterotrophic growth in fermenters, J. Biotechnol. 126 (2006) 499—507.

[15] U. Aguoruc, P.O. Okibe, Content and composition of lipid produced by *Chlorella vulgaris* for biodiesel production, Adv. Life Sci. Technol. 36 (2015) 96—100.

[16] S.M. Hamed, G. Klock, Improvement of medium composition and utilization of mixotrophic cultivation for green and blue green microalgae towards biodiesel production, Adv. Microbiol. 4 (3) (2014) 1—8.

[17] A. Abbaszaadeh, B. Ghobadian, M.R. Omidkhah, G. Najaf, Current biodiesel production technologies: a comparative review, Energy Convers. Manag. 63 (2012) 138—148.

[18] Y.C.L. Dennis, X. Wu, M.K.H. Leung, A review on biodiesel production using catalysed transesterification, Appl. Energy 87 (4) (2010) 1083—1095.

[19] Y. Chisti, Biodiesel from microalgae, Biotechnol. Adv. 25 (2007) 294—306.

[20] P. Spolaore, C. Joannis-Cassan, E. Duran, A. Isambert, Commercial applications of microalgae, J. Biosci. Bioeng. 101 (2006) 87—96.

[21] T.M. Mata, A.C. Melo, M. Simões, N.S. Caetano, Parametric study of a brewery effluent treatment by microalgae Scenedesmus obliquus, Bioresour. Technol. 107 (2012) 151—158.

[22] R.aI. Abou-Shanab, M.K. Ji, H.C. Kim, K.J. Paeng, B.H. Jeon, Microalgal species growing on piggery wastewater as a valuable candidate for nutrient removal and biodiesel production, J. Environ. Manag. 115 (2013) 257—264.

[23] B.S. Sekhon, Nanotechnology in agri-food production: an overview, Nanotechnol. Sci. Appl. 7 (2014) 31.

[24] E. Serrano, G. Rus, J. Garcia-Martinez, Nanotechnology for sustainable energy, Renew. Sustain. Energy Rev. 13 (9) (2009) 2373—2384.

[25] H. Ramsurn, R.B. Gupta, Nanotechnology in solar and biofuels, ACS Sustain. Chem. Eng. 1 (7) (2013) 779—797.

[26] S.C. Trindade, Nanotech biofuels and fuel additives, in: M.A. Dos Santos Bernardes (Ed.), Biofuels Engineering Process Technology Intech Open, 2011, pp. 103—104.

[27] X. Zhang, S. Yan, R. Tyagi, R. Surampalli, Biodiesel production from heterotrophic microalgae through transesterification and nanotechnology application in the production, Renew. Sustain. Energy Rev. 26 (2013) 216—223.

[28] A. Guldhe, P. Singh, F.A. Ansari, B. Singh, F. Bux, Biodiesel synthesis from microalgal lipids using tungstated zirconia as a heterogeneous acid catalyst and its comparison with homogeneous acid and enzyme catalysts, Fuel 187 (2017) 180—188.

[29] A. Guldhe, C.V. Moura, P. Singh, I. Rawat, E.M. Moura, Y. Sharma, F. Bux, Conversion of microalgal lipids to biodiesel using chromium-aluminum mixed oxide as a heterogeneous solid acid catalyst, Renew. Energy 105 (2017) 175—182.

[30] D.A. Kamel, H.A. Farag, N.K. Amin, A.A. Zatout, R.M. Ali, Smart utilization of jatropha (*Jatropha curcas* Linnaeus) seeds for biodiesel production: optimization and mechanism, Ind. Crops Prod. 111 (2018) 407—413.

[31] M. Di Serio, R. Tesser, M. Dimiccoli, F. Cammarota, M. Nastasi, E. Santacesaria, Synthesis of biodiesel via homogeneous Lewis acid catalyst, J. Mol. Catal. A Chem. 239 (1—2) (2005) 111—115.

[32] J. Marchetti, A. Errazu, Esterification of free fatty acids using sulfuric acid as catalyst in the presence of triglycerides, Biomass Bioenergy 32 (9) (2008) 892—895.

[33] B. Freedman, E.H. Pryde, T.L. Mounts, Variables affecting the yields of fatty esters from transesterified vegetable oils, J. Am. Oil Chem. Soc. 61 (1984) 1638—1643.

[34] K. Noiroj, P. Intarapong, A. Luengnaruemitchai, S. Jai-In, A comparative study of KOH/Al_2O_3 and KOH/NaY catalysts for biodiesel production via transesterification from palm oil, Renew. Energy 34 (2009) 1145—1150.

[35] R.A. Candeia, M.C.D. Silva, J.R. Carvalho Filho, M.G.A. Brasilino, T.C. Bicudo, I.M.G. Santos, et al., Influence of soybean biodiesel content on basic properties of biodiesel—diesel blends, Fuel 88 (2009) 738—743.

[36] M.K. Lam, K.T. Lee, A.R. Mohamed, Homogeneous, heterogeneous and enzymatic catalysis for transesterification of high free fatty acid oil (waste cooking oil) to biodiesel: a review, Biotechnol. Adv. 28 (4) (2010) 500—518.

[37] K. Narasimharao, A. Lee, K. Wilson, Catalysts in production of biodiesel: a review, J. Biobased Mater. Bioenergy 1 (3) (2007) 1—12.

[38] I.M. Atadashi, M.K. Aroua, a.R. Abdul Aziz, N.M.N. Sulaiman, The effects of catalysts in biodiesel production: a review, J. Ind. Eng. Chem. 19 (1) (2013) 14—26.

[39] Z. Helwani, M.R. Othman, N. Aziz, W.J.N. Fernando, J. Kim, Technologies for production of biodiesel focusing on green catalytic techniques: a review, Fuel Process. Technol. 90 (12) (2009) 1502—1514.

[40] G. Arzamendi, I. Campoa, E. Arguinarena, M. Sanchez, M. Montes, L.M. Gandia, Synthesis of biodiesel with heterogeneous NaOH/alumina catalysts: comparison with

homogeneous NaOH, Chem. Eng. J. 134 (2007) 123–130.

[41] A. MacArio, G. Giordano, L. Setti, A. Parise, J.M. Campelo, J.M. Marinas, D. Luna, Study of lipase immobilization on zeolitic support and transesterification reaction in a solvent free-system, Journal Biocatalysis and Biotransformation 25 (2007) 328–335.

[42] V. Minovska, E. Winkelhausen, S. Kuzmanova, Lipase Immobilized by different techniques on various support materials applied in oil hydrolysis, J. Serb. Chem. Soc. 70 (2005) 609–624.

[43] S.A. Ansari, Q. Husain, Potential applications of enzymes immobilized on/in nano materials: a review, Biotechnol. Adv. 30 (3) (2012) 512–523.

[44] K.R. Jegannathan, S. Abang, D. Poncelet, E.S. Chan, P. Ravindra, Production of biodiesel using immobilized lipase – a critical review, Crit. Rev. Biotechnol. 28 (2008) 253–264.

[45] P. Gupta, K. Dutt, S. Misra, S. Raghuwanshi, R.K. Saxena, Characterization of cross-linked immobilized lipase from thermophilic mould Thermomyces lanuginosa using glutaraldehyde, Bioresour. Technol. 100 (2009) 4074–4076.

[46] A. Gog, M. Roman, M. Toşa, C. Paizs, F.D. Irimie, Biodiesel production using enzymatic transesterification – current state and perspectives, Renew. Energy 39 (2012) 10–16.

[47] J.K. Poppe, R. Fernandez-Lafuente, R.C. Rodrigues, M.A.Z. Ayub, Enzymatic reactors for biodiesel synthesis: present status and future prospects, Biotechnol. Adv. 33 (2015) 511–525.

[48] S.S. Mahmod, F. Yusof, M.S. Jami, S. Khanahmadi, H. Shah, Development of an immobilized biocatalyst with lipase and protease activities as a multipurpose cross-linked enzyme aggregate (multi-CLEA), Process Biochem. 50 (2015) 2144–2157.

[49] E. Ibanez, M. Herrero, J.A. Mendiola, M.C. Puyana, Extraction and characterization of bioactive compounds with Health Benefits from marine resources: Macro and micro algae, Cyanobacteria, and Invertebrates, in: Marine Bioactive Compounds: Sources, Characterization and Applications, 2012, pp. 55–97 (Chapter 2).

[50] J. Jena, M. Nayak, H.S. Panda, N. Pradhan, C. Sarika, P.K. Panda, B.V.S.K. Rao, B.N. Rachapudi, Prasad, L.B. Sukla, Microalgae of Odisha coast as a potential source for biodiesel production, World Environ. 2 (1) (2012) 12–17.

[51] Y. Feng, C. Li, D. Zhang, Lipid production of Chlorella vulgaris cultured in artificial wastewater medium, Bioresour. Technol. 102 (1) (2001) 101–105.

[52] A. Converti, A.A. Casazza, E.Y. Ortiz, P. Perego, M.D. Borghi, Effect of temperature and nitrogen concentration on the growth and lipid content of Nannochloropsis oculata and Chlorella vulgaris for biodiesel production, Chem. Eng. Process 48 (6) (2009) 1146–1151.

[53] C.H. Liu, C.C. Huang, Y.W. Wang, D.J. Lee, J.S. Chang, Biodiesel production by enzymatic transesterification catalyzed by Burkholderia lipase immobilized on

hydrophobic magnetic particles, Appl. Energy 100 (2012) 41–46.

[54] C.H. Hsieh, W.T. Wu, Cultivation of microalgae for oil production with a cultivation strategy of urea limitation, Bioresour. Technol. 100 (17) (2001) 3921–3926.

[55] C. Yoo, S.Y. Jun, J.Y. Lee, C.Y. Ahn, H.M. Oh, Selection of microalgae for lipid production under high levels carbon dioxide, Bioresour. Technol. 101 (1) (2001) 71–74.

[56] S. Tamilarasan, R. Sahadevan, Ultrasonic assisted acid base transesterification of algal oil from marine macroalgae Caulerpa peltata: optimization and characterization studies, Fuel 128 (2014) 347–355.

[57] S. Siddiqua, A.A. Mamun, S.M.E. Babar, Production of biodiesel from coastal macroalgae (Chara vulgaris) and optimization of process parameters using Box Behnken design, Springerplus 4 (720) (2015) 1–11.

[58] A. Macarioa, F. Verria, U. Diazb, A. Cormab, G. Giordanoa, Pure silica nanoparticles for liposome/lipase system encapsulation: application in biodiesel production, Catal. Today 204 (2013) 148–155.

[59] S. Tamilarasan, N.N. Gandhi, S. Renganathan, Production of algal biodiesel from marine macroalgae Enteromorpha compressa by two step process: optimization and kinetic study, Bioresour. Technol. 128 (2013) 392–400.

[60] J.J. Milledge, B. Smith, P.W. Dyer, P. Harvey, Macroalgae-derived biofuel: a review of methods of energy extraction from seaweed biomass, Energies 7 (11) (2014) 7194–7222.

[61] X. Wang, X. Liu, C. Zhao, Y. Ding, P. Xu, Biodiesel production in packed-bed reactors using lipase–nanoparticle biocomposite, Bioresour. Technol. 102 (2011) 6352–6355.

[62] R. Tripathi, J. Singh, R.K. Bharti, I.S. Thakur, Isolation, purification and characterization of lipase from Microbacterium sp and its application in biodiesel production, Energy Procedia 54 (2014) 518–529.

[63] W. Xie, N. Ma, Immobilised lipase on Fe$_3$O$_4$ nanoparticles as biocatalyst for biodiesel production, Energy Fuel. 23 (2009) 1347–1353.

[64] D.T. Tran, C.L. Chen, J.S. Chang, Immobilization of Burkholderia sp. lipase on a ferric silica nanocomposite for biodiesel production, J. Biotechnol. 158 (3) (2012) 112–119.

[65] D.T. Tran, C.L. Chen, J.S. Chang, Effect of solvents and oil content on direct transesterification of wet oil-bearing microalgal biomass of Chlorella vulgaris ESP-31 for biodiesel synthesis using immobilized lipase as the biocatalyst, Bioresour. Technol. 135 (2013) 213–221.

[66] J.-Q. Lai, Z.-L. Hu, P.-W. Wang, Z. Yang, Enzymatic production of microalgal biodiesel in ionic liquid [BMIm][PF6], Fuel 95 (2012) 329–333.

[67] C.L. Teo, H. Jamauddin, N.A.M. Zain, A. Idris, Biodiesel production via lipase catalysed transesterification of microalgae lipids from Tetraselmis sp, Renew. Energy 68 (2014) 1–5.

[68] K. Kawakami, Y. Oda, R. Takahashi, Application of a Burkholderia cepacia lipase immobilized silica monolith to

batch and continuous biodiesel production with a stoichiometric mixture of methanol and crude Jatropha oil, Biotechnol. Biofuels 4 (42) (2014) 1−11.

[69] Y.-C. Chen, Immobilized microalga Scenedesmus quadricauda (Chlorophyta, Chlorococcales) for long-term storage and for application for water quality control in fish culture, Aquaculture 195 (1) (2001) 71−80.

[70] D.-T. Tran, K.-L. Yeh, C.-L. Chen, J.-S. Chang, Enzymatic transesterification of microalgal oil from *Chlorella vulgaris* ESP-31 for biodiesel synthesis using immobilized Burkholderia lipase, Bioresour. Technol. 108 (2012) 119−127.

[71] M. Odlare, E. Nehrenheim, V. Ribe, E. Thorin, M. Gavare, M. Grube, Cultivation of algae with indigenous species — potentials for regional biofuel production, Appl. Energy 88 (2011) 3280−3285.

[72] M. Raita, J. Arnthong, V. Champreda, N. Laosiripojana, Modification of magnetic nanoparticle lipase designs for biodiesel production from palm oil, Fuel Process. Technol. 134 (2015) 189−197.

[73] Y. Watanabe, Y. Shimada, A. Sugihara, Y. Tominaga, Conversion of degummed soybean oil to biodiesel fuel with immobilized *Candida antarctica* lipase, J. Mol. Catal. B Enzym. (2002) 151−155.

CHAPTER 17

Biosynthesis of Nanomaterials Using Algae

ASHIQUR RAHMAN • SHISHIR KUMAR • TABISH NAWAZ

INTRODUCTION

According to ISO/TS 80004 series of standards [1], from the International Organization for Standardization, nanomaterial is defined as a material with any external dimension in the nanoscale or having internal of external structure in the nanoscale; nanoscale is further defined as the length range approximately in the order 1–100 nm. The nanomaterials comprise both nanoobjects (which are discrete materials, e.g., nanoparticles, also known as NPs) and nanostructured materials (which have internal or external structure in the nanoscale, e.g., structured superhydrophobic surfaces). Nanomaterials are both engineered and naturally produced. Engineered nanomaterials are further classified as NPs (all three external dimensions in nanoscale, with longest and shortest axis not differing significantly), nanofibers (with two dimensions in nanoscale including nanorods and nanotubes), nanoplates, nanoribbons, nanocomposites, and nanofoam. Naturally occurring nanomaterials are lotus leaves, gecko's foot, some butterfly wings, shark skin, etc.

Nanomaterials exhibit unique chemical, physical, electronic, optical, thermal, mechanical, and biological properties that significantly differ from their bulk counterparts, mainly due to their small sizes, various shapes, and high specific surface areas. For instance, copper NPs smaller than 50 nm behave as a superhard material that does not exhibit the similar ductility and malleability as bulk copper [2]. The unique properties observed at nanoscale range have opened a floodgate for these materials' potential applications in numerous areas, including medicines, electronics, water treatment, optoelectronics, and catalysis, among many others. NPs made up of metals, semiconductors, or oxides have been used as quantum dots [3], chemical catalysts [4], adsorbents [5], drug delivery [6], and biosensors [7]. NPs are of scientific interest as they effectively bridge the gulf between bulk materials' properties at one end

and atomic/molecular structural properties at another end. They exhibit size-dependent properties such as surface plasmon resonance in the case of gold and silver NPs (AuNPs and AgNPs, respectively) [8] and superparamagnetism for magnetic NPs [9], able to confine their electrons and produce quantum effects (also known as quantum confinement) in semiconducting NPs [10].

Due to their unique properties and diverse range of applications, the production of engineered NPs has continuously shown an increasing trend in the past two decades. It is estimated that the production of NPs doubles every 3 years [11]. The use of silver NPs for their antimicrobial roles has pushed their projected market value from $0.79 billion in 2014 to $2.54 billion by 2022 [12]. The increase in production and applications of NPs has fueled the possibility of new kinds of contaminants entering the environment. In the face of lack of knowledge regarding their long-term implications on environmental health, many studies have been conducted to investigate their ecotoxicological effects. However, one way of addressing the issue is to develop methods of NP production that are more environmentally friendly and sustainable. This is more important when there are available evidences [13] that suggest that some of the toxic effects inherent in NPs can potentially come from the components utilized in their synthesis. For instance, surfactants such as sodium dodecylbenzene sulfonate [14] and sodium dodecyl sulfate [15], which are used as stabilizing agents in NP synthesis to prevent their agglomeration, have been reported to be the cause of toxicity in the environment.

NP synthesis is carried out by physical and chemical methods. The traditional and the most common approach is wet chemistry–based method. In chemical methods, precursor cations are first reduced by a reducing agent such as sodium borohydride or lithium aluminum hydride in a liquid medium. Subsequently, the NPs' growth is controlled and stabilized by stabilizing agents. The examples of chemical methods include chemical reduction [16], electrochemical [17], and photochemical techniques [18]. Among physical methods [19], the most common ones are attrition and pyrolysis. In attrition, the particles are first grinded by a size-reducing technique, and subsequently, air-classified and finally oxidized NPs are obtained. In pyrolysis, the precursor is burned by passing them through an air orifice at high pressure, followed by an air classification step for final recovery of oxidized NPs. The chemical methods are cheaper for high volume production, but their major drawbacks include contamination from precursor chemicals, application of toxic chemicals, and generation of hazardous by-products. The physical methods suffer from their high cost of production, low production rate, and high energy expenditure. Therefore, there exists credible need for developing environmentally safe methods for synthesizing nanomaterials. This has led researchers to focus on biological methods. The advantages include lower cost, nontoxic nature of ingredients used, and eco-friendliness [19]. Several studies report utilizing plant extract [20], bacteria [21], fungi [22], enzymes [23], and algae [19] to synthesize NP. Most of the biological methods of NP synthesis fall under green chemistry category. In green chemistry methods, the focus is on mitigating the toxic effects of producing a given chemical entity and minimizing the environmental footprints of production process.

In this chapter, a green chemistry-based approach on the biosynthesis of nanomaterials using algae has been reviewed. Special focus is placed on AgNP synthesis and microwave-assisted synthesis methods. Various factors that affect nanomaterial synthesis using algae are discussed in detail. The role of pH, temperature, incubation time, ionic strength, light intensity, and algal biomass has been extensively discussed. Also, the biosynthesis methodologies and the potential applications of as-produced NPs have been studied.

GREEN SYNTHESIS OF NANOMATERIALS

According to International Union of Pure and Applied Chemistry (IUPAC) [24], green chemistry is defined as "design of chemical products and processes that reduce or eliminate the use or generation of substance hazardous to humans, animals, plants, and the environment." The approach includes (1) preventing by-products and waste formation, (2) minimizing the use of potentially toxic chemicals, (3) reducing energy needs by carrying out the experiments at room temperature and pressure, and (4) utilizing renewable and natural resources. Biological methods by utilizing natural materials inherently satisfy criteria (2) and (4) stated above. The remaining criteria can be satisfied by virtue of developing methodology that meets them. In this work, available scientific literature is utilized to evaluate the present state-of-the-art methods in meeting all the criteria. The key focus of this work is on the application of algae, as a promoter of nanomaterials, in achieving the goals of green chemistry.

Why Green Synthesis

The significance of nanomaterials is evident considering the recent synergistic advances in materials science and biotechnology. This synergy has applications in the areas of electronics [25,26], cosmetics [27], coatings [28], packaging [29], biosensors [30], and biotechnology [31,32], to name a few [32]. Though nanomaterials have shown increasing importance in our daily life, their negative effects on the environment are well established [33]. Therefore, benign synthesis strategies should be considered to reduce the use of toxic reagents and solvents and lessen the generation of toxic by-products [33]. Accordingly, green chemistry, which values the design of reaction pathways with benign reagents, innocuous solvents, and a few by-products, is preferred to conventional strategies [33,34].

The use of sugars as reducing agents for the synthesis of inorganic NPs has shown promising results as they are cost-effective and harmless to the environment. Besides, these biogenic sources preclude the use of conventional reagents, such as sodium borohydride [32] and N,N-dimethylformamide [23,35−37], which further consolidates their position as the best choice for sustainable NP synthesis. Following the principles of "green chemistry," materials scientists have employed reducing sugars, such as glucose and hemicelluloses, to obtain AgNPs [19,23,35,36,38−40]. The as-produced NPs have found a large range of bio-applications, such as antimicrobial agents [41,42], catalysts [42], and theranostic vehicles [43] using dextran, chitosan, or marine biopolymers as the reducing agent. The use of NPs in the previously mentioned applications is dependent on not only their size and shape but also their stability [44,45]. To this end, NPs must be stabilized by a capping agent, for which starch is a suitable candidate as it is easily available [46], nontoxic (generally regarded as safe [GRAS] certified by the US Food and Drug Administration [FDA]) [37,47], and inexpensive [48].

Microwave-Assisted Synthesis of Metal Nanoparticles

Synthesis mechanisms often involve the use of a heat source to reduce the reaction time, improve kinetics, and achieve higher yields and selectivity [49]. Heating mantles are commonly used to provide the increased temperature condition for the reactions. However, it has been observed that conductive heating often leads to excessive thermal gradients and losses. As an alternative, microwave-assisted synthesis of NPs reduces time and energy requirements. Microwaves work on the principle of dipole rotation, where an alternating electric field causes a polar solvent, such as water, to continuously realign along the changing electric field, thus increasing molecular vibration, kinetic energy, and therefore the thermal energy in the reaction mixture. Additionally, laboratory-based microwaves have been observed to offer better tuning of process parameters by virtue of easily tuned heating rates and real-time process monitoring [50,51]. The ability to control various process variables aids in accurately establishing the reaction conditions and thus enabling the development of a process platform technology for the synthesis of a broad range of NPs.

Metallic NPs in the form of colloidal nanospheres have been of continuous research interest, since their intrinsic properties can be finely tuned by changing parameters, such as diameter, chemical composition, bulk structure, surface chemistry, and crystallinity. Up to now, they have found a broad range of applications in fields, such as drug delivery, biodiagnostics, combinatorial synthesis, and photonic bandgap crystals [32].

AgNPs

AgNPs have been extensively studied for many years owing to their excellent electronic and chemical properties that arise from the quantum size effect and large surface energy. In particular, AgNPs are used in optical imaging for biomedical diagnostics [52]. The surface plasmon resonance and large effective scattering cross-section of individual AgNPs make them ideal candidates for molecular labeling, where phenomenon, such as surface-enhanced Raman scattering, can be exploited. Microwave-assisted synthesis of AgNPs using glucose and starch was carried out by Raveendran et al. [53]. The use of starch in the process is twofold, acting as both a stabilizing agent and a template to produce sphere-like shapes with small particle diameters. Templates provide a constrained environment during the NP growth and hence have a significant effect on the size and shape of the as-produced NPs [54]. The reaction of AgNP formation is given in Eq. (17.1) as:

$$2AgNO_3 + C_6H_{12}O_6 + starch \xrightarrow{\Delta} 2Ag - starch$$
$$+ \text{ gluconic acid} \quad (17.1)$$

Carbon-Based Nanomaterials

The use of green method can also be applied to the synthesis of organic nanomaterials. The second-most popular nanomaterials after NPs, especially metal NPs, are nanotubes [55]. Carbon nanotubes (CNTs) find applications in a variety of applications from catalysis to drug delivery [56].

The extensive use of CNTs makes it significant to incorporate green methods for their synthesis. In the past, green synthesis methodologies have been used via hydrothermal treatment of soy milk to make CNTs wherein the CNTs were used as metal-free electrocatalysts for oxygen reduction [56]. The use of coconut oil as a precursor in a chemical vapor deposition (CVD) technique was made for synthesizing multiwalled CNTs in the size range of 80–90 nm [57]. Recent reports show the use of plants such as rose, walnuts, and garden grass in implementation of a CVD technique for the synthesis of CNTs. Therefore, the technique helped to overcome the challenges of CNT synthesis, which include the use of metal precursors, which in turn required high energy and cost as well as the possibility of metal contamination of the CNTs [58].

Nevertheless, the green synthesis of CNTs is still wrought with challenges. These challenges chiefly include the inconsistency in morphology that may arise when using biological substances for CNT synthesis.

Perspective on Green Synthesis of Nanomaterials

The state of nanomaterial synthesis highlighted here focuses on the methods for making metallic as well as organic nanomaterials using green synthesis methods. The chief nanomaterials being explored currently in industry and academia were discussed with respect to green synthesis methods [55]. These green methods make use of bio-based reagents for nanomaterial synthesis. To that extent, optimized strategies for metallic NP synthesis have been reported [59]. Furthermore, not just material usages but also the environmental impacts for these methods were established to further cement the green synthesis methods for metal NP synthesis [60].

Nevertheless, the utilization of plant-based biomolecules poses the vital questions of competing use of plants, with their application as food source taking center stage. Additionally, these crops also make use of arable land, which may lead to paucity of space for agriculture, to support and feed the human population worldwide. Thus, it is imperative to scout for sustainable and competition neutral sources of "green" precursors to successfully attain green synthesis of nanomaterials.

Green Chemistry Approach to Nanoparticle Synthesis: Role of Algae

Algae are unicellular or multicellular organisms found in different media such as freshwater, moist surfaces, and marine water [19]. They are either microscopic (microalgae) or macroscopic (macroalgae). Algae are classified into four different domains/kingdoms— Bacteria, Plantae, Chromista, and Protozoa. Totally, they constitute 15 divisions or phyla. More than half of these phyla have been reported for their ability to help biosynthesis of NPs, which is a new and developing field. The first work demonstrating the application of algae (*Chlorella vulgaris*) to prepare AuNPs appeared in 2007 [61]. Typically, an algal species synthesizes NPs by accumulating the cations within its cellular matrix and subsequently reducing them. It is claimed that bioremediation processes are the precursor of biosynthesis of NPs using algal sources [19]. Bioremediation and biosynthesis are basically the two aspects of the same process; for example, biosorption, where cations are adsorbed on the outer surface of the biomass, is like biosynthesis of NPs extracellularly and akin to bioaccumulation nanomaterials that are synthesized intracellularly. Biosynthesis typically commences when the metallic cations exceed a certain threshold limit. Algae-mediated biosynthesis typically proceeds via exposing cell cultures, cell extracts, or algal biomass to noble metal salt solutions. In some cases, for instance, in seaweeds [62], the reduction is catalyzed by biomolecules present in the cell walls, leading to nucleation of NPs, and other biomolecules act as surfactants and stabilize, direct, and control NPs' growth.

Algae are a valuable source of bioactive molecules, and the presence of these molecules can potentially enhance the medicinal and pharmaceutical properties of NPs as reported in certain seaweeds [62]. Also, in this technique, the use of additional chemical species to act as stabilizing agent is precluded since certain biomolecules produced by cells provide colloidal stability (the ability of NPs to form stable and homogenous suspension). Although, algal biosynthesis techniques are green and ecofriendly and pose low toxicity risks, they suffer from few limitations. The process kinetically is typically slower taking up to days to few weeks [19] as compared with wet chemical reduction. Moreover, the

yield is also variable [19], and the process control is an issue to be addressed. Also, nanomaterials cannot be synthesized beginning with concentration level higher than the toxicity level of any given species. Colloidal stability is also often an issue that needs some attention, because in certain cases high level of agglomeration has been noted. Therefore, to build large scale photobioreactors, further research is needed to address the issues of kinetics, yield, and cell viability. Also, a significant amount of research is needed to identify and establish the role of specific biomolecules responsible for the NP synthesis process. A comparative study on physiochemical properties of NPs synthesized by conventional methods and using algae is also needed.

All these discussions make it crucial to write about algae, especially as the resource of choice. In brief, these autotrophic polyphyletic organisms, which play a huge role in the natural ecological cycles and complex food webs, may come to our rescue even to serve the purpose of sustainable synthesis of nanomaterials. Algae not only grow in water and other simple-to-prepare simulated water media but can also often survive in harsh conditions of wastewater and yield the production of nanomaterials. Since these organisms do not necessarily need freshwater and survive well in brackish and nutrient-rich conditions, they totally seem to have an advantage over terrestrial crops as algae do not intensify the competition for arable space. On the contrary, they can be grown in much greater volume per unit area due to the possibility of placing parallel photobioreactor in an area of land. Besides this, algae can also be used with twofold application of wastewater remediation and NP synthesis in many cases. Hence, the use of algae as a green synthesis resource will be discussed in greater detail in this chapter.

NANOPARTICLE SYNTHESIS USING ALGAE

Biosynthesis of metallic NP using algae is a simple process. It incorporates exposing algal cell culture materials to different metal salt solutions and letting the bioactive components reduce the cations to form NPs [19]. Several methods have been explored. Nevertheless, there exist several methodologies that are conventionally implemented.

Biosynthesis of Nanoparticles Using Living Cultures

This is the simplest among the methods shown in Fig. 17.1. This method uses living cultures of algae to synthesize NPs without any intermediate steps in between. Generally, this refers to cultivating the cells along

FIG. 17.1 Various methodologies for the biosynthesis of nanomaterials using algae. (Adapted from S.A. Dahoumane, M. Mechouet, F.J. Alvarez, S.N. Agathos, C. Jeffryes, Microalgae: an outstanding tool in nanotechnology, Bionatura. 1 (2016), with permission from Revista Bionatura.)

with aqueous salt solutions where cells are expected to continue their metabolism, photosynthesis, and growth [19]. However, the age of culture and the concentration of salt solutions are two deciding factors in this regard [64]. Since its advent, the living cultures of more than half-dozens of algal phyla have shown their potential to promote the production of various NPs [63]. For example, AuNPs, AgNPs, and palladium NPs (PdNPs) have been produced suing cyanobacterial strains of *Anabaena flos-aquae, Calothrix pulvinata, and Leptolyngbya foveolarum*. The synthesis process can take place in two different pathways. When the synthesis takes place inside the cells, it is known as "intracellular synthesis." The intracellular formation can be often visualized with naked eye and further evidenced by optical and electron microscopes (as shown in Fig. 17.2). For an example, *Ulva intestinalis*, also known as sea lettuce, promoted AuNPs that was evident as the entire plant turned purple [65]. On the other hand, extracellular materials, secreted by algae to the surrounding media, can also promote the synthesis process that is coined as "extracellular synthesis." In the case of intracellular synthesis process, the NPs are often found to get released into the culture media where they become stabilized

[63]. However, between these two pathways, the extracellular synthesis is more common to occur, in general.

One of the advantages of this method is that the produced nanomaterials possess some unique and important features. When produced intracellularly, NPs show very narrow size distribution, in most of the cases [66]. In addition, as the NPs make their way from the cells to surrounding medium, they interact with the polysaccharides-based organic matrix present on the cell surface [67,68]. This entire process makes the separation of the NPs easier in two ways: (1) the biopolymers, acting as the capping agent, make the NPs colloidally stable and hence prevent any further change in the NP size and/or shape, and (2) stable NPs within the cells make them heavier than the surrounding media and hence promote their settling [63]. Moreover, the remaining cells, upon addition of fresh media, can stay viable and continue the cycle of biosynthesis.

Biosynthesis of Nanoparticles Using Extracted Biomolecules

Biomolecules are generally extracted by disrupting the algal cells obtained from their living cultures. In fact, this is the first ever reported method of alga-mediated

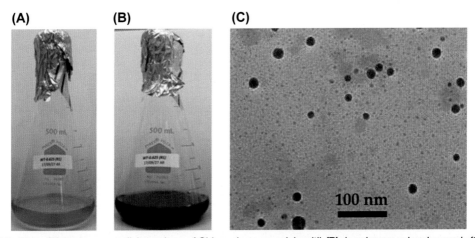

FIG. 17.2 **(A)** 21 days green living culture of *Chlamydomonas reinhardtii*; **(B)** deep brown color observed after 1 h from the addition of 0.625 mM AgNO$_3$ as the precursor; **(C)** the TEM micrograph of the produced AgNPs.

NP synthesis process. Biomolecules from *C. vulgaris* were extracted to synthesize AuNPs [19]. To elaborate this method, biomass of *C. vulgaris* was first lyophilized and then subjected to reverse-phase high-performance liquid chromatography (RP-HPLC). This process yielded the gold shape-directing protein (GSP) that was responsible for producing gold nanoplates. In a different study, proteins of both low (PLW) and high molecular weights (PHW) from the biomass of the same algae were obtained by using a dialysis membrane [19]. However, only PHW was successful in promoting AgNPs.

Although extracted biomolecules have the same NP synthesis potential as living cultures, there are some fundamental differences between these two methods. Unlike the previous method, the current method requires some energy input owing to the implementation of some harvesting techniques. In general, different techniques can be applied to extract and produce biomolecules of various forms. Some of the commonly used techniques include vortexing with glass beads and sonication [69]. Therefore, the biomolecules obtained by applying these techniques can be in the forms of fine powder [70], cell-free filtrate of disrupted cells [71], cell-free supernatant [69], etc. However, the aspect of intra- and extracellular synthesis does not apply in any of these forms [19].

Biosynthesis of Nanoparticles Using Cell-Free Supernatant

Cell-free supernatant generally refers to the method where algal cells are removed by centrifugation and the supernatant is used for NP biosynthesis. However, in some instances, the cells can be removed by filtration as well. For example, cell-free supernatants of various species of cyanobacteria and Chlorophyta can promote the synthesis of AgNPs [72].

Biosynthesis of Nanoparticles Using Whole Cells

This method implements centrifugation/filtration or simple washing of the cells to remove the growth media from the native culture. Later, the cells are simply resuspended in distilled water before they are exposed to precursor salt solutions. Therefore, this method is often considered the easiest among the methods that include harvesting [63]. Nonetheless, one major disadvantage of this method is that, being isolated from the culture media, the cells often lose their metabolic activity within a short time [63]. In some of the worst cases, the cells break out owing to the stressful environment by the distilled water. Algae that can successfully promote NP synthesis in this method include *Euglena gracilis* [70], *Euglena intermedia* [70], *Navicula minima* [73], etc.

FACTORS AFFECTING THE SYNTHESIS PROCESS

There are many factors that affect the synthesis process of NP synthesis using algae. Some of the major factors include temperature, pH of the reaction medium, incubation time, precursor ion concentration, illumination, and algae/biomass concentration. Fig. 17.3 portrays those factors.

The morphology and stability of NPs depend substantially on these factors. More importantly, the scalability of as-produced NPs strongly relies on the optimization of these parameters [19].

Temperature

In most cases, the NPs are synthesized at low temperature such as room temperature to which algae are cultured and grown. It gives a greater energy efficiency compared with other energy-intensive synthesis methods such as laser ablation and thermoreductive processes. Nevertheless, various temperatures, starting from room temperature to a temperature as high as 100°C, are often explored in the field of NP biosynthesis. In fact, algae can produce NPs of various size and shape at different temperatures. For example, aqueous suspension of harvested *Plectonema boryanum* whole cells synthesizes round-shaped AgNPs at 25°C and at 60°C while triangular platelets at 100°C [74].

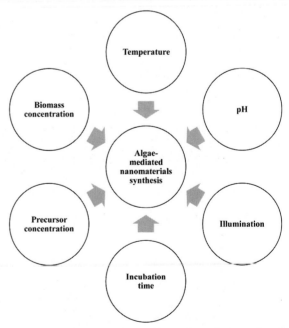

FIG. 17.3 Factors that affect the algae-mediated nanomaterials synthesis process.

Other examples of relatively high-temperature algal biosynthesis include AuNPs produced by seaweed *Sargassum muticum* at a temperature of 70°C and AgNPs produced by *Sargassum cinereum* at 100°C. In addition to their shape, the size of NPs largely depends on temperature: larger particle size is usually obtained at lower temperatures, while smaller particles tend to form at higher temperatures [75]. However, at high temperatures, biomolecules of algae may denature that leads to low or no NP formation [19]. Moreover, agglomeration of NPs is associated with high temperature. For example, 75-nm-size AgNPs produced at 65°C by *Cystophora moniliformis* tend to agglomerate with the increase in temperature and form clusters of particles with size ~2 μm at 95°C [76]. Hence, optimization of temperature is necessary. As an example, optimum synthesis of AuNPs by brown marine algae *Padina gymnospora* is found to take place at 75°C [77].

pH of the Reaction Medium

pH plays a very significant role in controlling the size and shape of algae-produced NPs [78]. Generally, algae-mediated NP synthesis process takes place at neutral pH. For example, AgNPs by *Spirulina platensis* and ZnO NPs by *Sargassum myriocystum* were produced at pH 7 and pH 8, respectively [79,80]. However, while many of the algae produce NPs at or close to neutral pH, there are instances of production at both acidic and basic pH. For example, biomass of brown algae *Fucus vesiculosus* produces AuNPs within optimum pH ranging from 4 to 9. However, maximum uptake of Au^{3+} occurs at pH 7 owing to the stable algal cells and optimal reactivity of their polysaccharides (i.e., the hydroxyl groups) at this pH [81]. Nonetheless, the fact that higher pH promotes NP formation whereas lower pH suppresses the process is well established. For instance, *Sargassum longifolium* produces anisotropic AgNPs at lower pH, while monodispersed and small-size particles at higher pH. In general, the reducing power of various functional groups reduces at lower pH owing to high H^+ concentration, and there have been instances when no particles were produced at low pH [19]. On the other hand, as the pH increases, the opposite happens that leads to less agglomeration and enhanced stability of the particles [20–22]. In many cases, this effect of pH toward NP morphology is primarily detected by a change in the absorption wavelength by the UV-Vis spectra. Generally, smaller particles absorb visible lights at a lower wavelength. For instance, absorption peaks for AuNPs produced by green algae *Rhizoclonium fontinale* plateau at 537, 529, and 517 nm for pH at 5, 7, and 9, respectively,

indicating smaller particle formation at higher pH [65,82]. However, the impact of pH on NP morphology can be species dependent. For instance, *S. longifolium* and *R. fontinale* yield maximum AuNPs and AgNPs, respectively, at respective pH of 8.4 and 9.0 [65,83], while optimum synthesis of AuNPs by brown marine algae *P. gymnospora* is found to take place at pH 10 [77]. Therefore, in the case of algae-mediated process, the tuning of pH is of high importance for producing good-quality NPs.

Incubation Time

In general, algae are known to take shorter time for synthesizing NPs than other biological entities [32,84]. Brown algae *Laminaria japonica* can intracellularly convert 90%–95% $HAuCl_4$ precursor (at 2 mM) to AuNPs within 10–20 min [85]. *S. cinereum* produces AgNPs with a high conversion in about 3 h only. However, longer reaction time for optimal yield is necessary in many cases. For instance, optimum synthesis of AuNPs by brown marine algae *P. gymnospora, Sargassum wightii,* and green freshwater algae *Prasiola crispa* takes place within 12 h. In addition, *S. longifolium* and *R. fontinale* take 64 and 72 h, respectively, for an optimal synthesis [65,83]. Apart from the time-dependent NP yield, the reaction time can also be an important factor in determining the shape of the particles. For example, AuNPs synthesized by *C. vulgaris* show predominance of nanoplates at longer reaction time [61]. *Klebsormidium flaccidum* promotes AuNPs that increase in size with the increase in incubation time [68]. Thus, proper optimization of time is crucial to optimize the algae-mediated NP synthesis processes.

Precursor Ion Concentration

It is well established that the precursor ion concentration is one of the most influencing parameters not only in terms of yield but also in terms of particle morphology. Often the impact of precursor concentration is so strong that the effect can be visualized by naked eyes. Significant changes in color are observed at different precursor ion concentrations that correspond NPs with different morphologies and yields [86]. Furthermore, the linear dependency of NP formation on the precursor concentration is well established [87] In fact, the effect of precursor concentration is respected even in the case of bimetallic NP synthesis. For example, *C. reinhardtii* produced bimetallic Ag-Au while maintaining the stoichiometric ratio of input silver and gold concentrations [88]. Overall, the amount of NP synthesis depends directly on the precursor concentration: the higher the concentration, the greater the yield. Nevertheless, in the case of metallic precursors, higher concentration often reduces the cell viability, owing to the cytotoxic effect from both cations and NPs, and therefore alters the yield and morphology of as-produced particles. For example, photosynthetic protist *E. gracilis* fails to process high concentrations (0.1 mM) of gold precursors [89]. However, the impact of toxicity depends on the species, as some species of algae can recover from toxicity at a concentration that is irrecoverably toxic to another species. For instance, in contrast to *E. gracilis* (cf. vide supra), *K. flaccidum* regained cell viability after few days when exposed to a same gold concentration of 0.1 mM. As additional examples, living cultures of *Nannochloropsis oculata and C. vulgaris* were only able to promote AgNP formation at 1 mM (lowest concentration) when exposed to $AgNO_3$ solutions at 1, 2, and 5 mM [90]. However, for algae being able to produce NPs, the precursor concentration needs to exceed a certain threshold. Hence, it is crucial to determine species-dependent precursor ion concentration to achieve recyclable and sustainable algae-based nanomaterial production platform. As an instance, optimum synthesis of AuNPs by *R. fontinale* is determined to take place at gold precursor concentration of 0.04 mM [82].

Illumination

Light works as a catalyst for algae-mediated biosynthesis reactions. Although there are instances when algae promote NP synthesis in the dark, the biosynthesis process, in general, is strongly light driven. It is worth noting that NP synthesis is often governed by photosynthetic pigments available in algae. For instance, fucoxanthin, a photosynthetic pigment of diatom *Amphora*-46, is held responsible for reducing Ag^+ into AgNPs [71]. Another example that corroborates the strong catalytic effect of light is when the extracellular polysaccharides of *Scenedesmus* sp., a green alga, fail to promote AgNPs in the absence of light [72]. However, as an exception, the cell-free extract of *C. reinhardtii*, a fresh water microalga, can promote AgNP formation; however, the reaction takes infeasibly long time ($\sim$ 13 days) to go into completion [69].

Algae/Biomass Concentration

In addition to precursor ion concentration, the biomass concentration plays a vital role in alga-mediated nanomaterials synthesis [91]. In fact, the size and shape of NPs are often substantially directed by the substrate concentration. For example, *C. vulgaris* promotes gold nanoplates

only at a relatively higher GSP concentration [61]. However, in contrast to ion concentration, the biomass concentration is always directly related to the NP yield.

Although these are the major factors affecting the biosynthesis of NP using algae, there are other factors (e.g., pressure, static, and stirring conditions) that are often taken into consideration.

APPLICATION OF ALGAE-MEDIATED NANOPARTICLES

Owing to their unique physical, chemical, and electrical properties, NPs find a myriad of application in different fields. Generally, chemical reactivity and the surface chemistry make them suitable for various applications [62]. However, the availability of hydrophilic surface groups, such as sulfate, carboxyl, and hydroxyl on algae-mediated NPs, gives them unique applicability [91]. In addition, as these algae themselves do not produce any toxin or harmful chemicals, these NPs can be effectively used in medical treatment. Owing to their sensitivity toward surface absorption of metals, these NPs can also be applied in imaging and sensing. Overall, algae-produced NPs are safe for direct utilization in different fields such as biomedicine, catalysis, and electronics [19]. Fig. 17.4 depicts these applications.

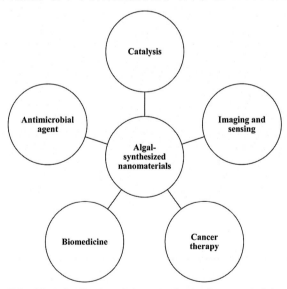

FIG. 17.4 Application of algae-synthesized nanomaterials.

AgNPs

AgNPs are well known for their antimicrobial effect. The AgNPs interact with the microbial cell membrane of the targeted harmful biological entities such as bacteria and fungi and cause substantial damage to the membrane. In addition, as the cellular uptake of the AgNP increases, the toxicological damage starts to get stronger [62]. These characteristics make them widely used for medical treatments including cancer therapy. For instance, AgNPs produced by *Padina tetrastromatica*, *Chlorella salina*, *Tetraselmis gracilis*, *Isochrysis galbana*, *Sargassum plagiophyllum*, *Chlorococcum humicola*, *Caulerpa racemose*, *Ulva fasciata*, *Microcoleus* sp., *S. cinereum*, and *diatom Amphora*-46 can effectively work against the growth of *Escherichia coli* and other bacterial pathogens. In addition to antibacterial activity, algae-based AgNPs are successful in inhibiting various fungal pathogens. In particular, AgNPs produced by *Gelidiella acerosa* and *S. longifolium* have strong antifungal activity against *Aspergillus fumigatus*, *Humicola insolens*, *Fusarium* sp., *Candida albicans*, *Trichoderma reesei*, etc. Moreover, AgNPs synthesized from *Enterococcus faecalis* can inhibit the biofilm produced by multidrug-resistant pathogens, while AgNPs synthesized from *Sargassum vulgare* have the ability to kill cancer cells such as human myeloblastic HL60 and HeLa [92]. Apart from their usefulness in medical treatment, algae-derived NPs have different other applications. For example, seaweed *Turbinaria conoides*–mediated AgNPs have unique properties that enable them to (1) form antibiological film against marine biofilm-forming bacteria and (2) effectively stabilize cotton fabrics that gives them the potential as materials for fabric strengthening and wound dressing [93–95].

AuNPs

Algae-mediated AuNPs have been significantly used in the fields of biosensor, pharmaceuticals, and electronics. Their unique size-dependent properties enable them to find versatile applications. Algae-mediated spherical AuNPs have very well-known antibacterial activity [62]. For instance, AuNPs synthesized from *Lemanea fluviatilis*, *S. platensis*, *Galaxaura elongata*, and *T. conoides* have antibacterial activity against various bacterial pathogens including *E. coli* [83,96–98]. Moreover, as-synthesized AuNPs are also used in DNA conjugation [99]. On the other hand, *K. flaccidum*–synthesized AuNPs can form "living" biohybrid material that has potential application in photosynthesis-based biosensors [68].

Metallic Alloy Nanoparticles

Besides metallic NPs, the metallic alloy NPs have many features that give them unique applicability in different fields. For instance, Ag-Au NPs produced by *Gracilaria* sp. can inhibit the growth of both gram-negative and gram-positive bacterial strains such as *Klebsiella pneumoniae* and *Staphylococcus aureus*, respectively [100].

Metal Oxide Nanoparticles

There has been an increasing interest in the application of algae-synthesized metal oxide NPs such as, ZnO, CuO, Cu_2O, and Fe_3O_4. In particular, the electronic, optical, and magnetic properties of these particles are phenomenal and make them of high industrial importance. Brown seaweed *S. muticum*–promoted Fe_3O_4 NPs have cytotoxicity and anticancer activity. They tend to enhance cell apoptosis and hence possess a great potential for the treatment of breast cancer, cervical cancer, liver cancer, and leukemia. In addition, the same seaweed-mediated ZnO NPs are found to have cytotoxicity over an elevated period, and hence they are potential in alternative chemotherapeutic treatment. Brown alga *Bifurcaria bifurcate*–produced CuO and Cu_2O NPs both exhibit antibacterial activity against both gram-negative and gram-positive bacterial strains such as *Enterobacter aerogenes* and *S. aureus*, respectively [101]. The red seaweed *Gracilaria gracilis*–promoted ZnO NPs have exceptional photocatalytic properties and can degrade aqueous solutions of phenol [102].

Other Nanoparticles

There are several algae-mediated NPs other than AgNPs, AuNPs, and their alloys. These include silver chloride (AgCl) NPs, cadmium sulfide (CdS) NPs, and PdNPs [62]. Although many of them are not well researched in terms of their specific applications, they do show properties similar to other metallic NPs. For example, *S. plagiophyllum*–synthesized AgCl NPs have bactericidal activity against *E. coli* [103]. *C. vulgaris*–mediated PdNPs have been used to catalyze the Mizoroki-Heck cross-coupling reaction to successfully produce substituted alkene.

To date, AgNPs and AuNPs are the most explored NPs synthesized by freshwater and marine algae. Table 17.1 summarizes some of these algae-produced NPs and their respective antimicrobial activity against a wide variety of pathogens.

TABLE 17.1

A Selection of Nanoparticles and their Antimicrobial Against Various Pathogens.

Nanoparticles	Algae	Antimicrobial Activities	References
Ag	*Padina tetrastromatica*	*Pseudomonas* sp., *Escherichia coli*, *Bacillus subtilis*, *Klebsiella planticola*	[104]
Ag	*Gracilaria corticata*	*Candida albicans*, *Candida glabrata*	[105]
Ag	*Gelidiella acerosa*	*Humicola insolens*, *Fusarium dimerum*, *Mucor indicus*, *Trichoderma reesei*	[106]
Ag	*Sargassum wightii* Grevilli	*Staphylococcus aureus*, *Bacillus rhizoids*, *Escherichia coli*, *Pseudomonas aeruginosa*	[107]
Ag	*Urospora* sp.	*S. aureus*, *E. coli*, *P. aeruginosa*, *Klebsiella pneumoniae*, *B. subtilis*	[108]
Ag	*Ulva lactuca*	*Bacillus* sp., *E. coli*, *Pseudomonas* sp.	[109]
Ag	*Enteromorpha flexuosa* Wulfen	*B. subtilis*, *Bacillus pumilus*, *Enterococcus faecalis*, *S. aureus*, *Staphylococcus epidermidis*, *E. coli*, *K. pneumoniae*, *P. aeruginosa*, *Aspergillus niger*, *C. albicans*, *Saccharomyces cerevisiae*	[110]
Ag	*S. wightii*	*P. aeruginosa*, *Vibrio cholerae*, *K. pneumoniae*, *S. aureus*, *E. coli*, *Staphylococcus pneumoniae*, *Salmonella typhi*	[111]
Ag	*Sargassum polycystum* C. Agardh	*P. aeruginosa*, *K. pneumoniae*, *E. coli*, *S. aureus*, MCF breast cancer	[112]
Ag and Au	*Turbinaria conoides*	*Salmonella* sp., *E. coli*, *Serratia liquefaciens*, *Aeromonas hydrophila*	[113]
AgCl	*Sargassum plagiophyllum*	*E. coli*	[103]
Au	*Stoechospermum marginatum*	*P. aeruginosa*, *Klebsiella oxytoca*, *E. faecalis*, *K. pneumoniae*, *Vibrio parahaemolyticus*, *V. cholerae*, *E. coli*, *Salmonella typhi*, *Salmonella paratyphi*, *Proteus vulgaris*	[114]
Au	*Padina pavonica*	*B. subtilis*, *E. coli*	[115]
Cu$_2$O and CuO	*Bifurcaria bifurcata*	*Enterobacter aerogenes*, *S. aureus*	[101]
ZnO	*Sargassum myriocystum*	*Staphylococcus mutans*, *Micrococcus luteus*, *V. cholerae*, *K. pneumoniae*, *Neisseria gonorrhoeae*	[72]

Adapted from D. Fawcett, J.J. Verduin, M. Shah, S.B. Sharma, G.E.J. Poinern, A Review of Current research into the biogenic synthesis of metal and metal oxide nanoparticles via marine algae and seagrasses, J Nanosci. 2017 (2017) with permission from Hindawi.

CONCLUSION

In this chapter, the green chemistry-based approach of nanomaterials production using algae has been discussed. Several biosynthesis methods, primarily classified as direct exploitation and harvested utilization, are explored thus far. Nevertheless, development of these methods, in terms of major parameters, is needed. Among various process parameters, the pH, temperature, incubation time, precursor concentration, and light intensity are found to play a significant role in the biosynthesis process. A vast majority of these nanomaterials find their potential application as antimicrobial agent. However, their extensive application may lie in biomedicine, cancer therapy, catalysis, and medical imaging. Further studies are encouraged to advance and extend the potential use of these nanomaterials.

REFERENCES

[1] T. ISO, 80004-1, Nanotechnologies-Vocabulary-Part 1: Core Terms, International Standards Organization, Geneva, Switzerland, 2007.

[2] M. Suganeswari, Nano particles: a novel system in current century, Int. J. Pharmaceut. Biol. Arch. 2 (2011).

[3] D.K. Yi, S.T. Selvan, S.S. Lee, G.C. Papaefthymiou, D. Kundaliya, J.Y. Ying, Silica-coated nanocomposites of magnetic nanoparticles and quantum dots, J. Am. Chem. Soc. 127 (2005) 4990–4991.

[4] H. Tsunoyama, H. Sakurai, N. Ichikuni, Y. Negishi, T. Tsukuda, Colloidal gold nanoparticles as catalyst for carbon- carbon bond formation: application to aerobic homocoupling of phenylboronic acid in water, Langmuir 20 (2004) 11293–11296.

[5] W. Wang, Y. Li, Q. Wu, C. Wang, X. Zang, Z. Wang, Extraction of neonicotinoid insecticides from environmental water samples with magnetic graphene nanoparticles as adsorbent followed by determination with HPLC, Anal. Methods 4 (2012) 766–772.

[6] A. Kumari, S.K. Yadav, S.C. Yadav, Biodegradable polymeric nanoparticles based drug delivery systems, Colloids Surf., B 75 (2010) 1–18.

[7] X. Luo, A. Morrin, A.J. Killard, M.R. Smyth, Application of nanoparticles in electrochemical sensors and biosensors, Electroanalysis 18 (2006) 319–326.

[8] P.K. Jain, X. Huang, I.H. El-Sayed, M.A. El-Sayed, Review of some interesting surface plasmon resonance-enhanced properties of noble metal nanoparticles and their applications to biosystems, Plasmonics 2 (2007) 107–118.

[9] A.-H. Lu, E.L. Salabas, F. Schüth, Magnetic nanoparticles: synthesis, protection, functionalization, and application, Angew. Chem. Int. Ed. 46 (2007) 1222–1244.

[10] L. Yanhong, W. Dejun, Z. Qidong, Y. Min, Z. Qinglin, A study of quantum confinement properties of photogenerated charges in ZnO nanoparticles by surface photovoltage spectroscopy, J. Phys. Chem. 108 (2004) 3202–3206.

[11] B.S. Prasad, P.S. Ganesh, Nanowaste: Need for Paradigms in Waste Management, American Academic Press, USA, 2018.

[12] C. Peyrot, K.J. Wilkinson, M. Desrosiers, S. Sauvé, Effects of silver nanoparticles on soil enzyme activities with and without added organic matter, Environ. Toxicol. Chem. 33 (2014) 115–125.

[13] S. Rebello, A.K. Asok, S. Mundayoor, M.S. Jisha, Surfactants: toxicity, remediation and green surfactants, Environ. Chem. Lett. 12 (2014) 275–287.

[14] G. Ramakrishna, H.N. Ghosh, Optical and photochemical properties of sodium dodecylbenzenesulfonate (DBS)-capped TiO_2 nanoparticles dispersed in nonaqueous solvents, Langmuir 19 (2003) 505–508.

[15] C.-H. Kuo, T.-F. Chiang, L.-J. Chen, M.H. Huang, Synthesis of highly faceted pentagonal-and hexagonal-shaped gold nanoparticles with controlled sizes by sodium dodecyl sulfate, Langmuir 20 (2004) 7820–7824.

[16] T.M.D. Dang, T.T.T. Le, E. Fribourg-Blanc, M.C. Dang, Synthesis and optical properties of copper nanoparticles prepared by a chemical reduction method, Adv. Nat. Sci. Nanosci. 2 (2011) 015009.

[17] T. Nawaz, S. Sengupta, C.-L. Yang, Silver recovery as Ag^0 nanoparticles from ion-exchange regenerant solution using electrolysis, J. Environ. Sci. (2018).

[18] M.L. Marin, K.L. McGilvray, J.C. Scaiano, Photochemical strategies for the synthesis of gold nanoparticles from Au (III) and Au (I) using photoinduced free radical generation, J. Am. Chem. Soc. 130 (2008) 16572–16584.

[19] S.A. Dahoumane, M. Mechouet, K. Wijesekara, C.D.M. Filipe, C. Sicard, D.A. Bazylinski, C. Jeffryes, Algae-mediated biosynthesis of inorganic nanomaterials as a promising route in nanobiotechnology – a review, Green Chem. 19 (2017) 552–587.

[20] S. Ahmed, M. Ahmad, B.L. Swami, S. Ikram, Green synthesis of silver nanoparticles using Azadirachta indica aqueous leaf extract, J. Radiat. Res. Appl. Sci. 9 (2016) 1–7.

[21] A.R. Shahverdi, S. Minaeian, H.R. Shahverdi, H. Jamalifar, A.-A. Nohi, Rapid synthesis of silver nanoparticles using culture supernatants of Enterobacteria: a novel biological approach, Process Biochem. 42 (2007) 919–923.

[22] I. Maliszewska, A. Juraszek, K. Bielska, Green synthesis and characterization of silver nanoparticles using Ascomycota fungi Penicillium nalgiovense AJ12, J. Clust. Sci. 25 (2014) 989–1004.

[23] D. Hebbalalu, J. Lalley, M.N. Nadagouda, R.S. Varma, Greener techniques for the synthesis of silver nanoparticles using plant extracts, enzymes, bacteria, biodegradable polymers, and microwaves, ACS Sustain. Chem. Eng. 1 (2013) 703–712.

[24] The IUPAC Green Chemistry Directory: Overview, IUPAC Green Chemistry Directory, 2018. http://www.incaweb.org/transit/iupacgcdir/overview.htm.

[25] T.M. Tolaymat, A.M. El Badawy, A. Genaidy, K.G. Scheckel, T.P. Luxton, M. Suidan, An evidence-based environmental perspective of manufactured silver nanoparticle in syntheses and applications: a systematic review and critical appraisal of peer-reviewed scientific papers, Sci. Total Environ. 408 (2010) 999–1006.

[26] S. Iravani, Green synthesis of metal nanoparticles using plants, Green Chem. 13 (2011).

[27] R.J. Aitken, M.Q. Chaudhry, A.B.A. Boxall, M. Hull, Manufacture and use of nanomaterials: current status in the UK and global trends, Occup. Med. 56 (2006) 300–306.

[28] A. Kumar, P.K. Vemula, P.M. Ajayan, G. John, Silver-nanoparticle-embedded antimicrobial paints based on vegetable oil, Nat. Mater. 7 (2008) 236.

[29] P.J.P. Espitia, N. de F.F. Soares, J.S. dos Reis Coimbra, N.J. de Andrade, R.S. Cruz, E.A.A. Medeiros, Zinc oxide nanoparticles: synthesis, antimicrobial activity and food packaging applications, Food Bioprocess. Technol. 5 (2012) 1447–1464.

[30] J. Lin, C. He, Y. Zhao, S. Zhang, One-step synthesis of silver nanoparticles/carbon nanotubes/chitosan film and its application in glucose biosensor, Sens. Actuators B Chem. 137 (2009) 768–773.

[31] O.V. Salata, Applications of nanoparticles in biology and medicine, J. Nanobiotechnol. 2 (2004) 3.

[32] K.N. Thakkar, S.S. Mhatre, R.Y. Parikh, Biological synthesis of metallic nanoparticles, Nanomed. Nanotechnol. 6 (2010) 257–262.

[33] J. Chen, J. Wang, X. Zhang, Y. Jin, Microwave-assisted green synthesis of silver nanoparticles by carboxymethyl cellulose sodium and silver nitrate, Mater. Chem. Phys. 108 (2008) 421–424.

[34] M.N. Nadagouda, R.S. Varma, Green synthesis of silver and palladium nanoparticles at room temperature using coffee and tea extract, Green Chem. 10 (2008) 859–862.

[35] N. Vigneshwaran, R.P. Nachane, R.H. Balasubramanya, P.V. Varadarajan, A novel one-pot 'green'synthesis of stable silver nanoparticles using soluble starch, Carbohydr. Res. 341 (2006) 2012–2018.

[36] V.K. Sharma, R.A. Yngard, Y. Lin, Silver nanoparticles: green synthesis and their antimicrobial activities, Adv. Colloid Interfac. 145 (2009) 83–96.

[37] P. Raveendran, J. Fu, S.L. Wallen, Completely "green" synthesis and stabilization of metal nanoparticles, J. Am. Chem. Soc. 125 (2003) 13940–13941.

[38] X. Gao, L. Wei, H. Yan, B. Xu, Green synthesis and characteristic of core-shell structure silver/starch nanoparticles, Matter Lett. 65 (2011) 2963–2965.

[39] P. Cheviron, F. Gouanvé, E. Espuche, Green synthesis of colloid silver nanoparticles and resulting biodegradable starch/silver nanocomposites, Carbohydr. Polym. 108 (2014) 291–298.

[40] J. Helmlinger, C. Sengstock, C. Gross-Heitfeld, C. Mayer, T.A. Schildhauer, M. Köller, M. Epple, Silver nanoparticles with different size and shape: equal cytotoxicity, but different antibacterial effects, RSC Adv. 6 (2016) 18490–18501.

[41] K.P. Bankura, D. Maity, M.M. Mollick, D. Mondal, B. Bhowmick, M.K. Bain, A. Chakraborty, J. Sarkar, K. Acharya, D. Chattopadhyay, Synthesis, characterization and antimicrobial activity of dextran stabilized silver nanoparticles in aqueous medium, Carbohydr. Polym. 89 (2012) 1159–1165.

[42] M. Venkatesham, D. Ayodhya, A. Madhusudhan, N.V. Babu, G. Veerabhadram, A novel green one-step synthesis of silver nanoparticles using chitosan: catalytic activity and antimicrobial studies, Appl. Nanosci. 4 (2014) 113–119.

[43] P. Manivasagan, S. Bharathiraja, M.S. Moorthy, Y.-O. Oh, H. Seo, J. Oh, Marine biopolymer-based nanomaterials as a novel platform for theranostic applications, Polym. Rev. 57 (2017) 631–667.

[44] L. Mulfinger, S.D. Solomon, M. Bahadory, A.V. Jeyarajasingam, S.A. Rutkowsky, C. Boritz, Synthesis and study of silver nanoparticles, J. Chem. Educ. 84 (2007) 322.

[45] M. Kitching, M. Ramani, E. Marsili, Fungal biosynthesis of gold nanoparticles: mechanism and scale up, Microb. Biotechnol. 8 (2015) 904–917.

[46] Global Starch – Food & Beverage, Report Linker, 2018. https://www.reportlinker.com/p04484177/Global-Starch-Food-Beverage.html.

[47] Foods, U.S Food and Drug Administration, 2018. https://www.fda.gov/Food/default.htm.

[48] F. He, D. Zhao, Preparation and characterization of a new class of starch-stabilized bimetallic nanoparticles for degradation of chlorinated hydrocarbons in water, Environ. Sci. Technol. 39 (2005) 3314–3320.

[49] N.N. Mallikarjuna, R.S. Varma, Microwave-assisted shape-controlled bulk synthesis of noble nanocrystals and their catalytic properties, Cryst. Growth Des. 7 (2007) 686–690.

[50] G.A. Kahrilas, L.M. Wally, S.J. Fredrick, M. Hiskey, A.L. Prieto, J.E. Owens, Microwave-assisted green synthesis of silver nanoparticles using orange peel extract, ACS Sustain. Chem. Eng. 2 (3) (2014) 367–376.

[51] S.V. Kumar, A.P. Bafana, P. Pawar, M. Faltane, A. Rahman, S.A. Dahoumane, A. Kucknoor, C.S. Jeffryes, Optimized production of antibacterial copper oxide nanoparticles in a microwave-assisted synthesis reaction using response surface methodology, Colloids Surf., A (2019).

[52] A. Pal, S. Shah, S. Devi, Microwave-assisted synthesis of silver nanoparticles using ethanol as a reducing agent, Mater. Chem. Phys. 114 (2009) 530–532.

[53] P. Raveendran, J. Fu, S.L. Wallen, A simple and "green" method for the synthesis of Au, Ag, and Au–Ag alloy nanoparticles, Green Chem. 8 (2006) 34–38.

[54] N.R. Jana, L. Gearheart, C.J. Murphy, Seed-mediated growth approach for shape-controlled synthesis of spheroidal and rod-like gold nanoparticles using a surfactant template, Adv. Mater. 13 (2001) 1389–1393.

[55] Nanomaterials Database | STATNANO, StatNano, 2018. https://statnano.com/nanomaterials.

[56] C. Zhu, J. Zhai, S. Dong, Bifunctional fluorescent carbon nanodots: green synthesis via soy milk and application as metal-free electrocatalysts for oxygen reduction, Chem. Commun. 48 (2012) 9367–9369.

[57] S. Paul, S.K. Samdarshi, A green precursor for carbon nanotube synthesis, N. Carbon Mater. 26 (2011) 85–88.

[58] N. Tripathi, V. Pavelyev, S.S. Islam, Synthesis of carbon nanotubes using green plant extract as catalyst: unconventional concept and its realization, Appl. Nanosci. 7 (2017) 557–566.

[59] S.V. Kumar, A.P. Bafana, P. Pawar, A. Rahman, S.A. Dahoumane, C.S. Jeffryes, High conversion synthesis of <10 nm starch-stabilized silver nanoparticles using microwave technology, Sci Rep-UK 8 (2018).

[60] A. Bafana, S.V. Kumar, S. Temizel-Sekeryan, S.A. Dahoumane, L. Haselbach, C.S. Jeffryes, Evaluating microwave-synthesized silver nanoparticles from silver nitrate with life cycle assessment techniques, Sci. Total Environ. 636 (2018) 936—943.

[61] J. Xie, J.Y. Lee, D.I. Wang, Y.P. Ting, Identification of active biomolecules in the high-yield synthesis of single-crystalline gold nanoplates in algal solutions, Small 3 (2007) 672—682.

[62] D. Fawcett, J.J. Verduin, M. Shah, S.B. Sharma, G.E.J. Poinern, A review of current research into the biogenic synthesis of metal and metal oxide nanoparticles via marine algae and seagrasses, J. Nanosci. 2017 (2017).

[63] S.A. Dahoumane, M. Mechouet, F.J. Alvarez, S.N. Agathos, C. Jeffryes, Microalgae: an outstanding tool in nanotechnology, Bionatura 1 (2016).

[64] J. Kaduková, O. Velgosová, A. Mražíková, R. Marcinčáková, The effect of culture age and initial silver concentration on biosynthesis of Ag nanoparticles, Nova Biotechnologica et Chimica 13 (2014) 28—37.

[65] D. Parial, H.K. Patra, A.K.R. Dasgupta, R. Pal, Screening of different algae for green synthesis of gold nanoparticles, Eur. J. Phycol. 47 (2012) 22—29.

[66] L.M. Rösken, S. Körsten, C.B. Fischer, A. Schönleber, S. van Smaalen, S. Geimer, S. Wehner, Time-dependent growth of crystalline Au⁰-nanoparticles in cyanobacteria as self-reproducing bioreactors: 1. Anabaena sp, J. Nano Res. 16 (2014).

[67] S.A. Dahoumane, C. Djediat, C. Yéprémian, A. Couté, F. Fiévet, T. Coradin, R. Brayner, Species selection for the design of gold nanobioreactor by photosynthetic organisms, J. Nano Res. 14 (2012).

[68] C. Sicard, R. Brayner, J. Margueritat, M. Hémadi, A. Couté, C. Yéprémian, C. Djediat, J. Aubard, F. Fiévet, J. Livage, Nano-gold biosynthesis by silica-encapsulated micro-algae: a "living" bio-hybrid material, J. Mater. Chem. 20 (2010) 9342—9347.

[69] I. Barwal, P. Ranjan, S. Kateriya, S.C. Yadav, Cellular oxido-reductive proteins of Chlamydomonas reinhardtii control the biosynthesis of silver nanoparticles, J. Nanobiotechnol. 9 (2011) 56.

[70] Y. Li, X. Tang, W. Song, L. Zhu, X. Liu, X. Yan, C. Jin, Q. Ren, Biosynthesis of silver nanoparticles using Euglena gracilis, Euglena intermedia and their extract, IET Nanobiotechnol. 9 (2014) 19—26.

[71] J. Jena, N. Pradhan, B.P. Dash, P.K. Panda, B.K. Mishra, Pigment mediated biogenic synthesis of silver nanoparticles using diatom Amphora sp. and its antimicrobial activity, J. Saudi Chem. Soc. 19 (2015) 661—666.

[72] V. Patel, D. Berthold, P. Puranik, M. Gantar, Screening of cyanobacteria and microalgae for their ability to synthesize silver nanoparticles with antibacterial activity, Biotechnol. Rep. 5 (2015) 112—119.

[73] N. Chakraborty, A. Banerjee, S. Lahiri, A. Panda, A.N. Ghosh, R. Pal, Biorecovery of gold using cyanobacteria and an eukaryotic alga with special reference to nanogold formation — a novel phenomenon, J. Appl. Phycol. 21 (2009) 145—152.

[74] M.F. Lengke, M.E. Fleet, G. Southam, Biosynthesis of silver nanoparticles by filamentous cyanobacteria from a silver(I) nitrate complex, Langmuir 23 (2007) 2694—2699.

[75] Y. Kang, P.M. Siegel, W. Shu, M. Drobnjak, S.M. Kakonen, C. Cordón-Cardo, T.A. Guise, J. Massagué, A multigenic program mediating breast cancer metastasis to bone, Cancer Cell 3 (2003) 537—549.

[76] T.N. Prasad, V.S.R. Kambala, R. Naidu, Phyconanotechnology: synthesis of silver nanoparticles using brown marine algae Cystophora moniliformis and their characterisation, J. Appl. Phycol. 25 (2013) 177—182.

[77] M. Singh, R. Kalaivani, S. Manikandan, N. Sangeetha, A.K. Kumaraguru, Facile green synthesis of variable metallic gold nanoparticle using Padina gymnospora, a brown marine macroalga, Appl. Nanosci. 3 (2013) 145—151.

[78] P.D. Shankar, S. Shobana, I. Karuppusamy, A. Pugazhendhi, V.S. Ramkumar, S. Arvindnarayan, G. Kumar, A review on the biosynthesis of metallic nanoparticles (gold and silver) using bio-components of microalgae: formation mechanism and applications, Enzym. Microb. Technol. 95 (2016) 28—44.

[79] K. Govindaraju, S.K. Basha, V.G. Kumar, G. Singaravelu, Silver, gold and bimetallic nanoparticles production using single-cell protein (Spirulina platensis) Geitler, J. Mater. Sci. 43 (2008) 5115—5122.

[80] S. Nagarajan, K.A. Kuppusamy, Extracellular synthesis of zinc oxide nanoparticle using seaweeds of gulf of Mannar, India, J. Nanobiotechnol. 11 (2013) 39.

[81] Y.N. Mata, E. Torres, M.L. Blazquez, A. Ballester, F. González, J.A. Munoz, Gold (III) biosorption and bioreduction with the brown alga Fucus vesiculosus, J. Hazard Mater. 166 (2009) 612—618.

[82] D. Parial, R. Pal, Biosynthesis of monodisperse gold nanoparticles by green alga Rhizoclonium and associated biochemical changes, J. Appl. Phycol. 27 (2015) 975—984.

[83] S. Rajeshkumar, C. Malarkodi, K. Paulkumar, M. Vanaja, G. Gnanajobitha, G. Annadurai, Algae mediated green fabrication of silver nanoparticles and examination of its antifungal activity against clinical pathogens, Int. J. (2014) 2014.

[84] P. Rauwel, S. Küünal, S. Ferdov, E. Rauwel, A review on the green synthesis of silver nanoparticles and their morphologies studied via TEM, Ann. Mater. Sci. Eng. 2015 (2015).

[85] G. Ghodake, D.S. Lee, Biological synthesis of gold nanoparticles using the aqueous extract of the brown algae Laminaria japonica, J. Nanoelectron. Optoelectron. 6 (2011) 268—271.

[86] A. Rahman, S. Kumar, A. Bafana, S.A. Dahoumane, C. Jeffryes, Biosynthetic conversion of Ag^+ to highly stable Ag^0 nanoparticles by wild type and cell wall deficient strains of *Chlamydomonas reinhardtii*, Molecules 24 (2019) 98.

[87] A. Rahman, S. Kumar, A. Bafana, S.A. Dahoumane, C. Jeffryes, Individual and combined effects of extracellular polymeric substances and whole cell components of *Chlamydomonas reinhardtii* on silver nanoparticle synthesis and stability, Molecules 24 (2019) 956.

[88] S.A. Dahoumane, K. Wijesekera, C.D.M. Filipe, J.D. Brennan, Stoichiometrically controlled production of bimetallic gold-silver alloy colloids using micro-alga cultures, J. Colloid Interface Sci. 416 (2014) 67−72.

[89] S.A. Dahoumane, C. Yéprémian, C. Djédiat, A. Couté, F. Fiévet, T. Coradin, R. Brayner, A global approach of the mechanism involved in the biosynthesis of gold colloids using micro-algae, J. Nano Res. 16 (2014).

[90] M. Mohseniazar, M. Barin, H. Zarredar, S. Alizadeh, D. Shanehbandi, Potential of microalgae and lactobacilli in biosynthesis of silver nanoparticles, Bioimpacts 1 (2011) 149.

[91] A. Rahman, S. Kumar, A. Bafana, J. Lin, S.A. Dahoumane, C. Jeffryes, A Mechanistic view of the light-Induced synthesis of silver nanoparticles using extracellular polymeric substances of *Chlamydomonas reinhardtii*, Molecules 24 (2019) 3506.

[91a] S. Ermakova, M. Kusaykin, A. Trincone, Z. Tatiana, Are multifunctional marine polysaccharides a myth or reality? Front Chem 3 (2015) 39.

[92] K. Govindaraju, K. Krishnamoorthy, S.A. Alsagaby, G. Singaravelu, M. Premanathan, Green synthesis of silver nanoparticles for selective toxicity towards cancer cells, IET Nanobiotechnol. 9 (2015) 325−330.

[93] J.M. Sheeba, S. Thambidurai, Extraction, characterization, and application of seaweed nanoparticles on cotton fabrics, J. Appl. Polym. Sci. 113 (2009) 2287−2292.

[94] F. LewisOscar, S. Vismaya, M. Arunkumar, N. Thajuddin, D. Dhanasekaran, C. Nithya, Algal nanoparticles: synthesis and biotechnological potentials, in: Algae-Organisms for Imminent Biotechnology, InTech, London, 2016.

[95] A. Schröfel, G. Kratošová, I. Šafařík, M. Šafaříková, I. Raška, L.M. Shor, Applications of biosynthesized metallic nanoparticles—a review, Acta Biomater. 10 (2014) 4023−4042.

[96] N. Abdel-Raouf, N.M. Al-Enazi, I.B. Ibraheem, Green biosynthesis of gold nanoparticles using *Galaxaura elongata* and characterization of their antibacterial activity, Arb J Chem 10 (2017) S3029−S3039.

[97] K.U. Suganya, K. Govindaraju, V.G. Kumar, T.S. Dhas, V. Karthick, G. Singaravelu, M. Elanchezhiyan, Blue green alga mediated synthesis of gold nanoparticles and its antibacterial efficacy against gram positive organisms, Mater. Sci. Eng. 47 (2015) 351−356.

[98] B. Sharma, D.D. Purkayastha, S. Hazra, M. Thajamanbi, C.R. Bhattacharjee, N.N. Ghosh, J. Rout, Biosynthesis of fluorescent gold nanoparticles using an edible freshwater red alga, *Lemanea fluviatilis* (L.) C. Ag. and antioxidant activity of biomatrix loaded nanoparticles, Bioproc. Biosyst. Eng. 37 (2014) 2559−2565.

[99] D. MubarakAli, J. Arunkumar, K.H. Nag, K.A. SheikSyedIshack, E. Baldev, D. Pandiaraj, N. Thajuddin, Gold nanoparticles from Pro and eukaryotic photosynthetic microorganisms—comparative studies on synthesis and its application on biolabelling, Colloids Surf. B 103 (2013) 166−173.

[100] C.M. Ramakritinan, E. Kaarunya, S. Shankar, A.K. Kumaraguru, Antibacterial effects of Ag, Au and bimetallic (Ag-Au) nanoparticles synthesized from red algae, in: Solid State Phenomena, Trans Tech Publ, 2013, pp. 211−230.

[101] Y. Abboud, T. Saffaj, A. Chagraoui, A. El Bouari, K. Brouzi, O. Tanane, B. Ihssane, Biosynthesis, characterization and antimicrobial activity of copper oxide nanoparticles (CONPs) produced using brown alga extract (*Bifurcaria bifurcata*), Appl. Nanosci. 4 (2014) 571−576.

[102] M. Francavilla, A. Pineda, A.A. Romero, J.C. Colmenares, C. Vargas, M. Monteleone, R. Luque, Efficient and simple reactive milling preparation of photocatalytically active porous ZnO nanostructures using biomass derived polysaccharides, Green Chem. 16 (2014) 2876−2885.

[103] T.S. Dhas, V.G. Kumar, V. Karthick, K.J. Angel, K. Govindaraju, Facile synthesis of silver chloride nanoparticles using marine alga and its antibacterial efficacy, Spectrochim. Acta 120 (2014) 416−420.

[104] S. Rajeshkumar, C. Kannan, G. Annadurai, Synthesis and characterization of antimicrobial silver nanoparticles using marine brown seaweed *Padina tetrastromatica*, Drug Invent. Today 4 (2012) 511−513.

[105] P. Kumar, S.S. Selvi, M. Govindaraju, Seaweed-mediated biosynthesis of silver nanoparticles using *Gracilaria corticata* for its antifungal activity against *Candida* spp, Appl. Nanosci. 3 (2013) 495−500.

[106] M. Vivek, P.S. Kumar, S. Steffi, S. Sudha, Biogenic silver nanoparticles by *Gelidiella acerosa* extract and their antifungal effects, Avicenna J. Med. Biotechnol. 3 (2011) 143.

[107] K. Govindaraju, V. Kiruthiga, V.G. Kumar, G. Singaravelu, Extracellular synthesis of silver nanoparticles by a marine alga, *Sargassum wightii* Grevilli and their antibacterial effects, J. Nanosci. Nanotechnol. 9 (2009) 5497−5501.

[108] J. Suriya, S.B. Raja, V. Sekar, R. Rajasekaran, Biosynthesis of silver nanoparticles and its antibacterial activity using seaweed *Urospora* sp, Afr. J. Biotechnol. 11 (2012) 12192−12198.

[109] N. Sangeetha, K. Saravanan, Biogenic silver nanoparticles using marine seaweed (*Ulva lactuca*) and evaluation of its antibacterial activity, J. Nanosci. Nanotechnol. 2 (2014) 99−102.

[110] M. Yousefzadi, Z. Rahimi, V. Ghafori, The green synthesis, characterization and antimicrobial activities of silver nanoparticles synthesized from green alga *Enteromorpha flexuosa* (wulfen), J. Agardh Matter Lett. 137 (2014) 1−4.

[111] N. Shanmugam, P. Rajkamal, S. Cholan, N. Kannadasan, K. Sathishkumar, G. Viruthagiri, A. Sundaramanickam, Biosynthesis of silver nanoparticles from the marine seaweed *Sargassum wightii* and their antibacterial activity against some human pathogens, Appl. Nanosci. 4 (2014) 881–888.

[112] N. Thangaraju, R.P. Venkatalakshmi, A. Chinnasamy, P. Kannaiyan, Synthesis of silver nanoparticles and the antibacterial and anticancer activities of the crude extract of *Sargassum polycystum* C Agardh, Nano Biomed Eng. 4 (2012) 89–94.

[113] S.R. Vijayan, P. Santhiyagu, M. Singamuthu, N. Kumari Ahila, R. Jayaraman, K. Ethiraj, Synthesis and characterization of silver and gold nanoparticles using aqueous extract of seaweed, *Turbinaria conoides*, and their antimicrofouling activity, Sci. World J. 2014 (2014).

[114] F.A.A. Rajathi, C. Parthiban, V.G. Kumar, P. Anantharaman, Biosynthesis of antibacterial gold nanoparticles using brown alga, *Stoechospermum marginatum* (kützing), Spectrochim. Acta 99 (2012) 166–173.

[115] G. Isaac, R.E. Renitta, Brown Algae mediated synthesis, characterization of gold nano particles using *Padina pavonica* and their antibacterial activity against human pathogens, Int J Pharm Tech Res 8 (2015) 31–40.

CHAPTER 18

Life Cycle Assessment and Techno-Economic Analysis of Algal Biofuel Production

DONGYAN MU • CHUNHUA XIN • WENGUANG ZHOU

INTRODUCTION

Overview of LCA Application in Algal Biofuel Production

Life cycle assessment (LCA) is "compilation and evaluation of inputs, outputs, and the potential environmental impacts of a product system throughout its life cycle." [1] A product's life cycle includes all production stages from raw material extraction, processing, manufacturing to product's end of life treatment. Therefore, LCA is also called a cradle-to-grave analysis. Based on its definition, LCA not only focuses on the life cycle of the product for analysis, but also includes life cycles of process inputs, infrastructure, equipment and labor in making the product. In addition, LCA not only focuses on one environmental impact, but often includes multiple impacts in different spaces and at varied time frames. In a word, LCA provides a full range of quantitative analysis of the environmental impacts of a product or a process.

Since the first project conducted by the Coca-Cola Company in the 1960s, LCA has been increasingly used by industries for environmental management and impact prevention. Especially when developing a new technology or process, LCA is a very effective quantitative tool for environmentally friendly design and manufacturing. Even for an existing product, LCA can help manufacturers reevaluate its environmental impacts and identify the most influential unit processes. By replacing those influential processes, manufactures could not only protect the environment, but also reduce the cost of energy use and waste treatment. Furthermore, the manufacturers can constrain suppliers and request them to pay attention to environmental protection by adopting energy-saving and environmentally friendly processes.

LCA has been applied in evaluating microalgae biofuels for more than 10 years. Early LCA study can be traced back to the article, "Life Cycle Assessment of Microalgae Biodiesel", published by Laurent Lardon in 2009 [2]. From that time on, researchers published many LCA studies on algal biofuels. Several major research groups and researchers in the U.S. include: Laura Brentner at Yale University [3], Liaw Batan at Colorado State University [4,5], Jason Quinn at Utah State University [6–10], Andres Clarens at the University of Virginia [11 13], Kyle Sander at Oregon State University [14], Group in the Argonne National Laboratory [15,16], Ryan Davis in the National Renewable Energy Laboratory [17], and Dongyan Mu at Kean University [18,19]. In addition, there were many LCA studies conducted by researchers across the world, such as the group of Xuming Ou [20] and Liang Sai [21] at Tsinghua University, China, Pierre Collet in INRIA, France [22–24], Peter K. Campbell in CSIRO, Australia [25], S. Jez at the University of Siena, Italy [26], Sarat Chandra Togarcheti in CSIR, India [27], Simone Souza in Brazil [28], and Edgard Gnansounou in Switzerland [29].

Overview of TEA of Microalgae-Based Biofuel Production

Development and commercialization of microalgae biofuels require consideration of social, environmental, and economic aspects. The techno-economic assessment (TEA) can examine the capital and operating costs and analyze relevant risks for various production processes and technologies, which is crucial to evaluate the economic feasibility of microalgae based biofuels and co-products. Presently, a number of TEA studies have been completed to analyze the economic feasibility of various microalgae-based biofuels. The key

Microalgae Cultivation for Biofuels Production. https://doi.org/10.1016/B978-0-12-817536-1.00018-7

factors that affect economic performance have been identified. Different solutions to reduce costs have been also evaluated by TEA.

In order to conduct a transparent assessment that enables the insistence in different studies, a framework has been proposed and successfully applied in many TEA studies for algal biofuel production [30,31]. In general, the framework includes four major steps:

(1) Market study: Investigate the market perspectives and external factors that affect the commercialization of the products for analysis.

(2) The process flow diagram, and mass and energy balance: Construct an excel-based spreadsheet model which is used throughout the whole process of techno-economic assessment.

(3) The economic assessment: Calculate economic investment criteria, such as the net present value (NPV) and the internal rate of return (IRR) through the comprehensive technological process to determine the economic feasibility.

(4) The sensitivity assessment: Examine the impacts of a variation of the input parameters on the economic output parameters.

In addition, the TEA modeling and cash flow calculations should be based on detailed process description including algae strain, algae cultivation method, cultivation equipment, method of separating algae and water, and method of algal biomass conversion. Before processing TEA, the system boundary should be identified, and the assumptions and conditions of process description should be given. Main parameters for algae cultivation, harvest and conversion should be listed. The capital and operating cost should be estimated based on process description and assumptions. The main products and their price should be identified.

LCA METHODOLOGY IN ALGAL BIOFUEL PRODUCTION

Conducting life cycle assessment (LCA) is a complex and time-consuming process, and usually with high level of uncertainty. Therefore, the International Standards Organization (ISO) has issued a series of standards, i.e., the ISO 14,000 series, in order to specify the process to conduct LCA. Although not a rigid requirement, current LCA practitioners generally follow the ISO standards to implement the life cycle assessment. In the ISO standard, the scope and depth of the evaluation is not stipulated, only the steps/phases to conduct LCA are specified shown in Fig. 18.1. This aims to ensure the transparency and repeatability of all projects, while allowing flexibility of project

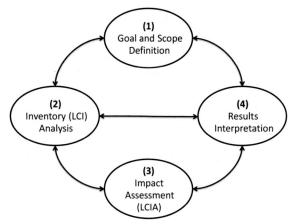

FIG. 18.1 Life cycle assessment phases.

execution. The session will discuss how to conduct LCA of biofuel production in each step defined in the ISO standard.

Goals and Scope Definition

The first step of LCA is to determine the research goals and scope for analysis. LCA could focus on analyzing a product, a process, or a service, or comparing different products and processes. The scope of a LCA project is then determined by identifying the goal for analysis. In this stage, it is also required to describe the functional units (FU) of the study. The FU is the "quantified performance of a product system for use as a reference unit" [1]. It is the product's function related unit used in quantifying all environmental impacts. In the case of biofuel production, it could be "per gallon/litter", or "per kilogram (kg) ", or "per megajoule (MJ)" of biofuel produced. Once the FU determined, all the impacts will be related to FU in the impact analysis stage (stage 3 in Fig. 18.1). In the goal and scope definition stage, it is also necessary to describe the system boundaries, data collection methods, and data quality requirements of the study. All of which are critical in determining the depth and breadth of the LCA and also the time and labor required for the LCA project.

For algal biofuel LCA, the research goal could be set as (1) analyzing environmental impacts of certain biofuel such as biodiesel, ethanol, crude oil, biogas, etc., (2) or comparing impacts of different biofuels in terms of production technologies and process integration pathways, (3) or comparing impacts of biofuels to non-renewable fuels. The life cycle of algal biofuel production is generally divided into three major stages with various unit processes.

- Stage 1 - Algae growth and collection: Algae farming, collection, separation, dewatering, and drying (sometimes not needed), etc.
- Stage 2 - Conversion of microalgae to biofuels: pretreatment (e.g., hydrolysis/homogenization), production of crude oil (e.g., lipid extraction/pyrolysis/gasification/liquefaction), crude oil refining (e.g., cracking), and algae residue recycling (e.g., anaerobic digestion/combustion).
- Stage 3 - Biofuel distribution and combustion: regular product transportation and distribution, and engine burning of biofuel.

The scope for analysis is determined by the goal of the study. When comparing microalgae biofuels to non-renewable energy, it is often chosen to evaluate the full life cycle. In this case, the functional unit could be set as the travel distance (kilometers) of vehicles in order to ensure a fair comparison. This is because the energy contained in per gallon/liter/kilogram of biofuels is lower than that of petroleum-based fuels, it would be better to set comparison based on the function provided by either the algal biofuel or the petroleum fuel, i.e., making cars move in certain distance. In contrast, when comparing different algal biofuel production technologies, the choice of research scope is more flexible. If the final products are the same, the LCA practitioner can exclude the biofuel use stage, because the emissions in this stage are the same and the conclusion will not be affected. In this case, the FU could be set as per kilogram or per megajoule (MJ) biofuel produced. Definitely, there are LCA studies only focusing on algae cultivation because they are targeted to compare various cultivation methods.

Life Cycle Inventory (LCI) Analysis

Following the goal and scope definition, the LCI includes data collection and model establishment to quantify relevant inputs and outputs of a product(s) for analysis [1]. LCI needs to collect all the inputs and outputs of unit processes, as well as materials and energy used for construction, equipment, land occupation, labor and so on. In particular, LCA needs to identify flows that are directly connected to the environmental systems, because those flows cause environmental impacts. It also requires relating of data to unit processes and relating data to the functional units. If there is more than one product yielded in the production processes, impacts should be allocated to various products. This stage is a very time consuming and laborious phase. Especially the data collection step which usually takes a long time.

Therefore, many LCA studies focus on collecting data of key processes, materials and components and neglecting others. In addition, as there are many commercial or free databases which include thousands of unit processes/products, the LCA practitioners can directly use the results in existing databases, and thus saving time to conduct analysis for each material and process involved in the boundary. Sometimes the goals and scope of the study must be revised based on the data availability and validity.

The inputs commonly involved in algal biofuel production are electricity, water, nutrients, and materials for infrastructure. Electricity is used for pumps, mixers, algae collection and dewatering processes. Water is used for replenishing water lost due to evaporation in cultivation facilities. Nutrients, including flue gas (containing carbon dioxide), nitrogen and phosphorus fertilizers, are used in the algae cultivation. The materials for infrastructure are concrete, steel, or fiberglass for culturing facilities applied. Lastly, the outputs of algal biofuel production include products (biofuels or bioenergy), co-products (heat, electricity, or animal feed, chemicals), wastes (algae residues, ash, or wastewater) and emissions (greenhouse gases (GHGs) and other gaseous pollutants). Many waste streams could be recycled and reused either in algal biofuel production or in other industries, which reduces environmental impacts.

In addition to direct inputs and outputs, the LCI needs to consider the upstream environmental impacts (i.e., indirect impacts) associated with the inputs at all stages of algal biofuel production. For conventional inputs, such as water, electricity, and fertilizers, their environmental impacts are often included in LCA databases. Therefore, there is no need to conduct LCA for those inputs separately. Once the inputs are determined for algal biofuel production, upstream impacts that are stored in LCA databases could be directly added into the inventory. Several commonly used databases include: EcoInvent, GREET Model (Greenhouse gases, Regulated Emissions, and Energy use in Transportation Model), US LCI, US-EI, BEES (Building for Environmental and Economic Sustainability), and EIO-LCA (Economic Input-Output Life Cycle Assessment), etc. Among those databases, the EcoInvent database contains algal biodiesel production with dry lipid extraction followed by transesterification. The GREET also includes emissions and energy use of microalgae biofuel production [32].

Another issue needs to be addressed in establishing LCI is when counting GHG emissions in algal biofuel production, the carbon in algae are usually treated as carbon fixed in algae. However, most carbon absorbed in the algae growing stage will be returned to the atmosphere during the biofuel production and use stages. Only carbon remained in the waste streams which are either dumped in the farmland, or buried in landfills, or discharged to local bodies of water, is carbon sequestered from atmosphere in algal biofuel production.

Life Cycle Impact Assessment (LCIA)

LCIA is a step for evaluating the potential environmental impacts by converting the LCI results into specific impact indicators. Conducting LCIA has to follow several substeps: First is to select impact categories for analysis. The major impact categories are divided into three general groups in terms of impacting subjects (shown in Fig. 18.2). Second is to assign the LCI results to different impact categories (classification). Third, the potential impact indicators are calculated (characterization). These three steps are mandatory for LCIA. Also, there are optional steps for LCIA, including relating the impact indicators to reference conditions (normalization), grouping, and weighting impacts. Currently, most LCIAs only perform three mandatory steps [1].

When LCIA of microalgae bioenergy is carried out, environmental impact categories can be selected according to project requirements. Many studies include only the most important environmental impacts and ignore other impacts. Of many impacts for analysis, the fossil fuel use is always included in algal biofuel LCIA, because the purpose to produce algal biofuel is to replace fossil fuels. It is necessary to examine if fossil fuel use of algal biofuels is less than producing petroleum-based fuels. Similarly, using algal biofuels is expected to reduce GHGs in the atmosphere. It is also necessary to study the GHG emissions of algal biofuels in order to evaluate whether algae biofuel is superior to traditional fossil energy. Other impacts in conducting algal biofuel LCIA include eutrophication (N), ozone depletion (O_3), acid rain (SO_2), respiration ($PM_{2.5}$), water use, and land use, etc.

LCA Results Interpretation

The final step of conducting LCA is to interpret of the LCIA results, identify major environmental issues in the product's life cycle, and evaluate results transparency, completeness, consistency, and uncertainty [1]. In this step, research is needed to determine which material or process or life stage has a significant impact on the environment through the sensitivity analysis. In addition, as the algal biofuel production involved many unit processes with multiple process parameters, the research results are influenced by many variables, and therefore, uncertainty analysis is especially important to evaluate precision and variability of LCA results. Currently, LCA uses either ranges or probability distributions of variables to determine uncertainty. In some cases, the Monte Carlo simulation is used to determine uncertainty in the results. Lastly, the results interpretation should include conclusions and limitations of the LCA studies and also give recommendations for reducing environmental impacts.

An LCA Case Study

A case study has been presented here to explain the LCA procedure. This study aims to conduct a cradle-to-grave LCA for the algal biodiesel production and use in a vehicle. The scope of analysis with unit processes is shown in Fig. 18.3. The algae cultivation modeled here is based on autotrophic algae grown in an open pond reactor, and the biodiesel production is based on wet lipid extraction, a widely studied technology in LCA. The waste reuse stage focuses on treating algae residue after lipid extraction. The popular way to reuse residue is to generate methane (CH_4) in anaerobic digestion, and then burn methane to generate electricity and heat for in-plant use. Lastly, the biodiesel is assumed to be used in a passage car with a diesel engine. In the flow diagram, material and energy flows are also included. The flows in black are either co-products or wastes that will be directly discharged into the environment. The flows in red are inputs that must be purchased outside the algal biofuel production plant.

Ecosystem Impacts	Human Impacts	Resource Depletion
Climate Change	Ozone Depletion	Fossil Fuel
Acid Rain	Smog	Freshwater
Eutrophication	Particulate Matter	Soil
Land Use Change	Carcinogens	Forest
Solid Waste	Toxicity	Grassland
Toxicity		Minerals

FIG. 18.2 Major environmental impact categories.

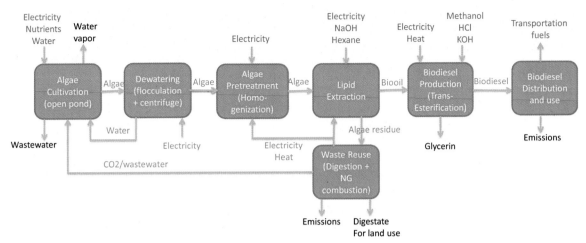

FIG. 18.3 LCA study scope for algal biofuel production.

Flows in blue are intermediate flows which only flow between unit processes. The intermediate flows do not cause environmental impacts. The functional unit selected is per mile vehicle driven.

After the goal and scope is determined, the life cycle inventory (LCI) should be established. The life cycle inventory includes all input and output flows and have to be converted to per mile vehicle driven. The LCI of the case study is included in Table 18.1. There are some assumptions when building the LCI. First, the study does not include the impacts of facilities such as tanks, machines, and buildings. Second, the LCA does not include impacts of digestate used as a soil amendment, because digestate's amendment effect has not been well studied yet. Third, the impacts of glycerin are not included because it is not sold as a product given the limited market demand. The negative values of inputs in LCI are either CO_2 extracted from the atmosphere or the electricity and heat generated which are treated as benefits to environment because they help substitute the same amount of electricity or heat produced from nonrenewable energy sources. The process calculation assumptions and parameters in process calculation could be referenced to the publication by Mu et al. [19].

The environmental impact analysis focuses on three major impacts including fossil fuel use, climate change (GHG emissions), and aquatic eutrophication. These three impacts are measured by MJ of energy use, kg CO_2 equivalent (eq.), and kg N equivalent (eq.). The impacts per unit of certain inputs are directly extracted from EcoInvent 3.0 database. The impacts of certain inputs of a process therefore have been

calculated by multiplying unit impacts and amount of inputs. For example, the GHG emissions of electricity generation are 0.746 kg CO_2 eq. per kWh, and the electricity use for the algae cultivation is 1.5576 kWh (shown in Table 18.1). Therefore the GHG emissions caused by electricity use in the cultivation stage are 1.16 kg CO_2. The impacts of inputs in all unit processes are then calculated and added into the total impacts. Lastly, the direct emissions from unit processes, e.g., the 1.3442 kg of CO2 eq. in waste reuse are also added into the total impacts. The impacts of unit processes and the total impacts are presented in Table 18.1.

The results show the algae cultivation stage cause major impacts in fossil fuel use because the electricity used in water pumping, mixing and dewatering. For the entire life cycle, the electricity used in each unit process is the major factor cause fossil fuel use. For the climate change impact, the waste reuse stage has the highest GHG emissions, which is caused by a large amount of CO_2 emission in the digestion process. In addition, the biodiesel combustion in the vehicle emits CO_2, making the biodiesel use another stage with high GHGs. Lastly for eutrophication, the algae cultivation stage has the highest impact because the wastewater is released in this stage. In order to reduce environmental impacts, the study suggests (1) selecting high efficient equipment to reduce electricity use, (2) recycling more CO_2 in algae residue digestion to feed into the algae culturing stage to reduce GHG emissions, and (3) improving the ability of algae to use nutrients in wastewater to reduce pollutants discharged into the environment.

TABLE 18.1
Life Cycle Inventory (per Mile Based) and Life Cycle Impact Assessment (LCIA) of The Case Study.

Input/Output Flows	Amount	Units	Fossil Fuel Use (MJ)	Climate Change (kgCO$_2$eq.)	Eutrophication (kg N eq.)
OPEN POND + DEWATERING					
Freshwater	0.2431	ton	0.4175	0.0405	6.85E-06
Electricity	1.5576	kWh	13.7779	1.1620	4.32E-03
PAM (flocculation)	0.0056	kg	0.4825	0.0159	1.37E-06
Urea (nutrient)	0.0381	kg	2.3817	0.1257	5.16E-05
DAP (nutrient)	0.0085	kg	0.7560	0.0623	3.22E-05
Wastewater	0.0070	m^3			0.0212
CO2 fixing	−1.6672	kg		−1.6672	
HOMOGENIZATION					
Electricity	0.2353	kWh	2.0812	0.1755	6.53E-04
Freshwater	0.0002	m^3	0.0003	0.0000	4.88E-09
LIPID EXTRACTION					
Electricity	0.0578	kWh	0.5110	0.0431	1.60E-04
Heat	0.7737	MJ	0.9852	0.0585	1.45E-05
Hexane	0.0051	kg	0.3067	0.0046	1.37E-05
NaOH	0.0026	kg	0.0344	0.0028	6.55E-07
TRANSESTERIFICATION					
Freshwater	0.0028	m^3	0.0049	0.0005	7.99E-08
Electricity	0.0101	kWh	0.0893	0.0075	2.80E-05
Heat	0.1907	MJ	0.2428	0.0144	3.57E-06
Methane	0.0131	kg	0.4839	0.0097	2.02E-06
HCl	0.0095	kg	0.1634	0.0125	2.70E-06
KOH	0.0017	kg	0.0442	0.0032	4.82E-07
Glycerin	−0.0136	kg	—	—	—
WASTE REUSE					
Electricity	−0.5662	kwh	−5.0086	−0.4224	−0.0016
Heat	−0.4682	MJ	−0.0016	−0.0016	−8.78E-06
Emissions			0	1.3442	0
CO$_2$	1.1331	kg			
CO	0.0002	kg			
CH$_4$	0.0051	kg			
N$_2$O	0.0003	kg			
NOx	0.0009	kg			
BIODIESEL DISTRIBUTION + USE					
Fossil fuel use	0.0755	MJ	0.0755		
Emissions				0.3271	0
CO$_2$	0.3234	kg			
CH$_4$	2.6E-06	kg			

Continued

TABLE 18.1

Life Cycle Inventory (per Mile Based) and Life Cycle Impact Assessment (LCIA) of The Case Study.—cont'd

Input/Output Flows	Amount	Units	Fossil Fuel Use (MJ)	Climate Change (kgCO$_2$eq.)	Eutrophication (kg N eq.)
N$_2$O	1.2E-05	kg			
NOx	1.2E-05	kg			
Total impacts			17.2334	1.2850	0.0249

RESULTS AND DISCUSSION FOR EXISTING LCA IN ALGAL BIOFUELS

Although many LCA studies have been conducted for microalgae bioenergy, their results varied significantly. For example, Sill and his colleagues [33] reviewed 15 LCA studies published between 2010 and 2012 and found out the energy input-output ratio (EROI) ranged from 0.09 to 4.3 (EROI = 1.0 indicates algae bioenergy production equals to non-renewable energy consumption), which indicates that existing LCA studies could not obtain a consistent conclusion that algae energy saves non-renewable energy. Similarly, most of the existing studies concluded that GHG emissions of algal biofuels were lower than equivalent fossil fuels, but emissions per functional unit of biofuel differed signif icantly among those studies [24]. Such a large difference in results was mainly due to the complexity of the microalgae bioenergy production and the lack of industrial or commercialized production. The scope and assumptions for those studies also varied. This section could only provide a general description of LCA results in previous studies.

Non-renewable energy (fossil fuel) use: 70%–90% of non-renewable energy is consumed in generating electricity and heat for microalgae cultivation and treatment. In addition, the fertilizers and facilities are major contributors to the fossil fuel use. Many LCA studies have shown that the life cycle non-renewable energy consumption of algae biofuels is even higher than that of petroleum gasoline and diesel. Therefore, they suggested using wastewater to grow algae in order to reduce fossil fuel use, because the wastewater was removed at the same time, which could be counted as a benefit that reduces the nonrenewable energy use in biofuel production. Besides, recycling waste streams such as algae residue or waste heat can reduce non-renewable energy consumption in algal biofuel production [34]. Of many biofuel production technologies, the hydrothermal liquefaction has less fossil fuel use than other biofuel production technologies because the hydrothermal liquefaction process does not require drying algae.

Climate change: The GHGs directly emitted from biofuel production and use stage. Also, the waste recycling processes release GHGs directly to the atmosphere. The indirect GHG emissions are related to electricity, heat, fertilizers and infrastructures. Most of LCA studies claimed the GHG emissions of algal biofuel were lower than petroleum-based fuels, especially when using flue gas from coal fire plants, the life cycle GHG emissions could be negative which indicates algal biofuel production sequestrates carbon in the atmosphere. In general, thermochemical conversion (pyrolysis, liquefaction, etc.) has higher life cycle GHG emissions than biochemical conversions (lipid extraction or fermentation), because in thermochemical conversion more carbon is converted to CO$_2$ because of high temperature operation conditions.

Eutrophication: The eutrophication impact of algal biofuel production is mainly caused by the direct discharge of nutrients (nitrogen and phosphorus) into natural bodies of water during microalgae cultivation stage. Those nutrients are stem from the excess fertilizers applied to grow algae. If sewage is used to grow algae, the eutrophication potential is generally negative due to the environmental credits obtained by removing nutrients from the sewage. The algae conversion to biofuel stage can also produce wastewater containing high nutrients, but recycling the wastewater in this stage can reduce nutrient discharges. The problem is many nutrients produced in this stage are difficult for algae to absorb. Only partial nutrients could be used. The rest of the nutrients are released into the environment, causing eutrophication [19].

Water use: Algal biofuel production consumes much more water than fossil fuels. Almost all the water is used to make up evaporation water loss in cultivation facilities. In addition, electricity and fertilizer production can lead to indirect water consumption. The use of sewage or saline water can reduce the water use. However, if the concentration of nutrients in sewage used is too high, such as animal manure, it needs to be diluted before being applied. The water consumption in using manure water is therefore much higher than the use of other sewage. Besides, the use of wastewater to raise algae needs to increase the frequency of cleaning, which increases the water use.

Land use: The land use for algal biofuel production mainly occurs in the microalgae cultivation stage. Algae cultivation ponds and facilities occupy land. However, since algae ponds can be built on abandoned or idle land, it is generally considered that there will be no negative impact on land resources. However, it is necessary to build a raceway open pond on relatively flat and open land. In contrast, photobioreactors are easier to disassemble and less demanding on the site. The majority of LCA studies did not count land use in the biofuel production stage, because the land occupation is relatively small comparing to algae growing.

Environmental Toxicity: Certain microalgae produce toxic and carcinogenic substances that can cause damage to aquatic organisms and ecosystems, but the toxic microalgae can be screen out in algae cultivation. The toxicity of biofuel production is also difficult to estimate. Many LCA studies did not conduct microalgae toxicity analysis, but some incorporated environmental toxicity in producing infrastructures and other inputs. Since the production of bioreactors requires a large amount of steel and PVC, the production of these materials emits a large amount of toxic and carcinogenic substances.

Acid rain/acidification: Acid rain is mainly caused by SO_2 emissions. Microalgae bioenergy has no direct SO_2 emission in the life cycle. Therefore, acidification is mainly stemmed from indirect SO_2 emissions from electricity and infrastructure [2]. In the case of open pond cultivation, acidification is mainly stem from the indirect SO_2 emissions associated with concrete, electricity and fertilizer production. If a bioreactor is used to grow algae, the main SO_2 emissions are indirect emissions from the production of steel and PVC [35]. In general, algal biofuel has higher acidification potential than conventional soy diesel and mineral diesel because more electricity and infrastructure is used in producing per unit of fuel.

Ozone depletion: Ozone depletion is caused by emissions of CFCs and similar substances. The algal biofuel production is free of direct such emissions. The ozone depletion potential is mainly caused by the indirect CFCs associated with electricity, fertilizer production, infrastructure materials including concrete, PVC, and steel [35]. According to research, microalgae biodiesel has higher ozone layer consumption than other biodiesel but lower than petroleum diesel [2].

Respiratory effects: The emission of particulate matter (PM), mainly PM2.5 causes respiratory effects. In algal biofuel production life cycle, direct PM emissions come from burning algal biomass to generate electricity and heat, and biofuel combustion in vehicles. Indirect PM is mainly caused by upstream impacts of process inputs [36].

Smog formation: NOx plays an important role in the formation of smog. In algal biofuel production, NOx emissions are mainly occurred in biofuel production stage. The biochemical conversion technologies, there are no direct NOx emissions due to the low process temperature. But in thermochemical conversion, 30% −40% of the nitrogen added into the system is released into the atmosphere [37]. The indirect emissions are mainly caused by electricity and building materials in algae cultivation stage.

RESULTS AND DISCUSSION OF TEA FOR ALGAL BIOFUELS

Many TEA studies have been conducted in order to evaluate the feasibility of algal biofuels. However, as the industrial production does not exist, the biofuel costs vary significantly by different TEA modeling assumptions including final products, cultivation systems, biomass productivity, oil content, production capacities and conversion technologies. Many studies indicated that the cost of producing bio-oil ranged from $5.2−$13/GJ [38−40]. In the U.S. Department of Energy (DOE) 2008 report [41], the algae biodiesel cost was estimated as $2.11/L (in 2008 USD), which was much higher than soybean biodiesel. The technology used in this DOE report was based on algae cultivation followed by a centrifuge for biomass harvesting, and then oil extraction followed by transesterification. Nagarajan and his colleagues [42] reviewed studies conducted before 2013 and claimed the final costs of algal biodiesel were in the range of $0.42−0.97/L (assuming biomass productivity of 30 g/m^2/day − 60 g/m^2/day with lipid content of 50%) with freshwater open pond cultivation, lipid extraction followed transesterification process.

Besides research done in the U.S., Mehmood Ali [43] reviewed research on costs of algal biodiesel to figure out economic feasibility of biodiesel production in Pakistan. The conclusion showed the microalgal biodiesel has promising prospects compared to the conventional diesel in Pakistan, and its large-scale cultivation should be considered if the high production cost of biodiesel can be reduced by using wastewater and marginal land of Pakistan in cultivation of microalgae as a feedstock to produce biodiesel. Also, the cost of algal biodiesel could compete with petroleum diesel if renewable energy such as solar power could be used in the harvesting and lipid extraction processes.

Vujadin Kovacevic et al. [44] conducted a cost-effectiveness analysis to analyze the algae energy production in the EU and compared the algae biodiesel with rapeseed biodiesel and fossil fuels in the EU-25 transportation sector from the aspects of private production, external and social costs and benefits of the fuels. The study showed that algal biodiesel has the lowest private cost for utilization in the EU road transportation and the break-even point of algal biodiesel is the lowest of the three when considering the CO_2 sequestration in algae growing as positive social costs and external costs. The study also suggested utilizing existing biodiesel facilities for algal biomass conversion and thus reducing necessary capital investment.

The TEA studies show the algae price is a major factor determining the finial biofuel price. The price is further determined by the algae yield, lipid content, nutrients, and cultivation and dewatering facilities. The higher yield and lipid content and the lower nutrient inputs will reduce cost accordingly. In addition, the open pond costs lower than PBRs in term of per algae produced [45], which is caused by relative lower material and energy uses in open ponds.

In order to reduce costs of algae, many studies proposed to utilize various wastes as free nutrient sources in algae production. Therefore, combining algae cultivation and wastewater treatment for biofuel production is recommended as the most feasible way for resource utilization. Biller et al. 2012 [47] claimed that utilizing primarily treated municipal wastewater helped to reduce the total cost during cultivation stage up to 75% by energy efficiency and optimum nutrient utilization. Xin et al. 2016 [46] conducted a comprehensive TEA to assess the feasibility of algal biofuel production with various types of wastewater. The estimated break-even selling price of the biooil was estimated at $2.23/gal with concentrate (wastewater collected in sludge dewatering in a secondary wastewater treatment plant) as algae feed. Nugroho Adi Sasongko [48] in Japan simulated algae cultivation with municipal wastewater and flue gas as nutrient sources. The study showed using wastewater reduced the nutrient cost from 1605.9 JPY/kg biomass to 160.6 JPY/kg biomass (90% drop). However, the unexpected presence of contaminants and suspended solids (SS) build up in wastewater may influence the feasibility of the proposed system.

Besides using wastewater, other nutrients sources were examined. The study by Barreiro et al. [49] showed using flue gases from the surrounding power plant and organic matter from hydrothermal processing co-products in algal biofuel production could provide another source of nutrients enhancement (medium potassium and sodium acetate) for microalgae growth. Fei et al. [50] proposed the use of volatile fatty acids (VFA) obtained from food waste to grow high lipid content microalgae species such as *protothecoides* through heterotrophic culturing. The TEA of this technology indicated the biofuel produced is cost competitive and very promising compared to other biofuels.

Drying algae for fuel production is another cost intensive process. The innovative drying strategies were proposed as well to reduce the cost. For example, algae blended with woody biomass have been proposed to reduce drying costs. Compared with general drying methods, blending 40% algae and 60% pine could obviously improve the drying effect and reduce cost while maintaining conversion facilities at full capacity [51].

Algae biomass conversion routes with various conversion technologies also affect TEA for algal biofuels production because they determine the algal biofuel yield. Traditional oil extraction followed by transesterification has lower cost than other conversion technologies when lipid content of algae is high (>25%). When lipid content is low, conversion technologies such as hydrothermal liquefaction (HTL) or pyrolysis followed by hydroupgrading is expected to yield more fuel, and therefore reduces the cost [46].

A better integration of various algae biomass conversion technologies and algae waste management could reduce the cost in algal biofuel production. For example, Delrue [52] developed an economic model of algal biodiesel production in an integration pathway with hybrid raceway/PBR cultivation, belt filter press dewatering, wet lipid extraction, oil hydrotreating and anaerobic digestion of residues. The result showed that the diesel costs between $2.52 and $4.35/L, which was better than a reference pathway. Jey-R S [53] also compared costs of four different routes of microalgal bioenergy conversion. The result indicated the supercritical gasification of algae produced the highest net energy and removed the highest CO_2, while integration digestion of the residuals after lipid extraction had the lowest net cost.

Most TEA studies showed microalgae biofuel systems have launched a new generation of biofuel production. However, microalgae biofuel has not been economically viable. Due to the significant initial capital cost and the high production cost, the algal biofuels have not been able to compete directly with the petroleum-based fuels so far.

POLICIES AND MARKET INTENSIVES FOR ALGAL BIOFUEL

Current EU policies do not concretely distinguish or support biofuels with better environmental performance.

However, there is an increasing momentum within governmental and especially non-governmental structures which could change that soon [44]. Biodiesel is exempt from fuel taxes in European countries such as Germany and France.

Many Asian countries set up energy policies and market intensives to promote biofuel production. Government of Pakistan has set of using 10% blended biodiesel by the year 2025 in Pakistan [34]. The incentives and investment subsidies were highly suggested to promote biofuels markets and attract private sector investment [54,55]. Similarly, Indonesia has established a policy of biofuel production by 2025 to achieve a 5% increase in basic energy supply. Malaysia has enacted a policy of introducing different proportions of blended biofuels and petroleum fuels in daily life [56]. Measures are also being taken to promote the development of biofuels. This policy mainly includes application of crops for transportation and factories, biofuel related machineries development, biofuel for trading purpose and biofuel for green atmosphere sustainability [57].

CONCLUSIONS

In conclusion, the feasibility of algal biofuel depends on its environmental and economic performance of production. Since the current industrial scale commercial microalgae application has not yet been developed, many production processes and parameters are still uncertain. This creates a lot of uncertainty in results of both LCA and TEA. How to improve accuracy and integrity is a huge challenge for LCA and TEA. It is also a problem to be solved in the future.

Many studies proposed to use wastewater and flue gas as a water and nutrient source to reduce both environmental impacts and operation costs. In addition, innovation on biofuel technologies was promoted to increase biofuel yields and therefore reduce environmental impacts and economic costs. In particular, the technology integration for algal biofuel production should avoid energy or emission intensive processes and include technologies to recycle and reuse waste streams created in biofuel production. Lastly, the biofuel production should focus on producing high value co-products to maximize return on investment.

The microalgae industry will become an independent industry in the next decade. It is inevitable that microalgae lipids will replace fossil fuels and become the mainstream fuels [58]. Several issues need to be paid attention to in transition of microalgae biofuel from laboratory to an actual industry. As the industrial scale production requires a large amount of microalgae biomass, algae processing, handling and storage will become a

major challenge. Land and weather could increase challenges in realizing algal biofuel commercialization. Using marginal lands and innovation on algae cultivation facilities and infrastructure are necessary. Lastly, in the biofuel production stage, the future project should address using existing refinery facilities and petroleum infrastructure, which could reduce impacts and costs significantly.

REFERENCES

[1] ISO, Environmental Management — Life Cycle Assessment — Principles and Framework, Assessment-Principles and Framework, ISO 14040) 2006, 2006.

[2] L. Lardon, A. Helias, B. Aialve, J.P. Steyer, O. Beranrd, Life cycle assessment of microalgae biodiesel, Environ. Sci. Technol. 43 (17) (2009) 6475–6481.

[3] L.B. Brentner, M.J. Eckelman, J.B. Zimmerman, Combinatorial life cycle assessment to inform process design of industrial production of algal biodiesel, Environ. Sci. Technol. 45 (16) (2011) 7060–7067.

[4] L. Batan, J. Quinn, B. Willson, T. Bradley, Supplementary information for net energy and greenhouse gas emissions evaluation of biodiesel derived from microalgae, Environ. Sci. Technol. 3539 (13) (2010) 1–17.

[5] L. Batan, J. Quinn, T. Bradley, B. Willson, Net energy and greenhouse gas emissions evaluation of biodiesel derived from microalgae, Environ. Sci. Technol. 45 (3) (2011) 1160.

[6] J.C. Quinn, R. Davis, The potentials and challenges of algae based biofuels: a review of the techno-economic, life cycle, and resource assessment modeling, Bioresour. Technol. 184 (2015) 444–452.

[7] J.Q. Quinn, A. Hanif, S. Sharvelle, T.H. Bradley, Microalgae to biofuels: life cycle impacts of methane production of anaerobically digested lipid extracted algae, Bioresour. Technol. 171 (2014) 37–43.

[8] J.C. Quinn, K.B. Catton, S. Johnson, T.H. Bradley, Geographical assessment of microalgae biofuels potential incorporating resource availability, Bioenergy Res. 6 (2) (2013) 591–600.

[9] J. Quinn, L. de Winter, T. Bradley, Microalgae bulk growth model with application to industrial scale systems, Bioresour. Technol. 102 (8) (2011) 5083–5092.

[10] J. Barlow, R. Sims, J. Quinn, Techno-economic and life-cycle assessment of an attached growth algal biorefinery, Bioresour. Technol. 220 (2016) 360–368.

[11] A.F. Clarens, H. Nassau, E.P. Resurreccion, M.A. White, L.M. Colosi, Environmental impacts of algae-derived biodiesel and bioelectricity for transportation, Environ. Sci. Technol. 45 (17) (2011) 7554–7560.

[12] A.F. Clarens, H. Nassau, E.P. Resurreccion, M.A. White, L.M. Colosi, Environmental life cycle comparison of algae to other bioenergy feedstocks, Environ. Sci. Technol. 44 (5) (2010) 1813–1819.

[13] X. Liu, B. Saydah, P. Eranki, L.M. Colosi, B. Greg Mitchell, J. Rhodes, A.F. Clarens, Pilot-scale data provide enhanced estimates of the life cycle energy and emissions profile of

algae biofuels produced via hydrothermal liquefaction, Bioresour. Technol. 148 (2013) 163−171.

[14] K. Sander, G.S. Murthy, Life cycle analysis of algae biodiesel, Int. J. Life Cycle Assess. 15 (7) (2010) 704−714.

[15] E.D. Frank, A. Elgowainy, J. Han, Z. Wang, Life cycle comparison of hydrothermal liquefaction and lipid extraction pathways to renewable diesel from algae, Mitig. Adapt. Strategies Glob. Change 18 (1) (2013) 137−158.

[16] E.D. Frank, J. Han, I. Palou-Rivera, A. Elgowainy, M.Q. Wang, Methane and nitrous oxide emissions affect the life-cycle analysis of algal biofuels, Environ. Res. Lett. 7 (1) (2012) 14030.

[17] R.E. Davis, D.B. Fishman, E.D. Frank, M.C. Johnson, S.B. Jones, C.M. Kinchin, R.L. Skaggs, E.R. Venetris, M.S. Wigmosta, Integrated evaluation of cost, emissions, and resource potential for algal biofuels at the national scale, Environ. Sci. Technol. 48 (10) (2014) 6035−6042.

[18] D. Mu, M. Min, B. Krohn, K.A. Mullins, R. Ruan, J. Hill, Life cycle environmental impacts of wastewater-based algal biofuels, Environ. Sci. Technol. 48 (19) (2014) 11696−11704.

[19] D. Mu, R. Ruan, M. Addy, S. Mack, P. Chen, Y. Zhou, Life cycle assessment and nutrient analysis of various processing pathways in algal biofuel production, Bioresour. Technol. 230 (2017) 33−42.

[20] X. Ou, X. Yan, X. Zhang, X. Zhang, Life-cycle energy use and greenhouse gas emissions analysis for bio-liquid jet fuel from open pond-based micro-algae under China conditions, Energies 6 (9) (2013) 4897−4923.

[21] S. Liang, M. Xu, T. Zhang, Life cycle assessment of biodiesel production in China, Bioresour. Technol. 129 (2013) 72−77.

[22] P. Collet, A. Hélias, L. Lardon, M. Ras, R.A. Goy, J.P. Steyer, Life-cycle assessment of microalgae culture coupled to biogas production, Bioresour. Technol. 102 (1) (2011) 207−214.

[23] P. Collet, D. Spinelli, L. Lardon, A. Hélias, J.P. Steyer, Life-cycle assessment of microalgal-based biofuels, in: A. Pandey, D.J. Lee, Y. Chisti, C.R. Soccol (Eds.), Biofuels from Algae, Elsevier, 2013, pp. 287−312.

[24] P. Collet, A. Hélias, L. Lardon, J.P. Steyer, O. Bernard, Recommendations for life cycle assessment of algal fuels, Appl. Energy 154 (2015) 1089−1102.

[25] P.K. Campbell, T. Beer, D. Batten, Life cycle assessment of biodiesel production from microalgae in ponds, Bioresour. Technol. 102 (1) (2011) 50−56.

[26] S. Jez, D. Spinelli, A. Fierro, A. Dibenedetto, M. Aresta, E. Busi, R. Basosi, Comparative life cycle assessment study on environmental impact of oil production from microalgae and terrestrial oilseed crops, Bioresour. Technol. 239 (2017) 266−275.

[27] S.C. Togarcheti, M.K. Mediboyina, V.S. Chauhan, S. Mukherji, S. Ravi, S.N. Mudliar, Life cycle assessment of microalgae-based biodiesel production to evaluate the impact of biomass productivity and energy source, Resour. Conserv. Recycl. 122 (2017) 286−294.

[28] S.P. Souza, A.R. Gopal, E.A. Joaquim, Life cycle assessment of biofuels from an integrated Brazilian algae-sugarcane biorefinery, Energy 81 (2015) 373−381.

[29] E. Gnansounou, J.K. Raman, Life cycle assessment of algae biodiesel and its co-products, Appl. Energy 161 (2016) 300−308.

[30] M. Van Deal, N. Marquez, P. Reumerman, L. Pelkmans, T. Kuppens, S. Van Passel, Development and techno-economic evaluation of a biorefinery based on biomass (waste) streams − case study in The Netherlands, Biofuels Bioprod. Bioref. 8 (5) (2014) 635−644.

[31] T. Kuppens, M. Van Deal, K. Vanreppelen, T. Thewys, J. Yperman, R. Carleer, S. Schreurs, S. Van Passel, Techno-economic assessment of fast pyrolysis for the valorization of short rotation coppice cultivated for phytoextraction, J. Clean. Prod. 88 (2015) 336−344.

[32] E.D. Frank, J. Han, I. Palou-Rivera, A. Elgowainy, M.Q. Wang, Life-Cycle Analysis of Algal Biofuel with the GREET Model, 2011.

[33] D.L. Sills, V. Paramita, M.J. Franke, M.C. Johnson, T.M. Akabas, C.H. Greene, J.W. Tester, Quantitative uncertainty analysis of life cycle assessment for algal biofuel production, Environ. Sci. Technol. 47 (2) (2012) 687−694.

[34] R. Slade, A. Bauen, Micro-algae cultivation for biofuels: cost, energy balance, environmental impacts and future prospects, Biomass Bioenergy 53 (0) (2013) 29−38.

[35] A.G. Silva, R. Carter, F.L.M. Merss, D.O. Corrêa, J.V.C. Vargas, A.B. Mariano, J.C. Ordonez, M.D. Scherer, Life cycle assessment of biomass production in microalgae compact photobioreactors, GCB Bioenergy 7 (2) (2015) 184−194.

[36] K. Soratana, W.F. Harper, A.E. Landis, Microalgal biodiesel and the renewable fuel standard's greenhouse gas requirement, Energy Policy 46 (2012) 498−510.

[37] P. Biller, B.K. Sharma, B. Kunwar, A.B. Ross, Hydroprocessing of bio-crude from continuous hydrothermal liquefaction of microalgae, Fuel 159 (2015) 197−205.

[38] S. Sarkar, A. Kumar, Large-scale biohydrogen production from bio-oil, Bioresour. Technol. 101 (19) (2010) 7350−7361.

[39] M. Ringer, V. Putsche, J. Scahill, Large-scale pyrolysis oil production: a technology assessment and economic analysis, Environ. Sci. 17 (2006) 21−33.

[40] G. Peacocke, A. Bridgwater, J. Brammer, Techno-economic assessment of power production production from the Wellman and BTG fast pyrolysis processes, in: A. Bridgwater, D. Boocock (Eds.), Science in Thermal and Chemical Biomass Conversion, 2006, p. 1785.

[41] Algae biofuels, E.E.R.E. U.S. Department of Energy, in: Growing America's Energy Future. Alternative Fuels Data Center, Engineering News-Record, ENR The McGraw-Hill Companies, Inc, Washington, DC, USA, 2012.

[42] S. Nagarajan, S.K. Chou, S. Cao, C. Wub, Z. Zhou, An updated comprehensive techno-economic analysis of algae biodiesel, Bioresour. Technol. 145 (2013) 150−156.

[43] M. Ali, Biological process −fuel, in: The News International, The News International, Karachi, Pakistan, 2016.

[44] V. Kovacevic, J. Wesseler, Cost-effectiveness analysis of algae energy production in the EU, Energy Policy 38 (10) (2010) 5749–5757.

[45] J.W. Richardson, M.D. Johnson, J.L. Outlaw, Economic comparison of open pond raceways to photo bio-reactors for profitable production of algae for transportation fuels in the Southwest, Algal Res. 1 (1) (2012) 93–100.

[46] C. Xin, M. Addy, J. Zhao, Y. Cheng, S. Cheng, D. Mu, Y. Liu, R. Ding, P. Chen, R. Ruan, Comprehensive techno-economic analysis of wastewater-based algal biofuel production: a case study, Bioresour. Technol. 211 (2016) 584–593.

[47] P. Biller, A.B. Ross, S.C. Skill, A. Lea-Langton, B. Balasundaram, C. Hall, et al., Nutrient recycling of aqueous phase for microalgae cultivation from the hydrothermal liquefaction process, Algal Res. 1 (1) (2012) 70–76.

[48] N.A. Sasongko, R. Noguchi, J. Ito, M. Demura, S. Ichikawa, M. Nakajima, Engineering study of a pilot scale process plant for microalgae-oil production utilizing municipal wastewater and flue gases: fukushima pilot plant, Energies 11 (7) (2018) 1693.

[49] D.L. Barreiro, W. Prins, F. Ronsse, ChemInform Abstract: hydrothermal liquefaction (HTL) of microalgae for biofuel production: state of the art review and future prospects, J. Biomass Bioenergy 53 (16) (2013) 113–127.

[50] Q. Fei, R. Fu, L. Shang, C.J. Brigham, H.N. Chang, Lipid production by microalgae *Chlorella protothecoides* with volatile fatty acids (VFAs) as carbon sources in heterotrophic cultivation and its economic assessment, Bioproc. Biosyst. Eng. 38 (4) (2015) 691–700.

[51] D. Bradley, M. Wahlen, S. Roni, G. Kara, L. Cafferty, M. Wendt, L. Tyler Westover, D.M. Stevens, D.T. Newby, Managing variability in algal biomass production through drying and stabilization of feedstock blends, Algal Research 24 (2017) 9–18.

[52] F. Delrue, P.A. Setier, C. Sahut, L. Cournac, A. Roubaud, G. Peltier, A.K. Froment, An economic, sustainability, and energetic model of biodiesel production from microalgae, Bioresour. Technol. 111 (2012) 191–200.

[53] J.-R.S. Venturaa, B. Yanga, Y.-W. Leeb, K. Leea, D. Jahnga, Life cycle analyses of CO_2, energy, and cost for four different routes of microalgal bioenergy conversion, Bioresour. Technol. 137 (6) (2013) 302–310.

[54] I. Mukherjee, B.K. Sovacool, Sustainability principles of the Asian Development Bank's (ADB's) energy policy: an opportunity for greater future synergies, Renew. Energy 48 (2012) 173–182.

[55] F. Braadbaart, I. Poole, H.D.J. Huisman, Fuel, Fire and Heat: an experimental approach to highlight the potential of studying ash and char remains from archaeological contexts, J. Archaeol. Sci. 39 (4) (2012) 836–847.

[56] M.A. Abu Hassan, M.H. Puteh, Pre-treatment of palm oil mill effluent (POME): a comparison study using Chitosan and Alum, MJCE 19 (2007) 128–141.

[57] K.R. Jalbuena, The Growing Popularity of Biofuels in Southeast Asia, 2012 [cited 2015 Jan] Available from: http:// oilprice.com/Alternative-Energy/Biofuels/The-Growing-Popularity-of-Biofuels-in-Southeast-Asia.html.

[58] L.Y. Batan, G.D. Graff, T.H. Bradley, Techno-economic and Monte Carlo probabilistic analysis of microalgae biofuel production system, Bioresour. Technol. 219 (2016) 45–52.

CHAPTER 19

Environmental Resilience by Microalgae

M.L. SEREJO • M. FRANCO MORGADO • D. GARCÍA • A. GONZÁLEZ-SÁNCHEZ •
H.O. MÉNDEZ-ACOSTA • A. TOLEDO-CERVANTES

INTRODUCTION

The increasing growth of the industrial sector (chemical and pharmaceutical, plastics, paper and pulp mills, textile, mills, agriculture, etc.) have drastically risen the release of toxic effluents into water bodies. In addition, the growing urbanization releases an extensive amount of domestic and municipal wastewaters. Although wastewater is often regulated being obligatory to remove most of the nutrients before discharging it into water bodies, accidental leaks and bad regulations, especially in developing nations, play a major role in causing environmental pollution. Conventional techniques such as chemical precipitation, aerated lagoons, activated sludge, electroflocculation, anaerobic digestion, membrane filtration, ion exchange, catalytic oxidation, and disinfection are energy-consuming, and some of them have large land requirements. In addition, processes devoted to organic matter oxidation are not efficient in removing nitrogen, phosphorus, and heavy metals. In this sense and despite wastewater environment can be particularly toxic to microalgae and cyanobacteria (both commonly referred to as microalgae), they can adapt to these conditions supporting a cost-effective wastewater treatment with final effluents containing very low levels of nutrients and other pollutants.

Wastewater treatment by microalgae involves three main processes or the combination of them: (1) bioaccumulation, where the compounds are taken up by cells; (2) biodegradation, in which the pollutants are converted into simpler molecules by enzymes (exoenzymes or endoenzymes); and (3) biosorption or bioadsorption, where microalgal biomass (live or dead) adsorbs compounds onto its surface. Several microalgae and cyanobacteria (*Chlorella, Scenedesmus, Spirulina, Botryococcus, Spirogyra, Chlamydomonas, Phormidium, Tetraselmis,* etc.) have been reported for the treatment of the most polluting wastewaters, including textile wastewater, wastewater containing heavy metals, wastewaters with pharmaceutical contaminants (PCs), petroleum wastewater, palm oil mill effluent, olive mill effluent, and distillery wastewaters. Nonetheless, the removal efficiencies of nutrients and other pollutants by microalgae mainly depend on the physical-chemical characteristics of the wastewaters, microalgae resilience and tolerance to pollutants, and environmental conditions.

The aim of this chapter is to show the resilience of some microalgae strains for treating highly polluted wastewaters, highlighting the main challenges during the wastewater treatment process and the current panorama for the produced algal biomass valorization.

AVAILABILITY OF NUTRIENTS IN WASTEWATERS FOR MICROALGAE GROWTH

Microalgae are able to grow in diverse water bodies such as freshwater, brackish water, seawater, and wastewater due to their ability to metabolize both organic and inorganic nutrients. However, microalgae growth is controlled by the carbon (C), nitrogen (N), and phosphorus (P) content in wastewater as well as the concentration of micronutrients. In this sense, Posadas et al. [1] concluded that, in the absence of inhibitory or recalcitrant compounds, the C/N/P ratio of the wastewater correlates with its biodegradability, the optimum biodegradability ratio being 100/18/2. Still, most research studies (domestic, industrial, agroindustrial effluents, etc.) operate under carbon limitation, and therefore, carbon dioxide (CO_2) addition is often needed to mitigate the carbon deficiency. Despite this fact, during microalgae-based wastewater treatment processes, the synergistic interactions between photosynthesis and heterotrophic/nitrifying metabolism support the oxidation of the organic matter and nitrogen. Table 19.1 shows the removal efficiencies of nutrients and other contaminants from different wastewaters. For efficient bioremediation of wastewater, proper physical, chemical, and biological conditions are needed.

Microalgae Cultivation for Biofuels Production. https://doi.org/10.1016/B978-0-12-817536-1.00019-9

TABLE 19.1
Removal Efficiencies of Nutrients and Other Contaminants From Different Wastewaters.

Wastewater	Color (%)	Heavy Metals (%)	TN (%)	TP (%)	References
Fish processing			>99	76	[24]
Centrate and domestic			70	85	[25]
Fish farm and domestic			83	94	[26]
Centrate			86	92	[8]
Domestic			81–82	61–67	[7]
Piggery		93–98	83–87	89–91	[27]
Textile	75–80		71–87	35–57	[13]

Temperature, pH, mixing, irradiance, dissolved oxygen concentration (DO), CO_2, and the hydraulic retention time (HRT) are key environmental parameters that govern the performance of microalgae growth during wastewater treatment in algal-bacterial photobioreactors. In this sense, the HRT determines the carbon and nutrient loads supplied to the system and therefore the biomass productivity. Thus, to avoid the overload and the possible toxic effect of any wastewater compounds, this parameter must be selected taking into account the photobioreactor configuration, type of wastewater, the environmental conditions, and the microalgae nutrient assimilation capacity.

Nowadays, microalgae-based wastewater treatment provides a good option for nutrient recovery in the form of biomass. The ability of microalgae-bacteria consortia to assimilate both organic and inorganic carbon, nutrients, and some metals results in high biomass productivities and therefore enhanced nutrient recoveries. Different microalgae species have been investigated for nutrient removal; for instance, Shriwastav et al. [2] studied the effects of variable nutrient levels (N and P) in wastewaters on growth, productivity, and nutrient uptake of *Chlorella sorokiniana* [2]. Evans et al. [3] explored the potential of *Chlorella vulgaris* to treat municipal primary settled wastewater and evaluated its efficiency in removing ammonia (N-NH_3), phosphates (P-PO_4^{3-}), and chemical oxygen demand (COD) under static culture conditions [3]. Similarly, Prandini et al. [4] used *Scenedesmus* spp. to enhance the nutrient removal (N–NH_3 and PO_4^{3-}) from swine wastewater digestate under autotrophic and mixotrophic conditions. Moreover, *Scenedesmus* species are recognized to be able to growth under mixotrophic conditions such as cheese whey permeate [5] and to assimilate amino acids, yeast extracts, and proteinaceous algal residues [6].

In this context, algal-bacterial processes have been successfully applied to treat domestic wastewater [7], centrates [8], vinasses [9], digestates livestock effluents [10], tetracycline polluted waters [11], toluene [12], textile wastewater [13], and piggery wastewater [14]. However, a recent study has shown that during the treatment of piggery wastewater, efficient removal of nitrogen and total phosphorous occurred regardless of the microalgae species inoculated (*Chlorella* sp., *Acutodesmus obliquus*, *Oscillatoria* sp.) [14].

Despite the abovementioned advantages, microalgae-based wastewater treatment exhibits some disadvantages for full-scale implementation such as high land requirements and poor control of microalgae population in open systems. Furthermore, the poor sedimentation properties of most species limit a cost-effective biomass harvesting process for nutrient recovery [15].

Nutrients Removal Mechanisms During Microalgae-Based Wastewater Treatment

Carbon (organic+inorganic)

Assimilation into biomass (biotic) and CO_2 stripping (abiotic) are the major mechanisms underlying carbon removal from wastewaters in open photobioreactors, being the contribution of CO_2 striping to total carbon removal influenced by the inorganic carbon concentration, pH, and the liquid-gas mass transfer coefficients according to CO_2 equilibrium (Eq. 19.1)

$$CO_2 + H_2O \leftrightarrow H_2CO_3 \leftrightarrow HCO_3^- + H^+ \leftrightarrow CO_3^{2-} + 2H^+$$

$$(19.1)$$

Carbon is fixed into microalgae biomass ($\sim$50% content) using the electrons released during the light-dependent water photolysis. Approximately 1.8 kg of

CO_2 is required per kg of microalgae produced (Eq. 19.2).

$$4.92\,CO_2 + 0.99\,H_2O + 0.15\,NO_3^-$$
$$+ 0.01\,PO_4^{3-} \rightarrow CH_{1.7}O_{0.4}N_{0.15}P_{0.01} + 1.5O_2 + 0.28\,H^+$$
$$(19.2)$$

Microalgae are able to sequester CO_2 from flue gases generated in power plants or biogas, contributing to significantly reduce greenhouse gas emissions, whose emissions have increased from 22 Gt in 1990 to 33 Gt in 2010, and it is expected to reach 41 Gt by 2030 [16]. CO_2 represents nowadays the most important greenhouse gas, which atmospheric concentration has increased up to 0.5% over the past decade. Microalgae-based wastewater treatment coupled to CO_2-contaminated gas stream abatement is a promising alternative to mitigate the environmental impact of industries. Some microalgae are also able to obtain carbon and energy from organic substrates in the absence of light (heterotrophic growth) or to simultaneous assimilate organic and inorganic carbon (mixotrophic growth). Thus, microalgae and bacteria interaction can support an efficient removal of organic and inorganic carbon, nutrients, heavy metals, recalcitrant compounds, and pathogens in a single-step process.

Nitrogen

One of the major problems associated with the continuous discharge of effluents into water bodies is the so-called eutrophication phenomenon, which is the enrichment of water resources with nutrients, mainly nitrogen and phosphorus. Nitrogen is essential for the construction of proteins, nucleic acids, chlorophyll, and other biomolecules, accounting for 7%−20% dry weight basis. Nitrogen in the ammonium form (N-NH_4^+) and organic nitrogen (urea) are typically present at high concentrations in domestic and livestock wastewaters, while the presence of nitrate (NO_3) is commonly found in industrial discharges. Ammonium at a concentration higher than 100 mg N−NH_4^+ L^{-1} and pH 7−8 might inhibit the photosynthetic activity in some microalgae species, increasing at high pHs based on the aqueous NH_4^+ equilibrium ($NH_4^+ + OH^- \rightarrow NH_3 + H_2O$, pKa = 9.25) [1]. Microalgae can uptake NO_3^- or NH_4^+ through the biological reactions in which the inorganic nitrogen is ultimately reduced to ammonium prior to being incorporated into amino acids (L-glutamine). Thereafter, nitrogen assimilation is carried out through the glutamine synthetase enzyme system, in which glutamate reacts with NH_4^+ (driven energetically by ATP) to form the amino acid glutamine [17].

Assimilative, dissimilatory (nitrification-denitrification), and abiotic mechanisms support nitrogen removals of 90%−98% during microalgae-based wastewater treatment [18]. However, nitrogen removal via assimilation is often limited by the concentration of total carbon (organic + inorganic) present in wastewater. NH_4^+ nitrification involves the aerobic ammonium oxidation into nitrite and nitrate, which demands up to 4.57 mg O_2/g N−NH_4^+ for complete oxidation. Nitrate and nitrite formed during nitrification can be reduced to N_2 in the presence of an organic and inorganic electron donor in the so-called heterotrophic or autotrophic denitrification, respectively [19]. During microalgae-based wastewater treatment, nitrification and denitrification occur simultaneously due to the occurrence of diffusional gradients between the inner part of the algal-bacterial flocs or biofilms and the culture broth. The nitrate assimilation pathway from nitrate to amino acids is relatively simple at the structural level. However, its regulation to ensure an efficient assimilation is complex [20]. Microalgae have an important role in converting inorganic nitrogen to its organic form by assimilation. For instance, García et al. [7] achieved a complete denitrification-nitrification process in a novel anoxic-aerobic photobioreactor during the treatment of real domestic wastewater. On the other hand, some cyanobacteria can fix atmospheric nitrogen into ammonia such as *Oscillatoria* sp., *Nostoc* sp., and *Anabaena* sp. Likewise, some diatoms, *Rhizosolenia* and *Hemiaulus*, which have cyanobacterial symbionts, are diazotrophic microorganisms, which can reduce N_2 to NH_4^+ by the nitrogenase enzyme complex [17]. Finally, the increase in pH and modification of metal ion speciation (Ca^{2+}, Mg^{2+}, and Fe^{2+}) in the cultivation broth induced by microalgae growth can also promote the abiotic removal of nitrogen and phosphorus.

Phosphorus

Phosphorus is another essential macronutrient for microalgae growth that contributes to the eutrophication of lakes and rivers. This element can be removed by assimilation into biomass (typically containing 0.5%−1% P dry weight basis) or by precipitation at high pHs. Phosphorus in wastewater can be in the form of orthophosphate, polyphosphate, pyrophosphate, metaphosphate, and organic molecules. However, soluble phosphate (P-PO_4^3) is required for microalgae growth since not all the phosphorus compounds are bioavailable [21]. The uptake rate of P-PO_4^3 is affected by environmental factors such as irradiation, pH, temperature, and ions availability

$(K^+, Na^+,$ and Mg^{2+}) [17]. The *enhanced biological phosphorus uptake* (or luxury uptake) involves phosphorous accumulation as a reserve when the external phosphorus concentration limits the growth or as an energy source [22]. Certain heterotrophic bacteria (polyphosphate-accumulating organisms [PAOs]) are capable of sequestering high levels of phosphorus as intracellular polyphosphate, the most well-known group of PAOs being *Candidatus accumulibacter phosphatis* [23].

During microalgae-based wastewater treatment, photosynthesis increases the pH of the cultivation broth to pH values above 8.0 in which phosphorus precipitation may occur in the presence of Ca^{+2} and Mg^{+2} via hydroxyapatite formation (Eq. 19.3). Surface adsorption, via hydrogen bonds formation with the extracellular polysaccharides secreted by microalgae, also contributes to phosphorous removal (Fig. 19.1).

$$3HPO_4^{2-} + 5\ Ca^{2+} + 4\ OH^- \rightarrow Ca_5(OH)(PO_4)_3 + 3H_2O$$
$$(19.3)$$

Micronutrients and Heavy Metals in Wastewaters

Heavy metals released into the environment by agro-industrial activities exhibit a significant hazard to both natural ecosystems and human health due to their toxicity and persistence in the food chain. In this sense, heavy metals can inhibit growth, photosynthesis, and promote morphological modifications in the

microalgae cell walls even at very low concentrations [28]. Nonetheless, the capacity of microalgae for heavy metal biosorption from aquatic systems has been suggested since the 1970s due to their negatively charged cell surface and their large cell surface to volume ratio that display ideal properties for intra- and/or extracellular adsorption of heavy metals [29]. The biosorption process includes transport across cell membrane, complexation, ion exchange, precipitation, and physical adsorption [30]. Thus, microalgae support removal of heavy metals from wastewater through a combination of passive and active mechanism (Fig. 19.1).

The most important metals are potassium, magnesium, sulfur, and calcium, among others such as silica, manganese, zinc, copper, iron, and cobalt. These micronutrients are required in trace amounts by microalgae and rarely limit growth when wastewaters are used as nutrient source. Potassium is an activator of several enzymes involved in photosynthesis and respiration. This element is commonly found in chemical fertilizers, and its exploitation causes air and water pollution [31]. Biofertilizers also keep the soil rich in all kinds of micro- and macronutrients via nitrogen fixation and potassium solubilization or mineralization [32]. In addition, there is high amount of potassium in specific wastes such as sugarcane processing, spent grains, yeast, and manure [31]. Magnesium is an essential micronutrient for microalgae accounting for 0.35%–0.7% of biomass content; however, a content as high as 7.5% can be found in some species. Magnesium is vital for cell

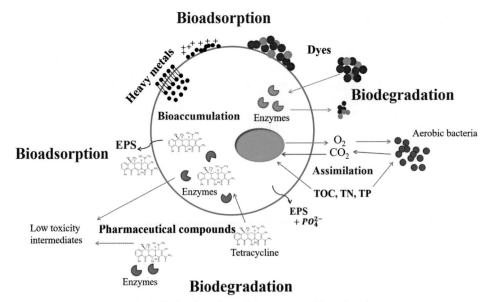

FIG. 19.1 Mechanisms for pollutants removal by microalgae.

processes such as ATP reactions for carbon fixation [17] and is a cofactor of many metabolic enzymes [33]. Normally, municipal wastewaters are rich in ammonium but deficient in magnesium, so magnesium supplementation is required. Other essential micronutrients by microalgae are sulfur, calcium, and iron since they constitute proteins, the cell wall, and chlorophyll, respectively [17]. Furthermore, iron is essential for the proteins that mediate electron transfer, *ferredoxin hydrogenase* [34].

TREATING AGRO-INDUSTRIAL AND INDUSTRIAL WASTEWATERS WITH MICROALGAE

Textile Wastewater

The textile industry is one of the most water-consuming industrial sectors and therefore releases large volume of wastewater. Textile wastewater (TWW) is characterized by its strong color and high amount of chemicals and auxiliaries (used during the dye solubilization process), which limit light absorption and severely deteriorate aquatic ecosystems. Furthermore, this kind of wastewater also presents significant concentrations of organic matter, nutrients, recalcitrant compounds, high salinity, heavy metals, and variable pH [35]. The composition of TWW depends mainly on the type of fabric, dyeing process, and chemicals/auxiliaries used during dyeing. Currently, more than 100,000 different textile dyes are used in several industrial processes [36]. Dyes are classified according to their functionalities in acid, basic, direct, disperse, metallic, reactive, sulfur, etc. [37]. There are several physicochemical and biological technologies to remove dyes depending on their concentrations and bioavailability. In general, physicochemical methods are effective for the removal of dyes but not for the complete mineralization of dyes and other recalcitrant compounds present in textile wastewaters. In this context, despite the fact that decolorization of TWW might be limited by its low biodegradability, microalgae cultivation for TWW treatment allows decolorization in the range of 13%–99%, while COD and total nitrogen (TN) reduction can reach up to 92% and 87%, respectively [13,36,38]. Some studies have reported different color and COD removal efficiencies; for instance, El-Kassas and Mohamed (2014) [39] reported ≈76% and 70%, respectively, while lower reductions of color (48%) and COD (62%) were obtained by Lim et al. [40], both treating real TWW using *C. vulgaris*. In another study, *Chlorella pyrenoidosa* was capable to decolorize 96% of diazo dye Direct Red-31, possible by the action of an azoreductase enzyme (Fig. 19.1),

while improving the water quality by reducing COD (83%), biochemical oxygen demand (BOD) (56%), sulfate (55%), phosphate (20%), and total dissolved solids (TDS) (84%) [41].

Mostly green microalgae of genera *Chlorella* have been reported for TWW treatment [42]. For instance, Aragaw and Asmare [38] reported the reduction of 83% of color, 92% of COD, 92% of BOD, and 89% of TDS by a microalgae consortium dominated by *Chlorella* sp. On the other hand, cyanobacteria *Anabaena* sp. and *Phormidium* sp. showed significant differences during the degradation of textile dyes, indigo and sulfur black. *Phormidium* sp. was able to decolorize the 91% of indigo dye and none of the sulfur black, while *Anabaena* sp. removed 72% and 91%, respectively [43]. Recently, a new photobioreactor configuration for the treatment of wastewaters with low C/N ratio based on promoting the nitrification-denitrification process was reported [13]. Dhaouefi et al. [13] observed effective disperse orange-3 and blue-1 color removals of ≈80% and 75%, respectively, in an anoxic-aerobic photobioreactor operated at an HRT of 10 days, using microalgae-bacteria consortium mainly composed by *A. obliquus*. Furthermore, the removal efficiencies of total organic carbon (TOC), TN, and total phosphorus (TP) of ≈48, 87% and 57%, respectively, were obtained. In general, microalgae cultivation for TWW treatment has been shown useful for dyes, organic matter, and nutrients removal by either dead biomass (biosorption) or living biomass (bioconversion) (Fig. 19.1). However, the efficient biodegradation of dyes depends on environmental factors such as temperature, pH, light, and nutrients availability that ultimately influence cell viability and enzymes activity, as well as the microalgae strain selection.

Wastewater Containing Heavy Metals

Heavy metals are one of the most hazardous environmental pollutants due to their persistence, toxicity, and biomagnification. Agro-industrial wastewaters frequently contain heavy metals, and their concentrations vary depending on the wastewater type and processing practices. Furthermore, different industries including metal plating, mining activities, smelting, battery and radiator manufactures, tanneries, chloralkali, petroleum refining, paint and pigment manufactures, pesticides, printing, and photographic produce wastewater containing high concentration of heavy metals such as chromium, copper, zinc, cadmium, nickel, lead, and mercury, among others [44]. Conventional technologies of heavy metal removal such as ion exchange, chemical precipitation, and reverse osmosis are ineffective and very expensive to treat low concentrations <100 mg/L [45]. Biosorption (with living or dead microorganisms) and bioaccumulation are the most studied biological removal mechanisms for heavy metals (Fig. 19.1) [46]. In this sense, microalgae remove heavy metals due to their large surface area and the presence of binding groups (i.e., carboxyl, sulfuryl, hydroxyl, phosphoryl, amine, carbohydrate, phosphate, etc.). Some reports suggest that dead microalgae can remove heavy metals ion similar to living cells, but involving different mechanisms. Passive uptake is a nonmetabolic, rapid, and reversible mechanism occurring in both living and nonliving cells, while the active uptake is a metabolism-dependent process, involving transport of metal ions across the cell membrane barrier and a subsequent accumulation inside the cell (Fig. 19.1) [44]. Doshi et al. [47] studied the potential of living and dead *Spirulina* sp. for the removal of chromium, nickel, and copper, finding the maximum uptake capacity (q_{max}, mg/g) with live *Spirulina* sp. ($\approx$333, 1378 and 389 mg/g, respectively). Similarly, live *Desmodesmus pleiomorphus* showed an uptake capacity of 85.3 mg/g for cadmium, while dead biomass allowed removing 58.6 mg/g [48,49].

Heavy metals removal by microalgae is affected by several factors, such as pH, concentration, nutrients availability, temperature, contact time, and the age of culture. The most important factor is pH since it affects the metal binding sites on algal biomass; however, its influence on the uptake capacity is metal specific. For instance, *C. vulgaris* cultured at pH 7.0 removed 99.4% of manganese (contact time of 3 h), while *Scenedesmus almeriensis* cultured at pH 9.5 removed 40.7% of arsenic with a contact time of 3 h and 38.6% of boron at pH 5.5 with only 10 min of contact time [50]. At this point, it is worth noticing that the heavy metal removal capacity of microalgae reported in the literature significantly varies from species to species and even among strains. This variation can be attributed to the different experimental conditions applied and the use of synthetic wastewaters. For instance, chromium reductions of $\approx$67%, 75%, and 100% were found using *Chlorococcum* sp., *Spirogyra condensate* plus *Rhizoclonium hieroglyphicum*, and *Chlorella miniata*, respectively, while other studies reported high removal of copper, cadmium, mercury, and zinc reaching up to 99%, 97%, 98%, and 85%, respectively, depending on the microalgae species and the environmental conditions [51–53]. Regarding bioremediation of real wastewaters, Ajayan et al. [54] observed removal of heavy metals from tannery wastewaters in the range of 81%–96%, 73%–98%, 75%–98%, and 65%–98% for chromium,

copper, lead, and zinc when culturing *Scenedesmus* sp. isolated from a local habitat. Palma et al. [55] studied the potential of microalgae biofilms (Chlorella-like microalga) for remediation of heavy metals in nickel refinery tailings waters. They found uptake capacities of 7.5, 0.5, 81.1, and 2.2 mg/g for nickel, cobalt, manganese, and strontium, respectively. Electroplating and galvanizing industry effluents were also treated using dried biomass of *C. vulgaris*, removing up to 81 mg/g of chromium (Cr^{+6}) [56]. Mohamed [57] studied the biosorption of zinc and cadmium by lyophilized and living cells of *Klebsiella pneumonia* KM609983. The maximum biosorption capacities for zinc were 243.9 and 232.4 mg/g, respectively, while those for cadmium were 227.3 and 212.8 mg/g, respectively, indicating a higher efficiency of lyophilized cells to heavy metals compared with living cells [57].

Pharmaceutical Contaminants

PCs are the excreted residue of diverse organic compounds consumed worldwide in agricultural activities, hospitals, industries, and domestic activities. PCs are emerging microcontaminants (ng/L to μg/L levels) that alter microbial communities contributing to the increase of antibiotic resistance, which is a major human health risk. Generally, PCs are resilient to biodegradation and have a long half-life time in the environment [58]. In addition, conventional biological wastewater treatment technologies are not designed to remove these emerging contaminants, often leading to its release into the environment. Therefore, more than 200 different PCs have been found in water bodies due to wastewater discharges [59], including antibiotics, analgesics, steroids, antidepressants, antipyretics, stimulants, antimicrobials, and disinfectants, among others [58]. For instance, 64 PCs were detected by Villar-Navarro et al. [60], in urban wastewater including acetaminophen (60−211 μg/L), ibuprofen (16−47 μg/L), and salicylic acid (13−43 μg/L), while azithromycin (5−3776 μg/L), erythromycin-H_2O (1−2009 μg/L), sulfadiazine (3−20 μg/L), and sulfamethazine (7−231 μg/L) were recorded in wastewaters from pharmaceutical industries [61]. Especially, attention has been paid to pharmaceutical compounds such as 17β-estradiol (E2), 17α-ethinylestradiol (EE2), diclofenac, and carbamazepine as priority emerging pollutants since they can cause adverse effects on aquatic ecosystems at very low levels [62].

The mechanisms of removal of PCs by microalgae include bioadsorption, bioaccumulation, and intracellular and extracellular biodegradation (Fig. 19.1) [58]. Other mechanisms possibly involved during the biological removal of PCs are indirect photodegradation, volatilization, hydrolysis, and conjugation [11,63,64]. Normally, when using microalgae cultures, an almost complete removal (99−100%) of ibuprofen, paracetamol, metoprolol, caffeine, triclosan, tetracycline, estradiol, diclofenac, 17a-ethynylestradiol, and salicylic acid [11,63,65−67] is reported with low removal for carbamazepine of ∼62% [63,64,68]. The biodegradation of PCs by microalgae mainly has been studied at laboratory scale using monocultures. For example, ibuprofen removals of 40%, 60%, 99%, and 100% were recorded for *Nannochloris* sp. [69], *Navicula* sp. [70], microalgae consortium (mainly *Chlorella* sp. and *Scenedesmus* sp.) [63], and *C. sorokiniana* [66], respectively. Nonetheless, removal of PCs in algal systems is a complex process due to daily and seasonable environmental conditions such as temperature, DO, and pH, and the competitive or symbiotic interaction between microalgae and bacteria. These factors can ultimately affect the removal mechanisms and influence their contribution to global removal efficiencies. Finally, different approaches have been proposed to increase the PCs removal efficiencies and flexibility of the process such as microalgae acclimation [71], cometabolism [68] and the integration of microalgae-based technologies to advanced oxidation processes [58]. In this context, Daneshvar et al. [72] have recently proposed combining different applications of microalgae to achieve different environmental goals at the same time: (1) microalgae cultivation in wastewater for nutrients removal; (2) microalgal lipid production as biodiesel feedstock; and (3) tetracycline removal using lipid-extracted biomass. This approach can reduce the cost of investment toward the environmental sustainability of the microalgae-based wastewater treatment technology.

Petroleum Wastewater

The extraction of petroleum and gas generates large quantities of wastewaters (also called produced water) approximately 7- to 10-fold the volume of oil. The composition of petroleum wastewaters (PWWs) depends on the crude quality (nature of the geological formation) and the operating condition in the refinery. These effluents contain petroleum compounds such as volatile aromatic fractions of the oil including benzene, toluene, ethylbenzene, and xylene isomers (BTEX), polycyclic aromatic hydrocarbons, organic acids, oils and greases, phenols, sulfides, heavy metals, and radionuclides, as well as a very high salt concentration [73].

Bioremediation of PWW is often limited by the low food to microorganism ratio (F/M) and biological inhibition due to toxic compounds. Therefore, a long period of acclimation and long retention time are needed during the PWW treatment. Some of these problems can be avoided by replacing the suspended biomass reactors with fixed biomass reactors. Furthermore, it is well known that aerobic heterotrophic bacteria together with cyanobacteria consortia mineralize petroleum in contaminated sites. Cyanobacteria play an indirect role in oil biodegradation by providing the oxygen needed for the breakdown of aliphatic and aromatic compounds. This synergistic relationship among photosynthetic microorganisms and bacteria can be exploited for the economic treatment of recalcitrant and toxic compounds, since they are much easier to degrade aerobically than anaerobically. In this sense, Abid et al. [74] showed the feasibility of *Hydrocabonoclastic* native microbial and *Spongiochloris* sp. microalgae cultured in airlift bioreactors interconnected through CO_2/O_2 recycling for petroleum wastewater treatment. CO_2 fixation rate of 3 g/L d, high COD removal (97%), and almost total petroleum hydrocarbon (TPH) removal (99%) were recorded during the PWW treatment by *Spongiochloris* sp. Zheng and Ke [75] investigated the feasibility of treating oilfield wastewater by an oil-degrading bacterial community and the microalgae *Scenedesmus obliquus* GH2 immobilized into Ca-alginate beads. The experiment was carried out by operating an algal-bacterial fluidized bed reactor at different organic loading rates. Results showed COD, N-NH₃, and oil removal efficiencies of ≈71, 61%, and 84%, respectively, at 16 h HRT (0.45−0.49 g-COD L d).

Other algal systems have been tested; for instance, Hodges et al. [76] compared the PWW treatment in a rotating algae biofilm reactor (RABR) versus suspended-biomass open pond lagoon reactors, obtaining maximum TN and TP removals up to 72% and 56%, and 14% and 19%, respectively. The authors also observed that filamentous cyanobacteria dominated in the RABR, while the open pond culture was mainly composed of green microalgae [76]. On the other hand, the single culture of the microalga *C. vulgaris* has been recently addressed by Madadi et al. [73], obtaining COD, BOD, TN, TP, and TPH removal efficiencies of 11%, 51%, 43%, 54%, and 7%, respectively. In this case, the authors observed that the addition of surfactants increased the COD, TP, and TPH removal up to 38%, 100%, and 27%.

Palm Oil Mill Effluent and Olive Mill Effluent

Palm oil mill effluent (POME) is an agro-industrial wastewater mixture generated from the whole palm oil production process. A world production of about

70 Mt of palm oil was estimated for 2018, which represents approximately $0.9-1.5 \, m^3$ of POME generated for each ton of crude palm oil [79]. The typical composition of the POME includes concentrations of COD of $15-100 \, g/L$, BOD of $10-44 \, g/L$, N-NH$_3$ of $4-80 \, mg/L$, and TN in the range of $180-1400 \, mg/L$, with pH values of $3.5-5.2$ [80]. Microalgae such as *Chlorella* and *Botryococcus* capable of assimilating both organic carbon and CO_2, performing respiration and photosynthesis, are suggested for treating this kind of wastewater that provides mixotrophic conditions. Table 19.2 shows the maximum removal efficiencies recorded during POME treatment by different microalgae species. The removal efficiencies of COD and nutrients depend on the microalgae cultivation parameters (light, pH, cosubstrate addition), the dilution rate of the POME, and the strain selection. For instance, Cheah et al. [81] observed COD, TN, and TP removal efficiencies of 64%, 92%, and 83%, respectively, using *C. sorokiniana* CY-1, while Cheirsilp et al. [82] recorded reductions up to 65%, 99%, and 64% using *Chlorella* sp. C-MR, respectively. Furthermore, Ahmad et al. [83] addressed the heavy metals biosorption from POME using Ca-alginate-immobilized *C. vulgaris*. They found Fe^{+2}, Mn^{+2}, and Zn^{+2} uptake capacities of ≈ 130, 116, and $105 \, mg/g$, respectively, at pH of 6.0, 300 min of contact time, and $25°C$ of temperature.

The olive mill effluent (OME) is an agro-industrial wastewater obtained from olive oil extraction industry in a proportion of $0.5-1.5 \, m^3$ of OME per ton of olives. In 2017, the production of olive oil accounted for approximately 2.6 Mt worldwide, being 97% produced only in Mediterranean countries [90]. OME has high concentrations of COD $(35-320 \, g/L)$, BOD $(30-132 \, g/L)$, and suspended solids $(1-9 \, g/L)$ and low pH $(2.2-5.9)$, depending on the production process applied. Furthermore, OME contains high concentrations of recalcitrant compounds such as lignins and tannins, and phenolic compounds (phenol, flavonoids or polyphenols) that have carcinogenic, genotoxic, and immunotoxic effects [91] at concentrations ranging from 0.6% to 5.5% [92].

Acclimated or adapted microorganisms are required to aerobically degrade the recalcitrant compounds in OME and to reduce its toxicity. In this sense, microalgae are able to degrade phenols [93], but fewer studies have focused on the OME treatment by microalgae. Papazi et al. [94] reported the biodegradation of tyrosol and hydroxytyrosol, two of the main phenolic compounds present in OME, by *S. obliquus*, being the appropriate culture conditions and toxicity level, control the key for a successful biodegradation. Hodaifa et al. [95]

also studied the effect of phenolic compounds on the growth of *S. obliquus* cultured in diluted OME. They observed that the deficiency of nutrients and the presence of fats and color in the OME limited the growth. Similarly, Di Caprio et al. [96] concluded that the biodegradation of phenols is significantly affected by the concentration of inorganic nutrients since they observed removal of higher phenols when the OME was supplemented with BG11 medium. Therefore, the low nutrients content (N, P, and micronutrients) and the dark color of this kind of wastewater (due to tannins presence) might prevent photosynthetic activity and phenol degradation when using microalgae. Nonetheless, technologies such as adsorption onto activated carbon or filtration could be combined with algal treatment to reduce the color, with an associated increase in cost and energy demand of the wastewater treatment process. In this sense, combination of wastewaters with high nutrient content such as piggery wastewaters, domestic wastewaters, and anaerobic effluents can be also addressed to compensate the low C/N of the OME.

Distillery Wastewaters

Fermentation industry is considered the major pollutant industry among all agro-processing industries based on the volume of wastewater generated. In addition, the production of alcohol has increased over the years due to its extensive applications and derived products such as acetaldehyde, acetic acid, and ethyl acetate to pharmaceuticals, foods, paints, and perfumery manufacturing industries [97]. Distillery wastewaters are dark brown effluents named *spentwash, stillage, slop, or vinasses*. These wastewaters are generated during the distillation step of ethanol, and their characteristics vary significantly depending on the feedstock raw material used, starch-based materials (e.g., rice, barley, wheat, maize, potato, fruits), or cellulose-based material (sugarcane molasses, agave), process efficiency, and water handling. In general, vinasses are characterized by low pH, dark brown color, extremely high COD and BOD content, $80-100$ and $40-50 \, kg/L$, respectively [98], high concentrations of K, P, S, Fe, Mn, Zn, Cu, heavy metals, polyphenolic compounds, and organic compounds such as polysaccharides, reduced sugars, lignin, proteins, melanoidin, waxes, etc. In this kind of wastewater, melanoidin is one of the major pollutants causing serious environmental and health problems. Physical treatment methods applied in most of the distilleries include screening, sedimentation, floatation, and air stripping, while the biological treatment consists in anaerobic digestion followed by aerobic treatment [99].

TABLE 19.2
Microalgae Cultivation Conditions and the Removal Efficiencies Recorded During the Raw Petroleum Wastewaters (PWW) Treatment.

Raw PWW Characteristics	Microalgae	Cultivation Conditions	Maximum Removal Efficiencies	References
BOD 0.2–1.5 g/L COD 0.6–2.3 g/L TPH 50–150 mg/L TN 4.4–18.0 mg/L TP 2.5–3.7 mg/L	*Chlorella vulgaris*	Batch; 15 days cultivation time; 20% PWW (v/v); surfactants addition (Tween 20 or Tween 60)	BOD 100% COD 38% TN 51% TP 100% TPH 27%	[73]
COD 2.9 g/L TPH 188 mg/L TN 64 mg/L TP 17 mg/L Phenol 14 mg/L	*Spongiochloris* sp.	Airlift photobioreactor; 100% PWW (v/v); 120 days cultivation time	COD 97% TPH 99%	[74]
BOD 0.1 g/L COD 0.3 g/L TDS 3.5 g/L TN 75–84 mg/L NH_3-N 53–68 mg/L TP 1.2–1.3 mg/L Oil 26–34 mg/L	*Scenedesmus obliquus* GH2	Fluidized bed photobioreactor; PWW + K_2HPO_4; 16 h HRT	COD 71% NH_3-N 61% Oil 84%	[75]
COD 0.2 g/L TN 25 mg/L TP 1.8 mg/L	Filamentous cyanobacteria	Continuous-flow RABR; 24 and 48 h HRT; 100% PWW (v/v)	24 h HRT: TN 72% TP 50% 48 h HRT: TN 71% TP 56%	[76]
COD 0.2 g/L TN 25 mg/L TP 1.8 mg/L	Green microalgae	Duplicate continuous-flow open pond lagoons; 32 h HRT; 100% PWW (v/v)	TN 14% TP 19%	[76]
COD 0.5 g/L TN 23 mg/L TP 1.9 mg/L	*C. vulgaris*	Batch; 13 days cultivation time	COD 71% TN 78% TP 100% Ca^{+2} 73% S^{-2} 70% Mg^{+2} 82% K^+ 54%	[77]
COD 1.3 g/L TDS 137.8 g/L Oil and grease 540 mg/L	*Nannochloropsis oculata*	Batch; 10% PWW (v/v)	Oil 89% COD 90%	[78]
COD 1.3 g/L TDS 137.8 g/L Oil and grease 540 mg/L	*Isochrysis galbana*	Batch; 10% PWW (v/v)	Oil 81% COD 72%	[78]

Distillery wastewater treatment by microalgae and cyanobacteria has focused on the major recalcitrant and toxic components. Kalavathi et al. [100] studied 13 marine cyanobacterial strains of *Oscillatoria,* *Phormidium, Spirulina,* and *Synechococcus* to determine their capacity to decolorize the anaerobically treated spentwash and their ability to use melanoidin as carbon and nitrogen source. They reported that the nonheterocystous form of *Oscillatoria boryana* BDU 92181 uses melanoidin as nitrogen and carbon source, leading to decolorization of the wastewater. The color removal was attributed to the production of hydrogen peroxide, hydroxyl anions, and oxygen during photosynthesis. In this context, *C. vulgaris* and the macrophyte *Lemna minuscula* have been also reported to be able to decolorize and remove organic matter from ethanol and citric acid production effluents [101]. Finally, efficient bioremediation of distillery wastewater using high-density semibatch culture of the acclimated *C. sorokiniana* was demonstrated by Solovchenko et al. [102]. They observed an almost complete removal of inorganic nutrients (nitrate (>95%), phosphate (77%), and sulfate (35%)) and organic matter (from 20 to 1.5 g/L) within 3–4 days.

Other Agro-industrial and Industrial Wastewaters

The use of microalgae cultures to treat several types of wastewaters has been shown earlier. Nonetheless, studies concerning to **tannery wastewaters** are still limited despite tanneries are a highly pollutant sector especially in terms of heavy metal release. The tannery process produces a complex wastewater that is highly saline with high organic matter content and other pollutants such as sulfides, chromium, oil and grease, dyes, surfactants, and pesticides (Fig. 19.2) [103,104]. Tannery wastewater has been studied for the cultivation of microalga *Scenedesmus* sp. under different wastewater concentrations (between 20% and 100%). *Scenedesmus* sp. was able to adapt to this nutrient source reaching a biomass concentration of 0.9 g/L while removing ∼85% of ammoniacal nitrogen (85.63%), ∼97% of phosphorus (96.78%), and ∼80% of COD at a tannery wastewater concentration of 88.4% [103]. Other studies have focused on the coculture of resilient microalgae to create a consortium that could effectively remediate these hazardous wastewaters. In this context, Das et al. [104] observed faster bioremediation of tannery wastewater with the *Chlorella* sp. and *Phormidium* sp. consortium than both of them individually. The consortium allowed TN and TP removals of 91.16% and 88%, respectively, while biosorption of chromium ranged from 90.17% to 94.45% [104]. Conversely, microalgae such as *Euglena proxima* and *Scenedesmus* sp. have shown high tolerance to heavy metals present in tannery wastewaters especially against Cr^{6+} (20 μg/mL) and Pb^{2+} (30 μg/mL) [54,105]. It is worth noticing that the toxic effects of chromium are valence dependent, and hexavalent chromium is highly soluble,

mutagenic, and carcinogenic, whereas trivalent form (Cr^{3+}) is less soluble and hence less bioavailable [104].

The brewery industry is another industrial sector that generates large volumes of highly polluted wastewater. The effluents derived from brewery industries are rich in organic compounds, such as proteins, phosphates, ammonia, and/or nitrate (Fig. 19.2). The fluctuant organic load of **brewery wastewaters** limits the selection of an appropriate treatment method since it has to be flexible enough to overcome the fluctuations while keeping the process economically viable. Regarding the potential of microalgae for brewery wastewaters treatment, there are few available. For instance, Marchão et al. [106] cultured the microalgae *S. obliquus* in a bubble column photobioreactor operated under batch and continuous regime. The most favorable condition for culturing the microalgae was a dilution rate of 0.26 d^{-1} with a biomass productivity of 0.2 g/d, achieving TN and COD removals of 97% and 74%, respectively. Raposo et al. [107] also studied the removal of nutrients and organic matter by using microalgae monocultures and consortium (microalgae *C. vulgaris* with cyanobacteria and bacteria), concluding that the consortium was more effective in reducing these wastewater contaminants. A new two-stage microalgae cultivation mode to increase lipid productivity while removing nutrients from brewery wastewaters was also evaluated by Farooq [108] using *Chlorella* spp. Finally, Travieso et al. [109] assessed the secondary treatment of brewery wastewater by culturing *C. vulgaris* SR/2. They observed a reduction of 98% of COD and BOD_5 while the treated effluent had a final quality suitable to be discharged to the environment [109].

On the other hand, the rapid increase in human population has risen the consumption of meat with the subsequent slaughterhouses and meat processing plants that generate a large volume of effluents. The consumption of water per slaughtered animal varies according to the animal, and thus, COD concentrations in **slaughterhouse wastewater** widely vary (Fig. 19.3). In general, slaughterhouse wastewaters contain high amounts of proteins, blood, fat, oil, feathers, lard, bones, hair, flesh, manure, heavy metals, disinfectants, cleaning agents, and pharmaceuticals used for veterinary purposes. Some studies have reported high removal efficiencies of TOC, TN, TP, and zinc when using slaughterhouse wastewater as a culture medium for microalgae. Taşkan [110] evaluated the performance of mixed algal species, achieving TOC, TN, and TP removal efficiencies of 89.6%, 70.2%, and 96.2%, respectively. Furthermore, microalgae collected from a storage lagoon containing

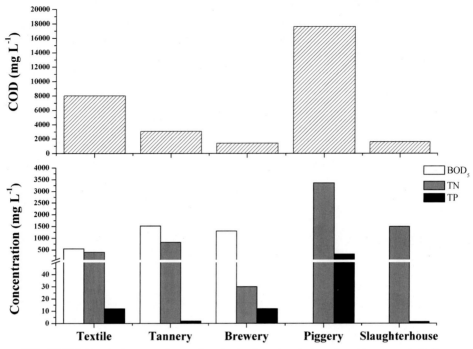

FIG. 19.2 Integrated process for wastewater treatment and microalgae biomass valorization.

aerobically treated swine manure were used for treating slaughterhouse wastewater in 75-L high rate algal ponds (HRAPs) [111]. Microalgae were able to adapt to the strong color and high organic content of the wastewater, with *Cyanophyta* species dominating over *Chlorophyta* and *Heterokontophyta* in the reactors. In this sense, the dynamics of microalgae population structure during wastewater treatment might vary according to its characteristics; however, *Chlorella* and *Scenedesmus* species are commonly the dominant species.

CHALLENGES IN TREATING WASTEWATERS WITH MICROALGAE

Among the challenges of treating agro-industrial and industrial wastewater applying microalgae-based processes are photobioreactor configuration, inorganic carbon availability, control of environmental parameters, and the downstream processing of microalgal biomass. Typically, the *photobioreactor configuration* used during the wastewater treatment processes is open system. The common configuration is the HRAP in which the factors that determine the algal biomass yield and therefore the nutrient removal performance are the incident solar radiation, geographic and climate localization, temperature, geometry, mass transfer, and

fluid dynamics, among others. This configuration is the simplest, and the construction and maintenance cost are low compared with closed photobioreactors. The basic configuration is a raceway pond of a half meter depth (normally operated at 30–40 cm liquid depth) with paddle wheels that keep the liquid in constant recirculation for mixing the nutrients and allow light penetration [112].

The coculture of microalgae and heterotrophic bacteria in this reactor configuration have proved to be effective in enhancing nutrients and other pollutants removal since microalgae provide the oxygen required by bacteria for the oxidation of organic matter. However, CO_2 released by bacteria might not be enough for microalgae growth, requiring other inorganic carbon sources [113]. In this context, some microalgae-based wastewater (domestic, industrial, agro-industrial effluents) treatment processes are operated under *carbon limitation*. CO_2 addition from flue gas or biogas could mitigate the abovementioned carbon deficiency. Thus, photosynthetic abatement of CO_2 coupled to nutrient removal from wastewaters can enhance the sustainability and economic viability of the effluents management. Furthermore, when biogas is used as CO_2 source, the algal symbiosis allows an integral upgrading (volatile organic compounds, hydrogen sulfur, and CO_2

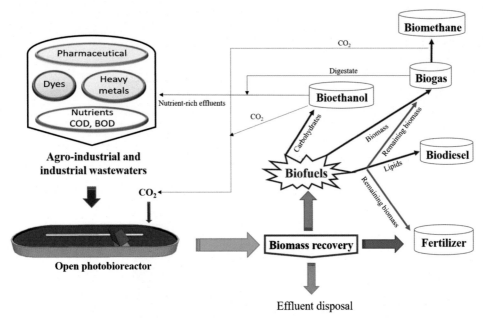

FIG. 19.3 Composition of different agro-industrial and industrial wastewaters.

removal) to biomethane suitable to be used as natural gas substitute [114]. The alkaline condition prevailing in these processes was recently identified as a key environmental parameter influencing biomethane quality [115]. In this sense, the *alkalinity* is also a key parameter that influences the kinetics of microalgae growth, phosphorous and nitrogen removal, and CO_2 absorption into the algal broth. Microalgae use CO_2 as carbon source in dissolved form; however, CO_2 is available in gas form, being its solubility affected by pH, salinity, pressure, and temperature. The solubility of CO_2 decreases when salinity and temperature in water increase, but in alkaline solutions, CO_2 dissociation equilibrium form HCO_3^-. Microalgae are capable to assimilate both CO_2 and HCO_3 of culture medium by diffusion, active transport, or using extracellular enzymes (carbonic anhydrase).

High algal biomass production is always desirable during the photobioreactor operation for wastewater treatment; however, the high *dissolved oxygen concentrations* reached in dense cultures can lead to the accumulation of reactive oxygen species such as superoxide anions ($O_2 \cdot^-$), hydroxyl radicals ($OH \cdot$), hydroxyl ions (OH^-), hydrogen peroxide (H_2O_2), and hypochlorite ions (OCl^-), which can induce detrimental effects in cells [116]. The increment of dissolved oxygen concentration at optimum irradiances can also induce photorespiration in microalgae cells (a competing

process to carboxylation by Rubisco enzyme). In this metabolic process, Rubisco that has oxygenase activity function reduces the net photosynthesis process up to 50%. A compromise between good irradiance supply for biomass production and low oxygen accumulation must be accomplished to guarantee the efficient wastewater treatment process. The maximum theoretical photosynthetic efficiency of solar energy conversion into biomass is around 12%, so the excess of photons can also lead to *photoinhibition* that reduces the photosynthetic yield by damaging key proteins in the photosynthetic apparatus. This phenomenon is common in open ponds since they are subjected to fluctuating environmental conditions. However, in high cell density cultures, self-shading prevents excessive light exposure but impacts on microalgae productivity. A good mixing is thus necessary to overcome the negative effects of self-shading and to prevent photoinhibition.

Another main issues prevailing in open ponds are contamination by other microorganisms or predators, high evaporation rates that concentrate the pollutants, high temperatures, limited diffusion of CO_2 from the atmosphere, large land requirements, and poor control over operating parameters. *Temperature* of algal broth becomes very important when microalgae are growth outdoors for wastewater treatment since circadian variations might promote temperatures between 10 and 45°C. The temperature influences both the CO_2

availability and the microalgae metabolic process. Microalgae are able to grow in a large range of temperature $15-35°C$ with optimal values at $20-25°C$. Nonetheless, there are several microalgae species capable of growing under extreme temperatures, i.e., *C. pyrenoidosa* (optimum growth $33°C$) that tolerates up to $45°C$ [117] and *Chlorogleopsis* sp. (thermophilic cyanobacteria) that grows at $50°C$ [118]. Strategies such as controlling the pH, temperature, alkalinity, and the dilution rate as well as nutrients supply coupled to CO_2 addition (flue gas or biogas) might counteract some of these problems [114,115].

Closed photobioreactors are proposed for better control of the operating parameters and to avoid contamination problems. Either if the light is provided artificially or by the sun, light limitation can occur at high biomass concentrations. In addition, oxygen accumulation up to 400% can be observed in this kind of systems due to the higher biomass concentration achieved compared with that reached in open ponds. The most common configurations of closed photobioreactors are bubble columns, flat plate reactors, and tubular photobioreactors. Recent studies have focused on digestate treatment in tubular photobioreactors [119] and agricultural runoff waste in hybrid (semi-closed) horizontal tubular photobioreactor [120]; however, for wastewater treatment, the use of closed photobioreactors is not widely accepted. Despite the technological advances in design and operation of closed photobioreactors, the *limited scalability*, high construction, operating and maintaining costs associated to the energy consumption for mixing the liquid broth hider its application to large scale [112]. Currently, the only large-scale microalgae cultivation systems are open ponds devoted to producing high-value products. McGinn et al. [121] have suggested that in the future the role of closed photobioreactor in the large-scale process will be as a repository of high-quality inoculum in the event of contamination or crash of the culture or as a component of a hybrid system.

Biomass concentrations typically achieved in closed photobioreactors are around $4-12$ g/L, while in open raceways, ponds treating domestic wastewaters low biomass concentrations are reported $(0.1-0.5$ g/L). Biomass concentrations can be improved by increasing the nutrient load to the system by means of operating at lower HRTs (typically $5-10$ days). Nevertheless, when increasing the nutrient load, it could cause washout of the photobioreactor. In this context, the decoupling of the HRT and biomass retention time (inversely related to microalgae productivity) is an innovative strategy for maximizing the biomass productivity during microalgae cultivation in high-strength wastewaters. The control of the biomass productivity via regulation of the biomass recovery rate maximizes the nutrient recovery from wastewaters and provides an alternative to conventional biomass recovery methods [122].

In this sense, microalgae harvesting or *biomass recovery* negatively affects the biomass valorization. To choose the adequate harvesting method, the final application of biomass and the cell wall characteristics must be considered. The main harvesting techniques applied for microalgae recovery are coagulation, flocculation, flotation, centrifugation, filtration, and a combination of these techniques [123]. Centrifugal separation of the microalgae depends on cell settling characteristics (cell size and density). This technique is fast and effective with high recovery efficiencies (>90%) but preferred for small-scale and laboratory purposes since it is expensive and high energy-consuming; it is only appropriate for recovery of high-valued products. Flotation is a gravity separation process in which air or gas bubbles are used to carry the suspended matter to the top of a liquid surface where they can be collected. The flotation depends on the type of collector (surfactant or flocculants), pH and ionic strength of the medium, type of bubble formation, recycling rate, air tank pressure, HRT, and particle floating rate. This technique is suitable at large-scale processes due to its low cost, low space requirement, and short operation time required; however, the addition of surfactants is needed. On the other hand, chemical coagulation of microalgae biomass is widely applied on large-scale processes since it does not damage the cells, has fewer energy requirements, and allows for high biomass recovery efficiencies. Microalgae have a negative surface charge that, by adding flocculants, agents can be neutralized to boost the agglomeration of microalgae. Chemical coagulants should be effective in low doses, cheap, and nontoxic to avoid water contamination for further treatment.

The recovery of biomass is critical for wastewater treatment processes since its economy viability is based on the valorization of the algal biomass. In this sense, some studies have reported that biomass harvesting represents among 20%–30% of the total cost of production [124]. According to Al Hattab et al. [125] and Singh and Patidar [123], criteria including dewatering efficiency, cost, toxicity, suitability for industrial scale, time, species specificity, reusability of media, and maintenance must be taken into account for choosing the proper harvesting technique. Table 19.3 shows the harvesting technique suggested according to microalgal biomass application. Another suggested approach for maximizing microalgae recovery is to promote the

TABLE 19.3
Maximum Removal Efficiencies Recorded During Palm Oil Mill Effluent Treatment by Different Microalgae Species.

| | MAXIMUM REMOVAL EFFICIENCIES | | |
Microalgae Species	BOD and COD	Macro- and Micronutrients	References
Spirulina platensis	COD 90%	N-NH$_3$ 87% TP 80% Cr 84% Fe 45% Mn 85% Zn 55%	[84]
Chlorella vulgaris	BOD 62% COD 51%	N-NH$_3$ 61% N-NH$_4^+$ 54% TP 84%	[85]
Scenedesmus dimorphus	BOD 87% COD 86%	N-NH$_3$ 100% N-NH$_4^+$ 92% TP 99%	[85]
Nannochloropsis oculata	BOD 97% COD 95%	TOC 75% TN 91%	[86]
Tetraselmis suecica	BOD 96% COD 94%	TOC 71% TN 79%	[86]
Chlamydomonas sp. UKM6	COD 29%	TN 73% N-NH$_4^+$ 100% TP 64%	[87]
Chlamydomonas incerta	COD 4%	Nitrate 13% N-NH$_3$ 4% P-PO$_4$ 66%	[88]
Chlorella sp. C-MR	COD 65%	TN 99% TP 64%	[82]
Chlorella sorokiniana	COD 91%	Nitrate 71% N-NH$_4^+$ 98% TP 64%	[89]
C. vulgaris		Fe^{+2} 26 mg/g Mn^{+2} 22 mg/g Zn^{+2} 19 mg/g	[83]
C. sorokiniana CY-1	COD 64%	TN 92% TP 83%	[81]

growth onto a surface, in which way the biomass is naturally concentrated and would be more easily harvested (Table 19.4).

CHALLENGES AND FUTURE PROSPECT OF THE END USE OF MICROALGAE CULTIVATED ON WASTEWATERS

Most published researches on agro-industrial and industrial wastewater treatment using microalgae are focused on the influence of operating conditions. However, process efficiency should be evaluated considering the biomass production and its biochemical profile for producing high-added value products, since the cost of algal biomass production would be covered by the wastewater treatment plant capital. Furthermore, the removal of nutrients from the wastewaters by microalgae assimilation cuts the cost of nutritional requirements, and the photosynthetically produced oxygen in turn reduces the need for aeration and its associated

TABLE 19.4
Harvesting Techniques According to Microalgae
Application.

Application of Microalgae	Harvesting Technique
Biofuels	Coagulation and flocculation
Human and animal food	Filtration and centrifugation
High-value products	Filtration and centrifugation
Water quality restoration	Filtration

costs [35]. In this way, new biotechnologies for wastewater treatment could become environmentally and economically feasible solutions to overcome the negative impact of industrial effluents. Generally, lipids (7%−23%), proteins (6%−71%), and carbohydrates (5%−64%) constitute the microalgae biomass with those proportions depending on algae species and growth conditions. In this sense, microalgae with high lipid content are suitable for biooil and biodiesel production, while biomass rich in sugars is better suited for bioethanol and biogas production. Furthermore, the wasted biomass (lipid-extracted) generated during the biodiesel production process can be used as feedstock for anaerobic digestion (Fig. 19.3). Thus, wastewater-based microalgal cultivation is an ideal platform for pollutants removal and renewable energy production.

Biogas production from microalgal biomass depends on the cell wall components and their chemical composition ($CO_{0.48}H_{1.83}N_{0.11}P_{0.01}$). Theoretical methane yields are 0.415, 0.851, and 1.014 L CH_4/g VS for carbohydrate, protein, and lipid, respectively, while for microalgae biomass, a wide range of values 0.4−0.8 L CH_4 gVS^{-1} have been reported [126]. Biogas production also depends on the microalgae cultivation parameters (that affect its macromolecular composition), biomass harvesting technique, pretreatment applied, and anaerobic digestion conditions (temperature, pH, inoculum to substrate ratio, reactor configuration, etc.). Time of microalgae harvesting is also important for the biogas production process since it affects the distribution of intracellular macromolecules [127]. In the past decades, two main approaches have been evaluated to produce biogas from microalgae biomass, anaerobic digestion of the whole biomass, and lipid-extracted biomass (Fig. 19.3). Codigestion with carbon-rich substrates has also been investigated to overcome the low C/N

ratio of microalgae biomass (C/N = 6−9) [128]. On the other hand, selecting easily biodegradable species of microalgae is crucial since this prevents the need for biomass pretreatment to break the cell structure [129]. However, the microalgae typically studied for wastewater treatment *Chlorella* and *Scenedesmus* have resistant trilaminar membrane-like structure containing nonhydrolyzable sporopollenin-like biopolymer-algaenan. The overall cell wall structure has a complex organization with three distinct layers: rigid internal microfibrillar, medial trilaminar, and external columnar. In addition, *Chlorella* and *Scenedesmus* have internal rigid cell walls either glucose-mannose type or glucosamine type. Thus, biomass pretreatments (thermal, mechanical, chemical, and biological methods) have been subject to high intense research in recent years. Passos et al. [130] concluded that the pretreatment techniques are species specific due to cell structure and the bioavailability of different organic compounds. Furthermore, the formation of recalcitrant compounds might hinder the anaerobic digestion process [130].

Biohydrogen and biomethane production by anaerobic digestion of algal biomass is a promising alternative for the valorization of biomass due to their high calorific values. These gases can be converted to heat and electricity in the wastewater treatment plants and/or used as the so-called biohythane gas [131]. Traditional anaerobic digestion process involves a one-stage process in which hydrogen is not detected since it is consumed during methanogenesis to produce methane. However, separation of the anaerobic digestion phases, acidogenic-methanogenic, allows optimizing both processes separately for higher recovery of hydrogen and methane. During the acidogenic phase, the macromolecules such as carbohydrates, proteins, and lipids are converted to volatile fatty acids (VFAs) and hydrogen, while in the methanogenic phase, the VFAs are subsequently converted into CO_2 and methane. At this point, it is important to highlight that the upgrading of the produced biogas by removing the CO_2, H_2S, and other volatile compounds and the valorization of digestate can also be achieved by microalgae cultures in a biorefinery approach (Fig. 19.3) [114,122]. Similar to the one-stage process, the H_2 and CH_4 productivities strongly depend on the biochemical composition and the operating parameter of the two-stage anaerobic digestion process. Jehlee et al. [132] studied the production of biohythane from hydrothermal pretreated *Chlorella* sp. biomass achieving 190.0 mL H_2 gVS^{-1} and 319.8 mL CH_4 gVS^{-1}. Lipid-extracted microalgal biomass residues have been also valorized for biohythane production by Yang et al. [133], producing 46 mL H_2 gVS^{-1} and 393 mL CH_4 gVS^{-1}.

Bioethanol is becoming an attractive alternative to gasoline, and their mix can be used in engines with no modification, providing a higher octane rating. The production of bioethanol involves saccharification of the algal biomass followed by yeast fermentation. Genera *Scenedesmus, Chlorella, Chlorococcum, Chlamydomonas,* and *Tetraselmis* from Chlorophyta division and *Synechococcus* among other cyanobacteria have high carbohydrate content and have been extensively studied as feedstock. In addition, carbohydrates in microalgae can exceed 60% of dry weight when nutritional stress strategies are applied (nitrogen, phosphorous, and iron starvation). Nonetheless, pretreatment of biomass to obtain the fermentable fraction is required being the acid pretreatment the best option in terms of cost-effectiveness and low energy consumption [134]. Ethanol yield of 0.47 g/g, which is 91% of the theoretical yield, can be reached by fermentation of *Chlorella* biomass hydrolyzates [135]. In addition, the produced CO_2 as a by-product of the fermentation process can be recycled as carbon sources for microalgae cultivation, reducing the greenhouse gas emissions (Fig. 19.3).

Biodiesel is a mixture of alkyl esters of fatty acids, produced by transesterification of triglycerides. Microalgae are capable of accumulating more than 50% of lipid per dried weight especially under stress conditions. Nonetheless, the extraction of lipids from microalgal cells after cultivation is required for biofuel production; the common extraction methods include physical, solvent and supercritical fluid, ultrasound, and/or microwave extractions [136]. Thus, biodiesel production includes two main steps: (1) lipids extraction, in which the lipid extraction techniques must be selected taking into account the species and growth stage, lipid content in biomass, and the harvest process, and (2) the transesterification process in which alkaline or acidic, homogenous, or heterogeneous chemical catalysts can be applied, being alkaline (NaOH and KOH) catalysts largely used [137]. Nonetheless, large-scale commercialization of algae-based biodiesel is still limited by the high costs for biomass production and lipid extraction.

In this sense, microalgae biomass conversion into biofuels involves multiple-step processes that still require further investigations for improving the competitiveness of the microalgae industry. At the moment, the commercial use of microalgae relies on specific markets of ***high-value products***, such as omega-3 fatty acids (DHA and EPA) and pigments (astaxanthin and lutein) [138]. However, for human usage, sanitary raw materials must be used, prohibiting the use of wastewater for microalgal biomass production. In addition, protein/peptide products derived from microalgae strains other than *Arthrospira* and *Chlorella* must be analyzed in a regulatory framework. In this context, the variability of nutrient concentration in agro-industrial and industrial wastewaters and the presence of toxic and recalcitrant compounds also prevent the monoalgal culture of microalgae. High ***microalgae diversity*** has been often found during the treatment of different types of wastewaters, with a higher microalgae diversity in open systems than in closed photobioreactors [139]. In addition, abiotic factors such as temperature and carbon availability influence the predominance of native microalgae in open ponds. In this sense, the unialgal culture of specific species cannot be guaranteed during long-term photobioreactor operation unless extreme cultivation conditions are applied as a selective pressure. Nonetheless, easy settleable species predominance and biomass productivity can be controlled by recycling the harvested algal biomass back to the system [140].

Finally, another alternative for the end use of microalgae cultivated on wastewaters is to recovery as slow nutrient release ***fertilizer*** to prevent the soil infertility, biodiversity loss, and eutrophication problems associated to the overuse of synthetic agrochemicals. In this context, Garcia-Gonzalez and Sommerfeld [141] evaluated the potential agricultural applications of the microalga *Acutodesmus dimorphus* as seed primer, foliar fertilizer, and soil biofertilizer. They found that seeds of Roma tomato treated with *A. dimorphus* culture triggered faster seed germination, while dry biomass enhanced plant growth. Microalgal bacterial flocs (from an outdoor raceway pond treating aquaculture wastewater) and the marine microalgae *Nannochloropsis oculata* were also studied as slow-release fertilizers for tomato cultivation obtaining similar results of that achieved with commercial fertilizers in terms of plant growth [142]. Maurya et al. [143] have also proposed using the lipid-extracted biomass of *Chlorella variabilis* and *Lyngbya majuscula* as nitrogen source during the fertilization of maize (*Zea mays*), finding that both can be employed to reduce the usage of chemical fertilizers. Furthermore, the technical feasibility of cultivating *Spirulina platensis* with fish water for agricultural fertilizers production was investigated by Wuang et al. [144]. The Spirulina-based fertilizer enhanced the growth of leafy vegetables and improved the germination of Chinese Cabbage and Kai Lan [144].

PROOF OF CONCEPT AND SUCCESSFUL CASES

The use of wastewater reduces the requirements of nutrients and freshwater for microalgae cultivation as feedstock for the production of biofertilizer/biofuels. However, studies focusing on real wastewaters are scarce, and most of them are reported at laboratory scale. Thus, there is a gap regarding the technoenvironmental analysis of microalgae-based wastewater processes since only a few pilot/demonstration plants are operating worldwide. Available information on biofuel production with microalgae cultivated in wastewater only involves biodiesel and biogas production by *Scenedesmus* and *Chlorella* biomass obtained through swine, brewery, slaughterhouse, municipal, and domestic wastewaters. In this context, recently, Uggetti et al. [120] published the start-up of a microalgae-based treatment system within a biorefinery concept to convert wastewater to bioproducts. The experimental plant located at the UPC site Agrópolis (Viladecans, Barcelona) is part of the project INCOVER: "Innovative Eco-technologies for Resource Recovery from Wastewater" (http://incover-project.eu) and coordinated by the AIMEN Technology Center (Spain). This is the first plant coupling microalgae-based wastewater treatment with the production of bioplastics, biomethane, and biofertilizers. Furthermore, a demonstrative facility of this technology was developed by a consortium of companies led by AQUALIA under the *All-gas project* funded by the European Union responding to the 2020 targets. The All-gas project objective was the large-scale demonstration of biofuel production from microalgae using urban wastewater as nutrient source. Since 2010, different research phases have been accomplished within the project including the algae-based wastewater treatment evaluation, impact of different wastewaters and operating strategies over process performance, biogas production from algal biomass, and the evaluation of the scale effect and hydrodynamic optimization. The construction of the industrial plant was finished in 2017 having a culture area of more than 2 ha. This is the world largest facility for the production of biofuel from microalgae coupled to wastewater treatment. The project claims to treat the wastewater generated by 5000 inhabitants per hectare while producing the energy to move 20 cars around 30,000 km/y (www.all-gas.eu/en). Data from these projects have not been published, but previous studies focusing on the life cycle assessment (LCA) of microalgal-based wastewater systems have already shown promising results. For instance, Garfi et al. [145] compared a conventional wastewater treatment plant against a high rate algal pond system. The LCA showed that the microalgae-based treatment (despite having a larger footprint) was the most environmentally friendly alternative. Similarly, Arashiro et al. [146] evaluated an HRAP system for wastewater treatment with two alternatives for biomass valorization, biogas and biofertilizer production. The HRAP coupled with biogas production has shown to be more environmentally friendly, while in terms of cost, the HRAP coupled with biofertilizer production seems to be more economically feasible. In this sense, another LCA study showed that the coproduction of energy (biogas) is the key aspect triggering the favorable life cycle performance of the system regardless of the biogas uses (heat and electricity or heat, electricity, and biomethane), since the specific option is conditioned by the impact categories prioritized by decision-makers [147].

ACKNOWLEDGMENTS

To Fondo Sectorial CONACyT-SENER, CEMIE-Bio project No. 247006.

REFERENCES

[1] E. Posadas, S. Bochon, M. Coca, M.C. García-González, P.A. García-Encina, R. Muñoz, Microalgae-based agro-industrial wastewater treatment: a preliminary screening of biodegradability, J. Appl. Phycol. 26 (2014) 2335–2345.

[2] A. Shriwastav, S.K. Gupta, F.A. Ansari, I. Rawat, F. Bux, Adaptability of growth and nutrient uptake potential of *Chlorella sorokiniana* with variable nutrient loading, Bioresour. Technol. 174 (2014) 60–66.

[3] L. Evans, S.J. Hennige, N. Willoughby, A.J. Adeloye, M. Skroblin, T. Gutierrez, Effect of organic carbon enrichment on the treatment efficiency of primary settled wastewater by *Chlorella vulgaris*, Algal Res. 24 (2017) 368–377.

[4] J.M. Prandini, M.L.B. da Silva, M.P. Mezzari, M. Pirolli, W. Michelon, H.M. Soares, Enhancement of nutrient removal from swine wastewater digestate coupled to biogas purification by microalgae *Scenedesmus* spp. Bioresour. Technol. 202 (2016) 67–75.

[5] J.M. Girard, M.L. Roy, M. Ben Hafsa, J. Gagnon, N. Faucheux, M. Heitz, R. Tremblay, J.S. Deschênes, Mixotrophic cultivation of green microalgae *Scenedesmus obliquus* on cheese whey permeate for biodiesel production, Algal Res. 5 (2014) 241–248.

[6] E. Barbera, E. Sforza, S. Kumar, T. Morosinotto, A. Bertucco, Cultivation of Scenedesmus obliquus in

liquid hydrolysate from flash hydrolysis for nutrient recycling, Bioresour. Technol. 207 (2016) 59—66.

[7] D. García, C. Alcántara, S. Blanco, R. Pérez, S. Bolado, R. Muñoz, Enhanced carbon, nitrogen and phosphorus removal from domestic wastewater in a novel anoxic-aerobic photobioreactor coupled with biogas upgrading, Chem. Eng. J. 313 (2017) 424—434.

[8] E. Posadas, D. Marín, S. Blanco, R. Lebrero, R. Muñoz, Simultaneous biogas upgrading and centrate treatment in an outdoors pilot scale high rate algal pond, Bioresour. Technol. 232 (2017) 133—141.

[9] M.L. Serejo, E. Posadas, M.A. Boncz, S. Blanco, P. García-Encina, R. Muñoz, Influence of biogas flow rate on biomass composition during the optimization of biogas upgrading in microalgal-bacterial processes, Environ. Sci. Technol. 49 (2015) 3228—3236.

[10] M. Franchino, V. Tigini, G.C. Varese, R. Mussat Sartor, F. Bona, Microalgae treatment removes nutrients and reduces ecotoxicity of diluted piggery digestate, Sci. Total Environ. 569 (2016) 40—45.

[11] Z.N. Norvill, A. Toledo-Cervantes, S. Blanco, A. Shilton, B. Guieysse, R. Muñoz, Photodegradation and sorption govern tetracycline removal during wastewater treatment in algal ponds, Bioresour. Technol. 232 (2017).

[12] R. Lebrero, R. Ángeles, R. Pérez, R. Muñoz, Toluene biodegradation in an algal-bacterial airlift photobioreactor: influence of the biomass concentration and of the presence of an organic phase, J. Environ. Manage. 183 (2016) 585—593.

[13] Z. Dhaoufi, A. Toledo-Cervantes, D. García, A. Bedoui, K. Ghedira, L. Chekir-Ghedira, R. Muñoz, Assessing textile wastewater treatment in an anoxic-aerobic photobioreactor and the potential of the treated water for irrigation, Algal Res. 29 (2018) 170—178.

[14] D. García, E. Posadas, S. Blanco, G. Acién, P. García-Encina, S. Bolado, R. Muñoz, Evaluation of the dynamics of microalgae population structure and process performance during piggery wastewater treatment in algal-bacterial photobioreactors, Bioresour. Technol. 248 (2018) 120—126.

[15] Y. Su, A. Mennerich, B. Urban, Municipal wastewater treatment and biomass accumulation with a wastewater-born and settleable algal-bacterial culture, Water Res. 45 (2011) 3351—3358.

[16] W. Bank, World development indicators: energy dependency, Efficiency and Carbon Dioxide Emission 2014 (2017). http://wdi.worldbank.org/table/3.8.

[17] G. Markou, D. Vandamme, K. Muylaert, Microalgal and cyanobacterial cultivation: the supply of nutrients, Water Res. 65 (2014) 186—202.

[18] C. Alcántara, P.A. García-Encina, R. Muñoz, Evaluation of the simultaneous biogas upgrading and treatment of centrates in a high-rate algal pond through C, N and P mass balances, Water Sci. Technol. 72 (2015) 150—157.

[19] B.E. Rittmann, P.L. McCarty, Environmental Biotechnology:Principles and Applications, Tata McGraw-Hill, New Delhi, 2012.

[20] E. Sanz-Luque, A. Chamizo-Ampudia, A. Llamas, A. Galvan, E. Fernandez, Understanding nitrate assimilation and its regulation in microalgae, Front. Plant Sci. 6 (2015).

[21] A.L. Gonçalves, J.C.M. Pires, M. Simões, A review on the use of microalgal consortia for wastewater treatment, Algal Res. 24 (2017) 403—415.

[22] N. Brown, A. Shilton, Luxury uptake of phosphorus by microalgae in waste stabilisation ponds: current understanding and future direction, Rev. Environ. Sci. Biotechnol. 13 (2014) 321—328.

[23] Z. Yuan, S. Pratt, D.J. Batstone, Phosphorus recovery from wastewater through microbial processes, Curr. Opin. Biotechnol. 23 (2012) 878—883.

[24] B. Riaño, D. Hernández, M.C. García-González, Microalgal-based systems for wastewater treatment: effect of applied organic and nutrient loading rate on biomass composition, Ecol. Eng. 49 (2012) 112—117.

[25] E. Posadas, P.A. García-Encina, A. Soltau, A. Domínguez, I. Díaz, R. Muñoz, Carbon and nutrient removal from centrates and domestic wastewater using algal-bacterial biofilm bioreactors, Bioresour. Technol. 139 (2013) 50—58.

[26] E. Posadas, A. Muñoz, M.-C. García-González, R. Muñoz, P. García-Encina, A case study of a pilot high rate algal pond for the treatment of fish farm and domestic wastewaters, J. Chem. Technol. Biotechnol. 90 (2014) 1094—1101.

[27] D. García, I. de Godos, C. Domínguez, S. Turiel, S. Bolado, R. Muñoz, A systematic comparison of the potential of microalgae-bacteria and purple phototrophic bacteria consortia for the treatment of piggery wastewater, Bioresour. Technol. (2019).

[28] K.P. Gopinath, M.N. Kathiravan, R. Srinivasan, S. Sankaranarayanan, Evaluation and elimination of inhibitory effects of salts and heavy metal ions on biodegradation of Congo red by *Pseudomonas* sp. mutant, Bioresour. Technol. 102 (2011) 3687—3693.

[29] T. Li, G. Lin, B. Podola, M. Melkonian, Continuous removal of zinc from wastewater and mine dump leachate by a microalgal biofilm PSBR, J. Hazard Mater. 297 (2015) 112—118.

[30] V. Javanbakht, S.A. Alavi, H. Zilouei, Mechanisms of heavy metal removal using microorganisms as biosorbent, Water Sci. Technol. 69 (2014) 1775—1787.

[31] D.J. Batstone, T. Hülsen, C.M. Mehta, J. Keller, Platforms for energy and nutrient recovery from domestic wastewater: a review, Chemosphere 140 (2015) 2—11.

[32] D. Bhardwaj, M.W. Ansari, R.K. Sahoo, N. Tuteja, Biofertilizers function as key player in sustainable agriculture by improving soil fertility, plant tolerance and crop productivity, Microb. Cell Fact. 13 (2014) 1—10.

[33] A. Masuda, Y. Toya, H. Shimizu, Metabolic impact of nutrient starvation in mevalonate-producing *Escherichia coli*, Bioresour. Technol. 245 (2017) 1634—1640.

[34] K. Chandrasekhar, Y.-J. Lee, D.-W. Lee, Biohydrogen production: strategies to improve process efficiency through microbial routes, Int. J. Mol. Sci. 16 (2015) 8266—8293.

[35] A.F. Mohd Udaiyappan, H. Abu Hasan, M.S. Takriff, S.R. Sheikh Abdullah, A review of the potentials, challenges and current status of microalgae biomass applications in industrial wastewater treatment, J. Water Process. Eng. 20 (2017) 8−21.

[36] T. Fazal, A. Mushtaq, F. Rehman, A. Ullah Khan, N. Rashid, W. Farooq, M.S.U. Rehman, J. Xu, Bioremediation of textile wastewater and successive biodiesel production using microalgae, Renew. Sustain. Energy Rev. 82 (2018) 3107−3126.

[37] S. Natarajan, H.C. Bajaj, R.J. Tayade, Recent advances based on the synergetic effect of adsorption for removal of dyes from waste water using photocatalytic process, J. Environ. Sci. 65 (2018) 201−222.

[38] T.A. Aragaw, A.M. Asmare, Phycoremediation of textile wastewater using indigenous microalgae, Water Pract. Technol. 13 (2018) 274−284.

[39] H.Y. El-Kassas, L.A. Mohamed, Bioremediation of the textile waste effluent by *Chlorella vulgaris*, Egypt J. Aquat. Res. 40 (2014) 301−308.

[40] S.L. Lim, W.L. Chu, S.M. Phang, Use of *Chlorella vulgaris* for bioremediation of textile wastewater, Bioresour. Technol. 101 (2010) 7314−7322.

[41] S. Sinha, R. Singh, A.K. Chaurasia, S. Nigam, Self-sustainable *Chlorella pyrenoidosa* strain NCIM 2738 based photobioreactor for removal of direct Red-31 dye along with other industrial pollutants to improve the water-quality, J. Hazard. Mater. 306 (2016) 386−394.

[42] V.V. Pathak, R. Kothari, A.K. Chopra, D.P. Singh, Experimental and kinetic studies for phycoremediation and dye removal by *Chlorella pyrenoidosa* from textile wastewater, J. Environ. Manage. 163 (2015) 270−277.

[43] P.M. Dellamatrice, M.E. Silva-Stenico, L.A.B. de Moraes, M.F. Fiore, R.T.R. Monteiro, Degradation of textile dyes by cyanobacteria, Brazilian J. Microbiol. 48 (2017) 25−31.

[44] K. Suresh Kumar, H.U. Dahms, E.J. Won, J.S. Lee, K.H. Shin, Microalgae − a promising tool for heavy metal remediation, Ecotoxicol. Environ. Saf. 113 (2015) 329−352.

[45] M.M. Montazer-Rahmati, P. Rabbani, A. Abdolali, A.R. Keshtkar, Kinetics and equilibrium studies on biosorption of cadmium, lead, and nickel ions from aqueous solutions by intact and chemically modified brown algae, J. Hazard. Mater. 185 (2011) 401−407.

[46] A.K. Zeraatkar, H. Ahmadzadeh, A.F. Talebi, N.R. Moheimani, M.P. McHenry, Potential use of algae for heavy metal bioremediation, a critical review, J. Environ. Manage. 181 (2016) 817−831.

[47] H. Doshi, A. Ray, I.L. Kothari, Bioremediation potential of live and dead Spirulina: spectroscopic, kinetics and SEM studies, Biotechnol. Bioeng. 96 (2007) 1051−1063.

[48] C.M. Monteiro, P.M.L. Castro, F.X. Malcata, Cadmium removal by two strains of *Desmodesmus pleiomorphus* cells, Water Air Soil Pollut. 208 (2010) 17−27.

[49] C.M. Monteiro, P.M.L. Castro, F.X. Malcata, Capacity of simultaneous removal of zinc and cadmium from contaminated media, by two microalgae isolated from a polluted site, Environ. Chem. Lett. 9 (2011) 511−517.

[50] R. Saavedra, R. Muñoz, M.E. Taboada, M. Vega, S. Bolado, Comparative uptake study of arsenic, boron, copper, manganese and zinc from water by different green microalgae, Bioresour. Technol. 263 (2018) 49−57.

[51] A.H. Sulaymon, A.A. Mohammed, T.J. Al-Musawi, Competitive biosorption of lead, cadmium, copper, and arsenic ions using algae, Environ. Sci. Pollut. Res. 20 (2013) 3011−3023.

[52] A. Singh, S.K. Mehta, J.P. Gaur, Removal of heavy metals from aqueous solution by common freshwater filamentous algae, World J. Microbiol. Biotechnol. 23 (2007) 1115−1120.

[53] A. Chan, H. Salsali, E. McBean, Heavy metal removal (copper and zinc) in secondary effluent from wastewater treatment plants by microalgae, ACS Sustain. Chem. Eng. 2 (2014) 130−137.

[54] K.V. Ajayan, M. Selvaraju, P. Unnikannan, P. Sruthi, Phycoremediation of tannery wastewater using microalgae Scenedesmus species, Int. J. Phytoremediation 17 (2015) 907−916.

[55] H. Palma, E. Killoran, M. Sheehan, F. Berner, K. Heimann, Assessment of microalga biofilms for simultaneous remediation and biofuel generation in mine tailings water, Bioresour. Technol. 234 (2017) 327−335.

[56] G. Sibi, Biosorption of chromium from electroplating and galvanizing industrial effluents under extreme conditions using *Chlorella vulgaris*, Green Energy Environ. 1 (2016) 172−177.

[57] R.M. Mohamed, Biosorption of zinc and cadmium by Klebsiella pneumonia KM609983 isolated from Sohag, Egypt. Glob. Adv. Res. J. Microbiol. 4 (2015) 2315−5116.

[58] J.Q. Xiong, M.B. Kurade, B.H. Jeon, Can microalgae remove pharmaceutical contaminants from water? Trends Biotechnol. 36 (2018) 30−44.

[59] S.R. Hughes, P. Kay, L.E. Brown, Global synthesis and critical evaluation of pharmaceutical data sets collected from river systems, Environ. Sci. Technol. 47 (2013) 661−677.

[60] E. Villar-Navarro, R.M. Baena-Nogueras, M. Paniw, J.A. Perales, P.A. Lara-Martín, Removal of pharmaceuticals in urban wastewater: high rate algae pond (HRAP) based technologies as an alternative to activated sludge based processes, Water Res. 139 (2018) 19−29.

[61] A. Bielen, A. Šimatović, J. Kosić-Vukšić, I. Senta, M. Ahel, S. Babić, T. Jurina, J.J. González Plaza, M. Milaković, N. Udiković-Kolić, Negative environmental impacts of antibiotic-contaminated effluents from pharmaceutical industries, Water Res. 126 (2017) 79−87.

[62] W. Brack, V. Dulio, M. Ågerstrand, I. Allan, R. Altenburger, M. Brinkmann, D. Bunke, R.M. Burgess, I. Cousins, B.I. Escher, et al., Towards the review of the

European Union Water Framework management of chemical contamination in European surface water resources, Sci. Total Environ. 576 (2017) 720−737.

[63] V. Matamoros, E. Uggetti, J. García, J.M. Bayona, Assessment of the mechanisms involved in the removal of emerging contaminants by microalgae from wastewater: a laboratory scale study, J. Hazard Mater. 301 (2016) 197−205.

[64] Z.N. Norvill, A. Shilton, B. Guieysse, Emerging contaminant degradation and removal in algal wastewater treatment ponds: identifying the research gaps, J. Hazard Mater. 313 (2016) 291−309.

[65] A. Hom-Diaz, M. Llorca, S. Rodríguez-Mozaz, T. Vicent, D. Barceló, P. Blánquez, Microalgae cultivation on wastewater digestate: β-estradiol and 17α-ethynylestradiol degradation and transformation products identification, J. Environ. Manage. 155 (2015) 106−113.

[66] A. de Wilt, A. Butkovskyi, K. Tuantet, L.H. Leal, T.V. Fernandes, A. Langenhoff, G. Zeeman, Micropollutant removal in an algal treatment system fed with source separated wastewater streams, J. Hazard Mater. 304 (2016) 84−92.

[67] V. Matamoros, R. Gutiérrez, I. Ferrer, J. García, J.M. Bayona, Capability of microalgae-based wastewater treatment systems to remove emerging organic contaminants: a pilot-scale study, J. Hazard Mater. 288 (2015) 34−42.

[68] J.Q. Xiong, M.B. Kurade, J.R. Kim, H.S. Roh, B.H. Jeon, Ciprofloxacin toxicity and its co-metabolic removal by a freshwater microalga Chlamydomonas mexicana, J. Hazard Mater. 323 (2017) 212−219.

[69] X. Bai, K. Acharya, Algae-mediated removal of selected pharmaceutical and personal care products (PPCPs) from Lake Mead water, Sci. Total Environ. (2017) 734−740, 581−582.

[70] T. Ding, M. Yang, J. Zhang, B. Yang, K. Lin, J. Li, J. Gan, Toxicity, degradation and metabolic fate of ibuprofen on freshwater diatom Navicula sp, J. Hazard Mater. 330 (2017) 127−134.

[71] J. Chen, F. Zheng, R. Guo, Algal feedback and removal efficiency in a sequencing batch reactor algae process (SBAR) to treat the antibiotic cefradine, PLoS One 10 (2015) 1−11.

[72] E. Daneshvar, M.J. Zarrinmehr, A.M. Hashtjin, O. Farhadian, A. Bhatnagar, Versatile applications of freshwater and marine water microalgae in dairy wastewater treatment, lipid extraction and tetracycline biosorption, Bioresour. Technol. 268 (2018) 523−530.

[73] R. Madadi, A.A. Pourbabaee, M. Tabatabaei, M.A. Zahed, M.R. Naghavi, Treatment of petrochemical wastewater by the green algae chlorella vulgaris, Int. J. Environ. Res. 10 (2016) 555−560.

[74] A. Abid, F. Saidane, M. Hamdi, Feasibility of carbon dioxide sequestration by Spongiochloris sp microalgae during petroleum wastewater treatment in airlift bioreactor, Bioresour. Technol. 234 (2017) 297−302.

[75] T. Zheng, C. wei Ke, Treatment of oilfield wastewater using algal−bacterial fluidized bed reactor, Sep. Sci. Technol. 52 (2017) 2090−2097.

[76] A. Hodges, Z. Fica, J. Wanlass, J. VanDarlin, R. Sims, Nutrient and suspended solids removal from petrochemical wastewater via microalgal biofilm cultivation, Chemosphere 174 (2017) 46−48.

[77] H. Znad, A.M.D. Al Ketife, S. Judd, F. AlMomani, H.B. Vuthaluru, Bioremediation and nutrient removal from wastewater by Chlorella vulgaris, Ecol. Eng. 110 (2018) 1−7.

[78] S.H. Ammar, H.J. Khadim, A.I. Mohamed, Cultivation of Nannochloropsis oculata and Isochrysis galbana microalgae in produced water for bioremediation and biomass production, Environ. Technol. Innov. 10 (2018) 132−142.

[79] I. Mundi. Palm oil production by country in 1000 MT. (2018).

[80] A. Ahmad, A. Buang, A.H. Bhat, Renewable and sustainable bioenergy production from microalgal co-cultivation with palm oil mill effluent (POME): a review, Renew. Sustain. Energy Rev. 65 (2016) 214−234.

[81] W.Y. Cheah, P.L. Show, J.C. Juan, J.S. Chang, T.C. Ling, Enhancing biomass and lipid productions of microalgae in palm oil mill effluent using carbon and nutrient supplementation, Energy Convers. Manag. 164 (2018) 188−197.

[82] B. Cheirsilp, J. Tippayut, P. Romprom, P. Prasertsan, Phytoremediation of secondary effluent from palm oil mill by using oleaginous microalgae for integrated lipid production and pollutant removal, Waste Biomass Valori. 8 (2017) 2889−2897.

[83] A. Ahmad, A.H. Bhat, A. Buang, Biosorption of transition metals by freely suspended and Ca-alginate immobilised with Chlorella vulgaris: kinetic and equilibrium modeling, J. Clean. Prod. 171 (2018) 1361−1375.

[84] A. Zainal, Z. Yaakob, M. Takriff, R. Renganathan, J. Ghani, Phycoremediation in anaerobically digested palm oil mill effluent using cyanobacterium, Spirulina platensis, J. Biobased Mater. Bioenergy 6 (2012) 1−6.

[85] K. Kamarudin, Z. Yaakob, R.R.M.S. Takriff, S. Tasirin, Bioremediation of palm oil mill effluents (POME) using Scenedesmus dimorphus and Chlorella vulgaris, Adv. Sci. Lett. 19 (2013) 2914−2918.

[86] S.M.U. Shah, A. Ahmad, M.F. Othman, M.A. Abdullah, Effects of palm oil mill effluent media on cell growth and lipid content of Nannochloropsis oculata and Tetraselmis suecica, Int. J. Green Energy 13 (2016) 200−207.

[87] G.T. Ding, Z. Yaakob, M.S. Takriff, J. Salihon, M.S. Abd Rahaman, Biomass production and nutrients removal by a newly-isolated microalgal strain Chlamydomonas sp in palm oil mill effluent (POME), Int. J. Hydrogen Energy 41 (2016) 4888−4895.

[88] H. Kamyab, S. Chelliapan, M.F.M. Din, R. Shahbazian-Yassar, S. Rezania, T. Khademi, A. Kumar, M. Azimi, Evaluation of Lemna minor and Chlamydomonas to

treat palm oil mill effluent and fertilizer production, J. Water Process. Eng. 17 (2017) 229–236.

[89] S. Haruna, S.E. Mohamad, H. Jamaluddin, Potential of treating unsterilized Palm Oil Mill Effluent (POME) using freshwater microalgae, Pakistan J. Biotechnol. 14 (2017) 221–225.

[90] International Olive Council World olive oil figures. (2018). http://www.internationaloliveoil.org/estaticos/view/131-world-olive-oil-figures

[91] D. Samanthakamani, N. Thangaraju, Potential of Freshwater Microalgae for Degradation of Phenol, 3, 2015, pp. 9–12.

[92] S. Dermeche, M. Nadour, C. Larroche, F. Moulti-Mati, P. Michaud, Olive mill wastes: biochemical characterizations and valorization strategies, Process Biochem. 48 (2013) 1532–1552.

[93] A. Papazi, K. Kotzabasis, Bioenergetic strategy of microalgae for the biodegradation of phenolic compounds-exogenously supplied energy and carbon sources adjust the level of biodegradation, J. Biotechnol. 129 (2007) 706–716.

[94] A. Papazi, A. Ioannou, M. Symeonidi, A.G. Doulis, K. Kotzabasis, Bioenergetic strategy of microalgae for the biodegradation of tyrosol and hydroxytyrosol, Z. Naturforsch C J Biosci. 72 (2017) 227–236.

[95] G. Hodaifa, M.E. Martínez, R. Órpez, S. Sánchez, Inhibitory effects of industrial olive-oil mill wastewater on biomass production of Scenedesmus obliquus, Ecol. Eng. 42 (2012) 30–34.

[96] F. Di Caprio, P. Altimari, F. Pagnanelli, Integrated biomass production and biodegradation of olive mill wastewater by cultivation of Scenedesmus sp, Algal Res. 9 (2015) 306–311.

[97] S. Ghosh Ray, M.M. Ghangrekar, Comprehensive review on treatment of high-strength distillery wastewater in advanced physico-chemical and biological degradation pathways, Int. J. Environ. Sci. Technol. (2018).

[98] N.K. Singh, and D.B. Patel. "Microalgae for bioremediation of distillery effluent," in Farming for Food and Water Security, ed. E. Lichtfouse (Dordrecht: Springer Netherlands), 83–109.

[99] A.K. Prajapati, P.K. Chaudhari, Physicochemical treatment of distillery wastewater—a review, Chem. Eng. Commun. 202 (2015) 1098–1117.

[100] D. Francisca Kalavathi, L. Uma, G. Subramanian, Degradation and metabolization of the pigment – melanoidin in distillery effluent by the marine cyanobacterium Oscillatoria boryana BDU 92181, Enzyme Microb. Technol. 29 (2001) 246–251.

[101] L.T. Valderrama, C.M. Del Campo, C.M. Rodriguez, L.E. De- Bashan, Y. Bashan, Treatment of recalcitrant wastewater from ethanol and citric acid production using the microalga Chlorella vulgaris and the macrophyte Lemna minuscula, Water Res. 36 (2002) 4185–4192.

[102] A. Solovchenko, S. Pogosyan, O. Chivkunova, I. Selyakh, L. Semenova, E. Voronova, P. Scherbakov, I. Konyukhov, K. Chekanov, M. Kirpichnikov, et al., Phycoremediation of alcohol distillery wastewater with a novel Chlorella sorokiniana strain cultivated in a photobioreactor monitored on-line via chlorophyll fluorescence, Algal Res. 6 (2014) 234–241.

[103] J.T. da Fontoura, G.S. Rolim, M. Farenzena, M. Gutterres, Influence of light intensity and tannery wastewater concentration on biomass production and nutrient removal by microalgae Scenedesmus sp, Process Saf. Environ. Prot. 111 (2017) 355–362.

[104] C. Das, N. Ramaiah, E. Pereira, K. Naseera, Efficient bioremediation of tannery wastewater by monostrains and consortium of marine Chlorella sp. and Phormidium sp, Int. J. Phytoremediation 20 (2018) 284–292.

[105] A. Rehman, Heavy metals uptake by Euglena proxima isolated from tannery effluents and its potential use in wastewater treatment, Russ. J. Ecol. 42 (2011) 44–49.

[106] L. Marchão, T. Silva, L. Gouveia, A. Reis, Microalgae-mediated brewery wastewater treatment: effect of dilution rate on nutrient removal rates, biomass biochemical composition, and cell physiology, J. Appl. Phycol. (2017) 1583–1595.

[107] M.F.D.J. Raposo, S.E. Oliveira, P.M. Castro, N.M. Bandarra, R.M. Morais, J.I. Brew, On the utilization of microalgae for brewery effluent treatment and possible applications of the produced biomass, J. Inst. Brew. 116 (2010) 285–292.

[108] W. Farooq, Y.C. Lee, B.G. Ryu, B.H. Kim, H.S. Kim, Y.E. Choi, J.W. Yang, Two-stage cultivation of two Chlorella sp. strains by simultaneous treatment of brewery wastewater and maximizing lipid productivity, Bioresour. Technol. 132 (2013) 230–238.

[109] L. Travieso, F. Benítez, E. Sánchez, R. Borja, M. Léon, F. Raposo, B. Rincón, Assessment of a microalgae pond for post-treatment of the effluent from an anaerobic fixed bed reactor treating distillery wastewater, Environ. Technol. 29 (2008) 985–992.

[110] E. Taşkan, Performance of mixed algae for treatment of slaughterhouse wastewater and microbial community analysis, Environ. Sci. Pollut. Res. 23 (2016) 20474–20482.

[111] D. Hernández, B. Riaño, M. Coca, M. Solana, A. Bertucco, M.C. García-González, Microalgae cultivation in high rate algal ponds using slaughterhouse wastewater for biofuel applications, Chem. Eng. J. 285 (2016) 449–458.

[112] H. Ting, L. Haifeng, M. Shanshan, Y. Zhang, L. Zhidan, D. Na, Progress in microalgae cultivation photobioreactors and applications in wastewater treatment: a review, Int. J. Agric. Biol. Eng. 10 (2017) 1–29.

[113] E. Posadas, M.D.M. Morales, C. Gomez, F.G. Acién, R. Muñoz, Influence of pH and CO_2 source on the performance of microalgae-based secondary domestic wastewater treatment in outdoors pilot raceways, Chem. Eng. J. 265 (2015) 239–248.

[114] M. Franco-Morgado, A. Toledo-Cervantes, A. González-Sánchez, R. Lebrero, R. Muñoz, Integral (VOCs, CO_2, mercaptans and H_2S) photosynthetic biogas upgrading using innovative biogas and digestate supply strategies, Chem. Eng. J. 354 (2018) 363–369.

[115] M. Del R. Rodero, E. Posadas, A. Toledo-Cervantes, R. Lebrero, R. Muñoz, Influence of alkalinity and temperature on photosynthetic biogas upgrading efficiency in high rate algal ponds, Algal Res. 33 (2017) 284–290.

[116] L. Peng, C.Q. Lan, Z. Zhang, Evolution, detrimental effects, and removal of oxygen in microalga cultures: a review Licheng, Environ. Prog. Sustain. Energy 32 (2013).

[117] S. Cao, X. Zhou, W. Jin, F. Wang, R. Tu, S. Han, H. Chen, C. Chen, G.J. Xie, F. Ma, Improving of lipid productivity of the oleaginous microalgae *Chlorella pyrenoidosa* via atmospheric and room temperature plasma (ARTP), Bioresour. Technol. 244 (2017) 1400–1406.

[118] E. Ono, J.L. Cuello, Carbon dioxide mitigation using thermophilic cyanobacteria, Biosyst. Eng. 96 (2007) 129–134.

[119] A. Anbalagan, A. Toledo-Cervantes, E. Posadas, E.M. Rojo, R. Lebrero, A. González-Sánchez, E. Nehrenheim, R. Muñoz, Continuous photosynthetic abatement of CO_2 and volatile organic compounds from exhaust gas coupled to wastewater treatment: evaluation of tubular algal-bacterial photobioreactor, J. CO_2 Util. 21 (2017).

[120] E. Uggetti, J. García, J.A. Álvarez, M.J. García-Galán, Start-up of a microalgae-based treatment system within the biorefinery concept: from wastewater to bioproducts, Water Sci. Technol. (2018) wst2018195.

[121] P.J. McGinn, K.E. Dickinson, S. Bhatti, J.C. Frigon, S.R. Guiot, S.J.B. O'Leary, Integration of microalgae cultivation with industrial waste remediation for biofuel and bioenergy production: opportunities and limitations, Photosynth. Res. 109 (2011) 231–247.

[122] A. Toledo-Cervantes, M.L. Serejo, S. Blanco, R. Pérez, R. Lebrero, R. Muñoz, Photosynthetic biogas upgrading to bio-methane: boosting nutrient recovery via biomass productivity control, Algal Res. 17 (2016) 46–52.

[123] G. Singh, S.K. Patidar, Microalgae harvesting techniques: a review, J. Environ. Manage. 217 (2018) 499–508.

[124] A.I. Barros, A.L. Gonçalves, M. Simões, J.C.M. Pires, Harvesting techniques applied to microalgae: a review, Renew. Sustain. Energy Rev. 41 (2015) 1489–1500.

[125] M. Al Hattab, A. Ghaly, A. Hammoud, Microalgae harvesting methods for industrial production of biodiesel: critical review and comparative analysis, J. Fundam. Renew. Energy Appl. 5 (2015) 1000154.

[126] B. Sialve, N. Bernet, O. Bernard, B. Sialve, N. Bernet, O. Bernard, Anaerobic digestion of microalgae as a necessary step to make microalgal biodiesel sustainable. To cite this version : HAL Id : hal-00854465, Biotechnol. Adv. 27 (2013) 409–416.

[127] A. Toledo-Cervantes, M. Morales, E. Novelo, S. Revah, Carbon dioxide fixation and lipid storage by *Scenedesmus obtusiusculus*, Bioresour. Technol. 130 (2013).

[128] J.L. Ramos-Suárez, A. Martínez, N. Carreras, Optimization of the digestion process of *Scenedesmus* sp. and Opuntia maxima for biogas production, Energy Convers. Manag. 88 (2014) 1263–1270.

[129] C. González-Fernández, B. Sialve, N. Bernet, J.P. Steyer, Thermal pretreatment to improve methane production

of Scenedesmus biomass, Biomass Bioenergy 40 (2012) 105–111.

[130] F. Passos, E. Uggetti, H. Carrère, I. Ferrer, Pretreatment of microalgae to improve biogas production: a review, Bioresour. Technol. 172 (2014) 403–412.

[131] A. Ghimire, G. Kumar, P. Sivagurunathan, S. Shobana, G.D. Saratale, H.W. Kim, V. Luongo, G. Esposito, R. Munoz, Bio-hythane production from microalgae biomass: key challenges and potential opportunities for algal bio-refineries, Bioresour. Technol. 241 (2017) 525–536.

[132] A. Jehlee, P. Khongkliang, W. Suksong, S. Rodjaroen, J. Waewsak, A. Reungsang, S. O-Thong, Bio-hythane production from *Chlorella* sp. biomass by two-stage thermophilic solid-state anaerobic digestion, Int. J. Hydrogen Energy 42 (2017) 27792–27800.

[133] Z. Yang, R. Guo, X. Xu, X. Fan, S. Luo, Hydrogen and methane production from lipid-extracted microalgal biomass residues, Int. J. Hydrogen Energy 36 (2011) 3465–3470.

[134] R. Harun, M.K. Danquah, Influence of acid pretreatment on microalgal biomass for bioethanol production, Process Biochem. 46 (2011) 304–309.

[135] N. Zhou, Y. Zhang, X. Wu, X. Gong, Q. Wang, Hydrolysis of Chlorella biomass for fermentable sugars in the presence of HCl and $MgCl_2$, Bioresour. Technol. 102 (2011) 10158–10161.

[136] W.Y. Cheah, T.C. Ling, P.L. Show, J.C. Juan, J.S. Chang, D.J. Lee, Cultivation in wastewaters for energy: a microalgae platform, Appl. Energy 179 (2016) 609–625.

[137] P. Verma, M.P. Sharma, Review of process parameters for biodiesel production from different feedstocks, Renew. Sustain. Energy Rev. 62 (2016) 1063–1071.

[138] O. Perez-Garcia, and Y. Bashan. "Microalgal heterotrophic and mixotrophic culturing for bio-refining: from metabolic routes to techno-economics," in Algal Biorefineries, eds. A. Prokop, R. K. Bajpai, M. E. Zappi (Cham: Springer International Publishing), 61–131.

[139] E. Posadas, P.A. García-Encina, A. Domínguez, I. Díaz, E. Becares, S. Blanco, R. Muñoz, Enclosed tubular and open algal-bacterial biofilm photobioreactors for carbon and nutrient removal from domestic wastewater, Ecol. Eng. 67 (2014) 156–164.

[140] J.B.K. Park, R.J. Craggs, A.N. Shilton, Recycling algae to improve species control and harvest efficiency from a high rate algal pond, Water Res. 45 (2011) 6637–6649.

[141] J. Garcia-Gonzalez, M. Sommerfeld, Biofertilizer and biostimulant properties of the microalga *Acutodesmus dimorphus*, J. Appl. Phycol. 28 (2016) 1051–1061.

[142] J. Coppens, O. Grunert, S. Van Den Hende, I. Vanhoutte, N. Boon, G. Haesaert, L. De Gelder, The use of microalgae as a high-value organic slow-release fertilizer results in tomatoes with increased carotenoid and sugar levels, J. Appl. Phycol. 28 (2016) 2367–2377.

[143] R. Maurya, K. Chokshi, T. Ghosh, K. Trivedi, I. Pancha, D. Kubavat, S. Mishra, A. Ghosh, Lipid extracted microalgal biomass residue as a fertilizer substitute for *Zea mays* L, Front. Plant Sci. 6 (2016) 1–10.

[144] S.C. Wuang, M.C. Khin, P.Q.D. Chua, Y.D. Luo, Use of Spirulina biomass produced from treatment of aquaculture wastewater as agricultural fertilizers, Algal Res. 15 (2016) 59−64.

[145] M. Garfí, L. Flores, I. Ferrer, Life cycle assessment of wastewater treatment systems for small communities: activated sludge, constructed wetlands and high rate algal ponds, J. Clean. Prod. 161 (2017) 211−219.

[146] L.T. Arashiro, N. Montero, I. Ferrer, F.G. Acién, C. Gómez, M. Garfí, Life cycle assessment of high rate algal ponds for wastewater treatment and resource recovery, Sci. Total Environ. (2018) 1118−1130, 622−623.

[147] A. Colzi Lopes, A. Valente, D. Iribarren, C. González-Fernández, Energy balance and life cycle assessment of a microalgae-based wastewater treatment plant: a focus on alternative biogas uses, Bioresour. Technol. 270 (2018) 138−146.

CHAPTER 20

Microalgae-based Remediation of Wastewaters

MYRSINI SAKARIKA • ELENI KOUTRA • PANAGIOTA TSAFRAKIDOU •
ANTONIA TERPOU • MICHAEL KORNAROS

INTRODUCTION

Microalgae are increasingly considered a green expedient for wastewater treatment. Key reasons [1,2] include (a) cost-effective treatment with (b) low energy requirements, that can be achieved through system optimization. Other prospects are (c) their low land requirements, (d) the residual organic and nutrient concentration is also low, and (e) they do not require further addition of chemicals. Furthermore, (f) sludge formation is lower than in conventional aerobic treatment, and (g) algal biomass can be used for biofuel or added-value product formation. Microalgae are efficient in removing nutrients, such as nitrogen (N) and phosphorus (P) as well as removing minerals and reducing the organic load. The latter includes the removal of phenolic and aromatic compounds. Nevertheless, intensive cultivation of microalgae might require large amounts of various nutrients that cannot be retrieved from the environment, as the utilization of nutrients is interwoven with their bioavailability. Microalgae are also capable of efficiently removing heavy metals (HMs), as well as emerging contaminants (ECs), even when they are contained in trace concentrations. A great advantage is that the use of microalgae for potentially human health-threatening compounds removal does not afflict the valorization of microalgal biomass for biofuel production (as would happen for applications such as fertilizers and feed products). The objective of this chapter is to highlight the challenges and opportunities of wastewater treatment using microalgae, as well as the potential valorization routes of the produced microalgal biomass (Fig. 20.1).

Availability of Nutrients in Wastewater for Microalgal Growth

The use of wastewater for microalgal growth is considered beneficial for minimizing the use of freshwater and reducing the cost of nutrient addition because nitrogen (N) and phosphorus (P) are removed from wastewater as bioresources for the production of biofuel or other added-value products [3]. Microalgae have been successfully cultivated in a variety of domestic and agroindustrial wastewaters (see Section 3). In case of lack of nutrients within a wastewater treatment plant a potentially significant source could be either raw or anaerobically digested agricultural wastes added in the influent of the system [4,5]. Although microalgae may have numerous advantages in wastewater treatment, their application is still challenged by the stringent discharge limits of nutrients [6]. Specifically, to achieve effective growth, microalgae require a basic nitrate (NO_3^-) concentration, which varies between 200 and 400 mg/L, in addition to other macronutrients such as phosphorous (P) and carbon (C) as well as micronutrients such as iron (Fe), manganese (Mn), zinc (Zn), sulfur (S), copper (Co), selenium (Se), potassium (K), and magnesium (Mg) [7−9].

Carbon

Microalgae assimilate inorganic carbon via photosynthesis. Through this process, solar energy is converted into chemical energy producing oxygen (O_2) as a byproduct, while in the second step chemical energy is used to assimilate carbon dioxide (CO_2) into sugars [9]. Microalgae are known to exist only in aqueous environments where the inorganic carbon required for their growth occurs in the form of CO_2, bicarbonate (HCO_3^-), and carbonate (CO_3^{2-}). Microalgae mainly metabolize inorganic carbon in the form of CO_2, while the enzyme carbonic anhydrase is required to convert HCO_3^- into CO_2 [10]. The most commonly used CO_2 source for microalgal growth is atmospheric air (CO_2 as 0.04% v/v, where v/v denotes volume of CO_2 per volume of air). However, about 2%−5% (v/v) of CO_2

Microalgae Cultivation for Biofuels Production. https://doi.org/10.1016/B978-0-12-817536-1.00020-5

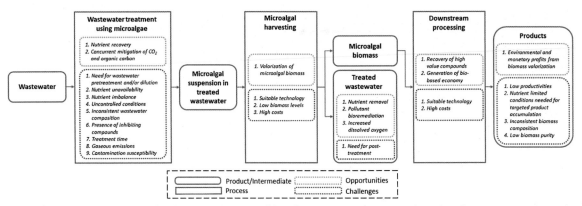

FIG. 20.1 Challenges and opportunities of wastewater treatment using microalgae.

in the supplied gas stream is usually required for sufficient microalgal biomass production [9]. As a matter of fact, it has been reported that 1 kg of microalgal biomass can use 1.83 kg of CO_2 but the theoretical yield can be predicted only when nitrogen source is defined [9]. The amount of CO_2 dissolved in wastewater might vary according to the pH. Specifically, at higher pH values (pH > 9) most of the inorganic carbon is in the form of carbonate (CO_3^{2-}), which cannot be assimilated by microalgae and as a result great care should be taken regarding the pH of wastewater [11]. In such a case, atmospheric CO_2 may be provided to the wastewater via forced aeration (air enriched by 1%–5% v/v CO_2) although this method has been proved economically unfeasible [9,12]. In case CO_2 becomes limited some microalgae can metabolize organic compounds such as sugars, organic acids, acetate, or glycerol as carbon source [9,13–16]. Scientific studies have reported that about 25%–50% of the algal carbon can be derived from heterotrophic utilization of organic carbon sources [14]. In fact, based on their metabolism, microalgae can be photoautotrophic, photoheterotrophic, mixotrophic, and heterotrophic while in several species the mode of carbon utilization can be shifted from autotrophy to heterotrophy, or even mixotrophy according to the variety and availability of the carbon source [14,16,17]. Many recent studies have targeted on the addition of various carbon sources along with wastewater for enhanced microalgal biomass production [16,18]. In addition, the use of food and agricultural waste along with wastewater for production of microalgae could be a beneficial approach from an economic point of view. Likewise, according to Chew et al. [13], the use of 25% of organic food waste along with inorganic residues can provide an enhanced *Chlorella*

vulgaris FSP-E biomass up to 11% compared with microalgal cultivation in a fully inorganic medium. Another approach is the use of mixed cultures of photosynthetic bacteria and microalgae, which could simultaneously remove all nutrients contained within synthetic wastewater [8]. Finally, temperature, illumination, pH, salinity, aeration, and substantial nutrients are the most important operating parameters to optimize for maximized CO_2 sequestration targeting sufficient microalgal biomass production.

Nitrogen

Nitrogen is the second most important macronutrient essential for microalgal growth and as such could directly influence their production yield. Nitrogen can be found in many forms within wastewater and is usually assimilated by microalgae as an inorganic substrate in the preference of ammonium (NH_4^+) > nitrate (NO_3^-) > nitrite (NO_2^-) > organic–N [19]. Many different wastewater streams contain high ammonium concentrations (100–9000 mg/L NH_4^+–N) and can be successfully used for production of microalgae [5]. Wastewater treatment plants present one of the greatest opportunities for ammonium recovery due to wastewater containing high concentrations of ammonium ions [5]. Ammonium shows higher bioavailability compared with nitrate or organic–N, and as a result, when it occurs in wastewater, alternative nitrogen sources are not assimilated [20]. High temperatures or high pH values could lead to an equilibrium of free ammonia (NH_3) and ammonium (NH_4^+) shifting toward ammonia production, and therefore inhibiting microalgal growth [20]. In some cases, this equilibrium can be used for the removal of excess nitrogen from wastewater as ammonia stripping and loss to the atmosphere can range between 17% and about 80% depending on wastewater pH [21]. Specifically, nitrogen is found in the form of ammonia at pH 10.5−11.5 and can be transferred from the liquid to the air through aeration. However, this process might result in high costs and can cause air pollution, and ammonia emissions may distort nitrogen recovery through microalgal treatment [22]. Some strains, on the other hand, have been reported to grow in high ammonia concentrations in a range up to 130 mg/L [8,23]. In addition to the inorganic nitrogen forms, urine $(CO(NH_2)_2)$ derived mainly from urban wastewaters can also be used as a nitrogen source, but only nitrification and stabilization could lead to high microalgae production yields [24]. When other nitrogen sources have become limited, amino acids can be assimilated by some species of cyanobacteria, but this method has been proved to be most cost-effective for microalgae growth [25]. As reported previously, ammonium recovery in wastewater seems preferable than other nitrogen forms. To this end, valorization of wastewater containing less desirable nitrogen entities seems feasible only by a combination of various wastewater sources (domestic wastewater, piggery wastewater, landfill leachate, and urine) that could lead to enhanced microalgal growth [5].

Phosphorus

Phosphorus (P) is a macronutrient essential for microalgal growth especially for the formation of DNA, RNA, adenosine triphosphate (ATP), and phospholipids in algal biomass [19]. Phosphorus is a nonrenewable resource, and thus its recycling is considered vital for future use. It can be retrieved from wastewater and can be assimilated by microalgae as an inorganic substrate preferably as orthophosphate (PO_4^{3-}). This substantial macronutrient should be present in wastewater at c.a. 40 mg/L PO_4−P to support efficient microalgal biomass production [23,26]. Nevertheless, nutrient uptake for microalgal growth depends on the associated concentration of nutrients inside the microalgal biomass. Because the molar N:P ratio in freshwater microalgal biomass ranges between 8:1 to 45:1 [19], phosphorus removal from wastewater is consistently lower than nitrogen due to the greater intracellular nitrogen content. There are also cases, for example, in Norway, in which phosphorous concentrations in the reject water from dewatering of anaerobically digested sludge are very low because all plants use metal salts for phosphorus precipitation and removal. As a result, this wastewater cannot be used as nutrient source for microalgal biomass production [23]. When inorganic phosphate is limited, many microalgal species can utilize organic phosphate sources converting it to orthophosphate using phosphatases located at the cell surface [27]. On the other hand, when inorganic phosphate appears in excess in wastewater, microalgae have the ability to store it within the cells in the form of acid-insoluble polyphosphate (volutin) granules [19]. Microalgae cultivated in wastewater may hence contain higher amounts of phosphorus than actually needed. These processes, which may lead to an extra uptake of phosphorus by microalgae, can be sufficient for prolonged growth in case phosphate concentration becomes limiting. As a result, the growth rate of microalgae may be immediately influenced by external concentration of phosphorus, in opposite to the responses to temperature, pH, or light [19,27].

Micronutrients

Micronutrients play important physiological roles in plant growth, and therefore they should be present in wastewater for efficient microalgal growth. Different species of microalgae require a range of micronutrients such as iron (Fe), manganese (Mn), zinc (Zn), sulfur (S), copper (Cu), selenium (Se), potassium (K), and magnesium (Mg), which might be considered essential. Micronutrient requirements usually differ among taxa, whereas in some cases their presence may have a toxic effect if accumulated in excess of cells' requirements [28]. As a result, the presence of different micronutrients in wastewater could be characterized as essential, required, replaceable, or toxic for microalgal growth [28].

Iron, zinc, and manganese are considered essential for microalgal growth as they take part in the oxygenic photosynthesis. Iron has also been reported as essential for the assimilation of nitrate and nitrite [29]. Generally, iron can be found in wastewater in two oxidation states; soluble (Fe^{2+}) and in a sparingly soluble oxidized form (Fe^{3+}). On the other hand, in oceanic water, the majority of dissolved iron (>99%) is found bound with organic complexes [30]. Nevertheless, microalgae have the ability to utilize both free and bound iron. A problem might occur when micronutrients such as iron are bound with other essential nutrients, inhibiting microalgal growth. For instance, it has been reported that when wastewater contains a lot of sediments, phosphate can be chemically bound to iron or aluminum resulting in phosphates' lower bioavailability [31]. Generally, microalgae can efficiently assimilate these micronutrients when found in adequate amounts according to each species. For example, according to Saavedra et al. [32] removal efficiency of manganese from contaminated river water (2.1 mg/L) was reported up to 99.4% by *C. vulgaris* at pH 7 within 3 h. Likewise, it has been reported that the monometallic effluents of zinc (150 mg/L) used for *Scenedesmus obliquus* cultivation supported an uptake of 22.3 mg/g [33]. However, zinc or other micronutrients could be inhibitory for the growth of microalgae if present at elevated concentrations, or as a result of antagonistic behaviors in the presence of other micronutrients [32]. For example, the copper uptake efficiency of *C. vulgaris* is decreased by 10% in the presence of zinc [34]. In multimetallic systems, biosorption might be another process taking place in parallel explaining the removal of metallic components from wastewater. The biosorption mechanisms in this case are very complex and are not yet fully understood.

Speciation in Contaminant Removal

As stated previously, microalgae use organic or inorganic carbon sources, nitrogen (N), phosphorus (P), and trace metals as nutrients for their growth, depending on the cultivation method [35]. Phycoremediation extends beyond the aforementioned chemicals, which are just a limited example of the substances that microalgal species can remove from the environment. Nowadays, the scientific concern is oriented toward the so-called EC such as pharmaceutical organics, endocrine disrupting compounds (EDCs), polycyclic aromatic hydrocarbons (PAHs), surfactants, personal care products (PCPs), pesticides, perfluorinated compounds, substituted phenolics, flame retardants, industrial additives, and nanomaterials, among others [36–39]. The biochemical routes that microalgae follow to remove these contaminants are still under investigation, but it is believed that cell adsorption, biodegradation, bioaccumulation, stripping (volatilization), and/or photodegradation are the possible mechanisms [36].

Numerous studies have been conducted regarding specific microalgal species and their removal capacity on certain contaminants. *Navicula* sp., which is a typical freshwater diatom, was used to degrade the antimicrobial agent Triclosan (5-chloro-2-(2,4-dichlorophenoxy)-phenol, TCS) in a synthetic medium. The presence of potassium permanganate ($KMnO_4$) as well as the pH value of the culture medium seem to influence the outcome, regarding the toxicity of TCS in this diatom. Researchers also identified the transformation products of the contaminant and suggested a possible biochemical pathway of TCS degradation in *Navicula* sp. [40]. The same research group investigated the biouptake, toxicity, and biotransformation of TCS in the freshwater algae *Cymbella* sp. and how humic acid (HA), as a typical constituent of dissolved organic matter found in aquatic environments, affects these mechanisms [41].

Natural organic matter and its role on the removal of bisphenol A by *Monoraphidium braunii* was the research subject of Gattullo et al. [42]. Organic matter affects algal growth rate and may hinder or promote xenobiotic removal under certain circumstances. *M. braunii* reached 48% removal of bisphenol A [42]. *Cymbella* sp. and *Scenedesmus quadricauda* were evaluated also in terms of their bioremediation efficiency regarding naproxen [40]. Wang et al. [43] studied the removal and metabolism of TCS by using *Chlorella pyrenoidosa*, *Desmodesmus* sp., and *S. obliquus*. They stated 99.7% removal by the three microalgae and detected that cellular uptake was the metabolic mechanism for the reduction of TCS by *C. pyrenoidosa*, whereas *Desmodesmus* sp. and *S. obliquus* cells use biotransformation to eliminate the contaminant. Previously published studies, referred to a 77.2% removal percentage by *C. pyrenoidosa* after its cultivation with 800 ng mL^{-1} TCS for 96 h [44].

C. vulgaris is a species thoroughly examined by many scientists for its capacity to remove various pollutants. Among six tested species, Xiong et al. [45] reported that *C. vulgaris* reached a 12% removal of the antibiotic levofloxacin (LEV) at 1 mg/L, which was the highest. Furthermore, acclimation of the microalgae and addition of 1% w/v sodium chloride seem to boost its removal efficiency [45]. The decrease of metal concentrations such as Zn, Cd, Cu, Zn, Fe, Al, Ni, and Mn ions [46–48], organic matter, nutrients [48], diclofenac [49], bisphenol A [50], diazinon [51], and enrofloxacin (ENR) [52] was also evaluated for *C. vulgaris* commercially available species, mutants [53], and/or local isolates [54].

Desmodesmus sp., *Chlorella sorokiniana*, *S. obliquus*, *Chlamydomonas mexicana*, *Chlamydomonas pitschmannii*, *Ourococcus multisporus*, *Micractinium resseri* are some other species that were tested for their removal capacity in comparison with *C. vulgaris* or in a consortium, regarding the abovementioned chemicals.

Liu et al. [55] reported the biodegradation and reduction of 17β-estradiol (E2) and diethylstilbestrol (DES) by *Raphidocelis subcapitata*. E2 and DES are two estrogens that can be found both in waste and natural water, and it is believed that their activity is linked to endocrine disruption in humans and may affect future generations due to their accumulation in the food chain. *R. subcapitata* was chosen because of its reported ability to remove and degrade also benzo(a)pyrene (BaP) through its dioxygenase enzyme system, parent PAHs by a cytochrome P-450 system, and via monooxygenase and dioxygenase enzymatic pathways remove phenanthrene, fluoranthene, and pyrene. In the mentioned study, *R. subcapitata* presented an efficient removal capacity for E2 and DES, achieving up to 88.5% removal through accelerated bioaccumulation and biodegradation. Parlade et al. [56] approached the same contaminant problem from a different angle. They investigated the removal of 17β-estradiol by an algal–bacterial consortium in indoor and outdoor conditions as well as the outcome of bioaugmentation with *Scenedesmus*. It was concluded that algal-based systems removed E2 totally, and the high diversity of algal community acts beneficially on the contaminant removal [56].

The importance of this field is reflected in various published reviews of the scientific community, which deals with the different species that have been tested and all the outstanding questions or research gaps that are still need to be unraveled [36,37,39,57].

The elucidation of metabolic pathways that microalgal species use to remove contaminants as well as the identification of transformation products is a valuable tool for risk assessment of the chemicals that can be found in natural waters. The screening of the capabilities of each species, regarding micropollutant elimination is crucial for an effective design of water treatment installations and protocols.

Treating Municipal, Agroindustrial and Industrial wastewater

This section focuses on the treatment of municipal wastewater effluent (MWE), palm oil mill effluent (POME), olive mill wastewater (OMW), textile wastewater (TW), HM-containing wastewater, petroleum wastewater (PW), and ECs. A typical composition of POME, OMW, TW, and PW is presented in Table 20.1, whereas the HM- and EC-containing wastewaters are not included because of their high variability depending on the origin.

Municipal wastewater

Existing literature focuses on treatment of different effluents of municipal wastewater treatment, for example, primary settling, denitrification tank, and secondary settling effluent; therefore, direct comparison is not possible.

This subsection focuses on the microalgal treatment of secondary effluent, further referred to as MWE. The physicochemical characteristics of this wastewater are substantially deviating from those of any (agro)industrial stream, due to the distinctively low concentrations (Table 20.1) [58]. MWE treatment with microalgae results in efficient removal of N and P, while it also offers the advantages of removing HMs, increasing the dissolved oxygen in the effluent and limiting the growth

of bacteria [2]. Nevertheless, the latter is mostly observed in attached growth systems rather than suspended. Gómez-Serrano et al. [63] cultivated *Muriellopsis* sp., *C. vulgaris*, *Chlorella fusca*, *Chlorella* sp., *Scenedesmus subspicatus*, and *Pseudokirchneriella subcapitata* in semicontinuous mode. The final biomass concentrations ranged between 0.7 and 1.7 g/L, while the N and P were totally eliminated from the cultivation medium (initial concentrations of 17.1 and 10 mg/L for TKN and TP, respectively). Shi et al. [59] used a twin-layer photobioreactor to immobilize *Halochlorella rubescens* for the treatment of MWE. Removal of 0.45 and 6.26 mg/L day was achieved for PO_4^-–P and NO_3^-–N, respectively (achieving 73.2% and 83.2% removal, respectively). The authors noted that this treatment achieved effluent values that meet the requirements of the European Water Framework Directive. Furthermore, Lv et al. [58] observed that the immobilization of microalgae as well as the establishment of a specific consortium enhances the nutrient/pollutant removal. Continuous cultivation of *Chlorella* and *Scenedesmus* sp. in MWE resulted in biomass concentrations varying between 0.1 and 1.3 g/L (at hydraulic retention time [HRT] 0.04–10 day). This resulted in TN removal of 36.0%–95.3% and TP removal of 40%–100%. It was noted that construction of microalgal consortia, apart from resulting in higher nutrient removals, also results in a more robust system that can face environmental fluctuations and hinders the invasion and prevalence of other species. Similarly, Wu et al. [64] found that the use of a coculture of *Scenedesmus* sp. LX1 and *Haematococcus pluvialis* resulted in better performance than the individual microalgae. It was concluded that the main mechanism responsible for this higher nutrient removal (7.2–7.5 mg/L residual concentration for monocultures compared with 3.3–5.0 mg/L for the mixed culture starting from 27.4 mg/L) was the interspecific nutrient competition between the two microalgae. Phosphorus was eliminated except from the case where *H. pluvialis* was the sole microalga grown. Finally, Wang et al. [2] summarized the mechanisms of N removal in MWE as: assimilation by microalgae (NH_4^+–N, NO_3^-–N, and NO_2^-–N and minor amounts of organic–N); loss as NH_3 due to increase in pH and temperature as well as mixing; and loss as N_2 due to bacterial nitrification–denitrification. On the other hand, P is removed through assimilation (for their anabolic and catabolic products); it is up-–taken to form polyphosphate granules (luxurious P uptake); or precipitated as $Ca_3(PO_4)_2$ and $Mg_3(PO_4)_2$ (in the presence of Ca^{2+} and Mg^{2+}, using the photosynthesis derived O_2, when pH > 8.5).

TABLE 20.1
Typical Characteristics of Raw Municipal Wastewater Effluent (MWE), Palm Oil Mill Effluent (POME), Olive Mill Wastewater (OMW), Textile Wastewater (TW), and Petroleum Wastewater (PW).

Parameters	Units	Municipal Wastewater Effluent (MWE)	Palm Oil Mill effluent (POME)	Olive Mill wastewater (OMW)	Textile Wastewater (TW)	Petroleum Wastewater (PW)
pH	–	8.40 ± 0.41	4.30 ± 0.21	5.13 ± 0.03	7.63 ± 5.34	7.82 ± 1.07
Alkalinity	g CaCO$_3$ L^{-1}	–	–	0.75 ± 0.10	–	–
Total solids (TS)	g L^{-1}	–	–	112 ± 1	5.86 ± 8.24	–
Volatile solids (VS)	g L^{-1}	–	–	82.3 ± 6.8	0.29 ± 0.34	0.22 ± 0.23
Total suspended solids (TSS)	g L^{-1}	–	19.6 ± 5.59	40.6 ± 0.6	0.09 ± 0.09	–
Volatile suspended solids (VSS)	g L^{-1}	–	28.7 ± 2.0	39.9 ± 1.1	–	–
Biochemical oxygen demand (BOD)	g O$_2$ L^{-1}	0.01 ± 0.01	0.35 ± 0.07	–	–	0.18 ± 0.20
Total chemical oxygen demand (COD)	g O$_2$ L^{-1}	0.05 ± 0.03	1.74 ± 0.54	169 ± 9	0.61 ± 0.54	0.39 ± 0.30
Soluble chemical oxygen demand (COD)	g O$_2$ L^{-1}	–	22.5 ± 2.8	89.2 ± 3.9	–	–
Total carbohydrates[a]	g L^{-1}	–	–	37.5 ± 1.7	–	–
Soluble carbohydrates[a]	g L^{-1}	–	–	26.0 ± 0.3	–	–
Phenolic compounds[b]	g L^{-1}	–	–	8.61 ± 0.98	–	–
Oil and grease	g L^{-1}	–	–	22.0 ± 4.0	–	0.69 ± 1.31
Total Kjeldahl nitrogen (TKN)	mg N L^{-1}	14.8 ± 5.6	–	735 ± 10	–	–
Total ammonium nitrogen (TAN)	mg N L^{-1}	9.17 ± 8.99	650 ± 212	86.1 ± 5.3	25.7 ± 35.6	19.6 ± 22.2
Total phosphorus	mg P L^{-1}	2.32 ± 3.19	–	513 ± 1	–	–
Soluble phosphorus	mg P L^{-1}	–	400 ± 283	309 ± 4	2.04 ± 2.79	–
Zinc (Zn)	mg/L^{-1}	–	–	–	1.52 ± 1.99	–
Manganese (Mn)	mg/L^{-1}	–	–	–	0.03 ± 0.02	–
Cadmium (Cd)	mg/L^{-1}	–	–	–	0.03 ± 0.03	–
Copper (Cu)	mg/L^{-1}	–	–	–	0.05 ± 0.07	–
Lead (Pb)	mg/L^{-1}	–	–	–	0.09 ± 0.01	–
Nickel (Ni)	mg/L^{-1}	–	–	–	0.03 ± 0.01	–
Arsenic (As)	mg/L^{-1}	–	–	–	<0.001	–
Iron (Fe)	mg/L^{-1}	–	–	–	0.14 ± 0.04	–
Reference	–	[58,59]	[60]	[60a]	[61]	[62]

"–" indicates unavailable values.
[a] in glucose equivalents.
[b] in syringic acid equivalents.

Palm oil mill effluent

POME is a high-strength wastewater, characterized by high concentrations of organic load (BOD and COD), organic N and P and other nutrients (Table 20.1). It consists of 95%–96% of water, 4%–5% of solids of which 50%–100% are suspended, and 0.6%–0.7% of oil [65]. Kamyab et al. [65] used *Chlamydomonas incerta* to treat POME in COD concentrations of 250 mg/L, 500 mg/L, and 1000 mg/L. The highest COD removal (67.4%) was observed at 250 mg/L in 28 days, gradually decreasing for the two higher concentrations (43.2% and 34.1%). The corresponding organic carbon removal was 56.2%, 54.7%, and 49.8%. POME concentrations higher than 500 mg/L resulted in slow growth of *C. incerta* due to substrate inhibition as indicated by the lower growth rates. Therefore, this wastewater requires pretreatment before microalgal remediation. Vairappan et al. [66] bioremediated 5% aerobically digested POME using *Isochrysis* sp. Biomass productivities of 69.0 and 91.7 mg/m^2 day were achieved with an indoor photobioreactor and an outdoor system, respectively. The outdoor system produced higher lipid content biomass (52.8% compared with 44.5% for the photobioreactor), a fact ascribed to stress induced by variations in irradiance and temperature during outdoor cultivation. The nutrient removals were comparable in both systems, achieving 42.2%, 84.4%, and 21.9% reductions of nitrate, phosphate, and BOD$_3$, respectively (corresponding to 3.55, 0.03, and 3.02 mg/L, respectively). Hadiyanto et al. [67] applied a two-stage treatment of anaerobically treated POME. The first stage (3–8 days) consisted of treatment with aquatic plants, hyacinth (*Eichhornia crassipes*), or water lily (*Nymphaea* sp.), whereas the second treatment stage (15 days) was performed using *Arthrospira platensis*. *A. platensis* achieved maximum growth rate of 0.412 day^{-1} while reducing the COD, N and P content by 50.8%, 96.5%, and 85.9% (reaching values between 400 and 790, 8.29–145, and 3.14–10.0 mg/L, respectively).

Olive mill wastewater

The biological treatment of OMW, the side-stream of olive oil production, is challenging due to its phytotoxic and antibacterial properties derived from the high (poly)phenolic content as well as the high organic load and nutrient content. OMW can be effectively treated through membrane filtration, resulting in a wastewater with a limited amount of polyphenols [68]. Nevertheless, a sustainable membrane-based treatment method requires the valorization of the concentrates, accounting to about 20% of the initial volume.

Cicci et al. [68] selected *Scenedesmus dimorphus* and *A. platensis* to treat OMW as well as ultra- and nanofiltration concentrates (all in 1:1 v/v). Both strains displayed higher polyphenol and phenol removals when treating ultrafiltration (UF) retentate (60.2%–72.7% and 72.7%–100%, respectively), while *A. platensis* eliminated phenolics during this treatment. OMW and UF retentate reduced microalgal growth due to the high polyphenol content, whereas nanofiltration concentrate promoted growth, as indicated by the lower specific growth rates (−59% to −30%, −124% to −26% and 53%–142% difference compared with the control for OMW, UF concentrate, and NF concentrate, respectively). The increase in growth in the latter case is ascribed to the reduced concentration of toxic compounds coupled with the presence of suitable soluble organic carbon. Hodaifa et al. [69] remediated physicochemically pretreated OMW using *C. pyrenoidosa*. Microalgal treatment contributed to 1.4% COD decrease, 8.0% phenolic compound (corresponding to 0.57 mg/L) as well as N removal of 19.2% (or 11.3 mg/L). The maximum specific growth rate and biomass productivity increased with the increase in concentration until 50% OMW (v/v), where the values of 0.07 h and 1.25 mg/L h were presented. Higher OMW concentrations resulted in drop of these parameters either due to inhibition or due to increased toxic effect. The low biomass productivities hinder the use of the biomass as a high-value product and therefore decrease the potential of economic viability. To this end, centrifuged OMW was supplied to *Scenedesmus* sp. through different strategies: (A) batch, providing 9% v/v; (B) fed-batch, with daily addition of 1% OMW; and (C) two-stage supply, consisting of photoautotrophic cultivation, followed by 9% OMW supplementation for heterotrophic growth [70]. The highest biomass concentration was achieved by the two-stage strategy (1.4 g/L compared with 0.9 g/L and 0.5 g/L during fed-batch and batch supply, respectively). Sugars were removed at the levels of 74.0%, 62.9%, and 61.0% during batch, fed-batch, and two-stage strategy, respectively. Phenol removal was reversely proportional to the concentration of phenolics (at concentrations >100 mg/L), reaching values of 66.0% in day 1 and gradually decreasing to 12.0% until day 7 of fed-batch cultivation. A slight phenol increase was observed during batch cultivation, as well as during the first days of the two-phase strategy, whereas 55% decrease was noted on day 12 of the heterotrophic phase. Therefore, three hypotheses were formed: (A) phenol degradation occurs after the depletion of highly biodegradable compounds, (B) phenol-tolerant microorganisms (bacteria or fungi) degraded the phenols, and (C) polyphenols were degraded from the beginning of the phase but this was not detected from the method used. On the other hand, Pinto et al. [71] did not observe high-molecular-weight phenol (e.g., lignins, tannins) degradation when treating OMW (1:10 v/v) with *Ankistrodesmus braunii* and *Scenedesmus quadricauda*. In contrast, low-molecular-weight phenols (<300 Da) were biotransformed during this short treatment, with hydroxytyrosol, catechol, and ferulic and sinapic acid completely removed in all cases. Experiments performed in the dark resulted in higher phenol removal efficiency (75%—100% compared with 35%—100% under light conditions), while light was found to result in high toxicity mainly due to autooxidation. The overall phenolic compound reduction after 5 days amounted to 12%. Finally, physicochemical analysis revealed that phenols were biotransformed into other aromatic compounds.

Textile wastewater

Textile industry generates various organic and inorganic streams that fluctuate from alkaline to acidic, the color of which can be orange, red, purple, blue, and black, presenting highly variable composition [72]. This wastewater, a typical composition of which is shown in Table 20.1, constitutes significant environmental threat, as apart from the nutrients and COD, also contains HMs. Biological treatment with microalgae seems promising, as these microbes degrade dyes to utilize them as N source [73]. Ghazal et al. [72] tested *Anabaena flos-aquae*, *Nostoc elepsosporum*, *Nostoc linckia*, *Anabaena variabilis*, and *C. vulgaris* for the treatment of TW. The highest color removal was noted by *N. elepsosporum*, followed by *C. vulgaris*, *A. variabilis*, *N. linckia*, and *A. flos-aquae* (50.8%—100% removal). The treatment increased the pH (from 9.5 up to 9.95) and DO (from 0.6 to 2.01—2.75 mg/L) due to photosynthesis. Furthermore, a drop was noted for EC (from 2.90 to 0.92—2.03 dS/m), COD (60.0%—98.9% removal at initial concentration of 430 mg/L) and BOD (80.0%—97.6% removal starting from 96 mg/L). The cell dry weight remained at low levels, with the best results presented by *N. elpsosporum* (11.5 mg/L), while the growth was 12.1%—19.4% reduced compared with the control. The best overall performance was noted by *N. elepsosporum*, achieving removals of 98.9% for COD, 97.6% for BOD, and 95.0%—99.0% for HMs, while color was totally removed. Moreover, *C. vulgaris* was used for TW bioremediation in high-rate algae ponds (HRAPs) [61]. NH_4—N was reduced by 44.4%—45.1%, PO_4—P by 33.1%—33.3% (achieving values of 3.43—3.57 mg N L and 4.76—64.23 mg P L)

and COD by 38.3%—62.3% (reaching values of 102—170 mg/L), while the color was 41.8%—50.0% removed through biosorption. Color removal decreased with the increase in initial color content. TW reduced the specific growth rate of *C. vulgaris* to 0.05 day^{-1} compared with 0.40 day^{-1} for the control pond (containing nutrient medium), while nutrient addition almost restored the specific growth rate to the control level (0.39 day^{-1}). Nevertheless, biomass concentrations remained low in all TW treatments (107—203 mg/L compared with 613 mg/L for the control). Nutrient deficiency was identified as the main limiting factor for *C. vulgaris* growth in TW. Finally, additional treatment with immobilized in alginate *Chlorella* cells would be suitable to further remove the color of TW because this approach creates a cell-dense system that can enhance dye adsorption.

Heavy metal—containing wastewater

The anthropogenic HM disposal into the environment causes detrimental effects, due to the inability of biodegradation that causes bioaccumulation. This provoked the penetration of HMs, with Cd, Cr, Cu, Hg, Pb, and Zn most commonly occurring into the food chain, causing environmental, public health, and economic impacts [74]. The main HM removal mechanisms in microalgae are cell incorporation and biosorption (passive adsorption process), while the great superiority of the microalgal HM remediation over conventional processes is that it has a great efficiency in removing HMs in trace levels [74,75]. Kumar et al. [74] categorized the factors affecting the HM remediation to biotic, that is, algal species used, tolerance, capacity, biomass concentration, size and volume of microalgal cell; and abiotic factors, that is, pH, ionic stress, salinity, temperature, metal speciation, and effect of combined metals. It was noted that the species of choice for HM removal are *Chlorella* and *Scenedesmus*, whereas the highest HM removals are generally presented at pH 5 [74]. This study summarized that *Chlamydomonas reinhardtii*, *S. platensis*, as well as several strains of *Scenedesmus* spp., *Tetraselmis* spp., and *Chlorella* spp. are efficient in removing Cd, taking it up in levels of 0.64—292 mg/g dry biomass. Co is more efficiently removed by *Spirogyra* spp. (12.8 mg/g) and *Oscillatoria angustissima* (15.3 mg/g). The most notable Cr^{6+} as well as Cu^{2+} uptake was performed by *Arthrospira* spp. reaching levels of 333 and 389 mg/g, respectively [76]. Immobilized in alginate *C. reinhardtii* presented the most prominent Hg^{2+} removal capacity of 107 mg/g as well as Pb^{2+} removal (381 mg/g) [74], while the highest Zn removal potential was presented by *Desmodesmus pleiomorphus* (360 mg/g) [77]. Finally,

A. flos-aquae, N. elepsosporum, N. linckia, A. variabilis, and *C. vulgaris* were able to remove Cr (60.4% −99.2%), Pb (73.0%−98.8%), Fe (73.1%−98.9%), Cu (60.4%−99.0%), Mo (60.8%−96.9%), and As (46.7%−95.0%) present in initial concentrations 2.60−95.2 mg/L [72].

Petroleum wastewater

Petroleum by-product contamination is a considerable polluting factor of surface- and groundwater occurring due to leakage of storage tanks, spills as well as incorrect disposal of PW. Compounds commonly found in petroleum by-products include benzene, toluene, ethylbenzene, and xylenes (BTEX). Takáčová et al. [78] investigated BTEX biodegradation using *Parachlorella kessleri.* The number of *P. kessleri* cells after 48 h was 13% lower compared with the control, whereas BTEX degradations of 39%−50% and 56%−64% were observed after 24 and 48h, respectively (initial BTEX concentration of 100 µg/L). Ansari et al. [79] used *Isochrysis* sp. for the treatment of crude oil and heavy duty marine diesel (HDMD), to assess the potential of degrading the water-soluble fractions of oceanic oil spill. Crude oil affected the growth due to oil toxicity. Specifically, concentrations of 30%−50% resulted in initial drop in cell density during early cultivation. HDMD had more toxic effect, notable even at 5%−10% concentration, while higher concentrations (20%) resulted in culture death. The higher toxicity of HDMD compared with crude oil is noted by the 7.4% decrease in maximum growth rate (0.24 day^{-1}) compared with the 58% decrease (0.55 day^{-1}) for crude oil, when both were supplemented at 10%. The lower toxicity of crude oil's water-soluble fractions is attributed to the lower concentration of polyaromatic hydrocarbons compared with HDMD. On the other hand, *Prototheca zopfii* degraded 10.7% motor oil and 41.4% crude oil [80]. More specifically, 10%−23% of the saturated aliphatic hydrocarbons as well as 10%−26% of the aromatic compounds contained in motor oil were removed. Regarding crude oil, the respective percentages were 38%−60% and 12%−41%. The higher removal concerning crude oil is ascribed to the higher content of alkanes (17% compared to 9% for motor oil) and the lower percentage of aromatics and cycloalkanes (35% and 45% for motor oil and 28% and 39% for crude oil). The degradation of cycloalkanes was inversely proportional to the number of rings, and the six-ring cycloalkanes did not get utilized whatsoever. Hodges et al. [81] compared two different systems for the treatment of petroleum refining wastewater. Specifically, they used rotating algae biofilm reactors (RABRs) as well as suspended-growth open pond lagoons,

containing mainly *Pseudanabaena, Oscillatoria,* and *Chroococcus.* N was reduced by 70.8%−72.4% (corresponding to 17.7−18.1 mg/L reduction), P was reduced by 50.0%−55.6% (0.90−1.00 mg/L) while the TSS concentration was 53.6%−61.3% lower (20.9−23.9 mg/L) through the RABR treatment. The open pond growth lagoon operated in 13.9% (3.47 mg/L) and 18.9% (0.34 mg/L) lower N and P content, while the TSS content increased by 46.9% (18.3 mg/L) due to the suspended growth of microalgae. The biomass productivity of the open pond amounted to 0.4 g m^2 day, whereas the RABR resulted in average biomass productivity of 4.11 g m^2 day while presenting the advantage of not requiring intensive biomass harvesting.

Emerging contaminants

ECs have recently penetrated natural environments. They are mostly found in municipal and pharmaceutical plant wastewater and landfills and derive from daily household products, cosmetics, and medicine [39]. Even though their concentration in aquatic environments is low (ranging from ng/L to µg/L), they can have deleterious ecological effects. As mentioned in section 17.2, pharmaceuticals, PCPs, EDCs, and pesticides are among the prime examples of ECs [39]. Microalgae have shown the potential to remove many types of ECs at levels of 9−24 µg/L, however, presenting difficulties in removing pesticides [39]. Ahmed et al. [39] identified the trend of removal: pharmaceuticals > PCPs > EDCs > pesticides. Furthermore, Wang et al. [44] found that *C. pyrenoidosa* can remove 50% of triclosan (antimicrobial agent) during the first 1 h of exposure when the initial concentration is 100−800 µg/L. The microalga removed 77.2% of triclosan after 96 h (initial concentration of 800 µg/L). Nevertheless, this compound negatively affected the microalga, as indicated by the disruption of the chloroplast. Matamoros et al. [82] tested the removal of 26 different ECs from urban wastewater using pilot scale HRAPs fed with 7−29 g of COD m^{-2} day^{-1}. The targeted compounds included pharmaceuticals and PCPs, fire retardants, surfactants, anticorrosive agents, pesticides, and plasticizers. The system was able to remove up to 90% of the most commonly occurring compounds (caffeine, acetaminophen, and ibuprofen), whereas the concentration of others was not reduced. Specifically, the authors of this study identified four different groups according to their removal efficiency: (A) > 90% biodegradation: caffeine, acetaminophen, ibuprofen, methyl dihydrojasmonate, and hydrocinnamic acid; (B) 60%−90% biodegradation: oxybenzone, ketoprofen, 5-methyl/benzotriazole, naproxen, galaxolide, tonalide, tributyl phosphate, triclosan, bisphenol A, and octylphenol; (C) 40%–60% biodegradation: diclofenac, benzotriazole, OH–benzothiazole, triphenyl phosphate, cashmeran, diazinon, celestolide, and atrazine; (D) < 30% biodegradation: carbamazepine, benzothiazole, methyl paraben, tris(2-chloroethyl) phosphate, and 2,4-D. Moreover, it was noted that the main mechanisms were biodegradation and photo-degradation. Xiong et al. [83] evaluated the biodegradation of carbamazepine using *C. mexicana* and *S. obliquus*. The growth of *C. mexicana* was 30% inhibited by the supplementation of 200 mg/L carbamazepine, whereas *S. obliquus* presented 97% inhibition. The maximum biodegradations achieved were 28% and 35% by *S. obliquus* and *C. mexicana* when 1 mg/L of the compound was supplemented, while 13% and 32% carbamazepine was biodegraded after 10 days of incubation (initial concentration of 25 mg/L). Batch reactors with *Chlorella* sp. and *Scenedesmus* sp. as prevalent species were used for the biodegradation of caffeine, ibuprofen, galaxolide, tributyl phosphate, 4-octylphenol, tris(2-chloroethyl) phosphate, and carbamazepine [84]. 4-octylphenol, galaxolide, and tributyl phosphate were removed up to 99% after 10 days due to air

stripping. However, caffeine and ibuprofen were biodegraded by 99% and 95%, respectively. The authors observed that the initial lag phase of 3 days for ibuprofen removal was eliminated to nondetectable time. Finally, Ahmed et al. [39] concluded that the combination of physicochemical treatment (e.g., ozonation, ultrasound treatment, UF) with biological treatments, such as microalgal treatment, will result in more effective EC removal. Summarizing, Table 20.2 illustrates the maximum EC removals by various microalgal species reported in recent literature.

Challenges in Wastewater Bioremediation

The challenges concerning the microalgal wastewater bioremediation need to be identified and elucidated, to advance the field of microalgal biotreatment. To this end, the following limitations were identified by Wang et al. [2]: (A) relatively long treatment time is required; (B) the separation of algae from the wastewater is cumbersome; and (C) reduced performance is observed when bacteria and zooplankton dominate. Additionally, (D) it is common to observe increase in organic compound and solid content due to the often

TABLE 20.2
Maximum Emerging Contaminant Removals by Various Microalgal Species.

Microalgal Species	Contaminant	Max. Removal (%)	References
Cymbella sp.	Triclosan (5–chloro–2–(2, 4–dichlorophenoxy)–phenol)–phenol	69	[41]
	Naproxen	97.1	[40]
Monoraphidium braunii	Bisphenol A	48	[42]
Scenedesmus quadricauda	Naproxen	58.8	[40]
Scenedesmus obliquus	Triclosan (5–chloro–2–(2, 4–dichlorophenoxy)–phenol)–phenol	99.7	[43]
	Diclofenac	79	[49]
Chlorella pyrenoidosa	Triclosan (5–chloro–2–(2, 4–dichlorophenoxy)–phenol)–phenol	69.3, 77.2	[43,44].
Chlorella vulgaris	Levofloxacin	12	[45]
	Nitrogen, phosphorus	87.7, 76.7	[48]
	Metal ions (Zn, Cd, Cu, Zn, Fe, Al, Ni, and Mn)	64.7–100	[46–48]
	Diclofenac	21.58	[49]
	Bisphenol A	23	[50]
	Diazinon	94	[51]
	Enrofloxacin	26	[52]
Desmodesmus sp.	Triclosan (5–chloro–2–(2, 4–dichlorophenoxy)–phenol)–phenol	92.9	[43]
Raphidocelis subcapitata	17β-Estradiol (E2)	88.5	[55]
	Diethylstilbestrol (DES)	71.8	[55]

not efficient harvesting of cells, or due to the intracellular products release from the lysis of the cells [2]. Even though membranes can shorten the HRT as well as retain the microalgal biomass, the costs and risk of fouling limit their full-scale implementations. Another approach is the stimulation of attached growth by introducing substratum for biofilm creation. In this case, the most common method of harvesting, mechanical scraping, is difficult to perform in full-scale installations. Furthermore, the reasons for the unpredictable biomass detachment need to be elucidated.

The characteristically low concentration of nutrients and organics in some wastewaters, for instance MWE, render its treatment challenging. Lv et al. [58] reported that the lack of organic carbon is one of the main challenges in the cultivation of microalgae in this effluent. The biggest fraction of organics contained in MWE are inert and therefore not degradable by microalgae [2]. Another challenge for treating MWE is the imbalanced concentration of nutrients and particularly the N:P ratio [2]. The nutrient-limiting conditions prevailing under the cultivation in MWE result in low biomass productivity, while also increasing the lipid and carbohydrate accumulation in the cell [63]. On the other hand, (agro)industrial wastewaters often need high dilution to be eligible for microalgal remediation. *Scenedesmus* sp. was used for the treatment of OMW, in concentration of 9% v/v during batch growth, and 1% v/v during fed-batch growth [70]. The lower growth during batch cultivation was ascribed to the exposure at high concentrations of growth-inhibiting compounds, as well as the dark color of the wastewater that does not permit light penetration. Therefore, special attention should be paid with wastewaters containing high amounts of inhibiting compounds, as for instance phenolics [69]. Di Caprio et al. [70] stated that a low OMW supply rate is suitable for efficient phenol removal from microalgae, for a low phenol as well as light variation. Moreover, wastewaters such as PW provoke oxidative stress, and therefore high dilution is required for microalgal treatment [85]. Cicci et al. [68] highlighted that apart from the compositional characteristics, the color of the wastewater hinders its treatment since a reduced fraction of light penetrates dark-colored wastewaters (e.g., OMW). This could be solved by cultivation in thin-layer bioreactors, nevertheless, the biomass productivity in such systems is yet to be determined. Substrate inhibition is a common reason for the lower growth of microalgae, even during heterotrophic conditions [86], where nutrient and organic load levels can generally be higher. For example, POME supplied in concentrations more than 500 mg/L inhibited the growth of *C. incerta* [65].

Furthermore, wastewaters are often nutrient poor and therefore do not favor the growth of microalgae [61,66]. To satisfy all the microalgal needs, Vairappan et al. [66] treated 5% POME, diluted in seawater and supplemented with 0.075% NPK fertilizer. The low percentage was selected due to (A) the high organic and nutrient load of POME, (B) the dark color that hinders the phototrophic cultivation, and (C) it is a freshwater-type wastewater that cannot be used for the cultivation of a marine microalga (*Isochrysis* sp.).

Furthermore, agroindustrial streams in many cases need extensive pretreatment before they become suitable for bioremediation using microalgae, mainly targeting suspended solids and COD reduction [68,69]. For instance, Hadiyanto et al. [67] used anaerobically digested POME, that was subsequently treated with aquatic plants prior to the treatment with *A. platensis*. For the latter, POME was filtered and used at 80% v/v, supplemented with N, P, and vitamin sources. Hodaifa et al. [69] proposed a complete treatment process composed of (A) flocculation—sedimentation to remove the biggest part of solids from the liquid, (B) photolysis to reduce the organic load and photodegrade the phenolic compounds, (C) microfiltration (0.2 μm) to remove the suspended particles and sterilize the wastewater, and finally (D) biotreatment of OMW using *C. pyrenoidosa*. The authors stated that this pretreatment was essential for ensuring microalgal growth in OMW. Similarly, Cicci et al. [68] used OMW that was filtered, flocculated through acidification, subsequently centrifuged and photocatalyzed. *S. platensis* appeared to be more sensitive to media composition than *Scenedesmus dimorphus*, as indicated by the high variation in specific growth rate under different media. The latter indicates that some strains are more sensitive than others, indicating that several experiments need to be performed before an optimal treatment is established.

Treatment efficiency is also affected by environmental conditions. For instance, when temperature deviates from the optimal range growth rates decline [87], having as a result the reduced pollutant removal in a given period of time. Similarly, increased temperature or pH can result in ammonia volatilization, resulting in the escape of the main nitrogen source [20]. Additionally, reduced nutrient bioavailability due to wastewater composition reduces the potential of wastewater bioremediation through microalgal treatment [87]. More specifically, nutrient speciation has an important role in the overall performance, as microalgae show preference in specific nutrient forms (for instance ammonium instead of organic—N) [19]. All these factors can lead to insufficient microalgal growth and therefore inefficient wastewater treatment. Special attention should be paid during wastewater treatment aiming at HM or EC removal. Kumar et al. [74] stated that there are several compounds that form complexes with HMs (e.g., amino acids, HAs, fulvic acid, EDTA) rendering them unavailable for microalgal treatment. Furthermore, there is a selectivity in EC removal, as for instance difficulties are presented in the elimination of EDCs and pesticides [39]. Moreover, EC removal efficiencies are season dependent as they get reduced with lower temperatures [39]. These compounds can present growth inhibition up to 97% [83] and therefore, extensive study is required to establish the suitable strains and cultivation conditions for EC biotreatment. Finally, it is also important to note that there are several phenomena occurring during the microalgal wastewater remediation, that remain yet to be identified. For instance, the biotreatment of OMW under illumination results in phenol autooxidation increasing therefore the toxicity of the wastewater [71].

End-use of Microalgae Cultivated in Wastewater

In modern societies, renewable sources of energy, dietary ingredients, and valuable chemicals have been in the limelight of scientific research, due to challenges concerning the increasing population, climate change,

water availability, and energy supply. In this regard, microalgae represent a promising feedstock for biofuels production, as well as unlimited number of added-value compounds, including lipids, proteins, carbohydrates, pigments, and numerous secondary metabolites used for cosmetics, pharmaceuticals, and other purposes, as shown in Fig. 20.2 [88,89].

From an environmental point of view, microalgal cultivation makes no use of arable land and enables consumption of sea water, wastewaters, and effluents, as well as flue gases contributing to recycling of nutrients, CO_2 mitigation, and substantial improvement in terms of cost of biomass production [90,91]. However, wastewater valorization in microalgal cultivation can be a demanding task that needs further investigation. Especially in the case where the produced biomass ends up in the food chain, wastewater pretreatment is needed to diminish concentration of several contaminants, including pathogens, HMs, and xenobiotic substances present in the biomass produced by specific types of effluents [6].

Biofuels

Microalgal biomass can be converted, through thermal, chemical, and biological processes to third-generation biofuels, including biodiesel, biogas, biohydrogen, bioethanol, and biocrude oil. In contrast to first- and

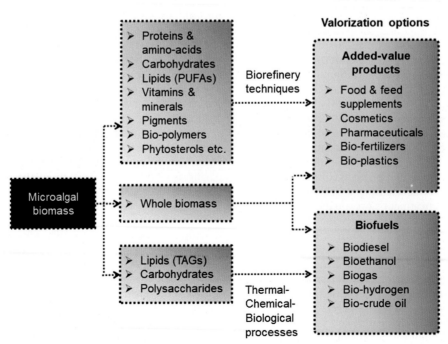

FIG. 20.2 Valorization options for microalgal biomass.

second-generation biofuels, produced from food crops and lignocellulosic biomass, agricultural and forestry residues, respectively, microalgal biofuels represent a sustainable alternative to fossil-based fuels, which can dominate the energy sector the following decades, provided that mass cultivation and downstream processing make substantial progress [92,93].

One of the most well-investigated valorization options for microalgae produced in wastewaters is biodiesel, associated with the lipid fraction of microalgal biomass. Typically, microalgae are characterized by 20%–50% intracellular lipids at dry weight basis, mainly consisting of C16–C18 fatty acids, which can be transesterified to high-quality biodiesel. Lipid content can reach up to 80% usually under nutrient-depleted conditions resulting in high lipid productivity. Concurrently, high biomass productivity can compensate for moderate lipid content inducing high biodiesel production [94]. Lipid productivity in wastewaters can be optimized through two-stage cultivation, where nutrients and COD are initially depleted maximizing biomass production, while lipid content is subsequently increased in the second stage characterized by nutrient starvation [95]. Low nitrogen concentration resulted in high lipid content (32.7%) of *C. vulgaris* cultivated in municipal wastewater [96]. Cultivation of *Nephroselmis* sp. with diluted industrial wastewater resulted in high specific growth rate (0.87 day^{-1}), higher lipid productivity (14.2 mg/L day) than that of the synthetic medium BBM and high percentage of palmitic and oleic acid for biodiesel production [97]. In addition, lipid content of *Chlorococcum* sp. RAP13 cultivated in dairy wastewater increased from 31% under mixotrophic conditions, to 42% under heterotrophic growth after the addition of 6% waste glycerol from biodiesel industry [98]. Microalgae produced from wastewater treatment can also be used as feedstock for biocrude oil production through hydrothermal liquefaction, a sustainable process that can be effectively applied in wet biomass, mixed cultures, and low-lipid microalgae, resulting in a potential jet fuel [92,99].

Carbohydrates represent another major component of microalgal biomass that can be further valorized for bioethanol production. In contrast to ethanol feedstock of former generations, microalgae are characterized by high concentration of the easily fermentable cellulose and starch, along with low concentration of hemicelluloses and absence of lignin [100]. With view to decreasing the cost of biomass production and materializing algal biofuels, carbohydrate-rich biomass can be cultivated in wastewater. *Nannochloropsis oculata* and *Tetraselmis suecica* were effectively grown in municipal

wastewater, and the produced biomass was fermented by *Saccharomyces cerevisiae*, after alkaline pretreatment, resulting in almost 4% bioethanol yield, despite the relatively low carbohydrate content [101].

In contrast to biodiesel and bioethanol production which make use of a single fraction of microalgal biomass, anaerobic digestion exploits the whole biomass produced toward biogas production [93] and can be applied without prior drying. Anaerobic digestion is highly determined by the rigidity of microalgal cell wall and biomass composition and can be performed after biomass pretreatment for cell permeability and lipid extraction or codigestion with other substrates for enhanced nutrient balance [102,103]. Biomass produced by a consortium of *C. vulgaris*, *S. obliquus*, and *C. reinhardtii* cultivated in swine wastewater under batch and semicontinuous mode of operation, resulted in methane yields up to 146 and 171 mL/g COD, respectively [104]. Methane yield of 346 and 415 mL/g VS was also recorded in case of *Chlorella kessleri* and *C. vulgaris*, which were effectively grown in municipal wastewater [105]. Another type of wastewater used for cultivation of microalgae intended for biogas production among others, is the effluent from anaerobic digesters, called digestate, in the framework of an integrated approach for zero-waste production of energy [106,107]. Lastly, clean energy, in the form of biohydrogen, is a highly promising option for microalgal technology and can be produced through direct biophotolysis or dark fermentation [92]. Carbohydrate-rich biomass produced during municipal wastewater treatment was fermented by *Enterobacter aerogenes*, resulting in hydrogen yield of 56.8 mL/g VS in case of *S. obliquus* [108]. In addition, an output of 11.7 mL/L hydrogen was recorded during cultivation of *Chlorella* sp. with crude glycerol, a by-product from biodiesel industry, under anaerobic conditions [109].

Added-value products
Despite the high potential of microalgal biomass for biofuels production, formation of added-value products should accompany energy applications, in the framework of a biorefinery approach, which will make scale-up of microalgae-based systems economically feasible [110]. To this end, utilization of wastewater and effluents to produce valuable compounds, including proteins, pigments, vitamins, and numerous bioactive substances represents an emerging and highly promising field of microalgal technology. Since ancient times, people have been using microalgae as a food and feed ingredient, while currently microalgae have been considered as a novel source of protein. This type of

protein is often called single-cell protein or microbial protein. In case effluents are used as cultivation media for microalgal production, several concerns arise in terms of safety and hygiene of the produced protein source [6,111]. Especially in the case of fecally contaminated wastewater, multiple pretreatment steps should be implemented before their use for microalgal biomass destined for feed and/or food applications, to enable public acceptance [112]. Wastewater characteristics and operational mode significantly affect biomass composition, with high levels of nitrogen, semicontinuous operation, and low hydraulic retention time inducing protein accumulation [104]. Biomass of a *Chlorella* strain (PY-ZU1) cultivated in swine digestate was considered suitable for feed applications, owing to its 46% protein content and low accumulation of HMs [113]. However, microalgae used for wastewater treatment usually coexist with different types of microorganisms, including bacteria, zooplankton, and debris, affecting total biomass quality. It has been found that microalgae grown on agroindustrial wastewaters demonstrate efficient nutrient removal capacity, although poorer biomass characteristics and lower than 30% algae abundance, when compared with purple phototrophic bacteria [114].

Microalgal biomass also constitutes a valuable source of pigments, mainly chlorophylls and carotenoids, which have been correlated with numerous health benefits, including prevention of cancer and cardiovascular diseases, antioxidant, antiinflammatory, and antidiabetic activity, as well as several applications as natural colorants, cosmetics, food, and feed supplements [115,116]. With view to alleviating environmental concerns and decreasing the upstream cost, POME was used to cultivate *Phaeodactylum tricornutum* for production of bioactive compounds, resulting in fucoxanthin productivity of 25.4 µg/L day, under 30% (v/v) effluent supplemented with urea [117]. In addition, carotenoid production of *Phormidium autumnale* at industrial scale can reach up to 108 ton year, under heterotrophic cultivation with slaughterhouse waste [118]. Besides pigments, several benefits and subsequent applications can stem from valuable biomass ingredients, including vitamins, minerals, polyunsaturated fatty acids, phenols, and phytosterols. Additionally, agriculture represents another area of application for microalgal biomass, due to the similarities and even enhanced characteristics compared to conventional fertilizers. Several microalgal species, including *Chlorella* and *Arthrospira* with high nutrient removal capacity, can return valuable nutrients and micronutrients, as well as soil organic carbon, substantially

improving soil quality and crop growth and coping with macro- and micronutrient deficiencies [119]. Lastly, high ash content of microalgae can be an advantageous characteristic, such in case of biomass of *Scenedesmus* and *Desmodesmus* sp. grown in effluent from a stabilization lagoon, which ash concentration close to 20% and low concentration of HMs can result in a high-quality digestate used as biofertilizer [120].

Concluding Remarks

In conclusion, microalgae-based wastewater treatment and biomass valorization represent a wide area of research and development with great potential. Even though having several bottlenecks, mainly in terms of economics, downstream processing and scale-up have to be handled, many strategies anticipated to meet these challenges exist. To this end, exploitation of wastewater and effluents for microalgal cultivation and bioproduct formation offer significant advantages both for the environment and modern societies' needs, thus it is expected to culminate in the near future.

REFERENCES

[1] J.P. Maity, J. Bundschuh, C.Y. Chen, P. Bhattacharya, Microalgae for third generation biofuel production, mitigation of greenhouse gas emissions and wastewater treatment: present and future perspectives — a mini review, Energy 78 (2014) 104—113.

[2] J.-H. Wang, T.-Y. Zhang, G.-H. Dao, X.-Q. Xu, X.-X. Wang, H.-Y. Hu, Microalgae-based advanced municipal wastewater treatment for reuse in water bodies, Appl. Microbiol. Biotechnol. 101 (2017) 2659—2675.

[3] S. Raghuvanshi, V. Bhakar, R. Chava, K.S. Sangwan, Comparative study using life cycle approach for the biodiesel production from microalgae grown in wastewater and fresh water, in: Procedia CIRP, Elsevier, 2018, pp. 568—572.

[4] E. Koutra, G. Grammatikopoulos, M. Kornaros, Microalgal post-treatment of anaerobically digested agroindustrial wastes for nutrient removal and lipids production, Bioresour. Technol. 224 (2017) 473—480.

[5] Y. Ye, H.H. Ngo, W. Guo, Y. Liu, S.W. Chang, D.D. Nguyen, H. Liang, J. Wang, A critical review on ammonium recovery from wastewater for sustainable wastewater management, Bioresour. Technol. 268 (2018) 749—758.

[6] G. Markou, L. Wang, J. Ye, A. Unc, Using agro-industrial wastes for the cultivation of microalgae and duckweeds: contamination risks and biomass safety concerns, Biotechnol. Adv 36 (2018) 1238—1254.

[7] H. Kamyab, F. Friedler, S. Chelliapan, S. Rezania, J. Sultan Yahya Petra, K. Lumpur, Bioenergy production and nutrients removal by green microalgae with

cultivation from agro-wastewater palm oil mill effluent (POME)-A review, Chem. Eng. Trans. 70 (2018).

[8] E.J. Olguín, Phycoremediation: key issues for cost-effective nutrient removal processes, in: Biotechnol. Adv., Elsevier, 2003, pp. 81−91.

[9] R. Verma, A. Srivastava, Carbon dioxide sequestration and its enhanced utilization by photoautotroph microalgae, Environ. Dev. 27 (2018) 95−106.

[10] M.A. Borowitzka, Chapter 3 − biology of microalgae, in: I.A. Levine, J. Fleurence (Eds.), Microalgae Heal. Dis. Prev., Academic Press, 2018, pp. 23−72.

[11] V.C. Eze, S.B. Velasquez-Orta, A. Hernández-García, I. Monje-Ramírez, M.T. Orta-Ledesma, Kinetic modelling of microalgae cultivation for wastewater treatment and carbon dioxide sequestration, Algal Res 32 (2018) 131−141.

[12] E. Becker, Microalgae — Biotechnology and Microbiology, Cambridge University Press, 1994.

[13] K.W. Chew, S.R. Chia, P.L. Show, T.C. Ling, S.S. Arya, J.S. Chang, Food waste compost as an organic nutrient source for the cultivation of Chlorella vulgaris, Bioresour. Technol. 267 (2018) 356−362.

[14] J. Hu, D. Nagarajan, Q. Zhang, J.-S. Chang, D.-J. Lee, Heterotrophic cultivation of microalgae for pigment production: a review, Biotechnol. Adv. 36 (2018) 54−67.

[15] W.B. Kong, H. Yang, Y.T. Cao, H. Song, S.F. Hua, C.G. Xia, Effect of glycerol and glucose on the enhancement of biomass, lipid and soluble carbohydrate production by Chlorella vulgaris in mixotrophic culture, Food Technol. Biotechnol. 51 (2013) 62−69.

[16] N. Poddar, R. Sen, G.J.O. Martin, Glycerol and nitrate utilisation by marine microalgae Nannochloropsis salina and Chlorella sp. and associated bacteria during mixotrophic and heterotrophic growth, Algal Res 33 (2018) 298−309.

[17] L. Lu, J. Wang, G. Yang, B. Zhu, K. Pan, Biomass and nutrient productivities of Tetraselmis chuii under mixotrophic culture conditions with various C:N ratios, Chin. J. Oceanol. Limnol. 35 (2017) 303−312.

[18] A. Otondo, B. Kokabian, S. Stuart-Dahl, V.G. Gude, Energetic evaluation of wastewater treatment using microalgae, Chlorella vulgaris, J. Environ. Chem. Eng. 6 (2018) 3213−3222.

[19] R. Whitton, F. Ometto, M. Pidou, P. Jarvis, R. Villa, B. Jefferson, Microalgae for municipal wastewater nutrient remediation: mechanisms, reactors and outlook for tertiary treatment, Environ. Technol. Rev. 4 (2015) 133−148.

[20] L. Leng, J. Li, Z. Wen, W. Zhou, Use of microalgae to recycle nutrients in aqueous phase derived from hydrothermal liquefaction process, Bioresour. Technol. 256 (2018) 529−542.

[21] G. Markou, D. Vandamme, K. Muylaert, Ammonia inhibition on Arthrospira platensis in relation to the initial biomass density and pH, Bioresour. Technol. 166 (2014) 259−265.

[22] F.M. Santos, J.C.M. Pires, Nutrient recovery from wastewaters by microalgae and its potential application as bio-char, Bioresour. Technol. 267 (2018) 725−731.

[23] A.K. Sahu, J. Siljudalen, T. Trydal, B. Rusten, Utilisation of wastewater nutrients for microalgae growth for anaerobic co-digestion, J. Environ. Manag. 122 (2013) 113−120.

[24] J. Coppens, R. Lindeboom, M. Muys, W. Coessens, A. Alloul, K. Meerbergen, B. Lievens, P. Clauwaert, N. Boon, S.E. Vlaeminck, Nitrification and microalgae cultivation for two-stage biological nutrient valorization from source separated urine, Bioresour. Technol. 211 (2016) 41−50.

[25] H. Kageyama, R. Waditee-Sirisattha, Mycosporine-like amino acids as multifunctional secondary metabolites in cyanobacteria: from biochemical to application aspects, in: Stud. Nat. Prod. Chem., Elsevier, 2019, pp. 153−194.

[26] A. Solovchenko, A.M. Verschoor, N.D. Jablonowski, L. Nedbal, Phosphorus from wastewater to crops: an alternative path involving microalgae, Biotechnol. Adv. 34 (2016) 550−564.

[27] S.T. Dyhrman, Nutrients and their acquisition: phosphorus physiology in microalgae, in: Physiol. Microalgae, Springer International Publishing, Cham, 2016, pp. 155−183.

[28] A. Quigg, Micronutrients, in: R.J. Borowitzka MA, J. Beardall (Eds.), Physiol. Microalgae, Springer International Publishing, Cham, 2016, pp. 211−231.

[29] A. Anbalagan, S. Schwede, C.F. Lindberg, E. Nehrenheim, Influence of iron precipitated condition and light intensity on microalgae activated sludge based wastewater remediation, Chemosphere 168 (2017) 1523−1530.

[30] D. Lannuzel, M. Grotti, M.L. Abelmoschi, P. van der Merwe, Organic ligands control the concentrations of dissolved iron in Antarctic sea ice, Mar. Chem. 174 (2015) 120−130.

[31] T.R. Devlin, A. Di Biase, V. Wei, M. Elektorowicz, J.A. Oleszkiewicz, Removal of soluble phosphorus from surface water using iron (Fe-Fe) and aluminum (Al-Al) electrodes, Environ. Sci. Technol. 51 (2017) 13825−13833.

[32] R. Saavedra, R. Muñoz, M.E. Taboada, M. Vega, S. Bolado, Comparative uptake study of arsenic, boron, copper, manganese and zinc from water by different green microalgae, Bioresour. Technol. 263 (2018) 49−57.

[33] C.M. Monteiro, S.C. Fonseca, P.M.L. Castro, F.X. Malcata, Toxicity of cadmium and zinc on two microalgae, Scenedesmus obliquus and Desmodesmus pleiomorphus, from Northern Portugal, J. Appl. Phycol. 23 (2011) 97−103.

[34] M.S. Rodrigues, L.S. Ferreira, J.C.M. de Carvalho, A. Lodi, E. Finocchio, A. Converti, Metal biosorption onto dry biomass of Arthrospira (Spirulina) platensis and Chlorella vulgaris: multi-metal systems, J. Hazard. Mater. 217−218 (2012) 246−255.

[35] D. Hoh, S. Watson, E. Kan, Algal biofilm reactors for integrated wastewater treatment and biofuel production: a review, Chem. Eng. J. 287 (2016) 466−473.

[36] S.P. Cuellar-Bermudez, G.S. Aleman-Nava, R. Chandra, J.S. Garcia-Perez, J.R. Contreras-Angulo, G. Markou, K. Muylaert, B.E. Rittmann, R. Parra-Saldivar, Nutrients utilization and contaminants removal. A review of two approaches of algae and cyanobacteria in wastewater, Algal Res 24 (2017) 438−449.

[37] Z.N. Norvill, A. Shilton, B. Guieysse, Emerging contaminant degradation and removal in algal wastewater treatment ponds: identifying the research gaps, J. Hazard Mater. 313 (2016) 291−309.

[38] Y. Wang, S.-H. Ho, C.-L. Cheng, W.-Q. Guo, D. Nagarajan, N.-Q. Ren, D.-J. Lee, J.-S. Chang, Perspectives on the feasibility of using microalgae for industrial wastewater treatment, Bioresour. Technol. 222 (2016) 485−497.

[39] M.B. Ahmed, J.L. Zhou, H.H. Ngo, W. Guo, N.S. Thomaidis, J. Xu, Progress in the biological and chemical treatment technologies for emerging contaminant removal from wastewater: a critical review, J. Hazard Mater. 323 (2017) 274−298.

[40] T. Ding, K. Lin, B. Yang, M. Yang, J. Li, W. Li, J. Gan, Biodegradation of naproxen by freshwater algae Cymbella sp. and Scenedesmus quadricauda and the comparative toxicity, Bioresour. Technol. 238 (2017) 164−173.

[41] T. Ding, K. Lin, L. Bao, M. Yang, J. Li, B. Yang, J. Gan, Biouptake, Toxicity and biotransformation of triclosan in diatom Cymbella sp. and the influence of humic acid, Environ. Pollut. 234 (2018) 231−242.

[42] C.E. Gattullo, H. Bährs, C.E.W. Steinberg, E. Loffredo, Removal of bisphenol A by the freshwater green alga Monoraphidium braunii and the role of natural organic matter, Sci. Total Environ. 416 (2012) 501−506.

[43] S. Wang, K. Poon, Z. Cai, Removal and metabolism of triclosan by three different microalgal species in aquatic environment, J. Hazard Mater. 342 (2018) 643−650.

[44] S. Wang, X. Wang, K. Poon, Y. Wang, S. Li, H. Liu, S. Lin, Z. Cai, Removal and reductive dechlorination of triclosan by Chlorella pyrenoidosa, Chemosphere 92 (2013) 1498−1505.

[45] J.Q. Xiong, M.B. Kurade, B.H. Jeon, Biodegradation of levofloxacin by an acclimated freshwater microalga, Chlorella vulgaris, Chem. Eng. J. 313 (2017) 1251−1257.

[46] M.A. Alam, C. Wan, X.Q. Zhao, L.J. Chen, J.S. Chang, F.W. Bai, Enhanced removal of Zn^{2+} or Cd^{2+} by the flocculating Chlorella vulgaris JSC-7, J. Hazard Mater. 289 (2015) 38−45.

[47] L. Rugnini, G. Costa, R. Congestri, L. Bruno, Testing of two different strains of green microalgae for Cu and Ni removal from aqueous media, Sci. Total Environ. 601−602 (2017) 959−967.

[48] F. Gao, C. Li, Z.H. Yang, G.M. Zeng, J. Mu, M. Liu, W. Cui, Removal of nutrients, organic matter, and metal from domestic secondary effluent through microalgae cultivation in a membrane photobioreactor, J. Chem. Technol. Biotechnol. 91 (2016) 2713−2719.

[49] C. Escapa, R.N. Coimbra, S. Paniagua, A.I. García, M. Otero, Comparative assessment of diclofenac removal from water by different microalgae strains, Algal Res 18 (2016) 127−134.

[50] M.K. Ji, A.N. Kabra, J. Choi, J.H. Hwang, J.R. Kim, R.A.I. Abou-Shanab, Y.K. Oh, B.H. Jeon, Biodegradation of bisphenol A by the freshwater microalgae Chlamydomonas mexicana and Chlorella vulgaris, Ecol. Eng. 73 (2014) 260−269.

[51] M.B. Kurade, J.R. Kim, S.P. Govindwar, B.H. Jeon, Insights into microalgae mediated biodegradation of diazinon by Chlorella vulgaris: microalgal tolerance to xenobiotic pollutants and metabolism, Algal Res 20 (2016) 126−134.

[52] J.Q. Xiong, M.B. Kurade, B.H. Jeon, Ecotoxicological effects of enrofloxacin and its removal by monoculture of microalgal species and their consortium, Environ. Pollut. 226 (2017) 486−493.

[53] J. Cheng, Q. Ye, K. Li, J. Liu, J. Zhou, Removing ethinylestradiol from wastewater by microalgae mutant Chlorella PY-ZU1 with CO_2 fixation, Bioresour. Technol. 249 (2018) 284−289.

[54] Q.T. Gao, Y.S. Wong, N.F.Y. Tam, Removal and biodegradation of nonylphenol by different Chlorella species, Mar. Pollut. Bull. 63 (2011) 445−451.

[55] W. Liu, Q. Chen, N. He, K. Sun, D. Sun, X. Wu, S. Duan, Removal and biodegradation of 17β-estradiol and diethylstilbestrol by the freshwater microalgae Raphidocelis subcapitata, Int. J. Environ. Res. Public Health 15 (2018) 452.

[56] E. Parladé, A. Hom-Diaz, P. Blánquez, M. Martínez-Alonso, T. Vicent, N. Gaju, Effect of cultivation conditions on β-estradiol removal in laboratory and pilot-plant photobioreactors by an algal-bacterial consortium treating urban wastewater, Water Res. 137 (2018) 86−96.

[57] J.Q. Xiong, M.B. Kurade, B.H. Jeon, Can microalgae remove pharmaceutical contaminants from water? Trends Biotechnol. 36 (2018) 30−44.

[58] J. Lv, J. Feng, Q. Liu, S. Xie, J. Lv, J. Feng, Q. Liu, S. Xie, Microalgal cultivation in secondary effluent: recent developments and future work, Int. J. Mol. Sci. 18 (2017) 79.

[59] J. Shi, B. Podola, M. Melkonian, Application of a prototype-scale twin-layer photobioreactor for effective N and P removal from different process stages of municipal wastewater by immobilized microalgae, Bioresour. Technol. 154 (2014) 260−266.

[60] H. Kamyab, M.F. Md Din, C.T. Lee, A. Keyvanfar, A. Shafaghat, M.Z.A. Majid, M. Ponraj, T.X. Yun, Lipid production by microalgae Chlorella pyrenoidosa cultivated in palm oil mill effluent (POME) using hybrid photo bioreactor (HPBR), Desalin. Water Treat. 55 (2015) 3737−3749.

[60a] K. Tsigkou, M. Sakarika, M. Kornaros, et al., Inoculum origin and waste solid content influence the biochemical methane potential of olive mill wastewater under mesophilic and thermophilic conditions, Biochem. Eng. J. 151 (2019) 107301.

[61] S.L. Lim, W.L. Chu, S.M. Phang, Use of *Chlorella vulgaris* for bioremediation of textile wastewater, Bioresour. Technol. 101 (2010) 7314−7322.

[62] B.H. Diya'uddeen, W.M.A.W. Daud, A.R. Abdul Aziz, Treatment technologies for petroleum refinery effluents: a review, Process Saf. Environ. Prot. 89 (2011) 95−105.

[63] C. Gómez-Serrano, M.M. Morales-Amaral, F.G. Acién, R. Escudero, J.M. Fernández-Sevilla, E. Molina-Grima, Utilization of secondary-treated wastewater for the production of freshwater microalgae, Appl. Microbiol. Biotechnol. 99 (2015) 6931−6944.

[64] Y.H. Wu, S.F. Zhu, Y. Yu, X.J. Shi, G.X. Wu, H.Y. Hu, Mixed cultivation as an effective approach to enhance microalgal biomass and triacylglycerol production in domestic secondary effluent, Chem. Eng. J. 328 (2017) 665−672.

[65] H. Kamyab, M.F.M. Din, A. Keyvanfar, M.Z.A. Majid, A. Talaiekhozani, A. Shafaghat, C.T. Lee, L.J. Shiun, H.H. Ismail, Efficiency of microalgae *Chlamydomonas* on the removal of pollutants from palm oil mill effluent (POME), Energy Procedia 75 (2015) 2400−2408.

[66] C.S. Vairappan, A.M. Yen, Palm oil mill effluent (POME) cultured marine microalgae as supplementary diet for rotifer culture, J. Appl. Phycol. 20 (2008) 603−608.

[67] M.C.,D.S. Hadiyanto, Phytoremediation of palm oil mill effluent (POME) by using aquatic plants and microalge for biomass production, J. Environ. Sci. Technol. 6 (2013) 79−90.

[68] A. Cicci, M. Stoller, M. Bravi, Microalgal biomass production by using ultra- and nanofiltration membrane fractions of olive mill wastewater, Water Res. 47 (2013) 4710−4718.

[69] G. Hodaifa, A. Malvis Romero, M. Halioui, S. Sánchez, Combined process for olive oil mill wastewater treatment based in flocculation, photolysis, microfiltration and microalgae culture, in: N. Khelifi, A. Kallel, M. Ksibi, H. Ben Dhia (Eds.), Recent Adv. Environ. Sci. From Euro-Mediterranean Surround. Reg. EMCEI 2017. Adv. Sci. Technol. Innov. (IEREK Interdiscip. Ser. Sustain. Dev., Springer, Cham, 2018, pp. 1127−1129.

[70] F. Di Caprio, P. Altimari, F. Pagnanelli, Integrated microalgae biomass production and olive mill wastewater biodegradation: optimization of the wastewater supply strategy, Chem. Eng. J. 349 (2018) 539−546.

[71] G. Pinto, A. Pollio, L. Previtera, M. Stanzione, F. Temussi, Removal of low molecular weight phenols from olive oil mill wastewater using microalgae, Biotechnol. Lett. 25 (2003) 1657−1659.

[72] F.M. Ghazal, E.-S.M. Mahdy, M.S.A. EL-Fattah, A. Elg, Y. EL-Sadany, N.M.E. Doha, The use of microalgae in bioremediation of the textile wastewater effluent, Nat. Sci. 3 (2018) 98−104.

[73] H.Y. El-Kassas, L.A. Mohamed, Bioremediation of the textile waste effluent by *Chlorella vulgaris*, Egypt, J. Aquat. Res. 40 (2014) 301−308.

[74] K. Suresh Kumar, H.-U. Dahms, E.-J. Won, J.-S. Lee, K.-H. Shin, Microalgae — a promising tool for heavy metal remediation, Ecotoxicol. Environ. Saf. 113 (2015) 329−352.

[75] H.V. Perales-Vela, J.M. Peña-Castro, R.O. Cañizares-Villanueva, Heavy metal detoxification in eukaryotic microalgae, Chemosphere 64 (2006) 1−10.

[76] H. Doshi, A. Ray, I.L. Kothari, Bioremediation potential of live and dead *Spirulina*: spectroscopic, kinetics and SEM studies, Biotechnol. Bioeng. 96 (2007) 1051−1063.

[77] C.M. Monteiro, A.P.G.C. Marques, P.M.L. Castro, F.X. Malcata, Characterization of *Desmodesmus pleiomorphus* isolated from a heavy metal-contaminated site: biosorption of zinc, Biodegradation 20 (2009) 629−641.

[78] A. Takáčová, M. Smolinská, M. Semerád, P. Matúš, Degradation of BTEX by microalgae *Parachlorella kessleri*, Pet. Coal. 57 (2015) 101−107.

[79] Z.A. Ansari, M.C. Saldanha, R. Rajkumar, Effects of petroleum hydrocarbons on the growth of a microalga, *Isochrysis sp.* (Chrysophyta), Indian J. Mar. Sci. 26 (1997) 372−376.

[80] J.D. Walker, R.R. Colwell, L. Petrakis, Degradation of petroleum by an alga, *Prototheca zopfii*, Appl. Microbiol. 30 (1975) 79−81.

[81] A. Hodges, Z. Fica, J. Wanlass, J. VanDarlin, R. Sims, Nutrient and suspended solids removal from petrochemical wastewater via microalgal biofilm cultivation, Chemosphere 174 (2017) 46−48.

[82] V. Matamoros, R. Gutiérrez, I. Ferrer, J. García, J.M. Bayona, Capability of microalgae-based wastewater treatment systems to remove emerging organic contaminants: a pilot-scale study, J. Hazard Mater. 288 (2015) 34−42.

[83] J.-Q.Q. Xiong, M.B. Kurade, R.A.I.I. Abou-Shanab, M.-K.K. Ji, J. Choi, J.O. Kim, B.-H.H. Jeon, Biodegradation of carbamazepine using freshwater microalgae *Chlamydomonas mexicana* and *Scenedesmus obliquus* and the determination of its metabolic fate, Bioresour. Technol. 205 (2016) 183−190.

[84] V. Matamoros, E. Uggetti, J. García, J.M. Bayona, Assessment of the mechanisms involved in the removal of emerging contaminants by microalgae from wastewater: a laboratory scale study, J. Hazard Mater. 301 (2016) 197−205.

[85] K. Ramadass, M. Megharaj, K. Venkateswarlu, R. Naidu, Toxicity of diesel water accommodated fraction toward microalgae, *Pseudokirchneriella subcapitata* and *Chlorella sp, MM3*, Ecotoxicol. Environ. Saf. 142 (2017) 538−543.

[86] M. Sakarika, M. Kornaros, Kinetics of growth and lipids accumulation in *Chlorella vulgaris* during batch heterotrophic cultivation: effect of different nutrient limitation strategies, Bioresour. Technol. 243 (2017) 356−365.

[87] A. Juneja, R.M. Ceballos, G.S. Murthy, Effects of environmental factors and nutrient availability on the biochemical composition of algae for biofuels production: a review, Energies 6 (2013) 4607−4638.

[88] M.A. Borowitzka, High-value products from microalgae-their development and commercialisation, J. Appl. Phycol. 25 (2013) 743–756.

[89] M. Koller, A. Muhr, G. Braunegg, Microalgae as versatile cellular factories for valued products, Algal Res 6 (2014) 52–63.

[90] T.-Y. Zhang, H.-Y. Hu, Y.-H. Wu, L.-L. Zhuang, X.-Q. Xu, X.-X. Wang, G.-H. Dao, Promising solutions to solve the bottlenecks in the large-scale cultivation of microalgae for biomass/bioenergy production, Renew. Sustain. Energy Rev. 60 (2016) 1602–1614.

[91] E.S. Salama, M.B. Kurade, R.A.I. Abou-Shanab, M.M. El-Dalatony, I.S. Yang, B. Min, B.H. Jeon, Recent progress in microalgal biomass production coupled with wastewater treatment for biofuel generation, Renew. Sustain. Energy Rev. 79 (2017) 1189–1211.

[92] B. Behera, A. Acharya, I.A. Gargey, N. Aly, P. Balasubramanian, Bioprocess engineering principles of microalgal cultivation for sustainable biofuel production, Bioresour. Technol. Reports. 5 (2019) 297–316.

[93] P. Vo Hoang Nhat, H.H. Ngo, W.S. Guo, S.W. Chang, D.D. Nguyen, P.D. Nguyen, X.T. Bui, X.B. Zhang, J.B. Guo, Can algae-based technologies be an affordable green process for biofuel production and wastewater remediation? Bioresour. Technol. 256 (2018) 491–501.

[94] S.-Y. Chiu, C.-Y. Kao, T.-Y. Chen, Y.-B. Chang, C.-M. Kuo, C.-S. Lin, Cultivation of microalgal *Chlorella* for biomass and lipid production using wastewater as nutrient resource, Bioresour. Technol. 184 (2015) 179–189.

[95] T. Cai, S.Y. Park, Y. Li, Nutrient recovery from wastewater streams by microalgae: status and prospects, Renew. Sustain. Energy Rev. 19 (2013) 360–369.

[96] M.K. Lam, M.I. Yusoff, Y. Uemura, J.W. Lim, C.G. Khoo, K.T. Lee, H.C. Ong, Cultivation of *Chlorella vulgaris* using nutrients source from domestic wastewater for biodiesel production: growth condition and kinetic studies, Renew. Energy 103 (2017) 197–207.

[97] M.K. Ji, H.S. Yun, B.S. Hwang, A.N. Kabra, B.H. Jeon, J. Choi, Mixotrophic cultivation of *Nephroselmis sp.* using industrial wastewater for enhanced microalgal biomass production, Ecol. Eng. 95 (2016) 527–533.

[98] S.B. Ummalyma, R.K. Sukumaran, Cultivation of microalgae in dairy effluent for oil production and removal of organic pollution load, Bioresour. Technol. 165 (2014) 295–301.

[99] W.T. Chen, Y. Zhang, J. Zhang, G. Yu, L.C. Schideman, P. Zhang, M. Minarick, Hydrothermal liquefaction of mixed-culture algal biomass from wastewater treatment system into bio-crude oil, Bioresour. Technol. 152 (2014) 130–139.

[100] C.Y. Chen, X.Q. Zhao, H.W. Yen, S.H. Ho, C.L. Cheng, D.J. Lee, F.W. Bai, J.S. Chang, Microalgae-based carbohydrates for biofuel production, Biochem. Eng. J. 78 (2013) 1–10.

[101] Z. Reyimu, D. Özçimen, Batch cultivation of marine microalgae *Nannochloropsis oculata* and *Tetraselmis suecica* in treated municipal wastewater toward bioethanol production, J. Clean. Prod. 150 (2017) 40–46.

[102] A.L. Marques, F.P. Pinto, O.Q.F. Araújo, M.C. Cammarota, Assessment of methods to pretreat microalgal biomass for enhanced biogas production, J. Sustain. Dev. Energy, Water Environ. Syst. 6 (2018) 394–404.

[103] J.C. Meneses-Reyes, G. Hernández-Eugenio, D.H. Huber, N. Balagurusamy, T. Espinosa-Solares, Oil-extracted *Chlorella vulgaris* biomass and glycerol bioconversion to methane via continuous anaerobic co-digestion with chicken litter, Renew. Energy 128 (2018) 223–229.

[104] B. Molinuevo-Salces, A. Mahdy, M. Ballesteros, C. González-Fernández, From piggery wastewater nutrients to biogas: microalgae biomass revalorization through anaerobic digestion, Renew. Energy 96 (2016) 1103–1110.

[105] M.P. Caporgno, A. Taleb, M. Olkiewicz, J. Font, J. Pruvost, J. Legrand, C. Bengoa, Microalgae cultivation in urban wastewater: nutrient removal and biomass production for biodiesel and methane, Algal Res 10 (2015) 232–239.

[106] L. Yang, X. Tan, B. Si, F. Zhao, H. Chu, X. Zhou, Y. Zhang, Nutrients recycling and energy evaluation in a closed microalgal biofuel production system, Algal Res 33 (2018) 399–405.

[107] E. Koutra, C.N. Economou, P. Tsafrakidou, M. Kornaros, Bio-based products from microalgae cultivated in digestates, Trends Biotechnol. 36 (2018) 819–833.

[108] A.P. Batista, L. Ambrosano, S. Graça, C. Sousa, P.A.S.S. Marques, B. Ribeiro, E.P. Botrel, P. Castro Neto, L. Gouveia, Combining urban wastewater treatment with biohydrogen production - an integrated microalgae-based approach, Bioresour. Technol. 184 (2015) 230–235.

[109] D. Sengmee, B. Cheirsilp, T.T. Suksaroge, P. Prasertsan, Biophotolysis-based hydrogen and lipid production by oleaginous microalgae using crude glycerol as exogenous carbon source, Int. J. Hydrogen Energy 42 (2017) 1970–1976.

[110] C.V. García Prieto, F.D. Ramos, V. Estrada, M.A. Villar, M.S. Diaz, Optimization of an integrated algae-based biorefinery for the production of biodiesel, astaxanthin and PHB, Energy 139 (2017) 1159–1172.

[111] M. van der Spiegel, M.Y. Noordam, H.J. van der Fels-Klerx, Safety of novel protein sources (insects, microalgae, seaweed, duckweed, and rapeseed) and legislative aspects for their application in food and feed production, Compr. Rev. Food Sci. Food Saf. 12 (2013) 662–678.

[112] W. Verstraete, P. Clauwaert, S.E. Vlaeminck, Used water and nutrients: recovery perspectives in a 'panta rhei' context, Bioresour. Technol. 215 (2016) 199–208.

[113] J. Cheng, J. Xu, Y. Huang, Y. Li, J. Zhou, K. Cen, Growth optimisation of microalga mutant at high CO_2 concentration to purify undiluted anaerobic digestion effluent

of swine manure, Bioresour. Technol. 177 (2015) 240−246.

[114] T. Hülsen, K. Hsieh, Y. Lu, S. Tait, D.J. Batstone, Simultaneous treatment and single cell protein production from agri-industrial wastewaters using purple phototrophic bacteria or microalgae − a comparison, Bioresour. Technol. 254 (2018) 214−223.

[115] G. Markou, E. Nerantzis, Microalgae for high-value compounds and biofuels production: a review with focus on cultivation under stress conditions, Biotechnol. Adv. 31 (2013) 1532−1542.

[116] A. Gille, U. Neumann, S. Louis, S.C. Bischoff, K. Briviba, Microalgae as a potential source of carotenoids: comparative results of an in vitro digestion method and a feeding experiment with C57BL/6J mice, J. Funct. Foods. 49 (2018) 285−294.

[117] M.M.A. Nur, W. Muizelaar, P. Boelen, A.G.J. Buma, Environmental and nutrient conditions influence fucoxanthin productivity of the marine diatom *Phaeodactylum tricornutum* grown on palm oil mill effluent, J. Appl. Phycol. (2018) 1−12.

[118] D.B. Rodrigues, É.M.M. Flores, J.S. Barin, A.Z. Mercadante, E. Jacob-Lopes, L.Q. Zepka, Production of carotenoids from microalgae cultivated using agroindustrial wastes, Food Res. Int. 65 (2014) 144−148.

[119] E. Alobwede, J.R. Leake, J. Pandhal, Circular economy fertilization: testing micro and macro algal species as soil improvers and nutrient sources for crop production in greenhouse and field conditions, Geoderma 334 (2019) 113−123.

[120] G.S. Diniz, A.F. Silva, O.Q.F. Araújo, R.M. Chaloub, The potential of microalgal biomass production for biotechnological purposes using wastewater resources, J. Appl. Phycol. 29 (2017) 821−832.

CHAPTER 21

Resource Recovery From Waste Streams Using Microalgae: Opportunities and Threats

NAIM RASHID • THINESH SELVARATNAM • WON-KUN PARK

INTRODUCTION

Microalgae are being considered as a convincing material to perform ecological services and respond to the sustainability challenges. Meritorious attributes of microalgae have shifted the focus of biorefinery from the traditional feedstock, that is, crop biomass to microalgae biomass. Microalgae return high crop turn-over, can use a wide variety of carbon sources, do not compete with food sources, and require less amount of water and nutrients to grow than the other plants [1,2]. The microalgae biomass does not contain lignin unlike wood biomass and possesses high biomolecules contents which are suitable for the production of value-added metabolites [3]. A wide variety of bioproducts can be obtained from microalgae biomass including biofuels (biodiesel, biohydrogen, and bioethanol), pharmaceutical ingredients, nutritional, feed and food supplements [1,4−7]. In the past few years, adequate studies have been dedicated to realizing the potential of microalgae for biofuel production [8,9]. However, the life cycle and technoeconomic analysis have revealed that microalgae-based biofuels are not cost-competitive in comparison with conventional petrochemical resources [8,10]. The high cost of microalgae biofuels is mainly because of biomass production and processing. According to an estimate, almost 70% of the total production cost is accounted only for cultivation and harvesting [11]. Cost of microalgal cultivation could be further classified into capital and operating cost. Capital cost includes land preparation, construction for the culture systems, and basic facilities. The operating cost includes raw material (nutrients and CO_2), utilities (water and power), labor, and others for regular expenses [12,13]. In the case of capital cost, it is usually fixed with the systemic design of the facility. Therefore, there is not much room for saving the cost.

However, in case of operating cost, one possible way to reduce the cost is to integrate microalgae cultivation system with wastewater industry. The water originated from waste streams is a rich source of nutrients, energy, and other bioresources. Wastewater can produce 6.5 MJ/kL of chemical energy, constituting 1% of total world energy [14]. The nutrients present in the form of carbon, nitrogen, and phosphorous can be turned into an economic opportunity by feeding them to microalgae [15]. Microalgae have the ability to utilize organic carbon present in wastewater and tailor it into biomass. The use of wastewater for microalgae cultivation in mixotrophy and heterotrophy cultivation mode can balance respiratory losses, improve energy budget, and give a boost to the biomass productivity [15,16].

Furthermore, according to Acien et al. [17] utilization of nutrients and water from wastewater and CO_2 from flue gas could decrease the portion of operation cost from 56% to 31% and reduce the total biomass production cost up to 35%. The recovery of bioresources such as heavy metals, rare earth metals, and value-added products can also compensate for the cost of the cultivation process [18−20]. Moreover, the wastewater would not need to go through the essential treatment step to meet the ecological and environmental regulations. Thus, the use of wastewater for microalgae cultivation would promote waste-free, carbon neutral, and environmentally sustainable technology. This chapter underlines the importance of integrating waste biorefinery with microalgae cultivation to reinforce the objectives of resource-efficient bioeconomy.

Microalgae Cultivation for Biofuels Production. https://doi.org/10.1016/B978-0-12-817536-1.00021-7

MECHANISM OF MICROALGAE FOR RESOURCE RECOVERY

Extensive studies have been carried out to realize the sustainability of wastewater biorefinery. The economics of wastewater industry is limited by high energy, operational cost, and resource sustainability. It triggered the research community to rethink the paradigm of wastewater treatment. New research activities are moving from the traditional "out of sight-out of mind" attitude to viewing the wastewaters as potential sources for resource recovery. The wastewaters originated from municipal, agricultural and industrial streams are enriched with high-value bioresources (Fig. 21.1). These bioresources can be recovered and turned into an economic opportunity, which will offset the cost of wastewater treatment and promote bio-based circular bioeconomy. Resource recovery from the wastewaters will help to limit the need for exploiting virgin resources and can improve the energy budget of the treatment processes. In this perspective, microalgae-based biotechnological processes offer a robust pathway to concentrate and transform resources from the wastewaters. The recoverable resources can be categorized into energy, nutrients (mainly carbon, nitrogen, phosphorous), and heavy metals. The following section will briefly discuss the mechanism of resource recovery by microalgae. A particular focus is only on the removal mechanism of the resources rather than the origins and sinks of the waste resources.

Carbon

Microalgae have the ability to fix atmospheric carbon dioxide, as well as the carbon, originated from the industrial waste gas streams through photosynthesis [21–23]. Microalgae can also use wastewaters containing organic loadings through the heterotrophic mechanism [24]. Mixotrophic microalgae strains can metabolize both organic and inorganic carbon and provide the opportunity to improve the overall sustainability of microalgae-based resource recovery pathways [25–27].

Nitrogen

Nitrogen plays a vital role in microbial growth. Nitrogen in the form of organic nitrogen can be found in a wide variety of biological substances such as energy transfer molecules (adenosine triphosphate [ATP], adenosine diphosphate [ADP]), genetic materials (DNA, RNA), chlorophylls, proteins, peptides, and enzymes [28]. Microalgae derive the necessary organic nitrogen from such sources including ammonia, ammonium, nitric acid, nitrite, and nitrate. Microalgae convert these inorganic nitrogen sources into organic nitrogen through assimilation [29–31]. Eukaryotic algae can perform assimilation when nitrogen is provided only in the form of ammonium, nitrite, and nitrate. The translocation of the inorganic nitrogen molecules occurs in the plasma membrane, which then followed by the reduction of oxidized nitrogen. The final step of the process is to incorporate the reduced nitrogen into amino acids [32]. Nitrate and nitrite molecules are reduced through nitrate and nitrite reductases. Afterward, the nitrite is reduced to ammonium by nitrite reductase and ferredoxin [31,32]. Ultimately, all forms of inorganic nitrogen are reduced to ammonium before being incorporated within the intracellular membrane as amino acids. Different algae strains showed a preference for one or more inorganic nitrogen sources [33,34]. Maestrini et al. [35] showed that algae prefer ammonium over nitrate, and nitrate consumption is suppressed until the complete ammonium uptake.

Phosphorus

Phosphorus is an essential element to support the growth of all living organisms being an integral part of lipids, proteins, phospholipids, and the intermediates of carbohydrate metabolism [36]. In microalgae cell growth and metabolism, inorganic phosphorus plays a vital role. During algal metabolism, phosphorus in the forms of dihydrogen phosphate and hydrogen phosphate is incorporated into organic compounds

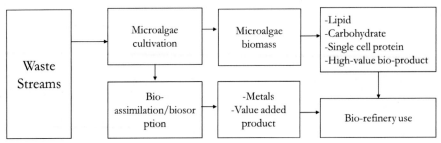

FIG. 21.1 The processes of resource recovery from waste stream using microalgae.

through the process of phosphorylation. The process involves the generation of ATP from ADP [37] accompanied by energy input. This energy input can be originated from the oxidation of respiratory substrates or sunlight. During the process, phosphates are transferred across the plasma membrane of the algal cells by an energized transport. Fully oxidized inorganic phosphates are the preferred sources of phosphorous for microalgal metabolism. However, several studies showed that some algal strains could use partially oxidized inorganic phosphate compounds and phosphorus found in organic esters [36,37].

Heavy Metals

Heavy metal contamination of the natural environment has increased ever since the industrial revolution, which poses an elevated risk to the ecosystem [38]. Current technologies often employ physiochemical approaches such as membrane separation, ion exchange, adsorption, chemical precipitation, and solvent extraction for heavy metal removal [39]. However, these methods are plagued with issues such as incomplete metal removal, high energy, and reagent requirements for produced sludge recovery/disposal, and expensive equipment requirements. The use of biomass-based metal adsorbing and removal system is viewed as an alternative to overcome some of the limitations of the physiochemical processes [40]. In recent years, algae-based systems have received increased interest among the researchers because of their potential to recover precious and rare earth metals [41−43]. The common mechanism of metal uptake by microalgae is often described as "biosorption" instead of accumulation. However, the biosorption describes well the process of metal ion sorption on the dead biomass, and it includes metal ion binding in both intracellular and extracellular ligands [44]. Heavy metal accumulation in microalgae occurs through two processes: passive initial rapid uptake followed by an active slower uptake [45]. Passive uptake is a metabolism-independent process in which heavy metal ions are adsorbed onto the algae surface in a short time. The metal sorption ability of algae differ from species to species, also varies within the strains [46−48]. Heavy metal biosorption by microalgae can be affected by various factors such as the metal ions concentration, algal biomass, temperature, pH, metabolic stage, and competing ions. During the active intake, metal ions transport across the algal cell membrane into the cytoplasm, and the process is metabolism dependent. Also, the heavy metal ion transport can occur through passive diffusion due to the metal-induced increase in permeability of the algal cell membrane [43]. Once the metal ions are inside the algal cell, they may precipitate or bind to intracellular compounds [49].

Overall, the corresponding resource removal mechanisms play a vital role in the subsequent recovery and reutilization of the resources. For example, thermochemical processes such as hydrothermal liquefaction of the algal biomass can use to recover (1) energy and carbon from the crude fraction (2) nutrients from the aqueous fraction of the process products [50−52].

BIOCHEMICALS RECOVERY FROM WASTE RESOURCES

Single-Cell Protein

Resource recovery in the form of the protein is an intriguing choice in modern waste biorefinery processes. Protein is an essential component of human and animal food. Traditionally, protein is obtained from plant or animal resources [25,53,54]. Owing to increasing population and diminishing agriculture resources, these protein sources are not sustainable. Also, their production imposes a negative impact on the world ecological system by producing greenhouse gases. In fact, plant growth requires high nitrogen input, provided in the form fertilizers. This process of nitrogen utilization is inherited with nutrient losses due to volatilization, run-off, and dissipation [55]. Additional conversion losses are involved while feeding plant-based protein to the animals, to obtain protein in the form of meat [56,57]. According to an estimate, only 17% of nitrogen input is translated into protein, and the rest is dedicated to processing loss [14]. Regardless of the low nitrogen conversion ratio of nitrogen into protein by plant and animal resources, they are being seriously threatened by urbanization and industrialization. In the future, the world carrying capacity will be challenged by the shortage of arable land, water, and food footprints [58,59]. These challenges have pushed the research interest to find an alternative, cheap, and sustainable protein sources. In this perspective, protein synthesis from microbial resources, called single-cell protein (SCP), can be an attracting choice. SCP offers distinct advantages over traditional protein sources. SCP production has high nutrient utilization efficiency and high crop turn-over with a lesser demand for land and water. In the SCP process, a wide variety of substrates can be employed, resulting in diverse protein composition. The major advantage of employing microorganism in protein synthesis is that they can recover nutrients from waste resources and can upcycle them into value-added bioproducts including SCP [14].

They can thrive in dilute as well as concentrated streams displacing the need for modulating water and nutrient inputs.

In the past, SCP has been produced by fungi, bacteria, and yeasts. Recently, microalgae have emerged as SCP source. The microalgae-based SCP contains essential amino acids and other essential nutrients required for animal and human growth. From sustainability and economic perspective, a standalone cultivation system of microalgae for SCP production is not affordable due to expensive carbon and nitrogen substrate requirements. Alternatively, the integration of microalga technology with wastewater biorefinery can be a sustainable model to promote bio-based economy [60]. Microalgae-based wastewater treatment can be proved superior to conventional systems. In conventional systems, oxygen is furnished through expensive aeration, whereas in the former system microalgae produce oxygen through photosynthesis.

Microalgae cultivation can be carried out in an open pond or closed photobioreactor system. A number of biotic and abiotic factors describe the system energetics to dictate the choice of a cultivation system [61]. It is reported that in the closed bioreactor system, up to 90% of nitrogen can be translated into biomass. Closed photo-bioreactor can return a homogenous composition, and thus, high-quality biomass. On the contrary, open pond cultivation system expectedly offers heterogeneous, and low-quality biomass due to high contamination risk by other microorganisms. The quality of SCP is affected by the provision of different carbon sources and light pattern defining into phototrophic, mixotrophic, and heterotrophic cultivation [62]. Phototrophic cultivation would result in relatively high-quality SCP due to better control on growth to maintain axenic culture. Mixotrophy and heterotrophic cultivation would return heterogenous biomass due to the inclusion of bacteria during cultivation to metabolize organic carbon. Heterotrophic cultivation, however, produces much higher biomass than the phototrophic and mixotrophic cultivation. Unfortunately, preferential studies have been carried out to investigate the effect of cultivation mode (based on carbon and light source) on protein quality. The effect of cultivation mode is widely studied in the perspective of lipid production; however, intended control of cultivation conditions to induce SCP might govern an entirely different mechanism of growth modeling. For example, lipid accumulation is enhanced under nitrogen- and sulfur-deprived conditions, but its impact on protein synthesis is not clearly known yet. Likewise, slow-growing microalgae accumulate more lipids, and thus, theoretically, fast-growing microalgae should return high SCP. Experimental investigations might redirect the cultivation focus. Thus, from SCP production standpoint, fast-growing microalgae can be the ultimate choice, which would significantly reduce the cost of cultivation (due to less growth time). Yet, the cost of SCP is higher than the traditional protein source. SCP can compete for the cost of agriculture and animal resource by exploiting the potential of wastewater resources and valorizing waste materials. Wastewater is a rich source of nitrogen and organic carbon and minerals, which can be assimilated by microalgae to produce SCP. Thus, wastewater is treated with negative energy input and resources are converted into a valuable bioproduct (SCP). Microalgae utilize inorganic nutrients from wastewater and convert them into organic materials through photosynthesis. Phototrophic microalgae in wastewater show relatively low resource recovery than the heterotrophs. Phototrophic microalgae produce organic carbon from inorganic resources and thus cannot use organic carbon present in the wastewater, whereas heterotrophic microalgae efficiency use organic carbon and turn it into SCP. Hulsen et al. (2018) showed that microalgae could efficiently remove nutrients from various wastewaters (livestock, food) and produce SCP. They also tested coculture of microalgae with purple phototrophic bacteria and concluded that coculture could result in the high quality of SCP than the monoculture. It encourages directing future research to focus on the coculture system. In a coculture system, a phototrophic microalga is grown with heterotrophic algae or bacteria in a symbiotic relationship. Phototrophic algae fix inorganic nutrients and produce oxygen, which is used by heterotrophic microorganisms to metabolize organic carbon. Coculture system is expected to be superior to the monoculture system, with an added advantage of bioflocculation. Recently, a number of studies have been carried out demonstrating the resource recovery through coculture system; however, they are mainly focused on microalgae—bacteria system. In nutrient recovery and bioremediation perspective, microalgae—bacteria can be a promising choice; however, for SCP production, it might not be appealing because it will produce heterogenous biomass. Future studies should be pledged onto microalgae—microalgae coculture system to reflect their potential for quantitative as well as qualitative estimation of SCP.

Feasibility of Single-cell protein

SCP production by microalgae is an ambitious biorefinery goal. The sustainability and large-scale production of SCP are restricted by various environmental, health, technical, and policy drivers. The use of microalgae-based SCP for food and feed has some safety hazards including toxins, nucleic acids, allergens, and indigestible fibers. The use of wastewater as a substrate for SCP synthesis can pose a negative impact on its quality due to high nutrient accumulation in the biomass, and its bad smell would limit its application as an edible protein for the humans. SCP application as food for the animal, aquaculture, and poultry has tremendous potential. Studies have shown that SCP meets environmental and food regulations. Regardless of the high quality and nutrition value of SCP, its application at large scale and potential to replace conventional protein sources is still questionable. Future research should be directed to realize the potential of SCP production combined with the objectives of climate change, environmental remediation, and waste biorefinery. Sustained funding can constitute a comprehensive research niche and can redirect the commercial focus of microalgae biorefinery from biofuels production to SCP synthesis.

Carbohydrate and lipids

The intense use of microalgae as a protein source is likely because they contain high protein contents and their targeted benefits to human health. Microalgae are composed of not only protein but also carbohydrates and lipids. The biochemical composition of microalgae biomass depends on nutrients input and culture conditions including abiotic and biotic factors [9,63]. Also, there are >30,000 identified species of microalgae, which offer diverse biomass composition [64,65]. Some microalgae species are rich in protein, whereas the others in lipids and carbohydrate contents. However, there is a trade-off between protein, carbohydrate, and lipid contents. The employment of microalgae systems for carbohydrate production also has an appeal. Waste streams contain a high concentration of carbon. Microalgae have the ability to concentrate carbon and accumulate in the form of carbohydrate [9,64,65].

Microalgae-based carbohydrate can be utilized as a feedstock for fermentation/anaerobic digestion process to produce biogas, bioethanol or butanol. Because microalgae do not contain lignin, hemicellulose unlike woody biomass, also some species do not even contain a cell wall, they can be easily digested after mild pretreatment process [66]. Microalgae-based carbohydrate can also be used for value-added bioproducts including food additives, cosmetics, textiles, and pharmaceutical products [67]. For example, fucose, which has high potentials in the cosmetic and pharmaceutical industry,

has been recognized as rare sugars in nature [68]. Hong et al. [69] show the production of fucose from lipid extracted algae. Also, there are exopolysaccharides released from microalgae, which serve as bioactive compounds to control microalgae and propagate the growth of other microorganisms [7]. For example, antiviral compounds from some cyanobacteria were able to inhibit the human immune deficiency virus and avian myeloblastosis virus and the sulfated polysaccharides from microalgae, *Gyrodinium impudicum* could inhibit Encephalomyocarditis virus [70,71], although their inhibitory mechanism is not exactly defined yet.

The use of microalgae for valuable bioproducts has tremendous potential to achieve biorefinery goals. However, the biomass needs to undergo a treatment process to remove nonfood contents. The treatment process is neither easy nor economically affordable. Thus, the use of microalgae through fermentation offers more benefits than the former one.

In addition to carbohydrates and proteins, microalgae have the ability to produce lipids. Lipids can be transformed into fatty acid methyl esters through esterification or transesterification. It shows similar properties as of diesel and thus can be used as fuel without modifications in a vehicle engine. Some lipids and fatty acids are recognized to have higher market values than biodiesel. In this context, polyunsaturated fatty acids (PUFAs) have been widely studied. PUFAs show high benefit to human health. In PUFAs, α-linolenic acid (ALA, 18:3, ω-3), eicosapentaenoic acid (EPA, 20:5, ω-3), and docosahexaenoic acid (DHA, 22:6 ω-3) play a vital role to cure cardiovascular and cardiac diseases by reducing total cholesterol and LD cholesterol [72]. Microalgae-based PUFAs are reported as superior to conventional fish oil due to their fine odor, quick absorption ability, and easy separation process [73]. DHA has the highest market pull because of various health benefits, for example, preventing arteriosclerosis, alleviating inflammatory conditions, asthma, depression, and rheumatoid arthritis [74–77]. According to an estimate, the retail price of DHA is 600–1000 times higher than the projected cost of algal biofuel (comparison conducted based on retail price of DHA, the cost for algal biodiesel with 50% margin, and density of biodiesel) [20,78,79].

Some lipophilic pigments (chlorophyll, carotenoids) also can be termed as valuable products. Chlorophyll and carotenoids play an important role in photosynthesis in plant cells. They show antioxidant properties and their antioxidant activity do a major role as a valuable product β-carotene; lutein and astaxanthin are also used as commercial products for health supplements [80]. The major products of microalgae biomass, including protein, carbohydrates, and lipids, are widely applied for various purposes. However, their commercial-scale application is not economically affordable. The price of microalgae biomass can be reduced to some extent by upcycling bioresources from waste streams. Still, microalgae biorefineries confront economic and sustainability challenges. These challenges can be overcome by extending the scope of microalgae application to pharmaceutical and nutritional coproducts.

RECENT DEVELOPMENTS AND OPPORTUNITIES FOR RESOURCE RECOVERY

The Use of Extremophile Microalgae

Over the past few decades, research activities were dedicated to investigating the potential of microalgae as a sustainable source of energy and bioproducts. However, most of the microalgae production facilities are geographically confined to the places that offer favorable growth conditions to the selected strains. Most microalgal strains have evolved under relatively mild climate and often not able to grow under extreme environmental conditions including temperature, pH, salinity, and the contaminants. The environmental conditions highly vary across the globe and thus provide the diverse potential of microalgae growth. In some areas, microalgae growth is fully or partially restricted due to extreme environmental conditions. The microalgal strains that can survive under extreme environmental conditions are termed as extremophiles. In addition, there are some strains that can withstand high levels of CO_2 or have the ability to grow in the environments with high levels of heavy metal presence. These organisms are called as polyextremophiles [81]. There are two extremophilic microalgal strains that have reached a stage of being a commercially traded product: (1) *Dunaliella*, green unicellular microalgae isolated from high saline water and (2) *Spirulina*, a filamentous cyanobacterium isolated in lakes with high pH in the range of 9−11 [82]. The most recent development in the algal research paradigm is to explore the use of extremophiles for resource recovery and the production of value-added bioproducts [41,83,84]. The following section will briefly discuss the potential of extremophiles in wastewater treatment to recover/remove heavy metals, high-value product generation, and food applications.

Wastewater Treatment

The use of municipal and industrial wastewater for algal growth is adequately studied over the years [85−87]. However, the use of extremophiles for their large-scale application is scarcely investigated. Extremophiles offer a unique platform for wastewater treatment through the possibility of growing algae in a wide temperature spectrum and thus enables the application in places where traditional algal strains underperform. Selvaratnam and Pegallapati [88] used acidophilic extremophile strain *Galdieria sulphuraria* to treat municipal wastewater to meet the regulatory drivers. The authors exploited the selected strains the ability to grow at low pH (2−4) to minimize the potential competition from the microorganisms present in municipal wastewater. They also evaluated the potential of carbon and nutrient recovery from the wastewater using the same extremophilic strain [26,52]. de-Bashan and Trejo [89] used *Chlorella sorokiniana*, heat and intense sunlight tolerant microalgae to treat wastewater and remove ammonium.

Heavy Metal Removal/Recovery

Heavy metal pollution stemming from acid mine drainage is the most aggravated environmental challenge in the mining industry worldwide [90]. In the United States alone, there are more than 500,000 abandoned mine sites and need intensive remediation [91]. The US Environmental Protection Agency estimated that the cleanup cost of selected 156 hard rock mine sites is $7 billion to $24 billion [92]. Finding economically viable ways to treat and recover metals from mine waste can provide not only a better environment but also an avenue to generate revenue. Algal-based biotechnological systems have received considerable attention as an ecologically safe method for accumulating heavy metals from mine wastes [41,42,93]. *Cyanidiales* group of extremophile algal strains, commonly found in hot sulfur springs at pH less than 5, showed a unique ability to grow in acid mine waste solutions enriched with sulfate anions [84,94]. Wood and Wang [93] used an extremophile, *Cyanidium caldarium* to demonstrate the removal of metal ions by precipitation from acid mine waters. Minoda and Sawada [41] have investigated the use of *G. sulphuraria* (074W) for recovery of rare earth metals. The study showed that Nd, Dy, and Cu were efficiently recovered (greater than 90%) from a solution containing a mixture of different metals under a semianaerobic heterotrophic condition at pH 2.5 using *G. sulphuraria* (074W). Similarly, Ju and Igarashi [42] demonstrated the effective and selective recovery of gold and palladium ions from wastewater using *G. sulphuraria*.

High-Value Product Generation

The extremophilic algal strains develop unique mechanisms to withstand the extreme environments.

Microalgal strains from cold climates tend to develop a distinguished membrane structure to protect the cells from freezing damage, whereas the strains from hot springs produce enzymes that are resistant to the elevated temperatures [81]. The typical example of this special mechanism is the accumulation of β-carotene as a protective agent against excess light or the accumulation of glycerol as an osmoregulant in *Dunaliella*. These mechanisms can be used to produce value-added bioproducts from the extremophiles. Fan and Vonshak [95] employed a similar phenomenon for the extraction of astaxanthin from the mass culture of *Haematococcus pluvialis*. Recent developments in the industry have enabled the researchers to develop several biotechnological applications such as (1) production of astaxanthin using green algae *Chlamydomonas nivalis* [96], (2) xanthophyll cycle pigments and vitamin E production using *Raphidonema* sp. [97], and (3) blue pigment phycocyanin production using red algae *G. sulphuraria* 074G [98] etc. The production of high-value products would promote resource efficient bioindustry.

Food Application

Microalgae can be used as a source of vitamins and fatty acids. Especially algal strains with high protein content might be useful to produce nutritious food products for both humans and animals [99]. The demonstrated ability of extremophilic algal growth in various wastewater streams can be used as a potential pathway to produce food sources. This pathway will enable to upcycle the nutrient and carbon recovery/reuse from the wastewater in a sustainable way. The extremophiles are less vulnerable to contamination than other microalgal species. Thus, they provide relatively pure and stable biomass composition, which is favorable for axenic food production.

Graziani and Schiavo [100] successfully employed *G. sulphuraria* in autotrophic and heterotrophic cultivation conditions to produce protein-rich as well as dietary fiber-rich biomass.

Recent studies on extremophile microalgal strains have provided an elegant solution to the waste industry to recover resources. It is being realized that extremophiles can tailor a sustainable biotechnology industry to support sustainable bioeconomy. However, further research should be carried out to identify other extremophile strains offering growth potential in diverse waste streams to maximize the utilization of waste resources.

Coculture system for resource recovery

In the last few years, the monoculture system is widely studied in microalgae biorefinery. Monoculture system is confronted with the high risk of contamination, which causes culture crash. Thus, the stability of monoculture system demands continuous monitoring and adopting aseptic techniques, which are not economically favorable for microalgae applications at industrial scale [101]. Other inherited challenges with the monoculture system include low biomass productivity, less diversity in biomass composition, inefficient metabolites exchange capacity, mass transfer, and high respiratory losses.

These issues have provided an opportunity to revisit the natural habitation of microalgae and establish another research niche that is closer to the natural ecosystem [101,102]. In a natural system, the microorganisms live in a community, communicate, and share the resources to benefit each other. In the community, they are exposed to either competitive or cooperative environment; however, they self-adjust to live cooperatively by identifying a friendly partner. This symbiotic relationship of the microbes envisions us to redirecting the microalgal technology toward the cocultivation system. In an intended (artificial) cocultivation system, two species of microalgae of similar characteristics are grown together [58]. They live in a symbiotic relationship, judiciously share the resources, produce the exudates and metabolites, and protect each other against unfavorable environmental (biotic or abiotic) conditions. Being living in a friendly environment, they return a high growth rate, biomass, and other biochemical productivity. Most of all, they produce extracellular polysaccharides, which cause self-flocculation in stationary phase, displacing the need of dewatering. In an intended cocultivation system, a phototrophic microalgae species is grown with either heterotrophic algae, yeast, or bacteria. Phototrophic algae turn inorganic carbon into organic carbon through photosynthesis, whereas the counterpart (heterotrophic algae) use this organic carbon and convert into carbon dioxide, which is used by phototrophic algae again to promote their growth. This system would offer enhanced growth and biomass productivity due to an efficient gas exchange (being produced within the cultivation matrix), show less dependency on external carbon (inorganic) supply to remove nutrients, and protection against invading microbes. This way, the cocultivation system is considered superior to the monoculture. Although cocultivation system provides a number of exciting benefits, in this chapter, we cite a few examples with regard to nutrients recovery only.

Rasouli et al. (2018) used coculture of microalgae—bacteria to recover nutrients from industrial wastewater [103]. They found that coculture could successfully remove organic carbon from the wastewater up to 91%, total nitrogen (TN) (67%) and total phosphorous (TP) (43%). Hulsen et al. (2018) employed coculture of microalgae—purple bacteria and achieved up to 74% of chemical oxygen demand (COD), 80% of NH_4—N, and 55% of PO_4—P [62]. Mujtaba et al. (2017) achieved 98%—100% of nitrogen, 92%—100% of phosphorus, and 94%—96% of COD by cocultivating *Chlorella* sp. with bacteria-containing activated sludge [104]. Zhou et al. (2017) achieved the maximum mean COD, TN, TP, and CO_2 removal efficiency can reach 68.29%, 61.75%, 64.21%, and 64.68%, respectively, by employing a coculture of microalgae—fungi [58]. Most of the studies are dedicated to microalgae—bacteria or microalgae—fungi cocultivation. Cocultivation with bacteria or fungi entails disadvantage of disturbing symbiotic balance due to different growth rates. Bacteria or fungi have much higher growth rates than the microalgae, and thus, outcompete microalgae, resulting in undesired biomass composition and biochemicals. Future research should be directed at investigating microalgae—microalgae cocultivation. Microalgae—microalgae cocultivation would be expectedly superior to bacteria- or fungi-based cultivation system because of various reasons. In microalgae—microalgae cultivation, both species would have similar biochemical characteristics and growth requirements including media composition, light pattern, bioreactor design, a carbon substrate, aeration, culture pH, and inoculation ratio. Thus, they would coexist in a cooperative relationship and are less likely to outcompete each other [105]. However, further optimization studies should be carried out to project the potential of microalgae—microalgae cultivation to draw a comparison with bacteria or fungi-based cocultivation. The typical factors in cocultivation would include identifying microalgae species offering a cooperative interaction with similar growth properties, optimizing inoculation ratio, studying the population dynamics and its impact on downstream bioprocessing. In nutrient recovery perspective, cocultivation of microalgae in wastewater should be carried out to realize the tolerance level of each counterpart (the microalga) and their ability for their reciprocal support (by producing oxygen and CO_2), valorizing the waste resources, and displacing the need for external resource supply [57,106]. Technoeconomic studies should be carried to ascertain the sustainability of microalgae cocultivation system in the perspective of bioremediation and ecological services.

Resource recovery using mixed wastewater

Microalgae have the ability to grow in diverse waste streams including municipal, agriculture, swine, and industry wastewaters. Their growth depends on wastewater composition because each wastewater contains a wide range of nutrients. Carbon, nitrogen, and phosphorous are major nutrients, which could be utilized by microalgae. The ratio of these nutrients displays a significant impact on microalgae growth. C:N:P ratio is a widely accepted abstraction to maintain nutrient concentration. Typically, Redfield ratio (41.1:7.2:1) is used to optimize C:N:P ratio [107]. The C:N:P ratio is specific to the microalgae strain, and also, their C:N:P ratios change over the course of cultivation time and conditions [108,109]. The disagreement of stoichiometry composition in the microalgae cell wall and the target waste stream can result in arrested microalgae growth and low nutrient removal efficiency. Thus, it is important to develop coherence between microalgae nutrient requirements and target waste streams. The nutrient requirements of microalgae growth can be satisfied if they are grown in the wastewater originated from a known source either municipal, agriculture, industry, or swine. The difficulty arises to grow microalgae in mixed wastewater. In an ideal wastewater-based microalgae cultivation system, minimum resource input and modifications in wastewater composition should be targeted. For example, secondary municipal wastewater is rich in phosphorous but has limited nitrogen concentration. To satisfy the N:P ratio, additional nitrogen should be supplemented to support microalgae growth, which will add cost to the cultivation system. Alternatively, phosphorous should be removed from the input material before supplementing it to the microalgae, which is also not desirable from the economical viewpoint. In this regard, the best approach would be to mix municipal wastewater with swine wastewater (being rich in nitrogen) and supplement it to the microalgae accordingly. Some waste streams offer concentrated wastewater, for example, biogas industry waste. High nutrient concentration can hamper microalgae growth. To surmount this difficulty, the wastewater can be simply diluted with water. However, it will pose a negative impact on water footprints and environmental sustainability. Future research should be directed to target microalgae cultivation in undiluted wastewater for efficient resource recovery, low energy budget, and water sustainability.

Limitations of using waste stream resources

As discussed, the utilization of waste and wastewater is an indispensable process to realize the economic sustainability of microalgae bio-refinery. Admittedly, it would reinforce the objectives of setting up sustainable waste biorefinery. However, there are some ethical issues to utilize waste resources for the production of high-value bioproducts. Wastewater is inherited with pathogens, toxic chemicals, heavy metals, and persistent organic pollutants too [110]. The use of these resources for microalgae feed would result in toxic biomass, and their subsequent use in pharmaceutical or nutritional products can pose a serious threat to human health. The toxicity can be curtailed by taking preventive pretreatment measures at upstream processing of the waste materials. However, the concern is beyond the toxicity. The public perception of the material input and preparation of these products can fuel a serious debate due to aesthetic and ethical issues. Moreover, it will urge government and civic organizations to inspect and take safety measures. Thus, so far, it turned out that the waste-stream-based microalgae biorefinery has not reached the technology-readiness level [110].

Alternatively, the use of clean waste sources for microalgae feed can be targeted. For example, organic waste originated from food industry is relatively clean, as it is obtained after peeling the food or processed to fit the product. The orange peel waste obtained during orange juice production can be a classic example of clean waste. Studies have shown that orange peel contains a reasonable amount of carbon, phosphorous, and other trace metals to support the growth of *Aurantiochytrium* sp. KRS101 [111]. Similarly, coffee waste and beer manufacturing wastes have been successfully applied for microalgae growth [112,113]. The use of clean organic waste resources for microalgae feed can promise a waste-free and carbon neutral bioindustry. Fig. 21.2 provides a summary of recent developments in microalgae bioprocessing.

Table 21.1 shows the composition of SCP obtained from different microbial biomass.

Potential Threats and Mitigation Strategies

Turning waste into an economic opportunity through microalgae is a promising approach; however, several issues threaten its application at large scale. The most staggering issue in microalga-based resource recovery system is the contamination and culture crash by the invading bacteria. The contamination would result in low biomass quality and complications in biomass purification. Traditionally, wastewater undergoes a necessary step of sterilization to control contamination

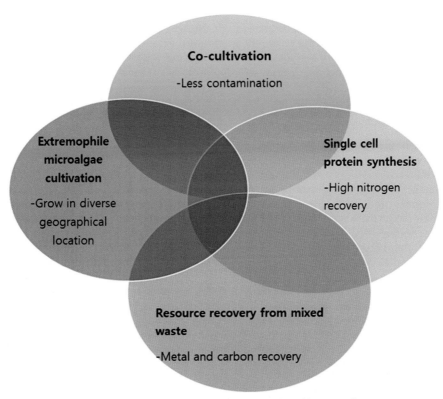

FIG. 21.2 Recent developments in microalgae bioprocessing.

TABLE 21.1
Single-Cell Protein (SCP) Composition of Various Microorganism Biomasses.

	Species Name	SCP Content, % of Biomass	References
Algae	Scenedesmus obliquus	30–50	[114]
	Euglena gracilis	50–70	[115]
	Chlorella sorokiniana	46–65	[116]
	Arthrospira maxima	60–71	[117]
	Aphanothece microscopica	42	[118]
Fungi	Kluyveromyces marxianus	59	[119]
	Yarrowia lipolytica Inulin	48–54	[120]
	Pleurotus florida	63	[121]
	Aspergillus flavus	10	[122]
	Candida tropicalis	56	[123]
Bacteria	Rhodopseudomonas palustris	55–65	[124]
	Cupriavidus necator	40–46	[125]
	Methylomonas sp.	69	[126]

before feeding it to microalgae. Unfortunately, sterilization is extremely expensive, and thus, its integration for a large volume of wastewater seems technically unrealistic. The use of extremophile microalgal species, which grow at low pH (<3.0) can limit the bacterial growth, as most of the bacteria cannot grow at this pH. The provision of chemicals to lower the pH of wastewater will still add significant cost to the system. Moreover, the use of only extremophiles will restrict the diversity of microalgal species to perform ecological services. Thus, extremophile cannot fully displace the need for sterilization. Recently, the use of a coculture system for contamination control is adequately reported. A number of microalgal species have been identified that can develop a symbiotic relationship with naturally born bacteria in wastewaters, offer protection against environmental changes, and equally share the resources to result in high biomass productivity. The potential of coculture in biomass productivity and biofuels production perspective has been realized in recent years; however, investigations on resource recovery should be carried out in future studies.

The basic notion of valorizing waste streams is to establish a sustainable bioindustry with minimum resource input. The ideal biorefinery approach would base upon putting minimum efforts in modifying waste resources and subsequently feeding them to the microalgae. In fact, waste streams contain concentrated carbon and other nutrients, which need several times

dilution to feed microalgae. The dilution causes high water footprints and thus threatens environmental and economic sustainability. Hence, microalgae cultivation in undiluted wastewaters should be targeted. Another approach can be employing the mixture of high-strength wastewater with low-strength wastewater for microalgae cultivation. Even so, the difficulty arises identifying wastewaters that can satisfy the nutrients stoichiometry demand of microalgae.

Microalgae-based resource recovery system mainly relies on photoautotrophs, which have the ability to produce oxygen through photosynthesis and metabolize the organic carbon and other contaminants during respiration. Photoautotrophs can displace the needs of costly mechanical aeration system, which is typically used in biological wastewater treatment. The use of only phototrophic microalgae for resource recovery has a number of fundamental issues. For example, their growth is contingent on the provision of light and seasonal variability. The sunlight duration highly changes year around and across the globe. Less light duration would lead to low photosynthetic efficiency, high respiratory losses, and high emission of greenhouse gases. Thus, relying only on phototrophic microalgae should not be desired. Rather, the use of polyculture for resource recovery should be targeted. In a polyculture system, various microorganisms live together in a niche. They offer high metabolic yield and resource recovery since some of them would be active under light and the others under a dark period. Polyculture produces composite biomass which is not favorable to obtain pure metabolites. Technology readiness is another major issue for microalgae-based resource recovery. Wastewater biorefinery is a long-known technology, and it has wide infrastructure around the world. The employment of microalgae in wastewater resource recovery would be a paradigm shift to redirect the research focus and to rebuilt or modify the existing infrastructure.

Conclusions and Future Perspectives

The use of microalgae for resource recovery from waste material is of paramount importance to drive the objectives of a sustainable bioindustry. It constitutes an interesting research focus to turn waste into an economic opportunity. It can tailor high energy budget and metabolites productivity with less material input. However, the realization of this technology at a commercial scale need to identify the challenges and devise a comprehensive strategy to address them. The diverse and unstable composition of waste streams is a real challenge to develop this technology and to integrate it with microalgae cultivation. Continued bioprospecting efforts are required to develop an agreement between wastewater composition and microalgae nutrients requirement. The use of undiluted wastewater should be encouraged to reduce water footprints and to conserve resources. Microalgae culture is highly vulnerable to contamination, which can be controlled to some extent by using extremophile species, and cocultivation system. In the wastewater system, polyculture return high resource recovery, biochemical productivity, and less reliance on carbon supplement. Redefining the scope of microalgae biorefinery from oil production to SCP synthesis, value-added bioproducts, and resource recovery can warrant establishing large-scale environmentally sustainable bioindustry.

REFERENCES

[1] G.L. Bhalamurugan, O. Valerie, L. Mark, Valuable bioproducts obtained from microalgal biomass and their commercial applications: a review, Environ. Eng. Res. 23 (2018) 229−241.

[2] Y. Chisti, J. Yan, Energy from algae: current status and future trends, Appl. Energy 88 (2011) 3277−3279.

[3] E.M. Trentacoste, A.M. Martinez, T. Zenk, The place of algae in agriculture: policies for algal biomass production, Photosynth. Res. 123 (2015) 305−315.

[4] A.K. Bajhaiya, J.Z. Moreira, J.K. Pittman, Transcriptional engineering of microalgae: prospects for high-value chemicals, Trends Biotechnol. 35 (2017) 95−99.

[5] M.M. El-Sheekh, M.M. Gharieb, G.W. Abou-El-Souod, Biodegradation of dyes by some green algae and cyanobacteria, Int. Biodeterior. Biodegrad. 63 (2009) 699−704.

[6] P. Das, S.S. Aziz, J.P. Obbard, Two phase microalgae growth in the open system for enhanced lipid productivity, De. Re Met. 36 (2011) 2524−2528.

[7] Y. Tang, J.N. Rosenberg, P. Bohutskyi, G. Yu, M.J. Betenbaugh, F. Wang, Microalgae as a feedstock for biofuel precursors and value-added products: green fuels and golden opportunities, Bioresources 11 (1) (2015) 2850−2885.

[8] T.M. Mata, A.A. Martins, N.S. Caetano, Microalgae for biodiesel production and other applications: a review, Renew. Sustain. Energy Rev. 14 (2010) 217−232.

[9] G. Markou, I. Angelidaki, D. Georgakakis, Microalgal carbohydrates: an overview of the factors influencing carbohydrates production, and of main bioconversion technologies for production of biofuels, Appl. Microbiol. Biotechnol. 96 (2012) 631−645.

[10] H.M. Amaro, A.C. Guedes, F.X. Malcata, Advances and perspectives in using microalgae to produce biodiesel, Appl. Energy 88 (2011) 3402−3410.

[11] C.V. García Prieto, F.D. Ramos, V. Estrada, M.A. Villar, M.S. Diaz, Optimization of an integrated algae-based biorefinery for the production of biodiesel, astaxanthin and PHB, Energy 139 (2017) 1159−1172.

[12] J. Li, D. Zhu, J. Niu, S. Shen, G. Wang, An economic assessment of astaxanthin production by large scale cultivation of Haematococcus pluvialis, Biotechnol. Adv. 29 (2011) 568−574.

[13] F.G. Acien, J.M. Fernandez, J.J. Magan, E. Molina, Production cost of a real microalgae production plant and strategies to reduce it, Biotechnol. Adv. 30 (2012) 1344−1353.

[14] D. Puyol, D.J. Batstone, T. Hülsen, S. Astals, M. Peces, J.O. Krömer, Resource recovery from wastewater by biological technologies: Opportunities, challenges, and prospects, Front. Microbiol. (2017) 7.

[15] S.K. Wang, X. Wang, H.H. Tao, X.S. Sun, Y.T. Tian, Heterotrophic culture of Chlorella pyrenoidosa using sucrose as the sole carbon source by co-culture with immobilized yeast, Bioresour. Technol. 249 (2018) 425−430.

[16] J. Wang, H. Yang, F. Wang, Mixotrophic cultivation of microalgae for biodiesel production: status and prospects, Biotechnol. Appl. Biochem. 172 (2014) 3307−3329.

[17] F.G. Acien Fernandez, C.V. Gonzalez-Lopez, J.M. Fernandez Sevilla, E. Molina Grima, Conversion of CO_2 into biomass by microalgae: how realistic a contribution may it be to significant CO_2 removal? Appl. Microbiol. Biotechnol. 96 (2012) 577−586.

[18] M.A. Kucuker, N. Wieczorek, K. Kuchta, N.K. Copty, Biosorption of neodymium on Chlorella vulgaris in aqueous solution obtained from hard disk drive magnets, PLoS One 12 (2017).

[19] Y. Cheng, W. Wang, W. Hua, W. Liu, P. Chen, R. Ruan, Bioremoval and recovery of metal ions by growing microalgae and via microwave assisted pyrolysis, Int. Agric. Eng. J. 25 (2016) 193−204.

[20] J. Li, Y. Liu, J.J. Cheng, M. Mos, M. Daroch, Biological potential of microalgae in China for biorefinery-based production of biofuels and high value compounds, N. Biotech. 32 (2015) 588–596.

[21] K.G. Zeiler, D.A. Heacox, S.T. Toon, K.L. Kadam, L.M. Brown, The use of microalgae for assimilation and utilization of carbon dioxide from fossil fuel-fired power plant flue gas, Energy Convers. Manag. 36 (1995) 707–712.

[22] I. Douskova, J. Doucha, K. Livansky, J. Machat, P. Novak, D. Umysova, et al., Simultaneous flue gas bioremediation and reduction of microalgal biomass production costs, Appl. Microbiol. Biotechnol. 82 (2009) 179–185.

[23] K.L. Kadam, Power plant flue gas as a source of CO_2 for microalgae cultivation: economic impact of different process options, Energy Convers. Manag. 38 (1997) S505–S510.

[24] X. Li, H. Xu, Q. Wu, Large-scale biodiesel production from microalga *Chlorella protothecoides* through heterotrophic cultivation in bioreactors, Biotechnol. Bioeng. 98 (2007) 764–771.

[25] A.P. Abreu, B. Fernandes, A.A. Vicente, J. Teixeira, G. Dragone, Mixotrophic cultivation of *Chlorella vulgaris* using industrial dairy waste as organic carbon source, Bioresour. Technol. 118 (2012) 61–66.

[26] T. Selvaratnam, A.K. Pegallapati, H. Reddy, N. Kanapathipillai, N. Nirmalakhandan, S. Deng, et al., Algal biofuels from urban wastewaters: maximizing biomass yield using nutrients recycled from hydrothermal processing of biomass, Bioresour. Technol. 182 (2015) 232–238.

[27] S.M. Henkanatte-Gedera, T. Selvaratnam, N. Caskan, N. Nirmalakhandan, W. Van Voorhies, P.J. Lammers, Algal-based, single-step treatment of urban wastewaters, Bioresour. Technol. 189 (2015) 273–278.

[28] B. Laura, G. Paolo, Algae Anatomy, Biochemistry, and Biotechnology, CRC Press, Boca Raton, 2005.

[29] T. Ramon, M. Adam, E. Robyn, L. Connie, Diversity of nitrogen assimilation pathways among microbial photosynthetic eukaryotes, J. Phycol. 51 (2015) 490–506.

[30] E. Sanz-Luque, A. Chamizo-Ampudia, A. Llamas, A. Galvan, E. Fernandez, Understanding nitrate assimilation and its regulation in microalgae, Front. Plant Sci. 6 (2015) 899.

[31] J.A. Hellebust, I. Ahmad, Regulation of nitrogen assimilation in green microalgae, Oceanography 6 (1989) 241–255.

[32] T. Cai, S.Y. Park, Y. Li, Nutrient recovery from wastewater streams by microalgae: status and prospects, Renew. Sustain. Energy Rev. 19 (2013) 360–369.

[33] M. Arumugam, A. Agarwal, M.C. Arya, Z. Ahmed, Influence of nitrogen sources on biomass productivity of microalgae Scenedesmus bijugatus, Bioresour. Technol. 131 (2013) 246–249.

[34] H. Kamyab, C. Tin Lee, M.F. Md Din, M. Ponraj, S.E. Mohamad, M. Sohrabi, Effects of nitrogen source on enhancing growth conditions of green algae to produce higher lipid, Desalination 52 (2014) 3579–3584.

[35] S.Y. Maestrini, J.-M. Robert, J.W. Leftley, Y. Collos, Ammonium thresholds for simultaneous uptake of ammonium and nitrate by oyster-pond algae, J. Exp. Mar. Biol. Ecol. 102 (1986) 75–98.

[36] M.M. Loera-Quezada, M.A. Leyva-Gonzalez, D. Lopez-Arredondo, L. Herrera-Estrella, Phosphite cannot be used as a phosphorus source but is non-toxic for microalgae, Plant Sci. 231 (2015) 124–130.

[37] M.E. Martínez, J.M. Jiménez, F. El Yousfi, Influence of phosphorus concentration and temperature on growth and phosphorus uptake by the microalga *Scenedesmus obliquus*, Bioresour. Technol. 67 (1999) 233–240.

[38] C.M. Monteiro, P.M.L. Castro, F.X. Malcata, Metal Uptake by Microalgae: Underlying Mechanisms and Practical Applications, Biotechnol Progr, 2012, pp. 299–311.

[39] H. Eccles, Treatment of metal-contaminated wastes: why select a biological process? Trends Biotechnol. 17 (1999) 462–465.

[40] B. Volesky, Z.R. Holan, Biosorption of heavy metals, Biotechnol. Prog. 11 (1995) 235–250.

[41] A. Minoda, H. Sawada, S. Suzuki, S.-i. Miyashita, K. Inagaki, T. Yamamoto, et al., Recovery of rare earth elements from the sulfothermophilic red alga *Galdieria sulphuraria* using aqueous acid, Appl. Microbiol. Biotechnol. 99 (2015) 1513–1519.

[42] X. Ju, K. Igarashi, S. Miyashita, H. Mitsuhashi, K. Inagaki, S. Fujii, et al., Effective and selective recovery of gold and palladium ions from metal wastewater using a sulfothermophilic red alga, Galdieria sulphuraria, Bioresour.Technol 211 (2016) 759–764.

[43] S.K. Mehta, J.P. Gaur, Use of algae for removing heavy metal ions from wastewater: progress and prospects, Crit. Rev. Biotechnol. 25 (2005) 113–152.

[44] Z. Aksu, Biosorption of heavy metals by microalgae in batch and continuous systems, in: Y.-S. Wong, N.F.Y. Tam (Eds.), Wastewater Treatment with Algae, Springer Berlin Heidelberg, Berlin, Heidelberg, 1998, pp. 37–53.

[45] S.S. Bates, A. Tessier, P.G.C. Campbell, J. Buffle, Zinc adsorption and transport by chlamydomonas varuiabilis and scenedesmus subspicatus (chlorophyceae) grown in semicontinuous culture, J. Phycol. 18 (1982) 521–529.

[46] S.K. Mehta, J.P. Gaur, Removal of Ni and Cu from single and binary metal solutions by free and immobilized *Chlorella vulgaris*, Eur. J. Protistol. 37 (2001) 261–271.

[47] E. Sandau, P. Sandau, O. Pulz, Heavy metal sorption by microalgae, Acta Biotechnol. 16 (1996) 227–235.

[48] G. Massaccesi, M.C. Romero, M.C. Cazau, A.M. Bucsinszky, Cadmium removal capacities of filamentous soil fungi isolated from industrially polluted sediments, in La Plata (Argentina), World J. Microbiol. Biotechnol. 18 (2002) 817–820.

[49] A. Malik, Metal bioremediation through growing cells, Env. Intl. 30 (2004) 261–278.

[50] T. Muppaneni, H.K. Reddy, T. Selvaratnam, K.P. Dandamudi, B. Dungan, N. Nirmalakhandan, et al., Hydrothermal liquefaction of *Cyanidioschyzon merolae* and the influence of catalysts on products, Bioresour. Technol. 223 (2017) 91–97.

[51] D.C. Elliott, T.R. Hart, A.J. Schmidt, G.G. Neuenschwander, L.J. Rotness, M.V. Olarte, et al., Process development for hydrothermal liquefaction of algae feedstocks in a continuous-flow reactor, Algal Res. 2 (2013) 445–454.

[52] T. Selvaratnam, H. Reddy, T. Muppaneni, F.O. Holguin, N. Nirmalakhandan, P.J. Lammers, et al., Optimizing energy yields from nutrient recycling using sequential hydrothermal liquefaction with *Galdieria sulphuraria*, Algal Res. 12 (2015) 74–79.

[53] S. Velea, F. Oancea, F. Fischer, Microalgae-based, Biofuels. Bioproducts. (2017) 45–65.

[54] E. Koutra, C.N. Economou, P. Tsafrakidou, M. Kornaros, Bio-based products from microalgae cultivated in digestates, Trends Biotechnol. 36 (2018) 819–833.

[55] T. Gervasi, V. Pellizzeri, G. Calabrese, G. Di Bella, N. Cicero, G. Dugo, Production of single cell protein (SCP) from food and agricultural waste by using *Saccharomyces cerevisiae*, Nat. Prod. Res. 32 (2018) 648–653.

[56] E.W. Becker, Micro-algae as a source of protein, Biotechnol. Adv. 25 (2007) 207–210.

[57] M.H. Haddadi, H.T. Aiyelabegan, B. Negahdari, Advanced biotechnology in biorefinery: a new insight into municipal waste management to the production of high-value products, Int. J. Environ. Sci. Technol. 15 (2018) 675–686.

[58] K. Zhou, Y. Zhang, X. Jia, Co-cultivation of fungal-microalgal strains in biogas slurry and biogas purification under different initial CO_2 concentrations, Sci. Rep. 8 (2018) 7786.

[59] S. Abinandan, S.R. Subashchandrabose, K. Venkateswarlu, M. Megharaj, Nutrient removal and biomass production: advances in microalgal biotechnology for wastewater treatment, Crit. Rev. Biotechnol. 38 (2018) 1244–1260.

[60] J. Lv, J. Guo, J. Feng, Q. Liu, S. Xie, Effect of sulfate ions on growth and pollutants removal of self-flocculating microalga *Chlorococcum* sp. GD in synthetic municipal wastewater, Bioresour. Technol. 234 (2017) 289–296.

[61] S. Carfagna, C. Bottone, P. Cataletto, M. Petriccione, G. Pinto, G. Salbitani, et al., Impact of sulfur starvation in autotrophic and heterotrophic cultures of the extremophilic Microalga *Galdieria phlegrea* (cyanidiophyceae), Plant Cell Physiol. 57 (2016) 1890–1898.

[62] T. Hülsen, K. Hsieh, Y. Lu, S. Tait, D.J. Batstone, Simultaneous treatment and single cell protein production from agri-industrial wastewaters using purple phototrophic bacteria or microalgae – a comparison, Bioresour. Technol. 254 (2018) 214–223.

[63] D. Cheng, D. Li, Y. Yuan, L. Zhou, X. Li, T. Wu, et al., Improving carbohydrate and starch accumulation in *Chlorella* sp. AE10 by a novel two-stage process with cell dilution, Biotechnol. Biofuels 10 (2017) 75.

[64] A. Demirbas, M. Fatih Demirbas, Importance of algae oil as a source of biodiesel, Energy Convers. Manag. 52 (2011) 163–170.

[65] C.-Y. Chen, X.-Q. Zhao, H.-W. Yen, S.-H. Ho, C.-L. Cheng, D.-J. Lee, et al., Microalgae-based carbohydrates for biofuel production, Biochem. Eng. J. 78 (2013) 1–10.

[66] J.H. Mussgnug, V. Klassen, A. Schluter, O. Kruse, Microalgae as substrates for fermentative biogas production in a combined biorefinery concept, J. Biotechnol. 150 (2010) 51–56.

[67] H.W. Yen, I.C. Hu, C.Y. Chen, S.H. Ho, D.J. Lee, J.S. Chang, Microalgae-based biorefinery–from biofuels to natural products, Bioresour. Technol. 135 (2013) 166–174.

[68] F. Freitas, V.D. Alves, C.A.V. Torres, M. Cruz, I. Sousa, M.J. Melo, et al., Fucose-containing exopolysaccharide produced by the newly isolated enterobacter strain A47 DSM 23139, Carbohydr. Polym. 83 (2011) 159–165.

[69] S.-B. Hong, J.-H. Choi, H. Park, N.-H.L. Wang, Y.K. Chang, S. Mun, Development of an efficient process for recovery of fucose in a multi-component mixture of monosugars stemming from defatted microalgal biomass, J. Ind. Eng. Chem. 56 (2017) 185–195.

[70] J.H. Yim, S.J. Kim, S.H. Ahn, C.K. Lee, K.T. Rhie, H.K. Lee, Antiviral effects of sulfated exopolysaccharide from the marine microalga *Gyrodinium impudicum* strain KG03, Marine. Biotechnol. (NY) 6 (2004) 17–25.

[71] A.F. Lau, J. Siedlecki, J. Anleitner, G.M. Patterson, F.R. Caplan, R.E. Moore, Inhibition of reverse transcriptase activity by extracts of cultured blue-green algae (cyanophyta), Planta Med. 59 (1993) 148–151.

[72] M.A. Briggs, K.S. Petersen, P.M. Kris-Etherton, Saturated fatty acids and cardiovascular disease: replacements for saturated fat to reduce cardiovascular risk, J. Healthc. Qual. 5 (2017) 29.

[73] E. Ryckebosch, C. Bruneel, K. Muylaert, I. Foubert, Microalgae as an alternative source of omega-3 long chain polyunsaturated fatty acids, Lipid Technol. 24 (2012) 128–130.

[74] C. von Schacky, Omega-3 fatty acids: antiarrhythmic, proarrhythmic or both? Curr. Opin. Clin. Nutr. Metab. Care 11 (2008) 94–99.

[75] A.P. Simopoulos, Omega-3 fatty acids in inflammation and autoimmune diseases, J. Acad. Nutr. Diet. 21 (2002) 495–505.

[76] Y. Osher, R.H. Belmaker, Omega-3 fatty acids in depression: a review of three studies, CNS Neurosci. Ther. 15 (2009) 128–133.

[77] E.A. Miles, P.C. Calder, Influence of marine n-3 polyunsaturated fatty acids on immune function and a systematic review of their effects on clinical outcomes in rheumatoid arthritis, Br. J. Nutr. 107 (Suppl. 2) (2012) S171–S184.

[78] S. Nagarajan, S.K. Chou, S. Cao, C. Wu, Z. Zhou, An updated comprehensive techno-economic analysis of algae biodiesel, Bioresour. Technol. 145 (2013) 150–156.

[79] E. Alptekin, M. Canakci, Determination of the density and the viscosities of biodiesel—diesel fuel blends, Renew. Energy 33 (2008) 2623—2630.

[80] S. Poonam, S. Nivedita, Industrial and biotechnological applications of algae: a review, J Adv. J Plant Physiol. (2017) 01—25.

[81] P. Varshney, P. Mikulic, A. Vonshak, J. Beardall, P.P. Wangikar, Extremophilic micro-algae and their potential contribution in biotechnology, Bioresour. Technol. 184 (2015) 363—372.

[82] M.A. Borowitzka, J.M. Huisman, The ecology of Dunaliella salina (chlorophyceae, Volvocales): effect of environmental conditions on aplanospore formation, Bot. Mar. (1993) 233.

[83] L. Sorensen, A. Hantke, N.T. Eriksen, Purification of the photosynthetic pigment C-phycocyanin from heterotrophic Galdieria sulphuraria, J. Sci. Food Agric. 93 (2013) 2933—2938.

[84] T. Selvaratnam, A.K. Pegallapati, F. Montelya, G. Rodriguez, N. Nirmalakhandan, W. Van Voorhies, et al., Evaluation of a thermo-tolerant acidophilic alga, Galdieria sulphuraria, for nutrient removal from urban wastewaters, Bioresour. Technol. 156 (2014) 395—399.

[85] I. Woertz, A. Feffer, T. Lundquist, Y. Nelson, Algae grown on dairy and municipal wastewater for simultaneous nutrient removal and lipid production for biofuel feedstock, J. Environ. Eng. 135 (2009) 1115—1122.

[86] J.B.K. Park, R.J. Craggs, A.N. Shilton, Wastewater treatment high rate algal ponds for biofuel production, Bioresour. Technol. 102 (2010) 35—42.

[87] W.J. Oswald, H.B. Gotaas, H.F. Ludwig, V. Lynch, Algae symbiosis in oxidation ponds: photosynthetic oxygenation, Sewage Ind. Wastes 25 (1953) 692—705.

[88] T. Selvaratnam, A. Pegallapati, F. Montelya, G. Rodriguez, N. Nirmalakhandan, P.J. Lammers, et al., Feasibility of algal systems for sustainable wastewater treatment, Renew. Energy 82 (2015) 71—76.

[89] L.E. de-Bashan, A. Trejo, V.A.R. Huss, J.-P. Hernandez, Y. Bashan, Chlorella sorokiniana UTEX 2805, a heat and intense, sunlight-tolerant microalga with potential for removing ammonium from wastewater, Bioresour. Technol. 99 (2008) 4980—4989.

[90] A. Akcil, S. Koldas, Acid Mine Drainage (AMD): causes, treatment and case studies, J. Clean. Prod. 14 (2006) 1139—1145.

[91] D.B. Johnson, K.B. Hallberg, Acid mine drainage remediation options: a review, Sci. Total Environ. 338 (2005) 3—14.

[92] T. Lovingood, Nationwide Identification of Hardrock Mining Sites, US EPA, 2004.

[93] J.M. Wood, H. K. Wang, Microbial resistance to heavy metals, Environ. Sci. Technol. 17 (1983), 582A-90A.

[94] I. Fukuda, Physiological studies on a thermophilic blue green alga, Cyanidium caldarium GEITLER, Bot. Mag. 71 (1958) 79—86.

[95] L. Fan, A. Vonshak, A. Zarka, S. Boussiba, Does astaxanthin protect Haematococcus against light damage? Z. Naturforschung C (1998) 93.

[96] D. Remias, U. Lutz-Meindl, C. Lütz, Photosynthesis, pigments and ultrastructure of the alpine snow alga Chlamydomonas nivalis, Eur. J. Phycol. 40 (2005) 259—268.

[97] T. Leya, A. Rahn, C. Lutz, D. Remias, Response of arctic snow and permafrost algae to high light and nitrogen stress by changes in pigment composition and applied aspects for biotechnology, FEMS Immunol. Med. Microbiol. 67 (2009) 432—443.

[98] J.K. Sloth, M.G. Wiebe, N.T. Eriksen, Accumulation of phycocyanin in heterotrophic and mixotrophic cultures of the acidophilic red alga Galdieria sulphuraria, Enzym. Microb. Technol. 38 (2006) 168—175.

[99] S.D. Varfolomeev, L.A. Wasserman, Microalgae as source of biofuel, food, fodder, and medicines, Biotechnol. Appl. Biochem. 47 (2011) 789—807.

[100] G. Graziani, S. Schiavo, M.A. Nicolai, S. Buono, V. Fogliano, G. Pinto, et al., Microalgae as human food: chemical and nutritional characteristics of the thermo-acidophilic microalga Galdieria sulphuraria, Food Funct 4 (2013) 144—152.

[101] L. Liu, J. Chen, P.-E. Lim, D. Wei, Dual-species cultivation of microalgae and yeast for enhanced biomass and microbial lipid production, J. Appl. Phycol. (2018) 1—11.

[102] N. Rashid, W.-K. Park, T. Selvaratnam, Binary culture of microalgae as an integrated approach for enhanced biomass and metabolites productivity, wastewater treatment, and bioflocculation, Chemosphere 194 (2018) 67—75.

[103] Z. Rasouli, B. Valverde-Pérez, M. D'Este, D. De Francisci, I. Angelidaki, Nutrient recovery from industrial wastewater as single cell protein by a co-culture of green microalgae and methanotrophs, Biochem. Eng. J. 134 (2018) 129—135.

[104] G. Mujtaba, K. Lee, Treatment of real wastewater using co-culture of immobilized Chlorella vulgaris and suspended activated sludge, Wat.Res. 120 (2017) 174—184.

[105] J.R. Benemann, I. Woertz, T. Lundquist, Autotrophic microalgae biomass production: from niche markets to commodities, Ind. Biotechnol. 14 (2018) 3—10.

[106] M. Kube, B. Jefferson, L. Fan, F. Roddick, The impact of wastewater characteristics, algal species selection and immobilisation on simultaneous nitrogen and phosphorus removal, Algal.Res 31 (2018) 478—488.

[107] A.C. Redfield, The biological control of chemical factors in the environment, Am. Sci. 46 (1958), 230A-21.

[108] R. Whitton, A. Le Mevel, M. Pidou, F. Ometto, R. Villa, B. Jefferson, Influence of microalgal N and P composition on wastewater nutrient remediation, Wat.Res. 91 (2016) 371—378.

[109] J. Liu, Z. Li, J.S. Guo, Y. Xiao, F. Fang, R.-c. Qin, et al., The effect of light on the cellular stoichiometry of Chlorella sp. in different growth phases: implications of nutrient drawdown in batch experiments, J. Appl. Phycol. 29 (2017) 123—131.

[110] G. Markou, L. Wang, J. Ye, A. Unc, Using agro-industrial wastes for the cultivation of microalgae and duckweeds:

contamination risks and biomass safety concerns, Biotechnol. Adv. 36 (2018) 1238–1254.

[111] W.-K. Park, M. Moon, S.-E. Shin, J.M. Cho, W.I. Suh, Y.K. Chang, et al., Economical DHA (Docosahexaenoic acid) production from *Aurantiochytrium* sp. KRS101 using orange peel extract and low-cost nitrogen sources, Algal. Res. 29 (2018) 71–79.

[112] J. Miller, N. Murphy, K. McCary, K. Scarlett, The effects of coffee grounds on cell growth and reproduction in algae in *Chlorella* sp, JIBI 4 (2016).

[113] B.G. Ryu, K. Kim, J. Kim, J.I. Han, J.W. Yang, Use of organic waste from the brewery industry for high-density cultivation of the docosahexaenoic acid-rich microalga, *Aurantiochytrium* sp. KRS101, Bioresour. Technol. 129 (2013) 351–359.

[114] V. T Duong, F. Ahmed, S. R Thomas-Hall, S. Quigley, E. Nowak, P. M Schenk, High protein- and high lipid-producing microalgae from northern Australia as potential feedstock for animal feed and biodiesel, Front. Biotechnol Bioeng. (2015) 53, https://doi.org/10.3389/fbioe.2015.00053, eCollection 2015.

[115] J. Rodriguez, A. Ferraz, R.F. P Nogueira, I. Ferrer, E. Esposito, N. Duran, Lignin biodegradation by the ascomycete *Chrysondia sitophila*, Biotechnol. Appl. Biochem. 62 (1997) 233–242.

[116] H. Safafar, P. U Norregaard, A. Ljubic, P. Moller, S. L Holdt, C. Jacobsen, Enhancement of protein and pigment content in two Chlorella species cultivated on industrial process water, J Mar Eng Technol 4 (2016) 84.

[117] M.A.C.L.D. Oliveira, M.P.C. Monteiro, P.G. Robbs, S.G. F Leite, Growth and chemical composition of Spirulina maxima and Spirulina platensis biomass at different temperatures, Aquacult. Int. 7 (1999) 261–275.

[118] L.Q. Zepka, E. Jacob-Lopez, R. Goldbeck, L. A Souza-Soares, M.I. Queiroz, Nutritional evaluation of single-cell protein produced by Aphanothece microscopica Nägeli, Bioresour. Technol. 101 (2010) 7107–7111.

[119] T. Aggelopoulos, K. Katsieris, A. Bekatorou, A. Pandey, I. M Banat, A. A Koutinas, Solid state fermentation of food waste mixtures for single cell protein, aroma volatiles and fat production, Food Chem. 145 (2014) 710–716.

[120] W. Cui, Q. Wang, F. Zhang, S.C. Zhang, Z.M. Chi, C. Madzak, Direct conversion of inulin into single cell protein by the engineered *Yarrowia lipolytica* carrying inulinase gene, Process Biochem. 46 (2011) 1442–1448.

[121] A. R Ahmadi, H. Ghoorchian, R. Hajihosaini, J. Khanifar, Determination of the amount of protein and amino acids extracted from the microbial protein (SCP) of lignocellulosic wastes, Pakistan Int. J. Biol. Sci. 13 (2010) 355–361.

[122] M.J.G. Valentino, L. S Ganado, J. R Undan, Single cell protein potential of endophytic fungi associated with bamboo using rice bran as substrate, Adv. Appl. Sci. Res. 7 (2016) 68–72.

[123] Y. Gao, D. Ki, Y. Liu, Production of single cell protein from soy molasses using *Candida tropicalis*, Ann. Microbiol. 62 (2012) 1165–1172.

[124] N. Kornochalert, D. Kantachote, S. Chaiprapat, Techkarnjanaruk, Use of *Rhodopseudomonas palustris* P1 stimulated growth by fermented pineapple extract to treat latex rubber sheet wastewater to obtain single cell protein, Ann. An. Microbiol. 64 (2014) 1021–1032.

[125] B. Kunasundari, V. Murugaiyah, G. Kaur, F.H.J. Maurer, S. Kumar, Revisiting the single cell protein application of *Cupriavidus necator* H16 and recovering bioplastic granules simultaneously, PLoS One 8 (2013), https://doi.org/10.1371/journal.pone.0078528.

[126] F. Yazdian, S. Hajizadeh, S.A. Shojaosadati, R. Khalilzadeh, M. Jahanshahi, M. Nosrati, Production of single cell protein from natural gas: parameter optimization and RNA evaluation, Iran J. Biotechno1. 3 (2005) 235–242.

FURTHER READING

[1] E.B. Kurbanoglu, O.F. Algur, Single-cell protein production from ram horn hydrolysate by bacteria, Bioresour. Technol. 85 (2002) 125–129.

Index

A

Accelerated solvent extraction, 137t–138t
Acetogenesis, 231t, 233
Acetogenic microorganisms, 232
Acid catalysts, 253
Acid-catalyzed biodiesel production, 90
Acid hydrolysis, 143t–145t
Acidogenesis, 153–154, 231t, 233
Acid rain/acidification, 288
Activator, 59–60, 59f
Active layer facing draw solution (AL-DS), 102
Active layer facing feed solution (AL-FS), 102
Acutodesmus dimorphus, 165
Adsorption immobilization, 256
Agar, 198–199
Airlift column photobioreactors, 22, 23f
Alcoholic biofuel production
  genomics, 216–218
  proteomics, 218
  transcriptomics, 218
Algae cultivation system
  challenges, 187–188
  heterotrophic and autotrophic microalgal culture system, 3
  open pond cultivation system, 3
  photobioreactor (PBR), 2–3
Algae growth factor
  biotic factors, 6
  components, 3–4
  environmental factors
    nutrients, 5–6
    pH measures, 4–5
    temperature, 6
  light intensity, 6
  mixing, 6
Algae-mediated nanoparticles
  AgNPs, 273
  applications, 272f
  AuNPs, 273
  metallic alloy nanoparticles, 273
  metal oxide nanoparticles, 273
Algaenan, 229–230
Alginates, 199
Algogenic organic matters (AOMs), 102–103
Alkali catalysts, 253–254

Alkali-catalyzed biodiesel production, 90
Aminoglycoside adenyltransferase (aadA)-encoding genes, 215
Ammonium/ammonia, 16, 318–319
  toxicity, 34–35
Amphiphilic lipids, 135, 140
Amplified ribosomal DNA restriction Analysis (ARDRA), 241
Anaerobic digestion (AD), 121–122, 154f, 165–166, 231t
  acidogenesis, 153–154
  biochemical processes, 232f
  community microbiological structure, 234f
  conditions and results, 230t
  drawbacks, 227
  hydrolysis, 153–154
  methane fuel combustion, 153–154
  methane production, 154
  methanogenesis, 153–154
  microalgae–bacteria interaction, 237–238
  microbes involved, 154
  molecular techniques, 238–242
    cloning and sequencing, 240–241
    experimental design, 240f
    fundamentals, 238–239
    nucleic acid extraction, 239
    "omics" techniques, 242
    polymerase chain reaction and amplification, 239–240
    restriction enzymes, 241
  organic matter, 227
  organic waste, 153
  pathways involved, 154f
  sewage microbiology community
    acetogenesis, 233
    acetogenic microorganisms, 232
    acidogenesis, 233
    archaeal communities, 235–237
    bacterial communities, 235
    homoacetogenic bacteria, 232
    hydrolysis, 232–233
    hydrolytic microorganisms, 232
    methanogenesis, 233–234
    microbial community, 234–235
    sulfur-reducing bacteria, 232
  stability process, 227
Animal feed
  amino acid profile, 176

Animal feed (*Continued*)
  aquaculture and shrimp feed, 177
  aquaculture feed, 191
  carbohydrate-active enzymes (CAZymes), 176–177
  carbohydrates, 176
  limitations, 176–177
  lipids, 176
  monogastric livestock, 177–178
  pigments and vitamins, 176
  poultry feed, 191
  protein, 176, 176t
  ruminants, 178
Animal-like algae, 1
Aquaculture and shrimp feed, 177
Aquaculture feed
  β-glucans (BGs), 203–204
  feeding trials, 204
  rotifers, 203
Aquaponics system, 43
Arachidonic acids, 199
Archaeal communities, 235–237, 236t
Astaxanthin, 39, 202
Asterarcys quadricellulare, 52
Attached cultivation system, 25, 26f
Autoclave, 142t–145t
Autoflocculation
  advantages and disadvantages, 114t
  calcium and phosphate precipitation, 116
  calcium-induced flocculation, 116
  laboratory and outdoor experiments, 116
  negatively charged algal cells, 115–116
  pH changes, 115
Autotrophic cultivation, 18, 36, 36f
Autotrophic microalgal culture system, 3
Auxin, 59–60
Axial vibration membrane (AVM), 105

B

Bacterial communities, 235
Basic-region/leucine zipper (bZIP), 221
Batch culture system, 2
Bead milling, 142t–145t

Note: Page numbers followed by "f" indicate figures and "t" indicates tables.

Benzene, toluene, ethylbenzene, and xylenes (BTEX), 325
β-glucans (BGs), 203–204
Bioactive compounds, 39, 191–192, 201t
  agar, 198–199
  alginates, 199
  carotenoids, 200
  carrageenan, 199
  chitin, 198
  extracellular polymeric substances (EPS), 179
  fucoidans, 198
  lipids, fatty acids, and sterols, 199
  marine algae, 197–198
  microalgal pigments, 178
  phenolic compounds, 199–200
  phycobiliproteins, 200
  polysaccharides, 198
  PUFA, 178–179
  terpenoids, 199
  tocopherols, 199
Bio-based solvent, 137t–138t
Biocatalysts, 254–255
Biochar, 154–155
Biochemical engineering approach, 212–213, 213f
Biodiesel
  acid-catalyzed biodiesel production, 90
  alkali-catalyzed biodiesel production, 90
  Chlorella SDEC-18 and *Scenedesmus* SDEC-8 cultivation, 165
  codigestion, 167
  enzymatic route, 189
  enzyme-catalyzed biodiesel production, 90
  enzyme-catalyzed process, 123–124
  free fatty acids (FFAs), 163–164
  glycerol, 124
  heterotrophic metabolism, 165
  for industrial products, 188–189
  lipase-immobilized nanocatalyst
    Burkholderia lipase, 258–260
    Chlorella prototothecoides, 256–257
    C. vulgaris cultivation, 257–258
    genetic engineering, 256–257
    lipase-immobilized magnetic NP catalyst, 258
    Thermomyces lanuginosus, 260
  lipid composition, 163–164
  microalgal lipid conversion
    biodiesel fuel properties, 89t
    fatty acid chains, 88
    green algae and diatoms, 88
    lipid extraction, 88
    neutral lipids, 88
    polar lipids, 88
    simultaneous lipid extraction and transesterification, 90–91
    transesterification reaction, 88–90, 89f
  microalgal strains, 165

Biodiesel (*Continued*)
  oil production, *Neochloris oleoabundans*, 164–165
  polyunsaturated fatty acids (PUFAs), 163–164
  production
    cell wall disruption techniques. *See* Cell wall disruption techniques
    microalgal lipid extraction strategies. *See* Microalgal lipid extraction strategies
  productivity, 83
  salinity stress, 165
  traditional sources and limitations, 6–7
  transesterification, 188–189
  transesterification reaction, 163–164
  two-stage anaerobic systems, 167–168
  vehicle fuel, 164–165
  wastewater, 308
Bioethanol, 153
  chemical methods, 172
  enzymatic methods, 172–173
  feedstock, 169
  fuel producing countries, 170t
  glucose, 173
  intracellular carbohydrate content, 169
  mechanical methods, 170–172
  productivity, 83
  separated hydrolysis and fermentation (SHF), 172–173
  simultaneous hydrolysis and fermentation (SSF), 172–173
  wastewater treatment, 308
Biofertilizers, 190–191
Bio film membrane photobioreactor (BMPBR), 98
Bioflocculation
  advantages and disadvantages, 114t
  coculturing process, 116–117
  filamentous fungi, 87
  flocculant-producing bacteria, 87–88
Biofouling, 102
Biofuel. *See also* Microalgae-based biofuel
  classification, 251
  classifications, 211
  first-, second-, and third-generation, 7
  life cycle assessment. *See* Life cycle assessment (LCA)
  microalgal source, 251–252
  nanomaterials. *See* Nanomaterials (NMs)
  policies and market intensives, 289–290
  techno-economic assessment (TEA), 281–282, 288–289
  wastewater treatment, 328–329
Biogas, 124
  AD, 189
  advantages, 229t

Biogas (*Continued*)
  algal biomass composition, 166
  anaerobic digestion. *See* Anaerobic digestion (AD)
  anaerobic digestion (AD), 165, 166f
  carbon dioxide removal, 69
  case study, 242–243
  crude biogas, 69
  disadvantages, 229t
  hydrolysis, 166–167
  hydrolytic bacteria, 189
  for industrial products, 189
  industrial wastewater, nutrient sources, 69
  integrated process, 69
  methane content, 69, 166
  methane yield, 167, 228
  microalgae cell wall, 231f
    biopolymers, 229–230
    composition, 231–232
    phylogenetic base factor, 229
    sporopollenin and algaenan polymers, 229–230
  microalgae cultivation scheme, 70f
  organic matter anaerobic digestion, 228
  organic matter hydrolysis, 228
  parameters, 165
  upgrading and lipid production
    biogas composition, 70–72
    CO_2 concentration, 70
    CO_2 content, 71–72
    cocultivations, 74–80
    CO_2 removal efficiency, 69
    gas flow rate, 72
    half-saturation constant, 70
    H_2S content, 70–71
    light intensity and photoperiods, 72–73
    light wavelength, 73
    maximum specific growth rate, 70
    methane content, 70
    microalgae cultivation, 71t, 75t–76t
    Monod model, 70
    nitrogen source and concentration, 73–74
    and wastewater treatment, 74
Biohydrogen, 151–152
  biophotolysis, 168–169
  dark fermentation (DF), 169, 170t
  for industrial products, 190–191
  production routes, 168f
  wastewater treatment, 307–308
Biomass conversion
  anaerobic digestion (AD), 153–154
  biophotolysis, 151–152
  fermentation, 152–153
  gasification, 157
  liquefaction process, 156–157
  pyrolysis, 154–155
  torrefaction, 155–156
  transesterification, 150–151, 151f
  transformation route, 150, 150f

Biomass water content, 133–134
Biooil, liquefaction process, 156–157
Biophotolysis, 168–169
  biohydrogen production, 151–152
  C. reinhardtii, 152
  direct biophotolysis, 151–152
  [FeFe]-hydrogenase, 152
  hydrogen gas, 151
  indirect biophotolysis, 152
  photosynthesis, 151
Bioplastic production, 192, 192f
Bligh and Dyer method, 137t–138t
Bligh-Dyer method, 134
Brewery wastewaters, 302
Bubble column photobioreactors, 22, 23f
Burkholderia lipase, 258–260
1,4-Butanediol, 219

C
Calcium-induced flocculation, 116
Calvin cycle, 14, 35–36
Candida antarctica lipase, 258–260
Capital cost, 337
Carbohydrate-active enzymes (CAZymes), 176–177
Carbohydrate metabolism, 54
Carbon, 317–318
  alkaline environments, 14–15
  assimilation, 31–33
  calcification, 14–15
  inorganic carbon, 14–15
  mass transfer rate, 15–16
  Na$_2$HCO$_3$ solubility and tolerance, 15, 15f
  organic carbon, 16
  passive uptake, 15
  pH, 14
  photosynthetic activity, 15
  production costs, 15
Carbonic anhydrase enzymes affect, 73
Carbon nanotubes (CNTs), 267
Carotenoids, 122, 178, 200
Carrageenan, 199
Caulerpa peltata, 256–257
Cellular chassis, 218–219
Cellulase, 172
Cellulose, 211
Cellulosomes, 219–221
Cell wall disruption techniques
  advantages and limitations, 141–145
  biomass pretreatment, 140
  cell disintegration, 140–141
  chemical methods, 140–141
  electroporation, 141
  expeller press crushes and breaks, 141
  Fenton's method, 141
  intracellular product release, 140
  mechanical methods, 140–141
  nonmechanical methods, 141
  process parameters and mode of action, 142t
  thermal methods, 140–141

Centrifugation systems, 44, 84–86, 112–113, 114t
Chaetoceros muelleri, 70–71
Chemical coagulation, 113–115, 114t
Chemical-enhanced backwashing, 106
Chemical extraction, 120
Chemical flocculation, 44, 113–115
  advantages and disadvantages, 114t
  flocculant types and dosages, 115
  microalgae flocs, 113–115
  negative surface charge, 113–115
  polymeric flocculants, 115
  pretreatments, 115
  theories, 115
Chemical oxygen demand (COD), 74
Chemical treatments, 142t
Chemical vapor deposition (CVD) technique, 267
Chitin, 198
Chlamydomonas reinhardtii, 190
Chloramphenicol acetyltransferase (cat), 215
Chlorella protothecoides, 52, 256–257
Chlorella pyrenoidosa, 103
Chlorella variabilis, 175
Chlorella vulgaris, 16, 138–139
Chlorococcum littorale, 190
Chlorophyll pigments, 73
Cloning, 240–241
Clostridium thermocellum, 167
Clostridium ultunense, 237
Clustered regularly interspaced short palindromic repeats/CRISPR-associated protein (CRISPR/CAS9), 221–222
Cocultivation
  activated sludge, 74–80
  filamentous fungi, 74, 79f
  process, 116–117
  resource recovery, 343–344
Column photobioreactor, 4t, 22, 41
Conventional solvent extraction, 136–138, 137t–138t
Copper nanoparticle, 265
Corallina elongata, 175
Cosolvents, 134–136
Covalent immobilization, 256
Cross-flow mode, 98–99
Cross-linking immobilization, 256
Culture systems
  open ponds. *See* Open ponds
  photobioreactors. *See* Photobioreactors (PBRs)
Cyanobacterial genetic engineering
  plasmid expression vectors, 215–216
  selectable markers, 215
  transformation, 214–215

D
Dark fermentation (DF), 169, 170t
Dead-end mode, 98
Denaturing gradient gel electrophoresis (DGGE) technique, 241–242

Derjaguin, Landau, Verwey, and Overbeek (DLVO) modeling, 118
Detergents, 143t–145t
Dewatering process, 101t
Diatomite dynamic membrane (DDM), 99–100
Diethylstilbestrol (DES), 320–321
Dinitrogen, 16–17
Direct biophotolysis, 151–152
Direct transesterification, 137t–138t, 139–140, 139f, 145–146, 151
Disc-stack centrifuge, 84–86
Dispersed air flotation, 117
Dissolved air flotation, 117
Dissolved flotation, 46
Distillery wastewaters, 300–302
Dry torrefaction, 155–156
Dynamic filtration system, 100

E
Electroflotation, 46
Electroflotation process, 86
Electrolytic flotation, 117
Electroporation, 141
Emulsifiers, 136
Endoenzymes, 232–233
Energy input-output ratio (EROI), 287
Entrapment immobilization, 256
Environmental resilience, 293–315
Environmental toxicity, 288
Enzymatic catalysis, 151
Enzymatic extraction, 120
Enzymatic treatments, 142t
Enzyme catalyst, 254–255
Enzyme-catalyzed biodiesel production, 90
Essential amino acids (EAAs), 176
Eutrophication, 287
Expeller/oil press, 142t
Expeller press, 137t–138t
Expeller press crushes and breaks, 141
Extracellular polymeric substances (EPS), 179
Extracted biomolecules, 269–270
Extremophile microalgae, 21, 342

F
Fatty acid–based biosurfactants, 136
Fatty acid methyl ester (FAME), 133–134, 150–151, 253
Fed-batch cultivation, 61, 62f
Feedstock material, 97
Fenton's method, 141, 143t–145t
Fermentation, 121
  bioethanol, 153
  bioethanol yield, 152
  carbohydrates concentration, 152
  pretreatment, 152, 153f
Fertilizer
  filamentous strains, 175
  lipid-extracted biomass, 175
  microalgal-bacterial flocs, 173–175
  micronutrients, 173
  mineralization tests, 173

Fertilizer (*Continued*)
  nutrient content, 174t
  nutrient release, 173
  plant growth and crop productivity, 175−176
  red marine algae, 175
  Spirulina-based fertilizers, 175
  unicellular microalgae, 175
Filtration, 86, 112, 114t
Flash pyrolysis, 154−155
Flat plate-airlift reactors, 4t
Flat plate bioreactor, 22, 23f, 41
Flocculant-producing bacteria, 87−88
Flocculation
  advantages and drawbacks, 114t
  bioflocculation
    filamentous fungi, 87
    flocculant-producing bacteria, 87−88
  chemical, 113−115
  mixotrophic culture, 86−87
  organic cationic polymers, 86−87
  preconcentration step, 86
Floch method, 120
Flotation, 46, 86
  advantages and disadvantages, 114t, 117
  application potential, 117
  Derjaguin, Landau, Verwey, and Overbeek (DLVO) modeling, 118
  dispersed air flotation, 117
  dissolved air flotation, 117
  electrolytic flotation, 117
  jet flotation, 117
  mechanism, 118
  parameters, 117−118
  principle, 117
  suspended air flotation, 117
Flux balance analysis (FBA), 216−218
Folch method, 134, 137t−138t
Forward osmosis (FO), 100−101, 101f
  draw solutes, 102
  fouling, 104
  illustration, 100−101, 101f
  operating conditions, 102
Fossil fuel combustion, 149
Foulants, 102
Fouling, membrane
  aeration and microbubbles, 105−106
  alkaline agents, 106
  axial vibration membrane (AVM), 105
  biofouling, 102
  cake layer formation, 102
  chemical-enhanced backwashing, 106
  concentration polarization, 102
  electropolarization, 105
  electrostatic interactions, 104
  feed pretreatments, 104
  flow hydrodynamics, 104
  forward osmosis process, 104
  foulants, 102
  internal fouling, 102

Fouling, membrane (*Continued*)
  irreversible, 102
  membrane orientation, 105
  membrane relaxation, 106
  minimizing cake formation, 105
  mitigation, 104−106
  operational conditions, 104
  physical and chemical cleanings, 105−106
  pressure-driven processes, 102−104
  reversible, 102
  surface modification, 105
  transmembrane pressure (TMP), 104
  ultrasonic cleaning, 105−106
  ultraviolet (UV)-initiated graft polymerization, 105
Free fatty acids (FFAs), 163−164
Freeze drying, 143t−145t
Fucoidans, 198
Fucoxanthin, 178
Fungal pellet-assisted bioflocculation, 87
Fungi-assisted sedimentation, 44−45

G
Gas flow rate, 72
Gasification, 121, 157
Genetic engineering
  flux balance analysis (FBA), 216
  systems biology research, 216
Genome editing, 221−222
Genome shuffling, 213−214
Global energy crisis, 129
Global warming, 211
Glutamate dehydrogenase (GDH) pathway, 33−34
Glutamine synthetase enzyme system, 16
Glutamine synthetase-glutamine oxoglutarate aminotransferase (GS-GOGAT) pathway, 33−35
Glycerol, 124, 189
Glycolysis pathway, 55t
Gold nanoparticle (AuNPs), 273
Gompertz kinetic model, 167
Gravity sedimentation, 44, 84
Gravity settling, 118
Greenhouse gases, Regulated Emissions, and Energy use in Transportation (GREET) model, 283
Green solvents approach, 138−139
Grinding, 143t−145t

H
Haematococcus pluvialis, 202
Harvesting process, 97
  biological method, 85f
  centrifugation, 84−86
  commercial systems, 84
  filtration, 84, 86
  flocculation, 86−87
  flotation process, 86
  gravity sedimentation, 84
  life cycle analysis, 84

Harvesting process (*Continued*)
  membrane technology
    fouling. *See* Fouling, membrane
    future prospects, 106
    osmotically driven membranes, 100−102
    pressure-driven membrane process. *See* Pressure-driven membrane process
  physical/chemical method, 84, 85f
  technologies
    centrifugation-based strategy, 44
    chemical flocculation, 44
    flotation, 46
    fungi-assisted sedimentation, 44−45
    nontoxic components-based flocculation, 45−46
    sedimentation, 44
    selection of, 44
Heavy metals, 339
  removal/recovery, 342
  wastewater, 324−325
Heterogeneous catalysts, 254
Heterotrophic cultivation, 18, 36−37
Heterotrophic metabolism, 165
Heterotrophic microalgal culture system, 3
High-pressure homogenization, 142t
High-value product generation, 342−343
Homoacetogenic bacteria, 232
Homogeneous catalysts, 254
Hydraulic power, 21
Hydrodynamic cavitation, 143t−145t
Hydrogen, 124
Hydrolysis, 153−154, 231t, 232−233
Hydrophilic organic foulants, 102
Hydrophobic organic foulants, 102
Hydroprocessing, 121
Hysteresis, 13

I
Indirect biophotolysis, 152
Inhibitor, 60
Inorganic carbon sources, 31−32
Inorganic nitrogen, 318−319
Integrative plasmids, 215−216
Interfacial tension, 136
Internal fouling, 102
International Standards Organization (ISO), 282
Ionic liquid extraction, 137t−138t
Ionized phosphorus, 35
Iron, 319−320
Irreversible fouling, 102
Isobutanol, 219−221

J
Jet flotation, 117

L
Lambert−Beer's law, 13
Land use, 288

Laurencia obtusa, 175
Life cycle assessment (LCA)
acid rain/acidification, 288
case study, 284–285, 286t–287t
climate change, 287
energy input-output ratio (EROI),
287
environmental toxicity, 288
eutrophication, 287
functional units (FU), 282
goals and scope, 282–283
International Standards Organization
(ISO), 282
land use, 288
non-renewable energy, 287
ozone depletion, 288
phases, 282f
product's life cycle, 281
respiratory effects, 288
results interpretation, 284
smog formation, 288
water use, 287–288
Life cycle impact assessment (LCIA),
284, 284f
Life cycle inventory (LCI) analysis
algae cultivation stage, 285
assumptions, 285
case study, 286t–287t
data collection, 283
electricity, 283
environmental impact analysis, 285
GREET model, 283
inputs and outputs, 283
net carbon sequestration, 283–284
Light intensity, 54–56, 72–73
Light wavelength, 73
Lipase-based reaction, 189
Lipase enzyme
chemical reaction system, 255
definition, 255
immobilization
adsorption, 256
covalent immobilization, 256
cross-linking immobilization, 256
entrapment immobilization, 256
immobilization, 255–256
nanocatalyst, 256–260
merits, 255
microorganisms, 3t
Lipid extraction process, 88, 129–130
Lipid synthesis, 37–39
Liquefaction process, 156–157
Livestock feed, 203
Lower-quality agar, 198–199
Lutein, 202

M

Magnesium, 17
Magnetically induced membrane
vibrating system (MMV), 100
Manganese, 319–320
Mass transfer mechanism, 136
Mechanical pressing, 143t–145t
Medium-quality agar, 198–199

Membrane bioreactor (MBR), 118
Membrane photobioreactors
(MPBRs), 24, 25f
Membrane technology
fouling. *See* Fouling, membrane
future prospects, 106
osmotically driven membranes,
100–102
pressure-driven membrane process.
See Pressure-driven membrane
process
Metabolic networks, 216–218
Metabolic production systems
autotrophic cultivation, 18
heterotrophic cultivation, 18
microalgae cultivation mode, 18, 19t
mixotrophic cultivation, 19
Metabolite channeling, 219–221
Metallic alloy nanoparticles, 273
Metal oxide nanoparticles, 273
Methanation, 157
Methane content, 69
Methane production, 154
Methanogenesis, 153–154, 231t,
233–234
Microalgae
batch culture system, 2, 2f
biology, 2
cell disruption techniques, 129–130
cell walls, 129–130
growth curve, 2f
lipid composition and distribution
classes of, 132f
lipid bodies, 131
nonpolar lipids, 130–131
physical and chemical properties,
130–131
polar lipids, 130–131
triacylglycerols (TAGs), 130–131
lipid content, 2, 3t
lipid extraction process, 129–130
lipid yield, 129
main components, 7–8
oil content and lipid productivity, 8t
peak growth phase, 2
potential applications, 131f
pretreatment approach, 129–130
processing, 129, 130f
solvent-based extraction, 129–130
valuable products, 131f
value-added components, 130–131
Microalgae–bacteria interaction,
237–238
Microalgae based biofuel
advantages, 149
biomass conversion
anaerobic digestion (AD),
153–154
biophotolysis, 151–152
fermentation, 152–153
gasification, 157
liquefaction process, 156–157
pyrolysis, 154–155
torrefaction, 155–156

Microalgae-based biofuel (*Continued*)
transesterification, 150–151, 151f
transformation route, 150, 150f
composition, 149
feedstock, 149
fossil fuel combustion, 149
high biomass and lipid productivity,
149
microalgae oil yield, 149
types, 149
Microalgae cultivation
animal-like algae, 1
applications, 1
autotrophic cultivation, 36, 36f
bioactive compounds, 39
carbon assimilation
flue gas, 33t
greenhouse effects, 32
inorganic carbon sources, 31–32
organic carbon sources, 32, 32t
technical problem, 32–33
waste carbon resources, 32–33
waste stream, 33, 33t
culture systems
open ponds, 20–21
photobioreactors, 21–25
heterotrophic cultivation, 36–37
light
Lambert–Beer's law, 13
photoinhibition, 13
and rate of photosynthesis, 13, 13f
solar radiation, 13–14
lipid synthesis, 37–39
metabolic production systems
autotrophic cultivation, 18
heterotrophic cultivation, 18
microalgae cultivation mode, 18,
19t
mixotrophic cultivation, 19
microalgal biomass, 1
mixotrophic cultivation, 37
net energy ratio (NER), 1
nitrogen assimilation
ammonium toxicity, 34–35
nitrogen utilization, 35
sources and assimilation pathways,
33–34
nutrients
calcium (Ca) content, 17–18
carbon, 14–16
iron (Fe), 18
magnesium, 17
nitrogen, 16–17
phosphorus, 17
sulfur, 17
pH, 18
phosphorus assimilation, 35–36
plant-like algae, 1
priorities, 1
protein synthesis, 39
renewable bioenergy, 1
research and industrial fields, 11
temperature
and light interdependence, 12

Microalgae cultivation (*Continued*)
    moderately elevated temperatures, 11–12
    optimal temperature range, 11
    specific growth rate, 12
    upper temperature, 11
Microalgae processing
    biomass conversion technologies
        anaerobic digestion, 121–122
        fermentation, 121
        gasification, 121
        hydroprocessing, 121
        pyrolysis, 122
        transesterification, 120–121
    extraction
        chemical, 120
        enzymatic, 120
        lipid extraction, 119
        physical method, 119–120
        solvent, 120
    harvesting and dewatering
        centrifugation, 112–113
        factors, 111–112
        filtration, 112
        flocculation, 113–117
        flotation, 117–118
        microalgae size, 111–112
        sedimentation process, 118
    microalgae exploitation, 111
    microalgal biomass, 111
    products
        biodiesel, 123–124
        biogas production, 124
        fatty acids, 122
        hydrogen, 124
        pigments and proteins, 122–123
        sulfated polysaccharides, 124
        syngas, 124
Microalgal-bacterial flocs (MaB-flocs), 173–175
Microalgal biomass
    aquaculture feed, 203–204
    astaxanthin, 202
    carbohydrate content, 83–84, 85t
    chlorophyll, 200–202
    cultivation stage, 97
    dried and concentrated microalgae, 205t
    essential fatty acids and carotenoid, 203t
    feed, 202
    feedstock material, 97
    genetic modification, 211–212
    harvesting process, 97
    lipid, 85t
    livestock feed, 203
    lutein, 202
    medium components and culture conditions, 83–84
    natural pigments, 202
    phycoerythrin, 202
    polyunsaturated long-chain fatty acids (PUFAs), 202
    production and processing stages, 97

Microalgal biomass (*Continued*)
    products, 83–84
    protein, 83–84, 85t
    small cell size, 97
Microalgal biorefinery, 164f, 188f
    animal feed, 176–178
    bioactive compounds, 178–179
    biodiesel, 163–165
    biogas, 165–168
    biohydrogen, 168–169
    fertilizer, 173–176
    industrial production, 163
    for industrial products
        animal feed, 191
        bioactive compounds, 191–192
        biodiesel, 188–189
        biofertilizers, 190–191
        biogas, 189
        biohydrogen, 190–191
        plastics, 192, 192f
        sugarcane ethanol industry, 190
    petroleum refinery, 163
Microalgal lipid extraction strategies
    advantages and limitations, 137t–138t
    biomass water content, 133–134
    cell wall structure and composition, 133
    challenges, 132–136
    direct transesterification, 139–140, 139f
    green solvents approach, 138–139
    key steps, 131, 133f
    lipid recovery, 132–133
    mass transfer mechanism, 136
    nonpolar solvents
        amphiphilic lipids, 135
        cosolvents, 135–136
        mechanism, 135
        micelles, 135
        neutral lipids extraction, 135–136
        wet biomass, 135, 135f
    physiological properties, 132–133
    polar solvent, 132
    solvent-based techniques, 132
    solvents, 132
    stable emulsion formation, 136
Microwave, 142t
Microwave-assisted extraction, 137t–138t, 267
Milling method, 88
Mixotrophic cultivation, 19, 37
    activator, 59–60, 59f
    advantages, 52
    advantages and disadvantages, 53t
    biofuel production, 61–63
    carbohydrate, 52
    carbohydrate metabolism, 54
    carbon emission, 52
    carbon source, 57–58
    fed-batch cultivation, 61, 62f
    inhibitor, 60
    light intensity, 54–56
    lipid and biomass productivity, 52

Mixotrophic cultivation (*Continued*)
    microalgae growth rate, 52
    microalgal cultivation mode, 51
    nitrogen source, 58
    parameters, 54
    temperature, 56–57
    two-stage cultivation, 60–61
Monod model, 70
Monogastric livestock, 177–178
Monoraphidium braunii, 320
Multiplex-automated genomic engineering (MAGE), 218–219
Municipal wastewater effluent (MWE), 321, 322t
Mutagenesis-based strain improvement, 213–214
Mutual cell shading, 72–73

N

Nannochloropsis gaditana lipids, 90
Nannochloropsis oceanica lipids, 90
Nannochloropsis oculata, 256–257
Nanomaterials (NMs)
    catalyst
        acid catalysts, 253
        alkali catalysts, 253–254
        biocatalysts, 254–255
        disadvantages, 253
        heterogeneous catalysts, 254
        homogeneous catalysts, 254
        lipase and immobilization, 255–256
    merits of, 252–253
    microalgal oil transesterification, 259t
Nanoparticle (NP), 265
    biosynthesis
        algae/biomass concentration, 272
        cell-free supernatant, 270
        extracted biomolecules, 269–270
        illumination, 272
        incubation time, 271
        living cultures, 268–269
        microwave-assisted synthesis, 267
        pH effect, 271
        precursor ion concentration, 271–272
        temperature, 270–271
        whole cells, 270
    chemical methods, 265–266
    green synthesis
        algae role, 268
        benign synthesis strategies, 266
        carbon nanotubes, 267
        green chemistry, 266
        plant-based biomolecules, 267–268
        silver nanoparticle, 267
        sugars, 266
    physical methods, 265–266
    silver nanoparticle, 267
Naphthalene acetic acid (NAA), 59
Natural organic matters (NOMs), 102–103

Negative pressure–driven bioreactor, 42–43, 42f
Neomycin phosphotransferase (npt)-encoding genes, 215
Net energy ratio (NER), 1
Netrium digitus, 179
Nitrate, 16
Nitrite, 16
Nitrogen, 318–319, 338
  ammonium/ammonia, 16
  dinitrogen, 16–17
  glutamine synthetase enzyme system, 16
  nitrate, 16
  nitrite, 16
  nitrite oxide, 16
  organic nitrogen, 17
Nitrogen starvation, 58
Non-essential amino acids (NEAAs), 176
Nonpolar lipids, 130–131
Non-renewable energy, 287
Nontoxic components-based flocculation, 45–46
Nozzle-type centrifuge, 84–86
Nucleic acid scaffolds, 219–221

O
Oil/expeller press, 143t–145t
Oleaginous microalgae, 83
  biogas upgrading and phytoremediation. *See* Biogas
  feedstock, 188
Olive mill effluent (OME), 300
Olive mill wastewater (OMW), 323–324
OMEGA bioreactor, 42
Open ponds, 3
  bacteria and microalgal symbiosis, 21
  engineering design, 21
  paddlewheel, 21
  raceway open pond
    configuration of, 20, 20f
      hydraulic diameter, flow conduit, 20–21
      light penetration depth, 20
      photon flux densities, 20–21
      Reynolds number (Re), 20–21
      turbulence, 20–21
  types, 20
Open raceway bioreactor, 40–41
Operating cost, 337
Organic carbon sources, 32, 32t
Organic matter anaerobic digestion, 228
Organic matter hydrolysis, 228
Organic nitrogen, 17
Organic phosphate, 319
Osmotically driven membranes
  applications and recent developments, 101t
  forward osmosis (FO), 100–101, 101f
    draw solutes, 102

Osmotically driven membranes (*Continued*)
    illustration, 100–101, 101f
    operating conditions, 102
Osmotic shock, 143t–145t
Oxidation, 143t–145t
Ozone depletion, 288
Ozone flotation process, 86

P
Palm oil mill effluent (POME), 299–300, 323
Paramylon, 203–204
Pathway scaffolding, 219–221
Pelvetia canaliculata, 257–258
Penicillium expansum lipase, 258–260
Petroleum wastewater, 325
Petroleum wastewaters (PWWs), 299
Pharmaceutical contaminants, 298–299
Phenolic compounds, 199–200
Phenotypic phase plane analysis, 216–218
Phlorotannins, 199–200
Phosphorus (P), 35–36, 319
Photobioreactors (PBRs), 2–3
  algal biofilm bioreactor
    biomass yield, 41–42
    critical control, 42
    microalgae biomass density, 41–42
  basic criteria, 21
  choice of, 21
  column photobioreactors, 22
  cultivation mode, 21
  flat-plate photobioreactors, 22
  limitations, 21
  membrane photobioreactors (MPBRs), 24, 25f
  OMEGA bioreactor, 42
  open raceway bioreactor, 40–41
  parameters, 39–40
  plastic bag photobioreactors, 24–25
  selection and design, 39–40
  solid-state photobioreactors, 25
  total incident intensity, 22
  traditional light radiative equation, 21–22
  tubular/flat plate/column bioreactor, 41
  tubular photobioreactor, 22–24
Photoinhibition, 13
Photoperiods, 72–73
Photosynthesis, 73
Photosynthetic photon flux density (PPFD), 72
Photosystem II (PSII), 168–169
Phycobilins, 200
Phycobiliproteins, 122, 200
Phycobilisomes, 200
Phycoerythrin, 202
Phytoplanktons, 2
Pinene, 219
Plant-like algae, 1

Plasmid expression vectors, 215–216
Plastic bag photobioreactors, 24–25, 25f
Polarization phenomenon, 112
Polar lipids, 130–131
Polar solvent, 132
Polyhydroxy butyrate (PHB), 192
Polymerase chain reaction (PCR), 239–240
Polymeric flocculants, 115
Polysaccharides, 198
Polyunsaturated fatty acids (PUFAs), 163–164, 202
Polyvinylidenefluoride (PVDF), 99
Poultry feed, 177, 191
Predators, 6
Pressure-driven membrane process, 99f
  bio film membrane photobioreactor (BMPBR), 98
  concentration method, 98
  cross-flow mode, 98–99
  dead-end mode, 98
  dynamic filtration system, 100
  fouling, 102–104
  membrane materials
    diatomite dynamic membrane (DDM), 99–100
    polymer materials, 99
    polyvinylidenefluoride (PVDF), 99
    selection, 99
    starch addition, 100
    steel-use-stainless (SUS), 100
  transmembrane pressure (TMP), 98–99
  ultrafiltration (UF) process, 99
Prokaryotic microcompartments, 219–221
Protein-based scaffolds, 219–221
Protein-rich biomass, 188
Protein synthesis, 39
Pulsed electric field, 142t–145t
Pyrolysis, 122
  advantages, 122
  biochar, 154–155
  biooil, 154–155
  different modes, 154t
  downstream techniques, 122
  Dunaliella tertiolecta, 154–155
  flash pyrolysis, 154–155
  heating rate, 154–155
  temperature, 154
Pyrosequencing, 240
Pyruvate-ferredoxin oxidoreductase (PFOR) enzyme, 152

Q
Quorum sensing (QS), 237–238

R
Red pigment astaxanthin, 202
Renewable energy, 97
Replicative plasmids, 215–216

Resource recovery
  capital cost, 337
  carbohydrate and lipids, 341–342
  coculture system, 343–344
  extremophile microalgae, 342
  food application, 343–345
  future perspectives, 347
  heavy metal removal/recovery, 342
  high-value product generation,
    342–343
  microalgal mechanism
    carbon, 338
    heavy metals, 339
    nitrogen, 338
    phosphorus, 338–339
  mixed wastewater, 344
  operating cost, 337
  processes, 338f
  resource limitations, 344–345
  single-cell protein, 339–342
  threats and mitigation strategies,
    345–347
  wastewater treatment, 342
Restriction enzymes, 241
Restriction terminal fragment length
  polymorphism (T-RFLP), 241
Reverse micelles, 135
Reversible fouling, 102
Reynolds number (Re), 20–21
Rhizoclonium fontinale, 271
Ribulose-1, 5-bisphosphate (RuBP),
  35–36
Ribulose-1,5-bisphosphate
  carboxylase oxygenase (Rubisco),
  11–12, 14
Rotating algae biofilm reactor (RABR),
  299
Ruminants, 178

S

Sabinene and terpinene, 219
Salinity stress, 165
Sargassum longifolium, 271
Sargassum myriocystum, 271
Scenedesmus obliquus, 100
Scenedesmus quadricauda, 258–260
Sedimentation, 44
Sedimentation process, 118
Separated hydrolysis and fermenta-
  tion (SHF), 172–173
Silver nanoparticle (AgNPs), 265, 267,
  273
Simultaneous hydrolysis and fermen-
  tation (SSF), 172–173
Single-cell protein, 339–342
Single-strand conformation polymor-
  phism (SSCP), 241
Slaughter-house wastewater, 303
Smog formation, 288
Solid-state photobioreactors, 25, 26f
Solvent extraction, 120, 132, 134
  Bligh-Dyer method, 134
  chloroform:methanol mixture, 134

Solvent extraction (*Continued*)
  conventional solvent extraction,
    136–138
  cosolvents, 134
  direct transesterification, 134
  methanol/biomass ratio, 134
  organic solvents, 134
  polar lipids, 134
  Soxhlet extraction, 134
Solvent-free extraction, 119–120
Sonication-assisted solvent extraction,
  88
Soxhlet extraction, 134, 137t–138t
Spirulina-based fertilizers, 175
Sporopollenin, 229–230
Steam explosion, 142t–145t
Steam reforming, 157
Steel-use-stainless (SUS), 100
Strain improvement, 214t
  biochemical engineering approach,
    212–213
  classical improvement techniques,
    213–214
  genome shuffling, 213–214
  graphical representation, 212f
  metabolic engineering, 213–214
  microorganism reproduction and
    survival, 212
  mutagenesis-based strain improve-
    ment, 213–214
  stress factors, 212
  synthetic biology
    1,4-butanediol, 219
    cellular chassis, 218–219
    definition, 218–219
    genome editing, 221–222
    pathway scaffolding, 219–221
    pinene and α-pinene production,
      219
    sabinene and terpinene, 219
    transcriptional factor engineering,
      221
Sugarcane ethanol industry, 190
Sulfated polysaccharides, 124
Sulfur, 17
Sulfur-reducing bacteria, 232
Supercritical fluid extraction, 88,
  137t–138t
Supercritical water gasification
  (SCWG), 157
Suspended air flotation, 117
Switchable solvents, 137t–138t
Syngas, 124, 157
Systems Biology Markup Language
  (SBML), 216–218

T

Tannery wastewaters, 302
Techno-economic assessment (TEA),
  281–282, 288–289
Terpenoids, 199
Textile wastewater (TWW), 297, 324
Thermomyces lanuginosus, 260

Tocopherols, 199
Torrefaction
  biochar production, 155
  dry torrefaction, 155–156
  wet torrefaction, 156
Transcriptional factor engineering,
  221
Transesterification, 88–90
  catalytic activity, 150–151
  challenges, 151
  chemical solvent, 151
  direct transesterification, 151
  enzymatic catalysis, 151
  enzymes, 121
  fatty acid methyl esters (FAMEs),
    150–151
  homogeneous acid and alkaline
    catalysts, 150–151
  reaction rate, 150–151
  steps involved, 120–121
  triglycerides, 150–151, 151f
Transmembrane pressure (TMP),
  98–99
Triacylglycerols (TAGs), 130–131
Triclosan, 320
Tubular bioreactor, 41
Tubular photobioreactor, 22–24, 24f
Tubular photobioreactors, 4t
Two-stage anaerobic systems,
  167–168
Two-stage cultivation, 60–61

U

Ultrafiltration (UF) process, 99
Ultrasonication, 142t
Ultrasound-assisted extraction,
  137t–138t
Ultraviolet (UV)-initiated graft
  polymerization, 105

V

Vacuum filtration, 84
Vegetable oils, 6–7
Vibratory microfiltration system, 100
Volatile fatty acids (VFAs), 233

W

Wastewater treatment
  biomass concentration, 305
  biomass recovery, 305
  brewery wastewaters, 302
  carbon limitation, 303–304
  challenges and opportunities, 318f,
    326–328
  contaminant removal, 320–321
  cultivated microalgae
    added-value products, 329–330
    biodiesel, 308
    bioethanol, 308
    biofuels, 328–329
    biogas production, 307
    biohydrogen and biomethane
      production, 307–308

Wastewater treatment (*Continued*)
  dissolved oxygen concentrations, 304
  distillery wastewaters, 300–302
  emerging contaminants, 325–326
  fertilizer, 308
  heavy metals, 296–298, 324–325
  limited scalability, 305
  microalgae diversity, 308
  municipal wastewater effluent
    (MWE), 321
  nutrients availability, 293–297
    carbon, 317–318
    micronutrients, 319–320
    nitrogen, 318–319
    phosphorus (P), 319

Wastewater treatment (*Continued*)
  nutrients removal mechanisms
    carbon, 294–295
    nitrogen, 295
    phosphorus, 295–296
  olive mill effluent (OME), 300
  olive mill wastewater (OMW),
    323–324
  palm oil mill effluent (POME),
    299–300, 306t, 323
  petroleum wastewaters (PWWs), 299,
    301t, 325
  pharmaceutical contaminants,
    298–299
  photobioreactor configuration, 303

Wastewater treatment (*Continued*)
  photoinhibition, 304
  proof of concept, 309
  removal efficiencies, 294t
  resource recovery, 342
  slaughter-house wastewater, 303
  successful cases, 309
  tannery wastewaters, 302
  temperature, 304–305
  textile wastewater (TWW), 297, 324
Water-gas shift (WGS), 157
Wet torrefaction, 156

Z
Zinc, 319–320

Printed in the United States
By Bookmasters